Solutions Manual

to accompany

PHYSICAL CHEMISTRY

Fourth Edition

Robert J. Silbey
Class of 1942 Professor of Chemistry
Massachusetts Institute of Technology

Robert A. Alberty
Professor Emeritus of Chemistry
Massachusetts Institute of Technology

Moungi G. Bawendi
Professor of Chemistry
Massachusetts Institute of Technology

WILEY

John Wiley & Sons, Inc.

Cover Illustration: Norm Christiansen

To order books or for customer service call 1-800-CALL-WILEY (225-5945).

ISBN 978-0-471-65802-3

Printed in the United States of America

10 9 8

Printed and bound by Malloy Lithographing, Inc

PREFACE

This Solutions Manual accompanies the text R. J. Silbey, R. A. Alberty, and M. G. Bawendi, PHYSICAL CHEMISTRY, Fourth Edition, John Wiley and Sons, Inc., Hoboken, NJ. The textbook has two types of problems at the end of each chapter. In addition to the usual problems that can be solved using a hand-held calculator, there are Computer Problems that require a personal computer with a mathematical application. The problems of the first type are divided into two nearly equivalent lists. The answers to problems in the first list are given in the back of the textbook, but the answers to problems in the second list are not. The answers to the Computer Problems are also not given in the textbook.

This Solutions Manual provides three types of information about the problems:

(1) Worked-out solutions are provided for problems in the first list of problems that can be solved using a hand-held calculator.

(2) Answers are given for problems in the second list that can be solved using a hand-held calculator.

(3) Solutions are provided in Mathematica™ (Wolfram Research, Inc., 100 Trade Center Drive, Champaign, Illinois 61820-7237) for all 170 problems that require a personal computer with a mathematical application. These solutions in digital form can be obtained from the web at http://www.wiley.com/college/silbey. In order to run these programs it is necessary to have Mathematica 5.0 installed in your computer. Each problem is in a separate Mathematica notebook.

The first section of the Solutions Manual contains the first two types of information, and the second section, which has been typed in Mathematica, contains the Mathematica programs and printouts. You can type the Mathematica programs in yourself or get them in digital form from the web, but in either case you are encouraged to modify the programs to make them apply to other substances, temperature ranges, values of physical properties, etc. The second section of this Solutions Manual also provides a ReadMe, an Introduction to Mathematica, and an Index of Mathematica Commands.

The solutions in Mathematica provide a kind of extension to the Examples in the textbook that can profitably be studied, whether you have a computer with a mathematical application or not. That is, an interesting problem is stated, an indication is given as to how to go about solving it, and the solution is given. In about 100 of the problems the solution is expressed by a 2D or 3D plot. You have already seen some of these plots in the textbook. These problems can also be solved with other mathematical programs, such as MathCad, MATLAB, and MAPLE.

Working problems is an important part of learning physical chemistry. Not all knowledge of physical chemistry is quantitative, but much of it is. Since physical chemistry utilizes physics and mathematics to predict and interpret chemical phenomena, there are many opportunities to use quantitative methods. In solving problems it is important to develop the habit of using units and canceling them to obtain the units for the answer because this helps prevent errors. In the first section of the Solutions Manual units are usually given, but in the Mathematica programs, units have been omitted as a simplification. Actually, Mathematica can work problems with units, and it can solve equations symbolically. These are things that you can try.

We are indebted to many physical chemists who have recommended problems and who have suggested improvements in this SOLUTIONS MANUAL. The following individuals made very useful suggestions as to how to improve the Mathematica™ solutions to Computer Problems: Ian Brooks (Wolfram Research), Carl W. David (U. Connecticut), Robert N. Goldberg (NIST), Mark R. Hoffmann (University of North Dakota), Andre Kuzniarek (Wolfram Research), W. Martin McClain (Wayne State University), Kathryn Tomasson (University of North Dakota), and Worth E. Vaughan (University of Wisconsin--Madison). We are also indebted to Peter Giunta who provided advice on word processing and Jennifer Yee who managed the production of the SOLUTIONS MANUAL at Wiley.

Robert J. Silbey
Robert A. Alberty
Moungi G. Bawendi

CONTENTS

FIRST SECTION: PROBLEMS THAT CAN BE SOLVED USING A HAND-HELD CALCULATOR

PART ONE: THERMODYNAMICS

PART TWO: QUANTUM CHEMISTRY

PART THREE: KINETICS

PART FOUR: MACROSCOPIC AND MICROSCOPIC STRUCTURES

SECOND SECTION: PROBLEMS THAT REQUIRE A PERSONAL COMPUTER WITH A MATHEMATICAL APPLICATION

PART ONE: THERMODYNAMICS

PART TWO: QUANTUM CHEMISTRY

PART THREE: KINETICS

PART FOUR: MACROSCOPIC AND MICROSCOPIC STRUCTURES

1

Thermodynamic State of a Gas

1.1 The intensive state of an ideal gas can be completely defined by specifying (1) T, P, (2) T, $\overline{V}$, or (3) P, $\overline{V}$. The extensive state of an ideal gas can be specified in four ways. What are the combinations of properties that can be used to specify the extensive state of an ideal gas? Although these choices are deduced for an ideal gas, they apply to real gases.

SOLUTION

The extensive state is determined by (1) P, V, T, (2) P, n, T, (3) P, V, n, or (4) V, n, T.

1.2 The ideal gas law also represents the behavior of mixtures of gases at low pressures. The molar volume of the mixture is the volume of a mole of the mixture. The partial pressure of gas i in a mixture is defined as $y_i P$, where y_i is its mole fraction, and P is the total pressure. Ten grams of N_2 is mixed with 5 g of O_2 and held at 25 °C at 0.750 bar. (a) What are the mole fractions of N_2 and O_2? (b) What are the partial pressures of N_2 and O_2? (c) What is the volume of the ideal mixture?

SOLUTION

(a)

$$n_{N_2} = \frac{(10\text{ g})}{(28.013\text{ g mol}^{-1})} = 0.357\text{ mol}$$

$$n_{O_2} = \frac{(5\text{ g})}{(32.000\text{ g mol}^{-1})} = 0.156\text{ mol}$$

$$y_{N_2} = \frac{(0.357\text{ mol})}{(0.513\text{ mol})} = 0.696$$

$$y_{O_2} = \frac{(0.156\text{ mol})}{(0.513\text{ mol})} = 0.304$$

(b)

$$P_{N_2} = (0.696)(0.750\text{ bar}) = 0.522\text{ bar}$$

$$P_{O_2} = (0.304)(0.750\text{ bar}) = 0.228\text{ bar}$$

(c) $V = nRT/P$

$$= \frac{(0.513 \text{ mol})(0.083145 \text{ L bar K}^{-1} \text{ mol}^{-1})(298.15 \text{ K})}{0.750 \text{ bar}}$$

$= 17.00$ L

1.3 A mixture of methane and ethane is contained in a glass bulb of 500 cm^3 capacity at 25 °C. The pressure is 1.25 bar, and the mass of gas in the bulb is 0.530 g. What is the average molar mass, and what is the mole fraction of methane?

SOLUTION

$$PV = nRT = \left(\frac{m}{M}\right)RT$$

$$M = \frac{mRT}{PV} = \frac{(0.530 \text{ g})(0.08315 \text{ L bar K}^{-1} \text{ mol}^{-1})(298 \text{ K})}{(1.25 \text{ bar})(0.500 \text{ L})}$$

$= 21.0$ g mol^{-1}

The molar mass of the mixture is a mole fraction weighted average of the molar masses of methane and ethane.

$$M = y_1 M_1 + y_2 M_2$$
$$21.0 = y_1\, 16.0 + (1 - y_1)30.0$$
$$14.0\, y_1 = 9.0$$
$$y_1 = \frac{9.0}{14.0} = 0.643$$

1.4 Nitrogen tetroxide is partially dissociated in the gas phase according to the reaction

$N_2O_4(g) = 2NO_2(g)$

A mass of 1.588 g of N_2O_4 is placed in a 500-cm^3 glass vessel at 298 K and dissociates to an equilibrium mixture at 1.0133 bar. (a) What are the mole fractions of N_2O_4 and NO_2? (b) What percentage of the N_2O_4 has dissociated? Assume the gases are ideal.

SOLUTION

There are two simultaneous equations because we know (1) the mass is equal to the sum of the masses of N_2O_4 and NO_2 and (2) the pressure of the mixture is equal to the sum of the partial pressures of the two gases.

$$m_{\text{total}} = 1.588 \text{ g} = m_{N_2O_4} + m_{NO_2} \qquad (1)$$

$$P_{\text{total}} = 1.0133 \text{ bar} = \frac{m_{N_2O_4}(0.08314 \text{ L bar K}^{-1} \text{ mol}^{-1})(298 \text{ K})}{(92 \text{ g mol}^{-1})(0.500 \text{ L})}$$

$$+\frac{m_{NO_2}(0.08314 \text{ L bar K}^{-1} \text{ mol}^{-1})(298 \text{ K})}{(46 \text{ g mol}^{-1})(0.500 \text{ L})}$$

$$= 0.5386\, m_{N_2O_4} + 1.0772\, m_{NO_2} \quad (2)$$

$m_{NO_2} = 1.588 - m_{N_2O_4}$

Substituting this in equation (2) yields

$1.0133 = 0.5386\, m_{N_2O_4\,4} + 1.01772(1.588 - m_{N_2O_4})$

Thus

$m_{N_2O_4} = 1.295$ g or 0.01407 mol

$m_{NO_2} = 0.293$ g or 0.006370 mol

$n = 0.02044$ mol

(a) $y_{N_2O_4} = \dfrac{0.01407 \text{ mol}}{0.02044 \text{ mol}} = 0.6884$

$y_{NO_2} = 1 - y_{N_2O_4} = 0.3116$

(b) $\% \text{ undissociated} = \dfrac{(0.01407 \text{ mol})(100)}{(1.588 \text{ g})/(92 \text{ g mol}^{-1})} = 81.51\ \%$

% dissociated = 100 - 81.51 = 18.49%

1.5 Although a real gas obeys the ideal gas law in the limit as $P \to 0$, not all of the properties of a real gas approach the values for an ideal gas as $P \to 0$. The second virial coefficient of an ideal gas is zero, and so $dZ/dP = 0$ at all pressures. But calculate dZ/dP for a real gas as $P \to 0$.

SOLUTION

$Z = 1 + B'P + C'P^2 + \ldots$

$$\frac{dZ}{dP} = B' + 2\, C'P + \ldots$$

so that

$$\left(\frac{dZ}{dP}\right)_{P \to 0} = B'$$

This shows that a real gas does not behave like an ideal gas in all respects as $P \to 0$.

*1.6 Show how the second virial coefficient of a gas and its molar mass can be obtained by plotting P/ρ versus P, where ρ is the density of the gas. Apply this method to the following data on ethane at 300 K.

P/bar	1	10	20
$\rho/10^{-3}$ g cm^{-3}	1.2145	13.006	28.235

SOLUTION

$P\bar{V} = RT + BP$
$PVM/m = RT + BP$
$P/\rho = RT/M + BP/M$
Thus the intercept of the plot of P/ρ versus P is RT/M, and the slope is B/M. Plot P/ρ = 823.38, 768.88, 708.35 bar/g cm^{-3} versus P = 1, 10, 20 bar. The intercept is 829.44 bar/g cm^{-3} = RT/M, and so the molar mass of ethane is 30.07 g mol^{-1}. The slope is -6.056 cm^3 g^{-1} = B/M, and so B = (-6.056 cm^3 g^{-1})(30.07 g mol^{-1}) = -183 cm^3 mol^{-1}. This problem can also be solved using a computer program that calculates the intercept and slope by the method of least squares.

*1.7 Calculate the second and third virial coefficients of hydrogen at 0 °C from the fact that the molar volumes at 50.7, 101.3, 202.6, and 303.9 bar are 0.4634, 0.2386, 0.1271, and 0.09004 L mol^{-1}, respectively.

SOLUTION

$P\bar{V}/RT = 1 + B/\bar{V} + C/\bar{V}^2 + ...$

P/bar	50.7	101.3	202.6	303.9
$\bar{V}$/L mol^{-1}	0.4634	0.2386	0.1271	0.09004
$P\bar{V}/RT$	1.035	1.064	1.134	1.205
$(1/\bar{V})$/mol L^{-1}	2.158	4.191	7.868	11.106

The following plot has been prepared using *Mathematica* ™. In this calculation, the point $1/\bar{V} = 0$, $P\bar{V}/RT = 1$ was included.

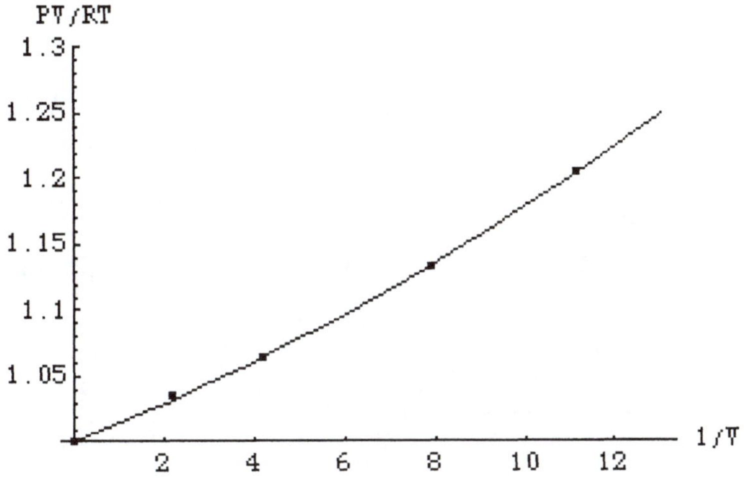

In the plot labels, V is the molar volume. In making the plot, the point $P\bar{V}/RT = 1$ at $1/\bar{V} = 0$ has been included in obtaining a quadratic fit. the use of Fit in *Mathematica*™ yields $PV/RT = 1.001 + 0.0135(1/V) + 0.00043(1/V)^2$. Thus the second virial coefficient B is 0.0135 L mol^{-1} and the third virial coefficient is $4.35\text{x}10^{-4}$ L^2 mol^{-2}..

*1.8 The critical temperature of carbon tetrachloride is 283.1 °C. The densities in g/cm^3 of the liquid ρ_1 and vapor ρ_v at different temperatures are as follows:

t /°C	100	150	200	250	270	280
ρ_1	1.4343	1.3215	1.1888	0.9980	0.8666	0.7634
ρ_v	0.0103	0.0304	0.0742	0.1754	0.2710	0.3597

What is the critical molar volume of CCl_4? It is found that the mean of the densities of the liquid and vapor does not vary rapidly with temperature and can be represented by

$$\frac{\rho_1 + \rho_v}{2} = A + Bt$$

where A and B are constants. The extrapolated value of the average density at the critical temperature is the critical density. The molar volume $\bar{V}_c$ at the critical point is equal to the molar mass divided by the critical density.

<u>SOLUTION</u>

The following plot has been prepared using *Mathematica* ™.

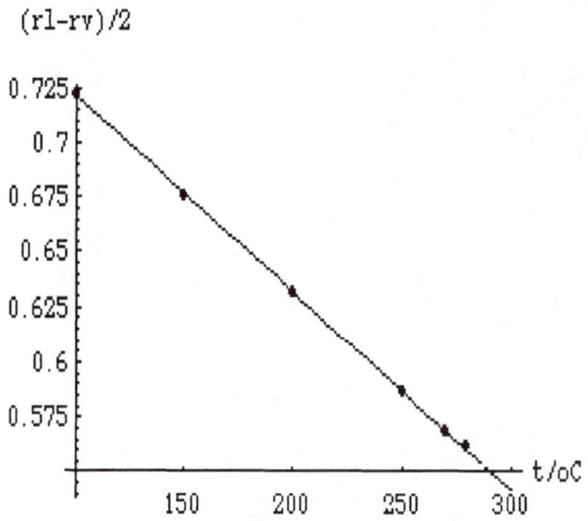

Extrapolating $\frac{\rho_1 + \rho_v}{2}$ to T_c we obtain

$\rho_c = 0.557 \text{ g cm}^{-3}$

$$\bar{V}_c = \frac{153.84 \text{ g mol}^{-1}}{0.557 \text{ g cm}^{-3}} = 276 \text{ cm}^3 \text{ mol}^{-1}$$

1.9 Show that for a gas of rigid spherical molecules, *b* in the van der Waals equation is four times the molecular volume times Avogadro's constant. If the molecular diameter of Ne is 0.258 nm (Table 17.4), approximately what value of *b* is expected?

SOLUTION

The molecular volume for a spherical molecule is

$$\frac{4}{3}\pi\left(\frac{d}{2}\right)^3 = \frac{\pi}{6}d^3$$

where *d* is the diameter. Since the center of a second spherical molecule cannot come within a distance *d* of the center of the first spherical molecule, the excluded volume per pair of molecules is $\frac{4}{3}\pi d^3$. The constant *b* in the van der Waals equation is the excluded volume per molecule times the Avogadro's constant

$$b = \left(\frac{4\pi}{6}d^3\right)N_A = \left(\frac{2}{3}\right)\pi d^3 N_A$$

For neon,

$$b = \left(\frac{2}{3}\right)\pi(0.258 \times 10^{-9} \text{ m})^3(6.02 \times 10^{23} \text{ mol}^{-1})$$
$$= 2.17 \times 10^{-5} \text{ m}^3 \text{ mol}^{-1}$$

Since 100 cm = 1 m, or 100 cm m^{-1} = 1, b can be expressed in $cm^3\ mol^{-1}$ by multiplying by $(100\ cm\ m^{-1})^3$.

$b = (2.17 \times 10^{-5}\ m^3\ mol^{-1})(100\ cm\ m^{-1})^3$

$= 21.7\ cm^3\ mol^{-1}$

1.10 What is the molar volume of n-hexane at 660 K and 91 bar according to (a) the ideal gas law and (b) the van der Waals equation? For n-hexane, $T_c = 507.7$ K and $P_c = 30.3$ bar

SOLUTION

(a) $$\bar{V} = \frac{RT}{P}$$

$$= \frac{(0.08314\ \text{L bar K}^{-1}\ \text{mol}^{-1})(660\ \text{K})}{91\ \text{bar}} = 0.603\ \text{L mol}^{-1}$$

(b) $$a = \frac{27\,R^2T_c^2}{64\,P_c} = \frac{(27)(0.08314\ \text{L bar K}^{-1}\ \text{mol}^{-1})^2(507.7\ \text{K})^2}{(64)(30.3\ \text{bar})}$$

$$= 24.81\ \text{L}^2\ \text{bar mol}^{-2}$$

$$b = \frac{RT_c}{8\,P_c} = \frac{(0.08314\ \text{L bar K}^{-1}\ \text{mol}^{-1})(507.7\ \text{K})}{8(30.3\ \text{bar})} = 0.174\ \text{L mol}^{-1}$$

$$P = RT/(\bar{V}_c - b) - a/\bar{V}^2$$

$$91\ \text{bar} = \frac{(0.08314)(660\ \text{K})}{(\bar{V} - 0.174)} - \frac{24.81}{\bar{V}^2}$$

Rather than solving a cubic equation, substituting successive values of $\bar{V}$ shows that $\bar{V} = 0.39$ L mol^{-1}.

1.11 Derive the expressions for van der Waals constants a and b in terms of the critical temperature and pressure; that is, derive equations 1.32 and 1.33 from 1.29 and 1.30.

SOLUTION

Equations 1.29 and 1.30 may be written

$$RT_c/(\bar{V}_c - b)^2 = 2a/(\bar{V}_c^3) \qquad (1)$$

$$2RT_c/(\bar{V}_c - b)^3 = 6a/\bar{V}_c^{4} \qquad (2)$$

Division of the first equation by the second yields

$\bar{V}_c = 3b$ (3)

Substitution of this expression in equation 1 yields

$T_c = 8a/27Rb$ (4)

Substitution of equations 3 and 4 in equation 1.31 yields

$P_c = a/27b^2$ (5)

Since there are three relations (equations 3-5) between the van der Waals constants and the critical constants, a and b may be expressed in terms of T_c and P_c or T_c and $\bar{V}_c$. Critical pressures are generally known more accurately than critical volumes, and so a and b are generally calculated using data on the critical temperature and pressure.

1.12 Calculate the second virial coefficient of methane at 300 K and 400 K from its van der Waals constants, and compare these results with Fig. 1.9.

SOLUTION

For methane $a = 2.283$ L^2 bar mol^{-2} and $b = 0.04278$ L mol^{-1}. From equation 1.28, the second virial coefficient at 300 K is

$$B = b - a/RT = 0.04278 - \frac{2.283}{(0.08314)(300)}$$

$$= -0.048 \text{ L mol}^{-1}$$

At 400 K, $B = -0.026$

Fig. 1.9 yields - 0.040 L mol^{-1} at 300 K and - 0.020 L mol^{-1} at 400 K.

*1.13 You want to calculate the molar volume of O_2 at 298.15 K and 50 bar using the van der Waals equation, but you don't want to solve a cubic equation. Use the first two terms of equation 1.26 to obtain an approximate solution. The van der Waals constants of O_2 are $a = 0.138$ Pa m^6 mol^{-1} and $b = 31.8 \times 10^{-6}$ m^3 mol^{-1}. What is the molar volume in L mol^{-1}?

SOLUTION

$$\frac{P\bar{V}}{RT} = 1 + \frac{1}{RT}\left(b - \frac{a}{RT}\right)P$$

$$\bar{V} = \frac{RT}{P}\left[1 + \frac{1}{RT}\left(b - \frac{a}{RT}\right)P\right]$$

$$= \frac{RT}{P} + b - \frac{a}{RT}$$

$$= \frac{(8.31451 \text{ Pa m}^3 \text{ mol}^{-1} \text{ K}^{-1})(298.15 \text{ K})}{50 \times 10^5 \text{ Pa}} + 31.8 \times 10^{-6} \text{ m}^3 \text{ mol}^{-1}$$

$$- \frac{0.138 \text{ Pa m}^6 \text{ m}^{-2}}{(8.314 \text{ Pa m}^3 \text{ K}^{-1} \text{ mol}^{-1})(298 \text{ K})}$$

$= (495.8 \times 10^{-6} + 31.8 \times 10^{-6} - 55.7 \times 10^{-6}) \text{ m}^3 \text{ mol}^{-1}$

$= 471.9 \times 10^{-6} \text{ m}^3 \text{ mol}^{-1}$

$= 0.4719 \text{ L mol}^{-1}$

If you have a computer program for solving polynomials, it can be used to check this approximation method.

1.14 The isothermal compressibility κ of a gas is defined in problem 1.17, and its value for an ideal gas is shown to be $1/P$. Use implicit differentiation of V with respect to P at constant T to obtain the expression for the isothermal compressibility of a van der Waals gas. Show that in the limit of infinite volume, the value for an ideal gas is obtained.

SOLUTION

The van der Waals equation can be written

$nRT = PV - nPb + n^2a/V - n^3ab/V^2$

Implicit differentiation with respect to P at constant T yields

$V + P(\partial V/\partial P)_T - nb - (n^2a/V^2)(\partial V/\partial P)_T + (2n^3ab/V^3)(\partial V/\partial P)_T = 0$

Solving for $(\partial V/\partial P)_T$ yields

$$(\partial V/\partial P)_T = \frac{nb - V}{P - n^2a/V^2 + 2n^3ab/V^3}$$

Thus the isothermal compressibility of a van der Waals gas is given by

$$\kappa = \frac{V - nb}{PV - n^2a/V + 2n^3ab/V^2}$$

When V is very large, this expression for the isothermal compressibility reduces to $\kappa = V/PV = 1/P$ as expected.

1.15 Calculate the second and third virial coefficients of O_2 from its van der Waals constants in Table 1.3.

SOLUTION

The van der Waals constants for O_2 are $a = 1.378 \text{ L}^2 \text{ bar mol}^{-2}$ and $b = 0.03183 \text{ L mol}^{-1}$.

We need the conversion factor between m^3 and L. Since a liter is 10^3 cm^3 and a cubic meter is 100^3 cm^3, the conversion factor is 10^3 L m^{-3}.

The second virial coefficient B is given by

$B = b - a/RT = .03183/10^3 - 1.378/(.08315 \text{x} 298.15 \text{x} 10^3) = -0.0000238 \text{ m}^3 \text{ mol}^{-1}$

The expected value is $-0.0000161 \text{ m}^3 \text{ mol}^{-1}$.

The third virial coefficient is given by
$C = b^2 = (.03183/10^3)^2 = 1.01\text{x}10^{-9}\ \text{m}^6\ \text{mol}^{-2}$.

The expected value is $1.2\text{x}10^{-9}\ \text{m}^6\ \text{mol}^{-2}$.

1.16 Calculate the critical properties for ethane using the van der Waals constants in Table 1.3.

SOLUTION

The critical properties for molecular oxygen are given by

$\bar{V}_c = 3b = 3(.0635\ \text{L mol}^{-1}) = 0.1914\ \text{L mol}^{-1}$

$$T_c = \frac{8a}{27Rb} = \frac{8(5.562\ \text{L}^2\ \text{bar mol}^{-2})}{27(0.038\ \text{L mol}^{-1})(0.083145\ \text{l bar K}^{-1}\ \text{mol}^{-1})} = 310.671\ \text{K}$$

$$P_c = \frac{a}{27b^2} = \frac{5.562\ \text{L}^2\ \text{bar mol}^{-2}}{27(0.0638\ \text{L mol}^{-1})^2} = 50.609\ \text{bar}$$

1.17 The cubic expansion coefficient α is defined by

$$\alpha = \frac{1}{V}\left(\frac{\partial V}{\partial T}\right)_P$$

and the isothermal compressibility is defined by

$$\kappa = -\frac{1}{V}\left(\frac{\partial V}{\partial P}\right)_T.$$

Calculate these quantities for an ideal gas.

SOLUTION

$V = nRT/P$

$$\left(\frac{\partial V}{\partial T}\right)_P = \frac{nR}{P}$$

$$\alpha = \frac{1}{T}$$

$$\left(\frac{\partial V}{\partial P}\right)_T = -\frac{nRT}{P^2}$$

$$\kappa = \frac{nRT}{VP^2} = \frac{1}{P}$$

1.18 What is the equation of state for a liquid for which the coefficient of cubic expansion α and the isothermal compressibility κ are constant?

SOLUTION

$$dV = \left(\frac{\partial V}{\partial T}\right)_P dT + \left(\frac{\partial V}{\partial P}\right)_T dP$$

$$= \alpha V dT - \kappa V dP$$

$$\int_{V_1}^{V_2} \frac{dV}{V} = \alpha \int_{T_1}^{T_2} dT - \kappa \int_{P_1}^{P_2} dP$$

$$\ln(V_2/V_1) = \alpha(T_2 - T_1) - \kappa(P_2 - P_1)$$

The indefinite integral and its exponential form are

$\ln V = \alpha T - \kappa P + \text{const.}$

$V = C\, e^{\alpha T} e^{-\kappa P}$

where C is a constant.

1.19 For a liquid the cubic expansion coefficient α is nearly constant over a narrow range of temperature. Derive the expression for the volume as a function of temperature and the limiting form for temperatures close to T_0.

SOLUTION

$$\int_{V_0}^{V} \frac{dV}{V} = \alpha \int_{T_0}^{T} dT$$

$$\ln \frac{V}{V_0} = \alpha\,(T - T_0)$$

$$V = V_0\, e^{\alpha(T - T_0)}$$

If $\alpha(T - T_0) << 1$,

$V = V_0[1 + \alpha(T - T_0)]$

1.20 (a) Derive the expressions for $(\partial P/\partial V)_T$ and $(\partial P/\partial T)_V$ for a gas that has the following equation of state:

$$P = \frac{nRT}{V - nb}$$

(b) Show that $(\partial^2 P/\partial V \partial T) = (\partial^2 P/\partial T \partial V)$. These are referred to as mixed partial derivatives.

SOLUTION

(a) $$\left(\frac{\partial P}{\partial V}\right)_T = \frac{-\,nRT}{(V - nb)^2}$$

$$\left(\frac{\partial P}{\partial T}\right)_V = \frac{nR}{V - nb}$$

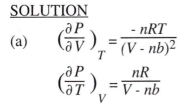

(b) $$\left(\frac{\partial^2 P}{\partial V \partial T}\right) = \frac{-nR}{(V - nb)^2}$$

$$\left(\frac{\partial^2 P}{\partial T \partial V}\right) = \frac{-nR}{(V - nb)^2}$$

1.21 Assuming that the atmosphere is isothermal at 0 °C and that the average molar mass of air is 29 g mol^{-1}, calculate the atmospheric pressure at 20,000 ft above sea level.

SOLUTION

$$h = (2.0 \text{ x } 10^4 \text{ ft})(12 \text{ in ft}^{-1})(2.54 \text{ cm in}^{-1})(10^{-2} \text{ m cm}^{-1})$$
$$= 6096 \text{ m}$$

$$P = P_o \, e^{-gMh/RT}$$

$$P = (1.013 \text{ bar}) \exp\left[\frac{-(9.8 \text{ m s}^{-2})(29 \text{ x } 10^{-3} \text{ kg mol}^{-1})(6096 \text{ m})}{(8.314 \text{ J K}^{-1} \text{ mol}^{-1})(273 \text{ K})}\right] = 0.472 \text{ bar}$$

1.22 Calculate the pressure and composition of air on the top of Mt. Everest assuming the atmosphere has a temperature of 0 °C independent of altitude (h = 29,141 ft). Assume that air at sea level is 20% O_2 and 80% N_2.

SOLUTION

$$h = (29{,}141 \text{ ft})(12 \text{ in ft}^{-1})(2.54 \text{ cm in}^{-1})(0.01 \text{ m cm}^{-1}) = 8{,}882 \text{ m}$$

For O_2, $P = (0.2 \text{ bar}) \exp(-9.8 \text{ x } 32 \text{ x } 10^{-3} \text{ x } 8882/(8.314 \text{ x } 273))$

$= 0.059$ bar

For N_2, $P = (0.8 \text{ bar}) \exp(-9.8 \text{ x } 28 \text{ x } 10^{-3} \text{ x } 8882/(8.314 \text{ x } 273))$

$= 0.274$ bar

The total pressure is 0.333 bar, $y_{O_2} = 0.177$ and $y_{N_2} = 0.823$.

1.23 (a) 98.1 kPa, (b) 9810 MPa, (c) 98.1 cm^2

1.24 (a) 0.02479 m^3, (b) 0.02560 m^3, (c) 0.7747, 0.1937, 0.0317

1.25 0.835

1.26 $B = y_1^2 B_{11} + 2y_1y_2 B_{12} + y_2^2 B_{22}$

1.27 276 $cm^3\ mol^{-1}$

1.28 - 1.96 x 10^{-3} bar^{-1}, - 0.60 x 10^{-4} bar^{-1}

1.29 $C = b^2$ and $D = b^3$

1.30 (a) 20.1 bar, (b) 1642 bar, (c) 2776 bar

1.31 99.8 bar, 152 bar

1.32 (a) 155 bar, (b) 88.6 bar

1.33 $V_2 = V_1 \exp[-\kappa(P_2 - P_1)]$

1.34 $\alpha = 1/T(1 + bP/RT)$

$\kappa = 1/P(1 + bP/RT)$

1.35 (a) $6.93\text{x}10^{-2}$ L mol^{-1}, (b) $8.62\text{x}10^{-2}$ L mol^{-1}.

1.36 0.843 bar

1.37 3.27×10^{-9} bar, $y_{O_2} = 0.015$, $y_{N_2} = 0.985$

1.38 $M = y_A M_A + y_B M_B$

2

First Law of Thermodynamics

2.1 How high can a person (assume a weight of 70 kg) climb on one ounce of chocolate, if the heat of combustion (628 kJ) can be converted completely into work of vertical displacement?

SOLUTION

$w = mgh$

$$h = \frac{w}{gm} = \frac{628\ 000\ \text{J}}{(9.806\ \text{m s}^{-2})(70\ \text{kg})} = 914\ \text{m}$$

2.2 A mole of sodium metal is added to water. How much work is done on the atmosphere by the subsequent reaction if the temperature is 25 °C?

SOLUTION

$$\text{Na(s)} + \text{H}_2\text{O(l)} = \text{NaOH(soln)} + \frac{1}{2}\ \text{H}_2\text{(g)}$$

$$\bar{V} = \frac{RT}{P} = \frac{(8.314\ \text{J K}^{-1}\ \text{mol}^{-1})(298\ \text{K})}{(101325\ \text{Pa})}$$
$$= 0.0245\ \text{m}^3\ \text{mol}^{-1}$$

$$w = -P\Delta\bar{V} = -(101\ 325\ \text{Pa})(0.0123\ \text{m}^3\ \text{mol}^{-1})$$
$$= -1.24\ \text{kJ mol}^{-1}$$

2.3 You want to heat 1 kg of water 10 °C, and you have the following four methods under consideration. The specific heat capacity of water is 4.184 J K^{-1} g^{-1}.

(a) You can heat it with a mechanical eggbeater that is powered by a 1 kg mass on a rope over a pulley. How far does the mass have to descend in the earth's gravitational field to supply enough work?

(b) You can send 1 ampere through a 100 Ω resistor. How long will it take?

(c) You can send the water through a solar collector that has an area of 1 m^2. How long will it take if the sun's intensity on the collector is 4 J cm^{-2} min^{-1}?

(d) You can make a charcoal fire. The heat of combustion of graphite is -393 kJ mol^{-1}. That is, 12 g of graphite will produce 393 kJ of heat when it is burned to CO_2(g) at constant pressure. How much charcoal will you have to burn?

SOLUTION:

(a) $(1000\text{ g})(4.184\text{ J K}^{-1}\text{ g}^{-1})(10\text{ K}) = 4.184 \times 10^4\text{ J}$

$mg\Delta h = (1\text{ kg})(9.8\text{ m s}^{-2})\Delta h$

$$\Delta h = \frac{4.184 \times 10^4}{9.8} = 4269\text{ m}$$

(b) $I^2Rt = 4.184 \times 10^4\text{ J} = (1\text{A})^2(100\ \Omega)t$

$$t = \frac{4.184 \times 10^4}{100} = 418.4\text{ s} = 6.97\text{ minutes}$$

(c) $4.184 \times 10^4\text{ J} = (4.0\text{ J cm}^{-2}\text{ min}^{-1})(100\text{ cm})^2\ t$

$$t = \frac{4.184 \times 10^4}{4 \times 10^4} = 1.05\text{ minutes}$$

(d) $4.184 \times 10^4\text{ J} = (393 \times 10^3\text{ J})\dfrac{m}{12}$

$$m = \frac{12(4.184 \times 10^4)}{393 \times 10^3} = 1.28\text{ grams}$$

2.4 Show that the differential df is inexact.

$df = dx - (x/y)dy$

Thus the integral $\int df$ depends on the path. However, we can define a new function g by

$dg = (1/y)df$

which has the property that dg is exact. Show that dg is exact, so that $\oint dg = 0$. In this case y is referred to as an integrating factor.

SOLUTION

To determine whether df is exact take the cross derivatives

$M(x,y) = 1;\ (\partial M/\partial y)_x = 0$

$N(x,y) = -x/y,\ \ (\partial N/\partial x)_y = -1/y$

Since $(\partial M/\partial y)_x \neq (\partial N/\partial x)_y$, df is inexact.

The new differential is given by

$dg = (1/y)\,dx - (x/y^2)dy$

Now take the cross derivatives.

$M = 1/y,\ (\partial M/\partial y)_x = -1/y^2$

$N = -x/y^2,\ (\partial N/\partial x)_y = -1/y^2$

Since $(\partial M/\partial y)_x = (\partial M/\partial x)_y$, dg is exact.

2.5 Show that the function $f(x,y)$ defined by

$df(x,y) = (x + 2y)dx - xdy$

is inexact. Test to see whether the integrating factor $1/x^3$ makes it an exact differential.

SOLUTION

The function is exact if

$$\frac{\partial M}{\partial y} = \frac{\partial N}{\partial x}$$

Taking the two mixed partial derivatives yields

$$\left[\frac{\partial (x + 2y)}{\partial y}\right]_x = 2$$

$$\left[\frac{\partial (-x)}{\partial x}\right]_y = -1$$

showing that the function $f(x,y)$ is inexact.

Multiplying by the integrating factor yields

$$\frac{df(x,y)}{x^3} = \frac{x + 2y}{x^3}\,dx - \frac{1}{x^2}\,dy$$

Taking the two mixed partial derivatives yields

$$\left[\frac{\partial \left(\frac{x + 2y}{x^3}\right)}{\partial y}\right]_x = \frac{2}{x^3}$$

$$\left[\frac{\partial \left(\frac{-1}{x^2}\right)}{\partial x}\right]_y = \frac{2}{x^3}$$

Since the mixed partial derivatives are equal, $f(x,y)/x^3$ is exact.

2.6 Show that the function defined by

$df(x,y) = (y^2 - xy)dx - x^2dy$

is inexact. Test the integrating factor $1/xy^2$ to see whether it produces an exact differential.

SOLUTION

The function is exact if

$$\frac{\partial M}{\partial y} = \frac{\partial N}{\partial x}$$

Taking the two mixed partial derivatives yields

$$\left[\frac{\partial (y^2 - xy)}{\partial y}\right]_x = 2y - x$$

$$\left[\frac{\partial (-x^2)}{\partial x}\right]_y = -2x$$

showing that the function $f(x,y)$ is inexact.

Multiplying by the integrating factor yields

$$\frac{df(x,y)}{xy^2} = \left(\frac{1}{x} - \frac{1}{y}\right)dx - \frac{x}{y^2}\,dy$$

Taking the partial derivatives yields

$$\left[\frac{\partial\left(\frac{1}{x}-\frac{1}{y}\right)}{\partial y}\right]_x = \frac{1}{y^2}$$

$$\left[\frac{\partial\left(\frac{x}{y^2}\right)}{\partial x}\right]_y = \frac{1}{y^2}$$

Since the mixed partial derivatives are equal, $f(x,y)/xy^2$ is exact.

2.7 What are the partial derivatives $(\partial z/\partial x)_y$ and $(\partial z/\partial y)_x$ of the following functions.

(a) z = xy, (b) z = x/y, (c) z = ln(xy), (d) z = ln(x/y), and (e) z = exp(xy).

SOLUTION

(a) $\left(\frac{\partial z}{\partial x}\right)_y = y;\quad \left(\frac{\partial z}{\partial y}\right)_x = x$

(b) $\left(\frac{\partial z}{\partial x}\right)_y = \frac{1}{y};\quad \left(\frac{\partial z}{\partial y}\right)_x = \frac{-x}{y^2}$

(c) $z = \ln x + \ln y;\ \left(\frac{\partial z}{\partial x}\right)_y = \frac{1}{x};\quad \left(\frac{\partial z}{\partial y}\right)_x = \frac{1}{y}$

(d) $z = \ln x - \ln y;\ \left(\frac{\partial z}{\partial x}\right)_y = \frac{1}{x};\quad \left(\frac{\partial z}{\partial y}\right)_x = \frac{-1}{y}$

(e) $\left(\frac{\partial z}{\partial x}\right)_y = ye^{xy};\ \left(\frac{\partial z}{\partial y}\right)_x = xe^{xy}$

2.8 (a) The surface tension of water at 25 °C is 0.072 N m^{-1}. How much work is required to form a surface 100 m by 100 m? (b) The force on a wire is due to a 75 kg person in the earth's gravitational field. If the wire stretches 1 m, how much work is done on the wire? (c) A gas expands 1 L against a constant external pressure of 1 bar. How much work is done on the gas?

SOLUTION

(a) $w = (0.072\ \text{N m}^{-1})(100\ \text{m})^2 = 720\ \text{J}$
(b) $w = (9.8\ \text{m s}^{-2})(75\ \text{kg})(1\ \text{m}) = 735\ \text{J}$

(c) $w = -P\Delta V = -(1\ \text{bar})(1\ \text{L}) = (10^5\ \text{Pa})(10^{-3}\ \text{m}^3) = -100\ \text{Pa m}^3 = -100\ \text{J}$

2.9 Over narrow ranges of temperature and pressure, the differential expression for the volume of a fluid as a function of temperature and pressure can be integrated to obtain

$V = K\,e^{\alpha T}\,e^{-\kappa P}$

(α and κ are defined in problem 1.17). Show that V is a state function.

SOLUTION

The test of a state function is that it forms exact differentials. For an exact differential the mixed second derivatives are equal.

$$\left(\frac{\partial V}{\partial T}\right)_P = K\,e^{-\kappa P}\,e^{\alpha T}\,\alpha$$

$$\left[\frac{\partial}{\partial P}\left(\frac{\partial V}{\partial T}\right)_P\right]_T = K\,e^{\alpha T}\,e^{-\kappa P}\,(-\kappa\alpha)$$

$$\left(\frac{\partial V}{\partial P}\right)_T = K\,e^{\alpha T}\,e^{-\kappa P}\,(-\kappa)$$

$$\left[\frac{\partial}{\partial T}\left(\frac{\partial V}{\partial P}\right)_T\right]_P = K\,e^{\alpha T}\,e^{-\kappa P}\,(-\kappa\alpha)$$

Thus V for a substance that can be described in this way is a state function.

2.10 One mole of nitrogen at 25 °C and 1 bar is expanded reversibly and isothermally to a pressure of 0.132 bar. (a) How much work w is done on the gas? (b) How much work is done on the gas if it is expanded against a constant pressure of 0.132 bar?

SOLUTION

(a) $w = RT\ln(P_2/P_1)$

$= (8.314\text{ J K}^{-1})(298.15\text{ K})\ln 0.132$

$= -5.03\text{ kJ mol}^{-1}$

(b) $\bar{V}_1 = RT/P_1 = \dfrac{(8.314)(298.15)}{10^5} = 0.0248\text{ m}^3\text{ mol}^{-1}$

$\bar{V}_2 = \dfrac{(8.314)(298.15)}{(0.132)(10^5)} = 0.188\text{ m}^3\text{ mol}^{-1}$

$w = -P\Delta\bar{V} = -(0.132 \times 10^5)(0.188 - 0.0248)$

$= -2.15\text{ kJ mol}^{-1}$

2.11 (a) Derive the equation for the work of reversible isothermal expansion of a van der Waals gas from V_1 to V_2. (b) A mole of CH_4 is expanded reversibly from 1 L to 50 L at 25 °C. Calculate the work in joules assuming (1) the gas is ideal, (2) the gas obeys the van der Waals equation. For CH_4 (g), $a = 2.283\text{ L}^2\text{ bar mol}^{-2}$, $b = 0.04278\text{ L mol}^{-1}$.

SOLUTION

(a) $w = -\int_{V_1}^{V_2} P\mathrm{d}V = -\int_{V_1}^{V_2} \frac{nRT\mathrm{d}V}{V-nb} + \int_{V_1}^{V_2} \frac{an^2\mathrm{d}V}{V^2}$

$= -nRT \ln \frac{V_2 - nb}{V_1 - nb} + an^2(\frac{1}{V_1} - \frac{1}{V_2})$

(b) $w = -nRT \ln \frac{V_2}{V_1}$

$= -(1\ \text{mol})(8.314\ \text{J K}^{-1}\ \text{mol}^{-1})(298\ \text{K}) \ln 50$

$= -9697\ \text{J}$

We have to be careful about units in adding these two terms. If we use $R = 8.314$ J K^{-1} mol^{-1} in the first term it will come out in J, as for the ideal gas. The second will come out in L bar. A convenient way to convert L bar to J is to note that

$$\frac{8.314\ \text{J K}^{-1}\ \text{mol}^{-1}}{0.08314\ \text{L bar K}^{-1}\ \text{mol}^{-1}} = 100\ \text{J/L bar}$$

$$w = (1\ \text{mol})(8.314\ \text{J K}^{-1}\ \text{mol}^{-1})(298\ \text{K}) \ln \frac{50 - 0.04278}{1 - 0.04278}$$

$$+ (100\ \text{J/L bar})(2.283\ \text{L}^2\ \text{bar mol}^{-2})(1\ \text{mol})^2\left(\frac{1}{1\ \text{L}} - \frac{1}{50\ \text{L}}\right)$$

$$= -9799\ \text{J} + 224\ \text{J} = -9575\ \text{J}$$

This is the work done on the gas. The work done on the surroundings is +9575 J. Less work is done by methane than an ideal gas because of intermolecular attractions.

2.12 Liquid water is vaporized at 100 °C and 1.013 bar. The heat of vaporization is 40.69 kJ mol^{-1}. What are the values of (a) w_{rev} per mole, (b) q per mole, (c) $\Delta\bar{U}$, and (d) $\Delta\bar{H}$.

SOLUTION

(a) Assuming that water vapor is an ideal gas and that the volume of liquid water is negligible,

$w = -P\Delta V = (8.314 \times 10^{-3}\ \text{kJ mol}^{-1})(273.15\ \text{K})$

$= -3.10\ \text{kJ mol}^{-1}$

(b) The heat of vaporization is 40.69 kJ mol^{-1}, and, since heat is absorbed, q has a positive sign.

$q = 40.69\ \text{kJ mol}^{-1}$

(c) $U = q + w$

$= (40.69 - 3.10)\ \text{kJ mol}^{-1}$

$= 37.59\ \text{kJ mol}^{-1}$

(d) $\Delta H = \Delta U + \Delta(PV) = \Delta U + P\Delta V$

$= \Delta U + RT$

$= 37.59 \text{ kJ mol}^{-1} + (8.314 \times 10^{-3} \text{ kJ K}^{-1} \text{ mol}^{-1})(373.15 \text{ K})$
$= 40.69 \text{ kJ mol}^{-1}$

2.13 An ideal gas is expanded reversibly and isothermally from 10 bar to 1 bar at 298.15 K. What are the values of (a) w per mole, (b) q per mole, (c) $\Delta \bar{U}$, and (d) $\Delta \bar{H}$. (e) If the ideal gas expands isothermally against a constant pressure of 1 bar, how much work is done on the gas?

SOLUTION

(a) $w = RT \ln(P_2/P_1)$
$= (8.314)(298.15) \ln (1/10)$
$= -5.70 \text{ kJ mol}^{-1}$

(b) $q = -w = 5.70 \text{ kJ mol}^{-1}$

(c) $\Delta \bar{U} = 0$ since an ideal gas

(d) $\Delta \bar{H} = \Delta \bar{U} + \Delta P\bar{V} = 0$

(e) $w = -P\Delta V = -P(V_f - V_i)$, where $V_i = RT/(10 \text{ bar})$, $V_f = RT/(1 \text{ bar})$

$$w = -(10^5 \text{ bar})(8.314 \text{ Pa m}^3 \text{ K}^{-1} \text{ mol}^{-1})(298 \text{ K})\left(\frac{1}{10^5 \text{ Pa}} - \frac{1}{10^6 \text{ Pa}}\right)$$
$= -2.23 \text{ J mol}^{-1}$

2.14 Calculate $\bar{H}^\circ(2000 \text{ K}) - \bar{H}^\circ(0 \text{ K})$ for H(g).

SOLUTION

$$\bar{H}^\circ(T_2) - \bar{H}^\circ(T_1) = \int_{T_1}^{T_2} \bar{C}_P^\circ \, dT$$

$$\bar{H}^\circ(2000 \text{ K}) - \bar{H}^\circ(0 \text{ K}) = \int_0^{2000} \frac{5}{2} R \, dT = \frac{5}{2} R(2000) = 41.572 \text{ kJ mol}^{-1}$$

Appendix C.3 yields $6.197 + 35.376 = 41.573 \text{ kJ mol}^{-1}$. (Note that for O(g) a slightly higher value is obtained because there is some absorption of heat by excitation to higher electronic levels.)

*2.15 The heat capacity of a gas may be represented by

$$\bar{C}_P = \alpha + \beta T + \gamma T^2$$

For N_2 $\alpha = 26.984 \text{ J K}^{-1} \text{ mol}^{-1}$, $\beta = 5.910 \times 10^{-3} \text{ J K}^{-2} \text{ mol}^{-1}$, and $\gamma = -3.377 \times 10^{-7} \text{ J K}^{-3} \text{ mol}^{-1}$. How much heat is required to heat a mole of N_2 from 300 K to 1000 K?

SOLUTION

$$q = \int_{T_1}^{T_2} (26.984 + 5.910 \times 10^{-3}\,T - 3.377 \times 10^{-7}\,T^2)\,dT$$

$$= 26.984(1000 - 300) + \frac{1}{2}(5.910 \times 10^{-3})(1000^2 - 300^2)$$

$$- \frac{1}{3}(3.377 \times 10^{-7})(1000^3 - 300^3)$$

$$= 21.468 \text{ kJ mol}^{-1}$$

2.16 (a) In a reversible adiabatic expansion of an ideal gas with $\gamma = C_P/C_V$ independent of temperature, the pressure and volume are related by
PV^γ = constant
Show that the work of adiabatic expansion from P_1,V_1 to P_2,V_2 is
$w = (P_2V_2 - P_1V_1)/(\gamma - 1)$
(b) Check this equation to be sure it gives the same amount of work as Example 2.17.

SOLUTION

(a)

$$w = -\int_{V_1}^{V_2} P\mathrm{d}V = -\int_{V_1}^{V_2} \frac{\text{const}}{V^\gamma}\mathrm{d}V = -\,\text{const}\,\frac{V_2^{1-\gamma} - V_1^{1-\gamma}}{1-\gamma}$$

$$= \frac{P_2V_2 - P_1V_1}{\gamma - 1}$$

(b) Example 2.17 for a gas with $\gamma = 5/3$ gives $V_1 = 22.7$ L, $V_2 = 45.4$ L, $P_1 = 1$ bar, and $P_2 = 0.315$ bar.

$w = ((45.4\text{x}0.315 - 22.7)/(2/3))(8.3145/0.083145) = -1260 \text{ J mol}^{-1}$

A ratio of gas constants is used to convert from energy units of L bar to joules.

2.17 Calculate the temperature increase and final pressure of helium if a mole is compressed adiabatically and reversibly from 44.8 L at 0 °C to 22.4 L.

SOLUTION

$$\gamma = \frac{\bar{C}_P}{\bar{C}_V} = \frac{(\frac{5}{2}R)}{(\frac{3}{2}R)} = \frac{5}{3}$$

$$\frac{T_1}{T_2} = \left(\frac{V_2}{V_1}\right)^{\gamma-1}$$

$T_2 = (273.25 \text{ K})(44.8 \text{ L}/22.4 \text{ L})^{2/3}$

$= 433.6 \text{ K or } 160.4\ °\text{C}$

Thus the temperature increase is 160.4 °C. The final pressure is given by

$$P = RT/\bar{V} = \frac{(0.08314 \text{ L bar K}^{-1} \text{ mol}^{-1})(433.6 \text{ K})}{22.4 \text{ L mol}^{-1}}$$

2.18 A mole of argon is allowed to expand adiabatically and reversibly from a pressure of 10 bar and 298.15 K to 1 bar. What is the final temperature, and how much work is done on the argon?

SOLUTION

$$\gamma = \bar{C}_P/\bar{C}_V = \left(\frac{5}{2} R\right) / \left(\frac{3}{2} R\right) = \frac{5}{3} \qquad (\gamma - 1)/\gamma = 2/5$$

$$\frac{T_1}{T_2} = \left(\frac{P_1}{P_2}\right)^{(\gamma - 1)/\gamma}$$

$T_2 = (298.15 \text{ K})(1/10)^{2/5} = 118.70 \text{ K}$

$$w = \int_{T_1}^{T_2} \bar{C}\, dT = \frac{3}{2} R(T_2 - T_1)$$

$$= \frac{3}{2}(8.314 \text{ J K}^{-1} \text{ mol}^{-1})(118.70 \text{ K} - 298.15 \text{ K}) = -2238 \text{ J mol}^{-1}$$

The maximum work that can be done on the surroundings is 2238 J mol^{-1}.

2.19 A tank contains 20 liters of compressed nitrogen at 10 bar and 25 °C. Calculate w when the gas is allowed to expand reversibly to 1 bar pressure (a) isothermally and (b) adiabatically.

SOLUTION

(a) For the isothermal expansion of 1 mol

$$w_{\text{rev}} = RT \ln \frac{P_2}{P_1}$$

$$= (8.314 \text{ J K}^{-1} \text{ mol}^{-1})(298.15 \text{ K}) \ln (1/10)$$

$$= -5708 \text{ J mol}^{-1}$$

There are

$(10 \text{ bar})(20 \text{ L})/(0.08314 \text{ L bar K}^{-1} \text{ mol}^{-1})(298 \text{ K}) = 8.07 \text{ mol}$. Therefore $w = -46.1$ kJ.

(b) For the adiabatic expansion we will assume that $\gamma = \bar{C}_P/\bar{C}_V$ has the value it has at room temperature. From Table C.3.

$$\gamma = 29.125/(29.125 - 8.314) = 1.399$$

$$\frac{T_1}{T_2} = \left(\frac{P_1}{P_2}\right)^{(\gamma - 1)/\gamma}$$

$$T_2 = (298.15 \text{ K})(1/10)^{0.285} = 154.7 \text{ K}$$

$$w = \int_{T_1}^{T_2} \bar{C}_V \, dT = \bar{C}_V(T_2 - T_1)$$

$$= (20.811 \text{ J K}^{-1} \text{ mol}^{-1})(154.7 \text{ K} - 298.15 \text{ K})$$

$$= -2.99 \text{ kJ mol}^{-1}$$

For 8.07 mol, $w = -24.1$ kJ.

2.20 An ideal monatomic gas at 298.15 K and 1 bar is expanded in a reversible adiabatic process to a final pressure of 1/2 bar. Calculate q per mole, w per mole, and $\Delta \bar{U}$.

SOLUTION

$$\frac{T_1}{T_2} = \left(\frac{P_1}{P_2}\right)^{(\gamma - 1)/\gamma}$$

$\bar{C}_V = \frac{3}{2}R$ $\bar{C}_P = \frac{5}{2}R$ $\gamma = \frac{5}{3}$

$$\frac{298.15}{T_2} = 2^{0.4}$$

$$T_2 = \frac{298.15}{2^{0.4}} = 226.0 \text{ K}$$

$q = 0$

$$w = \bar{C}_V(T_2 - T_1) = \frac{3}{2}R(226.0 - 298.15)$$

$$= -899.8 \text{ J mol}^{-1}$$

$$\Delta\bar{U} = q + w = -899.8 \text{ J mol}^{-1}$$

2.21 An ideal monatomic gas at 1 bar and 300 K is expanded adiabatically against a constant pressure of 1/2 bar until the final pressure is 1/2 bar. What are the values of q per mole, w per mole, $\Delta\bar{U}$, and $\Delta\bar{H}$? Given: $\bar{C}_V = (3/2)R$.

SOLUTION

For an adiabatic expression at constant pressure

$$\bar{C}_V \Delta T = -P_{op} \Delta\bar{V}$$

$$[(3/2)R](T_2 - 300) = -(\tfrac{1}{2}\text{ bar})(\bar{V}_2 - 300R/1\text{ bar})$$

$$= (-\tfrac{1}{2}\text{ bar})\left(\frac{RT_2}{0.5\text{ bar}} - \frac{300R}{1\text{ bar}}\right)$$

$T_2 = 240$ K

The work done on the gas is $w = \bar{C}_V\Delta T = (3/2)(8.314\text{ J K}^{-1}\text{ mol}^{-1})(-60\text{ K}) = -748\text{ J mol}^{-1}$. The value of q is zero since the process is adiabatic.

$$\Delta\bar{U} = q + w = -748\text{ J mol}^{-1}$$

$$\Delta\bar{H} = \int_{T_1}^{T_2} (5/2)\,R\,\mathrm{d}T = (5/2)(8.314)(-60) = -1247\text{ J mol}^{-1}$$

2.22 Derive the equation for calculating the work involved in a reversible, adiabatic pressure change of one mole of an ideal gas so that the work can be calculated from the initial temperature T_1, initial pressure P_1, and final pressure P_2.

SOLUTION

The first law shows that the differential of the work involved is given by $\mathrm{d}w = \bar{C}_V\mathrm{d}T$. Integrating yields

$$w = \bar{C}_V(T_2 - T_1) = \bar{C}_V T_1\left(\frac{T_2}{T_1} - 1\right)$$

Equation 2.79 shows that

$$\frac{T_2}{T_1} = \left(\frac{P_2}{P_1}\right)^{R/\bar{C}_P}$$

Substituting this expression into the equation for work yields

$$w = \bar{C}_V T_1\left[\left(\frac{P_2}{P_1}\right)^{R/\bar{C}_P} - 1\right]$$

2.23 Calculate $\Delta_r H^{\circ}_{298}$ for

$H_2(g) + F_2(g) = 2HF(g)$
$H_2(g) + Cl_2(g) = 2HCl(g)$
$H_2(g) + Br_2(g) = 2HBr(g)$
$H_2(g) + I_2(g) = 2HI(g)$

SOLUTION

(a) $2(-271.1) = -542.2\text{ kJ mol}^{-1}$

(b) $2(-92.31) = -184.62\text{ kJ mol}^{-1}$

(c) $2(-36.40) - 30.91 = -103.71\text{ kJ mol}^{-1}$

(d) $2(26.48) - 62.44 = -9.48\text{ kJ mol}^{-1}$

2.24 The following reactions might be used to power rockets.

(1) $H_2(g) + \frac{1}{2} O_2(g) = H_2O(g)$

(2) $CH_3OH(l) + 1\frac{1}{2} O_2(g) = CO_2(g) + 2H_2O(g)$

(3) $H_2(g) + F_2(g) = 2HF(g)$

(a) Calculate the enthalpy changes at 25 °C for each of these reactions per kilogram of reactants.

(b) Since the thrust is greater when the molar mass of the exhaust gas is lower, divide the heat per kilogram by the molar mass of the product (or the mole-fraction average molar mass in the case or reaction (2) and arrange the above reactions in order of effectiveness on the basis of thrust.

SOLUTION

(a) (1) $\Delta_r H = -\ 241.818\ \text{kJ mol}^{-1}$

$$= \frac{(-\ 241.818\ \text{kJ mol}^{-1})(100\ \text{g kg}^{-1})}{(18\ \text{g mol}^{-1})}$$

$$= -\ 13.4\ \text{MJ kg}^{-1}$$

(2) $\Delta_r H = -\ 393.509 + 2(-241.818) + 238.66$

$$= -\ 638.49\ \text{kJ mol}^{-1}$$

$$= \frac{(-\ 638.49\ \text{kJ mol}^{-1})(1000\ \text{g kg}^{-1})}{(80\ \text{g mol}^{-1})}$$

$$= -\ 7.98\ \text{MJ kg}^{-1}$$

(3) $\Delta_r H = 2(-\ 271.1) = -542.2\ \text{kJ mol}^{-1}$

$$= \frac{(-\ 542.2\ \text{kJmol}^{-1})(100\ \text{g kg}^{-1})}{(40\ \text{g mol}^{-1})} = -\ 13{,}600\ \text{kJ kg}^{-1}$$

(b) (1) $-\ 13.4/18 = -\ 0.744$

(2) $$\frac{-\ 7.98}{\frac{1}{3}(44 + 2 \times 18)} = -\ 0.299$$

(3) $-\ 13.6/20 = -\ 0.680$

(1) > (3) > (2)

2.25 Calculate $\Delta_r H^o$ for the dissociation
$O_2(g) = 2O(g)$
at 0, 298, and 3000 K. In Section 13.6 the enthalpy change for dissociation at 0 K will be found to be equal to the spectroscopic dissociation energy D_0.

SOLUTION

$\Delta_r H^o(0\ K) = 2(246.790) = 493.580\ kJ\ mol^{-1}$
$\Delta_r H^o(298\ K) = 2(249.173) = 498.346\ kJ\ mol^{-1}$
$\Delta_r H^o(3000\ K) = 2(256.741) = 513.482\ kJ\ mol^{-1}$

The spectroscopic dissociation energy of O_2 is given as 5.115 eV in Table 13.4. this can be converted to kJ mol^{-1} by multiplying by 96.485 kJ V^{-1} mol^{-1} to obtain 493.521 kJ mol^{-1}.

2.26 Methane may be produced from coal in a process represented by the following steps, where coal is approximated by graphite:
$2C(s) + 2H_2O(g) = 2CO(g) + 2H_2(g)$
$CO(g) + H_2O(g) = CO_2(g) + H_2(g)$
$CO(g) + 3H_2(g) = CH_4(g) + H_2O(g)$
the sum of the three reactions is
$2C(s) + 2H_2O(g) = CH_4(g) + CO_2(g)$
What is $\Delta_r H^o$ at 500 K for each of these reactions? Check that the sum of the $\Delta_r H^o$'s of the first three reactions is equal to $\Delta_r H^o$ for the fourth reaction. From the standpoint of heat balance would it be better to develop a process to carry out the overall reactions in three separate reactors, or in a single reactor?

SOLUTION

$$\Delta_r H^o_{500} = 2(-110.00) - 2(-243.83) = 267.66\ kJ\ mol^{-1}$$

$$\Delta_r H^o_{500} = -393.67 - (-110.00) - (243.83) = -39.84\ kJ\ mol^{-1}$$

$$\Delta_r H^o_{500} = -80.82 - 243.83 - (-110.00) = -214.65\ kJ\ mol^{-1}$$

$$\Delta_r H^o_{500} = -80.82 - 393.67 - 2(-243.83) = 13.17\ kJ\ mol^{-1}$$

Since the first reaction is very endothermic, there is an advantage in carrying the subsequent reactions out in the same reactor so that they can provide heat.

2.27 What is the heat of freezing water at -10 °C given that
$H_2O(1) = H_2O(s)$ $\qquad \Delta H^o\ (273\ K) = -6004\ J\ mol^{-1}$
Given: $\bar{C}_P(H_2O,1) = 75.3\ J\ K^{-1}\ mol^{-1}$ and $\bar{C}_P(H_2O,s) = 36.8\ J\ K^{-1}\ mol^{-1}$

SOLUTION

$$\Delta H^o = \Delta H^o(273\ K) + [\bar{C}_{P,H_2O(cr)} - \bar{C}_{P,H_2O(1)}] \times (263\ K - 273\ K)$$
$$= -6004\ J\ mol^{-1} + (-38.5\ J\ K^{-1}\ mol^{-1})(-10\ K)$$
$$= -5619\ J\ mol^{-1}$$

2.28 What is the enthalpy change for the vaporization of water at 0 °C? This value may be estimated from Appendix C.2 by assuming that the heat capacities of $H_2O(l)$ and $H_2O(g)$ are independent of temperature from 0 to 25 °C.

SOLUTION

$H_2O(l) = H_2O(g)$

$\Delta_{vap}H^o(298\ K) = -241.818 - (-385.830) = 44.011\ kJ\ mol^{-1}$

$\Delta_{vap}H_2^o = \Delta_{vap}H_1^o + \Delta_{vap}C_P(T_2 - T_1)$

$$\Delta_{vap}H^o(273\ K) = \Delta_{vap}H^o(298\ K) + [\bar{C}_P(H_2O,g) - \bar{C}_P(H_2O,l)](273 - 298)$$
$$= 44{,}011 + (33.577 - 75.291)(-25)$$
$$= 45{,}054\ J\ mol^{-1}$$

2.29 Calculate the standard enthalpy of formation of methane at 1000 K from the value at 298.15 K using the $\bar{H}^o - \bar{H}^o_{298}$ data in Appendix C.3.

SOLUTION

$$\begin{array}{lccccc} 1000\ K: & C(graphite) & + & 2H_2(g) & = & CH_4(g) \\ & \downarrow (\bar{H}^o_{1000} - \bar{H}^o_{298})_C & & \downarrow 2(\bar{H}^o_{1000} - \bar{H}^o_{298})_{H_2} & & \uparrow (\bar{H}^o_{1000} - \bar{H}^o_{298})_{CH_4} \\ 298\ K: & C(graphite) & + & 2H_2(g) & = & CH_4(g) \end{array}$$

The change in state in the reaction at 1000 K can be accomplished by cooling one mole of graphite and two moles of hydrogen to 298 K, converting them to methane, and then heating the methane to 1000 K.

$$\Delta_f H^o_{1000} = -(\bar{H}^o_{1000} - \bar{H}^o_{298})_C - 2(\bar{H}^o_{1000} - \bar{H}^o_{298})_{H_2} + \Delta_f\bar{H}^o(CH_4, 298\ K)$$
$$+ (\bar{H}^o_{1000} - \bar{H}^o_{298})_{CH_4}$$

$$= -11.795 - 2(20.680) - 74.873 + 38.179$$
$$= -89.849\ kJ\ mol^{-1}$$

2.30 For a diatomic molecule the bond energy is equal to the change in internal energy for the reaction

$X_2(g) = 2X(g)$

at 0 K. Of course, the change in internal energy and the change in enthalpy are the same at 0 K. Calculate the enthalpy of dissociation of $O_2(g)$ at 0 K. The enthalpy of formation of O(g) at 298.15 K is 249.173 kJ mol^{-1}. In the range 0-298 K the average value of the heat capacity of $O_2(g)$ is 29.1 J K^{-1} mol^{-1} and the average value of the heat capacity of O(g) is 22.7 J K^{-1} mol^{-1}. What is the value of the bond energy in electron volts? (When the changes in heat capacities in the range 0-298 K are taken into account, the enthalpy of dissociation at 0 K is 493.58 kJ mol^{-1}.)

SOLUTION

$$\Delta_r H_0^o = \Delta_r H_{298}^o - \Delta_r C_P(-298.15)$$
$$= 2(249.173) - (2 \times 22.7 - 29.1)(-298.15) \times 10^{-3}$$
$$= \Delta U_0^o = 493.486 \text{ kJ mol}^{-1}$$

$$\text{Bond energy for } O_2(g) = \frac{493486 \text{ J mol}^{-1}}{96485 \text{ C mol}^{-1}} = 5.115 \text{ eV}$$

2.31 One gram of liquid benzene is burned in a bomb calorimeter. The temperature before ignition was 20.826 ° C and the temperature after the combustion was 25.000 °C. This was an adiabatic calorimeter. The heat capacity of the bomb, the water around it, and the contents of the bomb after the combustion was 10,000 J K^{-1}. Calculate $\Delta_f H^o$ for $C_6H_6(1)$ at 298.15 K from these data. Assume that the water produced in the combustion is in the liquid state and the carbon dioxide produced in the combustion is in the gas state.

SOLUTION

$$C_6H_6(l) + 7\tfrac{1}{2}\, O_2(g) = 6CO_2(g) + 3H_2O(l)$$

$$\Delta U_{T_1} = -\int_{T_1}^{T_2} [C_V(R) + C_V(P)]dT$$
$$= -(10{,}000 \text{ J K}^{-1})(4.174 \text{ K})$$
$$= -41.74 \text{ kJ for 1 gram } C_6H_6$$
$$= -41.74 \text{ kJ/g} \times 78 \text{ g/mol} = -3255.7 \text{ kJ mol}^{-1}$$

$$q_P = q_V + (n_P - n_R)RT = q_v - 1.5\,RT$$

$$\Delta_c H(298) = -3255.7 - 1.5\,(8.314 \times 10^{-3})(298)$$
$$= -3259.4 \text{ kJ mol}^{-1}$$

$$\Delta_c H^o = 6\Delta_f H^o(CO_2) + 3\Delta_f H^o(H_2O,l) - \Delta_f H^o(C_6H_6,l)$$

$$-3259.4 = 6(-393.51) + 3(-285.83) - \Delta_f H^o(C_6H_6,l)$$

$$\Delta_f H^o = 41.4 \text{ kJ mol}^{-1} \text{ for } C_6H_6(l)$$

2.32 An aqueous solution of unoxygenated hemoglobin containing 5 g of protein (M = 64,000 g mol^{-1}) in 100 cm^3 of solution is placed in an insulated vessel. When enough molecular oxygen is added to the solution to completely saturate the hemoglobin, the temperature rises 0.031 °C. Each mole of hemoglobin binds 4 moles of oxygen. What is the enthalpy of reaction per mole of oxygen bound? The heat capacity of the solution may be assumed to be 4.18 J K^{-1} cm^{-3}.

SOLUTION

Oxygenated Hb(T_2)

$\Delta H = 0$ ↗ ↑ $\Delta H = C_P\Delta T$

Unoxygenated Hb(T_1) + O_2(T_1) → Oxygenated Hb(T_1)

$\Delta H(T_1)$

$$\Delta H(T_1) + C_P\Delta T = 0$$
$$\Delta H(T_1) = -(100\ \text{cm}^3)(4.18\ \text{J K}^{-1}\ \text{cm}^{-3})(0.031\ \text{K})$$
$$= -13.0\ \text{J}$$

$$\Delta_r H \text{ per mole of oxygen bound} = \frac{-13.0\ \text{J}}{(4 \times 5\ \text{g})/(64\,000\ \text{g mol}^{-1})}$$
$$= -41.6\ \text{kJ mol}^{-1}$$

2.33 Calculate the heat of hydration of Na_2SO_4(s) from the integral heats of solution of Na_2SO_4(s) and $Na_2SO_4 \cdot 10\ H_2O$(s) in infinite amounts of H_2O, which are -2.34 kJ mol^{-1} and 78.87 kJ mol^{-1}, respectively. Enthalpies of hydration cannot be measured directly because of the slowness of the phase transition.

SOLUTION

$Na_2SO_4(s) = Na_2SO_4(ai)$	$\Delta_r H^o = -2.43$ kJ mol^{-1}
$Na_2SO_4(ai) = Na_2SO_4 \cdot 10H_2O(s)$	$\Delta_r H^o = -78.87$ kJ mol^{-1}
$Na_2SO_4(s) + 10\ H_2O(1) = Na_2SO_4 \cdot 10\ H_2O(s)$	$\Delta_r H^o = -81.21$ kJ mol^{-1}

2.34 We want to determine the enthalpy of hydration of $CaCl_2$ to form $CaCl_2 \cdot 6H_2O$.
$CaCl_2(s) + 6H_2O(1) = CaCl_2 \cdot 6H_2O(s)$
We cannot do this directly for a couple of reasons: (1) reactions in the solid state are slow, and (2) there is a series of hydrates and so a mixture of different hydrates would probably be obtained. We can, however, determine the heats of solution of $CaCl_2$(s) and $CaCl_2 \cdot 6H_2O$(s) in water at 298 K and take the difference. The experimental heats of solution are as follows:
$CaCl_2(s) + Aq = CaCl_2(ai)$ $\Delta_r H = -81.33$ kJ mol^{-1}
$CaCl_2 \cdot 6H_2O(s) + Aq = CaCl_2(ai) + 6H_2O(1)$ $\Delta_r H = 15.79$ kJ mol^{-1}
What is the enthalpy of hydration of $CaCl_2$ to form $CaCl_2 \cdot 6H_2O$?

SOLUTION

Reverse the second reaction and add:
$CaCl_2(s) + 6H_2O(l) = CaCl_2 \cdot 6H_2O(s)$
$\Delta_r H = -81.33 - 15.79 = -97.12$ kJ mol^{-1}

2.35 The change in internal energy $\Delta_c U^o$ in the combustion of C_{60}(s) is -25 968 kJ mol^{-1} at 298.15 K (Kolesov, et al., J. Chem. Thermo., 28, 1121 (1996)). (a) What is the enthalpy of combustion $\Delta_c H^o$? (b) What is the enthalpy of formation

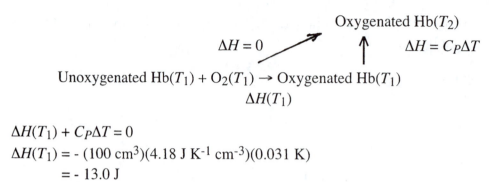

$\Delta_f H^o(C_{60}(s))$? (c) What is the enthalpy of vaporization of $C_{60}(s)$ to C(g) per mole of C(g)? (d) How does this compare with the enthalpy of vaporization of graphite and diamond to C(g)?

SOLUTION

(a) $C_{60}(s) + 60O_2(g) = 60CO_2(g)$

$\Delta_c H^o = \Delta_c U^o + \Delta n_g RT$

where n_g is the increase in the number of molecules of gas in the combustion reaction. Since $\Delta n_g = 0$, $\Delta_c H^o = -25\ 968$ kJ mol^{-1}.

(b) $\Delta_c H^o(C_{60},s) = 60\Delta_f H^o(CO_2,g) - \Delta_f H^o(C_{60},s)$

$-25\ 968 = 60(-393.509) - \Delta_f H^o(C_{60},s)$

$\Delta_f H^o(C_{60},s) = 2357$ kJ mol^{-1}

(c) $(1/60)C_{60}(s) = C(g)$

$\Delta_{vap} H^o = 716.68 - (1/60)(2357) = 677.4$ kJ mol^{-1}

(d) C(graphite) = C(g)

$\Delta_{vap} H^o = 716.68 - 0 = 716.68$ kJ mol^{-1}

C(diamond) = C(g)

$\Delta_{vap} H^o = 716.68 - 1.90 = 714.78$ kJ mol^{-1}

2.36 124.4, 497.6 kJ

2.37 2.91 kW-hr, 0.172, 4.4 kg

2.38 3.89 J

2.39 (a) No, (b) Yes

2.40 $\partial C_V/\partial \ln V \neq \partial(RT)/\partial T = R$

$\partial C_V/\partial \ln V = \partial R/\partial \ln T = 0$

2.42 (a) - 5.71, (b) -5.67 kJ mol^{-1}

2.43 (a) - 7.42 kJ mol^{-1} , (b) 7.42 kJ mol^{-1}, (c) 0, (d) 0

2.44 (a) 1.99, (b) - 23.30, (c) -21.31, (d) -23.30 kJ mol^{-1}

2.45 (a) - 1.72, (b) - 1.16 kJ mol^{-1}

2.46 98.098 kJ mol^{-1}

2.47 99.371 kJ mol^{-1}

2.48 (a) 570.2 K, (b) 9.48 bar, (c) 5.580 kJ mol^{-1}

2.49 (a) 0.496 bar; - 1739 J mol-1, (b) 0.310 bar; - 1390 J mol^{-1}

2.50 (a) 207.04 K, (b) 0.0435 m^3, (c) –825 J mol^{-1}, (d) –825 J mol^{-1}

2.51 (a) - 46.1 kJ, (b) -24.1 kJ

2.52 - 802.303, - 807.513 kJ mol^{-1}

2.53 The sum of the two enthalpies of reaction is more negative than the enthalpy change for the first reaction by the enthalpy of formation of methane.

2.54 12.2 kJ mol^{-1}

2.55 (a) - 119.9 kJ g^{-1}, (b) - 50.0 kJ g^{-1}, (c) - 19.9 kJ g^{-1}, (d) -45.2 kJ g^{-1}

2.56 206.11 kJ is liberated.

Burning CH_4 liberates 802.34 kJ.

Burning CO + $3H_2$ liberates 1008.45 kJ.

2.57 432.070, 435.998, 459.580 kJ mol^{-1}

2.58 (I) 214.63, - 97.94; (II) 116.69 kJ mol^{-1}
The disadvantage of the first process is the large demand for heat in the first step.

2.59 43.8 kJ mol^{-1}

2.60 459.579 kJ mol^{-1} (4.74551 eV)

2.61 1663.54 kJ mol^{-1}

2.62 0,340 mol graphite has to be burned.

2.63 About 2600 K

2.64 $\Delta_r U$ = - 5163 kJ mol^{-1}, $\Delta_r H$ = - 5168 kJ mol^{-1},
$\Delta_f H(C_{10}H_8)$ = 78 kJ mol^{-1}

2.65 (a) - 253.4, (b) - 249.7 kJ mol^{-1}

2.66 303 g

2.69 (a) - 1412.6 kJ mol^{-1}, - 1430.6 kJ mol^{-1}

(b) 11.5 kbar

3

Second and Third Laws of Thermodynamics

3.1 Show that $(\partial C_V/\partial V)_T = 0$ for an ideal gas, a gas following $P = nRT/(V - nb)$, and a van der Waals gas.

SOLUTION

$$C_V = \left(\frac{\partial U}{\partial T}\right)_V \qquad \text{(a)}$$

$$\left(\frac{\partial C_V}{\partial V}\right)_T = \frac{\partial^2 U}{\partial V \partial T} = \frac{\partial^2 U}{\partial T \partial V} = \frac{\partial}{\partial T}\left(\frac{\partial U}{\partial V}\right)_T \qquad \text{(b)}$$

It can be shown that

$$\left(\frac{\partial U}{\partial V}\right)_T = - P + T\left(\frac{\partial P}{\partial T}\right)_V \qquad \text{(c)}$$

This derivative is substituted in equation b to obtain

$$\left(\frac{\partial C_V}{\partial V}\right)_T = T\left(\frac{\partial^2 P}{\partial T^2}\right)_V \qquad \text{(d)}$$

For an ideal gas,

$P = nRT/V$

$$\left(\frac{\partial P}{\partial T}\right)_V = \frac{nR}{V}$$

$$\left(\frac{\partial^2 P}{\partial T^2}\right)_V = 0$$

For a gas following $P = nRT/(V - nb)$,

$$\left(\frac{\partial P}{\partial T}\right)_V = \frac{nR}{V - nb}$$

$$\left(\frac{\partial^2 P}{\partial T^2}\right)_V = 0$$

For a van der Waals gas,

$$P = \frac{nRT}{V - nb} + \frac{n^2 a}{V^2}$$

$$\left(\frac{\partial P}{\partial T}\right)_V = \frac{nR}{V - nb}$$

$$\left(\frac{\partial^2 P}{\partial T^2}\right)_V = 0$$

Thus $(\partial C_V/\partial V)_T = 0$ for all three cases.

3.2 Show that q_{rev} is not a state function for a gas obeying the equation of state $P(V - nb) = nRT$, but that q_{rev}/T is.

SOLUTION

Since for both gases $(\partial U/\partial V)_T = 0$, $dU = C_V dT$, and the second law yields

$$\bar{d}\, q_{rev} = C_V dT + P dV = C_V dT + \frac{nRT}{V - nb} dV$$

For such a gas it can be shown that C_V is independent of V. The test for exactness involves the calculation of the following two mixed partial derivatives:

$$\left(\frac{\partial C_V}{\partial V}\right)_T = 0$$

$$\left(\frac{\partial \left(\frac{nRT}{V - nb}\right)}{\partial T}\right)_V = \frac{nR}{V - nb}$$

Since these mixed partial derivatives are not equal, $\bar{d}\, q_{rev}$ is not a state function for a gas obeying this equation of state.

Use of the integrating factor $1/T$ yields

$$\frac{\bar{d} q_{rev}}{T} = \frac{C_V}{T} dT + \frac{nR}{V - nb} dV$$

The two mixed partial derivatives are given by

$$\left(\frac{\partial (C_V/T)}{\partial V}\right)_T = 0$$

$$\left(\frac{\partial [nR/(V - nb)]}{\partial T}\right)_V = 0$$

Since the mixed partial derivatives are equal, $\bar{d}\, q_{rev}/T$ is an exact differential and q_{rev}/T a state function for the gas.

3.3 Show that q_{rev} is not a state function for a gas obeying the van der Waals equation, but that q_{rev}/T is.

SOLUTION

For a van der Waals gas,

$$\left(\frac{\partial U}{\partial V}\right)_T = \frac{an^2}{V^2}$$

(see footnote in Section 2.6)

$$đ\, q_{rev} = C_V dT + \left(P + \left(\frac{\partial U}{\partial V}\right)_T\right) dV = C_V dT + \frac{nRT}{V - nb}\, dV$$

This is the same equation that was discussed in problem 3.2, and so the same proof applies if C_V is not a function of V, which is true if a is not a function of T.

3.4 An ideal gas initially at P_1,V_1,T_1 undergoes a reversible isothermal exapnsion to P_2,V_2,T_1. The same change in state of the gas can be accomplished by allowing it to expand adiabatically to P_3,V_2,T_2 and then heating it at constant volume to P_2,V_2,T_1. Show that the entropy change for the reversible isothermal expansion is the same as the sum of the entropy changes in the reversible adiabatic expansion and the reversible heating to P_2,V_2,T_1. This shows that ΔS is independent of path and is therefore a state function.

SOLUTION

For the reversible isothermal expansion, q_{rev} isgiven by

$$q_{rev} = nRT_1 \ln \frac{V_2}{V_1} \quad \text{(a)}$$

and the entropy change isgiven by

$$\Delta S(P_1,V_1,T_1\text{->}P_2,V_2,T_1) = \frac{q_{rev}}{T_1} = nR \ln \frac{V_2}{V_1} \quad \text{(b)}$$

For the reversible adiabatic expansion, $q_{rev} = 0$ and $\Delta S(P_1,V_1,T_1\text{-}$ $\text{>}P_3,V_2,T_2) = 0$ since the system is isolated. The final temperature T_2 for the reversible adiabatic expansion is given by

$$\int_{T_1}^{T_2} \frac{C_V dT}{T} = -nR \ln \frac{V_2}{V_1} \quad \text{(c)}$$

For the reversible heating from T_2 to T_1, q_{rev} is given by

$$q_{rev} = \int_{T_2}^{T_1} C_V dT \quad \text{(d)}$$

and the entropy change is given by

$$\Delta S(P_3,V_2,T_2\text{->}P_2,V_2,T_1) = \int_{T_2}^{T_1} \frac{C_V dT}{T} = nR \ln \frac{V_2}{V_1} \quad \text{(e)}$$

where the second term is obtained from equation c. Thus ΔS is the same by two paths, as expected for a state function.

3.5 Water is vaporized reversibly at 100 °C and 1.01325 bar. The heat of vaporization is 40.69 kJ mol^{-1}. (a) What is the value of ΔS for the water? (b) What is the value of ΔS for the water plus the heat reservoir at 100 °C?

SOLUTION

(a) $\Delta \bar{S}_{H_2O} = \dfrac{40\ 690 \text{ J mol}^{-1}}{373.15 \text{ K}} = 109.04 \text{ J K}^{-1} \text{ mol}^{-1}$

(b) $\Delta \bar{S}_{res} = \dfrac{-40\ 690 \text{ J mol}^{-1}}{373.15 \text{ K}} = -109.04 \text{ J K}^{-1} \text{ mol}^{-1}$

$\Delta \bar{S}_{syst} = 0$

This is necessarily true for a reversible process in an isolated system.

3.6 Assuming that CO_2 is an ideal gas, calculate ΔH^o and ΔS^o for the following process:
1 CO_2(g, 298.15 K, 1 bar) → 1 CO_2(g, 1000 K, 1 bar)
Given: $\bar{C}_P^o = 26.648 + 42.262 \times 10^{-3}\, T - 142.40 \times 10^{-7}\, T^2$ in J K^{-1} mol^{-1}

SOLUTION

$$\Delta \bar{H}^\circ = \int_{298\ 15}^{1000} C_P{}^\circ \mathrm{d}T = 26.648(1000 - 298.15)$$

$$+\left(\frac{42.262x10^{-3}}{2}\right)(1000^2 - 298.15^2) - \left(\frac{142.4x10^{-7}}{3}\right)(1000^3 - 298.15^3)$$

$= 33.34$ kJ mol^{-1}

$$\Delta \bar{S}^\circ = \int_{298.15}^{1000} \frac{C_P^\circ}{T} \mathrm{d}T = 26.648 \ln \frac{1000}{298.15}$$

$$+42.262x10^{-3}(1000 - 298.15) - \left(\frac{142.4x10^{-7}}{2}\right)(1000^2 - 298.15^2)$$

$= 55.42$ J K^{-1}mol^{-1}

3.7 The temperature of an ideal monatomic gas is increased from 300 K to 500 K. What is the change in molar entropy of the gas (a) if the volume is held constant and (b) if the pressure is held constant?

SOLUTION

(a) $\Delta \bar{S} = \bar{C}_V \ln \dfrac{T_2}{T_1} = \dfrac{3}{2}(8.314 \text{ J K}^{-1} \text{ mol}^{-1}) \ln \dfrac{500 \text{ K}}{300 \text{ K}}$
$= 6.371 \text{ J K}^{-1} \text{ mol}^{-1}$

(b) $\Delta\bar{S} = \bar{C}_P \ln\frac{T_2}{T_1} = \frac{5}{2}(8.314 \text{ J K}^{-1} \text{ mol}^{-1}) \ln\frac{500 \text{ K}}{300 \text{ K}}$
$= 10.618 \text{ J K}^{-1} \text{ mol}^{-1}$

3.8 Ammonia (considered to be an ideal gas) initially at 25 °C and 1 bar pressure is heated at constant pressure until the volume has trebled. Calculate (a) q per mole, (b) w per mole, (c) $\Delta\bar{H}$, (d) $\Delta\bar{U}$, and (e) $\Delta\bar{S}$.
Given: $\bar{C}_P = 25.895 + 32.999 \times 10^{-3}\, T - 30.46 \times 10^{-7}\, T^2$ in J K^{-1} mol^{-1}.

SOLUTION

(a) The higher temperature is 3(298 K) = 894 K.

$$q = \int_{T_1}^{T_2} \bar{C}_P \, dT$$

$$= \int_{298}^{894} \left(25.895 + 32.999 \times 10^{-3}\, T - 30.46 \times 10^{-7}\, T^2\right) dT$$

$$= (25.895)(596) + \frac{32.999 \times 10^{-3}}{2}(894^2 - 298^2)$$

$$- \frac{30.46 \times 10^{-3}}{3}(894^3 - 298^3) = 26.4 \text{ kJ mol}^{-1}$$

(b) $w = -P\Delta\bar{V} = -R(T_2 - T_1) = -(8.314 \text{ J K}^{-1} \text{ mol}^{-1})(596 \text{ K})$
$= -4.96 \text{ kJ mol}^{-1}$

(c) $\Delta\bar{H} = q_P = 26.4 \text{ kJ mol}^{-1}$

(d) $\Delta\bar{U} = q + w = 26.4 - 5.0 = 21.4 \text{ kJ mol}^{-1}$

(e) $$\bar{S} = \int_{298}^{894} \frac{\bar{C}_P}{T} \, dT$$

$$= \int_{298}^{894} \left(\frac{25.895}{T} + 32.999 \times 10^{-3} - 30.46 \times 10^{-7}\, T\right) dT$$

$$= 25.895 \ln\frac{894}{298} + 32.999 \times 10^{-3}\,(894 - 298)$$

$$- \frac{30.46 \times 10^{-7}}{2}(894^2 - 298^2) = 46.99 \text{ J K}^{-1} \text{ mol}^{-1}$$

3.9 Two blocks of the same metal are of the same size but are at different temperatures, T_1 and T_2. These blocks of metal are brought together and allowed to come to the same temperature. Show that the entropy change is given by

$$\Delta S = C_P \ln \left[\frac{(T_1 + T_2)^2}{4T_1T_2} \right]$$

if C_P is constant. How does this equation show that the change is spontaneous?

SOLUTION

If we designate the blocks as A and B, the total entropy change is given by

$$\Delta S = \int_{T_1}^{(T_1+T_2)/2} \frac{C_P}{T_A} \mathrm{d}T_A + \int_{T_1}^{(T_1+T_2)/2} \frac{C_P}{T_B} \mathrm{d}T_B$$

$$= C_P \ln\left(\frac{T_1 + T_2}{2T_1}\right) + C_P \ln\left(\frac{T_1 + T_2}{2T_2}\right)$$

$$= C_P \ln \left[\frac{(T_1 + T_2)^2}{4T_1T_2}\right]$$

Since this quantity is always positive, the change is spontaneous.

3.10 In the reversible isothermal expansion of an ideal gas at 300 K from 1 to 10 liters, where the gas has an initial pressure of 20.27 bar, calculate (a) ΔS for the gas and (b) ΔS for all systems involved in the expansion.

SOLUTION

(a) $$n = \frac{PV}{RT} = \frac{(20.27 \text{ bar})(1 \text{ L})}{(0.08314 \text{ L bar K}^{-1} \text{ mol}^{-1})(300 \text{ K})} = 0.812 \text{ mol}$$

$$\Delta S = nR \ln \frac{V_2}{V_1} = (0.812 \text{ mol})(8.314 \text{ J K}^{-1} \text{ mol}^{-1}) \ln \frac{10}{1} = 15.56 \text{ J K}^{-1}$$

(b) $\Delta S = 0$ since the process is carried out reversibly. The heat gained by the gas is equal to the heat lost by the heat reservoir, and both bodies are at the same temperature.

3.11 A mole of oxygen is expanded reversibly from 1 bar to 0.1 bar at 298 K. What is the change in entropy of the gas and what is the change in entropy for the gas plus the heat reservoir with which it is in contact?

SOLUTION

$$\Delta S(O_2) = -R \ln \frac{P_2}{P_1}$$

$$= -(8.314 \text{ J K}^{-1} \text{ mol}^{-1}) \ln 0.1 = 19.14 \text{ J K}^{-1} \text{ mol}^{-1}$$

The entropy of the gas and the reservoir taken together is not affected by a reversible process; therefore $\Delta S = 0$ for the whole system. Consequently the entropy of the reservoir decreases by 19.14 J K^{-1} mol^{-1}.

3.12 Three moles of an ideal gas expand isothermally and reversibly from 90 to 300 L at 300 K. (a) Calculate ΔU, ΔS, w, and q for this system. (b) Calculate $\Delta \bar{U}$, $\Delta \bar{S}$, w per mole, and q per mole. (c) If the expansion is carried out irreversibly by allowing the gas to expand into an evacuated container, what are the values of $\Delta \bar{U}$, $\Delta \bar{S}$, w per mole, and q per mole?

SOLUTION

(a) $\Delta U = 0$ kJ since the gas is ideal.

$\Delta S = nR \ln(V_2/V_1) = (3 \text{ mol})(8.314 \text{ J K}^{-1} \text{ mol}^{-1}) \ln(300 \text{ L}/90 \text{ L})$

$= 30.03 \text{ J K}^{-1}$

$w = - nRT \ln(V_2/V_1) = -(3 \text{ mol})(8.314 \text{ J K}^{-1} \text{ mol}^{-1})(300 \text{ K}) \ln(300/90)$

$= - 9.01 \text{ kJ}$

$q = - w = 9.01 \text{ kJ}$

(b) $\Delta \bar{U} = 0 \text{ kJ mol}^{-1}$

$\Delta \bar{S} = (30.03 \text{ J K}^{-1})/(3 \text{ mol}) = 10.01 \text{ J K}^{-1} \text{ mol}^{-1}$

$w = - 3.00 \text{ kJ mol}^{-1}$

$q = 3.00 \text{ kJ mol}^{-1}$

(c) $\Delta \bar{U} = 0 \text{ kJ mol}^{-1}$

$\Delta \bar{S} = 10.01 \text{ J K}^{-1} \text{ mol}^{-1}$

$w = 0 \text{ kJ mol}^{-1}$

$q = 0 \text{ kJ mol}^{-1}$

3.13 (a) A system consists of a mole of ideal gas that undergoes the following change in state

1 X(g, 298 K, 10 bar) = 1 X(g, 298 K,1 bar)

What is the value of $\Delta \bar{S}$ if the expansion is reversible? What is the value of $\Delta \bar{S}$ if the gas expands into a larger container so that the final pressure is 1 bar? (b) The same change in state takes place, but we now consider the gas plus the heat reservoir at 298 K to be our system. What is the value of $\Delta \bar{S}$ if the expansion is reversible? What is the value of $\Delta \bar{S}$ if the gas expands into a larger container so that the final pressure is 1 bar?

SOLUTION

(a) $w = RT \ln (10 \text{ bar}/1 \text{ bar}) = (8.314)(298) \ln 10 = 5705 \text{ J mol}^{-1} = q$

$$\Delta \bar{S} = \frac{5705 \text{ J mol}^{-1}}{298 \text{ K}} = 19.14 \text{ J K}^{-1} \text{ mol}^{-1}$$

This is $\Delta \bar{S}$, whether the expansion is reversible or irreversible.

(b) If the expansion is reversible, $\Delta\bar{S} = 0$, as it must be for any reversible process in an isolated system. The gas gains q and the surroundings lose q.

If the expansion is irreversible, $\Delta\bar{S}$ = 19.14 J K^{-1} mol^{-1}, which is positive, as it must be for any irreversible process in an isolated system.

3.14 An ideal gas at 298 K expands isothermally from a pressure of 10 bar to 1 bar. What are the value of w per mole, q per mole, $\Delta\bar{U}$, $\Delta\bar{H}$, and $\Delta\bar{S}$ in the following cases? (a) The expansion is reversible. (b) The expansion is free. (c) The gas and its surroundings form an isolated system, and the expansion is reversible. (d) The gas and its surroundings for an isolated system, and the expansion is free.
SOLUTION

	(a) Reversible	(b) Irreversible	(c) Isolated Rev.	(d) Isol. Irr.
w	- 5.71 kJ mol^{-1}0	0	0	
q	5.71 kJ mol^{-1}	0	0	0
$\Delta\bar{U}$	0	0	0	0
$\Delta\bar{H}$	0	0	0	0
$\Delta\bar{S}$	19.1 J K^{-1} mol^{-1}	19.1 J K^{-1} mol^{-1}	0	19.1 J K^{-1} mol^{-1}

$$w = -\int_{\bar{V}_2}^{\bar{V}_1} P\,d\Delta\bar{V} = RT\ln\frac{\bar{V}_2}{\bar{V}_1} = -RT\ln\frac{P_1}{P_2} = (8.314)(298)\ln 10 = -5.71 \text{ kJ mol}^{-1}$$

$$\Delta\bar{S} = \frac{5710 \text{ J mol}^{-1}}{298 \text{ K}} = 19.1 \text{ J K}^{-1} \text{ mol}^{-1}$$

3.15 An ideal monatomic gas is heated from 300 to 1000 K and the pressure is allowed to rise to 1 to 2 bar. What is the change in molar entropy?

SOLUTION

$$\Delta\bar{S} = \bar{C}_P \ln\left(\frac{T_2}{T_1}\right) - R\ln\left(\frac{P_2}{P_1}\right)$$

$$= \frac{5}{2}(8.314 \text{ J K}^{-1} \text{ mol}^{-1}) \ln\frac{1000 \text{ K}}{300 \text{ K}} - (8.314 \text{ J K}^{-1} \text{ mol}^{-1}) \ln\frac{2 \text{ bar}}{1 \text{ bar}}$$

$$= 19.26 \text{ J K}^{-1} \text{ mol}^{-1}$$

3.16 The purest acetic acid is often called glacial acetic acid because it is purified by fractional freezing at its melting point of 16.6 °C. A flask containing several moles of liquid acetic acid at 16.6 °C is lowered into an ice-water bath briefly. When it is removed it is found that exactly 1 mol of acetic acid has frozen. Given:

$\Delta_{fus}H$ (CH_3CO_2H) = 11.45 kJ mol^{-1} and $\Delta_{fus}H(H_2O)$ = 5.98 kJ mol^{-1}. (a) What is the change in entropy of the acetic acid? (b) What is the change in entropy of the water bath? (c) Now consider the water bath and acetic acid are in the same system. What is the entropy change for the combined system? Is the process reversible or irreversible? Why?

SOLUTION

(a) $$\Delta S_{aa} = \frac{\Delta H}{T_{fp}} = \frac{-11450 \text{ J}}{289.75 \text{ K}} = -39.52 \text{ J K}^{-1}$$

(b) $$\Delta S_{H_2O} = \frac{11450 \text{ J}}{273.15 \text{ K}}$$
$$= +41.92 \text{ J K}^{-1}$$

(c) $\Delta S_{syst} = 41.92 - 39.52 = 2.40 \text{ J K}^{-1}$

The process is irreversible because S increases in our isolated system. Once the freezing of acetic acid starts, it cannot be stopped by an infinitesimal change.

3.17 In Problem 2.21 an ideal monatomic gas at 1 bar and 300 K was expanded adiabatically against a constant pressure of 1/2 bar until the final pressure was 1/2 bar; a temperature of 240 K was reached. What is the value of $\Delta \bar{S}$ for this process?

SOLUTION

In order to calculate $\Delta \bar{S}$ we have to find a reversible path for the change in state which is

1 X(300 K, 1 bar) → 1 X(240 K, 1/2 bar)

This same change in state can be accomplished in two reversible steps.

Reversible isothermal expansion: 1 X(300 K, 1 bar) → 1 X(300 K, 1/2 bar)

$$\Delta \bar{S}_1 = R \ln 2 = 5.76 \text{ J K}^{-1} \text{ mol}^{-1}$$

Cooling process: 1 X(300 K, 1/2 bar) = 1 X(240 K, 1/2 bar)

$$\Delta \bar{S}_2 = \bar{C}_P \ln (T_2/T_1)$$
$$= (5/2)(8.314)\ln(240/300) = -4.64 \text{ J K}^{-1} \text{ mol}^{-1}$$

The entropy change for the reversible two-step process is

$$\Delta \bar{S} = \Delta \bar{S}_1 + \Delta \bar{S}_2 = 5.76 - 4.64 = 1.12 \text{ J K}^{-1} \text{ mol}^{-1}$$

Note that the entropy of the system increases in this irreversible expansion, even though $q = 0$.

3.18 Ten moles of H_2 and two moles of D_2 are mixed at 25 °C and 1 bar. What is the value of $\Delta \bar{S}^{\circ}$? Assume ideal gases.

SOLUTION

$$\Delta S^{o} = -R(n_1 \ln y_1 + n_2 \ln y_2)$$

$$= -(8.314 \text{ J K}^{-1} \text{ mol}^{-1})\left[(10 \text{ mol}) \ln \frac{10}{12} + (2 \text{ mol}) \ln \frac{2}{12}\right]$$

$$= 44.95 \text{ J K}^{-1}$$

3.19 (a) Write the expression for the entropy of a mixture of ideal gases A, B, and C at T and P using y_i for the mole fraction of gas i. (b) Now let us carry out the combination of terms contributing to the entropy of the system in two steps: First, imagine that gases A and C are mixed to form a mixture with mole fractions r_A and r_C within the A,C mixture, but that B remains unmixed. Write the equation for the entropy of the system with two terms, one for the $n_I = n_A + n_C$ moles of the A plus C mixture, which contributes pressure $P_I = P_A + P_C$, and the other $n_B\bar{S}_B$. (c) Second, imagine that B is mixed with mixture I, considered as one species, and show that this equation is the same as that obtained in (a).

SOLUTION

(a) $S = n_A\bar{S}_A + n_B\bar{S}_B + n_C\bar{S}_C$

$$= n_A\bar{S}_A^{o} + n_B\bar{S}_B^{o} + n_C\bar{S}_C^{o} - R(n_A \ln y_A + n_B \ln y_B + n_C \ln y_C) - (n_A + n_B + n_C)R\ln(P/P^{o})$$

(b) $S = [n_A\bar{S}_A + n_B\bar{S}_B] + n_C\bar{S}_C$

$$= [n_A\bar{S}_A^{o} + n_C\bar{S}_C^{o} - R(n_A \ln r_A + n_C \ln r_C) - (n_A + n_C)R\ln(P_I/P^{o})] + n_B\bar{S}_B$$

$$= n_I[r_A\bar{S}_A^{o} + r_C\bar{S}_C^{o} - R(r_A \ln r_A + r_C \ln r_C) - \text{R}\ln(P_I/P^{o})] + n_B\bar{S}_B$$

where $n_I = n_A + n_C$, $P_I = P_A + P_C$, and the quantity in brackets can be represented by $\bar{S}_I$. Now we can consider the system as being made up of n_I moles of I and n_B moles of B.

(c) $S = n_I\bar{S}_I + n_B\bar{S}_B = n_t[y_I\bar{S}_I + y_B\bar{S}_B]$

$$= n_t[y_I\bar{S}_I^{o} - y_I R\ln(P_I/P^{o}) + y_B\bar{S}_B - y_B R\ln(P_B/P^{o})]$$

$$= n_t[y_I\bar{S}_I^{o} + y_B\bar{S}_B^{o} - R(y_I \ln y_I + y_B \ln y_B) - R\ln(P/P^{o})]$$

Now if this equation is written out in terms of amounts of species, it can be rearranged to give the equation in (a).

3.20 One mole of A at 1 bar and one mole of B at 2 bar are separated by a partition and surrounded by a heat reservoir. When the partition is withdrawn, how much does the entropy change?

SOLUTION

The initial entropy $S_1 = \overline{S}_A^o - R\ln 1 + \overline{S}_B^o - R\ln 2$
The volume is $V = RT + RT/2 = 1.5RT$
The final pressure is $2RT/V = 4/3$ bar
The final entropy is

$S_2 = \overline{S}_A^o + \overline{S}_B^o - R(\ln 0.5 + \ln 0.5) - 2R\ln(4/3)$
The entropy of mixing is

$$\Delta_{mix}S = S_2 - S_1 = 3R\ln 2 - 2R\ln(4/3) = 12.51 \text{ J K}^{-1} \text{ mol}^{-1}$$

3.21 Use the microscopic point of view of Section 3.6 to show that for the expansion of amount n of an ideal gas by a factor of two, $\Delta S = nR \ln 2$. In this expression S is an extensive property.

SOLUTION

The probability that the molecules will all be found in one half of the container is
$\Omega'/\Omega = (1/2)^{N_A n}$
$\Delta S = k \ln(\Omega'/\Omega)$
$= kN_A n \ln(1/2)$
$= - kN_A n \ln 2$
where $R = N_A k$. For expansion from half the container to the full container,
$\Delta S = nR\ln 2$
ΔS has units of J K^{-1}, n has units of mol, and R has units of J K^{-1} mol^{-1}.

3.22 Calculate the change in molar entropy of aluminum which is heated from 600 °C to 700 °C. The melting point of aluminum is 660 °C, the heat of fusion is 393 J g^{-1}, and the heat capacities of the solid and liquid may be taken as 31.8 and 34.3 J K^{-1} mol^{-1}, respectively.

SOLUTION

$$\Delta\overline{S} = \int_{T_1}^{T_f} \frac{(\overline{C}_{P,s})}{T} dT + \frac{\Delta_{fus}H}{T_{fus}} + \int_{T_f}^{T_2} \frac{(\overline{C}_{P,l})}{T} dT$$

$$= \overline{C}_{P,s} \ln\frac{T_f}{T_1} + \frac{\Delta_{fus}H}{T_{fus}} + \overline{C}_{P,l} \ln\frac{T_2}{T_f}$$

$$= (31.8 \text{ J K}^{-1} \text{ mol}^{-1}) \ln \frac{933 \text{ K}}{873 \text{ K}} + \frac{(27 \text{ g mol}^{-1})(393 \text{ J g}^{-1})}{933 \text{ K}}$$

$$+ (34.3 \text{ J K}^{-1} \text{ mol}^{-1}) \ln \frac{973 \text{ K}}{933 \text{ K}} = 14.92 \text{ J K}^{-1} \text{ mol}^{-1}$$

3.23 Steam is condensed at 100 °C and the water is cooled to 0 °C and frozen to ice. What is the molar entropy change of the water? Consider that the average specific heat of liquid water is 4.2 J K^{-1} g^{-1}. The heat of vaporization at the boiling point and the heat of fusion at the freezing point are 2258.1 and 333.5 J g^{-1}, respectively.

SOLUTION

$$\Delta S = \frac{\Delta_{\text{vap}}H}{T_{\text{vap}}} + \int_{373 \text{ K}}^{273 \text{ K}} \frac{\bar{C}_P}{T} \, dT - \frac{\Delta_{\text{fus}}H}{T_{\text{fus}}}$$

$$= - \frac{(2258.1 \text{ J g}^{-1})(18.016 \text{ g mol}^{-1})}{373.15 \text{ K}}$$

$$+ (75.379 \text{ J K}^{-1} \text{ mol}^{-1}) \int_{373 \text{ K}}^{273 \text{ K}} d \ln T \quad \frac{(333.5 \text{ J g}^{-1})(18.016 \text{ g mol}^{-1})}{273.15 \text{ K}}$$

$$= - 154.4 \text{ J K}^{-1} \text{ mol}^{-1}$$

*3.24 Calculate the molar entropy of carbon disulfide at 25 °C from the following heat-capacity data and the heat of fusion, 4389 J mol^{-1}, at the melting point (161.11 K).

T/K	15.05	20.15	29.76	42.22	57.52	75.54	89.37
$\bar{C}_P$/J K^{-1} mol^{-1}	6.90	12.01	20.75	29.16	35.56	40.04	43.14
T/K	99.00	108.93	119.91	131.54	156.83	161-298	
$\bar{C}_P$/J K^{-1} mol^{-1}	45.94	48.49	50.50	52.63	56.62	75.48	

SOLUTION

$$\bar{S}° (298.15 \text{ K}) = \frac{\bar{C}_P(15.05 \text{ K})}{3} + \int_{15.05 \text{ K}}^{161.11 \text{ K}} \frac{\bar{C}_P}{T} \, dT + \frac{\Delta_{\text{fus}}H}{161.11 \text{ K}} + \int_{161.11 \text{ K}}^{298.15 \text{ K}} \frac{\bar{C}_P}{T}$$

dT

The first integral may be approximated by multiplying the average value of $\bar{C}_P/T$) for each temperature interval by the width of the interval. Thus the first contribution is

$$\frac{1}{2}\left(\frac{6.90}{15.05} + \frac{12.01}{20.15}\right)\left(20.15 - 15.05\right) = 2.69 \text{ J K}^{-1} \text{ mol}^{-1}$$

The first integral has the value 74.69 J K^{-1} mol^{-1}. thus

$$\bar{S}^\circ(298.15\text{ K}) = \frac{6.90}{3} + 74.69 + \frac{4389}{161.11} + 75.45 \ln\frac{298.15}{161.11} = 150.67\text{ J K}^{-1}\text{ mol}^{-1}$$

3.25 Ten grams of molecular hydrogen at 1 bar expands to triple the volume (a) isothermally and reversibly and (b) adiabatically and reversibly. In each case what are $\Delta S(H_2)$, ΔS(surr), and $\Delta S(H_2$ and surr)?

<u>SOLUTION</u>

(a) $\Delta S(H_2) = nR\ln(V_2/V_1)$

$= (10\text{ g}/2.016\text{ g mol}^{-1})(8.314\text{ J K}^{-1}\text{ mol}^{-1})\ln 3$

$= 45.31\text{ J K}^{-1}\text{ mol}^{-1}$

$\Delta S(\text{surr}) = -45.31\text{ J K}^{-1}\text{ mol}^{-1}$

$\Delta S(H_2\text{ and surr}) = 0\text{ J K}^{-1}\text{ mol}^{-1}$

(b) $\Delta S(H_2) = 0\text{ J K}^{-1}\text{ mol}^{-1}$ because q = 0

$\Delta S(\text{surr}) = 0\text{ J K}^{-1}\text{ mol}^{-1}$ because q = 0

$\Delta S(H_2\text{ and surr}) = 0\text{ J K}^{-1}\text{ mol}^{-1}$

3.26 Theoretically, how high could a gallon of gasoline lift an automobile weighing 2800 lb against the force of gravity, if it is assumed that the cylinder temperature is 2200 K and the exit temperature 1200 K? (Density of gasoline = 0.80 g cm^{-3}; 1 lb = 453.6 g; 1 ft = 30.48 cm; 1 L = 0.2642 gal. Heat of combustion of gasoline = 46.9 kJ g^{-1}.)

<u>SOLUTION</u>

$$q = \frac{(46.9 \times 10^3\text{ J g}^{-1})(1\text{ gal})(10^3\text{ cm}^3\text{ L}^{-1})(0.80\text{ g cm}^{-3})}{0.2642\text{ gal L}^{-1}}$$

$= 14.2 \times 10^7$ J

$$w = q\frac{T_2 - T_1}{T_2} = \frac{(14.2 \times 10^7\text{ J})(2200\text{ K} - 1200\text{ K})}{(2200\text{ K})}$$

$= 6.45 \times 10^7$ J

$w = mgh = (2800\text{ lb})(0.4536\text{ kg lb}^{-1})(9.8\text{ m s}^{-2})(0.3048\text{ m ft}^{-1})h$

$h = 17{,}000$ ft

3.27 (a) What is the maximum work that can be obtained from 100 J of heat supplied to a steam engine with a high-temperature reservoir at 100 °C if the condenser is at 20 °C? (b) If the boiler temperature is raised to 150 °C by the use of superheated steam under pressure, how much more work can be obtained?

<u>SOLUTION</u>

(a) $w = q \frac{T_2 - T_1}{T_2} = (1000 \text{ J}) \frac{80 \text{ K}}{373.1 \text{ K}} = 214 \text{ J}$

(b) $w = (1000 \text{ J}) \frac{130 \text{ K}}{423.1 \text{ K}} = 307 \text{ J}$ or 93 J more than (a)

3.28 The term adiabatic lapse rate used by meteorologists is the decrease in temperature with height that results from the adiabatic exapnsion of an air mass as it is pushed up a mountain by the wind. This adiabatic expansion is represented by

$P_0, V_0, T_0 \text{ ---> } P, V, T \qquad \Delta S_3 = 0$

where P_0, V_0, T_0 represents the sea level conditions. The calculation of the entropy change can be carried out in two steps (the first at constnt P, and the second at constant T)

$P_0, V_0, T_0 \text{ ---> } P_0, V^*, T \qquad \Delta S_1 = n\bar{C}_P \ln \frac{T}{T_0}$

$P_0, V^*, T \text{ ---> } P, V, T \qquad \Delta S_2 = nR \ln \frac{P_0}{P}$

We have seen in Section 1.11 that the pressure of the atmosphere drops off exponentially if temperature is independent of height h. (a) Since $\Delta S_1 + \Delta S_2 = \Delta S_3$, what is the expression for $\Delta T/\Delta h$? (b) If the temperature at the foot of a 14,000 foot mountain is 25 °C, what temperature would you expect at the summit from this adiabatic lapse rate? For this calculation the molar mass of air can be taken to be M = 29 g mol^{-1} and its heat capacity can be taken as 29.1 J K^{-1} mol^{-1}.

SOLUTION

(a) $n\bar{C}_P \ln \frac{T}{T_0} + nR \ln \frac{P_0}{P} = 0$

$$\ln \frac{T}{T_0} = \frac{R}{\bar{C}_P} \ln \frac{P}{P_0}$$

Differentiating this equation yields

$$\frac{\mathrm{d}T}{T} = \frac{R}{\bar{C}_P} \frac{\mathrm{d}P}{P}$$

since P_0 and T_0 are constants. Substituting equation 1.51 yields

$$\frac{\mathrm{d}T}{T} = \frac{R}{\bar{C}_P} \times \left(-\frac{Mg\mathrm{d}h}{RT}\right)$$

where g is the acceleration of gravity. Integration yields

$$\Delta T = -\frac{Mg\Delta h}{\bar{C}_P}$$

The adiabatic lapse rate is therefore

$$\frac{\Delta T}{\Delta h} = -\frac{(0.029 \text{ kg mol}^{-1})(9.8 \text{ m s}^{-2})}{29.1 \text{ J mol}^{-1} \text{ K}^{-1}} = -0.0098 \text{ K m}^{-1}$$

Thus the expected temperature at the summit of a 14,000 foot mountain is
$t/^{o}C = 25 - (0.0098\ ^{o}C\ m^{-1})(14,000\ feet)(12\ inch\ foot^{-1})(2.54\ cm\ inch^{-1})(0.01\ m\ cm^{-1}) = -16.8$

This calculation involves the assumption that the air is dry.

3.29 A single expansion is not a basis for an engine.

3.30 1.96 J K^{-1} mol^{-1}

3.31 126 J K^{-1} mol^{-1}

3.32 (a) 45.25, (b) 33.18 J K^{-1} mol^{-1}

3.33 1.82 J K^{-1} mol^{-1}

3.36 16.48 J K^{-1} mol^{-1}

3.37 (a) - 35.4, (b) 36.1, (c) 0.7 J K^{-1} mol^{-1} irreversible

3.38 9.13 J K^{-1} mol^{-1}

3.43 3.40 107.75 J K^{-1} mol^{-1}

3.41 21.0 x 10^6 J

3.42 0.114 kWh

4

Fundamental Equations of Thermodynamics

4.1 One mole of nitrogen is allowed to expand from 0.5 to 10 L. Calculate the change in entropy using (a) the ideal gas law and (b) the van der Waals equation.

SOLUTION

(a)
$$\begin{aligned}\Delta\bar{S} &= R\ln(\bar{V}_2/\bar{V}_1)\\ &= (8.314\ \text{J K}^{-1}\ \text{mol}^{-1})\ln(10/0.5)\\ &= 24.91\ \text{J K}^{-1}\ \text{mol}^{-1}\end{aligned}$$

(b)
$$\begin{aligned}\Delta\bar{S} &= R\ln\left[(\bar{V}_2 - b)/(\bar{V}_1 - b)\right]\\ &= (8.314\ \text{J K}^{-1}\ \text{mol}^{-1})\ln\left[\frac{(10 - 0.039)}{(0.5 - 0.039)}\right]\\ &= 25.55\ \text{J K}^{-1}\ \text{mol}^{-1}\end{aligned}$$

4.2 Derive the relation for $\bar{C}_P - \bar{C}_V$ for a gas that follows van der Waals' equation.

SOLUTION

$$\bar{C}_P - \bar{C}_V = T\bar{V}\alpha^2/\kappa = -T\left(\frac{\partial\bar{V}}{\partial T}\right)_P^2\left(\frac{\partial P}{\partial\bar{V}}\right)_T$$

The van der Waals equation can be written in the following form:

$$T = \frac{P\bar{V}}{R} - \frac{bP}{R} + \frac{a}{\bar{V}R} - \frac{ab}{R\bar{V}^2}$$

$$\left(\frac{\partial T}{\partial\bar{V}}\right)_P = \frac{P}{R} - \frac{a}{R\bar{V}^2} + \frac{2ab}{R\bar{V}^3}$$

$$\alpha^{-1} = \bar{V}\left(\frac{\partial T}{\partial\bar{V}}\right)_P = \frac{P\bar{V}}{R} - \frac{a}{R\bar{V}} + \frac{2ab}{R\bar{V}^2}$$

Eliminating P with the van der Waals equation yields

$$\alpha^{-1} = \frac{T\bar{V}}{\bar{V} - b} - \frac{2a}{R\bar{V}^2}(\bar{V} - b)$$

κ is calculated from

$$\left(\frac{\partial P}{\partial \bar{V}}\right)_T = -\frac{RT}{(\bar{V} - b)^2} + \frac{2a}{\bar{V}^3}$$

$$\bar{C}_P - \bar{C}_V = \frac{R}{1 - \frac{2a}{RT}\frac{(\bar{V} - b)^2}{\bar{V}^3}}$$

4.3 Earlier we derived the expression for the entropy of an ideal gas as a function of T and P. Now that we have the Maxwell relations, derive the expression for dS for any fluid.

SOLUTION

$$\begin{aligned} dS &= (\partial S/\partial T)_P\, dT + (\partial S/\partial P)_T\, dP \\ &= (C_P/T)\, dT - (\partial V/\partial T)_P\, dP \\ &= (C_P/T)\, dT - \alpha V dP \end{aligned}$$

4.4 What is the change in molar entropy of liquid benzene at 25 °C when the pressure is raised from 1 to 1000 bar? The coefficient of thermal expansion α is 1.237×10^{-3} K^{-1}, the density is 0.879 g cm^{-3}, and the molar mass is 78.11 g mol^{-1}.

SOLUTION

Equation 4.46 can be written as

$$\left(\frac{\Delta\bar{S}}{\Delta P}\right)_T = -\left(\frac{\partial \bar{V}}{\partial T}\right)_P = -\bar{V}\alpha$$

where α is defined by problem 1.17.

$$\begin{aligned} \Delta\bar{S} &= -\bar{V}\alpha\Delta P \\ &= -\frac{78.11 \text{ g mol}^{-1}}{0.879 \text{ g cm}^{-3}}(10^{-2} \text{ m cm}^{-1})^3 (1.237 \times 10^{-3} \text{ K}^{-1})(1000 \text{ bar}) \\ &= -10.99 \text{ J K}^{-1} \text{ mol}^{-1} \end{aligned}$$

4.5 Derive the expression for $\bar{C}_P - \bar{C}_V$ for a gas with the following equation of state.

$$[P + (a/\bar{V}^2)]\bar{V} = RT$$

SOLUTION

$$\bar{C}_P - \bar{C}_V = T\bar{V}\alpha^2/\kappa = -T\left(\frac{\partial \bar{V}}{\partial T}\right)_P^2\left(\frac{\partial P}{\partial \bar{V}}\right)_T$$

$$P = \frac{RT}{\bar{V}} - \frac{a}{\bar{V}^2}$$

$$\left(\frac{\partial P}{\partial \bar{V}}\right)_T = -\frac{RT}{\bar{V}^2} + \frac{2a}{\bar{V}^3}$$

$$\left(\frac{\partial \bar{V}}{\partial T}\right)_P = \frac{R\bar{V}}{2P\bar{V} - RT}$$

$$\bar{C}_P - \bar{C}_V = R\left(1 - \frac{2a}{\bar{V}RT}\right)^{-1}$$

4.6 What is the difference between the molar heat capacity of iron at constant pressure and constant volume at 25 °C? Given: $\alpha = 35.1 \times 10^{-6}\ \mathrm{K^{-1}}$, $\kappa = 0.52 \times 10^{-6}\ \mathrm{bar^{-1}}$, and the density is $7.86\ \mathrm{g\ cm^{-3}}$.

SOLUTION

$$\bar{V} = \frac{55.847\ \mathrm{g\ mol^{-1}}}{7.86\ \mathrm{g\ cm^{-3}}} = \frac{7.11\ \mathrm{cm^3\ mol^{-1}}}{(10^2\ \mathrm{cm\ m^{-1}})^3} = 7.11 \times 10^{-6}\ \mathrm{m^3\ mol^{-1}}$$

$$\bar{C}_P - \bar{C}_V = \alpha^2 T\bar{V}/\kappa$$

$$= \frac{(35.1 \times 10^{-6}\ \mathrm{K^{-1}})^2(298\ \mathrm{K})(7.11 \times 10^{-6}\ \mathrm{m^3\ mol^{-1}})(10^5\ \mathrm{Pa\ bar^{-1}})}{(0.52 \times 10^{-6}\ \mathrm{bar^{-1}})}$$

$$= 0.51\ \mathrm{J\ K^{-1}\ mol^{-1}}$$

4.7 In equation 1.26 we saw that the compressibility factor of a van der Waals gas can be written as

$$Z = 1 + \frac{1}{RT}\left(b - \frac{a}{RT}\right) P + \cdots$$

where terms in P^2 and higher are negligible. (a) To this degree of approximation, derive the expression for $(\partial \bar{H}/\partial P)_T$ for a van der Waals gas. (b) Calculate $(\partial \bar{H}/\partial P)_T$ for $CO_2(g)$ in $\mathrm{J\ bar^{-1}\ mol^{-1}}$ at 298 K. Given: $a = 3.640\ \mathrm{L^2\ bar\ mol^{-2}}$ and $b = 0.04267\ \mathrm{L\ mol^{-1}}$.

SOLUTION

(a) $(\partial \bar{H}/\partial P)_T = \bar{V} - T(\partial \bar{V}/\partial T)_P$

$$P\bar{V} \cong RT + (b - a/RT)P$$

$$\bar{V} \cong RT/P + b - a/RT$$

$$(\partial \bar{V}/\partial T)_P \cong R/P + a/RT^2$$

$$(\partial \bar{H}/\partial P)_T \cong \bar{V} - RT/P - a/RT$$
$$\cong b - 2a/RT$$

(b) $$(\partial \bar{H}/\partial P)_T \cong 0.04267 \text{ L mol}^{-1} - \frac{2(3.64 \text{ L}^2 \text{ bar mol}^{-2})}{(0.08314 \text{ L bar K}^{-1} \text{ mol}^{-1})(298 \text{ K})}$$
$$\cong -0.251 \text{ L mol}^{-1}$$
$$\cong \frac{-0.251 \text{ L mol}^{-1}}{0.01 \text{ L bar J}^{-1}} = -25 \text{ J bar}^{-1} \text{ mol}^{-1}$$

Since $R = 8.314$ J K^{-1} mol^{-1} = 0.08314 L bar K^{-1} mol^{-1}.
Therefore, 1 J = 0.01 L bar or 0.01 L bar J^{-1} = 1.

4.8 Derive the expression for $(\partial U/\partial \bar{V})_T$ (the internal pressure) for a gas following the virial equation with $Z = 1 + B/\bar{V}$.

SOLUTION

$$P = \frac{RT}{\bar{V}} + \frac{BRT}{\bar{V}^2}$$

$$\left(\frac{\partial P}{\partial T}\right)_{\bar{V}} = \frac{R}{\bar{V}} + \frac{BR}{\bar{V}^2} + \frac{RT}{\bar{V}^2}\left(\frac{\partial B}{\partial T}\right)_{\bar{V}}$$

$$(\partial \bar{U}/\partial \bar{V})_T = T\left(\frac{\partial P}{\partial T}\right)_{\bar{V}} - P$$
$$= \frac{RT^2}{\bar{V}^2}\left(\frac{\partial B}{\partial T}\right)_{\bar{V}}$$

4.9 In Section 3.4 we calculated that the enthalpy of freezing water at - 10 °C is -5619 J mol^{-1}, and we calculated that the entropy of freezing water is - 20.54 J K^{-1} mol^{-1} at - 10 °C. What is the Gibbs energy of freezing water at - 10 °C?

SOLUTION

$$\Delta G^\circ = \Delta H^\circ - T\Delta S^\circ$$
$$= -5619 \text{ J mol}^{-1} - (263.15 \text{ K})(-20.54 \text{ J K}^{-1} \text{ mol}^{-1})$$
$$= -213.9 \text{ J mol}^{-1}$$

This is negative, as expected for a spontaneous process at constant T and P. If the water was at -10 °C in an isolated system, the temperature would rise, but part of the water would freeze. In this case the increase in order due to the crystallization of part of the ice is more than compensated for by the increase in disorder of the system as a whole by the rise in temperature.

4.10 (a) Integrate the Gibbs-Helmholtz equation to obtain an expression for ΔG_2 at temperature T_2 in terms of ΔG_1 at T_1, assuming ΔH is independent of temperature.

(b) Obtain an expression for ΔG_2 using the more accurate approximation that $\Delta H = \Delta H_1 + (T - T_1)\Delta C_P$ where T_1 is an arbitrary reference temperature.

SOLUTION

(a) Using equation 4.63

$$\int_{\Delta G_1/T}^{\Delta G_2/T} d(\Delta G/T) = -\int_{T_1}^{T_2} (\Delta H/T^2)\, dT$$

$$\frac{\Delta G_2}{T_2} - \frac{\Delta G_1}{T_1} = -\Delta H\left(\frac{1}{T_1} - \frac{1}{T_2}\right)$$

$$\Delta G_2 = \Delta G_1 T_2/T_1 + \Delta H[1 - (T_2/T_1)]$$

(b)

$$\int d(\Delta G/T) = -\int \frac{\Delta H_1}{T^2}\, dT + \Delta C_P \int \frac{(T - T_1)}{T^2}\, dT$$

$$\frac{\Delta G_2}{T_2} - \frac{\Delta G_1}{T_1} = -\Delta H\left(\frac{1}{T_1} - \frac{1}{T_2}\right) + \Delta C_P \ln\left(\frac{T_2}{T_1}\right) - T_1\Delta C_P\left(\frac{1}{T_1} - \frac{1}{T_2}\right)$$

$$\Delta G_2 = \Delta G_1 T_2/T_1 + (\Delta H_1 + T_1\Delta C_P)\left(1 - \frac{T_2}{T_1}\right) + T_2\Delta C_P \ln\left(\frac{T_2}{T_1}\right)$$

4.11 When a liquid is compressed its Gibbs energy is increased. To a first approximation the increase in molar Gibbs energy can be calculated using $(\partial \bar{G}/\partial P)_T = \bar{V}$, assuming a constant molar volume. What is the change in the molar Gibbs energy for liquid water when it is compressed to 1000 bar?

SOLUTION

$$\int_{\bar{G}_1}^{\bar{G}_2} d\bar{G} = \int_{P_1}^{P_2} \bar{V}\, dP$$

$$\Delta \bar{G} = \bar{V}\ \Delta P = (18 \times 10^{-6}\ \text{m}^{-3}\ \text{mol}^{-1})(999\ \text{bar})(10^5\ \text{Pa bar}^{-1})$$
$$= 1.8\ \text{kJ mol}^{-1}$$

4.12 An ideal gas is allowed to expand reversibly and isothermally (25 °C) from a pressure of 1 bar to a pressure of 0.1 bar. (a) What is the change in molar Gibbs energy? (b) What would be the change in molar Gibbs energy if the process occurred irreversibly?

SOLUTION

(a)

$$\left(\frac{\partial \bar{G}}{\partial P}\right)_T = \bar{V} = \frac{RT}{P}$$

$$\Delta \bar{G} = RT \ln \frac{P_2}{P_1} = (8.314\ \text{J K}^{-1}\ \text{mol}^{-1})(298.15\ \text{K}) \ln 0.1$$
$$= -5708\ \text{J mol}^{-1}$$

(b) $\Delta\bar{G} = -5708 \text{ J mol}^{-1}$ because G is a state function and depends only on the initial state and the final state.

4.13 The standard entropy of $O_2(g)$ at 1 bar is listed in Appendix C.2 as 205.138 J K^{-1} mol^{-1} at 298 K, and the standard Gibbs energy of formation is listed as 0 kJ mol^{-1}. Assuming O_2 is an ideal gas, what will be the molar entropy and molar Gibbs energy of formation at 100 bar?

SOLUTION

$$\bar{S} = \bar{S}^{\circ} - R \ln \frac{P}{P^{\circ}} = 205.137 - 8.314 \ln 100 = 166.848 \text{ J K}^{-1} \text{ mol}^{-1}$$

$$\Delta_f G^{\circ} = 0 + RT \ln (P/P^{\circ}) = (8.314 \times 10^{-3})(298) \ln 100 = 11.41 \text{ kJ mol}^{-1}$$

4.14 Helium is compressed isothermally and reversibly at 100 °C from a pressure of 2 to 10 bar. Calculate (a) q per mole, (b) w per mole, (c) $\Delta\bar{G}$, (d) $\Delta\bar{A}$, (e) $\Delta\bar{H}$, (f) $\Delta\bar{U}$, and (g) $\Delta\bar{S}$, assuming helium is an ideal gas.

SOLUTION

(a) $\Delta\bar{U} = q + w = 0$

$$q = -w = -RT \ln \frac{P_2}{P_1}$$

$$= -(8.314 \text{ J K}^{-1} \text{ mol}^{-1})(373.15 \text{ K}) \ln \frac{10 \text{ bar}}{2 \text{ bar}} = -4993 \text{ J mol}^{-1}$$

(b) $w = -q = 4993 \text{ J mol}^{-1}$

(c) $$\Delta\bar{G} = \int_{P_1}^{P_2} V dP = RT \ln \frac{P_2}{P_1} = 4993 \text{ J mol}^{-1}$$

(d) $\Delta\bar{A} = w_{max} = 4993 \text{ J mol}^{-1}$

(e) $\Delta\bar{H} = \Delta\bar{U} + \Delta(P\bar{V}) = 0$

(f) $\Delta\bar{U} = 0$

(g) $$\Delta\bar{S} = \Delta\bar{H} - \frac{\Delta\bar{G}}{T} = 0 - \frac{4933 \text{ J mol}^{-1}}{373.15 \text{ K}} = -13.38 \text{ J K}^{-1} \text{ mol}^{-1}$$

4.15 Toluene is vaporized at its boiling point, 111 °C. The heat of vaporization at this temperature is 361.9 J g^{-1}. For the vaporization of toluene, calculate (a) w per mole, (b) q per mole, (c) $\Delta\bar{H}$, (d) $\Delta\bar{U}$, (e) $\Delta\bar{G}$, and (f) $\Delta\bar{S}$.

SOLUTION

(a) Assuming that toluene vapor is an ideal gas and that the volume of the liquid is negligible, the work on the toluene is

$$w = -P\Delta\bar{V} = -RT = -(8.314 \text{ J K}^{-1} \text{ mol}^{-1})(384 \text{ K})$$

$= -3193 \text{ J mol}^{-1}$

(b) $q_P = \Delta \bar{H} = (361.9 \text{ J g}^{-1})(92.13 \text{ g mol}^{-1}) = 33{,}340 \text{ J mol}^{-1}$.

(c) $\Delta\bar{H} = 33{,}340 \text{ J mol}^{-1}$

(d) $\Delta\bar{U} = q + w = 33{,}340 - 3193 = 30{,}147 \text{ J mol}^{-1}$

(e) $\Delta\bar{G} = 0$ because the evaporation is reversible at the boiling point and 1 atm.

(f) $\Delta\bar{S} = \frac{q_{\text{rev}}}{T} = \frac{33\,340 \text{ J mol}^{-1}}{384 \text{ K}} = 86.8 \text{ J K}^{-1} \text{ mol}^{-1}$

4.16 If the Gibbs energy varies with temperature according to

$G/T = a + b/T + c/T^2$

How will the enthalpy and entropy vary with temperature? Check that these three equations are consistent.

SOLUTION

$G = aT + b + c/T$

$\left(\frac{\partial G}{\partial T}\right)_P = - S = a - c/T^2$

$S = - a + c/T^2$

$\left[\frac{\partial (G/T)}{\partial T}\right]_P = - H/T^2 = - b/T^2 - 2c/T^3$

$H = b + 2c/T$

$G = H - TS = aT + b + c/T$

4.17 Calculate the change in molar Gibbs energy $\bar{G}$ when supercooled water at -3 °C freezes at constant T and P. The density of ice at -3 °C is 0.917 x 10^3 kg m^{-3}, and its vapor pressure is 475 Pa. The density of supercooled water at -3 °C is 0.9996 kg m^{-3} and its vapor pressure is 489 Pa.

SOLUTION

$\bar{V}_s = \frac{(18.015 \times 10^{-3} \text{ kg mol}^{-1})}{(0.917 \times 10^3 \text{ kg m}^{-3})}$

$= 1.965 \times 10^{-5} \text{ m}^3 \text{ mol}^{-1}$

$\bar{V}_l = \frac{(18.015 \times 10^{-3} \text{ kg mol}^{-1})}{(0.9996 \times 10^3 \text{ kg m}^{-3})}$

$= 1.802 \times 10^{-5} \text{ m}^3 \text{ mol}^{-1}$

Since the actual process is irreversible, the calculation uses the following reversible isothermal path:

$H_2O(l, 270.15\text{ K}, 10^5\text{ Pa})$ $\qquad$ $H_2O(s, 270.15\text{ K}, 10^5\text{ Pa})$

$\downarrow \Delta \bar{G}_1 = \int_{10^5}^{489} \bar{V}_l\, dP = -1.7\text{ J mol}^{-1}$ $\qquad$ $\uparrow \Delta \bar{G}_5 = \int_{475}^{10^5} \bar{V}_s\, dP = -1.9\text{ J mol}^{-1}$

$H_2O(l, 270.15\text{ K}, 489\text{ Pa})$ $\qquad$ $H_2O(s, 270.15\text{ K}, 475\text{ Pa})$

$\downarrow \Delta \bar{G}_2 = 0$ $\qquad$ $\uparrow \Delta \bar{G}_4 = 0$

$H_2O(g, 270.15\text{ K}, 489\text{ Pa})$ $=$ $H_2O(g, 270.15\text{ K}, 475\text{ Pa})$

$$\Delta \bar{G}_3 = RT \ln(475/489) = -65.2\text{ J mol}^{-1}$$

$$\begin{aligned}\Delta \bar{G} &= \Delta \bar{G}_1 + \Delta \bar{G}_2 + \Delta \bar{G}_3 + \Delta \bar{G}_4 + \Delta \bar{G}_5 \\ &= -1.7 + 0 - 65.2 + 1.9 \\ &= -65.0\text{ J mol}^{-1}\end{aligned}$$

4.18 Calculate the molar Gibbs energy of fusion when supercooled water at -3 °C freezes at constant T and P. The molar enthalpy of fusion of ice is 6000 J mol^{-1} at 0 °C. The heat capacities of water and ice in the vicinity of the freezing point are 75.3 and 38 J K^{-1} mol^{-1}, respectively.

SOLUTION

Since the actual process in irreversible, the calculation uses the following reversible isobaric path.

$H_2O(l, 273.15\text{ K}, 1\text{ bar}) \rightarrow H_2O(s, 273.15, 1\text{ bar})$

$\uparrow \qquad\qquad \downarrow$

$H_2O(l, 270.15\text{ K}, 1\text{ bar}) \rightarrow H_2O(s, 270.15, 1\text{ bar})$

To obtain $\Delta_{fus}G$ (270.15 K), we have to calculate $\Delta_{fus}H$ (270.15 K) and $\Delta_{fus}S$ (270.15 K) separately and use

$\Delta_{fus}G = \Delta_{fus}H - 270.15\, \Delta_{fus}S.$

$$\begin{aligned}\Delta_{fus}H(270.15\text{ K}) &= \Delta_{fus}H(273.15\text{ K}) + \int_{273.15}^{270.15} \Delta_{fus}C_P\, dT \\ &= -6000 + (38 - 75.3)(-3) \\ &= -5888\text{ J mol}^{-1}\end{aligned}$$

$\Delta_{fus}S(270.15\text{ K}) = -6000/273.15 + \Delta_{fus}C_P \ln(270.15/273.15)$

$$= - 21.56 \text{ J mol}^{-1}$$

$$\Delta_{fus}G = \Delta_{fus}H - T\Delta_{fus}S = - 5888 - (270.15)(- 21.56)$$
$$= - 63.6 \text{ J mol}^{-1}$$

4.19 At 298.15 K and a particular pressure, a real gas has a fugacity coefficient ø of 2.00. At this pressure, what is the difference in the chemical potential of this real gas and an ideal gas?

SOLUTION

$$\mu(\text{ideal}) = \mu^o + RT\ln(P/P^o)$$
$$\mu(\text{real}) = \mu^o + RT\ln(\phi P/P^o)$$
$$\mu(\text{real}) - \mu(\text{ideal}) = RT\ln\phi = (8.314 \text{ J K}^{-1} \text{ mol}^{-1})(298.15 \text{ K})\ln 2$$
$$= 1.72 \text{ kJ mol}^{-1}$$

4.20 As shown in Example 4.6, the fugacity of a van der Waals gas is given by a fairly simple expression if only the second virial coefficient is used. To this degree of approximation derive the expressions for $\bar{G}$, $\bar{S}$, $\bar{A}$, $\bar{U}$, $\bar{H}$, and $\bar{V}$.

SOLUTION

$$\bar{G} = \bar{G}^o + RT\ln\left(\frac{f}{P^o}\right) = \bar{G}^o + RT\ln\left(\frac{P}{P^o}\right) + \left(b - \frac{a}{RT}\right)P$$

$$\bar{S} = -\left(\frac{\partial \bar{G}}{\partial T}\right)_P = \bar{S}^o - R\ln\left(\frac{P}{P^o}\right) + \frac{aP}{RT^2}$$

$$\bar{A} = \bar{G} - P\left(\frac{\partial \bar{G}}{\partial P}\right)_T = \bar{G}^o + RT\ln\left(\frac{P}{P^o}\right) - RT$$

$$\bar{U} = \bar{G} - T\left(\frac{\partial \bar{G}}{\partial T}\right)_P - P\left(\frac{\partial \bar{G}}{\partial P}\right)_T = \bar{G}^o + T\bar{S}^o - RT - \frac{aP}{RT}$$
$$= \bar{H}^o - RT - \frac{aP}{RT} = \bar{U}^o - \frac{aP}{RT}$$

Since the second term is small, we can use the ideal gas law to obtain

$$\bar{U} = \bar{U}^o - (a/\bar{V})$$

Note that this agrees with the earlier result (Section 4.3) that

$$\left(\partial\bar{U}/\partial\bar{V}\right)_T = a/\bar{V}^2$$

$$\bar{H} = \bar{U} + P\bar{V} = \bar{U}^o + \frac{aP}{RT} + RT + bP - \frac{aP}{RT}$$

$$= \bar{H}^{\circ} + (b - \frac{2a}{RT})P$$

$$\bar{V} = \left(\frac{\partial \bar{G}}{\partial P}\right)_T = \frac{RT}{P} + b - \frac{a}{RT}$$

4.21 A mole of a van der Waals gas is expanded isothermally from V_1 to V_2. Derive the expressions for the changes in Helmoltz energy and internal energy.

SOLUTION

$dA = -SdT - PdV$

Since T is constant and

$P = RT/(V - b) - a/V^2$

$dA = -(RT/(V - b) - a/V^2)dV$

Integrating from V_1 to V_2 yields

$\Delta A = -RT\ln((V_2\text{-}b)/(V_1\text{-}b)) - a(1/V_2 - 1/V_1)$

The Gibbs-Helmholtz equation

$$\Delta U = -T^2\left(\frac{\partial(\Delta A / T)}{\partial T}\right)_V$$

can be used to calculate the change in internal energy in the isothermal expansion.

$\Delta U = a(1/V_2 - 1/V_1)$

4.22 A one-component system has three natural variables, as is evident from

$dU = TdS - PdV + \mu dn$

W have seen how three additional potentials can be defined by making Legendre transforms and how a complete Legendre transform yields the Gibbs-Duhem equation, which in a certain sense is like a thermodynamic potential but has the value zero. The total number of thermodynamic potentials for a system with D natural variables is 2^D. This is the number of Legendre transforms that can be made taking all possible pairs of conjugate variables, two pairs, three pairs, ...Since $2^3 = 8$, there are three more thermodynamic potentials for this system that can be defined by Legendre transforms. Write their fundamental equations; let's call them X, Y, and Z.

SOLUTION

$X(S,V,\mu) = U - \mu n$

$dX = TdS - PdV - nd\mu$

$Y(T,V,\mu) = U - TS - \mu n$

$\mathrm{d}Y = -S\mathrm{d}T - P\mathrm{d}V - n\mathrm{d}\mu$

$Z(S,P,\mu) = U + pV - \mu n$

$\mathrm{d}Z = T\mathrm{d}S + V\mathrm{d}P - n\mathrm{d}\mu$

This is all the thermodynamic potentials that can be defined for a system with three natural variables. Are any of these three additional thermodynamic potentials really useful? Actually, $Y(T,V,\mu)$ is the thermodynamic potential that corresponds with the grand canonical ensemble of Gibbs (see Chapter16).

4.23 Using the relation derived in Example 4.7, calculate the fugacity of $H_2(g)$ at 100 bar at 298 K.

SOLUTION

$$f = P\exp\left[\left(b - \frac{a}{RT}\right)\frac{P}{RT}\right]$$
$$= (100\text{ bar})\exp\left\{[0.02661 - 0.2476/(0.08314)(298)]100/(0.08314)(298)\right\}$$
$$= 106.9\text{ bar}$$

4.24 Show that if the compressibility factor is given by $Z = 1 + BP/RT$ the fugacity is given by $f = Pe^{Z-1}$. If Z is not very different from unity, $e^{Z-1} = 1 + (Z - 1) + \cdots \cong Z$ so that $f = PZ$. Using this approximation, what is the fugacity of $H_2(g)$ at 50 bar and 298 K using its van der Waals constants?

SOLUTION

Substituting this expression for the compressibility factor in equation 4.78,

$$\frac{f}{P} = \exp\left[\int_0^P (B/RT)\,\mathrm{d}P\right]$$
$$= \exp(BP/RT)$$
$$= \exp(Z - 1)$$

If Z is not very different from unity, $f = PZ$.
For $H_2(g)$ at 50 bar and 298 K,

$$Z = 1 + (b - a/RT)\,P/RT$$
$$= 1 + \left(0.02661\text{ L mol}^{-1} - \frac{0.02476\text{ L}^2\text{ bar mol}^{-2}}{0.08314\text{ L bar K}^{-1}\text{ mol}^{-1}}\right)\frac{50\text{ bar}}{(0.08314)(298)}$$
$$= 1.0335$$

$f = (50\text{ bar})(1.0335) = 5.17\text{ bar}$

*4.25 Calculate the partial molar volume of zinc chloride in 1-molal $ZnCl_2$ solution using the following data.

% by weight of $ZnCl_2$	2	6	10	14	18
Density/g cm^{-3}	1.0167	1.0532	1.0891	1.1275	1.1665

SOLUTION

Consider a kilogram of solution that is 2% by weight $ZnCl_2$, so that it contains 20 g of $ZnCl_2$ and 980 g of H_2O. Since there are (20 g/136.28 g mol^{-1}) mol for 980 g H_2O, the molality is given by m($ZnCl_2$) = 20 x 1000/136.58 x 980 = 0.14975 mol kg^{-1}. The volume of solution containing 1000 g of H_2O is given by

$$\frac{1020.4 \text{ g}}{1.0167 \text{ g cm}^{-3}} = 1003.6 \text{ cm}^3$$

Wt %	Molality	Volume Containing 1000 g of H_2O
2	0.1497 m	1003.2
6	0.4683	1010.1
10	0.8152	1020.2
14	1.194	1031.3
18	1.610	1045.5

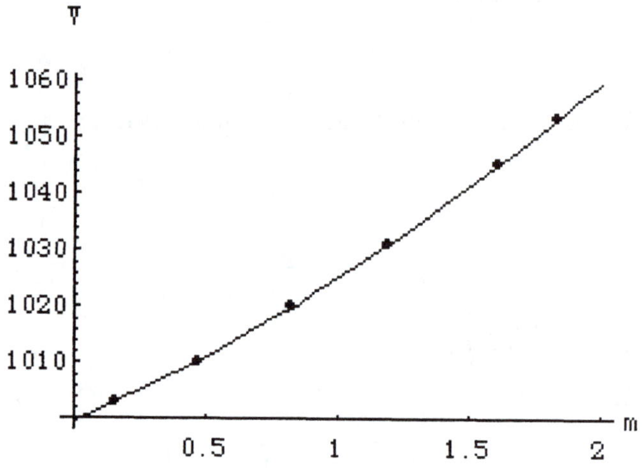

The slope of this plot at $m = 1$ molar can be obtained by drawing a tangent at $m = 1$ and calculating its slope. This yields 29.3 cm^3 mol^{-1}, and so this is the partial molar volume of $ZnCl_2$. *Mathematica* ™ was used to obtain a quadratic fit of the data. This yielded

$V = 999.706 + 21.1601\, m + 4.4639\, m^2$

so that the slope is 21.1601 + (2)(4.4639)m. At m = 1 molal, this yields 30.06 cm^3 mol^{-1}.

4.26 Calculate $\Delta_{mix}G$ and $\Delta_{mix}S$ for the formation of a quantity of air containing 1 mol of gas by mixing nitrogen and oxygen at 298.15 K. Air may be taken to be 80% nitrogen and 20% oxygen by volume.

SOLUTION

$$\begin{aligned}\Delta_{mix}G &= RT(y_1 \ln y_1 + y_2 \ln y_2)\\ &= (8.314 \text{ J K}^{-1} \text{ mol}^{-1})(298.15 \text{ K})(0.8 \ln 0.8 + 0.2 \ln 0.2)\\ &= -1239 \text{ J mol}^{-1}\end{aligned}$$

$$\Delta_{mix}S = -R(y_1 \ln y_1 + y_2 \ln y_2) = 4.159 \text{ J K}^{-1} \text{ mol}^{-1}$$

4.27 A mole of gas A is mixed with a mole of gas B at 1 bar and 298 K. How much work is required to separate these gases to produce a container of each at 1 bar and 298 K?

SOLUTION

The mixture can be separated by diffusion through a perfect semipermeable membrane. The highest partial pressure of each that can be reached is 1/2 bar. These gases then have to be compressed to 1 bar.

$$w = (2 \text{ mol}) \int_{1/2}^{1} \bar{V} \, dP = (2 \text{ mol}) \int_{1/2}^{1} (RT/P) \, dP = (2 \text{ mol}) RT \ln 2$$

$$= 2(8.314)(298) \ln 2$$

$$= 3.4 \text{ kJ}$$

The actual process will require more work.

4.28 The fundamental equation for the enthalpy is given by equation 4.20. Show that the fundamental equation for the internal energy can be obtained by using the inverse Legendre transform $U = H - PV$. This is an example of what is meant by saying that there is no loss of information in making a Legendre transform.

SOLUTION

$$dU = dH - PdV - VdP$$

Substitute equation 4.20 in this to obtain

$$dU = TdS + VdP + \Sigma \mu_i dn_i - PdV - VdP$$

$$= TdS - PdV + \Sigma \mu_i dn_i$$

4.29 Derive

$$C_P = -T\left(\frac{\partial^2 G}{\partial T^2}\right)$$

SOLUTION

$$H = -T^2\left(\frac{\partial(G/T)}{\partial T}\right)_P$$

$$\left(\frac{\partial(G/T)}{\partial T}\right)_P = -\frac{G}{T^2} + \frac{1}{T}\left(\frac{\partial G}{\partial T}\right)_P$$

Multiply this by $-T^2$ to obtain

$$H = G - T\left(\frac{\partial G}{\partial T}\right)_P$$

$$C_P = \left(\frac{\partial H}{\partial T}\right)_P = -T\left(\frac{\partial^2 G}{\partial T^2}\right)_P$$

4.30 In studying statistical mechanics we will find (see Table 16.1) that for a monatomic ideal gas, the molar Gibbs energy is given by

$$\bar{G} = -T \ln \frac{T^{5/2}}{P}$$

where numerical constants have been omitted so that only the functional dependence on the natural variables of $\bar{G}$, that is T and P, is shown. Derive the corresponding equations for $\bar{S}$, $\bar{H}$, $\bar{V}$, $\bar{U}$, and $\bar{A}$.

SOLUTION

The fundamental equation for $\bar{G}$ is

$$\mathrm{d}\bar{G} = -\bar{S}\,\mathrm{d}T + \bar{V}\,\mathrm{d}P$$

Since

$$\bar{G} = -\frac{5}{2}T \ln T + T \ln P$$

the molar entropy is given by

$$\bar{S} = -\left(\frac{\partial \bar{G}}{\partial T}\right)_P = \frac{5}{2} + \frac{5}{2}\ln T - \ln P = \frac{5}{2} + \ln \frac{T^{5/2}}{P}$$

and the molar volume is given by

$$\bar{V} = \left(\frac{\partial \bar{G}}{\partial P}\right)_T = \frac{T}{P}$$

The expressions for the molar enthalpy, internal energy, and Helmholtz energy can be calculated from the following Legendre transforms:

$$\bar{H} = \bar{G} - T\bar{S} = \frac{5}{2}T$$

$$\bar{U} = \bar{H} - P\bar{V} = \frac{3}{2}T$$

$$\bar{A} = \bar{G} - \mathrm{P}\bar{V} = -\frac{5}{2}T\ln T + T\ln P - T = -T\ln\frac{T^{5/2}}{P} - T$$

4.31 Statistical mechanics shows that for a monatomic ideal gas, the molar Gibbs energy is given by

$$\bar{G} = -\frac{5}{2}T\ln T + T\ln P$$

where the numerical factors have been omitted so that only the functional dependence on the natural variables, T and P, is shown. If we want to treat the thermodynamics of an ideal monatomic gas at specified T and $\bar{V}$ without losing any information, we cannot simply replace P with $T/\bar{V}$ and use

$$\bar{G} = -\frac{5}{2}T\ln T + T\ln\frac{T}{\bar{V}}$$

even though this relation is correct. If we want to treat the thermodynamics of an ideal monatomic gas at specified T and $\bar{V}$ without losing any information, we have to use the following Legendre transform to define the molar Helmholtz energy $\bar{A}$.

$$\bar{A} = \bar{G} - P\bar{V}$$

Use the expression for $\bar{A}$ obtained in this way to calculate $\bar{S}$, $\bar{V}$, $\bar{H}$, and $\bar{U}$ for an ideal monatomic gas as a function of T and $\bar{V}$. Show that these expressions agree with the expressions obtained in the preceding problem.

SOLUTION

Substitute the expressions for $\bar{G}$ and P into the Legendre transform to obtain

$$\bar{A} = -\frac{3}{2}T\ln T - T\ln\bar{V} - T$$

The fundamental equation for the molar Helmholtz energy is

$$d\bar{A} = -\bar{S}\,dT - P\,d\bar{V}$$

Thus the molar entropy is given by

$$\bar{S} = -\left(\frac{\partial \bar{A}}{\partial T}\right)_{\bar{V}} = \frac{5}{2} + \frac{3}{2}\ln T + \ln \bar{V} = \frac{5}{2} + \ln T^{3/2}\bar{V}$$

and the molar volume is given by

$$P = -\left(\frac{\partial \bar{A}}{\partial \bar{V}}\right)_T = \frac{T}{\bar{V}}$$

The values of the molar enthalpy and molar internal energy are given by the following Legendre transforms:

$$\bar{H} = \bar{A} + T\bar{S} + P\bar{V} = \frac{5}{2}T$$

$$\bar{U} = \bar{A} + T\bar{S} = \frac{3}{2}T$$

Substituting $\bar{V} = T/P$ in these expressions yields the same equations obtained in the previous problem for these thermodynamic properties. The moral is that we can change the variables in the expression for a thermodynamic property, but we cannot differentiate this expression to obtain other thermodynamic properties unless the variables in the expression are the natural variables for the thermodynamic potential.

4.32 We already know enough about the thermodynamics of a monatomic ideal gas to express *V*, *U*, and *S* in terms of the natural variables of *G*, namely *T*, *P*, and *n*.

$$V = nRT/P$$

$$U = \frac{3}{2}nRT$$

$$S = nR\left\{\frac{\bar{S}^{\circ}}{R} + \ln\left[\left(\frac{T}{T^{\circ}}\right)^{5/2}\left(\frac{P^{\circ}}{P}\right)\right]\right\}$$

The last equation is the Sackur-Tetrode equation, where $\bar{S}^{\circ}$ is the molar entropy at the standard temperature T°(298.15 K) and standard pressure P°(1 bar). The Gibbs energy $G(T,P,n)$ of the ideal monatomic gas can be calculated by using the Legendre transform

$$G = U + PV - TS$$

The fundamental equation for *G* is

$$dG = -S\,dT + V\,dP + \mu\,dn$$

Show that the correct expressions for *S*, *V*, and μ are obtained by using the partial derivatives of *G* indicated by this fundamental equation.

SOLUTION

$$G = \frac{3}{2}nRT + nRT - nRT\left\{\frac{\bar{S}^{\circ}}{R} + \ln\left[\left(\frac{T}{T^{\circ}}\right)^{5/2}\left(\frac{P^{\circ}}{P}\right)\right]\right\}$$

$$S = -\left(\frac{\partial G}{\partial T}\right)_{P,n}$$

$$= -\frac{3}{2}nRT - nR + nR\left\{\frac{\bar{S}^{\circ}}{R} + \ln\left[\left(\frac{T}{T^{\circ}}\right)^{5/2}\left(\frac{P^{\circ}}{P}\right)\right]\right\} + nRT\frac{\mathrm{d}}{\mathrm{d}T}\left(\frac{5}{2}\ln T\right)$$

$$= nR\left\{\frac{\bar{S}^{\circ}}{R} + \ln\left[\left(\frac{T}{T^{\circ}}\right)^{5/2}\left(\frac{P^{\circ}}{P}\right)\right]\right\}$$

$$V = \left(\frac{\partial G}{\partial P}\right)_{T,n} = nRT/P$$

$$\mu = \left(\frac{\partial G}{\partial n}\right)_{T,P} = RT\left(\frac{5}{2} - \bar{S}^{\circ}\right) + RT\ln\left[\left(\frac{T}{T^{\circ}}\right)^{5/2}\left(\frac{P}{P^{\circ}}\right)\right]$$

Thus the correct expressions are obtained for S and V and the equation for the chemical potential is consistent with $\mu = \mu^{\circ} + RT\ln(P/P^{\circ})$.

4.34 $\mathrm{d}S = (C_V/T)\mathrm{d}T + (\alpha/\kappa)\mathrm{d}V$

4.35 - 0.0252 J K^{-1} mol^{-1}

4.39 88.2 J mol^{-1}

4.40 (a) 4993, (b) 3655 J mol^{-1}

4.41 168.97 J K^{-1} mol^{-1}, 0, -11.42 kJ mol^{-1}

4.42 (a) 6820 J mol^{-1} (b) 6072 J mol^{-1} (c) 75.9 J K^{-1} mol^{-1} (d) 0

4.43 (a) - 5229 J mol^{-1}, (b) 5229 J mol^{-1}, (c) 0, (d) - 5229 J mol^{-1}, (e) 19.14 J K^{-1} mol^{-1}, (f) 0, (g) 0, (h) 0, (i) - 5229 J mol^{-1}, (j) 19.14 J K^{-1} mol^{-1}, (k) 0, (l) 19.14 J K^{-1} mol^{-1}

4.44 3100 J mol^{-1}, - 40,690 J mol^{-1}, - 37.6 kJ mol^{-1}, - 34.5 kJ mol^{-1}, 3.1 kJ mol^{-1}, - 109.0 J K^{-1} mol^{-1}, 0 kJ mol^{-1}

4.45 (a) - 6.754 kJ mol^{-1}, (b) 0, (c) 0, (d) -6.754 kJ mol^{-1}, (e) 22.51 J K^{-1} mol^{-1}

4.46 0, 19.16 J K^{-1} mol^{-1}

4.47 0, -13.38 J K^{-1} mol^{-1}, 4991 J mol^{-1}

4.48 33.34 kJ mol^{-1}, 73.52 J K^{-1} mol^{-1}, 0

4.49 - 2121 J mol^{-1}

4.50 $\Delta\bar{G} = RT\ln(P_2/P_1) + (b - a/RT)(P_2 - P_1)$

4.52 0.79 x 10^{-3} m^3 kg^{-1}

4.53 (a) 0.2033×10^{-3} m^{-3} kg^{-1} (b) 19.36×10^{-6} m^3 mol^{-1}

4.54 - 4731 J, 15.88 J K^{-1}

4.55 $\Delta_{vap}S = 109.3$ J K^{-1} mol^{-1}

$\Delta_{vap}H = 39.9$ kJ mol^{-1}

$\Delta_{vap}G = - 3.6$ kJ mol^{-1}

5

Chemical Equilibrium

5.1 For the reaction $N_2(g) + 3H_2(g) = 2NH_3(g)$, $K = 1.60 \times 10^{-4}$ at 400 °C. Calculate (a) $\Delta_r G^o$ and (b) $\Delta_r G$ when the pressures of N_2 and H_2 are maintained at 10 and 30 bar, respectively, and NH_3 is removed at a partial pressure of 3 bar. (c) Is the reaction spontaneous under the latter conditions?

SOLUTION

(a) $\Delta_r G^o = - RT \ln K$

$$= - (8.314 \text{ J K}^{-1} \text{ mol}^{-1})(673 \text{ K}) \ln 1.60 \times 10^{-4} = 48.9 \text{ kJ mol}^{-1}$$

(b) $$\Delta_r G = \Delta_r G^o + RT \ln \frac{(P_{NH_3}/P^o)^2}{(P_{N_2}/P^o)(P_{H_2}/P^o)^3}$$

$$= 48.9 + (8.314 \times 10^{-3})(673) \ln \frac{3^2}{10(30)^3} = - 8.78 \text{ kJ mol}^{-1}$$

(c) Yes

5.2 At 1:3 mixture of nitrogen and hydrogen was passed over a catalyst at 450 °C. It was found that 2.04% by volume of ammonia was formed when the total pressure was maintained at 10.13 bar. Calculate the value of K for $\frac{3}{2}$ $H_2(g) + \frac{1}{2}$ $N_2(g) = NH_3(g)$ at this temperature.

SOLUTION

At equilibrium $P_{H_2} + P_{N_2} + P_{NH_3} = 10.13$ bar

$P_{NH_3} = (10.13 \text{ bar})(0.0204) = 0.207$ bar

$P_{H_2} + P_{N_2} = 10.13 \text{ bar} - 0.207 \text{ bar} = 9.923$ bar

$P_{H_2} = 3P_{N_2}$ because this initial ratio is not changed by reaction

$$P_{N_2} = \frac{9.923 \text{ bar}}{4} = 2.481 \text{ bar}$$

$$P_{H_2} = \frac{3}{4}(9.923 \text{ bar}) = 7.442 \text{ bar}$$

$$K = \frac{(P_{NH_3}/P^o)}{(P_{H_2}/P^o)^{3/2}(P_{N_2}/P^o)^{1/2}}$$

$$= \frac{0.207}{7.442^{3/2}\ 2.482^{1/2}} = 6.47 \times 10^{-3}$$

5.3 At 55 °C and 1 bar the average molar mass of partially dissociated N_2O_4 is 61.2 g mol^{-1}. Calculate (a) ξ and (b) K for the reaction $N_2O_4(g) = 2NO_2(g)$. (c) Calculate ξ at 55 °C if the total pressure is reduced to 0.1 bar.

SOLUTION

(a) $$\xi = \frac{M_1 - M_2}{M_2} = \frac{92.0 - 61.2}{61.2} = 0.503$$

(b) $$K = \frac{4\xi^2(P/P^o)}{1 - \xi^2} = \frac{4(0.503)^2}{1 - 0.503^2} = 1.36$$

(c) $$\frac{\xi^2}{1 - \xi^2} = \frac{K}{4(P/P^o)} = \frac{1.36}{4(0.1)}$$

$\xi = 0.879$

(Note that ξ is the dimensionless extent of reaction.)

5.4 A 1 liter reaction vessel containing 0.233 mol of N_2 and 0.341 mol of PCl_5 is heated to 250 °C. The total pressure at equilibrium is 29.33 bar. Assuming that all the gases are ideal, calculate K for the only reaction that occurs.
$PCl_5(g) = PCl_3(g) + Cl_2(g)$

SOLUTION

	PCl_5	=	PCl_3	+	Cl_2	
initial	0.341		0		0	
eq.	0.341 - ξ		ξ		ξ	total = 0.341 + ξ

$$K = \frac{\xi^2(P/P^o)^2}{(0.341 + \xi)(0.341 - \xi)}$$

$$= \frac{\xi^2(P/P^o)}{0.341^2 - \xi^2} \text{ where } P = P_{PCl_5} + P_{PCl_3} + P_{Cl_2}$$

Calculate the equilibrium pressure of the reacting gases by subtracting the partial pressure of nitrogen from the total pressure.

$$P = 29.33 \text{ bar} - \frac{(0.233 \text{ mol})(0.08314 \text{ L bar K}^{-1}\text{mol}^{-1})(523 \text{ K})}{1 \text{ L}}$$

$$= 19.20 \text{ bar} = \frac{(0.341 + \xi)(0.08314)(523)}{1}$$

$\xi = 0.1005$

$$K = \frac{(0.1005)^2(19.2)}{0.341^2 - 0.1005^2} = 1.83$$

5.5 An evacuated tube containing 5.96×10^{-3} mol L^{-1} of solid iodine is heated to 973 K. The experimentally determined pressure is 0.496 bar. Assuming ideal gas behavior, calculate K for $I_2(g) = 2I(g)$.

SOLUTION

$$P = \frac{n}{V} RT$$

$$0.496 \text{ bar} = (1 + \xi)(5.96 \times 10^{-3} \text{ L}^{-1}) \text{x} (0.08314 \text{ L bar K}^{-1} \text{ mol}^{-1})(973 \text{ K})$$

$$\xi = \frac{0.496 \text{ bar}}{(5.96 \times 10^{-3} \text{ mol L}^{-1})(0.08314 \text{ L bar}^{-1} \text{ K}^{-1} \text{ mol}^{-1})(973 \text{ K})} - 1$$

$= 0.0288$

$$K = \frac{4\xi^2(P/P^o)}{1 - \xi^2} = \frac{4(0.0288)^2(0.496)}{1 - 0.0287^2} = 1.64 \times 10^{-3}$$

5.6 Nitrogen trioxide dissociates according to the reaction
$N_2O_3(g) = NO_2(g) + NO(g)$
When one mole of $N_2O_3(g)$ is held at 25 °C and 1 bar total pressure until equilibrium is reached, the extent of reaction is 0.30. What is $\Delta_r G^o$ for this reaction at 25 °C?

SOLUTION

	$N_2O_3(g)$	=	$NO_2(g)$	$+ NO(g)$	
init.	1		0	0	
eq.	$1 - \xi$		ξ	ξ	total = $1 + \xi$
mole fr.	$\frac{1 - \xi}{1 + \xi}$		$\frac{\xi}{1 + \xi}$	$\frac{\xi}{1 + \xi}$	

$$K = \frac{\xi^2 P^2 (1+\xi)}{(1+\xi)^2(1-\xi)P} = \frac{\xi^2 P}{1-\xi^2} = \frac{0.30^2}{0.91} = 0.099$$

$\Delta_r G^o = - RT \ln K = - 8.314\ (298) \ln 0.099 = 5.73$ kJ mol^{-1}

*5.7 For the reaction
$2HI(g) = H_2(g) + I_2(g)$
at 698.6 K, $K = 1.83 \times 10^{-2}$. (a) How many grams of hydrogen iodide will be formed when 10 g of iodine and 0.2 g of hydrogen are heated to this temperature in a 3 L vessel? (b) What will be the partial pressures of H_2, I_2, and HI?

SOLUTION

(a) Pressures due to reactants prior to reaction:

$$P_{I_2} = \frac{(10\text{ g})(0.08314\text{ L bar K}^{-1}\text{ mol}^{-1})(698.6\text{ K})}{(254\text{ g mol}^{-1})(3\text{ L})} = 0.762\text{ bar}$$

$$P_{H_2} = \frac{(0.2)(0.08314\text{ L bar K}^{-1}\text{ mol}^{-1})(698.6\text{ K})}{2 \text{ x } 3} = 1.936\text{ bar}$$

$$K = \frac{(0.762 - x)(1.936 - x)}{(2x)^2} = 1.83 \text{ x } 10^{-2}$$

$$x = 0.730\text{ bar}$$

(b) $P_{H_2} = 1.936 - 0.730 = 1.206$ bar

$P_{I_2} = 0.762 - 0.730 = 0.032$ bar

$P_{HI} = 1.460$ bar

5.8 Express K for the reaction
$CO(g) + 3H_2(g) = CH_4(g) + H_2O(g)$
in terms of the equilibrium extent of reaction ξ when one mole of CO is mixed with one mole of hydrogen.

SOLUTION

	CO +	$3H_2$ =	CH_4 +	H_2O	
initial	1	1	0	0	
equilibrium	$1 - \xi$	$1 - 3\xi$	ξ	ξ	total = $2 - 2\xi$

$$K = \frac{\left(\frac{\xi}{2 - 2\xi}\right)^2\left(\frac{P}{P^o}\right)^2}{\left(\frac{1 - \xi}{2 - 2\xi}\right)\left(\frac{1 - 3\xi}{2 - 2\xi}\right)^3\left(\frac{P}{P^o}\right)^4}$$

$$K = \frac{(\xi)^2(2 - 2\xi)^2}{(1 - \xi)(1 - 3\xi)^3(P/P^o)^2}$$

5.9 What are the percentage dissociations of $H_2(g)$, $O_2(g)$, and $I_2(g)$ at 2000 K and a total pressure of 1 bar?

SOLUTION

$H_2(g) = 2H(g)$

$\Delta_r G^o = 2(106{,}760 \text{ J mol}^{-1})$

$= - RT \ln K$

$= - (8.3145 \text{ J K}^{-1} \text{ mol}^{-1})(2000 \text{ K}) \ln K$

$$K = 2.65 \times 10^{-6} = \frac{4\xi^2}{1 - \xi^2}$$

$$\xi = \left(\frac{K}{K+4}\right)^{1/2} = \left(\frac{2.65 \times 10^{-6}}{4}\right)^{1/2} = 0.000814 \text{ or } 0.0814\%$$

$O_2(g) = 2O(g)$

$\Delta_r G^o = 2(121{,}552 \text{ J mol}^{-1})$

$K = 4.48 \times 10^{-7}$ $\qquad \xi = 0.0335\%$

$I_2(g) = 2I(g)$

$\Delta_r G^o = 2(- 29{,}410 \text{ J mol}^{-1})$

$K = 34.37$ $\qquad \xi = 94.6\%$

5.10 In order to produce more hydrogen from "synthesis gas" ($CO + H_2$) the water gas shift reaction is used.

$$CO(g) + H_2O(g) = CO_2(g) + H_2(g)$$

Calculate K at 1000 K and the equilibrium extent of reaction starting with an equimolar mixture of CO and H_2O.

SOLUTION

$\Delta G^o = - 395{,}886 - (- 200{,}275) - (- 192{,}590) = -3021 \text{ J mol}^{-1}$

$= - (8.314 \text{ J K}^{-1} \text{ mol}^{-1})(1000 \text{ K}) \ln K$

$$K = 1.44 = \frac{P_{H_2} P_{CO_2}}{P_{CO} P_{H_2O}} = \frac{\xi^2}{(1 - \xi)^2}$$

$\xi = 0.545$

(Note that this reaction is exothermic so that there will be a larger extent of reaction at lower temperatures. In practice this reaction is usually carried out at about 700 K.)

5.11 Calculate the extent of reaction ξ of 1 mol of $H_2O(g)$ to form $H_2(g)$ and $O_2(g)$ at 2000 K and 1 bar. (Since the extent of reaction is small, the calculation may be simplified by assuming that $P_{H_2O} = 1$ bar.)

SOLUTION

$H_2O(g)$	=	$H_2(g)$ +	$\frac{1}{2} O_2(g)$
init. 1		0	0

eq. $1 - \xi$ ξ $\frac{1}{2}\xi$

$\Delta_r G^o = 135{,}528 \text{ J mol}^{-1}$

$= - (8.1315 \text{ J K}^{-1} \text{ mol}^{-1})(2000 \text{ K}) \ln K$

$K = 2.887 \times 10^{-4}$

$$K = \frac{\left(\frac{P_{H_2}}{P^o}\right)\left(\frac{P_{O_2}}{P^o}\right)^{1/2}}{\left(\frac{P_{H_2O}}{P^o}\right)} ; \frac{P_{O_2}}{P^o} = \frac{P_{H_2}}{2P^o}$$

$$\left(\frac{P_{H_2}}{P^o}\right)\left(\frac{P_{H_2}}{2P^o}\right)^{1/2} = \frac{1}{\sqrt{2}}\left(\frac{P_{H_2}}{P^o}\right)^{3/2}$$

$$\xi = \left(\frac{P_{H_2}}{P^o}\right) = (\sqrt{2}\ K)^{2/3} = 0.0055$$

5.12 At 500 K CH_3OH, CH_4 and other hydrocarbons can be formed from CO and H_2. Until recently the main source of the CO mixture for the synthesis of CH_3OH was methane.

$CH_4(g) + H_2O(g) = CO(g) + 3H_2(g)$

When coal is used as the source, the "synthesis gas" has a different composition.

$C(\text{graphite}) + H_2O(g) = CO(g) + H_2(g)$

Suppose we have a catalyst that catalyzes only the formation of CH_3OH. (a) What pressure is required to convert 25% of the CO to CH_3OH at 500 K if the "synthesis gas" comes from CH_4? (b) If the synthesis gas comes from coal?

SOLUTION

(a) $CO + 2H_2 = CH_3OH$

Initial 1 3 0

Equil. $1 - \xi$ $3 - 2\xi$ ξ Total = $4 - 2\xi$

$$K = \frac{y_{CH_3OH}}{y_{CO}y_{H_2}^2P^2} = \frac{\xi(4 - 2\xi)^2}{(1 - \xi)(3 - 2\xi)^2P^2}$$

$\Delta_r G^o = - 134.27 - (- 155.41) = 22.14 \text{ kJ mol}^{-1}$

$K = 6.188 \times 10^{-3}$

$$P = \left[\frac{\xi(4 - 2\xi)^2}{K(1 - \xi)(3 - 2\xi)^2}\right]^{1/2}$$

$$= \left[\frac{(0.25)(3.5)^2}{6.188 \times 10^{-3}(0.75)(2.5)^2}\right]^{1/2} = 10.3 \text{ bar}$$

(b) $CO + 2H_2 = CH_3OH$

Initial	1	1	0	
Equil.	$1 - \xi$	$1 - 2\xi$	ξ	Total = $2 - 2\xi$

$$K = \frac{\xi(2 - 2\xi)^2}{(1 - \xi)(1 - 2\xi)^2P^2}$$

$$P = \left[\frac{(0.25)(1.5)^2}{K(0.75)(0.5)^2}\right]^{1/2} = 22.0 \text{ bar}$$

5.13 Many equilibrium constants in the literature were calculated with a standard state pressure of 1 atm (1.01325 bar). Show that the corresponding equilibrium constant with a standard pressure of 1 bar can be calculated using

$K(\text{bar}) = K(\text{atm})(1.01325)^{\Sigma\nu_i}$

where the ν_i are the stoichiometric numbers of the gaseous reactants.

SOLUTION

$K(\text{bar}) = \prod [P_i/(1 \text{ bar})]^{\nu_i}$

$K(\text{atm}) = \prod [P_i/(1 \text{ atm})]^{\nu_i}$

$$\frac{K(\text{bar})}{K(\text{atm})} = \prod_i \left(\frac{1 \text{ atm}}{1 \text{ bar}}\right)^{\nu_i} = \left(\frac{1 \text{ atm}}{1 \text{ bar}}\right)^{\Sigma\nu_i}$$

$$= 1.01325^{\Sigma\nu_i}$$

5.14 Older tables of chemical thermodynamic properties are based on a standard state pressure of 1 atm. Show that the corresponding $\Delta_f G_i^\circ$ with a standard state pressure of 1 bar can be calculated using

$\Delta_f G_j^\circ(\text{bar}) = \Delta_f G_j^\circ(\text{atm}) - (0.109 \times 10^{-3} \text{ kJ K}^{-1} \text{ mol}^{-1})\, T\, \Sigma\nu_i$

where the ν_i are the stoichiometric numbers of the gaseous reactants and products in the formation reaction.

SOLUTION

$$\Delta_f G_j^o(\text{bar}) = -RT\ln\{\prod_i [P_i/(1\ \text{atm})]^{\nu_i}\}$$

$$= -RT\ln\{\prod_i [P_i/(1\ \text{bar})]^{\nu_i}[(1\ \text{atm})/(1\ \text{bar})]^{\Sigma\nu_i}\}$$

$$= -RT\ln\{\prod_i [P_i/(1\ \text{atm})]^{\nu_i}\} - RT\ln[(1\ \text{atm})/(1\ \text{bar})]^{\Sigma\nu_i}$$

$$= \Delta_f G_j^o(\text{atm}) - RT\ln(1.01325)^{\Sigma\nu_i}$$

5.15 Show that the equilibrium mole fractions of *n*-butane and *iso*-butane are given by

$$y_n = \frac{e^{-\Delta_f G_n{}^o/RT}}{e^{-\Delta_f G_n{}^o/RT} + e^{-\Delta_f G_{iso}{}^o/RT}}$$

$$y_{iso} = \frac{e^{-\Delta_f G_{iso}{}^o/RT}}{e^{-\Delta_f G_n{}^o/RT} + e^{-\Delta_f G_{iso}{}^o/RT}}$$

SOLUTION

4C(graphite) + $5H_2$(g) = C_4H_{10}(g,n)

$$K_n = P_n/P_{H_2}^5 = e^{-\Delta_f G_n{}^o/RT}$$

4C(graphite) + $5H_2$(g) = C_4H_{10}(g,iso)

$$K_{iso} = P_{iso}/P_{H_2}^5 = e^{-\Delta_f G_{iso}{}^o/RT}$$

$$y_n = \frac{P_n}{P_n + P_{iso}} = \frac{P_{H_2}^5 e^{-\Delta_f G_n{}^o/RT}}{P_{H_2}^5\left(e^{-\Delta_f G_n{}^o/RT} + e^{-\Delta_f G_{iso}{}^o/RT}\right)}$$

$$= \frac{e^{-\Delta_f G_n{}^o/RT}}{e^{-\Delta_f G_n{}^o/RT} + e^{-\Delta_f G_{iso}{}^o/RT}}$$

*5.16 Calculate the molar Gibbs energy of butane isomers for extents of reaction of 0.2, 0.4, 0.6, and 0.8 for the reaction

n-butane = *iso*-butane

at 1000 K and 1 bar. At 1000 K,
$\Delta_f G^o$ (*n*-butane) = 270 kJ mol^{-1}

$\Delta_f G^o$ (*iso*-butane) = 276.6 kJ mol^{-1}

Make a plot and show that the minimum corresponds with the equilibrium extent of reaction.

SOLUTION

n-butane = *iso*-butane

1	0	moles at $t = 0$
$1 - \xi$	ξ	moles at equilibrium; $n_T = 1$ mol

$$G = n_n\mu_n + n_i\mu_i = n_n\mu_n^o + n_nRT\ln(x_n) + n_i\mu_i^o + n_iRT\ln(x_i)$$

$$= (1 - \xi)\,\Delta_f G_n^o + (1 - \xi)RT\ln(1 - \xi) + \xi\Delta_f G_i^o + \xi RT\ln(\xi)$$

$$= 270 + (276.6 - 270)\,\xi + RT[(1 - \xi)\ln(1 - \xi) + \xi\ln(\xi)]$$

$$G(\xi) = 270 + 6.6\xi + 8.314[(1 - \xi)\ln(1 - \xi) + \xi\ln(\xi)] \text{ kJ mol}^{-1}$$

$G(0.2) = 267.2$ $\quad$ $G(0.6) = 268.4$

$G(0.4) = 267.0$ $\quad$ $G(0.8) = 271.1$ $\quad$ in kJ/mol

$$\Delta G^o = \Delta_f G_i^o - \Delta_f G_n^o = 6.6 \text{ kJ mol}^{-1} = -RT\ln K$$

$$K = \frac{\xi}{1 - \xi} = \exp\left[\frac{-\Delta_r G^o}{RT}\right] = 0.4521\ (T = 10^3\text{ K})$$

$\xi = 0.311$

$G(0.311) = 266.9$ kJ/mol corresponds to the minimum of the following graph of $G(\xi)$ versus extent of reaction

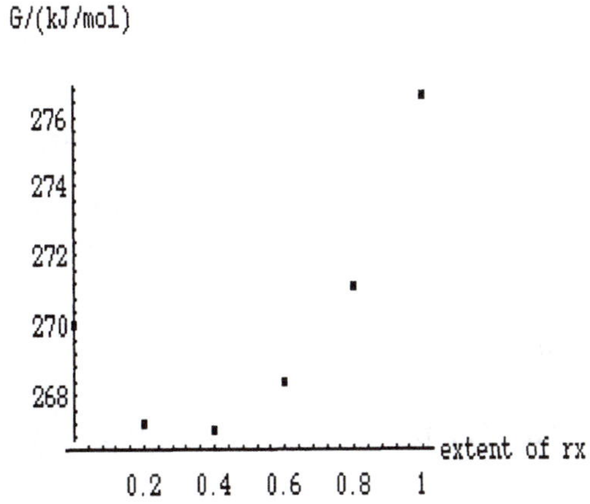

5.17 In the synthesis of methanol by $CO(g) + 2H_2(g) = CH_3OH(g)$ at 500 K, calculate the total pressure required for a 90% conversion to methanol if CO and H_2 are initially in a 1:2 ratio. Given: $K = 6.09 \times 10^{-3}$.

SOLUTION

$$CO(g) + 2H_2(g) = CH_3OH(g)$$

initial moles 1 2 0

equil. moles 0.1 0.2 0.9 total 1.2

$$K = \frac{(P_{CH_3OH}/P^o)}{(P_{CO}/P^o)(P_{H_2}/P^o)^2} = 6.09 \times 10^{-3}$$

$$= \frac{\frac{0.9}{1.2}\frac{P}{P^o}}{\frac{0.1}{1.2}\frac{P}{P^o}\left(\frac{0.2}{1.2}\frac{P}{P^o}\right)^2}$$

$$\frac{P}{P^o} = \sqrt{\frac{(0.9)(1.2)^2}{(0.1)(0.04)(6.09 \times 10^{-3})}} = 231$$

$P = 231$ bar = total pressure for 90% conversion to CH_3OH

5.18 At 1273 K and at a total pressure of 30.4 bar the equilibrium in the reaction $CO_2(g) + C(s) = 2CO(g)$ is such that 17 mole % of the gas is CO_2. (a) What percentage would be CO_2 if the total pressure were 20.3 bar? (b) What would be the effect on the equilibrium of adding N_2 to the reaction mixture in a closed vessel until the partial pressure of N_2 is 10 bar? (c) At what pressure of the reactants will 25% of the gas be CO_2?

SOLUTION

(a) $P_{CO_2} = (30.4 \text{ bar})(0.17) = 5.2$ bar

$P_{CO} = (30.4 \text{ bar})(0.83) = 25.2$ bar

$$K = \frac{(25.2)^2}{5.2} = 122$$

Let x = mole fraction CO_2

$$K = \frac{[20.3(1 - x)]^2}{20.3x} = 122$$

$x = 0.127$

Percentage CO_2 at equilibrium = 12.7%

(b) No effect for ideal gases because the partial pressures of the reactants are not affected.

(c) $$K = \frac{[0.75(P/P^o)]^2}{0.25(P/P^o)}$$

$P = 54$ bar

5.19 When alkanes are heated up, they lose hydrogen and alkenes are produced. For example,

$C_2H_6(g) = C_2H_4(g) + H_2(g) \quad K = 0.36$ at 1000 K

If this is the only reaction that occurs when ethane is heated to 1000 K, at what total pressure will ethane be (a) 10% dissociated and (b) 90% dissociated to ethylene and hydrogen?

SOLUTION

$$C_2H_6 = C_2H_4 + H_2$$

Init.	1	0	0	
Equil.	$1 - \xi$	ξ	ξ	total = $1 + \xi$

$$K = \frac{\xi^2(P/P^o)}{(1 + \xi)(1 - \xi)} = \frac{\xi^2(P/P^o)}{1 - \xi^2}$$

(a) $$0.36 = \frac{0.1^2(P/P^o)}{1 - 0.1^2} = \frac{0.01(P/P^o)}{0.99}$$

$$\frac{P}{P^o} = \left[\frac{(0.99)(0.36)}{0.01}\right] = 35.6$$

(b) $$0.36 = \frac{0.9^2(P/P^o)}{1 - 0.9^2} = \frac{0.81(P/P^o)}{0.19}$$

$$P = \left[\frac{(0.19)(0.36)}{0.81}\right] = 0.084 \text{ bar}$$

5.20 At 2000 °C water is 2% dissociated into oxygen and hydrogen at a total pressure of 1 bar.

(a) Calculate K for $H_2O(g) = H_2(g) + \frac{1}{2} O_2(g)$

(b) Will the extent of reaction increase or decrease if the pressure is reduced? (c) Will the extent of reaction increase or decrease if argon gas is added, holding the total pressure equal to 1 bar? (d) Will the extent of reaction change if the pressure is raised by addition of argon at constant volume to the closed system containing partially dissociated water vapor? (e) Will the extent of reaction increase or decrease if oxygen is added while holding the total pressure constant at 1 bar?

SOLUTION

(a)	$H_2O(g)$	$=$	$H_2(g)$	$+ \frac{1}{2} O_2(g)$
initial	1		0	0
equilibrium	$1 - \xi$		ξ	$\xi/2$ total = $1 + \xi/2$

$$P_{H_2O} = \frac{1-\xi}{1+\frac{\xi}{2}} P \qquad P_{H_2} = \frac{\xi}{1+\frac{\xi}{2}} P \qquad P_{O_2} = \frac{\xi/2}{1+\frac{\xi}{2}} P$$

$$K = \frac{\left[\frac{\xi/2}{1+\xi/2}\frac{P}{P^o}\right]^{1/2}\left[\frac{\xi}{1+\xi/2}\frac{P}{P^o}\right]}{\frac{1-\xi}{1+\xi/2}\frac{P}{P^o}} = \frac{\xi^{3/2}(P/P^o)^{1/2}}{\sqrt{2}(1+\xi/2)^{1/2}(1-\xi)}$$

$$= \frac{0.02^{3/2}\ 1^{1/2}}{\sqrt{2}\ (1.01)^{1/2}\ (0.98)} = 2.03 \times 10^{-3}$$

(b) If the total pressure is reduced, the extent of reaction will increase because the reaction will produce more molecules to fill the volume.

(c) If argon is added at constant pressure, the extent of reaction will increase because the partial pressure due to the reactants will decrease.

(d) If argon is added at constant volume, the extent of reaction will not be changed because the partial pressure due to the reactants will not change.

(e) If oxygen is added at constant total pressure, the extent of reaction of H_2O will decrease because the reaction will be pushed to the left.

5.21 At 250 ^oC PCl_5 is 80% dissociated at a pressure of 1.013 bar, and so $K = 1.80$. What is the extent of reaction at equilibrium after sufficient nitrogen has been added at constant pressure to produce a nitrogen partial pressure of 0.9 bar? The total pressure is maintained at 1 bar.

SOLUTION

$$K = \frac{0.8^2\ (1.013)}{1 - 0.64} = 1.80$$

$$K = \frac{\xi^2(0.1)}{1 - \xi^2}$$

$\xi = 0.973$ or 97.3% dissociated.

5.22 The following exothermic reaction is at equilibrium at 500 K and 10 bar.

$CO(g) + 2H_2(g) = CH_3OH(g)$

Assuming the gases are ideal, what will happen to the amount of methanol at equilibrium when (a) the temperature is raised, (b) the pressure is increased, (c) an inert gas is pumped in at constant volume, (d) an inert gas is pumped in at constant pressure, and (e) hydrogen gas is added at constant pressure?

SOLUTION

(a) $n(CH_3OH)$ decreases because heat is involved.

(b) $n(CH_3OH)$ increases because there are fewer molecules of gas in the product.

(c) No effect.

(d) $n(CH_3OH)$ decreases because the volume increases.

(e) $n(CH_3OH)$ increases because there is more of a reactant. Note: the effect of the addition of CO is more complicated.

5.23 The following reaction is nonspontaneous at room temperature and endothermic.
$3C(graphite) + 2H_2O(g) = CH_4(g) + 2CO(g)$
As the temperature is raised, the equilibrium constant will become equal to unity at some point. Estimate this temperature using data from Appendix C.3.

SOLUTION

At 1000 K

$\Delta_r G^o = 19.492 + 2(-200.275) - 2(-192.590) = 4.122 \text{ kJ mol}^{-1}$

$= -(8.3145 \times 10^{-3} \text{ kJ K}^{-1} \text{ mol}^{-1})(1000 \text{ K}) \ln K$

$K = 0.609$

$\Delta_r H^o = -89.849 + 2(-111.983) - 2(-247.857) = 181.899 \text{ kJ mol}^{-1}$

$$\ln \frac{K_2}{K_1} = \frac{\Delta H^o}{R}\left(\frac{1}{T_1} - \frac{1}{T_2}\right)$$

$$\ln \frac{1}{0.609} = \frac{181\,899 \text{ J mol}^{-1}}{8.3145 \text{ J K}^{-1} \text{ mol}^{-1}}\left(\frac{1}{1000 \text{ K}} - \frac{1}{T_2}\right)$$

$$T_2 = \frac{1}{\frac{1}{1000 \text{ K}} - \frac{8.3145}{181899} \ln \frac{1}{0.609}} = 1023 \text{ K}$$

5.24 The measured density of an equilibrium mixture of N_2O_4 and NO_2 at 15 °C and 1.013 bar is 3.62 g L^{-1}, and the density at 75 °C and 1.013 bar is 1.84 g L^{-1}. What is the enthalpy change of the reaction $N_2O_4(g) = 2NO_2(g)$?

SOLUTION

At 15 °C

$$M = \frac{RT}{P}\frac{g}{V} = \frac{(0.08314)(288)(3.62)}{1.013} = 85.57 \text{ g mol}^{-1}$$

$$\xi = \frac{M_1 - M_2}{M_2} = \frac{92.01 - 85.57}{85.57} = 0.0753$$

$$K = \frac{4\xi^2 P}{1 - \xi^2} = \frac{4(0.0753)^2(1.013)}{1 - 0.0753^2} = 0.0231$$

At 75 °C

$$M = \frac{(0.08314)(348)(1.84)}{1.013} = 52.55 \text{ g mol}^{-1}$$

$$\xi = \frac{92.01 - 52.55}{52.55} = 0.751 \qquad K = 5.24$$

$$\Delta_r H = \frac{RT_1T_2}{(T_2 - T_1)} \ln\frac{K_2}{K_1} = \frac{(8.314)(288)(348)}{60} \ln\frac{5.24}{0.0231}$$

$$= 75 \text{ kJ mol}^{-1}$$

5.25 Calculate K_c for the reaction in problem 5.19 at 1000 K and describe what it is equal to.

SOLUTION

$$K_c = K_P\left(\frac{P^o}{c^oRT}\right)^{\Sigma\nu_i}$$

$$= 0.36 \frac{1 \text{ bar}}{(1 \text{ L mol}^{-1})(0.08314 \text{ L bar K}^{-1} \text{ mol}^{-1})(238 \text{ K})}$$

$$= 0.0145$$

$$= \frac{\left(\frac{[C_2H_4]}{c^o}\right)\left(\frac{[H_2]}{c^o}\right)}{\frac{[C_2H_6]}{c^o}}$$

where [ ] indicates concentrations in moles per liter and $c° = 1 \text{ mol L}^{-1}$.

*5.26 The equilibrium constant for the reaction
$N_2(g) + 3H_2(g) = 2NH_3(g)$
is 35.0 at 400 K when partial pressures are expressed in bars. Assume the gases are ideal. (a) What is the equilibrium volume when 0.25 mol N_2 is mixed with 0.75 mol H_2 at a temperature of 400 K and a pressure of 1 bar? (b) What is the equilibrium composition and equilibrium pressure if this mixture is held at a constant volume of 33.26 L at 400 K?

SOLUTION

(a)	N_2	+	$3H_2$	=	$2NH_3$	
initial	0.25		0.75		0	
equil.	0.25 - ξ		0.75 - 3ξ		2ξ	total = 1 - 2ξ

$$K_P = \frac{(2\xi)^2(1 - 2\xi)^2}{(0.25 - \xi)(0.75 - 3\xi)^3} = 35.0$$

The method of successive approximations or the use of an equation solver yields $\xi = 0.1652$.

$P(N_2)/P^o = (0.25 - \xi)/(1 - 2\xi) = 0.1266$

$P(H_2)/P^o = 3(0.1266) = 0.3798$

$P(NH_3)/P^o = 2\xi/(1 - 2\xi) = 0.493$

Since the total pressure is 1 bar, these numbers also give the equilibrium mole fractions.

$V = (1 - 2\xi)RT/P$

$= (0.6696 \text{ mol})(0.8314 \text{ L bar K}^{-1} \text{ mol}^{-1})(400 \text{ K})/(1 \text{ bar})$

$= 22.27 \text{ L}$

When *Mathematica*™ is used to calculate the extent of reaction, four solutions are obtained; two of them are not satisfactory because they are imaginary and one is 0.334 mol, which is impossible because the extent of reaction has to be less than 0.25 mol. The fourth solution is correct.

(b) In order to calculate the equilibrium composition at constant volume, it is convenient to use K_c.

$K_c = (P^o/c^oRT)^{\Sigma\nu_i} K_P$

$$= \left[\frac{1 \text{ bar}}{(1 \text{ mol L}^{-1})(0.08314 \text{ L bar K}^{-1} \text{ mol}^{-1})(400 \text{ K})}\right]^{-2} 35.0$$

$= 3.871 \times 10^4$

$K_c = [c(NH_3)/c^o]^2/[c(N_2)/c^o][c(H_2)/c^o]^3$

$= (2\xi)^2\ 33.26^2/(0.25 - \xi)(0.75 - 3\xi)^3$

$= 3.871 \times 10^4$

The method of successive approximation or the use of an equation solver yields $\xi = 0.151$.

$n(N_2) = 0.25 - \xi = 0.0990$

$n(H_2) = 0.75 - 3\xi = 0.2970$

$n(NH_3) = 2\xi = 0.302$

$n(\text{total}) = 0.698$

$$P = \frac{(0.698 \text{ mol})RT}{33.26 \text{ L}} = 0.698 \text{ bar}$$

5.27 Show that to a first approximation the equation of state of a gas that dimerizes to a small extent is given by

$$\frac{P\bar{V}}{RT} = 1 - K_c/\bar{V}$$

SOLUTION

$2A(g) = A_2(g)$

$K_c = [A_2]/[A]^2 = n_{A_2}V/n_A^2$

assuming an ideal gas mixture.

n_0 = Total number of moles of A $= n_A + 2n_{A_2} = n_A + 2K_c n_A^2/V$ (a)

This can be solved for n_A using the quadratic formula for $ax^2 + bx + c = 0$:

$$x = \frac{-b \pm (b^2 - 4ac)^{1/2}}{2a}$$

This yields

$$n_A = n_0(1 - 2K_c n_0/V) \qquad \text{(b)}$$

when the approximation $(1 + x)^{1/2} = 1 + x/2 - x^2/8 + ...$ is used.

The gas law indicates that

$$\frac{PV}{RT} = n_A + n_{A_2} = n_A + K_c n_A^2/V = n_A(1 + K_c n_A/V) \qquad \text{(c)}$$

Substituting equation b in equation c yields

$$\frac{PV}{RT} = n_0(1 - 2K_c n_0/V)(1 + K_c n_A/V) \qquad \text{(d)}$$

Since n_A is just a little smaller than n_0, it can be replaced by n_0 in a term that is small compared to unity. Thus equation d can be written as

$$\frac{PV}{RT} = n_0(1 - 2K_c/\bar{V})(1 + K_c/\bar{V}) \qquad \text{(e)}$$

When this is multiplied out ignoring higher order terms and the n_0 is moved to the left-hand side of the equation, we obtain

$$\frac{P\bar{V}}{RT} = 1 - K_c/\bar{V}$$

5.28 Water vapor is passed over coal (assumed to be pure graphite in this problem) at 1000 K. Assuming that the only reaction occurring is the water gas reaction

$C(graphite) + H_2O(g) = CO(g) + H_2(g) \qquad K = 2.52$

calculate the equilibrium pressures of H_2O, CO, and H_2 at a total pressure of 1 bar. [Actually the water gas shift reaction

$CO(g) + H_2O(g) = CO_2(g) + H_2(g)$

occurs in addition, but it is considerably more complicated to take this additional reaction into account.]

SOLUTION

$$K = \frac{(P_{CO}/P^o)(P_{H_2}/P^o)}{(P_{H_2O}/P^o)} = \frac{x^2}{y} = \frac{x^2}{1 - 2x} = 2.52$$

$2x + y = 1 \qquad x^2 = 2.52 - 5.04x$

$x^2 + 5.04x - 2.52 = 0$

$$x = \frac{-5.04 \pm \sqrt{5.04^2 + 4(2.52)}}{2} = 0.458$$

$$= \frac{P_{CO}}{P^o} = \frac{P_{H_2}}{P^o}$$

$$1 - 2x = 0.084 = \frac{P_{H_2O}}{P^o}$$

$P_{H_2O} = 0.084$ bar $\qquad P_{CO} = 0.458$ bar $\qquad P_{H_2} = 0.458$ bar

5.29 What is the equilibrium partial pressure of NO in air at 1000 K at a pressure of 1 bar?

$(1/2)N_2(g) + (1/2)O_2(g) = NO(g)$

SOLUTION

The equilibrium constant can be calculated using data in Appendix C.3. Since the standard reaction Gibbs energy is 77.772 kJ mol^{-1}, the equilibrium constant is given by

$K = \exp(-77772/(8.3145\text{x}1000) = 8.663\text{x}10^{-5}$

$$= \frac{(P_{NO}/P^\circ)}{(P_{N_2}/P^\circ)^{1/2}(P_{O_2}/P^\circ)^{1/2}} = \frac{(P_{NO}/P^\circ)}{(0.80)^{1/2}(0.20)^{1/2}}$$

$P_{NO} = 3.465\text{x}10^{-5}$ bar

5.30 Starting with the fundamental equation for G in the form

$dG = - SdT + VdP + \Delta_r G d\xi$

derive equations for calculating $\Delta_r S$, $\Delta_r V$, and $\Delta_r H$ from experimental data on $\Delta_r G$ for a chemical reaction as a function of T and P.

SOLUTION

This fundamental equation has three Maxwell relations, and we need the following two:

$$-\left(\frac{\partial S}{\partial \xi}\right)_{T,P} = \left(\frac{\partial \Delta_r G}{\partial T}\right)_{\xi,P} = -\Delta_r S$$

$$\left(\frac{\partial V}{\partial \xi}\right)_{T,P} = \left(\frac{\partial \Delta_r G}{\partial P}\right)_{\xi,T} = \Delta_r V$$

The corresponding Gibbs-Helmholtz equation can be derived as follows:

Substitute $H/T - G/T$ for S in the fundamental equation. Now there is the following Maxwell relation:

$$-\left(\frac{\partial(H/T - G/T)}{\partial\xi}\right)_{T,P} = \left(\frac{\partial\Delta_r G}{\partial T}\right)_{\xi,P}$$

$$-\frac{1}{T}\left(\frac{\partial H}{\partial\xi}\right)_{T,P} + \frac{1}{T}\left(\frac{\partial G}{\partial\xi}\right)_{T,P} = \left(\frac{\partial\Delta_r G}{\partial T}\right)_{\xi,P}$$

$$-\Delta_r H + \Delta_r G = T\left(\frac{\partial\Delta_r G}{\partial T}\right)_{\xi,P}$$

$$\Delta_r H = \Delta_r G - T\left(\frac{\partial\Delta_r G}{\partial T}\right)_{\xi,P} = -T^2\left(\frac{\partial(\Delta_r G/T)}{\partial T}\right)_{\xi,P}$$

5.31 What is $\Delta_r S^o$ (298 K) for

$$H_2O(1) = H^+(ao) + OH^-(ao)$$

Why is this change negative and not positive?

SOLUTION

$\Delta_r S^o = - 10.75 - 69.92 = - 80.67$ J K^{-1} mol^{-1}

The ions polarize neighboring water molecules and attract them. For this reason the product state is more ordered than the reactant state.

5.32 Mercuric oxide dissociates according to the reaction $2HgO(s) = 2Hg(g) + O_2(g)$. At 420 °C the dissociation pressure is 5.16×10^4 Pa, and at 450 °C it is 10.8×10^4 Pa. Calculate (a) the equilibrium constants, and (b) the enthalpy of dissociation per mole of HgO.

SOLUTION

(a) $P_{Hg} = 2P_{O_2}$ $\quad P_{Hg} = \frac{2}{3}P$ $\quad P_{O_2} = \frac{1}{3}P$

$$K_{420} = P_{Hg}^2 P_{O_2} = \left(\frac{2}{3}\right)^2\left(\frac{1}{3}\right)P^3$$

$$= \left(\frac{4}{27}\right)\left(\frac{5.16 \times 10^4\ \text{Pa}}{1.013 \times 10^5\ \text{Pa}}\right)^3 = 0.0196$$

$$K_{450} = \left(\frac{4}{27}\right)\left(\frac{10.8 \times 10^4\ \text{Pa}}{1.013 \times 10^5\ \text{Pa}}\right)^3 = 0.1794$$

(b) $$\Delta_r H^o = \frac{RT_1T_2}{T_2 - T_1}\ln\frac{K_2}{K_1}$$

$$= \frac{(8.314\ \text{J K}^{-1}\ \text{mol}^{-1})(693\ \text{K})(723\ \text{K})}{30\text{K}}\ln\frac{0.1794}{0.0196}$$

$= 308$ kJ mol^{-1} for the reaction as written

$= 154$ kJ mol^{-1} of HgO(s)

5.33 The decomposition of silver oxide is represented by
$2Ag_2O(s) = 4Ag(s) + O_2(g)$
Using data from Appendix C.2 and assuming $\Delta_r C_P = 0$ calculate the temperature at which the equilibrium pressure of O_2 is 0.2 bar. This temperature is of interest because Ag_2O will decompose to yield Ag at temperatures above this value if it is in contact with air.

SOLUTION

$\Delta_r H^o = -2(-31.05) = 62.10$ kJ mol^{-1}

$\Delta_r S^o = 4(42.55) + 205.138 - 2(121.3) = 132.7$ J K^{-1} mol^{-1}

$\Delta_r G = \Delta G^o + RT \ln P_{O_2}$

$0 = \Delta H^o - T\Delta S^o + RT \ln 0.2$

$\Delta_r H^o = T(\Delta S^o - R \ln 0.2)$

$$T = \frac{\Delta_r H^o}{\Delta_r S^o - R \ln 0.2} = \frac{62100 \text{ J mol}^{-1}}{(132.7 - 8.314 \ln 0.2) \text{ J K}^{-1} \text{ mol}^{-1}}$$

$= 425$ K

5.34 The dissociation of ammonium carbamate takes place according to the reaction
$(NH_2)CO(ONH_4)(s) = 2NH_3(g) + CO_2(g)$
When an excess of ammonium carbamate is placed in a previously evacuated vessel, the partial pressure generated by NH_3 is twice the partial pressure of the CO_2, and the partial pressure of $(NH_2)CO(ONH_4)$ is negligible in comparison. Show that

$$K = \left(\frac{P_{NH_3}}{P^o}\right)^2\left(\frac{P_{CO_2}}{P^o}\right) = \frac{4}{27}\left(\frac{P}{P^o}\right)^3$$

where P is the total pressure.

SOLUTION

$P = P_{NH_3} + P_{CO_2} = 3P_{CO_2}$ since $P_{NH_3} = 2P_{CO_2}$

$P_{CO_2} = \frac{P}{3}$ $\qquad P_{NH_3} = \frac{2}{3}P$

$$K = \left(\frac{P_{NH_3}}{P^o}\right)^2\left(\frac{P_{CO_2}}{P^o}\right) = \left(\frac{2}{3}\frac{P}{P^o}\right)^2\left(\frac{1}{3}\frac{P}{P^o}\right) = \frac{4}{27}\left(\frac{P}{P^o}\right)^3$$

5.35 At 1000 K methane at 1 bar is in the presence of hydrogen. In the presence of a sufficiently high partial pressure of hydrogen, methane does not decompose to form graphite and hydrogen. What is this partial pressure?

SOLUTION

$CH_4(g) = C(\text{graphite}) + 2H_2(g)$

$\Delta G^o = - RT \ln K = - 19.46 \text{ kJ mol}^{-1}$

$$K = 10.39 = \frac{\left(\frac{P_{H_2}}{P^o}\right)^2}{\frac{P_{CH_4}}{P^o}}$$

$P_{H_2} = P^o[(10.39)(1)]^{1/2} = 3.2 \text{ bar}$

5.36 For the reaction

$Fe_2O_3(s) + 3CO(g) = 2Fe(s) + 3CO_2(g)$

the following values of K are known.

t/°C	250	1000
K	100	0.0721

At 1120 °C for the reaction $2CO_2(g) = 2CO(g) + O_2(g)$, $K = 1.4 \times 10^{-12}$. What equilibrium partial pressure of O_2 would have to be supplied to a vessel at 1120 °C containing solid Fe_2O_3 just to prevent the formation of Fe?

SOLUTION

$$\ln\left(\frac{K_2}{K_1}\right) = \frac{\Delta_r H^o(T_2 - T_1)}{RT_1T_2}$$

$$\Delta_r H^o = \frac{(8.314)(523)(1273) \ln (0.0721/100)}{(750)}$$

$$= -53.4 \text{ k J mol}^{-1}$$

$$\ln\left(\frac{K_{1393}}{K_{1273}}\right) = \frac{(53\,400)(120)}{8.314\,(1393)(1273)}$$

$$= - 0.4346$$

$$K_{1393} = (0.0721) \exp(- 0.4346)$$

$$= 0.0467 = \frac{P_{CO_2}^3}{P_{CO}^3}$$

For $2CO_2(g) = 2CO(g) + O_2$

$$K_{1393} = \frac{(P_{CO}/P^o)^2(P_{O_2}/P^o)}{(P_{CO}/P^o)^2}) = 1.4 \times 10^{-12}$$

$$\frac{P_{O_2}}{P^o} = \frac{(1.4 \times 10^{-12})(P_{CO_2}/P^o)^2}{(P_{CO}/P^o)}$$

$= (1.4 \times 10^{-12})(0.0467)^{2/3}$

$= 1.82 \times 10^{-13}$

$P_{O_2} = 1.82 \times 10^{-13}$ bar

5.37 When a reaction is carried out at constant pressure, the entropy change can be used as a criterion of equilibrium by including a heat reservoir as part of an isolated system containing the reaction chamber. Show that - $\Delta_r G/T$ is the global increase in entropy for the reaction system plus heat reservoir.

SOLUTION

If we write $-\Delta_r G/T = -\Delta_r H/T + \Delta_r S$, the first term on the right is the increase in entropy of the heat reservoir as a result of the reaction and the second term on the right is the increase in entropy in the reaction system. Therefore, $-\Delta_r G/T$ is the "global" increase in entropy for the system plus surroundings. The advantage in using the Gibbs energy as the criterion for equilibrium is that we do not have to think about the entropy changes in the surroundings.

5.38 The effect of temperature on K_P is given by equation 5.38, and the effect of temperature on K_c is given by

$$\Delta U^o = RT^2\left(\frac{\partial \ln K_c}{\partial T}\right)_V$$

Is it possible for a gas reaction to have K_P increase with increasing temperature, but K_c decreases with increasing T. If so, what has to be true?

SOLUTION

$\Delta H^o = \Delta U^o + \Delta \nu RT$
where $\Delta \nu$ is the number of moles of gaseous products minus the number of moles of gaseous reactants. If ΔH^o is positive, ΔU^o for the reaction must be negative for this to happen. This occurs when $\Delta \nu$ is sufficiently positive so that $\Delta \nu RT > |\Delta U^o|$.

5.39 Calculate the partial pressure of CO_2(g) over $CaCO_3$(calcite) - CaO(s) at 500 °C using the equation in Example 5.11 and data in Appendix C.3.

SOLUTION

$CaCO_3$(calcite) = CaO(s) + CO_2(g)

$\Delta_r H^o = -635.09 + (-393.51) - (-1206.92) = 178.32$ kJ mol^{-1}

$\Delta_r C_P^o = 42.80 + 37.11 - 81.88 = -1.97$ J K^{-1} mol^{-1}

$\Delta_r S^o = 39.75 + 213.74 - 92.9 = 160.6$ J K^{-1} mol^{-1}

Substituting in the equation in Example 5.11,

$$\ln K_{773} = -\frac{178\,320}{(8.314)(773.15)} + \frac{160.6}{8.314} + \frac{1.97}{8.314}\left(1 - \frac{298.15}{773.15} - \ln\frac{773.19}{298.15}\right)$$

$$K_{773} = \frac{P_{CO_2}}{P^o} = 20 \times 10^{-5}$$

Table 5.1 gives $\frac{P_{CO_2}}{P^o} = 9.2 \times 10^{-5}$ bar

5.40 The NBS Tables contain the following data at 298 K:

	$\Delta_f H^o$/kJ mol^{-1}	$\Delta_f G^o$/kJ mol^{-1}
$CuSO_4(s)$	-771.36	-661.8
$CuSO_4 \cdot H_2O(s)$	-1085.83	-918.11
$CuSO_4 \cdot 3H_2O(s)$	-1684.31	-1399.96
$H_2O(g)$	-241.818	-228.572

(a) What is the equilibrium partial pressure of H_2O over a mixture of $CuSO_4(s)$ and $CuSO_4 \cdot H_2O(s)$ at 25 °C?
(b) What is the equilibrium partial pressure of H_2O over a mixture of $CuSO_4 \cdot H_2O(s)$ and $CuSO_4 \cdot 3H_2O(s)$ at 25 °C?
(c) What are the answers to (a) and (b) if the temperature is 100 °C and ΔC_P^o is assumed to be zero?

SOLUTION

(a) $CuSO_4 \cdot H_2O(s) = CuSO_4(s) + H_2O(g)$

$\Delta G^o = -228.572 - 661.8 + 918.11$

$= 27.7$ kJ mol^{-1}

$$K = \exp\left(\frac{-\Delta G^o}{RT}\right) = 1.4 \times 10^{-5} = \frac{P_{H_2O}}{P^o}$$

$P_{H_2O} = 1.4 \times 10^{-5}$ bar

(b) $CuSO_4 \cdot 3H_2O(s) = CuSO_4 \cdot H_2O(s) + 2H_2O(g)$

$\Delta G^o = 2(-228.572) - 918.11 + 1399.46$

$= -24.71$ kJ mol^{-1}

$$P_{H_2O} = \exp\left[\frac{-24\,710}{(2)(8.314)(298)}\right]$$

$= 6.83 \times 10^{-3}$ bar

(c) For the first reaction

$$\ln\frac{K_{100}}{1.4 \times 10^{-5}} = \frac{72\,650(75)}{8.314(298)(373)}$$

$\Delta H^o = -241.818 - 771.36 + 1085.83 = 72.65$ kJ mol^{-1}

$$\ln \frac{K_{100}}{1.4 \times 10^{-5}} = 5.896$$

$K_{100} = 363.6(1.4 \times 10^{-5}) = 0.0051$ bar

For the second reaction

$$\Delta H^o = 2(-241.818) - 1085.83 + 1684.31$$
$$= 114.84 \text{ kJ mol}^{-1}$$

$$\ln \frac{K_{100}}{4.7 \times 10^{-5}} = \frac{114\,840(75)}{8.314(298)(373)} = 9.32$$

$K_{100} = 0.52 \text{ bar}^2 = P^2_{H_2O}$

$P_{H_2O} = 0.72$ bar

5.41 One micromole of CuO(s) and 0.1 μmole of Cu(s) are placed in a 1 L container at 1000 K. Determine the identity and quantity of each phase present at equilibrium if $\Delta_f G^o$ of CuO is -66.66 kJ mol^{-1} and that of Cu_2O is -77.94 kJ mol^{-1} at 1000 K. (From H. F. Franzen, *J. Chem. Ed.* **65**, 146 (1988).)

SOLUTION

$Cu(s) + CuO(s) = Cu_2O(s)$

$\Delta_r G^o = -77.94 - (-66.66) = -11.28$ kJ mol^{-1}

Therefore, this reaction goes to completion to the right. The two solids are in equilibrium with O_2(g).

$$2CuO(s) = Cu_2O(s) + \frac{1}{2}O_2(g)$$

$\Delta_r G^o = -77.94 - 2(-66.66) = 55.38 \text{ kJ mol}^{-1} = -RT \ln P^{1/2}_{O_2}$

$$P_{O_2} = \exp\left[-\frac{2(55\,380)}{(8.314)(1000)}\right] = 1.64 \times 10^{-6}$$

The amount of O_2(g) at equilibrium is

$$n_{O_2} = \frac{PV}{RT} = \frac{(1.64 \times 10^{-6})(1)}{(0.08314)(1000)} = 1.97 \times 10^{-8} \text{ mol}$$

Thus the amounts at equilibrium are essentially

$n_{CuO} = 0.9 - 4(0.02) = 0.82$ μmol

$n_{Cu_2O} = 0.1 + 2(0.02) = 0.14$ μmol

$n_{O_2} = 0.02$ μmol

5.42 For the heterogeneous reaction

$CH_4(g) = C(s) + 2H_2(g)$

derive the expression for the extent of reaction in terms of the equilibrium constant and the applied pressure, where the extent of reaction when graphite is in equilibrium with the gas mixture. Is this the same expression (equation 5.33) that was obtained for the reaction $N_2O_4(g) = 2NO_2(g)$?

SOLUTION

$CH_4(g) \;=\; C(s) + 2H_2(g)$

$1 - \xi \qquad \xi \qquad 2\xi$

$\frac{1-\xi}{1+\xi}P \qquad \frac{2\xi}{1+\xi}P$

$$K = \frac{4\xi^2(P/P^o)}{1-\xi^2}$$

$$\xi = \left(\frac{K}{4(P/P^o)+K}\right)^{1/2}$$

This is the same as equation 5.33, but it only applies when graphite is in equilibrium with the gas.

5.43 Calculate the equilibrium extent of the reaction $N_2O_4(g) = 2NO_2(g)$ at 298.15 K and a total pressure of 1 bar if the $N_2O_4(g)$ is mixed with an equal volume of $N_2(g)$ before the reaction occurs. As shown by Example 5.4, $K = 0.143$. Do you expect the same equilibrium extent of reaction as in example? If not do you expect a larger or smaller equilibrium extent of reaction?

SOLUTION

If there is initially 1 mol of N_2O_4, the total amount of gas at equilibrium is $2 + \xi$. Thus the expression for the equilibrium constant is

$$K = \frac{4\xi^2(P/P^o)}{(2+\xi)(1-\xi)}$$

where P is the total pressure. When the total pressure is 1 bar, the equilibrium extent of reaction obtained by solving this quadratic equation with the formula

$$\xi = \frac{-b \pm (b^2 - 4ac)^{1/2}}{2a}$$

is 0.249.

This equilibrium extent of reaction is smaller than that in Example 5.3 because the partial pressure of $N_2O_4(g)$ plus $NO_2(g)$ is larger than 0.5 bar. The partial pressure of $N_2(g)$ was initially 0.5 bar, but it is less than this in the equilibrium mixture because of the expansion of the reaction mixture during the reaction at a constant pressure of 1 bar.

5.44 (a) A system contains $CO(g)$, $CO_2(g)$, $H_2(g)$, and $H_2O(g)$. How many chemical reactions are required to describe chemical changes in this system? Give an example. (b) If solid carbon is present in the system in addition, how many independent chemical reactions are there? Give a suitable set.

SOLUTION

(a)

	CO	CO_2	H_2	H_2O
C	1	1	0	0
O	1	2	0	2
H	0	0	2	2

To perform a Gaussian elimination, subtract the first row from the second, and divide the third row by 2.

1	1	0	0
0	1	0	1
0	0	1	1

Subtract the second row from the first.

1	0	0	-1
0	1	0	1
0	0	1	1

The rank of the ***A*** matrix is 3, and so the number of independent reactions is
$R = N$ - rank $\boldsymbol{A} = 4 - 3 = 1$
where N is the number of species.
The stoichiometric numbers for a suitable reaction is obtained by changing the sign of the numbers in the last column, and extending the vector with a 1; that is, 1, -1, -1, 1. These are the stoichiometric numbers for the species across the top of ***A*** .
$CO - CO_2 - H_2 + H_2O = 0$
$CO_2 + H_2 = CO + H_2O$

(b)

	CO	CO_2	H_2	H_2O	C
C	1	1	0	0	1
O	1	2	0	1	0
H	0	0	2	2	0

Subtract the first row from the second and divide the third row by 2.

1	1	0	0	1
0	1	0	1	-1
0	0	1	1	0

Subtract the second row from the first

1	0	0	-1	2
0	1	0	1	-1
0	0	1	1	0

Rank ***A*** = 3 and $R = N$ - rank $\boldsymbol{A} = 5 - 3 = 2$. To obtain a suitable set of reactions, change the signs in the last two columns and put an identity matrix at the bottom.

1	-2
-1	1
-1	0
1	0
0	1

These two independent reactions are
$CO - CO_2 - H_2 + H_2O = 0$
$-2CO + CO_2 + C = 0$
or
$CO_2 + H_2 = CO + H_2O$
$2CO = CO_2 + C$

*5.45 For a closed system containing C_2H_2, H_2, C_6H_6, and $C_{10}H_8$, use a Gaussian elimination to obtain a set of independent chemical reactions. Perform the matrix multiplication to verify $\boldsymbol{A\nu} = \mathbf{0}$.

SOLUTION

	C_2H_2	H_2	C_6H_6	$C_{10}H_8$
C	2	0	6	10
H	2	2	6	8

1	0	3	5
1	1	3	4

1	0	3	5
0	1	0	-1

$\boldsymbol{\nu}$ =

C_2H_2	-3	-5
H_2	0	1
C_6H_6	1	0
$C_{10}H_8$ 0	1	

$3C_2H_2 = C_6H_6$

$5C_2H_2 = C_{10}H_8 + H_2$

$$\begin{pmatrix} 2 & 0 & 6 & 10 \\ 2 & 2 & 6 & 8 \end{pmatrix} \begin{pmatrix} -3 & -5 \\ 0 & 1 \\ 1 & 0 \\ 0 & 1 \end{pmatrix} = \begin{pmatrix} 0 & 0 \\ 0 & 0 \end{pmatrix}$$

5.46 The reaction A + B = C is at equilibrium at a specified T and P. Derive the fundamental equation for G in terms of components by eliminating μ_C.

SOLUTION

The fundamental equation for G is

$$dG = -SdT + VdP + \mu_A dn_A + \mu_B dn_B + \mu_C dn_C \quad (1)$$

When the reaction is at equilibrium,

$$\mu_A + \mu_B = \mu_C \quad (2)$$

Eliminating μ_C from equation 1 yields

$$dG = -SdT + VdP + \mu_A(dn_A + dn_C) + \mu_B(dn_B + dn_C) \quad (3)$$

This equation is written in terms of 2 components rather than 3 species because $C = N - R = 3 - 1 = 2$. The two components can be referred to as the A,C pseudoisomer group with amount

$$n_I' = n_A + n_C \quad (4)$$

and the B component with amount

$$n_B' = n_B + n_C \quad (5)$$

Thus at chemical equilibrium the fundamental equation can be written as

$$dG = -SdT + VdP + \mu_A dn_I' + \mu_B dn_B' \quad (6)$$

Note that the chemical potentials of the components are the same as the chemical potentials of two of the species.

*5.47 The article C. A. L. Figueiras, J. of Chem. Educ., 69, 276 (1992) illustrates an interesting problem you can get into in trying to balance a chemical equation. Consider the following reaction without stoichiometric numbers:

$ClO_3^- + Cl^- + H^+ = ClO_2 + Cl_2 + H_2O$

There is actually an infinite number of ways to balance this equation. The following steps in unraveling this puzzle can be carried out using a personal computer with a program like *Mathematica* ™, which can do matrix operations. Write the conservation matrix **A** and determine the number of components. How many independent reactions are there for this system of six species? What are the stoichiometric numbers for a set of independent reactions? These steps show that chemical change in this system is represented by two chemical reactions, not one.

SOLUTION

The conservation matrix **A** for this system is

	ClO_3^-	Cl^-	H^+	H_2O	ClO_2	Cl_2
H	0	0	1	2	0	0
O	3	0	0	1	2	0
Cl	1	1	0	0	1	2
charge	-3	-1	+1	0	0	0

Row reduction of this matrix yields

	ClO_3^-	Cl^-	H^+	H_2O	ClO_2	Cl_2
ClO_3^-	1	0	0	0	5/6	1/3
Cl^-	0	1	0	0	1/6	5/3
H^+	0	0	1	0	1	2
H_2O	0	0	0	1	-1/2	-1

This indicates that there are 4 components. Thus $R = N_s - C = 6 - 4 = 2$. The last two columns with changed signs and augmented by a 2x2 unit matrix at the bottom give the stoichiometric numbers of two independent reactions. Another way to obtain a set of independent reactions is to calculate the null space of the **A** matrix. The null space $\boldsymbol{v}$ is

	ClO_3^-	Cl^-	H^+	H_2O	ClO_2	Cl_2
rx 1	-1	-5	-6	3	0	3
rx 2	-5	-1	-5	3	6	0

5.48 A chemical reaction system contains three species: C_2H_4 (ethylene), C_3H_6 (propene), and C_4H_8 (butene). (a) Write the **A** matrix. (b) Row reduce the **A** matrix, (c) How many components are there? (d) Derive a set of independent reactions from the **A** matrix.

SOLUTION

(a)

	C_2H_4	C_3H_6	C_4H_8
$\boldsymbol{A}$ =	2	3	4
	4	6	8

(b) Multiplying the first row by 2 and subtracting it from the second row, and dividing by 2 yields

$\boldsymbol{A}$ =	1	3/2	2
	0	0	0

(c) There is one component because there is one independent row.

(d) Taking the last two columns, changing the sign, and appending a 2x2 matrix below it yields

		rx 1	rx 2
$\boldsymbol{\nu}$ =	C_2H_4	-3/2	-2
	C_3H_6	1	0
	C_4H_8	0	1

Thus two independent reactions are

rx 1: $1.5C_2H_4 = C_3H_6$

rx 2: $2C_2H_4 = C_4H_8$

If the columns in the $\boldsymbol{A}$ matrix are put in another order, a different set of independent reactions will be obtained, but they will also be suitable.

5.49 How many degrees of freedom are there for the following systems, and how might they be chosen?

(a) $CuSO_4 \cdot 5H_2O$(cr) in equilibrium with $CuSO_4$(cr) and H_2O(g).

(b) N_2O_4 in equilibrium with NO_2 in the gas phase.

(c) CO_2, CO, H_2O, and H_2 in chemical equilibrium in the gas phase.

(d) The system described in (c) is made up with stoichiometric amounts of CO and H_2.

SOLUTION

(a) $C = N_s - R = 3 - 1 = 2$

$F = C - p + 2 = 2 - 3 + 2 = 1$

Only the temperature or pressure may be fixed.

(b) $C = N_s - R = 2 - 1 = 1$

$F = C - p + 2 = 1 - 1 + 2 = 2$

Temperature and pressure may be fixed.

(c) $C = N_s - R = 4 - 1 = 3$

$F = C - p + 2 = 3 - 1 + 2 = 4$

Temperature, pressure, and two mole fractions may be fixed.

(d) $C = N_s - R - s = 4 - 1 - 2 = 1$

$F = C - p + 2 = 1 - 1 + 2 = 2$

Only temperature and pressure may be fixed if $n_c(C)/n_c(H)$ and $n_c(C)/n_c(O)$ are both fixed.

5.50 Graphite is in equilibrium with gaseous H_2O, CO, CO_2, H_2, and CH_4. How many degrees of freedom are there? What degrees of freedom might be chosen for an equilibrium calculation?

SOLUTION

$C = N_s - R = 6 - 3 = 3$

$F = C - p + 2 = 3 - 2 + 2 = 3$

The degrees of freedom chosen might be T, P, $n_c(H)/n_c(O)$.

5.51 A gaseous system contains CO, CO_2, H_2, H_2O, and C_6H_6 in chemical equilibrium. (a) How many components are there? (b) How many independent reactions? (c) How many degrees of freedom are there?

SOLUTION

(a)

	CO	CO_2	H_2	H_2O	C_6H_6
C	1	1	0	0	6
O	1	2	0	1	0
H	0	0	2	2	6

Subtract the first row from the second and divide the third row by 2 to obtain

1	1	0	0	6
0	1	0	1	-6
0	0	1	1	3

Subtract the second row from the first to obtain

1	0	0	-1	12
0	1	0	1	-6
0	0	1	1	3

The rank of this matrix is 3, and so there are 3 independent components.

(b) The stoichiometric numbers of 2 independent reactions are given by the last 2 columns.

$$CO_2 + H_2 = H_2O + CO$$

$$12CO + 3H_2 = C_6H_6 + 6CO_2$$

If the species are arranged in a different order in the matrix, a different pair of independent equations will be obtained.

(c) $F = C - p + 2 = 3 - 1 + 2 = 4$

These four degrees of freedom can be taken to be T, P, $n_c(C)/n_c(O)$ and $n_c(C)/n_c(H)$. Alternatively, the mole fractions of 2 species and the temperature and pressure may be specified. The two equilibrium constant expressions provide two relations between 5 mole fractions, 4 of which are independent since $\Sigma y_i = 1$. If 2 mole fractions are known, the other two can be calculated from the two simultaneous equations.

5.52 0.696

5.53 0.0166

5.54 0.351 bar

5.55 3.74 bar

5.56 0.803, 1.84

5.57 $K = (2\xi)^2(4 - 2\xi)^2/(1 - \xi)(3 - 2\xi)^3(P/P^o)^2$

5.58 26.3

5.59 (a) 3.81×10^{-2}, (b) 0.348

5.60 (a) 0.0788, (b) 0.0565

5.61 0.0273, 0.0861

5.64 (a) 16.69 kJ mol^{-1}, (b) 0.787 bar

5.65 1.84×10^6

5.66 30.1 bar

5.67 (a) 0.5000, 0.4363, 0.0637

(b) 0.5481, 0.3946, 0.0574

(c) When additional N_2 is added, the equilibrium shifts so that the mole fraction of N_2 is reduced below what it otherwise would have been.

5.68 (a) Yield of CH_4 will increase. (b) Yield of CH_4 will decrease. (c) Mole fraction CH_4 computed without including N_2 will increase.

5.69 (a) No, (b) 0.286

5.70 (a) 0.0050 bar, (b) 0.0220 bar

5.72 225.1 kJ mol^{-1}

5.73 (a) 0.1852, 0.3775, 0.6284, (b) 60.7 kJ mol^{-1}, (c) 0.320, (d) 0.371

5.75 $K_P = 0.0024$, $K_c = 4.15$

5.76 - 5.082, 0.070 J K^{-1} mol^{-1}

5.77 98.752, 113.526, 122.521 J K^{-1} mol^{-1}

5.78 (a) - 33.0, (b) - 77.4 J K^{-1} mol^{-1}

5.79 56.6 J K^{-1} mol^{-1}

5.80 Hydrogen dissolves as atoms.

5.81 Fe_2O_3

5.82 7.56 bar

5.83 16.06 kJ mol^{-1}

5.84 (a) 6.66×10^{-3}, (b) 12.1 kJ mol^{-1}

5.85 (a) 57.3 kJ mol^{-1}, (b) 5.82 kJ mol^{-1}, (c) 154.7 J K^{-1} mol^{-1}

5.86 - 14.8 kJ mol^{-1}

5.87 $K_1 = 2.5$, $K_2 = 0.61$, $K_3 = 0.35$, $y_{CH_4} = 0.27$, $y_{CO} = 0.27$, $y_{H_2O} = 0.46$

5.88 Two, for example, $2C_2H_6 = 2CH_4 + C_2H_4$

5.90 (a) $F = N - R - p + 2 = 3 - 2 - 1 + 2 = 2$. T and P can be chosen.

(b) $F = 3 - 2 - 1 + 2 = 2$. T and P can be chosen.

6

Phase Equilibrium

6.1 The boiling point of hexane at 1 atm is 68.7 °C. What is the boiling point at 1 bar? Given: The vapor pressure of hexane at 49.6 °C is 53.32 k Pa.

SOLUTION

$$\ln\frac{P_2}{P_1} = \frac{\Delta_{vap}H(T_2 - T_1)}{RT_1T_2}$$

$$\Delta_{vap}H = \frac{RT_1T_2}{(T_2 - T_1)}\ln\left(\frac{P_2}{P_1}\right)$$

$$= \frac{(8.3145)(322.8)(341.9)}{19.1}\ln\left(\frac{101.325\text{ k Pa}}{53.32\text{ k Pa}}\right) = 30{,}850\text{ J mol}^{-1}$$

$$\ln\left(\frac{101.325}{100}\right) = \frac{30850}{8.3145}\left(\frac{1}{T_1} - \frac{1}{341.9}\right)$$

$$T_1 = \frac{1}{\frac{8.3145}{30850}\ln\left(\frac{101.325}{100}\right) + \frac{1}{341.9}} = 341.5\text{ K}$$

Thus the boiling point is reduced 0.4 °C to 68.3 °C.

6.2 What is the boiling point of water 2 miles above sea level? Assume the atmosphere follows the barometric formula (equation 1.54) with $M = 0.0289$ kg mol^{-1} and $T = 300$ K. Assume the enthalpy of vaporization of water is 44.0 kJ mol^{-1} independent of temperature.

SOLUTION

h = (2 miles)(5280 ft/mile)(12 in/ft)(2.54 cm/in)(0.01 m/cm) = 3219 m

$P = P_o \exp(-gMh/RT)$

$$= (1.01325\text{ bar})\exp\left[-\frac{(9.8\text{ m s}^{-2})(0.0289\text{ kg mol}^{-1})(3219\text{ m})}{(8.314\text{ J K}^{-1}\text{ mol}^{-1})(300\text{ K})}\right] = 0.703\text{ bar}$$

$$\ln\left(\frac{P_2}{P_1}\right) = \Delta_{vap}H(T_2 - T_1)/RT_2T_2$$

$$\ln\left(\frac{1.01325}{0.703}\right) = \frac{(44\,000)(373 - T_1)}{(8.314)(373)T_1}$$

$$\frac{373 - T_1}{T_1} = 2.576 \times 10^{-2}$$

$T_1 = 364$ K = 90.6 °C

6.3 The barometric equation 1.46 and the Clausius-Clapeyron equation 6.8 can be used to estimate the boiling point of a liquid at a higher altitude. Use these equations to derive a single equation to make this calculation. Use this equation to solve problem 6.2.

SOLUTION

The two equations are

$\ln(P_o/P) = gMh/RT_{atm}$

$\ln(P_o/P) = \Delta_{vap}H(T_o - T)/RTT_o$

where P_o is the atmospheric pressure at sea level and P is the atmospheric pressure at a higher altitude. T_o is the boiling point at sea level, and T is the boiling point at the higher altitude. Setting the right-hand sides of these equations equal to each other yields

$$\frac{1}{T} = \frac{1}{T_o} + \frac{gMh}{T_{atm}\Delta_{vap}H}$$

Substituting the values from problem 6.2 yields

$$\frac{1}{T} = \frac{1}{373.15} + \frac{(9.8 \text{ m s}^{-2})(0.0289 \text{ kg mol}^{-1})(3.219 \times 10^3 \text{ m})}{(298.15 \text{ K})(44{,}000 \text{ J mol}^{-1})}$$

$T = 363.58$ K = 90.43 °C

6.4 Liquid mercury has a density of 13.690 g cm^{-3}, and solid mercury has a density of 14.193 g cm^{-3}, both being measured at the melting point, -38.87 °C, at 1 bar pressure. The heat of fusion is 9.75 J g^{-1}. Calculate the melting points of mercury under a pressure of (a) 10 bar and (b) 3540 bar. The observed melting point under 3540 bar is -19.9 °C.

SOLUTION

$$\frac{\Delta T}{\Delta P} = \frac{T(V_l - V_s)}{\Delta_{fus}H} = \left[\frac{(234.3 \text{ K})\left(\frac{1}{13.690} - \frac{1}{14.193}\right) \text{cm}^3 \text{ g}^{-1}}{9.75 \text{ J g}^{-1}}\right](10^{-2} \text{ m cm}^{-1})^3$$

$= 6.22 \times 10^{-8}$ K Pa^{-1}

(a) $\Delta T = (6.22 \times 10^{-8} \text{ K Pa}^{-1})(9 \times 100{,}000 \text{ Pa}) = 0.056$ K

$t = -38.87 + 0.06 = -38.81$ °C

(b) $\Delta T = (6.22 \times 10^{-8} \text{ K Pa}^{-1})(3539 \times 100{,}000 \text{ Pa}) = 22.0$ K

$t = -38.87 + 22.0 = -16.9$ °C

6.5 From the $\Delta_f G^o$ of Br_2(g) at 25 °C, calculate the vapor pressure of Br_2(l). The pure liquid at 1 bar and 25 °C is taken as the standard state.

SOLUTION

$Br_2(l) = Br_2(g) \qquad K = \frac{P_{Br_2}}{P^o}$

$\Delta G^o = -RT \ln\left(\frac{P_{Br_2}}{P^o}\right)$

$$\frac{P_{Br_2}}{P^o} = \exp\left[-\frac{3110 \text{ J mol}^{-1}}{(8.314 \text{ J K}^{-1} \text{ mol}^{-1})(298.15 \text{ K})}\right]$$

$P_{Br_2} = 0.285$ bar

6.6 Calculate ΔG^o for the vaporization of water at 0 °C using data in Appendix C.2 and assuming that ΔH^o for the vaporization is independent of temperature. Use ΔG^o to calculate the vapor pressure of water at 0 °C.

SOLUTION

$H_2O(l) = H_2O(g)$

$\Delta G^o(298.15 \text{ K}) = -228.572 + 237.129 = 8.577 \text{ kJ mol}^{-1}$

$\Delta H^o(298.15 \text{ K}) = -241.818 + 285.830 = 44.012 \text{ kJ mol}^{-1}$

Using the equation in problem 4.10(a)

$$\Delta G_2 = \frac{\Delta G_1 T_2}{T_1} + \Delta H\left(1 - \frac{T_2}{T_1}\right)$$

$$\Delta G(0\ ^\circ\text{C}) = 8.577\left(\frac{273.15}{298.15}\right) + 44.012\left(1 - \frac{273.15}{298.15}\right)$$

$$= 7.858 + 3.690 = 11.548 \text{ kJ mol}^{-1}$$

$$= -RT \ln\left(\frac{P}{P^o}\right)$$

$$P = P_o \exp\left(\frac{-11.548}{8.3145 \times 10^{-3} \times 273.15}\right)$$

$$= 6.190 \times 10^{-3} \text{ bar}$$

6.7 The change in Gibbs energy for the conversion of aragonite to calcite at 25 °C is -1046 J mol^{-1}. The density of aragonite is 2.93 g cm^{-3} at 15 °C and the density of calcite is 2.71 g cm^{-3}. At what pressure at 25 °C would these two forms of $CaCO_3$ be in equilibrium?

SOLUTION

aragonite = calcite $\qquad \Delta G^o = -1046 \text{ J mol}^{-1}$

$$\Delta \overline{V} = \frac{100.09 \text{ g mol}^{-1}}{2.71 \text{ g cm}^{-3}} - \frac{100.09 \text{ g mol}^{-1}}{2.93 \text{ g cm}^{-3}}$$

$$= 2.77 \text{ cm}^3 \text{ mol}^{-1} = 2.77 \times 10^{-6} \text{ m}^3 \text{ mol}^{-1}$$

$$\left(\frac{\partial \Delta G}{\Delta P}\right)_T = \Delta \overline{V}; \quad \int dG = \Delta \overline{V} \int dP$$

$\Delta G_2 - \Delta G_1 = \Delta \overline{V} (P - 1)$

$0 + 1046 \text{ J mol}^{-1} = (2.77 \times 10^{-6} \text{ m}^3 \text{ mol}^{-1})(P - 1)$

$P = 3780$ bar

*6.8 *n*-Propyl alcohol has the following vapor pressures:

t/°C	40	60	80	100
P/kPa	6.69	19.6	50.1	112.3

Plot these data so as to obtain a nearly straight line, and calculate (a) the enthalpy of vaporization, (b) the boiling point at 1 bar, and (c) the boiling point at 1 atm.

SOLUTION:

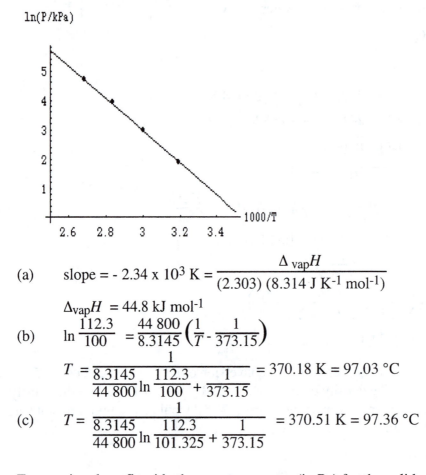

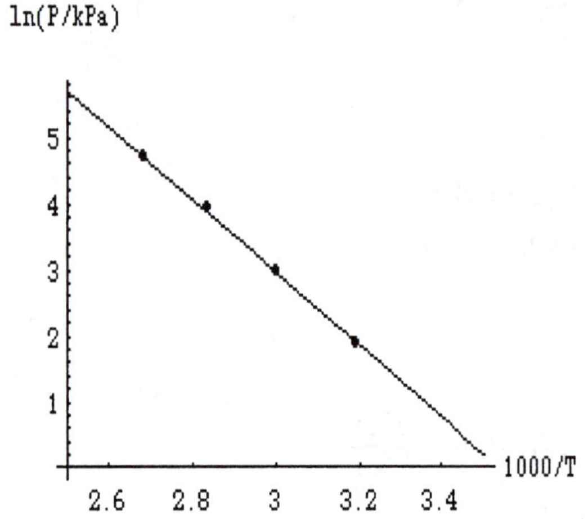

(a) $\text{slope} = -2.34 \times 10^3 \text{ K} = \dfrac{\Delta_{\text{vap}}H}{(2.303)(8.314 \text{ J K}^{-1} \text{ mol}^{-1})}$

$\Delta_{\text{vap}}H = 44.8 \text{ kJ mol}^{-1}$

(b) $\ln \dfrac{112.3}{100} = \dfrac{44\,800}{8.3145}\left(\dfrac{1}{T} - \dfrac{1}{373.15}\right)$

$T = \dfrac{1}{\dfrac{8.3145}{44\,800} \ln \dfrac{112.3}{100} + \dfrac{1}{373.15}} = 370.18 \text{ K} = 97.03\ °\text{C}$

(c) $T = \dfrac{1}{\dfrac{8.3145}{44\,800} \ln \dfrac{112.3}{101.325} + \dfrac{1}{373.15}} = 370.51 \text{ K} = 97.36\ °\text{C}$

6.9 For uranium hexafluoride the vapor pressures (in Pa) for the solid and liquid are given by

$\ln P_s = 29.411 - 5893.5/T$

$\ln P_l = 22.254 - 3479.9/T$

Calculate the temperature and pressure of the triple point.

SOLUTION

$$29.411 - \frac{5893.5}{T} = 22.254 - \frac{3479.9}{T}$$

$T = 337.2$ K = 64.0 °C

$P = e^{29.411 - 5893.5/337.2} = 152.2$ kPa

6.10 The heats of vaporization and of fusion of water are 2490 J g^{-1} and 33.5 J g^{-1} at 0 °C. The vapor pressure of water at 0 °C is 611 Pa. Calculate the sublimation pressure of ice at -15 °C, assuming that the enthalpy changes are independent of temperature.

SOLUTION

$\Delta_{fus}H$

liquid ←------- solid

↓ $\Delta_{vap}H$ ↙ $\Delta_{sub}H$

vapor

$$\Delta_{sub}H = \Delta_{fus}H + \Delta_{vap}H$$

$$= 33.5 + 2490 = 2824 \text{ J g}^{-1}$$

$$\ln \frac{P_2}{P_1} = \frac{\Delta_{sub}H\,(T_2 - T_1)}{RT_1T_2}$$

$$P_2 = P_1 \exp\left[\frac{\Delta_{sub}H(T_2 - T_1)}{RT_1T_2}\right]$$

$$= (611 \text{ Pa}) \exp\left[\frac{(2824 \times 18 \text{ J mol}^{-1})(-15 \text{ K})}{(8.314 \text{ J K}^{-1} \text{ mol}^{-1})(273.15 \text{ K})(258.15 \text{ K})}\right]$$

$$= 166 \text{ Pa}$$

6.11 The sublimation pressures of solid Cl_2 are 352 Pa at -112 °C and 35 Pa at -126.5 °C. The vapor pressures of liquid Cl_2 are 1590 Pa at -100 °C and 7830 Pa at -80 °C. Calculate (a) $\Delta_{sub}H$, (b) $\Delta_{vap}H$, (c) $\Delta_{fus}H$, and (d) the triple point.

SOLUTION

(a) $$\Delta_{sub}H = \frac{RT_1T_2}{T_2 - T_1} \ln \frac{P_2}{P_1}$$

$$= \frac{(8.314 \text{ J K}^{-1} \text{ mol}^{-1})(161.15 \text{ K})(146.65 \text{ K})}{14.5 \text{ K}} \ln \frac{352}{35}$$

$$= 31.4 \text{ kJ mol}^{-1}$$

(b) $$\Delta_{vap}H = \frac{(8.314 \text{ J K}^{-1} \text{ mol}^{-1})(173.15 \text{ K})(193.15 \text{ K})}{20 \text{ K}} \ln \frac{7830}{1590}$$

$$= 22.2 \text{ kJ mol}^{-1}$$

(c) $\Delta_{fus}H = \Delta_{sub}H - \Delta_{vap}H = 31.4 - 22.2 = 9.2$ kJ mol^{-1}

(d) For the solid

$$\ln P = \ln 352 + \frac{31\,400}{8.314}\left(\frac{1}{161.15} - \frac{1}{T}\right) = 29.300 - \frac{3777}{T}$$

For the liquid

$$\ln P = \ln 1590 + \frac{9200}{8.314}\left(\frac{1}{173.15} - \frac{1}{T}\right) = 13.762 - \frac{1107}{T}$$

At the triple point

$$29.300 - \frac{3777}{T} = 13.762 - \frac{1107}{T} \qquad T = \frac{2670}{15.538} = 172 \text{ K}$$

$T = 172$ K - 273.15 K = - 101.2 °C

6.12 The vapor pressure of solid benzene, C_6H_6, is 299 Pa at -30 °C and 3270 Pa at 0 °C, and the vapor pressure of liquid C_6H_6 is 6170 Pa at 10 °C and 15,800 Pa at 30 °C. From these data, calculate (a) the triple point of C_6H_6, and (b) the enthalpy of fusion of C_6H_6.

SOLUTION

Calculate the enthalpy of sublimation of C_6H_6.

$$\Delta_{sub}H = \frac{(8.314)\,(253.15)(223.15)}{30} \ln \frac{3270}{299}$$

$$= 44{,}030 \text{ J mol}^{-1}$$

Express sublimation pressures as a function of T.

$$\ln P_{sub} = -\frac{\Delta_{sub}H}{RT} + \frac{\Delta_{sub}S}{R}$$

$$\text{At 273.15 K, } \ln 3270 = -\frac{44\,030}{RT} + \frac{\Delta_{sub}S}{R}$$

$$\ln P_{sub} = -\frac{44\,030}{RT} + 27.481$$

Calculate the enthalpy of vaporization of C_6H_6

$$\Delta_{vap}H = \frac{(8.314)(283.15)(303.15)}{20} \ln \frac{15800}{6170}$$

$$= 33{,}550 \text{ J mol}^{-1}$$

Express vapor pressures as a function of T.

$$\ln P_{vap} = -\frac{\Delta_{vap}H}{RT} + \frac{\Delta_{vap}S}{R}$$

$$\text{At 303.15 K, } \ln 15{,}800 = -\frac{33\,550}{8.314(303.15)} + \frac{\Delta_{vap}S}{R}$$

$$\ln P_{vap} = -\frac{33\,550}{8.314\,T} + 22.979$$

(a) At the triple point, $\ln P_{sub} = \ln P_{vap}$

$$-\frac{44\,030}{RT} + 27.481 = -\frac{33\,550}{RT} + 22.979$$

$T = 279.99$ K = 6.85 °C

$$P = \exp\left(-\frac{44\,030}{8.314 \times 279.99} + 27.481\right) = 5249 \text{ Pa}$$

(b) $\Delta_{fus}H = \Delta_{sub}H - \Delta_{vap}H = 44.03 - 33.55 = 10.48$ kJ mol^{-1}

6.13 The surface tension of toluene at 20 °C is 0.0284 N m^{-1}, and its density at this temperature is 0.866 g cm^{-3}. What is the radius of the largest capillary that will permit the liquid to rise 2 cm?

SOLUTION

$$\gamma = \frac{1}{2} h \rho g r$$

$$r = \frac{2\gamma}{\rho g h} = \frac{2(0.284 \text{ N m}^{-1})}{(0.866 \times 10^3 \text{ kg m}^{-3})(9.80 \text{ m s}^{-2})(0.02 \text{ m})}$$

$$= 3.35\text{x}10^{-2} \text{ cm}$$

6.14 Mercury does not wet a glass surface. Calculate the capillary depression if the diameter of the capillary is (a) 0.1 mm, and (b) 2 mm. The density of mercury is 13.5 g cm^{-3}. The surface tension of mercury at 25 °C is 0.520 N m^{-1}.

SOLUTION

(a) $$h = \frac{2\gamma}{g\rho r}$$

$$= \frac{(2)(0.520 \text{ N m}^{-1})}{(9.8 \text{ m s}^{-2})(13.5\text{x}10^3 \text{ kg m}^{-3})(0.05\text{x}10^{-3} \text{ m})} = 15.7 \text{ cm}$$

(b) $h = (15.7 \text{ cm})(0.05 \text{ mm})/(1 \text{ mm}) = 7.86 \text{ mm}$

6.15 If the surface tension of a soap solution is 0.05 N m^{-1}, what is the difference in pressure across the film for (a) a soap bubble of 2 mm in diameter and (b) a bubble 2 cm in diameter.

SOLUTION

(a) $$\Delta P = \frac{2\gamma}{R} = \frac{4(0.05 \text{ N m}^{-1})}{0.001 \text{ m}} = 200 \text{ Pa}$$

(b) $$\Delta P = \frac{4(0.050 \text{ n m}^{-1})}{0.01 \text{ m}} = 20 \text{ Pa}$$

*6.16 From tables giving $\Delta_f G^o$, $\Delta_f H^o$, and $\overline{C}_P^o$ for $H_2O(1)$ and $H_2O(g)$ at 298 K, calculate (a) the vapor pressure of $H_2O(1)$ at 25 °C and (b) the boiling point at 1 atm.

SOLUTION

(a) $H_2O(1) = H_2O(g)$

$$\Delta G^o = -228.572 - (-237.129) = 8.557 \text{ kJ mol}^{-1}$$

$$\Delta G^o = -RT \ln\left(\frac{P}{P^o}\right)$$

$$\frac{P}{P^o} = \exp\frac{8557 \text{ J mol}^{-1}}{(8.314 \text{ J K}^{-1} \text{ mol}^{-1})(298.15 \text{ K})} = 3.168 \times 10^{-2}$$

$$P = (3.168 \times 10^{-2})(10^5 \text{ Pa}) = 3.17 \times 10^3 \text{ Pa}$$

(b) ΔH^o (298.15 K) = - 214.818 - (- 285.830) = 44.012 kJ mol^{-1}

In the absence of data on the dependence of $\bar{C}_P$ on T we will calculate $\Delta H^o(T)$ from

$$\Delta H^o(T) = 44\,012 + \int_{298.15}^{T} (33.577 - 75.291)\ \mathrm{d}T$$

$$= 44\,012 - 41.714\,(T - 298.15)$$

$$\left[\frac{\partial\left(\frac{\Delta G^o}{T}\right)}{\partial T}\right]_P = -\frac{\Delta H^o(T)}{T^2}$$

$$\frac{\Delta G^o}{T} = -\int\left[\frac{-44\,012}{T^2} - \frac{41.714}{T} + \frac{(41.714)(298.15)}{T^2}\right]\mathrm{d}T$$

$$= \frac{44\,012}{T} + 41.714 \ln T + \frac{(41.714)(298.15)}{T} + I$$

where I is an integration constant to be evaluated from ΔG^o(298.15 K).

$$\frac{8590}{298.15} = \frac{44.012}{298.15} + 41.714 \ln 298.15 + \frac{(41.714)(298.15)}{(298.15)} + I$$

I = - 398.191

At the boiling point $\Delta G^o = 0$ and so we need to solve the following equation by successive approximations.

$$0 = \frac{44012}{T} + 41.714 \ln T + \frac{(41.714)(298.15)}{T} - 398.191$$

Trying T = 373 K RHS = 0.163

Trying T = 374 K RHS = - 0.130

Thus, the standard boiling point calculated in this way is close to 373.5 K.

6.17 What is the maximum number of phases that can be in equilibrium in one-, two-, and three-component systems?

SOLUTION

$F = C - p + 2$

$0 = 1 - p + 2, p = 3$

$0.= 2 - p + 2, p = 4$

$0 = 3 - p + 2, p = 5$

6.18 The vapor pressure of water at 25 °C is 23.756 mm Hg. What is the vapor pressure of water when it is in a container with an air pressure of 100 bar, assuming the dissolved gases do not affect the vapor pressure. The density of water is 0.99707 g/m^3.

SOLUTION

$RT \ln (P/P_o) = \bar{V}_L(P - P_o)$

$(8.3145 \text{ J K}^{-1} \text{ mol}^{-1})(298.15 \text{ K}) \ln (P/23.756)$

$= (18.015 \text{ g mol}^{-1}/0.99707 \text{ g m}^{-3})(10^{-2} \text{ m/cm})^3(99 \times 10^5 \text{ Pa})$

$P/23.756 = \exp 0.07289$

$P = 25.552$ mm Hg = 3406 Pa

6.19 A binary liquid mixture of A and B is in equilibrium with its vapor at constant temperature and pressure. Prove that $\mu_A(g) = \mu_A(l)$ and $\mu_B(g) = \mu_B(l)$ by starting with
$G = G(g) + G(l)$
and the fact that $dG = 0$ when infinitesimal amounts of A and B are simultaneously transferred from the liquid to the vapor.

SOLUTION

$dG = dG(l) + dG(g) = 0$ (1)

since the transfer is made at equilibrium.

$dG(l) = \mu_A(l)\, dn_A(l) + \mu_B(l)\, dn_B(l)$ (2)

$dG(g) = \mu_A(g)\, dn_A(g) + \mu_B(g)\, dn_B(g)$ (3)

since $dn_A(l) = -\, dn_A(g)$ and $dn_B(l) = -\, dn_B(g)$, substituting equation 2 and 3 into equation 1 yields

$dG = [\mu_A(g) - \mu_A(l)]\, dn_A(g) + [\mu_B(g) - \mu_A(l)]\, dn_B(g) = 0$ (4)

Since $dn_A(g)$ and $dn_B(g)$ are independently variable, both terms in equation 4 have to be zero for dG to be zero. This yields the desired expression for components A and B.

6.20 Ethanol and methanol form very nearly ideal solutions. At 20 °C, the vapor pressure of ethanol is 5.93 kPa, and that of methanol is 11.83 kPa. (a) Calculate the mole fractions of methanol and ethanol in a solution obtained by mixing 100 g of each. (b) Calculate the partial pressures and the total vapor pressure of the solution. (c) Calculate the mole fraction of methanol in the vapor.

SOLUTION

(a) $$x_{CH_3OH} = \frac{100/32}{100/46 + 100/32} = 0.590$$

$$x_{C_2H_5OH} = \frac{100/46}{100/46 + 100/32} = 0.410$$

(b) $P_{CH_3OH} = x_{CH_3OH} P^*_{CH_3OH}$

$= (0.590)(11{,}830 \text{ Pa}) = 6980 \text{ Pa}$

$P_{C_2H_5OH} = x_{C_2H_5OH} P^*_{C_2H_5OH}$

$= (0.410)(5930 \text{ Pa}) = 2430 \text{ Pa}$

$P_{total} = 2430 \text{ Pa} + 6980 \text{ Pa} = 9410 \text{ Pa}$

(c) $y_{CH_3OH,vapor} = \dfrac{6980 \text{ Pa}}{9410 \text{ Pa}} = 0.742$

6.21 One mole of benzene (component 1) is mixed with two moles of toluene (component 2). At 60 °C the vapor pressures of benzene and toluene are 51.3 and 18.5 kPa, respectively. (a) As the pressure is reduced, at what pressure will boiling begin? (b) What will be the composition of the first bubble of vapor?

SOLUTION

(a) $P = P_2^* + (P_1^* - P_2^*)x_1$

$= 18.5 \text{ kPa} + (51.3 \text{ kPa} - 18.5 \text{ kPa})(0.333)$

$= 29.4 \text{ kPa}$

(b) $y_1 = \dfrac{x_1 P_1^*}{P_2^* + (P_1^* - P_2^*)x_1}$

$= \dfrac{(0.333)(51.3 \text{ kPa})}{18.5 \text{ kPa} + 32.8 \text{ kPa}\,(0.333)}$

$= 0.581$

6.22 The vapor pressures of benzene and toluene have the following values in the temperature range between their boiling points at 1 bar:

t/°C	79.4	88	94	100	110.0
$P^*_{C_6H_6}$/bar	1.000	1.285	1.526	1.801	
$P^*_{C_7H_8}$/bar		0.508	0.616	0.742	1.000

(a) Calculate the compositions of the vapor and liquid phases at each temperature and plot the boiling point diagram. (b) If a solution containing 0.5 mole fraction benzene and 0.5 mole fraction toluene is heated, at what temperature will the first bubble of vapor appear and what will be its composition?

SOLUTION

(a) At 88 °C $x_B(1.285) + (1 - x_B)\,0.508 = 1$

$x_B = 0.633$

$y_B = (0.633)(1.285)/1 = 0.814$

At 94 °C $x_B(1.526) + (1 - x_B)\,0.616 = 1$

$x_B = 0.422$

$y_B = (0.4222)(1.526)/1 = 0.644$

At 100 °C $x_B(1.801) + (1 - x_B)\,0.742 = 1$

$x_B = 0.244$

$y_B = (0.244)(1.801)/1 = 0.439$

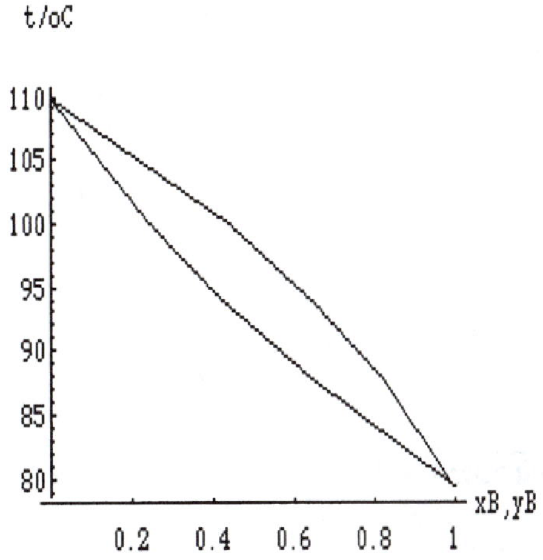

The liquid curve is below the vapor curve. These curves were plotted by using *Mathematica* ™ to fit them with quadratic equations.

(b) The fit of the liquid data with $t = a + bx_B + cx_B^2$ yields $t = 109.9 - 42.72x_B + 12.29x_B^2$. At $x_B = 0.5$, $t = 91.6$ °C. Fit of the vapor data yields $t = 109.9 - 15.13y_B - 15.13y_B^2$. Setting the temperature in this equation equal to the bubble point temperature 91.6 °C yields $y_B = 0.72$, by use of the quadratic formula. This is the composition at which the first bubble of vapor will appear.

6.23 At 1.013 bar pressure propane boils at -42.1 °C and *n*-butane boils at -0.5 °C; the following vapor-pressure data are available.

t/°C	-31.2	-16.3
P/kPa(propane)	160.0	298.6
P/kPa(n-butane)	26.7	53.3

Assuming that these substances form ideal binary solutions with each other, (a) calculate the mole fractions of propane at which the solution will boil at 1.013 bar pressure at -31.2 and -16.3 °C. (b) Calculate the mole fractions of propane in the equilibrium vapor at these temperatures. (c) Plot the temperature-mole fraction diagram at 1.013 bar, using these data, and label the regions.

SOLUTION

(a) At -31.2 °C, $P_P = x_P/60.0$ kPa, $P_B = (1 - x_P)$ 26.7 kPa

$P = 101.3$ kPa $= x_P\ 160.0 + (1 - x_P)\ 26.7$

$= 26.7 + 133.3\ x_P$

$x_P = 0.560$

At -16.3 °C, $P_P = x_P$ 298.6 kPa, $P_B = (1 - x_P)$ 53.3 kPa

$P = 101.3$ kPa $= x_P\ 298.6 + (1 - x_P)\ 53.3$

$= 53.3 + 245.3\ x_P$

$x_P = 0.196$

(b) At -31.2 °C,

$$y_P = \frac{(0.560)(160.0)}{(0.560)(160.0) + (0.440)(26.7)} = 0.884$$

At -16.3 °C

$$y_P = \frac{(0.196)(298.6)}{(0.196)(298.6) + (0.804)(53.3)} = 0.577$$

(c)

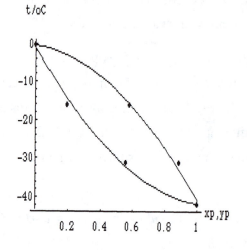

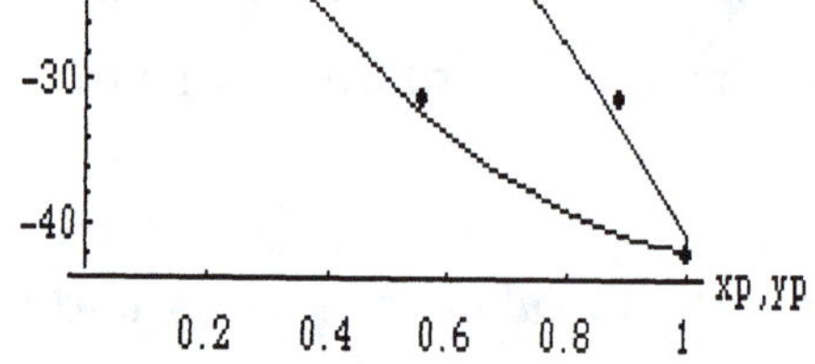

6.24 The following table gives mole % acetic acid in aqueous solutions and in the equilibrium vapor at the boiling point of the solution at 1.013 bar.

B.P., °C	118.1	113.8	107.5	104.4	102.1	100.0
Mole % of acetic acid						
In liquid	100	90.0	70.0	50.0	30.0	0
In vapor	100	83.3	57.5	37.4	18.5	0

Calculate the minimum number of theoretical plates for the column required to produce an initial distillate of 28 mole % acetic acid from a solution of 80 mole % acetic acid.

SOLUTION

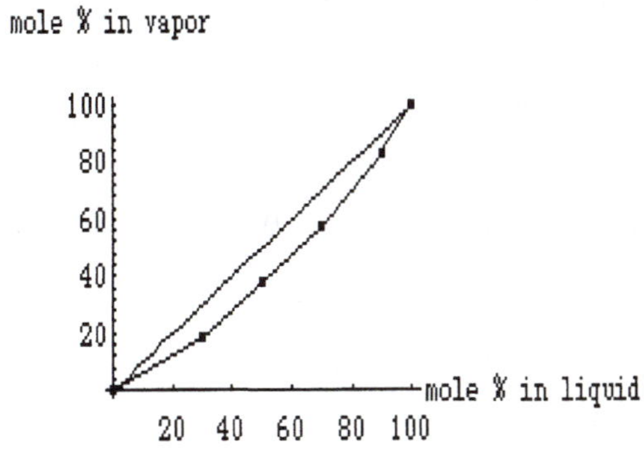

Since there are four steps, three theoretical plates are required in the column. The distilling pot counts as one plate.

6.25 If two liquids (1 and 2) are completely immiscible, the mixture will boil when the sum of the two partial pressures exceeds the applied pressure: $P = P_1^* + P_2^*$. In the vapor phase the ratio of the mole fractions of the two components is equal to the ratio of their vapor pressures.

$$\frac{P_1^*}{P_2^*} = \frac{x_1}{x_2} = \frac{m_1 M_2}{m_2 M_1}$$

where m_1 and m_2 are the masses of components 1 and 2 in the vapor phase, and M_1 and M_2 are their molar masses. The boiling point of the immiscible liquid system naphthalene-water is 98 °C under a pressure of 97.7 kPa. The vapor pressure of water at 98 °C is 94.3 kPa. Calculate the weight percent of naphthalene in the distillate.

SOLUTION

$$\frac{m_1}{m_2} = \frac{P_1^* M_1}{P_2^* M_2} = \frac{(97.7 - 94.3)(128)}{(94.3)(18)} = 0.256$$

$$\text{Weight percent naphthalene} = \frac{0.256}{1.256} = 0.204 \text{ or } 20.4\%$$

6.26 A regular binary solution is defined as one for which

$$\mu_1 = \mu_1^{\text{o}} + RT \ln x_1 + w x_2^2$$

$$\mu_2 = \mu_2^{\text{o}} + RT \ln x_2 + w x_1^2$$

Derive $\Delta_{\text{mix}}G$, $\Delta_{\text{mix}}S$, $\Delta_{\text{mix}}H$, and $\Delta_{\text{mix}}V$ for the mixing of x_1 moles of component 1 with x_2 moles of component 2. Assume that the coefficient w is independent of temperature.

SOLUTION

$$G = \sum_i n_i\mu_i = x_1\mu_1^{\text{o}} + x_1 RT \ln x_1 + w x_1 x_2^2 + x_2\mu_2^{\text{o}} + x_2 RT \ln x_2 + w x_2 x_1^2$$

$$G^{\text{o}} = \sum_i n_i\mu_i^{\text{o}} = x_1\mu_1^{\text{o}} + x_2\mu_2^{\text{o}}$$

$$\Delta_{\text{mix}}G = RT(x_1 \ln x_1 + x_2 \ln x_2) + w x_1 x_2$$

$$\Delta_{\text{mix}}S = -(\partial\Delta_{\text{mix}}G/\partial T)_P = -R(x_1 \ln x_1 + x_2 \ln x_2)$$

$$\Delta_{\text{mix}}H = \Delta_{\text{mix}}G + T\Delta_{\text{mix}}S = w x_1 x_2$$

$$\Delta_{\text{mix}}V = (\partial\Delta_{\text{mix}}G/\partial P)_T = 0$$

6.27 From the data given in the following table construct a complete temperature-composition diagram for the system ethanol-ethyl acetate for 1.013 bar. A solution containing 0.8 mole fraction of ethanol, EtOH, is distilled completely at 1.013 bar. (a) What is the composition of the first vapor to come off? (b) That of the last drop of liquid to evaporate? (c) What would be the values of these quantities if the distillation were carried out in a cylinder provided with a piston so that none of the vapor escapes?

x_{EtOH}	y_{EtOH}	B.P., °C	x_{EtOH}	y_{EtOH}	B.P., °C
0	0	77.15	0.563	0.507	72.0
0.025	0.070	76.7	0.710	0.600	72.8
0.100	0164	75.0	0.833	0.735	74.2
0.240	0.295	72.6	0.942	0.880	76.4

0.360	0.398	71.8	0.982	0.965	77.7
0.462	0.462	71.6	1.000	1.000	78.3

SOLUTION

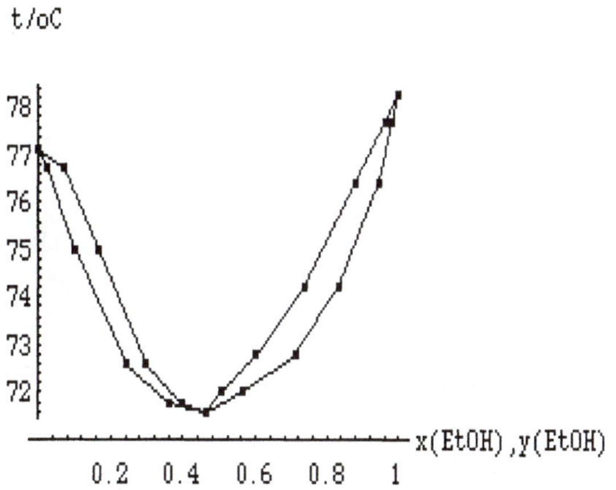

The composition of the liquid is given by the lower curve, and the composition of the vapor is given by the upper curve.

(a) The liquid with 0.8 mole fraction ethanol will start to boil at 73.7 °C. The vapor in equilibrium with it is 0.69 mole fraction in ethanol.

(b) As the boiling continues, more and more of the EtAc will be removed. As the mole fraction of EtOH increases, the boiling point will also increase. Finally, the very last vapors will be pure EtOH.

(c) The first vapor to come off will be the same as in part (a). However, when all of the mixture is in the vapor state, the mole fraction of the vapor state will be 0.800, simply because that is the total mole fraction of EtOH in the entire mixture. Thus, to find the composition of the last drop to evaporate into the cylinder, simply note that liquid ($y_{EtOH} = 0.9$) is in equilibrium with vapor which has 0.800 EtOH in it.

6.28 The Henry law constants for oxygen and nitrogen in water at 0 °C are 2.54×10^4 bar and 5.45×10^4 bar, respectively. Calculate the lowering of the freezing point of water by dissolved air with 80% N_2 and 20% O_2 by volume at 1 bar pressure.

SOLUTION

$$x_{N_2} = \frac{P_{N_2}}{K_{N_2}} = \frac{0.8 \text{ bar}}{5.45 \times 10^4 \text{ bar}} = 1.47 \times 10^{-5}$$

$$x_{O_2} = \frac{P_{O_2}}{K_{O_2}} = \frac{0.2 \text{ bar}}{2.54 \times 10^4 \text{ bar}} = 7.87 \times 10^{-6}$$

The mole fraction of dissolved air is $2.26 \times 10^{-5} = \dfrac{m}{m + \dfrac{1000}{18.02}}$

$m_{air} = 1.25 \times 10^{-3}$ mol/1000 g solvent

$\Delta T_{fp} = K_{fp}\, m = -(1.86)(1.25 \times 10^{-3}) = -0.00233$ K

6.29 Use the Gibbs-Duhem equation to show that if one component of a binary liquid solution follows Raoult's law, the other component will too.

SOLUTION

$$x_1 d\mu_1 + x_2 d\mu_2 = 0$$

If $\mu_1 = \mu_1^o + RT \ln x_1$

$$d\mu_1 = \frac{RT}{x_1} dx_1$$

Using the first equation

$$d\mu_2 = -\frac{x_1}{x_2} d\mu_1 = -\frac{RT}{x_2} dx_1$$

Since $x_1 + x_2 = 1$, $\quad dx_2 = -dx_1$

$$d\mu_2 = \frac{RT}{x_2} dx_2 = RT\, d \ln x_2$$

$$\mu_2 = \text{const} + RT \ln x_2$$

If $x_2 = 1$, $\quad$ const $= \mu_2^o$

$$\mu_2 = \mu_2^o + RT \ln x_2$$

6.30 The following data on ethanol-chloroform solutions at 35 °C were obtained by G. Scatchard and C. L. Raymond [*J. Am. Chem. Soc.* **60**, 1278 (1938)]:

$x_{EtOH,liq}$	0	0.2	0.4	0.6	0.8	1.0
$y_{EtOH,vap}$	0.0000	0.1382	0.1864	0.2554	0.4246	1.0000
Total P/kPa	39.345	40.559	38.690	34.387	25.357	13.703

Calculate the activity coefficients of ethanol and chloroform based on the deviations from Raoult's law.

SOLUTION

$x_{EtOH} = 0.2 \qquad \gamma_{EtOH} = \dfrac{y_{EtOH} P}{x_{EtOH} P^*_{EtOH}} = \dfrac{(0.1382)(40.559)}{(0.2)(13.703)} = 2.04$

$x_{EtOH} = 0.4 \qquad = \dfrac{(0.1864)(38.690)}{(0.4)(13.703)} = 1.316$

$x_{EtOH} = 0.6 \qquad = \dfrac{(0.2554)(34.387)}{(0.6)(13.703)} = 1.065$

$x_{EtOH} = 0.8$ $= \dfrac{(0.4246)(25.357)}{(0.8)(13.703)} = 0.982$

$x_{EtOH} = 1.0$ $= \dfrac{(1.000)(13.703)}{(1.000)(13.703)} = 1.000$

$x_{CHCl_3} = 1$ $\gamma_{CHCl_3} = \dfrac{y_{CHCl_3}P}{x_{CHCl_3}P^*_{CHCl_3}} = \dfrac{(1)(39.345)}{(1)(39.345)} = 1$

$x_{CHCl_3} = 0.8$ $= \dfrac{(0.8618)(40.559)}{(0.8)(39.345)} = 1.111$

$x_{CHCl_3} = 0.6$ $= \dfrac{(0.8136)(38.690)}{(0.6)(39.345)} = 1.333$

$x_{CHCl_3} = 0.4$ $= \dfrac{(0.7446)(34.387)}{(0.4)(39.345)} = 1.627$

$x_{CHCl_3} = 0.2$ $= \dfrac{(0.5754)(25.357)}{(0.2)(39.345)} = 1.854$

6.31 Show that the equations for the bubble-point line and dew-point line for nonideal solutions are given by

$$x_1 = \frac{P - \gamma_2 P_2^*}{\gamma_1 P x_1^* - \gamma_2 P_2^*}$$

$$y_1 = \frac{P\gamma_1 P_1^* - \gamma_1\gamma_2 P_1^* P_2^*}{P\gamma_1 P_1^* - P\gamma_2 P_2^*}$$

SOLUTION

$$P = P_1 + P_2 = x_1\gamma_1 P_1^* + (1 - x_1)\gamma_2 P_2^*$$

$$= \gamma_2 P_2^* + x_1(\gamma_1 P_1^* - \gamma_2 P_2^*)$$

$$x_1 = \frac{P - \gamma_2 P_2^*}{\gamma_1 P_1^* - \gamma_2 P_2^*}$$

$$y_1 = \frac{P_1}{P_1 + P_2} = \frac{x_1\gamma_1 P_1^*}{\gamma_2 P_2^* + x_1(\gamma_1 P_1^* - \gamma_2 P_2^*)}$$

Substituting the expression for x_1 yields the desired expression for y_1.

6.32 A regular binary solution is defined as one for which

$$\mu_1 = \mu_1^{\circ} + RT \ln x_1 + wx_2^2$$

$$\mu_2 = \mu_2^{\circ} + RT \ln x_2 + wx_1^2$$

Derive the expressions for the activity coefficients of γ_1 and γ_2 in terms of w.

SOLUTION

$$\mu_1 = \mu_1^{\circ} + RT \ln \gamma_1 x_1$$

$$\mu_1 = \mu_1^{\circ} + RT \ln x_1 + wx_2^2$$

$$= \mu_1^{\circ} + RT \ln x_1 + RT \ln e^{wx_2^2/RT}$$

$$= \mu_1^{\circ} + RT \ln (e^{wx_2^2/RT}) x_1$$

Therefore,

$$\gamma_1 = e^{wx_2^2/RT} \qquad \gamma_2 = e^{wx_1^2/RT}$$

6.33 The expressions for the activity coefficients of the components of a regular binary solution were derived in the preceding problem. Derive the expression for γ_1 in terms of the experimentally measured total pressure P, the vapor pressures of the two components, and the composition of the solution for the case that the deviations from ideality are small.

SOLUTION

$$\gamma_1 = \exp(wx_2^2/RT) \qquad \gamma_2 = \exp(wx_1^2/RT)$$

$$P = P_1 + P_2 = \gamma_1 x_1 P_1^* + \gamma_2 x_2 P_2^*$$

$$= x_1 P_1^* \exp\left(\frac{wx_2^2}{RT}\right) + x_2 P_2^* \exp\left(\frac{wx_1^2}{RT}\right)$$

If the exponentials are not far from unity, the exponentials may be expanded to obtain

$$P = x_1 P_1^* \left(1 + \frac{wx_2^2}{RT}\right) + x_2 P_2^* \left(1 + \frac{wx_1^2}{RT}\right)$$

$$\frac{w}{RT} = \frac{P - (x_1 P_1^* + x_2 P_2^*)}{x_1 x_2 \, [P_1^* + x_1 (P_2^* - P_1^*)]}$$

Thus from a measurement of the total pressure of a mixture of the two components, the activity coefficients of the components can be calculated over the entire concentration range using the first two equations given above. This is a particular example of the general situation that γ_1 and γ_2 can be calculated from the total pressure as a function of x_1.

6.34 Using the data in Problem 6.75, calculate the activity coefficients of water (1) and *n*-propanol (2) at 0.20, 0.40, 0.60, and 0.80 mole fraction of *n*-propanol, based on deviations from Henry's law and considering water to be the solvent.

<u>SOLUTION</u>

The data at the lowest mole fractions of *n*-propanol are used to calculate the Henry law constant for *n*-propanol in water.

x_2	$K_2 = P_2/x_2$
0.02	33.5
0.05	28.8
0.10	17.6

The intercept indicates that $K_2 = 37$ kPa.

At

$x_2 = 0.2$ $\quad \gamma_2' = \frac{P_2}{x_2 K_2} = \frac{1.81}{0.2(37)} = 0.24$

$x_2 = 0.4$ $$= \frac{1.89}{0.4(37)} = 0.13$$

$x_2 = 0.6$ $$= \frac{2.07}{0.6(37)} = 0.093$$

$x_2 = 0.8$ $$= \frac{2.37}{0.8(37)} = 0.080$$

$x_2 = 0.2$ $$\gamma_1 = \frac{P_1}{x_1 P_1^*} = \frac{2.91}{(0.8)(3.17)} = 1.15$$

$x_2 = 0.4$ $$= \frac{2.89}{(0.6)(3.17)} = 1.52$$

$x_2 = 0.6$ $$= \frac{2.65}{(0.4)(3.17)} = 2.09$$

$x_2 = 0.8$ $$= \frac{1.79}{(0.2)(3.17)} = 2.82$$

6.35 If 68.4 g of sucrose ($M = 342$ g mol^{-1}) is dissolved in 1000 g of water: (a) What is the vapor pressure at 20 °C? (b) What is the freezing point? The vapor pressure of water at 20 °C is 2.3149 kPa.

SOLUTION

(a) $$x_2 = \frac{\frac{68.4}{342}}{\frac{68.4}{342} + \frac{1000}{18}} = 3.59 \times 10^{-3}$$

$$\frac{P_1^* - P_1}{P_1^*} = x_2 \qquad \frac{2.3149 - P_1}{2.3149} = 3.59 \times 10^{-3}$$

$$P_1 = 2.3149(1 - 3.59 \times 10^{-3}) = 2.3066 \text{ kPa}$$

(b) $\Delta T_f = K_f m = 1.86(0.2) = -0.372$

$T_f = -0.372$ °C

6.36 The protein human plasma albumin has a molar mass of 69,000 g mol^{-1}. Calculate the osmotic pressure of a solution of this protein containing 2 g per 100 cm^3 at 25 °C in (a) pascals and (b) millimeters of water. The experiment is carried out using a salt solution for solvent and a membrane permeable to salt as well as water.

SOLUTION

$$\pi = \frac{cRT}{M} = \frac{(20 \times 10^{-3} \text{ kg L}^{-1})(10^3 \text{ L m}^{-3})(8.314 \text{ J K}^{-1} \text{ mol}^{-1})(298.15 \text{ K})}{69 \text{ kg mol}^{-1}}$$

$= 719$ Pa

$$h = \frac{\pi}{dg} = \frac{719\ \text{Pa}}{(1\ \text{g cm}^{-3})(10^{-3}\ \text{kg g}^{-1})(10^{2}\ \text{cm m}^{-1})(9.8\ \text{m s}^{-2})}$$

$= 7.34 \times 10^{-2}$ m = 73.4 mm of water

*6.37 The following osmotic pressures were measured for solutions of a sample of polyisobutylene in benzene at 25 °C.

c /kg m^{-3}	5	10	15	20
π /Pa	49.5	101	155	211

Calculate the number average molar mass from the value of π/c extrapolated to zero concentration of the polymer.

SOLUTION

c/kg m^{-3}	5	10	15	20
(π/c)/(Pa/kg m^{-3})	9.90	10.1	10.3	10.6

$$\frac{\pi}{c} = \frac{RT}{M} + Bc$$

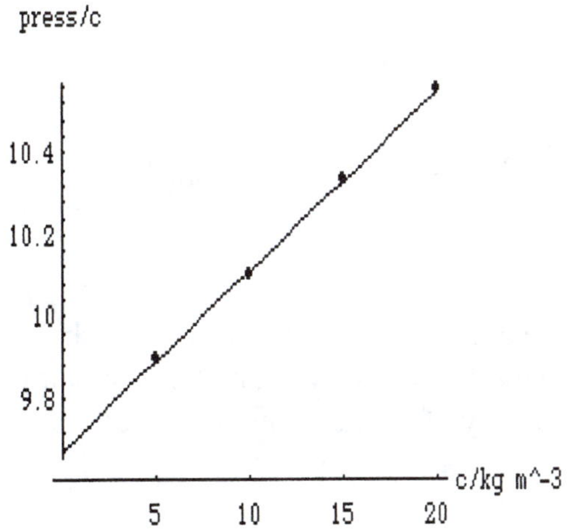

$$M = \frac{RT}{9.7\ \text{Pa/kg m}^{-3}} = \frac{(8.314\ \text{J K}^{-1}\ \text{mol}^{-1})(298\ \text{K})}{9.7\ \text{Pa m}^{3}\ \text{kg}^{-1}} = 255\ \text{kg mol}^{-1}$$

6.38 Calculate the osmotic pressure of a 1 mol L^{-1} sucrose solution in water from the fact that at 30 °C the vapor pressure of the solution is 4.1606 kPa. The vapor pressure of water at 30 °C is 4.2429 kPa. The density of pure water at this temperature (0.99564 g cm^{-3}) may be used to estimate V_1 for a dilute solution. To do this problem, Raoult's law is introduced into equation 6.74.

SOLUTION

$\bar{V}_1\pi = - RT \ln x_1$

Substituting Raoult's law $P_1 = x_1 P_1^o$

$$\pi = -\frac{RT}{\bar{V}_1} \ln \frac{P_1}{P_1^o}$$

$$\bar{V}_1 = \frac{18.02 \text{ g mol}^{-1}}{0.99564 \text{ g cm}^{-3}} = 18.10 \text{ cm}^3 \text{ mol}^{-1} = 0.01810 \text{ L mol}^{-1}$$

$$\pi = \frac{(0.08314 \text{ L bar K}^{-1} \text{ mol}^{-1})(303.15 \text{ K})}{0.0180 \text{ L mol}^{-1}} \ln \frac{4.2429 \text{ kPa}}{4.1606 \text{ kPa}} = 27.3 \text{ bar}$$

6.39 Calculate the solubility of *p*-dibromobenzene in benzene at 20 °C and 40 °C assuming ideal solutions are formed. The enthalpy of fusion of *p*-dibromobenzene is 13.22 kJ mol^{-1} at its melting point (86.9 °C).

SOLUTION

At 20 °C

$$\ln x_2 = -\frac{\Delta_{fus}H_2(T - T_{2f})}{RTT_{2f}}$$

$$= \frac{(13\ 220 \text{ J mol}^{-1})(-66.9 \text{ K})}{(8.314 \text{ J K}^{-1} \text{ mol}^{-1})(360.1 \text{ K})(293.2 \text{ K})}$$

$x_2 = 0.365$

At 40 °C

$$\ln x_2 = \frac{(13\ 220 \text{ J mol}^{-1})(-46.9 \text{ K})}{(8.314 \text{ J K}^{-1} \text{ mol}^{-1})(360.1 \text{ K})(313.2 \text{ K})}$$

$x_2 = 0.516$

6.40 Calculate the solubility of naphthalene at 25 °C in any solvent in which it forms an ideal solution. The melting point of naphthalene is 80 °C, and the enthalpy of fusion is 19.19 kJ mol^{-1}. The measured solubility of naphthalene in benzene is $x_1 = 0.296$.

SOLUTION

$$-\ln x_1 = \Delta_{fus}H_1(T_{0,1} - T)/RTT_{0,1}$$

$$\ln x_1 = \frac{-(19\ 190 \text{ J mol}^{-1})(353 \text{ K} - 298 \text{ K})}{(8.314 \text{ J K}^{-1} \text{ mol}^{-1})(353 \text{ K})(298 \text{ K})}$$

$x_1 = 0.297$

6.41 The addition of a nonvolatile solute to a solvent increases the boiling point above that of the pure solvent. The elevation of the boiling point is given by

$$\Delta T_b = \frac{R(T_{bA})^2 M_A m_B}{\Delta_{vap}H_A^o} = K_b m_B$$

where T_{bA} is the boiling point of the pure solvent and M_A is its molar mass. The derivation of this equation parallels that of equation 6.82 very closely, and so it is not given. What is the elevation of the boiling point when 0.1 mol of nonvolatile solute is added to 1 kg of water? The enthalpy of vaporization of water at the boiling point is 40.6 kJ mol^{-1}.

SOLUTION

$$K_b = \frac{(8.314\ \text{J K}^{-1}\ \text{mol}^{-1})(393.1\ \text{K})^2(0.018\ 01\ \text{kg mol}^{-1})}{40\ 600\ \text{J mol}^{-1}}$$

$$= 0.513\ \text{K kg mol}^{-1}$$

$$\Delta T_b = (0.513\ \text{K kg mol}^{-1})(0.1\ \text{mol kg}^{-1}) = 0.0513\ \text{K}$$

6.42 The NBS Tables of Chemical Thermodynamic Properties list $\Delta_f G^o$ for I_2 in $C_6H_6{:}x$ as 7.1 kJ mol. The x indicates that the standard state for I_2 in C_6H_6 is on the mole fraction scale. What is the solubility of I_2 in C_6H_6 at 298 K on the mole fraction scale? A chemical handbook lists the solubility as 16.46 g I_2 in 100 cm^3 of C_6H_6. Are these solubilities consistent?

SOLUTION

I_2 (cr) + [C_6H_6] = I_2 in C_6H_6

$\Delta G^o = -RT \ln x$

$7.1 = -(8.314 \times 10^{-3})(298) \ln x$

$x = 0.0569$ where x is the equilibrium mole fraction of I_2

The mole fraction calculated from the chemical handbook is

$$\frac{\dfrac{16.46}{253.8}}{\dfrac{16.46}{253.8} + \dfrac{(100)(0.8787)}{78.12}} = 0.0545$$

so the values are consistent.

6.43 The following cooling curves have been found for the system antimony-cadmium.

Cd Wt. %	0	20	37.5	47.5	50	58	70	93	100
First break in curve, °C	--	550	461	--	419	--	400	--	--
Constant temperature, °C	630	410	410	410	410	439	295	295	321

Construct a phase diagram, assuming that no breaks other than these actually occur in any cooling curve. Label the diagram completely and give the formula of any compound formed. How many degrees of freedom are there for each area and at each eutectic point?

SOLUTION

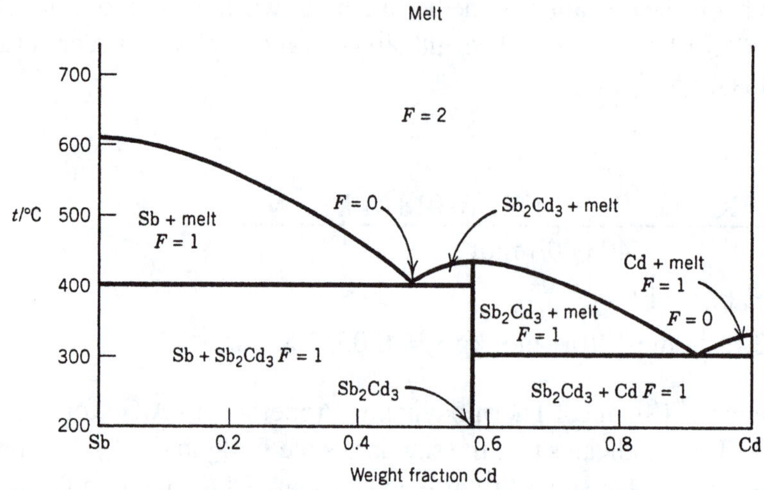

6.44 The phase diagram for magnesium-copper at constant pressure shows that two compounds are formed: $MgCu_2$ that melts at 800 °C, and Mg_2Cu that melts at 580 °C. Copper melts at 1085 °C, and Mg at 648 °C. The three eutectics are at 9.4% by weight Mg (680 °C), 34% by weight Mg (560 °C), and 65% by weight Mg (380 °C). Construct the phase diagram. How many degrees of freedom are there for each area and at each eutectic point?

SOLUTION

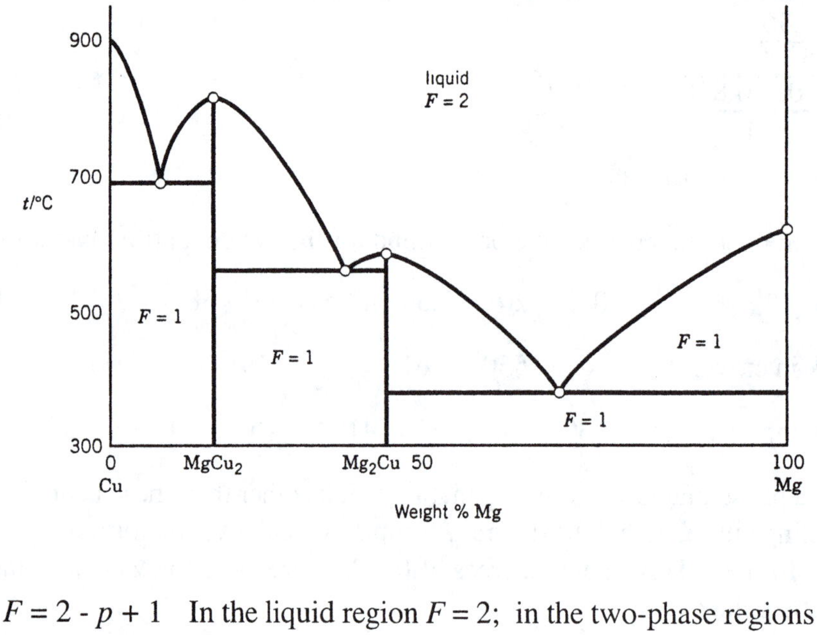

$F = 2 - p + 1$ In the liquid region $F = 2$; in the two-phase regions $F = 1$; and at the eutectic point $F = 0$.

6.45 The Gibbs-Duhem equation in the form

$$\left(\frac{\partial M}{\partial T}\right)_{P,x} \mathrm{d}T + \left(\frac{\partial M}{\partial P}\right)_{T,x} \mathrm{d}P - \sum (x_i \mathrm{d}M_i) = 0$$

applies to any molar thermodynamic property M in a homogeneous phase. If this is applied to G^E, it may be shown that if the vapor is an ideal gas.

$$x_1 \frac{\mathrm{d}\ln(y_1P)}{\mathrm{d}x_1} + x_2 \frac{\mathrm{d}\ln(y_2P)}{\mathrm{d}x_1} = 0 \qquad \text{(constant } T\text{)}$$

Show that this can be rearranged to the coexistence equation

$$\frac{\mathrm{d}P}{\mathrm{d}y_1} = \frac{P(y_1 - x_1)}{y_1(1 - y_1)}$$

Thus if P versus y_1 is measured, there is no need for measurements of x_1.

SOLUTION

$$x_1 \mathrm{d}\ln y_1 + x_1 \mathrm{d}\ln P + x_2 \mathrm{d}\ln y_2 + x_2 \mathrm{d}\ln P = 0$$

$$\mathrm{d}\ln P = -\frac{x_1}{y_1}\mathrm{d}y_1 - \frac{x_2}{y_2}\mathrm{d}y_2 = \left(-\frac{x_1}{y_1} + \frac{x_2}{y_2}\right)\mathrm{d}y_1$$

$$= \frac{x_2y_1 - x_1y_2}{y_1y_2}\mathrm{d}y_1 = \frac{y_1 - x_1}{y_1(1 - y_1)}\mathrm{d}y_1$$

6.46 For a solution of ethanol and water at 20 °C that has 0.2 mole fraction ethanol, the partial molar volume of water is 17.9 cm^3 mol^{-1} and the partial molar volume of ethanol is 55.0 cm^3 mol^{-1}. What volumes of pure ethanol and water are required to make a liter of this solution? At 20 °C the density of ethanol is 0.789 g cm^{-3} and the density of water is 0.998 g cm^{-3}.

SOLUTION

Component 1 is water and component 2 is ethanol.

$V = n_1\bar{V}_1 + n_2\bar{V}_2 = 4n_2\bar{V}_1 + n_2\bar{V}_2 = n_2(4\bar{V}_1 + \bar{V}_2)$

1000 cm^3 = n_2[(4)(17.9 cm^3 mol^{-1}) + 55.0 cm^3 mol^{-1}]

n_2 = 7.90 mol = w_2/46.07 g mol^{-1}

w_2 = 363.9 g of ethanol

Volume of pure ethanol = 363.9 g/0.789 g cm^{-3} = 461 cm^3

$n_1 = 4n_2$ = 31.50 = w_1/18.016 g mol^{-1}

w_1 = 569.3 g of water

Volume of pure water = 569.3 g/0.998 g cm^{-3} = 570 cm^3

Thus there is a shrinkage of 31 cm^3 when these amounts of ethanol and water are mixed at 20 °C.

6.47 Since the average entropy of vaporization at the standard boiling point (at 1 atm) is 88 J K^{-1} mol^{-1} (see Table 6.2), the vapor pressure of a liquid can be estimated using

$$\Delta_{vap}S = \Delta_{vap}H/T_b = 88 \text{ J K}^{-1} \text{ mol}^{-1}$$

where T_b is the temperature at the boiling point. This equation is often referred to as Trouton's rule. Estimate the vapor pressure of benzene at 25 °C from the fact that its boiling point is 80.1 °C.

SOLUTION

The vapor pressure of benzene at 80.1 °C is 1.013 bar. The enthalpy of vaporization at this temperature is therefore given by $\Delta_{vap}H = T_b\Delta_{vap}S$ = (353.1 K)(88 J K^{-1} mol^{-1}) = 31.1 kJ mol^{-1}. Equation 6.14 can be used to calculate the vapor pressure at 25 °C.

$$\ln\frac{1.013 \text{ bar}}{P_1} = \frac{(31.1x10^3)(55.1)}{(8.3145)(298.2)(353.1)}$$

$$P_1 = 0.143 \text{ bar}$$

6.48 Calculate the solubility of bismuth in an ideal solution at 150 °C and 200 °C and compare the results with Fig. 6.17. The enthalpy of fusion of bismuth at its melting point (273 °C) is 10.5 kJ mol^{-1}.

SOLUTION

Assuming that the enthalpy of fusion is independent of temperature, we can use equation 6.79. At 150 °C

$$x_{Bi} = \exp\left[-\frac{\Delta_{fus}H_{Bi}}{R}\left(\frac{1}{T} - \frac{1}{T_{fus}}\right)\right]$$

$$= \exp\left[-\frac{(10.5x10^3 \text{ J mol}^{-1})}{(8.3145 \text{ J K}^{-1} \text{ mol}^{-1})}\left(\frac{1}{433 \text{ K}} - \frac{1}{546 \text{ K}}\right)\right]$$

$$= 0.510$$

At 200 °C

$$x_{Bi} = \exp\left[-\frac{(10.5x10^3 \text{ J mol}^{-1})}{(8.3145 \text{ J K}^{-1} \text{ mol}^{-1})}\left(\frac{1}{473 \text{ K}} - \frac{1}{546 \text{ K}}\right)\right]$$

$$= 0.700$$

The solubility increases with temperature, as we would expect for an endothermic process.

6.49 The molar volume of a binary solution is given by

$$\overline{V} = x_1\overline{V}_1 + x_2\overline{V}_2$$

This kind of additive equation applies to other thermodynamic properties at constant T and P as well. A convenient way to treat the data on the molar volume or other thermodynamic property of a solution is to fit it to a function (for example, a function of x_2) and then calculate the molar volumes of the substances involved by differentiation of the polynomial. Show that

$$\overline{V}_1 = \overline{V} - x_2\left(\frac{\partial \overline{V}}{\partial x_2}\right)$$

$$\overline{V}_2 = \overline{V} + (1 - x_2)\left(\frac{\partial \overline{V}}{\partial x_2}\right)$$

<u>SOLUTION</u>

The three partial molar volumes are defined by

$$\overline{V} = \frac{V}{n_1 + n_2}, \overline{V}_1 = \left(\frac{\partial V}{\partial n_1}\right)_{n_2}, \overline{V}_2 = \left(\frac{\partial V}{\partial n_2}\right)_{n_1}$$

So

$$\overline{V}_1 = \left(\frac{\partial (n_1 + n_2)\overline{V}}{\partial n_1}\right)_{n_2} = \overline{V} + (n_1 + n_2)\left(\frac{\partial x_2}{\partial n_1}\right)_{n_2}\left(\frac{\partial V}{\partial x_2}\right)$$

But

$$x_2 = \frac{n_2}{n_1 + n_2} \text{ and } \left(\frac{\partial x_2}{\partial n_1}\right)_{n_2} = -\frac{n_2}{(n_1 + n_2)^2}$$

This yields

$$\overline{V}_1 = \overline{V} - x_2\left(\frac{\partial \overline{V}}{\partial x_2}\right)$$

Similarly,

$$\overline{V}_2 = \overline{V} + (1 - x_2)\left(\frac{\partial \overline{V}}{\partial x_2}\right)$$

6.50 Derive the Gibbs-Duhem equation for the volume of a binary solution and show that if the partial molar volume for substance 1 can be determined as a function of x_2, the partial molar volume of substance 2 can be calculated by integrating the relation obtained from the Gibbs-Duhem equation.

SOLUTION

The total differential of the equation for the molar volume of a binary solution is

$$\mathrm{d}\overline{V} = x_1\mathrm{d}\overline{V}_1 + x_2\mathrm{d}\overline{V}_2 + \overline{V}_1\mathrm{d}x_1 + \overline{V}_2\mathrm{d}x_2$$

but we know that

$$\mathrm{d}\overline{V} = \overline{V}_1\mathrm{d}x_1 + \overline{V}_2\mathrm{d}x_2$$

since $\overline{V}$ is a state property and $\mathrm{d}\overline{V}$ is an exact differential. Subtracting the second equation from the first yields the Gibbs-Duhem equation

$$0 = x_1\mathrm{d}\overline{V}_1 + x_2\mathrm{d}\overline{V}_2$$

Thus

$$\mathrm{d}\overline{V}_1 = -\frac{x_2}{x_1}\mathrm{d}\overline{V}_2$$

This equation can be integrated from $x_2 = 0$ to $x_2 = a$:

$$\overline{V}_1(a) = \overline{V}_1^* - \int_0^a \frac{x_2}{x_1}\mathrm{d}\overline{V}_2$$

6.51 Calculate the partial molar volumes of water and glycerol in solutions at 20 °C.

The molar volumes are given as a function of the molar volume of glycerol in the following table:

x_2	$\overline{V}$ /(cm^3 mol^{-1})
0	18.05
0.0212	19.18
0.0466	20.53
0.1153	24.18
0.2269	30.21
0.4390	41.82
0.6923	55.87
1.000	73.02

(a) Fit these data to $\overline{V} = A + Bx_2 + Cx_2^2$, where x_2 is the mole fraction of glycerol. (b) Calculate the two partial molar volumes as a function of x_2. (c) Show that the molar volumes in the above table can be calculated using the partial molar volumes. (d) Show that the partial molar volume of water can be calculated by using the function for the partial molar volume of glycerol by using the Gibbs-Duhem equation. (See problems 6.49 and 6.50.)

SOLUTION

(a) These data are fit by the following polynomial:

$$\overline{V} = 18.023 + 53.57x_2 + 1.45x_2^2$$

(b)

$$\overline{V}_1 = \overline{V} - x_2\left(\frac{\partial \overline{V}}{\partial x_2}\right)$$

$$= 18.023 - 1.45x^2$$

$$\overline{V}_2 = \overline{V} + (1 - x_2)\left(\frac{\partial \overline{V}}{\partial x_2}\right)$$

$$= 71.60 + 2.90x_2 - 1.45x_2^2$$

(c)

$$\overline{V} = (1 - x_2)\overline{V}_1 + x_2\overline{V}_2$$
$$= 18.023 + 53.57x_2 + 1.45x_2^{\ 2}$$

(d)

$$\overline{V}_1 = \overline{V}_1^* - \int_0^{x_2} \frac{x_2}{x_1} \mathrm{d}\overline{V}_2$$
$$= \overline{V}_1^* - \int_0^{x_2} \frac{x_2}{1 - x_2}(2.90 - 2.90x_2)\mathrm{d}x_2$$
$$= 18.023 - 1.45x_2$$

6.52 The freezing point is lowered to -5 °C under the skates, and so the answer is yes.

6.53 2.66 x 10^{-6} bar

6.54 4.9 K

6.55 53.0 kJ mol^{-1}, 28.4 kJ mol^{-1}, 158.6 J K^{-1} mol^{-1}

6.56 383.38 K, -0.4 °C

6.57 96 °C

6.58 27.57 kJ mol^{-1}

6.59 0.0773 Pa

6.60 (a) 0.123 14 bar, (b) 0.125 110 bar

6.61 (a) 38.1 kJ mol^{-1}, (b) 3.78 kPa

6.62 (a) 50.91 kJ mol^{-1}, (b) 50.14 Pa K^{-1}, 44.20 Pa K^{-1}, (c) P_{ice} = 361 Pa, P_{liq} = 390 Pa

6.63 (a) - 0.46 °C, (b) - 0.59 °C

6.64 - 8.59 kJ mol^{-1}

6.65 4

6.66 (a) 1, (b) 2

6.67 166.5 Pa

6.68 (a) $y_{EtBr_2} = 0.802$, (b) $x_{EtBr_2} = 0.425$

6.69 (a) $y_{CHCl_3} = 0.635$, (b) 20.91 kPa

6.70 (a) $y_B = 0.722$, $P = 69.55$ kPa, (b) 0.536

6.71 (a) $x_{ClB} = 0.591$; (b) $y_{ClB} = 0.731$; (c) 1.081 bar

6.72 $x_B = 0.240$, $y_B = 0.434$

6.73 571 g of $H_2O(g)$

6.74 5.293 J K^{-1} mol^{-1}, -1577 J mol^{-1}

6.75 $y_{n\text{-Pr}} = 0.406$

6.76 $x_{C_6H_6} = 0.55$; pure C_6H_6 can be obtained by distillation provided $x_{C_6H_6} > 0.55$

6.77 (a) $y_{Pr} = 0.37$, (b) $y_{Pr} = 0.59$

6.78 33.0% O_2, 67.0% N_2

6.79 $\gamma_{acetone} = 1.67$, $\gamma_{CS_2} = 1.38$

6.80

x_1	0	0.2	0.4	0.6	0.8	1
γ_1	-	0.58	0.70	0.84	0.96	1.00
γ_2	1.00	0.98	0.89	0.74	0.61	-

6.81 - 1311 J mol^{-1}

6.82 11.17 kJ mol^{-1}

6.83 (a) K = 19.9 kPa (b) γ'_{CHCl_3} = 1.13, 1.37, 1.65, 1.88, 1.96

6.84 γ_1 = 3.12, 1.63, 1.19, 1.02

γ'_2 = 0.314, 0.417, 0.571, 0.772

6.86 92.4 g mol^{-1}

6.88 122 kg mol^{-1}

6.89 x_A = 0.108. The solution is not ideal.

6.90 x_{Cd} = 0.842

7

Electrochemical Equilibrium

7.1 How much work is required to bring two protons from an infinite distance of separation to 0.1 nm? Calculate the answer in joules using the protonic charge 1.602×10^{-19} C. What is the work in kJ mol^{-1} for a mole of proton pairs?

SOLUTION

$$\text{Potential } \phi = \frac{Q_2}{4\pi\varepsilon_o r}$$

$$= \frac{(1.602 \times 10^{-19}\ \text{C})(0.89875 \times 10^{10}\ \text{N m}^2\ \text{C}^{-2})}{10^{-10}\ \text{m}}$$

$$= 14.398\ \text{J C}^{-1}$$

$$Q_1\phi = (1.602 \times 10^{-19}\ \text{C})(14.398\ \text{J C}^{-1})$$

$$= 2.307 \times 10^{-18}\ \text{J}$$

$$= (2.307 \times 10^{-18}\ \text{J})(6.022 \times 10^{23}\ \text{mol}^{-1})(10^{-3}\ \text{kJ J}^{-1})$$

$$= 1389.3\ \text{kJ mol}^{-1}$$

7.2 How much work in kJ mol^{-1} can in principle be obtained when an electron is brought to 0.5 nm from a proton?

SOLUTION

$$w = \frac{Q_1 Q_2}{4\pi\varepsilon_o r}$$

$$= \frac{(0.8988 \times 10^{10}\ \text{N m}^2\ \text{C}^{-2})(1.602 \times 10^{-19}\ \text{C})^2(6.022 \times 10^{23}\ \text{mol}^{-1})}{(5 \times 10^{-10}\ \text{m})(10^3\ \text{J kJ}^{-1})}$$

$$= 277.8\ \text{kJ mol}^{-1}$$

7.3 A small dry battery of zinc and ammonium chloride weighing 85 g will operate continuously through a 4-Ω resistance for 450 min before its voltage falls below 0.75 V. The initial voltage is 1.60 V, and the effective voltage over the whole life of the battery is taken to be 1.00 V. Theoretically, how many kilometers above the earth could this battery be raised by the energy delivered under these conditions?

SOLUTION

$I = \frac{E}{R} = \frac{1\text{ V}}{4\ \Omega} = 0.25\text{ A}$

Power = $I^2R = (0.25\text{ A})^2(4\ \Omega) = 0.25$ watt

Work = $(0.25\text{ W})(450 \times 60\text{ s}) = 6.75 \times 10^3$ J

$= (0.085\text{ kg})(9.80\text{ m s}^{-2})h$

$h = \frac{6.75 \times 10^3\text{ J}}{(0.085\text{ kg})(9.80\text{ m s}^{-2})} = 8103\text{ m}$

$= \frac{(8103 \times 10^5\text{ cm})}{(2.54\text{ cm in}^{-1})(12\text{ in ft}^{-1})(5280\text{ ft mile}^{-1})} = 5.04$ miles

7.4 (a) The mean ionic activity coefficient of 0.1 molar HCl(aq) at 25 °C is 0.796. What is the activity of HCl in this solution? (b) The mean ionic activity coefficient of 0.1 molar H_2SO_4 is 0.265. What is the activity of H_2SO_4 in this solution? Assume complete dissociation.

SOLUTION

$a_{A_{\nu_+}B_{\nu_-}} = \gamma_\pm^{\nu_\pm} m^{\nu_\pm} (\nu_+^{\nu_+}\nu_-^{\nu_-})$, where $\nu_\pm = \nu_+ + \nu_-$

(a) $a_{HCl} = (0.796)^2(0.1)^2 = 0.00634$

(b) $a_{H_2SO_4} = (0.265)^3(0.1)^3\ 2^2 1^1 = 7.44 \times 10^{-5}$

7.5 The solubility of Ag_2CrO_4 in water is 8.00×10^{-5} mol kg^{-1} at 25 °C, and its solubility in 0.04 mol kg^{-1} $NaNO_3$ is 8.84×10^{-5} mol kg^{-1}. What is the mean ionic activity coefficient of Ag_2CrO_4 in 0.04 mol kg^{-1} $NaNO_3$?

SOLUTION

$Ag_2CrO_4(s) = 2Ag^+ + CrO_4^{2-}$

$K_{sp} = a_{Ag^+}^2\ a_{CrO_4^{2-}} = 4\gamma_\pm^3\ m^3$

When Ag_2CrO_4 is dissolved in H_2O, the ionic strength is so low that the Debye-Huckel theory can be used.

$\log \gamma_\pm = Az_+z_-I^{1/2} = -\ 0.509 \times 2 \times (3m)^{1/2} = -1.58 \times 10^{-2}$

$\gamma_\pm \approx 0.964$

This is the mean ionic activity coefficient when Ag_2CrO_4 is dissolved in water. The thermodynamic value of the solubility product in water is given by

$K_{sp} = 4 \times (0.964)^3 \times (8.00 \times 10^{-5})^3 = 1.83 \times 10^{-12}$

Now we can use the solubility in 0.04 mol kg^{-1} $NaNO_3$ to calculate the mean ionic activity coefficient of Ag_2CrO_4 in 0.04 mol kg^{-1} $NaNO_3$.

$K_{sp} = 1.83 \times 10^{-12} = 4(8.84 \times 10^{-5})^3 \gamma_{\pm}^3$

$\gamma_{\pm} = 0.872$

7.6 A solution of NaCl has an ionic strength of 0.24 mol kg^{-1}. (a) What is its molality? (b) What molality of Na_2SO_4 would have the same ionic strength? (c) What molality of $MgSO_4$?

SOLUTION

(a) $I = \frac{1}{2}(m_1 z_1^2 + m_2 z_2^2)$

$0.24 = \frac{1}{2}(1^2 + 1^2)\ m \qquad m = 0.24 \text{ mol kg}^{-1}$

(b) $0.24 = \frac{1}{2}(2m \times 1^2 + m \times 2^2) \qquad m = 0.08 \text{ mol kg}^{-1}$

(c) $0.24 = \frac{1}{2}(2^2 + 2^2)\ m \qquad m = 0.06 \text{ mol kg}^{-1}$

7.7 Using the limiting law calculate the mean ionic activity coefficients at 25 °C in water of the following electrolytes at 10^{-3} m: (a) NaCl, (b) $CaCl_2$, (c) $LaCl_3$.

SOLUTION

(a) $I = 10^{-3}$

$\gamma_{\pm} = 10^{A z_+ z_- I^{1/2}}$

$= 10^{-0.509(10^{-3})^{1/2}} = 0.964$

(b) $I = \frac{1}{2}(0.001 \times 2^2 + 0.002) = 3 \times 10^{-3}$

$\gamma_{\pm} = 10^{-2(0.509)(3 \times 10^{-3})^{1/2}} = 0.880$

(c) $I = \frac{1}{2}(0.001 \times 3^2 + 0.003) = 6 \times 10^{-3}$

$\gamma_{\pm} = 10^{-3(0.509)(6 \times 10^{-3})^{1/2}} = 0.762$

7.8 Estimate the electromotive force of the cell
Zn(s)|$ZnCl_2$(aq,0.02 mol kg^{-1})|AgCl(s)|Ag(s)
at 25 °C using the Debye-Hückel theory.

SOLUTION

Right: $AgCl(s) + e^- = Ag(s) + Cl^-$ $\qquad E^o = 0.222$ V

Left: $\frac{1}{2}\ Zn^{+2} + e^- = \frac{1}{2}\ Zn(s)$ $\qquad E^o = -0.763$ V

$$\text{AgCl(s)} + \frac{1}{2}\text{ Zn(s)} = \text{Ag(s)} + \frac{1}{2}\text{ Zn}^{+2} + \text{Cl}^- \qquad E^o = 0.985\text{ V}$$

$$E = E^o - \frac{RT}{F}\ln[a(\text{ZnCl}_2)^{1/2}] = E^o - \frac{RT}{2F}\ln a(\text{ZnCl}_2)$$

$$= E^o - \frac{RT}{2F}\ln[4(m/m^o)^3\gamma_{\pm}{}^3]$$

where $m^o = 1$ mol kg^{-1}

$$\log\gamma_{\pm} = (0.509\text{ mol}^{-1/2}\text{ kg}^{-1/2})(-2)\sqrt{0.06\text{ mol kg}^{-1}}$$

$$\gamma_{\pm} = 0.563$$

$$E = 0.985\text{ V} - \frac{(8.314\text{ J K}^{-1}\text{ mol}^{-1})(298\text{ K})}{2(96\,485\text{ C mol}^{-1})}\ln[4(0.02)^3(0.563)^3]$$

$$= 1.140\text{ V}$$

7.9 The cell Pt|H_2(1 bar)|HBr(m)|AgBr|Ag has been studied by H. S. Harned, A. S. Keston, and J. G. Donelson [*J. Am. Chem. Soc.* **58**, 989 (1936)]. The following table gives the electromotive forces obtained at 25 °C.

m /mol kg^{-1}	0.01	0.02	0.05	0.10
E/V	0.3127	0.2786	0.2340	0.2005

Calculate (a) E^o and (b) the activity coefficient for a 0.10 mol kg^{-1} solution of hydrogen bromide.

SOLUTION

(a) Plot $E' = E + 0.1183\log m - 0.0602\sqrt{m}$ versus m. A more linear plot is obtained by ignoring the measurement at 0.1 molar because the deviations from the Debye-Hückel theory are the largest there.
The intercept at $m = 0$ is $E^o = 0.0710$ V.

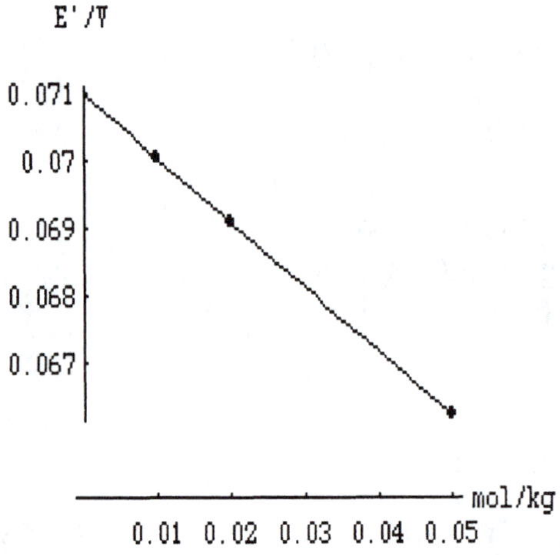

The intercept shows that the standard electromotive force is 0.0710 V.

(b) $E = 0.0710\ \text{V} - (0.05915\ \text{V}) \log \gamma_{\pm}^2 m^2$

$0.2005\ \text{V} = 0.0710\ \text{V} - (0.1183\ \text{V}) \log \gamma_{\pm} - (0.1183\ \text{V}) \log 0.1$

$$\log \gamma_{\pm} = -\frac{0.2005 - 0.1183 - 0.0710}{0.1183}$$

$\gamma_{\pm} = 0.804$

7.10 Design cells without liquid junction that could be used to determine the activity coefficients of aqueous solutions of (a) NaOH and (b) H_2SO_4. Give the equations relating electromotive force to the mean ionic activity coefficient at 25 °C.

SOLUTION

(a) Na|NaOH(m)|H_2|Pt

$H_2O + e^- = \frac{1}{2} H_2 + OH^-$ $\quad E_R^o$

$Na^+ + e^- = Na$ $\quad E_L^o$

$H_2O + Na = NaOH + \frac{1}{2} H_2$ $\quad E^o = E_R^o - E_L^o$

At 25 °C, $E = E^o - 0.0591 \log\left(m^2 \gamma_{\pm}^2 P_{H_2}{}^{1/2}\right)$

(b) Pt|H_2|H_2SO_4(m)|Ag_2SO_4|Ag

$Ag_2SO_4 + 2e^- = SO_4^{2-} + 2Ag$ $\quad E_R^o$

$2H^+ + 2e^- = H_2$ $\quad E_L^o$

$H_2 + Ag_2SO_4 = H_2SO_4 + 2Ag$ $\quad E^o = E_R^o - E_L^o$

At 25°C

$$E = E^o - \frac{0.0591}{2} \log\left(\frac{4m^3\gamma_{\pm}^3}{P_{H_2}}\right)$$

7.11 The electromotive force of the cell

Pb(s)|$PbSO_4$(s)|$Na_2SO_4 \cdot 10H_2O$(sat)|Hg_2SO_4(s)|Hg(1)

is 0.9647 V at 25 °C. The temperature coefficient is 1.74×10^{-4} V K^{-1}. (a) What is the cell reaction? (b) What are the values of $\Delta_r G$, $\Delta_r S$, and $\Delta_r H$?

SOLUTION

(a) $Pb(s) + Hg_2SO_4(s) = PbSO_4(s) + 2Hg(1)$

(b) $\Delta_r G = -|\nu_e| F E = -(2)(96\,485\ \text{C mol}^{-1})(0.9647\ \text{V}) = -186.16\ \text{kJ mol}^{-1}$

$$\Delta_r S = |\nu_e| F \left(\frac{\partial E}{\partial T}\right)_P = (2)(96\,485\ \text{C mol}^{-1})(1.74 \times 10^{-4}\ \text{V K}^{-1})$$

$= 33.58 \text{ J K}^{-1} \text{ mol}^{-1}$

$$\Delta_r H = -|\nu_e|FE + |\nu_e|FT\left(\frac{\partial E}{\partial T}\right)_P$$

$= -2(96\,485 \text{ C mol}^{-1})(0.9647 \text{ V}) + 2(96\,485 \text{ C mol}^{-1})$
$\times (298.15 \text{ K})(1.74 \times 10^{-4} \text{ V K}^{-1})$

$= -176.15 \text{ kJ mol}^{-1}$

7.12 For the galvanic cell

$$H_2(1 \text{ bar})|HCl(ai)|Cl_2(1 \text{ bar})$$

the standard electromotive force at 298.15 K is 1.3604 V and $(\partial E^\circ/\partial T)_P =$ -1.247×10^{-3} V K^{-1}. (a) For the cell reaction, what are the values of $\Delta_r G^\circ$, $\Delta_r H^\circ$, $\Delta_r S^\circ$? (b) For Cl^-(ao) what are the values of $\Delta_f G^\circ$, $\Delta_f H^\circ$, and $\bar{S}^\circ$?

SOLUTION

(a) $\frac{1}{2} Cl_2(g) + e^- = Cl^-(aq)$

$H^+(aq) + e^- = \frac{1}{2} H_2(g)$

$\frac{1}{2} Cl_2 + \frac{1}{2} H_2 = H^+(aq) + Cl^-(aq)$

$\Delta_r G^\circ = -|\nu_e|FE^\circ = -(96{,}485 \text{ C mol}^{-1})(1.3604 \text{ V}) = -131.260 \text{ kJ mol}^{-1}$

$\Delta_r H^\circ = \Delta_r G^\circ + T\Delta_r S^\circ$

$= -131.2604 + (298.15)(-120.3 \times 10^{-3} \text{ kJ K}^{-1} \text{ mol}^{-1})$

$= -167.127 \text{ kJ mol}^{-1}$

$$\Delta_r S^\circ = |\nu_e|F\left(\frac{\partial E^\circ}{\partial T}\right)_P = (96{,}485 \text{ C mol}^{-1})(-1.237 \times 10^{-3} \text{ V K}^{-1})$$

$= -120.3 \text{ J K}^{-1} \text{ mol}^{-1}$

(b) Since $\Delta_f G^\circ[H^+(ao)] = 0$,
$\Delta_f G^\circ[Cl^-(ao)] = -131.260 \text{ kJ mol}^{-1}$
Since $\Delta_f H^\circ[H^+(ao)] = 0$,
$\Delta_f H^\circ[Cl^-(ao)] = -167.127 \text{ kJ mol}^{-1}$

Since $\bar{S}^\circ[H^+(ao)] = 0$

$$\Delta_r S^\circ = \bar{S}^\circ[Cl^-(ao)] - \frac{1}{2}\bar{S}^\circ[H_2(g)] - \frac{1}{2}\bar{S}^\circ[Cl_2(g)]$$

$$-120.3 = \bar{S}^\circ[Cl^-(ao)] - \frac{1}{2}(130.684) - \frac{1}{2}(223.066)$$

$\bar{S}^\circ[Cl^-(ao)] = 56.6 \text{ J K}^{-1} \text{ mol}^{-1}$

7.13 In problem 4.10 two equations were derived for calculating ΔG at another temperature if it is known at one. Compare the values of $\Delta_r G^o$ (323 K) and K_w calculated with these equations for

$H_2O(1) = H^+(ao) + OH^-(ao)$

<u>SOLUTION</u>

	$\Delta_f G^o$(298 K)	$\Delta_f H^o$ (298 K)	C_P^o(298 K)
$H_2O(1)$	- 237.129	- 285.830	75.291
$H^+(ao)$	0	0	0
OH^-	-157.144	-229.994	-148.5

$\Delta_r G^o$ = - 157.244 + 237.129 = 79.885 kJ mol^{-1}

$\Delta_r H^o$ = - 229.994 + 285.830 = 55.836 kJ mol^{-1}

$\Delta_r C_P$ = - 148.5 - 75.291 = - 223.8 J K^{-1} mol^{-1}

(a) $$\Delta_r G_2^o = \Delta_r G_1^o\, T_2/T_1 + \Delta_r H(1 - T_2/T_1)$$

$$\Delta_r G^o(323) = \frac{(79.885)(323.15)}{298.15} + 55.836\left(1 - \frac{323.15}{298.15}\right)$$

$$= 81.902 \text{ kJ mol}^{-1}$$

$K_w = 5.69 \times 10^{-14}$

(b) $$\Delta_r G_2 = \Delta_r G_1\, T_2/T_1 + (\Delta_r H_1 - T_1 \Delta_r C_P)(1 - T_2/T_1) + T_2\, \Delta_r C_P \ln (T_2/T_1)$$

$$= \frac{(79.885)(323.15)}{298.15} + [55.836 + (298.15)(-\ 0.2238)]$$

$$\times \left(1 - \frac{323.15}{298.15}\right) + (323.15)(-\ 0.2238) \ln\left(\frac{323.15}{298.15}\right)$$

$$= 81.673 \text{ kJ mol}^{-1}$$

$K_w = 6.28 \times 10^{-14}$

This value is more accurate because the variation of $\Delta_r H^o$ with temperature is taken into account.

7.14 Calculate E^o for the half cell $OH^-|H_2|Pt$ at 25 °C using the value of the ion product for water, which is 1.006 x 10^{-14} (Section 8.1).

<u>SOLUTION</u>

For the cell $Pt|H_2|OH^-|H^+|H_2|Pt$

$$H^+ + e^- = \frac{1}{2}H_2 \qquad \overset{o}{E}_R = 0$$

$$H_2O + e^- = OH^- + \frac{1}{2}H_2 \qquad \overset{o}{E}_L$$

$$H^+ + OH^- = H_2O \qquad E^o = \overset{o}{E}_R - \overset{o}{E}_L = -\overset{o}{E}_L$$

$$E = -\overset{o}{E}_L = \frac{RT}{F}\ln\frac{1}{a_{H^+}a_{OH^-}}$$

$$\overset{o}{E}_L = \frac{RT}{F}\ln K_w$$

$$= \frac{(8.3145\ K^{-1}\ mol^{-1})(298.15\ K)}{(96\ 485\ C\ mol^{-1})}\ln(1.006 \times 10^{-14})$$

$$= 0.828\ V$$

7.15 What are the values of $\Delta_r G^o$ and K for the following reactions at 298 K from Appendix C.2?

(a) $Cu(s) + Zn^{2+}(ao) = Cu^{2+}(ao) + Zn(s)$

(b) $H_2(g) + Cl_2(g) = 2HCl(ai)$

(c) $Ca^{2+}(ao) + CO_3^{2-}(ao) = CaCO_3(s, calcite)$

(d) $\frac{1}{2}Cl_2(g) + Br^-(ao) = \frac{1}{2}Br_2(g) + Cl^-(ao)$

(e) $Ag^+(ao) + Fe^{2+}(ao) = Fe^{3+}(ao) + Ag(s)$

SOLUTION

(a) $\Delta_r G^o = 65.49 + 147.06 = 212.55\ kJ\ mol^{-1}$
$K = 5.79 \times 10^{-38}$

(b) $\Delta_r G^o = 2(-131.228) = -262.46$
$K = 9.56 \times 10^{45}$

(c) $\Delta_r G^o = -1128.79 + 553.58 + 527.81 = -47.40$
$K = 2.01 \times 10^{8}$

(d) $\Delta_r G^o = \frac{1}{2}(3.110) - 131.228 + 103.96 = -25.71$
$K = 3.20 \times 10^{4}$

(e) $\Delta_r G^o = -4.7 - 77.107 + 78.90 = -2.9$
$K = 3.23$

7.16 Derive the equation giving the effect of ionic strength at 298.15 K on (a) $\Delta_f S_i^o$ of an ionic species and (b) E^o for a cell reaction according to the extended Debye-Hückel equation.

SOLUTION

(a) The standard entropy of formation of an ion at a specified ionic strength is given by

$$\Delta_f S_i^o(I) = (\Delta_f H_i^o(I) - \Delta_f G_i^o(I))/T$$

Substituting equations 7.79 and 7.80 yields

$$\Delta_f S_i^o(I) = \Delta_f S_i^o(I = 0) + 14.7319\, z_i^2 I^{1/2}/(1 + B\, I^{1/2})$$

at 298.15 K, where the standard entropies of formation have the units J K^{-1} mol^{-1}.

(b) The standard electromotive force for a cell reaction at a specified ionic strength is given by

$$E^o(I) = -\frac{\Delta_r G^o(I)}{|\nu_e|F}$$

where F is the Faraday constant (96.485 kJ mol^{-1}). Substituting the expression for the standard Gibbs energy of reaction for the cell reaction from equation 8. yields

$$E^o(I) = E^o(I{=}0) + \frac{0.03021}{|\nu_e|}\, z_i^2 I^{1/2}/(1 + B\, I^{1/2})$$

7.17 From standard electrode potentials in Table 7.2 what are the standard Gibbs energies of formation at 25 °C for Cl^-(ao), OH^-(ao) and Na^+(ao)?

SOLUTION

For H_2(g)|HCl(ai)|Cl_2(g)

R $\quad \frac{1}{2} Cl_2 + e^- = Cl^-$ $\qquad E^o = 1.3604$ V

L $\quad H^+ + e^- = \frac{1}{2} H_2$ $\qquad E^o = 0$

$\frac{1}{2} Cl_2 + \frac{1}{2} H_2 = H^+(ao) + Cl^-(ao)$ $\qquad E^o = 1.3604$ V

$\Delta_r G^o = -|\nu_e|FE^o = -(96{,}485\ C\ mol^{-1})(1.3604\ V) = -131.258\ kJ\ mol^{-1}$

This is $\Delta_f G^o[Cl^-(ao)]$ since $\Delta_f G^o\ [H^+(ao)]$ is zero by convention. Appendix C.2 gives - 131.228 kJ mol^{-1}.

For H_2(g)|H^+(ao) | | OH^-(ao)|O_2(g)

R $\frac{1}{4}O_2 + \frac{1}{2}H_2O(1) + e^- = OH^-(ao)$ $E^o = 0.401$ V

L $H^+(ao) + e^- = \frac{1}{2}H_2$ $E^o = 0$

$\frac{1}{4}O_2 + \frac{1}{2}H_2 + \frac{1}{2}H_2O = H^+(ao) + OH^-(ao)$ $E^o = 0.401$ V

$\Delta_r G^o = -(96{,}485\ \text{C mol}^{-1})(0.401\ \text{V}) = -38.69\ \text{kJ mol}^{-1}$

$= \Delta_f G^o[OH^-(ao)] - \frac{1}{2}(237.129) = -157.26\ \text{kJ mol}^{-1}$

Appendix C.2 gives -157.244 kJ mol^{-1}.

For $Na(cr)|Na^+(ao)\,||\,H^+(ao)|H_2(g)$

R $H^+ + e^- = \frac{1}{2}H_2$ $E^o = 0$

L $Na^+ + e^- = Na$ $E^o = -2.714$ V

$Na(cr) + H^+(ao) = Na^+(ao) + \frac{1}{2}H_2(g)$ $E^o = 2.714$ V

$\Delta_r G^o = -(96{,}485\ \text{C mol}^{-1})(2.174\ \text{V}) = -261.86\ \text{kJ mol}^{-1} = \Delta_f G^o[Na^+(ao)]$
Appendix C.2 gives - 261.905 kJ mol^{-1}.

7.18 According to Table 7.2 what are the equilibrium constants for the following reactions at 25 °C?

(a) $H^+(ao) + Li(s) = Li^+(ao) + \frac{1}{2}H_2(g)$

(b) $2H^+(ao) + Pb(s) = Pb^{2+}(ao) + H_2(g)$

(c) $3H^+(ao) + Au(s) = Au^{3+}(ao) + \frac{3}{2}H_2(g)$

SOLUTION

(a) R $H^+ + e^- = \frac{1}{2}H_2$ $E^o = 0$

L $Li^+ + e^- = Li$ $E^o = -3.045$ V

$H^+ + Li = Li^+ + \frac{1}{2}H_2$ $E^o = 3.045$ V

$K = \exp(|\nu_e|FE^{o}/RT)$
$= \exp[(96{,}485 \text{ C mol}^{-1})(3.045 \text{ V})/(8.314 \text{ J K}^{-1} \text{ mol}^{-1})(298 \text{ K})]$
$= 3.1 \times 10^{51}$

(b) R $2H^+ + 2e^- = H_2$ $E^o = 0$

L $Pb^{2+} + 2e^- = Pb$ $E^o = -0.126$ V

$2H^+ + Pb = Pb^{2+} + H_2$ $E^o = 0.126$

$K = \exp[2(96{,}485)(0.126)/(8.314)(298)] = 1.8 \times 10^4$

(c) R $3H^+ + 3e^- = \frac{3}{2} H_2$ $E^o = 0$

L $Au^{3+} + 3e^- = Au$ $E^o = 1.50$ V

$3H^+ + Au = Au^{3+} + \frac{3}{2} H_2$ $E^o = -1.50$ V

$K = \exp[3(96{,}485)(-1.50)/(8.314)(298)] = 7.8 \times 10^{-77}$

The striking resistance of gold to corrosion by acid is evident.

7.19 Use Appendix C.2 to calculate the standard electrode potential for $Cl^-|AgCl(s)|Ag$ at 90 °C if $\Delta_r C_P^o = 0$.

SOLUTION

For

$AgCl(s) + e^- = Ag(s) + Cl^-$

$\Delta G^o = -131.228 - (-109.789) = -21.439 \text{ kJ mol}^{-1} = -FE^o$

$E^o = -(-21{,}439 \text{ J mol}^{-1})/(96{,}485 \text{ C mol}^{-1}) = 0.2222$ V

$\Delta H^o = -167.159 - (-127.068) = -40.091 \text{ kJ mol}^{-1}$

$$\left[\frac{\partial(\Delta G/T)}{\partial T}\right]_P = -\frac{\Delta H}{T^2}$$

$$\int d\left(\frac{\Delta G}{T}\right) = -\Delta H \int \frac{dT}{T^2}$$

$$\frac{\Delta G_2}{T_2} - \frac{\Delta G_1}{T_1} = -\Delta H\left[-\frac{1}{T_2} + \frac{1}{T_1}\right]$$

$$\frac{\Delta G^o(90\ ^oC)}{363.15} + \frac{21.439}{298.15} = 40.091\left[-\frac{1}{363.15} + \frac{1}{298.15}\right]$$

$\Delta G^o(90\ ^oC) = -\ 17.373$ kJ mol^{-1}

$E^o(90\ ^oC) = -\ (-\ 17.373$ J mol^{-1})/(98,485 C mol^{-1}) = 0.1801 V

7.20 The phase rule for an electrochemical cell is $F = C - p + 3$. (a) Why is this so? (b) Calculate the number of degrees of freedom of the following reaction considered as a chemical reaction.

$H_2(g) + 2AgCl(s) = HCl(aq) + 2Ag(s)$

(c) Calculate the number of degrees of freedom for the following electrochemical reaction.

$H_2(g) + 2AgCl(s) + 2e^-(Pt_R) = HCl(aq) + 2Ag(s) + 2e^-(Pt_L)$

SOLUTION

(a) The fundamental equation for G for an electrochemical cell has an additional term ϕdQ, where ϕ is electric potential and Q is electric charge. Therefore, there is an additional independent variable beyond T and P.

(b) The chemical reaction involves 5 species, and the reaction is assumed to be at equilibrium. Therefore,
$C = 5$ (including water) $- 1 = 4$, $p = 4$, $F = 4 - 4 + 2 = 2$ (T,P)

(c) For the electrochemical reaction there are 7 species, and so
$C = 7$ (including water) $- 1 = 6$, $p = 6$, $F = 6 - 6 + 3 = 3$ (T,P,E or $T,P,n_{HCl}/n_{H_2O}$)

7.21 The NBS Tables have entries for H_2CO_3(ao), HCO_3^-(ao), and CO_3^{2-}(ao), where ao means not dissociated and ai means completely ionized, with a note explaining that the table is based on the convention that for

$$A(aq) + nH_2O(l) = A{\cdot}nH_2O(aq) \quad \text{(a)}$$

$\Delta_r G^o = \Delta_r H^o = \Delta_r S^o$. This means that for the reaction

$$CO_2(aq) + H_2O(l) = H_2CO_3(aq) \quad \text{(b)}$$

the equilibrium constant is taken equal to unity. The reason for this convention is that in dilute solutions in water it is impossible to determine the equilibrium constants for these reactions by varying the concentration of water. Thus the properties for H_2CO_3(ao) apply to the sum $H_2CO_3(aq) + CO_2(aq)$, where these are interpreted as species. (a) To see how this works, calculate K_1 and K_2 for carbonic acid at 298.15 K and zero ionic strength. (b) Since the hydration of $CO_2(aq)$ in the neutral pH range is slow (half life about 1 second), it has been possible to determine the equilibrium constant K_h for equation b.

$$K_h = \frac{[H_2CO_3(aq)]}{[CO_2(aq)]} = 2.6\text{x}10^{-3} \quad \text{(c)}$$

Given this information, calculate K_a for $H_2CO_3(aq)$:

$$H_2CO_3(aq) = H^+ + HCO_3^-(ao) \quad \text{(d)}$$

$$K_a = \frac{[H^+][HCO_3^-(ao)]}{[H_2CO_3(aq)]} \qquad \text{(e)}$$

SOLUTION

(a) $$K_1 = \frac{[H^+][HCO_3^-(ao)]}{[H_2CO_3(ao)]} = \frac{[H^+][HCO_3^-(ao)]}{[H_2CO_3(aq)] + [CO_2(aq)]}$$

$\Delta_r G^o = -586.77 + 623.08 = 36.31 \text{ kJ mol}^{-1}$

$K_1 = \exp[-36310/(8.3145)(298.15)] = 4.35\text{x}10^{-7}$

$$K_2 = \frac{[H^+][CO_3^{2-}(ao)]}{[HCO_3^-(ao)]}$$

$\Delta_r G^o = -527.81 + 586.77 = 58.96 \text{ kJ mol}^{-1}$

$K_2 = \exp[-58960/(8.3145)(298.15)] = 4.69\text{x}10^{-11}$

(b) $$K_1 = \frac{[H^+][HCO_3^-(ao)]}{[H_2CO_3(aq)](1 + [CO_2(aq)]/[H_2CO_3(aq)])}$$

$$= \frac{[H^+][HCO_3^-(ao)]}{[H_2CO_3(aq)](1 + 1/K_h])}$$

$$= K_a/(1 + 1/K_h)$$

$K_a = K_1(1 + 1/K_h) = (4.35\text{x}10^{-7})(1 + 1/(2.6\text{x}10^{-3})) = 1.68\text{x}10^{-4}$

Thus the acid dissociation constant for $H_2CO_3(aq)$ is about what you would expect from the dissociation constants for acetic acid and formic acid.

7.22 At 25 °C the standard electrode potential for the $Ag^+|Ag$ electrode is 0.7991 V, and the solubility product for AgI is 8.2 x 10^{-17}. What is the standard electrode potential for $I^-|AgI|Ag$?

SOLUTION

$Ag|Ag^+ || I^-|AgI|Ag$

$AgI + e^- = Ag + I^-$ $\qquad E^o_R$

$Ag^+ + e^- = Ag$ $\qquad E^o_L = 0.7991$ V

$AgI = Ag^+ + I^-$ $\qquad E^o = E^o_R - 0.7991$ V

$$E^{\circ}_{R} - 0.7991 = \frac{RT}{F} \ln K_{sp}$$
$$= \frac{(8.314)(298)}{96\,485} \ln 8.2 \times 10^{-17}$$
$$E^{\circ}_{R} = -0.152 \text{ V, the } E^{\circ} \text{ for } I^{-}|AgI|Ag$$

7.23 Using data from Appendix C.2 calculate the solubility of AgCl(cr) in water at 298.15 K. The salt is completely dissociated in the aqueous phase.

SOLUTION

$AgCl(s) = Ag^{+}(ao) + Cl^{-}(ao)$

$\Delta_r G^{\circ} = 77.107 - 131.228 + 109.789 = 55.668 \text{ kJ mol}^{-1}$

$K = \exp(-55{,}668/8.3145 \times 298.15) = 1.767 \times 10^{-10}$

$= (a_{Ag^+})(a_{Cl^-}) = m^2$ γ**Error!**$_{)} \approx m^2$

$m = 1.33 \times 10^{-5} \text{ mol kg}^{-1}$

7.24 Calculate standard electrode potentials at 25 °C for the following electrodes using Appendix C.2.

(a) Li^{+} (ao)|Li(s)

(b) F^{-}(ao)|F_2(g)

(c) Pb^{2+}(ao)|PbO_2(s)|Pb

SOLUTION

(a) $Li^{+} + e^{-} = Li$

$\Delta_r G^{\circ} = -(293.31) = 293.31 \text{ kJ mol}^{-1}$

$E^{\circ} = -\Delta_r G^{\circ}/|\nu_e|F = -(293{,}310 \text{ J mol}^{-1})/(96{,}485 \text{ C mol}^{-1}) = -3.040 \text{ V}$

(b) $\frac{1}{2} F_2(g) + e^{-} = F^{-}$

$\Delta_r G^{\circ} = -278.79 \text{ kJ mol}^{-1}$

$E^{\circ} = (278{,}790 \text{ J mol}^{-1})/(96.485 \text{ C mol}^{-1}) = 2.889 \text{ V}$

(c) $\frac{1}{2} PbO_2 + 2H^{+} + e^{-} = \frac{1}{2} Pb^{2+} + H_2O$

$$\Delta_r G^o = -237.129 + \frac{1}{2}(-24.43) - \frac{1}{2}(-217.33) = -140.68 \text{ kJ mol}^{-1}$$

$$E^o = (140{,}680 \text{ J mol}^{-1})/(96{,}485 \text{ C mol}^{-1}) = 1.458 \text{ V}$$

7.25 Using Appendix C.2 calculate the values of $\Delta_r G^o$, $\Delta_r H^o$, $\Delta_r S^o$, and $\Delta_r C_P^o$ at 25 °C for the electrode reaction for the $Na^+|Na$ electrode.

SOLUTION

The shorthand notation for the electrode reaction is $Na^+ + e^- = Na$, but we must remember that we are really talking about the cell for which the cell reaction is

$$Na^+ + \frac{1}{2} H_2(g) = Na(s) + H^+$$

$$\Delta_r G^o = -(-261.905) = 261.905 \text{ kJ mol}^{-1}$$

$$\Delta_r H^o = -(-240.12) = 240.12 \text{ k J mol}^{-1}$$

$$\Delta_r S^o = 51.21 - 59.0 - \frac{1}{2}(130.684) = -73.132 \text{ J K}^{-1} \text{ mol}^{-1}$$

$$\Delta_r C_P^o = 28.24 - 46.4 - \frac{1}{2}(28.824) = -32.572 \text{ J K}^{-1} \text{ mol}^{-1}$$

According to Table 7.2

$$\Delta_r G^o = -|\nu_e| F E^o = -(96{,}485 \text{ C mol}^{-1})(-2.714 \text{ V}) = 261.860 \text{ kJ mol}^{-1}$$

7.26 The standard electrode potentials E^o in the earlier literature are based on a standard state pressure of 1 atm. Show that when the bar is used as the standard state pressure, standard electrode potentials E^o(atm) need to be corrected to E^o(bar) using
$E^o(\text{bar}) = E^o(\text{atm}) + (0.000169 \text{ V})\Delta\nu$
where $\Delta\nu$ is the increase in the number of gaseous molecules as the cell reaction (including hydrogen) proceeds as written.

SOLUTION

$$H_2(g) + Ox = 2H^+ + Red$$

We have already seen (problem 5.63)

$$\Delta G^o(\text{bar}) = \Delta G^*(\text{atm}) - [RT \ln (P^*/P^o)]\, \Delta\nu$$

$$E^o(\text{bar}) = -\Delta G^o(\text{bar})/|\nu_e|F = E^*(\text{atm}) + [RT|\nu_e|F \ln(P^*/P^o)]\Delta\nu$$

7.27 Calculate $\Delta_f S^o$ for Na^+(ao) at 298.15 K from
(a) $\Delta_f G^o(Na^+)$ and $\Delta_f H^o(Na^+)$
(b) $\bar{S}^o(Na^+)$

SOLUTION

(a) $$\Delta_f S^o(Na^+) = \frac{\Delta_f H^o(Na^+) - \Delta_f G^o(Na^+)}{T}$$

$$= \frac{-240\ 120 + 261\ 910}{298.15} = 73.08 \text{ J K}^{-1} \text{ mol}^{-1}$$

(b) $Na(s) = Na^+(ao) + e^-$

$$\Delta_f S^o(Na^+) = \bar{S}^o(Na^+) + \bar{S}^o(e^-) - \bar{S}^o[Na(s)]$$

$$= 59.0 + \tfrac{1}{2}(130.684) - 51.21 = 73.13 \text{ J K}^{-1} \text{ mol}^{-1}$$

7.28 When a hydrogen electrode and a normal calomel electrode are immersed in a solution at 25 °C a potential of 0.664 V is obtained. Calculate (a) the pH and (b) the hydrogen-ion activity.

SOLUTION

(a) $$E = E^o - \frac{RT}{|\nu_e|F} \ln a_{H^+} = E^o + 0.0591 \text{ pH}$$

$$\text{pH} = \frac{E - E^o}{0.0591} = \frac{0.664 - 0.2802}{0.0591} = 6.49$$

(b) $a_{H^+} = 10^{-6.49} = 3.24 \times 10^{-7}$

7.29 Calculate the equilibrium constant at 25 °C for the reaction
$2H^+ + D_2(g) = H_2(g) + 2D^+$
from the electrode potential for $D^+|D_2|Pt$, which is - 3.4 mV at 25 °C.

SOLUTION

$Pt|D_2, D^+ || H^+, H_2|Pt$

$E = 0 - (-0.0034) = 0.0034$ V

$K = \exp[|\nu_e|FE^o/RT]$

$$= \exp\left[\frac{2(96500 \text{ C mol}^{-1})(0.0034 \text{ V})}{(8.314 \text{ J K}^{-1} \text{ mol}^{-1})(298 \text{ K})}\right] = 1.30$$

7.30 A water electrolysis cell operated at 25 °C consumes 25 kWh/lb of hydrogen produced. Calculate the cell efficiency using $\Delta_r G^o$ for the decomposition of water.

SOLUTION

$$H_2O(l) = H_2(g) + \frac{1}{2} O_2(g)$$

$\Delta_r G^o = - 237.129 \text{ kJ mol}^{-1}$

The electrical energy used per mole of H_2 produced is

$(25 \text{ kWh/lb})(10^3 \text{ W kW}^{-1})(3600 \text{ s hr}^{-1})(454 \text{ g lb})^{-1}(2\text{g mol}^{-1}) = 396{,}475 \text{ J mol}^{-1}$

$$\text{Efficiency} = \frac{237.129}{396.475} = 0.60$$

7.31 Calculate E^o at 25 °C for fuel cells utilizing the reactions

(a) $C_2H_6(g) + 3\frac{1}{2} O_2(g) = 2CO_2(g) + 3H_2O(l)$

(b) $C_2H_4(g) + 3O_2(g) = 2CO_2(g) + 2H_2O(l)$

Catalysts have not yet been developed to make these fuel cells possible.

SOLUTION

$\Delta_r G^o$ can be calculated for each reaction using Appendix C.2. The problem is to calculate the number of electrons involved. The number of electrons involved can be determined by looking at the formation of H_2O.

$O_2 + 4H^+ + 4e^- = 2H_2O$

Thus two electrons are involved for each O, so that 14 electrons are involved in the first reaction and 12 in the second.

(a) $\Delta_r G^o = 2(- 394.36) + 3(- 237.13) + 32.82 = - 1467.3 \text{ kJ mol}^{-1}$

$$E^o = \frac{1467.3 \text{ kJ mol}^{-1}}{(14)(96.485 \text{ kC mol}^{-1})} = 1.086 \text{ V}$$

(b) $\Delta_r G^o = 2(- 394.36) + 2(- 237.13) + 68.15 = - 1194.8 \text{ kJ mol}^{-1}$

$$E^o = \frac{1194.8 \text{ kJ mol}^{-1}}{(12)(96.485 \text{ kC mol}^{-1})} = 1.032 \text{ V}$$

7.32 (a) When methane is oxidized completely to $CO_2(g)$ and $H_2O(1)$ at 25 °C, how much electrical energy can be produced using a fuel cell, assuming no electrical losses? What is the electromotive force of the fuel cell? (b) When one mole of methane is oxidized completely in a Carnot engine that operates between 500 K and 300 K, how much electrical energy can be produced, assuming that the mechanical energy can be converted completely into electrical energy?

SOLUTION

(a) R $2\ O_2 + 8\ e^- + 8\ H^+ = 4\ H_2O$

L $CO_2 + 8\ H^+ + 8\ e^- = CH_4 + 2\ H_2O$

$2\ O_2 + CH_4 = CO_2 + 2\ H_2O$

$\Delta_r G^o = -\ 394.359 + 2(-\ 237.129) - (50.72) = -\ 817.90\ \text{kJ mol}^{-1}$

$= -\ |\nu_e| F E^o$

$$E^o = \frac{817.90 \times 10^3\ \text{J mol}^{-1}}{8(96\ 485\ \text{C mol}^{-1})} = 1.0596\ \text{V}$$

(b) $\Delta_r H^o = -\ 393.509 + 2(-\ 285.830) - (-\ 74.81)$

$= -\ 890.4\ \text{kJ mol}^{-1}$

$$|w| = |q| \frac{(T_1 - T_2)}{T_1} = 890.4 \frac{200}{500} = 356\ \text{kJ mol}^{-1}$$

Thus the fuel cell would produce over twice as much electrical energy.

7.33 Calculate the electromotive force of Li(l)|LiCl(l)|Cl_2(g) at 900 K for $P_{Cl_2} = 1$ bar. This high-temperature battery is attractive because of its high electromotive force and low atomic masses. Lithium chloride melts at 883 K and lithium at 453.69 K. [The $\Delta_f G^o$ for LiCl(l) at 900 K in JANAF Thermochemical Tables is - 335.140 kJ mol^{-1}.]

SOLUTION

$$\text{Li(l)} + \frac{1}{2}\ Cl_2\text{(g)} = \text{LiCl(l)}$$

$\Delta_r G^o = -\ 335{,}140\ \text{J mol}^{-1} = -\ FE^o$

$$E^o = \frac{-\ 335140\ \text{J mol}^{-1}}{-\ 96485\ \text{C mol}^{-1}} = 3.474\ \text{V}$$

7.34 A membrane permeable only by Na^+ is used to separate the following two solutions:

α	0.10 mol kg^{-1} NaCl	0.05 mol kg^{-1} KCl
β	0.05 mol kg^{-1} NaCl	0.10 mol kg^{-1} KCl

What is the membrane potential at 25 °C and which solution has the highest positive potential?

SOLUTION

$$\phi^\beta - \phi^\alpha = -\frac{RT}{z_iF}\ln\frac{a_i^\beta}{a_i^\alpha}$$

$$= -\frac{(8.314 \text{ J K}^{-1}\text{ mol}^{-1})(298 \text{ K})}{96\,485 \text{ C mol}^{-1}}\ln\frac{0.05}{0.10} = 0.018 \text{ V}$$

The β phase is more positive because of the diffusion of Na^+ from α to β. Since the ionic strengths of the two solutions are the same, the activity coefficients of Na^+ in α and β are very nearly the same.

7.35 Since Table 7.2 does not give $E^\circ[Fe^{3+},Fe(s)]$, calculate it from other data in the table.

SOLUTION

$Fe^{2+} + 2e^- = Fe(s)$ $\quad \Delta_r G^\circ = -2(96485 \text{ C mol}^{-1})(-0.440 \text{ V}) = 84.907 \text{ J mol}^{-1}$

$Fe^{3+} + e^- = Fe^{2+}$ $\quad \Delta_r G^\circ = -2(96485 \text{ C mol}^{-1})(0.771 \text{ V}) = -74390 \text{ J mol}^{-1}$

Adding these reactions yields

$Fe^{3+} + 3e^- = Fe(s)$ $\quad \Delta_r G^\circ = 10517 \text{ J mol}^{-1} = -3(96485 \text{ C mol}^{-1})E^\circ$

$E^\circ = -0.036$ V

Note that although you can add standard Gibbs energies of reactions, you cannot add electrode potentials of half reactions involving different numbers of electrons.

7.36 In an electrolysis experiment, 0.1575 g of copper is plated on the cathode from a solution of copper sulfate when a current of 0.400 amperes is passed for 1200 s. (a) Calculate the value of the Faraday constant. (b) Given that the charge on an electron is $1.602\text{x}10^{-19}$ C, calculate the Avogadro constant. (This experiment is described by C. A. Seiglie, J. Chem. Ed. 80, 668 (2003).)

SOLUTION

(a) $F = qM/nm$, where q is the charge transferred in coulombs, M is the molar mass of copper, n is the number of electrons involved in plating one atom of copper, and m is the mass of copper plated.

$$F = \frac{(0.400x1200 \text{ C})(63.54 \text{ g mol}^{-1})}{2(0.1575 \text{ g})}$$

$$= 96.8x10^3 \text{ C mol}^{-1}$$

(b) $N_A = F/e = (96.9\text{x}10^3 \text{ C mol}^{-1})/(1.602\text{x}10^{-19} \text{ C}) = 6.04\text{x}10^{23} \text{ mol}^{-1}$

7.37 $5.759 \text{ x } 10^9 \text{ V m}^{-1}$

7.38 (a) 463 kJ mol^{-1} (b) 46.3 kJ mol^{-1} (c) 5.78 kJ mol^{-1}

7.39 $a = 4\, m^3 \gamma_\pm^3$

7.40 (a) $\dfrac{a_{\text{LiCl}}^{1/2}}{m}$ (b) $\dfrac{a_{\text{AlCl}_3}^{1/4}}{27^{1/4}\, m}$ (c) $\dfrac{a_{\text{MgSO}_4}^{1/2}}{m}$

7.41 (a) 0.1 (b) 0.3 (c) 0.4 (d) 0.4

7.42 0.834

7.43 $K = a(\mathrm{HCl})/(P_{\mathrm{H_2}}/P^{\mathrm{o}})^{1/2} = 5.701 \times 10^3$

7.44 $\mathrm{p}K = 1.018\sqrt{I} + \log\dfrac{m_1}{m_2} + \dfrac{(E - E^{\mathrm{o}})F}{2.303\,RT} + \log m_3$

7.45 (a) $2\mathrm{AgCl} + \mathrm{Zn} = 2\mathrm{Ag} + \mathrm{ZnCl_2}(0.555\ \mathrm{mol\ kg^{-1}})$; (b) - 195.9 kJ $\mathrm{mol^{-1}}$; (c) - 77.6 J $\mathrm{K^{-1}\ mol^{-1}}$; (d) - 219.0 kJ $\mathrm{mol^{-1}}$

7.46 - 130.318 kJ $\mathrm{mol^{-1}}$, - 125 J $\mathrm{K^{-1}\ mol^{-1}}$, - 167.580 kJ $\mathrm{mol^{-1}}$

7.47 (a) The 10.02% electrode is negative; (b) ΔH = - 2913 J $\mathrm{mol^{-1}}$; (c) 0.030462 V

7.48 791.885 kJ $\mathrm{mol^{-1}}$, 55.835 kJ $\mathrm{mol^{-1}}$, -80.668 J $\mathrm{K^{-1}\ mol^{-1}}$, 1.008×10^{-14}

7.49 - 109.805 kJ $\mathrm{mol^{-1}}$

7.50 (a) $Fe^{3+} + Cu^{+} = Fe^{2+} + Cu^{2+}$; (b) 0.620 V: (c) - 59.82 kJ $\mathrm{mol^{-1}}$; (d) - 58.69 kJ $\mathrm{mol^{-1}}$; (e) - 58.83 kJ $\mathrm{mol^{-1}}$

7.52 - 131.258 kJ $\mathrm{mol^{-1}}$

7.53 - 0.36 V

7.54 1.322×10^{-5} mol $\mathrm{kg^{-1}}$

7.56 (a) 4.405, (b) 4.400 V

7.56 $H_2O(l) = H^+(ao) + OH^-(ao)$

$K_w = 1.003 \times 10^{-14}$

7.58 Ag|Ag^+ | | Br^-|AgBr|Ag, $K = 10^{-11.85}$

7.59 1.34×10^{-5} mol $\mathrm{L^{-1}}$

7.60 - 0.17 mV, - 0.34 mV

7.61 - 0.4009 V

7.62 - 108.86 kJ $\mathrm{mol^{-1}}$

7.63 - 744.49 kJ $\mathrm{mol^{-1}}$

7.64 1.239 V

7.65 (a) 2.62 (b) 2.399×10^{-3}

7.66 (a) 237.129 kJ $\mathrm{mol^{-1}}$ (b) 285.83 kJ $\mathrm{mol^{-1}}$ (c) 50.8 K

7.67 1.172 V

7.68 15.3

7.69 (a) Right: $Cu^+ + e^- = Cu(s)$

Left: $Cu^{2+} + e^- = Cu^+$

(b) $2Cu^+ = Cu^{2+} + Cu(s)$; (c) 0.368 V; (d) $K = a(Cu^{2+})/a(Cu^+)^2 = 1.66 \times 10^6$

7.70 (a) Right: $Cu^+ + e^- = Cu(s)$

Left: $\frac{1}{2} Cu^{2+} + e^- = \frac{1}{2} Cu(s)$

(b) $Cu^+ = \frac{1}{2} Cu(s) + \frac{1}{2} Cu^{2+}$ 0.182 V

(c) $K = 1.192 \times 10^3$, which has to be squared to be compared with K in the preceding problem.

8

Thermodynamics of Biochemical Reactions

8.1 Show that the slope of the titration curve of a monobasic weak acid is given by

$$\frac{d\alpha}{dpH} = \frac{2.303\,K[H^+]}{(K + [H^+])^2}$$

where α is the degree of neutralization.

SOLUTION

	HA	=	H^+	+	A^-
Initial	1		0		0
eq	$1 - \alpha$		$[H^+]$		α

$$K = \frac{[H^+]\alpha}{1 - \alpha}$$

$$\alpha = \frac{K}{K + [H^+]}$$

$$\frac{d\alpha}{d[H^+]} = -\frac{K}{(K + [H^+])^2}$$

$$\frac{dpH}{d[H^+]} = -\frac{1}{2.303}\frac{d\ln[H^+]}{d[H^+]} = \frac{-1}{2.303[H^+]}$$

$$\frac{d\alpha}{dpH} = \frac{d\alpha}{d[H^+]}\frac{d[H^+]}{dpH} = \frac{2.303\,K[H^+]}{(K + [H^+])^2}$$

8.2 A liter of 0.10 M solution of a monoprotic weak acid is titrated with a sufficiently concentrated NaOH solution that there is not a significant change in volume. Derive the expression for the amount n of NaOH that would have to be added to reach a specified pH, assuming that the ionic strength can be taken as zero.

SOLUTION

Three types of equations must be satisfied in such a titration:

(1) The equilibrium constant expression for the weak acid must be satisfied:

$$K = \frac{[\mathrm{H^+}][\mathrm{A^-}]}{[\mathrm{HA}]} = \frac{10^{-\mathrm{pH}}[\mathrm{A^-}]}{c_0 - [\mathrm{A^-}]} = 10^{-\mathrm{p}K}$$

where c_0 is the initial concentration of the weak acid. The concentration of the base form of the weak acid at a specified pH is given by

$$[\mathrm{A^-}] = \frac{c_0 10^{-\mathrm{p}K}}{10^{-\mathrm{pH}} + 10^{-\mathrm{p}K}}$$

(2) The ion product of water must be satisfied.

$$K_\mathrm{w} = 10^{-14} = [\mathrm{H^+}][\mathrm{OH^-}] = 10^{-\mathrm{pH}}[\mathrm{OH^-}]$$

(3) The solution must remain electrically neutral:

$$[\mathrm{Na^+}] + [\mathrm{H^+}] = [\mathrm{A^-}] + [\mathrm{OH^-}]$$

$$n + 10^{-\mathrm{pH}} = [\mathrm{A^-}] + 10^{-14} / 10^{-\mathrm{pH}}$$

where n is the amount of NaOH added to a liter of solution. Now we have two expressions for [A⁻] that can be set equal to each other. Solving this expression for n yields

$$n = \frac{c_0 10^{-\mathrm{p}K} + 10^{-14} + 10^{-14+\mathrm{pH}-\mathrm{p}K} - 10^{-2\mathrm{pH}} - 10^{-\mathrm{pH}-\mathrm{p}K}}{10^{-\mathrm{pH}} + 10^{-\mathrm{p}K}}$$

This equation can be used to plot titration curves for monoprotic weak acids (see Computer Problem 8.I).

8.3 According to Appendix C.2 what are the values of $\Delta_r G^o$, $\Delta_r H^o$, and $\Delta_r S^o$ at 298 K for

$H_2O(1) = H^+(ao) + OH^-(ao)$

Show that the same value of $\Delta_r S^o$ is obtained from $\Delta_r G^o$ and $\Delta_r H^o$ as by using $\Delta_r S^o = \Sigma\, v_i \overline{S}_i^{\,o}$. Calculate K_w at 298 K.

SOLUTION

$\Delta_r G^o$ = - 157.244 + 2237.129 = 79.885 kJ mol^{-1}

$\Delta_r H^o$ = - 229.994 + 285.830 = 55.836 kJ mol^{-1}

$\Delta_r S^o$ = - 10.75 - 69.91 = - 80.66 J K^{-1} mol^{-1} or

$$\Delta_r S^o = \frac{\Delta H^o - \Delta G^o}{T} = \frac{(55\ 836 - 79\ 885)\ \text{J mol}^{-1}}{298.15\ \text{K}}$$

$= - 80.66$ J K^{-1} mol^{-1}

$K_w = \exp[- 79.885/(8.3143 \times 10^{-3})(298.15)]$

$= 1.010 \times 10^{-14}$

Note the use of the value of R used in constructing Appendix C.2.

8.4 For the acid dissociation of acetic acid, $\Delta_r H^o$ is approximately zero at room temperature in H_2O. For the acidic form of aniline, which is approximately as strong an acid as acetic acid, $\Delta_r H^o$ is approximately 21 kJ mol^{-1}. Calculate $\Delta_r S^o$ for each of the following reactions.

$CH_3CO_2H = H^+ + CH_3CO_2^-$ pK = 4.75

$C_6H_5NH_3^+ = H^+ + C_6H_5NH_2$ pK = 4.63

How do you interpret these entropy changes? What compensates for the increase in entropy expected from the increase in number of molecules in the balanced chemical reaction?

SOLUTION

For acetic acid:

$\Delta_r G^o = - RT \ln K$

$= RT\ 2.303\ \text{p}K$

$= (8.314$ J K^{-1} mol^{-1})(298 K)(2.303)(4.75) = 27.1 kJ mol^{-1}

$\Delta_r S^o = (\Delta_r H^o - \Delta_r G^o)/T = - (27.1 \times 10^3$ J mol^{-1})/(298 K) = - 91 J K^{-1} mol^{-1}

This increase in order is due to the hydration of the ions that are formed.

For aniline:

$\Delta_r G^o = (8.314$ J K^{-1} mol^{-1})(298 K)(2.303)(4.63)

$= 26.4$ kJ mol^{-1}

$\Delta_r S^o = [(21 - 26.4) \times 10^3$ J mol^{-1}]/(298 K)

$= - 18$ J K^{-1} mol^{-1}

The entropy change is much smaller than for acetic acid because there is no change in the number of ions.

8.5 Estimate pK_3 and pK_2 for H_3PO_4 at 25 °C and 0.1 mol L^{-1} ionic strength. The values at zero ionic strength are pK_3 = 2.148 and pK_2 = 7.198.

SOLUTION

$$\text{p}K_I = \text{p}K_{I=0} - \frac{(2n + 1)AI^{1/2}}{1 + I^{1/2}}$$

$A = 0.509$ at 25 °C

n is defined by $HA^{-n} = H^+ + A^{-(n+1)}$

For pK_3 of H_3PO_4, $n = 0$

$$pK_3 = 2.148 \frac{(0.509)(0.1)^{1/2}}{1 + (0.1)^{1/2}} = 2.148 - 0.122 = 2.026$$

For pK_2 of H_3PO_4, $n = 1$

$$pK_2 = 7.198 - \frac{(3)(0.509)(0.1)^{1/2}}{1 + (0.1)^{1/2}} = 7.198 - 0.369 = 6.831$$

8.6 In a strong acid solution, the amino acid histidine binds three protons. The acid dissociation constants numbered from the weakest acid dissociation are 6.92×10^{-10}, 1.00×10^{-6}, and 1.51×10^{-2} at 25 °C. Calculate the concentrations of the four forms of histidine (His^-, HisH, $HisH_2^+$, and $HisH_3^{2+}$) in a 0.1 M solution of histidine at pH 7, assuming these constants apply at the ionic strength of the solution.

SOLUTION

The equations derived for H_3PO_4 can be used.

$$K_1 = 6.92 \times 10^{-10}$$

$$K_1K_2 = 6.92 \times 10^{-16}$$

$$K_1K_2K_3 = 1.045 \times 10^{-17}$$

$$1 + [H^+]/K_1 + [H^+]^2/K_1K_2 + [H^+]^3/K_1K_2K_3$$

$$= 1 + 144.5 + 14.45 + 9.6 \times 10^{-5} = 159.95$$

$[His^-] = 0.1(1)/159.95 = 0.00063$ M

$[HisH] = 0.1(144.5)/159.95 = 0.0903$ M

$[HisH_2^+] = 0.1(14.45)/159.95 = 0.00903$ M

$[HisH_3^{2+}] = 0.1(9.6 \times 10^{-5})/159.95 = 6 \times 10^{-8}$ M

8.7 The pK for the dissociation of $CaATP^{2-}$ at 25 °C in 0.2 mol L^{-1} (n-propyl)$_4$NCl is 3.60. The pK for

$$HATP^{3-} = H^+ + ATP^{4-}$$

is 6.95. Calculate the apparent pK of this ATP ionization when ATP is titrated in a solution 0.1 mol L^{-1} $CaCl_2$. Assume that the Ca^{2+} concentration is much larger than the total ATP concentrations.

SOLUTION

$$K_{app} = K_{ATP}(1 + [Ca^{2+}]/K)$$

$$pK_{app} = pK_{ATP} - \log(1 + [Ca^{2+}]/K)$$

$$= 6.95 - \log(1 + 0.1/10^{-3.6}) = 4.35$$

8.8 To illustrate what we mean by a component in a solution at a specified pH, consider a very simple system, namely, a monoprotic weak acid HA and its salt in aqueous solution. Write the fundamental equation for G for this system and use the equilibrium expression in terms of chemical potentials for the acid dissociation to write the fundamental equation in terms of two components, the hydrogen component and the A component. The cation of the salt can be omitted from the fundamental equation because there are always enough cations for charge balance.

SOLUTION

$$dG = - SdT + VdP + \mu(H^+)dn(H^+) + \mu(A^-)dn(A^-) + \mu(HA)dn(HA) \quad (1)$$

At chemical equilibrium

$$\mu(H^+) + \mu(A^-) = \mu(HA) \quad (2)$$

We use this equation to eliminate n(HA) from equation 1.

$$dG = - SdT + VdP + \mu(H^+)dn(H^+) + \mu(A^-)dn(A^-) + [\mu(H^+) + \mu(A^-)]dn(HA)$$
$$= - SdT + VdP + \mu(H^+)[dn(H^+) + dn(HA)] + \mu(A^-)[dn(A^-) + dn(HA)] \quad (3)$$

This form of the fundamental equation applies at chemical equilibrium, and it can be written in terms of the hydrogen component with amount

$$n'(H^+) = n(H^+) + n(HA) \quad (4)$$

and the A component with amount

$$n'(A) = n(A^-) + n(HA) \quad (5)$$

Thus equation 1 can be written in terms of the two components at chemical equilibrium

$$dG = - SdT + VdP + \mu(H^+)dn'(H^+) + \mu(A^-)dn'(A) \quad (6)$$

rather than three species.

8.9 Since we have been dealing with dilute solutions, we have assumed that the chemical potential of a species is given by

$$\mu_i = \mu_i^o + RT\ln(c_i/c^o) \quad (a)$$

and then later in equation 8.55, we assumed that the transformed chemical potential μ_i' of a reactant made up of two species, for example HPO_4^{2-} + $H_2PO_4^-$, is given by

$$\mu_i' = \mu_i'^o + RT\ln((c_1 + c_2)/c^o) \quad (b)$$

at a specified pH. This looks reasonable, but it is a good idea to write out the mathematical steps. The transformed Gibbs energy of a reactant that is made up of two species is given by

$$G' = n_1\mu_1' + n_2\mu_2' \quad (c)$$

The amounts of the two species can be replaced with $n_1 = r_1n'(P_i)$ and $n_2 = r_2n'(P_i)$, where n' is the amount of inorganic phosphate and r_1 and r_2 of are the equilibrium mole fractions of HPO_4^{2-} and HPO_4^-. Thus equation c can be rewritten as

$$G' = n'(P_i)\{r_1\mu_1'^o + r_2\mu_2'^o + RT[r_1\ln(c_1/c^o) + r_2\ln(c_2/c^o)]\} \quad \text{(d)}$$

The last term looks a lot like an entropy of mixing, and so we add $RT\ln([P_i]/c^o)$ and subtract $(r_1 + r_2)RT\ln[(c_1 + c_2)/c^o)]$, which are equal. Show that this leads to

$$G' = n'(P_i)\{\mu'^o(P_i) + RT\ln([P_i]/c^o)\} = n'(P_i)\mu_i'(P_i) \quad \text{(e)}$$

where

$$\mu'^o(P_i) = r_1\mu_1'^o + r_2\mu_2'^o + RT\ln(r_1\ln r_1 + r_2\ln r_2) \quad \text{(f)}$$

This confirms equation b and shows that the standard transformed chemical potential of a reactant at a specified pH is equal to a mole fraction average transformed chemical potential for the two species plus an entropy of mixing. In making numerical calculations, the standard transformed chemical potentials are replaced by standard transformed Gibbs energies of formation.

SOLUTION

Adding and subtracting the terms described in the problem to equation d yields

$$G' = n'(P_i)\{r_1\mu_1'^o + r_2\mu_2'^o + RT\ln(r_1\ln r_1 + r_2\ln r_2) + RT\ln([P_i]/c^o)\}$$

The terms involving r_1 and r_2 do not depend on the total phosphate concentration, and so they make up the standard transformed chemical potential $\mu_i'^o$ of the reactant at the specified pH. Thus, the standard transformed chemical potential of inorganic phosphate at specified pH is given by

$$\mu'^o(P_i) = r_1\mu_1'^o + r_2\mu_2'^o + RT\ln(r_1\ln r_1 + r_2\ln r_2)$$

Another way of writing this relation is given in equation 8.79, which can be written

$$\mu'^o(P_i) = -RT\ln\sum_{i=1}^{N_I}\exp(-\mu_i'^o/RT)$$

8.10 Write out the equations for calculating the standard transformed Gibbs energy of formation and standard transformed enthalpy of formation of a partially neutralized weak acid (HA) at a specified pH.

SOLUTION

The standard transformed Gibbs energy of formation of reactant A at a specified pH is given by

$$\Delta_f G'^o(A) = -RT\ln\{\exp[-\Delta_f G^o(A^-)/RT] + \exp[-(\Delta_f G^o(HA) - \Delta_f G^o(H^+) - RT\ln([H^+]/c^o))/RT]\}$$

The mole fractions of the two species in the pseudoisomer group are given by

$$r(A^-) = \frac{\exp[-\Delta_f G^o(A^-)/RT]}{\exp[-\Delta_f G'^o/RT]}$$

$$r(\text{HA}) = \frac{\exp[-(\Delta_f G^o(\text{HA}) - \Delta_f G^o(\text{H}^+) - RT\ln([\text{H}^+]/c^o))/RT]}{\exp[-\Delta_f G'^o/RT]}$$

The standard transformed enthalpy of formation is given by

$$\Delta_f H'^o = r(\text{A}^-)\Delta_f H^o(\text{A}^-) + r(\text{HA})[\Delta_f H^o(\text{HA}) - \Delta_f H^o(\text{H}^+)]$$

8.11 Will 0.01 mol L^{-1} creatine phosphate react with 0.01 mol L^{-1} adenosine diphosphate to produce 0.04 mol L^{-1} creatine and 0.02 mol L^{-1} adenosine triphosphate at 25 °C, pH 7, pMg 4? What concentration of ATP can be formed if the other reactants are maintained at the indicated concentration?

SOLUTION

Creatine P + H_2O = Creatine + P $\quad\quad \Delta_r G'^o = -43.5$ kJ mol^{-1}

ADP + P = ATP + H_2O $\quad\quad \Delta_r G'^o = 39.8$ kJ mol^{-1}

Creatine P + ADP = Creatine + ATP $\quad\quad \Delta_r G'^o = -3.7$ kJ mol^{-1}

$$\Delta_r G' = \Delta_r G'^o + RT\ln\frac{[\text{Creatine}][\text{ATP}]}{[\text{Creatine P}][\text{ADP}]}$$

$$= -3700 + (8.314)(298)\ln\frac{(0.04)(0.02)}{(0.01)(0.01)} = 1340 \text{ J mol}^{-1}$$

Therefore the answer to the first question is no.

$$K = e^{-\Delta_r G'^o/RT} = e^{3700/(8.1314)(298)} = 4.5$$

$$= \frac{[\text{Creatine}][\text{ATP}]}{[\text{Creatine P}][\text{ADP}]}$$

If the reactants are maintained at the indicated concentrations,

$$[\text{ATP}] = \frac{4.5\,[\text{Creatine P}][\text{ATP}]}{[\text{Creatine}]} = \frac{4.5\,(0.01)(0.01)}{0.04}$$

$$= 1.1 \times 10^{-2} \text{ mol L}^{-1}$$

8.12 The cleavage of fructose 1,6-diphosphate (FDP) to dihydroxyacetone phosphate (CHP) and glyceraldehyde 3-phosphate (GAP) is one of a series of reactions most organisms use to obtain energy. At 37 °C and pH 7, $\Delta G'^o$ for the reaction FDP = DHP + GAP is 23.97 kJ mol^{-1}. What is $\Delta_r G'^o$ in an erythrocyte in which [FDP] = 3×10^{-6} mol L^{-1}, [DHP] = 138×10^{-6} mol L^{-1}, and [GAP] = 18.5×10^{-6} mol L^{-1}?

SOLUTION

FDP + H_2O = DHP + GAP

$$\Delta G' = \Delta G'^o + RT\ln\frac{[\text{DHP}][\text{GAP}]}{[\text{FDP}]}$$

$$= 23{,}970 + (8.314)(310)\ln\frac{(138 \times 10^{-6})(18.5 \times 10^{-6})}{(3 \times 10^{-6})} = 5770 \text{ J mol}^{-1}$$

8.13 How many grams of ATP have to be hydrolyzed to ADP to lift 100 lb 100 ft if the available Gibbs energy can be converted into mechanical work with 100% efficiency? It is assumed that [ATP] = [ADP] = [P] = 0.01 mol L^{-1} and that $\Delta G'^{o}$ is - 39.8 kJ mol^{-1} at 25 °C.

SOLUTION

$$w = mgh$$
$$= \left(\frac{100\ \text{lb}}{2.2\ \text{lb/kg}}\right)(9.8\ \text{ms}^{-2})(100\ \text{ft})(12\ \text{in/ft})(2.54\ \text{cm/in})(0.01\ \text{m/cm})$$
$$= 1.358 \times 10^4\ \text{J}$$
$$\Delta_r G' = \Delta_r G'^{o} + RT \ln \frac{[\text{ADP}][\text{P}]}{[\text{ATP}]}$$

for ATP + H_2O = ADP + P

$$\Delta_r G' = -39{,}800\ \text{J} + (8.314)(298)\ln[(0.01)^2/0.01] = -51{,}210\ \text{J mol}^{-1}$$
$$= -(51{,}210\ \text{J mol}^{-1})/(507.2\ \text{g mol}^{-1}) = -101\ \text{J g}^{-1}$$
$$\text{Mass of ATP required} = \frac{1.358 \times 10^4\ \text{J}}{101\ \text{J g}^{-1}} = 135\ \text{g}$$

8.14 Biochemistry textbooks give $\Delta_r G'^{o}$ = - 20.1 kJ mol^{-1} for the hydrolysis of ethyl acetate at pH 7 and 25 °C. Experiments in acid solution show that

$$\frac{[CH_3CH_2OH][CH_3CO_2H]}{[CH_3CO_2CH_2CH_3]} = 14$$

where the equilibrium concentrations are in moles per liter. What is the value of $\Delta G'^{o}$ obtained from this equilibrium constant? The pK of acetic acid = 4.60 at 25 °C.

SOLUTION

$$K' = \frac{[CH_3CH_2OH]([CH_3CO_2H] + [CH_3CO_2^-])}{[CH_3CO_2CH_2CH_3]}$$
$$= \frac{[CH_3CH_2OH][CH_3CO_2H]}{[CH_3CO_2CH_2CH_3]}\left(1 + \frac{[CH_3CO_2^-]}{[CH_3CO_2H]}\right)$$
$$= 14\ \text{mol L}^{-1}\left(1 + \frac{K_{CH_3CO_2H}}{[H^+]}\right)$$

At pH 7

$$K' = 14\left(1 + \frac{10^{-4.6}}{10^{-7}}\right) = 3530$$
$$\Delta_r G'^{o} = -RT \ln K = -(8.314\ \text{J K}^{-1}\ \text{mol}^{-1})(298.15\ \text{K}) \ln 3540$$
$$= -20.3\ \text{kJ mol}^{-1}$$

8.15 Fumarase catalyzes the reaction fumarate + H_2O = L-malate. At 25 °C and pH 7, $K' = 4.4$ = [L-malate]/[fumarate].

What is the value of K' at pH 4?

Given: For fumaric acid $K_1 = 10^{-4.18}$

For L-malic acid $K_1 = 10^{-4.73}$

SOLUTION

$$\frac{[\text{L-malate}]}{[\text{fumarate}]} = \frac{[\text{M}] + [\text{HM}]}{[\text{F}] + [\text{HF}]} = \frac{[\text{M}][1 + [\text{H}^+]/K_{1\text{M}}]}{[\text{F}][1 + [\text{H}^+]/K_{1\text{F}}]}$$

$$= 4.4\,\frac{[1 + 10^{-4}/10^{-4.73}]}{[1 + 10^{-4}/10^{-4.18}]} = 4.4\,\frac{6.37}{2.51} = 11.2$$

*8.16 Given $\Delta_r G^o = 49.4$ kJ mol^{-1} for

$$ATP^{4-} + H_2O = AMP^{2-} + P_2O_4^{7} + 2H^+$$

calculate $\Delta G'^o$ at pH 7 and 25 °C and 0.2 mol L^{-1} ionic strength. The pK values that are needed are: for ATP, pK_1 = 6.95; for ADP, pK_1 = 6.88; for AMP, pK_1 = 6.45; for pyrophosphate, pK_1 = 8.95 and pK_2 = 6.12.

SOLUTION

$$ATP^{4-} \quad + \quad H_2O \quad = \quad AMP^{2-} \quad + \quad P_2O_7^{4-} \quad + \quad 2H^+$$

$\downarrow\uparrow 10^{-6.95}$ (ATP) $\quad\downarrow\uparrow 10^{-6.45}$ (AMP) $\quad\downarrow\uparrow 10^{-8.95}$ ($P_2O_7^{4-}$)

$HATP^{3-}$ $\quad HAMP^{1-}$ $\quad HP_2O_7^{3-}$

$\downarrow\uparrow 10^{-6.12}$

$H_2P_2O_7^{2-}$

$$K' = \frac{([AMP^{2-}] + [HAMP^-])([P_2O_7^{4-}] + [HP_2O_7^{3-}] + [H_2P_2O_7^{2-}])}{([ATP^{4-}] + [HATP^{3-}])}$$

$$= \frac{[AMP^{2-}][P_2O_7^{4-}]\left(1 + \dfrac{[HAMP^-]}{[AMP^{2-}]}\right)\left(1 + \dfrac{[HP_2O_7^{3-}]}{[P_2O_7^{4-}]} + \dfrac{[H_2P_2O_7^{2-}]}{[P_2O_7^{4-}]}\right)}{[ATP^{4-}]\left(1 + \dfrac{[HATP^{3-}]}{[ATP^{4-}]}\right)}$$

$$= \frac{[AMP^{2-}][P_2O_7^{4-}][H^+]^2}{[ATP^{4-}]} \cdot \frac{\left(1 + \dfrac{[H^+]}{K_{1AMP}}\right)\left(1 + \dfrac{[H^+]}{K_{1PP}} + \dfrac{[H^+]^2}{K_{1PP}K_{2PP}}\right)}{[H^+]^2\left(1 + \dfrac{[H^+]}{K_{1ATP}}\right)}$$

$$= e^{-49,400/(8.314)(298)} \cdot \frac{\left(1 + \frac{10^{-7}}{10^{-6.45}}\right)\left(1 + \frac{10^{7}}{10^{-8.95}} + \frac{(10^{-7})^2}{10^{-8.95}\,10^{-6.12}}\right)}{(10^{7})^2\left(1 + \frac{10^{-7}}{10^{-6.95}}\right)}$$

$$= 1.53 \times 10^{7}$$

$$\Delta_r G'^{o} = -RT \ln K' = -(8.314 \text{ J K}^{-1} \text{ mol}^{-1})(298 \text{ K}) \ln (1.53 \times 10^{7})$$

$$= -41.0 \text{ kJ mol}^{-1}$$

8.17 Calculate the enthalpy of ionization for $H_2PO_4^- = H^+ + HPO_4^{2-}$ at (a) $I = 0$ and (b) $I = 0.25$ M, given that CODATA shows

	$\Delta_f H^o$/kJ mol^{-1}
H^+	0
HPO_4^{2-}	-1299.0
$H_2PO_4^-$	-1302.6

SOLUTION

(a) $\Delta_f H^o(I=0) = 0 - 1299.0 + 1302.6 = 3.6 \text{ kJ mol}^{-1}$

(b) $$\Delta_f H^o(I) = \Delta_f H^o(I=0) + \frac{1.4775\, z_i^2\, I^{1/2}}{1 + 1.6\, I^{1/2}}$$

$$\Delta_f H^o(H^+, I=0.25 \text{ M}) = 0 + \frac{1.4775(0.25)^{1/2}}{1 + 1.6(0.25)^{1/2}} = 0.41 \text{ kJ mol}^{-1}$$

$$\Delta_f H^o(HPO_4^{2-}, I=0.25 \text{ M}) = -1299.0 + \frac{1.4775(4)(0.25)^{1/2}}{1 + 1.6(0.25)^{1/2}} = -1297.36 \text{ kJ mol}^{-1}$$

$$\Delta_f H^o(H_2PO_4^-, I=0.25 \text{ M}) = -1302.6 + 0.41 = -1302.19 \text{ kJ mol}^{-1}$$

The enthalpy of ionization is given by

$$\Delta_r H^o(I=0.25 \text{ M}) = 0.41 - 1297.36 + 1302.19 = 5.24 \text{ kJ mol}^{-1}$$

This calculation can be made more simply by use of

$$\Delta_r H^o(I) = \Delta_r H^o(I=0) + \frac{1.4775\, (\Sigma\, z_i^2)\, I^{1/2}}{1 + 1.6\, I^{1/2}}$$

$$= 3.60 + (0.41)(1^2 + 2^2 - 1^2) = 5.24 \text{ kJ mol}^{-1}$$

8.18 If n molecules of a ligand A combine with a molecule of protein to form PA_n without intermediate steps, derive the relation between the fractional saturation Y and the concentration of A.

SOLUTION

$$P + nA = PA_n$$

$$K = \frac{[P][A]^n}{[PA_n]}$$

$$[P]_0 = [P] + [PA_n]$$

$$K = \frac{([P]_0 - [PA_n])[A]^n}{[PA_n]} \text{, or } [P]_0 [A]^n = [PA_n] (K + [A]^n)$$

$$Y = \frac{[PA_n]}{[P]_0} = \frac{1}{1 + K/[A]^n} = \frac{[A]^n}{1 + [A]^n/K}$$

This equlibrium represents a cooperative effect in that as soon as one ligand molecule is bound, the other $(n - 1)$ ligand molecules are also bound.

8.19 A protein M can bind two molecules of a ligand L, which is a gas. The macroscopic equilibrium constants, written in terms of the partial pressures of the ligand, are defined by

$M + L = ML \qquad K_1 = [ML]/[M]P_L$

$ML + L = ML_2 \qquad K_2 = [ML_2]/[ML]P_L$

Assume that the two binding sites are different and that ML can be distinguished from LM. How are the microscopic dissociation constants

$M + L = ML \qquad K_1{}^* = [ML]/[M]P_L$

$M + L = LM \qquad K_2{}^* = [LM]/[M]P_L$

$ML + L = LML \qquad K_3{}^* = [LML]/[ML]P_L$

$LM + L = LML \qquad K_4{}^* = [LML]/[LM]P_L$

related to the macroscopic dissociation constants K_1 and K_2? How many of the microscopic dissociation constants are independent? If there is a relation between them, what is it?

SOLUTION

$$[ML] = K_1{}^*[M]P_L$$

$$[LM] = K_2{}^*[M]P_L$$

$$[LML] = K_3{}^*[ML]P_L = K_1{}^*K_2{}^*[M]P_L{}^2 \text{ or } K_2{}^*K_4{}^*[M]P_L{}^2$$

$$K_1 = \frac{[ML] + [LM]}{[M]P_L} = K_1{}^* + K_2{}^*$$

$$K_2 = \frac{[LML]}{([ML] + [LM])P_L} = \frac{1}{\frac{1}{K_3^*} + \frac{1}{K_4^*}}$$

Since there are 5 species and two components (protein and ligand), the number of independent equilibria is 3; $C = N - R$ is $2 = 5 - 3$. Since there are two paths from M to LML,

$K_1^*K_3^* = K_2^*K_4^*$

8.20 Since it is difficult to determine the values of the four dissociation constants in equation 8.84, the empirical Hill equation

$$Y = \frac{1}{1 + \frac{K_h}{P_{O_2}^h}}$$

is frequently used to characterize binding. Show that the Hill coefficient h may be obtained by plotting log $[Y/(1 - Y)]$ versus P_{O_2}.

SOLUTION

$$\frac{Y}{1 - Y} = \frac{P_{O_2}^h}{K_h}$$

$$\log \frac{Y}{1 - Y} = -\log K_h + h \log P_{O_2}.$$

For a variety of binding systems relatively linear Hill plots are obtained for values of Y in the range 0.1 to 0.9, but deviations usually occur at the extremes unless $h = 1$. At the extremes the plot usually approaches a slope of unity.

8.21 Hemoglobin is made up of two alpha chains and two beta chains, and so it can be represented by $(\alpha\beta)_2$. Hemoglobin dissociates into $\alpha\beta$ subunits. The association constant K' for the reaction $2\alpha\beta = (\alpha\beta)_2$ depends on the partial pressure of molecular oxygen, but at relatively high concentration of molecular oxygen at pH 7 and 21.5 °C, $K' = 9.47 \times 10^5$, when molar concentrations are used. If a solution is 0.0025 M in hemoglobin (that is, $(\alpha\beta)_2$), what are the concentrations of the dimer and tetramer at equilibrium? What if the hemoglobin solution is 0.00025 M?

SOLUTION

	$2\alpha\beta$	$=$ $(\alpha\beta)_2$
initial conc. (mol/L)	0	0.0025
equil. conc	x	$0.0025 - x/2$

$$K' = 9.47 \times 10^5 = \frac{0.0025 - x/2}{x^2}$$

Use of the quadratic formula yields

$$x = \frac{-b + (b^2 - 4ac)^{1/2}}{2a} = 0.511 \times 10^{-5}\ \text{M} = [\alpha\beta]$$

$[(\alpha\beta)_2] = 0.0025 - (0.511 \times 10^{-5})/2 = 0.00247$ M (1.04% dissociated)

If the initial concentration of hemoglobin is 0.00025 M, $[\alpha\beta] = 1.60\times10^{-5}$ M and

$[(\alpha\beta)_2] = 0.000242$ M (3.2% dissociated)

*8.22 The percent saturation of a sample of human hemoglobin was measured at a series of oxygen partial pressures at 20 °C, pH 7.1, 0.3 mol L^{-1} phosphate buffer, and 3×10^{-4} mol L^{-1} heme.

P_{O_2}/Pa	Percent Saturation
393	4.8
787	20
1183	45
2510	78
2990	90

Calculate the values of h and K_h in the Hill equation. (See problem 8.20.)

SOLUTION

$\log P$	$\log\left(\frac{Y}{1-Y}\right)$
2.594	-1.31
2.896	-0.602
3.074	-0.087
3.400	+0.550
3.476	+0.954

$$\log\left(\frac{Y}{1-Y}\right) = -\log K_h + n \log P$$

n = slope = 2.4 $\qquad K_h = 4 \times 10^7$

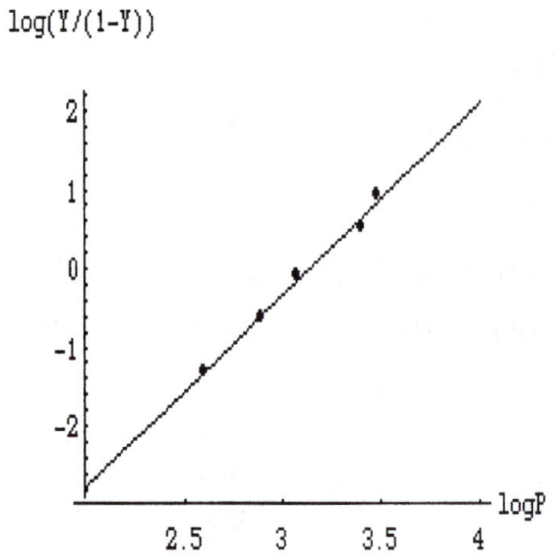

8.23 Use fundamental equations and Gibbs-Duhem equations to determine the number F of degrees of freedom and number D of variables required to define the extensive state of a system at constant T and P and with the reaction A + B = C at equilibrium from the following two points of view: (a) Start with the fundamental equation for G in terms of species and derive the form in terms of components to obtain D. Then derive the Gibbs-Duhem equation to obtain F. State the criterion for spontaneous change. (b) Define a transformed Gibbs energy G' to introduce μ_B as an intensive variable. Use the corresponding fundamental equation to obtain D and the Gibbs-Duhem equation to obtain F. State the criterion for spontaneous change.

SOLUTION

(a) The fundamental equation for G in terms of species is

$$dG = -SdT + VdP + \mu_A dn_A + \mu_B dn_B + \mu_C dn_C \quad (1)$$

This form of the fundamental equation can be used to derive the equilibrium relation $\mu_A + \mu_B = \mu_C$. Substituting this relation into equation 1 to eliminate μ_C yields

$$dG = -SdT + VdP + \mu_A dn_{cA} + \mu_B dn_{cB} \quad (2)$$

where $n_{cA} = n_A + n_B$ and $n_{cB} = n_B + n_C$ are amounts of components. This indicates that the criterion for spontaneous change is $(dG)_{T,P,n_{cA},n_{cB}} \leq 0$, and so $D = 4$. The corresponding Gibbs-Duhem equation is

$$0 = -SdT + VdP - n_{cA}d\mu_A - n_{cB}d\mu_B \quad (3)$$

Since only three of these intensive variables are independent, $F = 3$, and they can be taken as T, P, and μ_B.

(b) In order to use μ_B as an independent intensive variable, a transformed Gibbs energy G' is defined with the Legendre transform

$$G' = G - n_{cB}\mu_B \quad (4)$$

where the product of conjugate variables is subtracted. The differential of this equation is

$$dG' = dG - n_{cB}d\mu_B - \mu_B dn_{cB} \quad (5)$$

Substituting equation 2 yields

$$dG' = - SdT + VdP + \mu_A dn_{cA} - n_{cB} d\mu_B \qquad (6)$$

The differential terms on the right-hand side of equation 6 indicate that the natural variables for G' are T, P, n_{cA}, and μ_B, and so $D = 4$. The criterion for equilibrium is $(dG')_{T,P,n_{cA},\mu_B} \le 0$. The corresponding Gibbs-Duhem equation has already been given as equation 3, and so $F = 3$. This is in agreement with the phase rule, which gives $F = 3 - 1 - 1 + 2 = 3$. The relation $D = F + p = 3 + 1 = 4$ is also satisfied. Thus making the Legendre transform has not changed F or D, but it has indicated that different choices can be made of independent variables.

8.24 The equation for ln$K$ when $\Delta_r C_P{}^o$ is constant (see Example 5.11) is often used in a different form in treating the equilibrium constant for the thermal denaturation of a protein, which can be represented by N = D and K_D = [D]/[N]. The reason for this is that $\Delta_r C_P{}^o$ is generally quite large and nearly independent of temperature. The denaturation reaction is often highly cooperative, so that the reaction can be treated in terms of two states. Show that

$$\Delta_r G_D^o(T) = \Delta_r H^\circ(T_g)(1 - \frac{T}{T_g}) - \Delta_r C_P^o \left\{ T_g - T + T \ln\left(\frac{T}{T_g}\right) \right\}$$

where T_g is the temperature at which $K_D = 1$. This equation was first derived by W. J. Bectel and J. A. Schellman, Biopolymers, 26, 1858-1877 (1987).

SOLUTION

Multiplying the equation in Example 5.11 by $-RT$ yields

$$\Delta_r G^\circ(T) = \Delta_r H^\circ(298) - T\Delta_r S^\circ(298) + T\Delta_r C_P^o \left\{ 1 - \frac{298}{T} - \ln\left(\frac{T}{298}\right) \right\}$$

Since the 298 K is arbitrary, we can choose the reference temperature to be T_g at which K_D $(T_g) = 1$.

$$\Delta_r G^\circ(T) = \Delta_r H^\circ(T_g) - T\Delta_r S^\circ(T_g) + T\Delta_r C_P^o \left\{ 1 - \frac{T_g}{T} - \ln\left(\frac{T}{T_g}\right) \right\}$$

At this temperature $\Delta_r S^\circ(T_g) = \Delta_r H^\circ(T_g)/T_g$ so that

$$\Delta_r G^\circ(T) = \Delta_r H^\circ(T_g)(1 - \frac{T}{T_g}) - \Delta_r C_P^o \left\{ T_g - T + T \ln\left(\frac{T}{T_g}\right) \right\}$$

According to this equation, the dependence of K_D on temperature can be represented by two constants, $\Delta_r H^\circ(T_g)$ and $\Delta_r C_P{}^o$ Evaluating these constants for a reaction makes it possible to calculate $\Delta_r G^\circ(T)$, $\Delta_r H^\circ(T)$, and $\Delta_r S^\circ(T)$ for any temperature in the range in which $\Delta_r C_P^o$ is independent of temperature (see Computer Problem 8.M).

8.25 Given that the transformed Gibbs energy G', transformed enthalpy H', and transformed entropy S' of a system are given by

$$G' = G - n_c(\mathrm{H})G_m(\mathrm{H}^+)$$

$$H' = H - n_c(\mathrm{H})H_m(\mathrm{H}^+)$$

$$S' = S - n_c(\mathrm{H})S_m(\mathrm{H}^+)$$

Show that $G' = H' - TS'$.

SOLUTION

Substitute $G = H - TS$ and $G_m(\mathrm{H}^+) = H_m(\mathrm{H}^+) - TS_m(\mathrm{H}^+)$

in the definition of the transformed Gibbs energy to obtain

$$G' = H - TS - n_c(\mathrm{H}) \{H_m(\mathrm{H}^+) - TS_m(\mathrm{H}^+)\}$$

$$= H - n_c(\mathrm{H})H_m(\mathrm{H}^+) - T\{S - n_c(\mathrm{H})S_m(\mathrm{H}^+)\}$$

$$= H' - TS'$$

This shows that the Legendre transform that brings in the transformed Gibbs energy of a system also brings in a corresponding set of other thermodynamic potentials.

8.26 Solutions of two single -tranded DNAs are mixed and a double stranded DNA is formed:

5'-A-G-C-T-G-3'

+ = 5'-A-G-C-T-G-3'
3'-T-C-G-A-C-5'

5'-C-A-G-C-T-3'

Use of the parameters of J. SantaLucia, Jr., H. T.Allawi, and P. A.Seneviratne (Biochemistry, **35**, 3555 {1996) for 298 K, pH 7, and 1 M NaCl yields $\Delta_r G^o$ = -20.5 kJ mol^{-1} and $\Delta_r H^o$ = -128.4 kJ mol^{-1} for this reaction (G. G. Hammes, *Thermodynamics and Kinetics for the Biological Sciences*, Wiley, New York, 2000). (a) Calculate the equilibrium constant for this reaction at 298 K and the equilibrium fraction of single strands when the initial concentrations of the two single strands are 10^{-4} M. (b) Calculate the equilibrium constant for this reaction at 313 K and the equilibrium fraction of single strands when the initial concentrations of the two single strands are 10^{-4} M.

SOLUTION

(a) At 298 K,

$K = \exp\{20.5/(8.31451*0.298)\} = 3919$

$K = x/(c-x)^2$

Solution of this quadratic equation for x yields $2.32\text{x}10^{-5}$ M. The equilibrium fraction of single strands is

$f_S = 1 - x/c = 0.769$

(b) The equilibrium constant at 313K can be calculated using

$$\ln\frac{K_{313}}{3919} = \frac{-128{,}400(313-298)}{8.31451x298x313}$$

$K_{313} = 327$

Solution of the quadratic equation for x yields $3.07\text{x}10^{-6}$ M. The equilibrium fraction of single strands is

$f_S = 1 - x/c = 0.969$

8.27 $K_w = 1.148\text{x}10^{-15}$, pH = 7.47

8.28 $5.301 \text{ x } 10^{-14}$, 6.638

8.29 - 10.75 J K^{-1} mol^{-1}

8.30 - 0.25, 27.15 kJ mol^{-1}, - 92.1 J K^{-1} mol^{-1}, $1.752 \text{ x } 10^{-5}$

8.31 $[H^+] = 1.321 \text{ x } 10^{-3}$ M, $[A^-] = 1.321 \text{ x } 10^{-3}$ M, [HA] = 0.09868 M

8.32 pK = 6.33

8.33 $K = 80.19$, $\Delta_r H^o = 1.08$ kJ mol^{-1}, $\Delta_r S^o = 40.08$ J K^{-1} mol^{-1}

No effect of ionic strength

8.35 $P_o = 1/D$, $P_A = K_A[A]/D$, $P_B = K_B[B]/D$, and $P_{AB} = K_A K_B[A][B]/D$.

8.36 3.14

8.37 (a) $K_1 = K/3$, $K_2 = K$, $K_3 = 3K$

(b) $f_{-3} = 1/8, f_{-2} = 3/8, f_{-1} = 3/8, f_0 = 1/8$

8.38 1.3

8.39 0.53

8.40 - 42.3 kJ mol^{-1}

8.41 0.038 mol L^{-1}

8.42 - 29.8 kJ mol^{-1}

8.43 - 29.8 kJ mol^{-1}

8.44 2.74 x 10^4

8.45 (a) $K' = K_1(1 + K_{CH}/[H^+])/(1 + K_{AH}/[H^+])$

(b) $K_2K_{CH}/K_1K_{AH} = 1$

8.46 0.158 mol L^{-1}

8.48 64 ppm

9

Quantum Theory

9.1 A detector is exposed to a monochromatic source of radiation for 40 ms and indicates that the power level is 10 μW. If 10^9 photons are incident on the detector in this time, what is the frequency of the radiation? What type of electromagnetic radiation is this?

SOLUTION

$\varepsilon = h\nu$

$$\frac{(10 \times 10^{-6}\ \text{J s}^{-1})(40 \times 10^{-3}\ \text{s})}{10^9} = (6.626 \times 10^{-34}\ \text{J s})\nu$$

$\nu = 6.037 \times 10^{17}\ \text{s}^{-1}$

$\lambda = c/\nu = 4.966 \times 10^{-10}$ m

X-rays

9.2 Calculate the energy per photon and the number of photons emitted per second from (a) a 100-W yellow light bulb ($\lambda = 550$ nm) and (b) a 1-kW microwave source ($\lambda = 1$ cm).

SOLUTION

(a) $$\varepsilon_{\text{photon}} = \frac{hc}{\lambda}$$

$$= \frac{(6.626 \times 10^{-34}\ \text{J s})(2.998 \times 10^8\ \text{m s}^{-1})}{550 \times 10^{-9}\ \text{m}} = 3.61 \times 10^{-19}\ \text{J}$$

$$\#\text{photons/s} = \frac{100\ \text{W}}{3.61 \times 10^{-19}\ \text{J}} = 2.77 \times 10^{20}\ \text{photons/s}$$

(b) $$\varepsilon_{\text{photon}} = \frac{hc}{\lambda}$$

$$= \frac{(6.626 \times 10^{-34}\ \text{J s})(2.998 \times 10^{8}\ \text{m s}^{-1})}{10^{-2}\ \text{m}} = 1.99 \times 10^{-23}\ \text{J}$$

$$\#\text{photons/s} = \frac{1000\ \text{W}}{1.99 \times 10^{-23}\ \text{J}} = 5.03 \times 10^{25}\ \text{photons/s}$$

9.3 In the photoelectric effect an electron is emitted from a metal as the result of absorption of a photon of light. Part of the energy of the photon is required to release the electron from the metal; this energy ϕ is called the work function or binding energy. The kinetic energy of the ejected electron is given by

$$\frac{1}{2}\ mv^2 = h\nu - \phi$$

where m and v are the mass and velocity of the electron. For the 100 face of silver metal (see Chapter 23) the velocity of electrons emitted using 200 nm photons is 7.42×10^5 m s^{-1}. Calculate the work function of this face in eV.

SOLUTION

$$\frac{1}{2}\ mv^2 = \frac{1}{2}(9.10 \times 10^{-31}\ \text{kg})(7.42 \times 10^{5}\ \text{m s}^{-1})^2$$

$$= 2.50 \times 10^{-19}\ \text{J}$$

$$\phi = \frac{(6.626 \times 10^{-34}\ \text{J s})(2.998 \times 10^{8}\ \text{m s}^{-1})}{200 \times 10^{-9}\text{m}} - 2.50 \times 10^{-19}\ \text{J}$$

$$\phi = 7.432 \times 10^{-19}\ \text{J}$$

$$= \frac{7.432 \times 10^{-19}\ \text{J}}{1.602 \times 10^{-19}\ \text{J/eV}} = 4.64\ \text{eV}$$

9.4 Photoelectron spectroscopy utilizes the photoelectron effect to measure the binding energy of electrons in molecules and solids, by measuring the kinetic energy of the emitted electrons and using the relation in problem 9.3 between kinetic energy, wavelength, and binding energy. One variant of photoelectron spectroscopy is X-ray Photoelectron Spectroscopy (XPS). If the X-ray wavelength is 0.2 nm, calculate the velocity of electrons emitted from molecules in which the binding energies are 10 eV, 100 eV, and 500 eV.

SOLUTION

$$\frac{1}{2}\ mv^2 = \frac{hc}{\lambda} - \phi$$

$$v = \left[\frac{2}{m}\left(\frac{hc}{\lambda} - \phi\right)\right]^{1/2}$$

$$v_{10\text{eV}} = 4.668 \times 10^{7}\ \text{m s}^{-1}$$

$v_{100eV} = 4.634 \text{ x } 10^7 \text{ m s}^{-1}$

$v_{500eV} = 4.473 \text{ x } 10^7 \text{ m s}^{-1}$

9.5 Electrons are accelerated by a 1000-V potential drop. (a) Calculate the de Broglie wavelength. (b) Calculate the wavelength of the X-rays that could be produced when these electrons strike a solid.

SOLUTION

(a) $Ee = \frac{1}{2} mv^2$

$$v = \left(\frac{2Ee}{m}\right)^{1/2} = \left[\frac{2(1000 \text{ V})(1.602 \text{ x } 10^{-19} \text{ C})}{9.110 \text{ x } 10^{-31} \text{ kg}}\right]^{1/2}$$

$$= 1.875 \text{ x } 10^7 \text{ m s}^{-1}$$

$$\lambda = \frac{h}{mv} = \frac{6.626 \text{ x } 10^{-34} \text{ J s}}{(9.110 \text{ x } 10^{-31} \text{ kg})(1.875 \text{ x } 10^7 \text{ m s}^{-1})} = 0.0387 \text{ nm}$$

(b) $Ee = \frac{hc}{\lambda}$ $\qquad \lambda = \frac{hc}{Ee}$

$$\lambda = \frac{(6.626 \text{ x } 10^{-34} \text{ J s})(2.998 \text{ x } 10^8 \text{ m s}^{-1})}{(1000 \text{ V})(1.602 \text{ x } 10^{-19} \text{ C})} = 1.24 \text{ nm}$$

9.6 An ultraviolet photon ($\lambda = 58.4$ nm) from a helium gas discharge tube is absorbed by a hydrogen molecule which is at rest. Since momentum is conserved, what is the velocity of the hydrogen molecule after absorbing the photon? What is the translational energy of the hydrogen molecule in J mol^{-1}?

SOLUTION

$$p = \frac{h}{\lambda} = \frac{6.626 \text{ x } 10^{-34} \text{ J s}}{58.4 \text{ x } 10^{-9} \text{ m}} = 1.135 \text{ x } 10^{-26} \text{ kg m s}^{-1} = mv$$

$$mv = \frac{2(1.0079 \text{ x } 10^{-3} \text{ kg mol}^{-1})}{6.022 \text{ x } 10^{23} \text{ mol}^{-1}} v$$

$$v = \frac{(1.135 \text{ x } 10^{-26} \text{ kg m s}^{-1})(6.022 \text{ x } 10^{23} \text{ mol}^{-1})}{2(1.0079 \text{ x } 10^{-3} \text{ kg mol}^{-1})} = 3.39 \text{ m s}^{-1}$$

$$E = \frac{1}{2} mv^2 N_A = 0.012 \text{ J mol}^{-1}$$

9.7 What is the de Broglie wavelength of an oxygen molecule at room temperature? Compare this to the average distance between oxygen molecules in a gas at 1 bar at room temperature.

SOLUTION

$\frac{1}{2}\,mv^2 = \frac{3}{2}\,kT \qquad v = \left(\frac{3kT}{m}\right)^{1/2}$

$$\lambda = \frac{h}{mv} = \frac{h}{(3mkT)^{1/2}}$$

$$= \frac{6.63 \times 10^{-34}\ \mathrm{J\ s}}{[(3)(5.31 \times 10^{-26}\ \mathrm{kg})(1.38 \times 10^{-23}\ \mathrm{J\ K^{-1}})(298\ \mathrm{K})]^{1/2}}$$

$= 0.0259$ nm

where $m = \frac{(32 \times 10^{-3}\ \mathrm{kg\ mol^{-1}})}{(6.022 \times 10^{23}\ \mathrm{mol^{-1}})} = 5.31 \times 10^{-26}$ kg

To calculate the average distance l between molecules we calculate the length of the side of the cube containing one molecule.

$$\bar{V} = \frac{RT}{P} = \frac{(8.314\ \mathrm{J\ K^{-1}\ mol^{-1}})(298\ \mathrm{K})}{(10^5\ \mathrm{Pa})}$$

$= 0.0247\ \mathrm{m^3\ mol^{-1}}$

$$l^3 = \frac{(0.0247\ \mathrm{m^3\ mol^{-1}})}{(6.022 \times 10^{23}\ \mathrm{mol^{-1}})}$$

$l = 3.46$ nm

Since the de Broglie wavelength is much shorter than the average distance between molecules, translational motion can be treated classically.

9.8 What is the de Broglie wavelength of a thermal neutron at 300 K?

SOLUTION

$$E = \frac{3}{2}\,kT = \frac{1}{2}\,m_n v^2 = \frac{p^2}{2m_n}$$

$$p = \sqrt{3kTm_n}$$

$$= \sqrt{3(1.381 \times 10^{-23}\ \mathrm{J\ K^{-1}})(300\ \mathrm{K})(1.675 \times 10^{-27}\ \mathrm{kg})}$$

$= 4.563 \times 10^{-24}\ \mathrm{kg\ m\ s^{-1}}$

$$\lambda = \frac{h}{p} = \frac{6.626 \times 10^{-34}\ \mathrm{J\ s}}{4.563 \times 10^{-24}\ \mathrm{kg\ m\ s^{-1}}} = 1.452 \times 10^{-10}\ \mathrm{m}$$

$= 0.1452$ nm

9.9 Calculate the de Broglie wavelengths of
(a) a 1 g bullet with velocity 300 m s^{-1}
(b) a 10^{-6} g particle with velocity 10^{-6} m s^{-1}
(c) a 10^{-10} g particle with velocity 10^{-10} m s^{-1}
(d) a H_2 molecule with energy of (3/2) kT at $T = 20$ K.

SOLUTION

(a) $\lambda = \frac{h}{p} = \frac{6.626 \text{ x } 10^{-34} \text{ J s}}{10^{-3} \text{ kg x } 300 \text{ m s}^{-1}} = 2.21 \text{ x } 10^{-33} \text{ m}$

(b) $\lambda = \frac{6.626 \text{ x } 10^{-34} \text{ J s}}{10^{-9} \text{ kg x } 10^{-6} \text{ m s}^{-1}} = 6.626 \text{ x } 10^{-19} \text{ m}$

(c) $\lambda = \frac{6.626 \text{ x } 10^{-34} \text{ J s}}{10^{-13} \text{ kg x } 10^{-10} \text{ m s}^{-1}} = 6.626 \text{ x } 10^{-10} \text{ m}$

(d) $m_{H_2} = \frac{2 \text{ x } 1.008 \text{ x } 10^{-3} \text{ kg mol}^{-1}}{N_A} = 3.348 \text{ x } 10^{-27} \text{ kg}$

$$\frac{1}{2} m_{H_2} v^2 = \frac{3}{2} kT$$

$$v = \left(\frac{3kT}{m_{H_2}}\right)^{1/2} = \left[\frac{3(1.38 \text{ x } 10^{-23})(20)}{3.348 \text{ x } 10^{-27}}\right]^{1/2} = 497.3 \text{ m s}^{-1}$$

$$\lambda = \frac{6.626 \text{ x } 10^{-34} \text{ J s}}{(3.348 \text{ x } 10^{-27} \text{ kg})(497.3 \text{ m s}^{-1})} = 3.980 \text{ x } 10^{-10} \text{ m}$$

9.10 The lifetime of a molecule in a certain electronic state is 10^{-10} s. What is the uncertainty in energy of this state? Give the answer in J and in J mol^{-1}.

SOLUTION

$$\Delta E \geq \frac{\hbar}{2\Delta t} = \frac{6.626 \text{ x } 10^{-34} \text{ J s}}{4\pi(10^{-10} \text{ s})} = 5.27 \text{ x } 10^{-25} \text{ J}$$

$$= (5.27 \text{ x } 10^{-25} \text{ J})(6.022 \text{ x } 10^{23} \text{ mol}^{-1}) = 0.318 \text{ J mol}^{-1}$$

9.11 Show that the function $f = 8e^{5x}$ is an eigenfunction of the operator d/dx. What is the eigenvalue?

SOLUTION

$\frac{df}{dx} = \frac{d}{dx} 8e^{5x} = 40 \, e^{5x} = 5f$. Thus the eigenvalue is 5.

9.12 What are the results of operating on the following functions with the operators d/dx and d^2/dx^2: (a) e^{-ax^2}, (b) cos bx, (c) e^{ikx} ? Which functions are eigenfunctions of these operators? What are the corresponding eigenvalues?

SOLUTION

(a) $$\frac{de^{-ax^2}}{dx} = -2axe^{-ax^2}$$

$$\frac{d^2}{dx^2}e^{-ax^2} = 4ax^2e^{-ax^2} - 2ae^{-ax^2}$$

(b) $$\frac{d}{dx}\cos bx = -b\sin bx$$

$$\frac{d^2}{dx^2}\cos bx = -b^2\cos bx$$

(Thus $\cos bx$ is an eigenfunction of $\frac{d^2}{dx^2}$ with eigenvalue $-b^2$)

(c) $$\frac{d}{dx}e^{ikx} = ik^2e^{ikx}$$

$$\frac{d^2}{dx^2}e^{ikx} = -k^2e^{ikx}$$

[Thus e^{ikx} is an eigenfunction of both $\frac{d}{dx}$ and $\frac{d^2}{dx^2}$ with eigenvalues ik and $(ik)^2 = -k^2$, respectively.]

9.13 Show that the operators for the x coordinate and for the momentum in the x direction p_x do not commute. Calculate the operator representing the commutator of x and p_x.

SOLUTION

$$\hat{x} = x \text{ and } \hat{p} = \frac{\hbar}{i}\frac{\partial}{\partial x}$$

$$\hat{x}\hat{p}_x f(x) = x \cdot \frac{\hbar}{i} f'(x) \text{ where } f'(x) \text{ is } \frac{\partial f(x)}{\partial x}.$$

$$\hat{p}_x\hat{x} f(x) = \frac{\hbar}{i}\frac{\partial}{\partial x}[xf'(x)] = \frac{\hbar}{i}[f(x) + xf'(x)]$$

Since these operators do not commute, we cannot specify precise values of both x and p_x. The relationship between the uncertainties in these two variables is given by the Heisenberg uncertainty principle.

$$(\hat{x}\hat{p}_x - \hat{p}_x\hat{x})f(x) = x\frac{\hbar}{i} f'(x) - \frac{\hbar}{i}[f(x) + xf'(x)]$$

$$= -\frac{\hbar}{i} f(x)$$

$$(\hat{x}\hat{p}_x - \hat{p}_x\hat{x}) = -\frac{\hbar}{i} = i\hbar$$

9.14 For a particle in a one-dimensional box, the ground state wave function is $\phi = (2/a)^{1/2} \sin(\pi x/a)$.
(a) What is the probability that the particle is in the right half of the box?
(b) What is the probability that the particle is in the middle third of the box?

SOLUTION

(a) $\frac{1}{2}$, by symmetry

(b) $$\text{probability} = \int_{a/3}^{2a/3} \psi^2 dx = \frac{2}{a} \int_{a/3}^{2a/3} \sin^2\left(\frac{\pi x}{a}\right) dx$$

$$= \frac{2}{a} \left[\frac{1}{2} x - \frac{a}{4\pi} \sin\left(\frac{2\pi x}{a}\right)\right]_{a/3}^{2a/3}$$

$$= \frac{2}{a} \left[\frac{a}{3} - \frac{a}{6} - \frac{a}{4\pi} \sin \frac{2\pi 2a}{3a} + \frac{a}{4\pi} \sin \frac{2\pi a}{3a}\right]$$

$$= \frac{2}{a} \left[\frac{a}{6} - \frac{a}{4\pi} \sin \frac{4\pi}{3} + \frac{a}{4\pi} \sin \frac{2\pi}{3}\right]$$

$$= \frac{1}{3} - \frac{1}{2\pi} \sin \frac{4\pi}{3} + \frac{1}{2\pi} \sin \frac{2\pi}{3}$$

$$= \frac{1}{3} - \frac{1}{2\pi}\left(-\frac{\sqrt{3}}{2}\right) + \frac{1}{2\pi}\left(\frac{\sqrt{3}}{2}\right) = \frac{1}{3} + \frac{\sqrt{3}}{2\pi} = 0.609$$

9.15 (a) Calculate the energy levels for $n = 1$, 2, and 3 for an electron in a potential well of width 0.25 nm with infinite barriers on either side. The energies should be expressed in kJ mol^{-1}. (b) If an electron makes a transition from $n = 2$ to $n = 1$ what will be the wavelength of the emitted radiation ?

SOLUTION

(a) For $n = 1$,

$$E = \frac{n^2h^2}{8ma^2} = \frac{(6.626 \times 10^{-34})^2}{8(9.109 \times 10^{-31} \text{ kg})(2.5 \times 10^{-10} \text{ m})^2}$$

$$= 9.640 \times 10^{-19} \text{ J}$$

$$= 580.5 \text{ kJ/mol}$$

For $n = 2$,

$E = 2322$ kJ/mol

For $n = 3$,

$E = 5225$ kJ/mol

(b) $\Delta E = (4 - 1)(9.640 \times 10^{-19})\text{ J} = 2.892 \times 10^{-18}\text{ J}$

$$\Delta E = \frac{hc}{\lambda}$$

$$\text{Thus } \lambda = \frac{(6.626 \times 10^{-34})(2.998 \times 10^{8})}{2.892 \times 10^{-18}}\text{ m}$$

$$= 6.87 \times 10^{-8}\text{ m} = 68.7\text{ nm}$$

9.16 For a helium atom in a one-dimensional box calculate the value of the quantum number of the energy level for which the energy is equal to $(3/2)kT$ at 25°C (a) for a box 1 nm long, (b) for a box 10^{-6} m long and (c) for a box 10^{-2} m long.

SOLUTION

(a) $E = \frac{3}{2}kT = \frac{3}{2}(1.38 \times 10^{-23}\text{ J K}^{-1})(298\text{ K}) = 6.17 \times 10^{-21}\text{ J}$

$$n = \frac{a}{n}(8mE)^{1/2}$$

$$= \frac{10^{-9}\text{ m}}{6.63 \times 10^{-34}\text{ J s}}\left[\frac{8(4.003 \times 10^{-3}\text{ kg/mol})(6.17 \times 10^{-21}\text{ J})}{6.02 \times 10^{23}\text{ mol}^{-1}}\right]^{1/2}$$

$= 27.3$ or approximately 27

(b) $n = 2.7 \times 10^{4}$

(c) $n = 2.7 \times 10^{8}$

9.17 Show that the wave functions for a particle in a one-dimensional box are orthogonal using

$$\sin\alpha \sin\beta = \frac{1}{2}\cos(\alpha - \beta) - \frac{1}{2}\cos(\alpha + \beta)$$

SOLUTION

$$\psi_n(x) = \left(\frac{2}{a}\right)^{1/2} \sin\frac{n\pi x}{a} \qquad n = 1,2,\ldots$$

$$\frac{2}{a}\int_0^a \sin\frac{n\pi x}{a}\sin\frac{m\pi x}{a}\,dx = \frac{1}{a}\int_0^a \cos\frac{(n-m)\pi x}{a}\,dx - \frac{1}{a}\int_0^a \cos\frac{(n+m)\pi x}{a}\,dx$$

$= 0$ if $m \neq n$

because $n - m$ and $n + m$ are both integers. The integrals are over complete cycles of $\cos\frac{(\text{integer})\pi x}{a}$ which are equal to zero.

9.18 Calculate the degeneracies of the first three levels for a particle in a cubical box.

SOLUTION

$E = \frac{h^2}{8ma^2}(n_x^2 + n_y^2 + n_z^2)$ where n_x, n_y, n_z are 1, 2, 3 . . .

n_x	n_y	n_x	$E / (\frac{h^2}{8ma^2})$	
1	1	1	3	Degeneracy = 1
2	1	1		
1	2	1	6	Degeneracy = 3
1	1	2		
2	2	1		
2	1	2	9	Degeneracy = 3
1	2	2		

9.19 (a) Calculate $\Delta x \Delta p_x$ for a particle in a linear box for n = 1, 2, and 3 using the equation in Example 9.13. Compare these values with the minimum product of uncertainties from the Heisenberg uncertainty principle. (b) What is the uncertainty in x for a particle in a 0.2 nm box when its quantum number is unity? In a 2 nm box?

SOLUTION

(a) $\Delta x \Delta p_x = \frac{\hbar}{2}\left(\frac{\pi^2 n^2}{3} - 2\right)^{1/2}$

Since we want to compare with the Heisenberg uncertainty principle, we might as well express these products in terms of $\hbar$.

For n = 1, $\Delta x \Delta p_x = 0.568\hbar$

For n = 2, $\Delta x \Delta p_x = 1.67\hbar$

For n = 3, $\Delta x \Delta p_x = 2.63\hbar$

For the Heisenberg uncertainty principle, $\Delta x \Delta p_x \geq 0.500\hbar$

(b) $\Delta x = \frac{a}{2\pi n}\left(\frac{\pi^2 n^2}{3} - 2)\right)^{1/2}$

For a = 0.2 nm, Δx = 0.03615 nm.

For a = 2 nm, Δx = 0.3615 nm.

9.20 Calculate the standard deviation for the x coordinate of a harmonic oscillator at

$v = 1$. Since $< x > = 0$, it is only necessary to calculate $< x^2 >$.

SOLUTION

The wave function for the harmonic oscillator at $v = 1$ is

$$\psi_1(x) = \left(\frac{4\alpha^3}{\pi}\right)^{1/4} x e^{-\alpha x^2/2}$$

The mean value of x^2 is given by

$$< x^2 > = \int_{-\infty}^{\infty} \psi_1(x) x^2 \psi_1(x) dx = \left(\frac{4\alpha^3}{\pi}\right)^{1/2} \int_{-\infty}^{\infty} x^4 e^{-\alpha x^2} dx$$

The value of this integral is given in Table 17.1, and so.

$$< x^2 > = \frac{3}{2\alpha}$$

Since $< x > = 0$,

$$\sigma_x^2 = < x^2 > - < x >^2 = < x^2 >$$

$$\sigma_x = \left(\frac{3}{2\alpha}\right)^{1/2}$$

where $\alpha = (k\mu/\hbar^2)^{1/2}$.

9.21 Calculate the standard deviation for the momentum p of a harmonic oscillator at $v = 1$. Since $< p > = 0$, it is only necessary to calculate $< p^2 >$.

SOLUTION

The wave function for the harmonic oscillator at $v = 1$ is

$$\psi_1(x) = \left(\frac{4\alpha^3}{\pi}\right)^{1/4} x e^{-\alpha x^2/2}$$

The mean value of p^2 is given by

$$< p^2 > = \int_{-\infty}^{\infty} \psi_1(x) \hat{p}^2 \psi_1(x) dx$$

where

$$\hat{p}^2 = -\hbar^2 \frac{d^2}{dx^2}$$

Substituting these relations into the equation for $< p^2 >$ yields

$$< p^2 > = \int_{-\infty}^{\infty} dx \left(\frac{4\alpha^3}{\pi}\right)^{1/4} x e^{-\alpha x^2/2} \left\{ -\hbar^2 \frac{d^2}{dx^2} \left[\left(\frac{4\alpha^3}{\pi}\right)^{1/4} x e^{-\alpha x^2/2}\right]\right\}$$

$$= -\left(\frac{4\alpha^3}{\pi}\right)^{1/2}\hbar^2\left[-3\alpha\int_{-\infty}^{\infty} dx\, x^2 e^{-\alpha x^2} + \alpha^2\int_{-\infty}^{\infty} dx\, x^4 e^{-\alpha x^2}\right]$$

The values of these two integrals are given in Table 17.1. This yields

$$<p^2> = \frac{3}{2}\alpha\hbar^2$$

Since $<p> = 0$ for a harmonic oscillator,

$$\sigma_p^2 = <p^2> - <p>^2 = <p^2>$$

$$\sigma_p = \left(\frac{3\alpha}{2}\right)^{1/2}\hbar$$

where $\alpha = (k\mu/\hbar^2)^{1/2}$.

9.22 Using the results of the two previous problems, calculate $\sigma_x\sigma_p$ and compare it with the Heisenberg uncertainty principle.

SOLUTION

$$\sigma_x\sigma_p = \left(\frac{3}{2\alpha}\right)^{1/2}\left(\frac{3\alpha}{2}\right)^{1/2}\hbar = 3\frac{\hbar}{2}$$

This is in agreement with the Heisenberg uncertainty principle $\sigma_x\sigma_p \geq \hbar/2$.

9.23 The fundamental vibration frequency of $^{12}C^{16}O$ is 2169.814 cm^{-1}. Calculate the force constant.

SOLUTION

The reduced mass is given by

$$\mu = \frac{(12.011\text{x}10^{-3}\text{ kg mol}^{-1})(15.9994\text{x}10^{-3}\text{ kg mol}^{-1})}{(12.011\text{x}10^{-3}\text{ kg mol}^{-1}+15.994\text{x}10^{-3}\text{ kg mol}^{-1})(6.022367\text{x}10^{23}\text{ mol}^{-1})}$$

$$= 1.139\ 19\text{x}10^{-26}\text{ kg}$$

$$k = (2\pi c\tilde{\nu})^2\mu$$

$$= (2\pi(2.99792\text{x}10^{10}\text{ cm s}^{-1})(2169.814\text{ cm}^{-1}))^2(1.13919\text{x}10^{-26}\text{ kg})$$

$$= 1903.01\text{ N m}^{-1}$$

Note that kg s^{-2} is equivalent to N m^{-1}, which is a more appropriate unit for a "force" constant.

9.24 Using data from the previous problem, calculate the fundamental vibration frequency for $^{12}C^{18}O$, assuming the force constant is the same.

SOLUTION

$$\tilde{\nu} = \frac{1}{2\pi c}\left(\frac{k}{\mu}\right)^{1/2}$$

$$\mu = \frac{(12.011\text{x}10^{-3}\text{ kg mol}^{-1})(17.999159\text{x}10^{-3}\text{ kg mol}^{-1})}{(12.011\text{x}10^{-3}\text{ kg mol}^{-1}+17.999159\text{x}10^{-3}\text{ kg mol}^{-1})(6.022367\text{x}10^{23}\text{ mol}^{-1})}$$

$$= 1.196\ 18\text{x}10^{-26}\text{ kg}$$

$$\tilde{\nu} = \frac{1}{2\pi(2.99792\text{x}10^{10}\text{ cm s}^{-1})}\left(\frac{1903.01\text{ N m}^{-1}}{1.196\ 18\text{x}10^{-26}\text{ kg}}\right)^{1/2} = 2117.5\text{ cm}^{-1}$$

9.25 Calculate the root-mean-square displacement of the nuclei of $^{12}C^{16}O$ in the $v = 0$ state and compare it with the equilibrium bond length of 112.832 pm.

SOLUTION

$$< x^2 >^{1/2} = \frac{\hbar^{1/2}}{(4\mu k)^{1/4}}$$

$$= \frac{(1.054\text{x}10^{-34}\text{ J s})^{1/2}}{[4(1.139\text{x}10^{-26}\text{ kg})(1903\text{ kg s}^{-2})]^{1/4}}$$

$$= 3.364\text{x}10^{-12}\text{ m} = 3.364\text{ pm}$$

Thus the mean square displacemment in the ground state is 2.98 % of the equilibrium bond length.

9.26 (a) Later in Table 13.4, we will find that the following molecules have the indicated vibrational frequencies: $^{35}Cl_2$(560 cm^{-1}), $^{39}K^{35}Cl$(281 cm^{-1}), $^{1}H_2$(4401 cm^{-1}). What are the force constants for these molecules if we treat them as harmonic oscillators? (b) Assuming the force constant for $^{37}Cl_2$ is the same as for $^{35}Cl_2$, predict the fundamental vibrational frequency of $^{37}Cl_2$.

SOLUTION

(a) $k = \mu(2\pi\nu)^2 = \mu(2\pi c\tilde{\nu})^2$

$^{35}Cl_2$:

$$k = \frac{17.50\text{ x }10^{-3}\text{ kg mol}^{-1}}{6.022\text{ x }10^{23}\text{ mol}^{-1}}(2\pi\text{ x }2.998\text{ x }10^{8}\text{ m s}^{-1}\text{ x }5.60\text{ x }10^{+4}\text{ m}^{-1})^2$$

$$= 323.3\text{ N m}^{-1}$$

$^{39}K\ ^{35}Cl$:

$$k = \frac{18.446 \text{ x } 10^{-3} \text{ kg mol}^{-1}}{6.022 \text{ x } 10^{23} \text{ mol}^{-1}} (2\pi \text{ x } 2.998 \text{ x } 10^{8} \text{ m s}^{-1} \text{ x } 2.81 \text{ x } 10^{4} \text{ m}^{-1})^{2}$$

$= 85.81 \text{ N m}^{-1}$

$^1H_2$:

$$k = \frac{0.5039 \text{ x } 10^{-3}}{6.022 \text{ x } 10^{23}} (2\pi \text{ x } 2.998 \text{ x } 10^{8} \text{ x } 4.40 \text{ x } 10^{5})^{2}$$

$= 575.1 \text{ N m}^{-1}$

(b) $\quad \dfrac{\tilde{\nu}_{37}}{\tilde{\nu}_{35}} = \sqrt{\left(\dfrac{\mu_{35}}{\mu_{37}}\right)} = \left(\dfrac{17.5}{18.5}\right)^{1/2} = 0.9726 \Rightarrow \tilde{\nu}_{37} = 545 \text{ cm}^{-1}$

9.27 Check the normalizations of ψ_0 and ψ_1 for the harmonic oscillator and show they are orthogonal.

SOLUTION

$$\int_{-\infty}^{\infty} \psi_0^2 \, dx = \left(\frac{a}{\pi}\right)^{1/2} \int_{-\infty}^{\infty} e^{-ax^2} \, dx = \left(\frac{a}{\pi}\right)^{1/2} \frac{\pi^{1/2}}{\alpha^{1/2}} = 1$$

$$\text{since} \int_{0}^{\infty} e^{-a^2x^2} \, dx = \frac{1}{2\alpha} \pi^{1/2}$$

$$\int_{-\infty}^{\infty} \psi_1^2 \, dx = 2\left(\frac{\alpha^3}{\pi}\right)^{1/2} \int_{-\infty}^{\infty} x^2 e^{-ax^2} \, dx = 4\left(\frac{\alpha^3}{\pi}\right)^{1/2} \int_{0}^{\infty} x^2 e^{-ax^2} \, dx$$

$$= 4\left(\frac{\alpha^3}{\pi}\right)^{1/2} \frac{\pi^{1/2}}{2^2\alpha^{3/2}} = 1 \text{ since} \int_{0}^{\infty} x^2 e^{-ax^2} \, dx = \frac{\pi^{1/2}}{4a^{3/2}}$$

To check orthogonality

$$\int_{-\infty}^{\infty} \psi_0\psi_1 dx = 2^{1/2}\left(\frac{a}{\pi}\right)^{1/2} \int_{-\infty}^{\infty} x \, e^{-ax^2} \, dx = 0$$

since x is an odd function and e^{-ax^2} is an even function.

9.28 Substitute the v =1 eigenfunction for the harmonic oscillator into the Schrödinger equation for the harmonic oscillator, and obtain the expression for the eigenvalue (energy).

SOLUTION

$\psi_1 = c \, x \, e^{-ax^2} \quad a = \dfrac{\pi}{h} (km)^{1/2}$

$$\frac{d\psi_1}{dx} = c(e^{-ax^2} - 2ax^2e^{-ax^2})$$

$$\frac{d^2\psi_1}{dx^2} = c\{-6axe^{-ax^2} + 4a^2x^3e^{-ax^2}\} = \{-6a + 4a^2x^2\}\,\psi_1$$

Substituting into the Schrödinger equation

$$\left(-\frac{\hbar}{2m}\frac{d^2}{dx^2} + \frac{1}{2}\,k\,x^2\right)\psi_1 = E\psi_1$$

the terms proportional to x^2 cancel, so

$$E = \frac{\hbar^2}{2m} \cdot 6a = \frac{h^2}{8\pi^2 m} \cdot \frac{6\pi}{m}\,(km)^{1/2} = \frac{3}{4}\frac{h}{\pi}\left(\frac{k}{m}\right)^{1/2}$$

$$= \frac{3}{2}\,h\left[\frac{1}{2\pi}\left(\frac{k}{m}\right)^{1/2}\right] = \frac{3}{2}\,h\nu_0$$

9.29 In the vibrational motion of HI, the iodine atom essentially remains stationary because of its large mass. Assuming that the hydrogen atom undergoes harmonic motion and that the force constant k is 317 N m^{-1}, what is the fundamental vibrational frequency ν_0? What is ν_0 if H is replaced by D?

SOLUTION

$$m = \frac{1.0078 \times 10^{-3}\ \text{kg mol}^{-1}}{6.022 \times 10^{23}\ \text{mol}^{-1}} = 1.674 \times 10^{-27}\ \text{kg}$$

$$\nu_0 = \frac{1}{2\pi}\sqrt{\frac{k}{m}} = \frac{1}{2\pi}\sqrt{\frac{317\ \text{kg s}^{-2}}{1.674 \times 10^{-27}\text{kg}}} = 6.93 \times 10^{13}\ \text{s}^{-1}$$

If H is replaced by D:

$$\nu_0 = \left(\frac{1}{\sqrt{2}}\right)(6.93 \times 10^{13}\ \text{s}^{-1}) = 4.90 \times 10^{13}\ \text{s}^{-1}$$

9.30 What are the expectation values for $< x >$ and $< x^2 >$ for a quantum mechanical harmonic oscillator in the $\upsilon = 1$ state? What is the standard deviation Δx?

SOLUTION

$$\langle x \rangle = \int \psi^*(x)\, x\, \psi_1(x)\, dx$$

$$= \left(\frac{4\alpha^3}{\pi}\right)^{1/2} \int_{-\infty}^{\infty} x^3\, e^{-ax^2}\, dx = 0 \quad \text{since the integrand is odd}$$

$$\langle x^2 \rangle = \left(\frac{4\alpha^3}{\pi}\right)^{1/2} \int_{-\infty}^{\infty} x^4\, e^{-ax^2}\, dx = \frac{1}{\alpha}\left(\frac{4}{\pi}\right)^{1/2} \int_{-\infty}^{\infty} z^4 e^{-z^2}\, dz$$

$$= \frac{1}{\alpha}\left(\frac{4}{\pi}\right)^{1/2}\frac{3\pi^{1/2}}{4} = \frac{3}{2\alpha} = \frac{3}{2}\left(\frac{\hbar^2}{k\mu}\right)^{1/2}$$

$$\Delta x = [\langle x^2\rangle - \langle x\rangle^2]^{1/2} = \left[\frac{3}{2}\left(\frac{\hbar^2}{k\mu}\right)^{1/2}\right]^{1/2}$$

9.31 $^{12}C^{16}O$ is an example of a stiff diatomic molecule, and it has a vibration frequency of 2170 cm^{-1}. (a) What is the value of the force constant k? (b) What is the value of the standard deviation Δx in the internuclear distance? (c) What is the standard deviation Δp_x of the momentum of the vibrational motion? (d) Check that the product $\Delta x \Delta p_x$ yields $\hbar/2$ in accordance with the Heisenberg uncertainty principle.

SOLUTION

(a) $\mu = m_1 m_2/(m_1 + m_2)$

$$= \frac{(12 \text{ x } 10^{-3} \text{ kg mol}^{-1})(15.995 \text{ x } 10^{-3} \text{ kg mol}^{-1})}{(27.995 \text{ x } 10^{-3} \text{ kg mol}^{-1})(6.022 \text{ x } 10^{23} \text{ mol}^{-1})}$$

$$= 1.1285 \text{ x } 10^{-26} \text{ kg}$$

$$k = (2\pi\nu)^2\mu = [2\pi(2.998 \text{ x } 10^{10} \text{ cm s}^{-1})(2170 \text{ cm}^{-1})]^2(1.1285 \text{ x } 10^{26} \text{ kg})$$

$$= 1886 \text{ N m}^{-1}$$

(b)

$$\alpha = \frac{(k\mu)^{1/2}}{\hbar} = \frac{[(1886 \text{ N m}^{-1})(1.1285x10^{-26} \text{ kg})]^{1/2}}{1.0546x10^{-34} \text{ J s}}$$

$$= 4.374x10^{22} \text{ m}^{-2}$$

$$\Delta x = 1/(2\alpha)^{1/2} = 1/[2(4.374x10^{22} \text{ m}^{-2})]^{1/2} = 3.381x10^{-12} \text{ m}$$

(c) $\langle p^2\rangle = \hbar^2\alpha/2 = (1.0546 \text{ x } 10^{-34} \text{ J s})^2(4.374 \text{ x } 10^{22} \text{ m}^{-2})/2$

$$= 2.432 \text{ x } 10^{-45} \text{ kg}^2 \text{ m}^2 \text{ s}^{-2}$$

$$\Delta p_x = \langle p^2\rangle^{1/2} = 1.560 \text{ x } 10^{-23} \text{ kg m s}^{-1}$$

(d) $\Delta x \Delta p_x = (3.381 \text{ x } 10^{-12} \text{ m})(1.560 \text{ x } 10^{-23} \text{ kg m s}^{-1}) = 5.273 \text{ x } 10^{-35} \text{ J s}$

$\hbar/2 = 5.273 \text{ x } 10^{-35} \text{ J s}$

9.32 Use information in Example 9.21 to calculate the frequency and wavenumber of the radiation required to take the $H^{35}Cl$ molecule from J = 1 to J = 2.

SOLUTION

For $J = 2$, $E_2 = (\hbar /2I)(2 \times 3) = 12.618 \times 10^{-22}$ J

$h\nu = E_2 - E_1 = 12.618 \times 10^{-22}\text{ J} - 4.206 \times 10^{-22}\text{ J} = 8.412 \times 10^{-22}$ J

$\nu = (8.412 \times 10^{-22}\text{ J})/(6.626 \times 10^{-34}\text{ J s}) = 1.344 \times 10^{12}\text{ s}^{-1}$

$\tilde{\nu} = (1.344 \times 10^{-12}\text{ s}^{-1})/(2.998 \times 10^{10}\text{ cm s}^{-1}) = 44.82\text{ cm}^{-1}$

9.33 Use the Schrö dinger equation for a rigid rotor in three dimensions to calculate the rotational energy when the wavefunction is given by the spherical harmonic Y_1^0. What is the magnitude of the angular momentum?

SOLUTION

$$Y_1^0 = \left(\frac{3}{4\pi}\right)^{1/2} \cos\theta$$

Equation 9.142 for this wavefunction is

$$-\frac{\hbar^2}{2I}\left[\frac{1}{\sin\theta}\frac{\partial}{\partial\theta}\left(\sin\theta\frac{\partial}{\partial\theta}\right)\right]\left(\frac{3}{4\pi}\right)^{1/2}\cos\theta = E\,Y_1^0$$

$$\left(\frac{3}{4\pi}\right)^{1/2}\frac{\hbar^2}{2I}\left[\frac{1}{\sin\theta}\frac{\partial}{\partial\theta}\left(\sin^2\theta\right)\right] = E\,Y_1^0$$

$$\frac{\hbar^2}{I}\left(\frac{3}{4\pi}\right)^{1/2}\cos\theta = E\,Y_1^0$$

Therefore,

$$E = \frac{\hbar^2}{I}$$

in agreement with equation 9.165.

The angular momentum is $[l(l + 1)]^{1/2}\hbar = 2^{1/2}\hbar$.

9.34 What are the reduced mass and moment of inertia of $^{23}Na^{35}Cl$? The equilibrium internuclear distance R_e is 236 pm. What are the values of E for the states with $J = 1$ and $J = 2$?

SOLUTION

$$\mu = \frac{1}{\frac{1}{m_1} + \frac{1}{m_2}} = \frac{m_1 m_2}{m_1 + m_2}$$

$$= \frac{(22.98977 \times 10^{-3}\text{ kg mol}^{-1})(34.96885 \times 10^{-3}\text{ kg mol}^{-1})}{[(22.98977 + 34.96885) \times 10^{-3}\text{ kg mol}^{-1}](6.022137 \times 10^{23}\text{ mol}^{-1})}$$

$= 2.30328 \times 10^{-26}$ kg

$I = \mu R_e^2 = (2.303 \times 10^{-26}\text{ kg})(236 \times 10^{-12}\text{ m})^2$

$= 1.283 \times 10^{-45}\text{ kg m}^2$

$$E = \frac{\hbar^2}{2I} J(J+1) = \frac{(6.626 \times 10^{-34}\text{ J s})^2(J(J+1))}{8\pi^2(1.283 \times 10^{-45}\text{ kg m}^2)}$$

$= 8.668 \times 10^{-24}$ J for $J = 1$

$= 2.600 \times 10^{-23}$ J for $J = 2$

9.35 The commutator of two operators, $\hat{A}$ and $\hat{B}$, is defined as the operator $\hat{A}\hat{B} - \hat{B}\hat{A}$. Using the definitions of $\hat{L}_x$, $\hat{L}_y$, and $\hat{L}_z$ given in equation 9.164-9.166, find the commutator of $\hat{L}_x$ with $\hat{L}_y$ and $\hat{L}_x$ with $\hat{L}_z$.

SOLUTION

$$\hat{L}_x = -i\hbar\left(y\frac{\partial}{\partial z} - z\frac{\partial}{\partial y}\right)$$

$$\hat{L}_y = -i\hbar\left(z\frac{\partial}{\partial x} - x\frac{\partial}{\partial z}\right)$$

$$\hat{L}_z = -i\hbar\left(x\frac{\partial}{\partial y} - y\frac{\partial}{\partial x}\right)$$

$$\therefore \hat{L}_x\hat{L}_y = -\hbar^2\left(y\frac{\partial}{\partial z} - z\frac{\partial}{\partial y}\right)\left(z\frac{\partial}{\partial x} - x\frac{\partial}{\partial z}\right)$$

$$\hat{L}_y\hat{L}_x = -\hbar^2\left(z\frac{\partial}{\partial x} - x\frac{\partial}{\partial z}\right)\left(y\frac{\partial}{\partial z} - z\frac{\partial}{\partial y}\right)$$

so,

$$\hat{L}_x\hat{L}_y - \hat{L}_y\hat{L}_x = -\hbar^2\left[y\frac{\partial}{\partial z}\left(z\frac{\partial}{\partial x}\right) + zx\frac{\partial}{\partial y}\frac{\partial}{\partial z} - zy\frac{\partial}{\partial x}\frac{\partial}{\partial z} - x\frac{\partial}{\partial z}\left(z\frac{\partial}{\partial y}\right)\right]$$

$$= -\hbar^2\left(y\frac{\partial}{\partial x} - x\frac{\partial}{\partial y}\right) = i\hbar\hat{L}_z$$

Similarly

$$\hat{L}_x\hat{L}_z - \hat{L}_z\hat{L}_x = -i\hbar\hat{L}_y$$

9.36 The $^{12}C^{16}O$ molecule has an equilibrium bond distance of 112.8 pm. Calculate (a) the reduced mass and (b) the moment of inertia. Calculate (c) the wavelength of the photon emitted when the molecule makes the transition from $J = 1$ to $J = 0$ using equation 9.165 for the energy levels.

SOLUTION

(a) reduced mass

$$\mu = \frac{m_1 m_2}{m_1 + m_2} = \frac{(12)(15.995)}{(12 + 15.995)}\frac{10^{-3}}{6.022 \times 10^{23}}$$

$\mu = 1.139 \times 10^{-26}$ kg

(b) moment of inertia

$I = (1.139 \times 10^{-26})(112.8 \times 10^{-12})^2$ kg m^2

$= 1.449 \times 10^{-46}$ kg m^2

(c) $$\Delta E = \frac{\hbar^2}{2I}[J(J+1) - 0(0+1)] = \frac{\hbar^2}{I} = \frac{h^2}{4\pi^2 I}$$

$= 7.675 \times 10^{-23}$ J

$$\lambda = \frac{hc}{\Delta E} = 2.58 \times 10^{-3} \text{ m}$$

9.37 (a) The distribution of wavelengths from a certain star peaks in the visible at $\lambda = 600$ nm. Assuming the distribution obeys the Planck distribution law, use Wien's displacement law to estimate the temperature of the star. (b) A metal bar is heated to "red-heat" so that its radiation peaks at $\lambda =$ 800 nm. Estimate the temperature of the bar.

SOLUTION

(a) $$T = \frac{2.898 \times 10^{-3} \text{ K m}}{600 \times 10^{-9} \text{ m}} = 4830 \text{ K}$$

(b) $$T = \frac{2.898 \times 10^{-3} \text{ K m}}{800 \times 10^{-9} \text{ m}} = 3623 \text{ K}$$

9.38 (a) Derive the value of the constant in the Wien displacement law (equation 9.198) in terms of h, c, and k. (b) If, from experiment, the values of h and c were measured to be 6.6×10^{-34} J s^{-1} and 3.0×10^8 m s^{-1}, and the value of the constant in the Wien displacement law were measured to be 2.9×10^{-3} K/m, find the value of k from (a). Since R is measured to be 8.3 J K^{-1} mol^{-1}, you can also calculate the Avogadro's constant.

SOLUTION

(a) $$\rho_\lambda(\lambda,T) = \frac{8\pi hc}{\lambda^5}\frac{1}{e^{hc/\lambda kT} - 1}$$

Differentiate w.r.t. λ, and set = 0:

$(x = hc/\lambda kT)$

$$5 = \frac{x_{max}e^{x_{max}}}{e^{x_{max}} - 1} = \frac{x_{max}}{1 - e^{-x_{max}}}$$

Solve by successive approximation: $x_{max} = 4.965$

Thus

$$\lambda_{max}T = \frac{hc}{4.965k}$$

(b) $$k = \frac{hc}{(4.965)(2.9 \times 10^{-3}\ \text{K/m})} = \frac{(6.6 \times 10^{-34})(3 \times 10^{8})}{(4.965)(2.9 \times 10^{-3})}$$

$$= 1.4 \times 10^{-23}\ \text{J K}^{-1}$$

$$N_A = R/k = \frac{8.3\ \text{J K}^{-1}\ \text{mol}^{-1}}{1.4 \times 10^{-23}\ \text{J K}^{-1}} = 5.9 \times 10^{23}\ \text{mol}^{-1}$$

9.39 $3.49\text{x}10^{20}$, $1{,}764\text{x}10^{18}$

9.41 $\nu_{max} = (5.87 \times 10^{10}\ \text{s}^{-1}\ \text{K}^{-1})\ T$

9.42 (a) $1.76 \times 10^{14}\ \text{s}^{-1}$ (1700 nm), $5.88 \times 10^{14}\ \text{s}^{-1}$ (5.10 nm), (b) 9.66, 290 nm

9.43 (a) 0.0388 nm, (b) 9.052×10^{-4} nm

9.44 0.1303 nm

9.45 1.325×10^{-27} kg m s^{-1}, 0.0113 m s^{-1}

9.46 5.38×10^{-26} J, 1.6×10^{-5} %

9.48 (a) - $6a$, (b) $1/kx$

9.49 $E = n^2/8\ ml^2$

9.50 145.1, 580.5, 1306.1 kJ mol^{-1}

9.51 1, 2, 2, 2

9.52 $(km)^{1/2}\ \pi/k$

9.53 (a) $P = 2 \int_{x_{class}}^{\infty} (\alpha/\pi)^{1/2} e^{-\alpha x^2} dx$

where $x_{class} = \hbar^{1/2}/(\mu k)^{1/4}$

$\alpha = (k\mu/\hbar^2)^{1/2}$

(b) 16% outside the classically allowed region

9.54 $E = (n_x + 1/2) h\nu_x + (n_y + 1/2)h\nu_y + (n_z + 1/2)h\nu_z$

$E_0 = h\nu_x/2 + h\nu_y/2 + h\nu_z/2$

9.55 $E = (n_x + n_y + n_z + 3/2) h\nu$

The degeneracies are 1, 3, and 6.

9.59 1.240×10^{-27} kg, 2.827×10^{-46} kg m^2.

$E_0 = 0$, $E_1 = 3.92 \times 10^{-23}$ J, $E_2 = 11.76 \times 10^{-23}$ J

9.61 $N = (1/a)^{1/2}$

10

Atomic Structure

10.1 Using data from Appendix C.3 at 0 K, what is the ionization energy of H(g)?

$H(g) = H^+(g) + e^-$

SOLUTION

H(g) at 0 K $\Delta_f H^\circ = 216.037\ \text{kJ mol}^{-1}$

$H^+(g)$ at 0 K $\Delta_f H^\circ = 1528.085\ \text{kJ mol}^{-1}$

$e^-(g)$ at 0 K $\Delta_f H^\circ = 0$

$\Delta H^\circ = 1312.048\ \text{J mol}^{-1}$

$$= \frac{1312.048\ \text{J mol}^{-1}}{(1.602\ 17733 \times 10^{-19}\ \text{C})(6.022\ 136\ 7 \times 10^{23}\ \text{mol}^{-1})}$$

$= 13.598\ 42\ \text{eV}$

10.2 How much energy in eV and kJ mol^{-1} is required to remove electrons from the following orbitals in an H atom? (a) 3d, (b) 4f, (c) 4p, (d) 6s.

SOLUTION

$$\Delta E = \frac{13.6058\ \text{eV}}{n^2} = \frac{(13.6058\ \text{V})(96\ 485\ \text{C mol}^{-1})}{(10^3\ \text{J kJ}^{-1})n^2}\ \frac{\text{kJ}}{\text{mol}}$$

(a) $n = 3$ $\Delta E = 1.512\ \text{eV} = 145.9\ \text{kJ/mol}$

(b) $n = 4$ $\Delta E = 0.8504\ \text{eV} = 82.05\ \text{kJ/mol}$

(c) $n = 4$ same as (b)

(d) $n = 6$ $\Delta E = 0.3779\ \text{eV} = 36.47\ \text{kJ/mol}$

10.3 Calculate the ground state ionization potentials for He^+, Li^{2+}, Be^{3+}, B^{4+}, and C^{5+}.

SOLUTION

$E = (13.61 \text{ eV})Z^2$

	Z	E/eV		Z	E/eV
He^+	2	54.44	B^{4+}	5	340.25
Li^{2+}	3	122.49	C^{5+}	6	489.96
Be^{3+}	4	217.76			

10.4 Since H and D have different reduced masses, they also have slightly different energy levels. Calculate (a) the ionization potentials and (b) the wavelengths of the first line in the Balmer series for these atoms.

SOLUTION

$$\mu_H = \frac{m_e m_p}{m_e + m_p} = 9.104\ 431 \times 10^{-31} \text{ kg}$$

$$\mu_D = \frac{m_e m_D}{m_e + m_D} = 9.106\ 908 \times 10^{-31} \text{ kg}$$

(a) $$E_i(H) = h_c R_H = hcR_\infty \frac{\mu_H}{m_e} = 2.178\ 718 \times 10^{-18} \text{ J}$$

$$= 13.598\ 48 \text{ eV}$$

$$E_i(D) = hcR_\infty \frac{\mu_D}{m_e} = 2.179\ 311 \times 10^{-18} \text{ J}$$

$$= 13.602\ 18 \text{ eV}$$

% diff = 2.7×10^{-2} %

(b) $$\frac{1}{\lambda_H} = R_H \left(\frac{1}{2^2} - \frac{1}{3^2}\right) = \frac{R_H}{7.2} \ ; \ \frac{1}{\lambda_D} = \frac{R_D}{7.2}$$

$\lambda_H = 656.470$ nm

$\lambda_D = 656.291$ nm

10.5 Since we know the expression for the wave function for the hydrogen atom in its ground state, show that

$$\hat{H}\,\psi_{100} = -\frac{\hbar^2}{2m_e a_0^2}\,\psi_{100}$$

and express the ground state energy in Hartrees.

SOLUTION

$$\psi_{100} = \frac{1}{\pi^{1/2} a_0^{3/2}}\, e^{-r/a_0}$$

The Bohr radius is

$$a_0 = \frac{\hbar^2(4\pi\varepsilon_0)}{m_e e^2}$$

The Hamiltonian for the hydrogen atom in its ground state is

$$\hat{H} = -\frac{\hbar^2}{2m_e}\left[\frac{1}{r^2}\frac{\partial}{\partial r}\left(r^2\frac{\partial}{\partial r}\right)\right] - \frac{e^2}{4\pi\varepsilon_0 r}$$

since no angular coordinates are involved in ψ_{100}. Therefore,

$$\hat{H}\ \psi_{100} = -\frac{\hbar^2}{2m_e}\left[\frac{1}{r^2}\frac{\partial}{\partial r}\left(r^2\frac{\partial \psi_{100}}{\partial r}\right)\right]\psi_{100} - \frac{e^2\psi_{100}}{4\pi\varepsilon_0 r}$$

The differentiations can be done in the following steps:

$$A = \frac{\partial \psi_{100}}{\partial r} = -\frac{e^{-r/a_0}}{\pi^{1/2}a_0^{5/2}}$$

$$B = r^2 A = -\frac{r^2 e^{-r/a_0}}{\pi^{1/2}a_0^{5/2}}$$

$$C = \frac{\partial B}{\partial r} = \frac{r^2 e^{-r/a_0}}{\pi^{1/2}a_0^{7/2}} - \frac{2re^{-r/a_0}}{\pi^{1/2}a_0^{5/2}}$$

$$D = \frac{1}{r^2}\ C = \frac{e^{-r/a_0}}{\pi^{1/2}a_0^{7/2}} - \frac{2e^{-r/a_0}}{r\pi^{1/2}a_0^{5/2}}$$

Thus

$$\hat{H}\ \psi_{100} = -\frac{\hbar^2}{2m_e}\left[\frac{e^{-r/a_0}}{\pi^{1/2}a_0^{7/2}} - \frac{2e^{-r/a_0}}{r\pi^{1/2}a_0^{5/2}}\right] - \frac{e^2\psi_{100}}{4\pi\varepsilon_0 r}$$

Taking ψ_{100} out of the first two terms yields

$$\hat{H}\ \psi_{100} = -\frac{\hbar^2}{2m_e}\left[\frac{1}{a_0^2} - \frac{2}{ra_0}\right]\psi_{100} - \frac{e^2\psi_{100}}{4\pi\varepsilon_0 r}$$

Using the expression for a_0 given above to eliminate ε_0 from the last term yields

$$\hat{H}\ \psi_{100} = -\frac{\hbar^2}{2m_e}\left[\frac{1}{a_0^2} - \frac{2}{ra_0} + \frac{2}{ra_0}\right]\psi_{100} = -\frac{\hbar^2}{2m_e a_0}\ \psi_{100}$$

Thus the energy of the ground state of the hydrogen atom is

$$E = -\frac{\hbar^2}{2m_e a_0^2} = -\frac{E_h}{2}$$

where E_h is the Hartree energy, which is given by

$$E_h = \frac{\hbar^2}{m_e a_0^2}$$

10.6 The muon is an elementary particle with a negative charge equal to the charge of the electron and a mass approximately 200 times the electron mass. The muonium atom is formed from a proton and a muon. Calculate the reduced mass, the Rydberg constant, and the formula for the energy levels for this atom. What is the most probable radius of the 1s orbital for this atom?

SOLUTION

$$\mu_M = \mu_{muonium} = \frac{m_{muon} m_{prot}}{m_{muon} + m_{prot}} = \frac{200\, m_e m_{prot}}{200\, m_e + m_{prot}} = 1.646\,397 \text{ x } 10^{-28} \text{ kg}$$

$$R_M = R_\infty \frac{\mu_M}{m_e} = 1.983\,35 \text{ x } 10^{-9} \text{ m}^{-1}$$

$$E = -\frac{hcR_M}{n^2} = \frac{-3.9425 \text{ x } 10^{-16} \text{ J}}{n^2} = \frac{-2461}{n^2} \text{ eV}$$

$$R_{mp} = a_M = a_0 m_e/\mu_M = \frac{(52.918 \text{ pm})(9.1094 \text{ x } 10^{-31} \text{ kg})}{1.6463 \text{ x } 10^{-28} \text{ kg}}$$

$$= 0.2928 \text{ pm}$$

*10.7 A hydrogenlike atom has a series of spectral lines at $\lambda = 26.2445$ nm, $\lambda = 19.4404$ nm, $\lambda = 17.3578$ nm, and $\lambda = 16.4028$ nm. What is the atomic number Z of this atom? What is the formula for this spectral series (i.e., n_1 and n_2 in eq. 10.16)?

SOLUTION

$$\lambda^{-1} = R_\infty Z^2 \left(\frac{1}{n_1^2} - \frac{1}{n_2^2}\right) \; ; \; n_2 = n_1 + m$$

$$= 1.097\,373\,2 \text{ x } 10^7 Z^2 \left(\frac{1}{n_1^2} - \frac{1}{n_2^2}\right) \text{m}^{-1} \left(10^{-9}\frac{\text{nm}^{-1}}{\text{m}^{-1}}\right)$$

$$= 0.010\,973\,732\, Z^2 \left(\frac{1}{n_1^2} - \frac{1}{n_2^2}\right) \text{nm}^{-1}$$

Expt: $\lambda^{-1} = 0.038\,103\,2 \text{ nm}^{-1}$
$= 0.051\,439\,3 \text{ nm}^{-1}$
$= 0.057\,611\,0 \text{ nm}^{-1}$
$= 0.060\,967\,52 \text{ nm}^{-1}$

By trial and error substitution or by graphing these numbers versus $\frac{1}{(n_1 + m)^2}$ for $m = 1, 2, 3, \cdots$ and looking for the intercept for $n_1 + m = \infty$ gives $\frac{Z^2}{n_1^2}$; the slope gives Z^2. This leads to $Z = 5$, $n_1 = 2$ and

$$\lambda^{-1} = 25(0.010\ 973\ 732)\left(\frac{1}{4} - \frac{1}{n_2^2}\right)$$

$n_2 = 2 + m,\ m = 1,2,\cdots$

10.8 Calculate the wavelengths in μm of the first three lines of the Paschen series for atomic hydrogen.

SOLUTION

$$\lambda = \frac{1}{R}\frac{n_1^2 n_2^2}{(n_2^2 - n_1^2)} = \frac{1}{R}\frac{9\, n_2^2}{(n_2^2 - 9)}$$

For $n_2 = 4$ $\quad \lambda = \dfrac{(9)(16)}{(109677.58\ \text{cm}^{-1})(7)} = 1.8756\ \mu\text{m}$

For $n_2 = 5$ $\quad \lambda = \dfrac{(9)(25)}{(109677.58\ \text{cm}^{-1})(16)} = 1.2822\ \mu\text{m}$

For $n_2 = 6$ $\quad \lambda = \dfrac{(9)(36)}{(109677.58\ \text{cm}^{-1})(27)} = 1.0941\ \mu\text{m}$

10.9 Calculate the wavelength of light emitted when an electron falls from the n = 100 orbit to the n = 99 orbit of the hydrogen atom. Such species are known as high Rydberg atoms. They are detected in astronomy and are more and more studied in the laboratory.

SOLUTION

$$E_{100} - E_{99} = -\ (1.0967 \text{ x } 10^7\ \text{m}^{-1})(10^{-2}\ \text{m cm}^{-1})\left(\frac{1}{100^2} - \frac{1}{99^2}\right)$$

$$= 0.2227\ \text{cm}^{-1}$$

$\lambda = 1/\tilde{\nu} = 4.49$ cm

10.10 Calculate the Rydberg constant R_H for a hydrogen atom and the Rydberg constant R_D for a deuterium atom given the value for R_∞ given in the Appendix.

SOLUTION

R is proportional to the reduced mass,

$$\mu = \frac{m_{nucl} m_e}{m_{nucl} + m_e}$$

where m_{nucl} is the mass of the nucleus and m_e is the mass of the electron. If the nucleus is infinitely massive, $\mu = m_e$. Therefore,

$$\frac{R_H}{R_\infty} = \frac{m_H m_e}{(m_H + m_e)m_e}$$

and $$R_H = \frac{m_H R_\infty}{m_H + m_e}$$

$$= \frac{(1.672\,623\,1 \text{ x } 10^{-27}\text{ kg})(10\,973\,731.534\text{ m}^{-1})}{(1.672\,623\,1 \text{ x } 10^{-27} + 9.109 \text{ x } 10^{-31})\text{ kg}}$$

$$= 10\,967\,758.56\text{ m}^{-1}$$

For deuterium,

$$R_D = \frac{(3.343\,586\,0 \text{ x } 10^{-27}\text{ kg})R_\infty}{3.44\,4969 \text{ x } 10^{-27}\text{ kg}}$$

$$= 10\,970\,742.75\text{ m}^{-1}$$

10.11 Check the normalization of the hydrogen atomic wavefunction ψ_{100}.

SOLUTION

$$\int_0^\infty \mathrm{d}r\, r^2 \int_0^\pi \mathrm{d}\theta \sin\theta \int_0^{2\pi} \mathrm{d}\phi\, \psi_{100}^*\, \psi_{100} = \int_0^{2\pi} \mathrm{d}\phi \int_0^\pi \mathrm{d}\theta \sin\theta \int_0^\infty \mathrm{d}r\, r^2 \left[\frac{1}{\pi}\left(\frac{Z}{a_0}\right)^3 \mathrm{e}^{-2\sigma}\right]$$

$$= 4\left(\frac{Z}{a_0}\right)^3 \int_0^\infty \mathrm{d}r\, \mathrm{e}^{-2\sigma} r^2$$

$$= 4\int_0^\infty \mathrm{e}^{-2\sigma}\sigma^2\, \mathrm{d}\sigma = 1$$

since $\sigma = Zr/a_0$, so that $\mathrm{d}r = a_0\mathrm{d}\sigma/Z$ and $r^2 = a_0^2\sigma^2/Z^2$. The last integral is given in Appendix D.2.

10.12 Show that ψ_{100} and ψ_{200} for the hydrogen atom are orthogonal.

SOLUTION

$$\int_0^\infty \mathrm{d}r\, r^2 \int_0^\pi \mathrm{d}\theta \sin\theta \int_0^{2\pi} \mathrm{d}\phi\, \psi_{100}^*\, \psi_{200}$$

$$= \int_0^{2\pi} d\phi \int_0^{\pi} d\theta \sin\theta \int_0^{\infty} dr\, r^2 \left[\frac{1}{\pi^{1/2}}\left(\frac{Z}{a_0}\right)^{3/2} e^{-\sigma}\right]\left[\frac{1}{(32\pi)^{1/2}}\left(\frac{Z}{a_0}\right)^{3/2}(2-\sigma)\, e^{-\sigma/2}\right]$$

$$= \frac{1}{2^{1/2}}\left(\frac{Z}{a_0}\right)^3 \int_0^{\infty} dr\, r^2 (2-\sigma)\, e^{-3\sigma/2}$$

Since $\sigma = Zr/a_0$, $dr = a_0 d\sigma/Z$, and $r^2 = a_0^2\sigma^2/Z^2$ so that this expression becomes

$$= \frac{1}{2^{1/2}} \int_0^{\infty} (2\sigma^2 - \sigma^3)\, e^{-3\sigma/2}\, d\sigma = 0$$

The last two integrals are given in Appendix D.2.

10.13 In a hydrogen-like atom in the 1s state, there is a difference between the average distance $< r >$ between the electron and the nucleus and the most probable distance between the electron and the nucleus. (a) Derive the expression for the average distance $< r >$ between the electron and the nucleus. (b) Derive the expression for the most probable distance r_{mp}.

<u>SOLUTION</u>

(a) The average distance can be calculated like any average property.

$$< r > = \int_0^{\infty} \psi^* r \psi\, 4\pi r^2 dr = \int_0^{\infty} dr\, 4\pi r^3 \frac{Z^3}{\pi a_0^3}\, e^{-2Zr/a_0} = \frac{3}{2}\frac{a_0}{Z}$$

Integration over the ϕ and θ parts of the volume element yields 4π.

(b) The most probable value of r is the value that yields the maximum value of $4\pi r^2\psi^2$.

$$\frac{d(4\pi r^2\psi^2)}{dr} = 0$$

This yields $r_{mp} = a_0/Z$.

10.14 The spin functions α and β cannot be expressed in terms of spherical harmonics, but they can be represented as column matrices:

$$\alpha = \begin{bmatrix} 1 \\ 0 \end{bmatrix} \text{ and } \beta = \begin{bmatrix} 0 \\ 1 \end{bmatrix}$$

The spin operator can be represented by the following Pauli matrix;

$$\hat{S}_z = \frac{1}{2}\begin{bmatrix} 1 & 0 \\ 0 & -1 \end{bmatrix}$$

Show that $\hat{S}_z\alpha = \frac{1}{2}\alpha$ and $\hat{S}_z\beta = -\frac{1}{2}\beta$.

SOLUTION

If you are not familiar with matrix multiplication, see the Appendix.

$$\frac{1}{2}\begin{bmatrix} 1 & 0 \\ 0 & -1 \end{bmatrix}\begin{bmatrix} 1 \\ 0 \end{bmatrix} = \frac{1}{2}\begin{bmatrix} 1 \\ 0 \end{bmatrix}$$

$$\frac{1}{2}\begin{bmatrix} 1 & 0 \\ 0 & -1 \end{bmatrix}\begin{bmatrix} 0 \\ 1 \end{bmatrix} = \frac{1}{2}\begin{bmatrix} 0 \\ -1 \end{bmatrix} = -\frac{1}{2}\begin{bmatrix} 0 \\ 1 \end{bmatrix}$$

10.15 The spin parts of the wavefunctions for the first excited state of helium are $\psi_a = \alpha(1)\alpha(2)$, $\psi_b = \alpha(1)\beta(2) + \alpha(2)\beta(1)$, and $\psi_c = \beta(1)\beta(2)$. Since $\hat{S}_z\psi = \hbar M_S\psi$, what are the M_S values for the three wavefunctions?

SOLUTION

Note that

$\hat{S}_z = \hat{S}_{z1} + \hat{S}_{z2}$,

$\hat{S}_z\alpha = +\frac{1}{2}\hbar\,\alpha$,

and $\hat{S}_z\beta = -\frac{1}{2}\hbar\,\beta$

Therefore,

$\hat{S}_z\,\psi_a = (\hat{S}_{z1} + \hat{S}_{z2})[\alpha(1)\alpha(2)] = \frac{1}{2}\hbar\,\alpha(1)\alpha(2) + \frac{1}{2}\hbar\,\alpha(1)\alpha(2) = \hbar\,\alpha(1)\alpha(2)$ so that $M_S = 1$

$\hat{S}_z\,\psi_b = (\hat{S}_{z1} + \hat{S}_{z2})[\alpha(1)\beta(2) + \alpha(2)\beta(1)]$

$= \frac{1}{2}\hbar\,\alpha(1)\beta(2) - \frac{1}{2}\hbar\,\alpha(2)\beta(1) - \frac{1}{2}\hbar\,\alpha(1)\beta(2) + \frac{1}{2}\hbar\,\alpha(2)\beta(1) = 0$ so that $M_S = 0$

$\hat{S}_z\,\psi_c = (\hat{S}_{z1} + \hat{S}_{z2})[\beta(1)\beta(2)] = -\frac{1}{2}\hbar\,\beta(1)\beta(2) - \frac{1}{2}\hbar\,\beta(1)\beta(2) = -\hbar\,\beta(1)\beta(2)$ so that $M_S = -1$

10.16 What are the degeneracies of the following orbitals for hydrogen-like atoms? (a) $n = 1$, (b) $n = 2$, and (c) $n = 3$.

SOLUTION

(a) $n = 1 \quad l = 0 \quad m = 0 \quad m_s = \pm\ 1/2$
degeneracy = 2

(b) $n = 2 \quad l = 0 \quad m = 0 \quad m_s = \pm\ 1/2$
$l = 1 \quad m = -1, 0, +1 \quad m_s = \pm\ 1/2$
degeneracy = 8

(c) $n = 3 \quad l = 0 \quad m = 0 \quad m_s = \pm\ 1/2$
$l = 1 \quad m = -1, 0, +1 \quad m_s = \pm\ 1/2$
$l = 2 \quad m = -2, -1, 0, +1, +2 \quad m_s = \pm\ 1/2$
degeneracy = 18
[Note that the degeneracy is $2n^2$.]

10.17 Show that for a 1s orbital of a hydrogen-like atom the most probable distance from proton to electron is a_o/Z. Find the numerical values for C^{5+} and B^{3+}.

SOLUTION

$$\psi_{1s} \propto e^{-Zr/a_o}$$

Probability density of $p(r) \propto r^2 \psi_{1s}^2 = r^2 e^{-2Zr/a_o}$.
Setting the derivative of the probability density equal to zero,

$$r^2 e^{-2Zr/a_o}\left(\frac{-2Z}{a_o}\right) + 2re^{-Zr/a_o} = 0 \qquad r = \frac{a_o}{Z}$$

$$r(C^{5+}) = \frac{a_o}{5} = 105.8 \text{ pm}$$

$$r(B^{3+}) = \frac{a_o}{3} = 176.4 \text{ pm}$$

10.18 Find the values of r for which the hydrogenlike atom 2s and 3s wave functions are equal to zero (these are the radial nodes). Compare to Fig. 10.5.

SOLUTION

2s $$\psi_{2s} = \frac{1}{4\sqrt{2\pi}}\left(\frac{Z}{a_o}\right)^{1/2}\left(2 - \frac{Zr}{a_o}\right)e^{-Zr/a_o}$$

$$r_{NODE} = \frac{2a_o}{Z}$$

3s $$\psi_{3s} = \frac{1}{81\sqrt{3\pi}}\left(\frac{Z}{a_o}\right)^{1/2}\left[27 - 18\frac{Z}{a_o} + 2\left(\frac{Zr}{a_o}\right)^2\right]$$

$$\left(\frac{Zr}{a_o}\right)_{NODE} = \frac{1}{4}\left[18 \pm \sqrt{(18)^2 - 8(27)}\right]$$

$$r_{NODE} = 7.098\frac{a_o}{Z}, 1.902\frac{a_o}{Z}$$

10.19 What is the average distance from an orbital electron to the nucleus for a 2s and 2p electron in (a) H and (b) Li^{2+}?

SOLUTION

(a) $$<r>_{n,l} = \frac{n^2a_o}{Z}\left\{1 + \frac{1}{2}\left[1 - \frac{l(l+1)}{n^2}\right]\right\}$$

For H

$$<r>_{2,0} = \frac{2^2(52.92 \text{ pm})}{1} \cdot \frac{3}{2} = 317.5 \text{ pm}$$

$$<r>_{2,1} = \frac{2^2(52.92 \text{ pm})}{1}\left\{1 + \frac{1}{2}\left[1 - \frac{2}{4}\right]\right\} = 264.6 \text{ pm}$$

(b) For Li^{2+}

$$<r>_{2,0} = \frac{2^2(52.92 \text{ pm})}{3} \cdot \frac{3}{2} = 105.8 \text{ pm}$$

$$<r>_{2,1} = \frac{2^2(52.92 \text{ pm})}{3}\left\{1 + \frac{1}{2}\left[1 - \frac{2}{4}\right]\right\} = 88.2 \text{ pm}$$

10.20 Calculate the expectation value $<r>$ of the radius of a 2s orbital and a 2p orbital for a hydrogenlike atom. Is this the result that you expected?

SOLUTION

For a 2s orbital, equation 10.30 yields

$$<r>_{2,0} = \frac{2^2a_o}{Z}\left(1 + \frac{1}{2}\right) = 6a_o/Z$$

For a 2p orbital,

$$<r>_{2,1} = \frac{2^2a_o}{Z}\left[1 + \frac{1}{2}\left(1 - \frac{2}{4}\right)\right] = 5a_o/Z$$

Thus the average distance is a little less for the 2p orbital. How is this explained?

10.21 In the laboratory there is a limit to the number of lines that can be observed in the spectrum of a hydrogenlike atom because of pressure broadening. As atoms are excited to higher quantum numbers, their effective radii increase and because of crowding, excited atoms contact nearby atoms and do not act independently. However, in interstellar space emissions from hydrogen atoms at extremely low pressures with very high quantum numbers can be detected because of the large volumes per atom.

Radio astronomers have detected hydrogen atoms undergoing the transition $n = 253$ to 252 (D. B. Clark, *J. Chem. Ed.* 68, 454 (1991)). At what frequency and wavelength was this observation made and what is the expectation value for the radius of the emitting hydrogen atom?

SOLUTION

Equation 10.16 yields

$$\nu = (1.0968 \times 10^{-7}\ \mathrm{m^{-1}})(2.998 \times 10^{8}\ \mathrm{m\ s^{-1}})\left(\frac{1}{252^2} - \frac{1}{253^2}\right)$$

$$= 408.5\ \mathrm{MHz}$$

$$\lambda = (2.998\times10^{8}\ \mathrm{m\ s^{-1}})/(408.5 \times 10^{6}\ \mathrm{s^{-1}}) = 0.7339\ \mathrm{m}$$

Equation 10.30 yields

$$\langle r \rangle_{253,0} = 1.5n^2a_o = 1.5(253^2)(52.92 \times 10^{-12}\ \mathrm{m}) = 0.00508\ \mathrm{mm}$$

10.22 Calculate the expectation value of the distance between the nucleus and the electron of a hydrogen-like atom in the $2p_z$ state using equation 10.29. Show that the same result is obtained using equation 10.30.

SOLUTION

$$\langle r \rangle = \int \psi_{21}^{*}\ r\ \psi_{21}\mathrm{d}\tau = \int \psi_{21}^{2}\ r\ \mathrm{d}\tau$$

$$\psi_{21} = \frac{1}{\sqrt{4}\ 2\pi}\left(\frac{Z}{a_o}\right)^{3/2}\left(\frac{Zr}{a_o}\right)\cos\theta \exp(-Zr/2a_o)$$

$$\langle r \rangle = \frac{1}{16(2\pi)\left(\frac{Z}{a_o}\right)^5}\int_0^{2\pi}\int_0^{\pi}\int_0^{\infty} r^3\exp(-Zr/a_o)\cos^2\theta\, r^2\sin\theta\, \mathrm{d}\theta\mathrm{d}\phi\mathrm{d}r$$

The element of volume in spherical polar coordinates is

$\mathrm{d}\tau = r^2\sin\theta\, \mathrm{d}\theta\mathrm{d}\phi\mathrm{d}r$

and the ranges of the coordinates are

$0 < \theta < \pi \qquad 0 < \phi < 2\pi \qquad 0 < r < \infty$

as may be seen in Figure 9.8. When the integration is carried out using

$\int_0^{\infty} x^n e^{-bx}\mathrm{d}x = n!/b^{n+1}$ we obtain $\langle r \rangle = 5a_o/Z$

Equation 10.30 yields

$$\langle r \rangle = \frac{4a_o}{Z}\left[1 + \frac{1}{2}\left(1 - \frac{1}{2}\right)\right] = \frac{5a_o}{Z}$$

10.23 It can be shown that a linear combination of two eigenfunctions belonging to the same degenerate level is also an eigenfunction of the Hamiltonian with the same energy. In terms of mathematical formulas, if $\hat{H}\ {}_1\psi_1 = E_1\psi_1$ and $\hat{H}\ \psi_2 = E_1\psi_2$, then $\hat{H}(c_1\psi_1 + c_2\psi_2) = E_1(c_1\psi_1 + c_2\psi_2)$. The wave function $c_1\psi_1 + c_2\psi_2$ still needs to be normalized. Equation 10.8 yields the following expressions for the 2p eigenfunctions

$$\psi_{2p_{+1}} = b\ \mathrm{e}^{-Zr/2a}\ r \sin\theta\ \mathrm{e}^{\mathrm{i}\phi}$$

$$\psi_{2p_{-1}} = b\ \mathrm{e}^{-Zr/2a}\ r \sin\theta\ \mathrm{e}^{-\mathrm{i}\phi}$$

$$\psi_{2p_0} = c\ \mathrm{e}^{-Zr/2a}\ r \cos\theta$$

Use this information to find the real functions for the 2p orbitals that are given in Table 10.1. Given:

$$\mathrm{e}^{\mathrm{i}\phi} = \cos\phi + \mathrm{i}\sin\phi$$

$$\mathrm{e}^{-\mathrm{i}\phi} = \cos\phi - \mathrm{i}\sin\phi$$

SOLUTION

ψ_{2p_x} and ψ_{2p_y} are defined as the following linear combinations

$$\psi_{2p_x} = \frac{\psi_{2p_{+1}} + \psi_{2p_{-1}}}{\sqrt{2}}$$

$$\psi_{2p_x} = \frac{\psi_{2p_{+1}} + \psi_{2p_{-1}}}{\mathrm{i}\sqrt{2}}$$

Substituting in these expressions we obtain

$$\psi_{2p_x} = \frac{b\ \mathrm{e}^{-Zr/2a}\ r \sin\theta\ (\mathrm{e}^{\mathrm{i}\phi} + \mathrm{e}^{-\mathrm{i}\phi})}{\sqrt{2}}$$

$$= \frac{2b\ \mathrm{e}^{-Zr/2a}\ r \sin\theta \cos\phi}{\sqrt{2}}$$

$$\psi_{2p_y} = \frac{b\ \mathrm{e}^{-Zr/2a}\ r \sin\theta\ (\mathrm{e}^{\mathrm{i}\phi} - \mathrm{e}^{-\mathrm{i}\phi})}{\mathrm{i}\sqrt{2}}$$

$$= \frac{2b\ \mathrm{e}^{-Zr/2a}\ r \sin\theta \sin\phi}{\sqrt{2}}$$

10.24 For a hydrogen-like atom, what is the magnitude of the orbital angular momentum, and what are the possible values of L_z for electrons in the 2p and 3d orbitals?

SOLUTION

2p:

$n = 2 \quad l = 1 \quad m = -1, 0, +1$

$$L = [l(l+1)]^{1/2}\ \hbar$$

$$L = \frac{2^{1/2}(6.626x10^{-34}\ \text{J s})}{2\pi} = 1.491x10^{-14}\ \text{J s}$$

$$L_z = m\hbar = \frac{m(6.626x10^{-34}\ \text{J s})}{2\pi}$$

$$= \begin{cases} +1.054\text{x}10^{-34}\ \text{J s} \\ 0 \\ -1.054\text{x}10^{-34}\ \text{J s} \end{cases}$$

3d: $\quad n = 3 \quad l = 2 \quad m = -2, -1, 0, +1, +2$

$$L = \sqrt{6}\ \hbar = 2.583 \text{ x } 10^{-34} \text{ J s}$$

$$L_z = \begin{cases} -2\hbar = -2.109 \text{ x } 10^{-34} \text{ J s} \\ -1\hbar = -1.055 \text{ x } 10^{-34} \text{ J s} \\ 0\hbar = 0 \\ +1\hbar = 1.055 \text{ x } 10^{-34} \text{ J s} \\ +2\hbar = 2.109 \text{ x } 10^{-34} \text{ J s} \end{cases}$$

10.25 What is the magnitude of the angular momentum for electrons in 3s, 3p and 3d orbitals? How many radial and angular nodes are there for each of these orbitals?

SOLUTION

	3s	3p	3d
$L = \sqrt{l(l+1)}\ \hbar$	0	$\sqrt{2}\hbar$	$\sqrt{6}\hbar$
Radial nodes	2	1	0
Angular nodes	0	1	2

10.26 How many angular, radial, and total nodes are there for the following hydrogenlike wave functions? (a) 1s, (b) 2s, (c) 2p, (d) 3p, and (e) 3d.

SOLUTION

The number of angular nodes is l.

The number of radial nodes is $n - l - 1$

The total number of nodes is $n - 1$.

	1s	2s	2p	3p	3d
n	1	2	2	3	3
l	0	0	1	1	2
angular	0	0	1	1	2
radial	0	1	0	1	0
total	0	1	1	2	2

10.27 The antisymmetric spin function for 2 electrons is $N[\alpha(1)\beta(2) - \alpha(2)\beta(1)]$. Derive the value for the normalization constant N.

SOLUTION

Normalization requires that

$$1 = N^2 \int [\alpha(1)\beta(2) - \alpha(2)\beta(1)]^2 d\sigma$$

$$= N^2 \int \{[\alpha(1)\beta(2)]^2 - 2[\alpha(1)\beta(2)\alpha(2)\beta(1)] + [\alpha(2)\beta(1)]^2\} d\sigma$$

The first and last terms in the integral each contribute 1 because

$$\int \alpha^* \alpha d\sigma = \int \beta^* \beta d\sigma = 1$$

The middle term is equal to zero because

$$\int \alpha^* \beta d\sigma = \int \alpha \beta^* d\sigma = 0$$

Thus $1 = 2N^2$, and $N = 2^{-1/2}$.

10.28 Using equation 10.55, calculate the difference in energy between the two spin angular momentum states of a hydrogen atom in the 1s orbital in a magnetic field of 1 Tesla. What is the wavelength of radiation emitted when the electron spin "flips"? In what region of the electromagnetic spectrum is this?

SOLUTION

$$\Delta E = g_e \mu_B B = (2.0023)\left(\frac{e\hbar}{2m_e}\right)(1\ \text{T})$$

$$= (2.0023)(9.274 \times 10^{-24}\ \text{J T}^{-1})(1\ \text{T})$$

$$= 1.8569 \times 10^{-23}\ \text{J}$$

$$\lambda = \frac{hc}{\Delta E} = \frac{6.626 \times 10^{-34} \times 2.998 \times 10^{8}}{1.8569 \times 10^{-23}}\ \text{m} = 1.0698 \times 10^{-2}\ \text{m} = 1.07\ \text{cm}$$

Microwave

10.29 For the wavefunction

$$\psi = \begin{vmatrix} \psi_A(1) & \psi_A(2) \\ \psi_B(1) & \psi_B(2) \end{vmatrix}$$

show that (a) the interchange of two columns changes the sign of the wave function, (b) the interchange of two rows changes the sign of the wave function, and (c) the two electrons cannot have the same spin orbital.

SOLUTION

(a) The interchange of two columns yields

$$\begin{vmatrix} \psi_A(2) & \psi_A(1) \\ \psi_B(2) & \psi_B(1) \end{vmatrix} = \psi_A(2)\ \psi_B(1) - \psi_A(1)\ \psi_B(2)$$

This is the wavefunction for the atom with the two electrons interchanged, and it is the negative of the function given in the problem, as required by the anti-symmetrization principle.

(b) The interchange of two rows yields

$$\psi = \begin{vmatrix} \psi_B(1) & \psi_B(2) \\ \psi_A(1) & \psi_A(2) \end{vmatrix} = \psi_A(2)\ \psi_B(1) - \psi_A(1)\ \psi_B(2)$$

which is the negative of the wave function given in the problem. Thus the interchange of two columns or two rows is equivalent to the interchange of two electrons between orbitals.

(c) If two electrons are represented by the same spin orbitals, the determinant

$$\begin{vmatrix} \psi_A(1) & \psi_A(2) \\ \psi_A(1) & \psi_A(2) \end{vmatrix} = 0$$

is in accord with the principle that two electrons in an atom cannot have four identical quantum numbers.

10.30 What are the electron configurations for H^-, Li^+, O^{2-}, F^-, Na^+, and Mg^{2+}?

SOLUTION

$1s^2$, $1s^2$, $1s^22s^22p^6$, $1s^22s^22p^6$, $1s^22s^22p^6$, $1s^22s^22p^6$

10.31 How many electrons can enter the following sets of atomic orbitals:

1s, 2s, 2p, 3s, 3p, 3d?

SOLUTION

There are $2l + 1$ values of m for n given l, and each of these orbitals may be occupied by 2 electrons so $2(2l + 1)$ electrons can enter a set of orbitals.

1s: $l = 0$, $2(2l + 1) = 2$

2s: $l = 0$, $2(2l + 1) = 2$

2p: $l = 1$, $2(2l + 1) = 6$

3s: $l = 0$, $2(2l + 1) = 2$

3p: $l = 1$, $2(2l + 1) = 6$

3d: $l = 2$, $2(2l + 1) = 10$

10.32 $2.740\ 03 \times 10^{14}\ s^{-1}$, $1.094\ 11 \times 10^{-4}$ cm

10.33 $4.052\ 27 \times 10^{-4}$, $2.625\ 87 \times 10^{-4}$ nm

10.34 3.40 V, 1.259

10.35 1.84, 2.26, 2.777

10.36 $-2.903\ E_h$, $-7.478\ E_h$, $-14.668\ E_h$

10.38 0.7542 eV

10.39 ^3P_o

10.40 (b) and (d)

10.41 (a) 3p, 4p,... (b) 4s, 5s,... or 4d, 5d,... (c) 4p, 5p,... or 4f, 5f,....

10.43 $k = \frac{1}{\pi^{1/2}}\left(\frac{Z}{a_o}\right)^{3/2}$

10.44 13.598 V

10.45 -2.626 kJ mol^{-1}. This is half the classical value.

10.46 $n_1 \approx 10$

10.48 656.172 12, 656.163 58 nm

10.49 1.98×10^{-7} m

10.50 $6\ a_o$

10.51 (a) 243 nm, (b) 6.8 eV

10.52 $\theta = 0, \pi$, $r = 2a_o$ are the most probable positions

10.53 $n \approx 8$

10.56 17.639 pm

10.57 0.7144, 0.6615, 0.5557 nm

10.58 $\Delta E = 5.788 \times 10^{-4}$ eV

$kT = 0.0259$ eV

10.59

	4s	4p	4d	4f
$\lvert L \rvert / \hbar$	0	$\sqrt{2}$	$\sqrt{6}$	$\sqrt{12}$

radial nodes	3	2	1	0
angular nodes	0	1	2	3

10.63 Sc [Ar]$3d4s^2$

Sc^+ [Ar]3d4s

Sc^{2+} [Ar]3d

Sc^{3+} [Ar] = $1s^22s^22p^63s^23p^6$

11

Molecular Electronic Structure

11.1 Given that the equilibrium distance in H_2^+ is 106 pm and that of H_2 is 74.1 pm, calculate the internuclear repulsion energy in both cases at R_e. Using $D_e(H_2^+) = 2.79$ eV and $D_e(H_2) = 4.78$ eV, calculate E_{el} at R_e for both.

SOLUTION

$$H_2^+ : \frac{e^2}{4\pi\varepsilon_o R_e} = \frac{(1.602 \times 10^{-19})^2}{4\pi(8.854 \times 10^{-12})(1.06 \times 10^{-10})} = 2.176 \times 10^{-18} \text{ J} = 13.6 \text{ eV}$$

$$H_2: \frac{e^2}{4\pi\varepsilon_o R_e} = 19.4 \text{ eV}$$

Given: $D_e(H_2^+) = 2.79 \text{ eV} = E_{H_2^+}(\infty) - E_{H_2^+}(R_e)$

$$\therefore E_{H_2^+}(R_e) = E_{H_2^+}(\infty) - 2.79 \text{ eV}$$

$$E_{el,H_2^+}(\infty) = -13.6 \text{ eV}$$

and,

$$E_{el,H_2^+}(R_e) = -13.6 \text{ eV} + E_{el,H_2^+}(\infty) - 2.79 \text{ eV} = -30.0 \text{ eV}$$

Similarly,

$$E_{el,H_2}(R_e) = E_{H_2}(\infty) - 4.78 - 19.4 \text{ eV} = -51.4 \text{ eV}$$

11.2 Given the equilibrium dissociation energy D_e for N_2 in Table 11.2 and the fundamental vibration frequency 2331 cm^{-1}, calculate the spectroscopic dissociation energy D_o in kJ mol^{-1}.

SOLUTION

$$\frac{1}{2}h\nu_o = \frac{(6.626 \times 10^{-34}\text{ J s})(2.998 \times 10^8\text{ m s}^{-1})(2.331 \times 10^5\text{ m}^{-1})}{2(1000\text{ J kJ}^{-1})} \times (6.022 \times 10^{23}\text{ mol}^{-1})$$

$$= 13.93\text{ kJ mol}^{-1}$$

$$D_o = D_e - \frac{1}{2}h\nu_o = 955.42 - 13.93 = 941.49\text{ kJ mol}^{-1}$$

This is the thermodynamic dissociation energy at absolute zero. The standard change in enthalpy for dissociation at 0 K calculated from Appendix A.3 is 941.57 kJ mol^{-1}.

11.3 Derive the values of the normalization constants given in equations 11.11 and 11.12.

SOLUTION

$$\psi_g = C(1s_A + 1s_B)$$

$$\int \psi_g^2\, d\tau = C^2 \int \left[1s_A^2 + 2(1s_A 1s_B) + 1s_B^2\right] d\tau$$

$$1 = C^2\left[1 + 2\int 1s_A 1s_B\, d\tau + 1\right]$$

$$C = \frac{1}{(2 + 2S)^{1/2}}, \text{ where } S = \int 1s_A 1s_B\, d\tau$$

$$\psi_u = C'(1s_A - 1s_B)$$

$$\int \psi_u^2\, d\tau = (C')^2 \int \left[1s_A^2 - (2 1s_A 1s_B) + 1s_B^2\right] d\tau$$

$$= (C')^2\,[2 - 2S]$$

$$C' = \frac{1}{(2 - 2S)^{1/2}}$$

11.4 Plot ψ_g and ψ_u versus distance along the internuclear axis for H_2^+ in the ground state without worrying about normalization (R_e = 106 pm).

SOLUTION

The values of these wave functions along the line through the two nuclei are given by

$$\psi_g = C\left(e^{-z/a_o} + e^{-(z - R_e)/a_o}\right)$$

$$\psi_u = C'\left(e^{-z/a_o} - e^{-(z - R_e)/a_o}\right)$$

where z is the distance measured from nucleus A toward nucleus B, and C and C' are the normalization constants that we are going to ignore.

11.5 The overlap integral S for the H_2^+ molecule can be evaluated as

$$S = [1 + R/a_0 + R^2/3a_0^2]\, e^{-R/a_0}$$

Plot this as a function of R/a_0. At what value of R is it a maximum? At what value of R is it a minimum? What is the value of S at the equilibrium separation, $R = 106$ pm?

SOLUTION

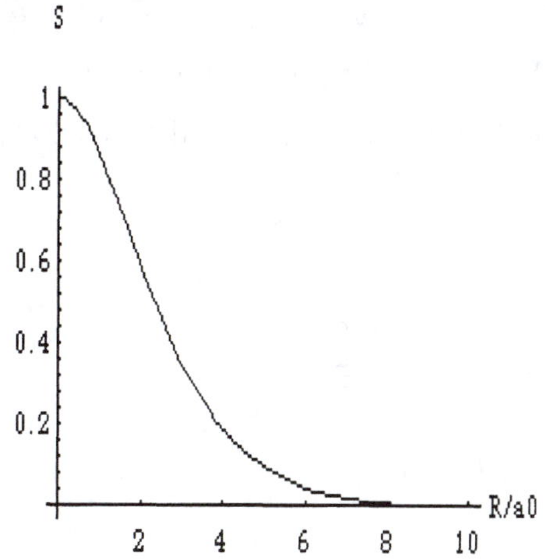

Maximum at $R = 0$

Minimum at $R = \infty$

$S(106 \text{ pm}) = 0.766$

11.6 Express the four valence bond wave functions for H_2 as Slater determinants.

SOLUTION

$$\psi_1 = \frac{1}{2^{1/2}} \begin{vmatrix} 1s_A(1)\alpha(1) & 1s_A(2)\alpha(2) \\ 1s_B(1)\beta(1) & 1s_B(2)\beta(2) \end{vmatrix} - \frac{1}{2^{1/2}} \begin{vmatrix} 1s_A(1)\beta(1) & 1s_A(2)\beta(2) \\ 1s_B(1)\alpha(1) & 1s_B(2)\alpha(2) \end{vmatrix}$$

$$\psi_2 = \frac{1}{2^{1/2}} \begin{vmatrix} 1s_A(1)\alpha(1) & 1s_A(2)\alpha(2) \\ 1s_B(1)\alpha(1) & 1s_B(2)\alpha(2) \end{vmatrix}$$

ψ_3 is the same as ψ_2 with β replacing α. ψ_4 is obtained from ψ_1 by replacing the minus sign with a plus sign.

11.7 Show that the hybrid orbitals for tetravalent carbon given by equations 11.57 to 11.60 are orthogonal.

SOLUTION

$\int \psi_{sp^3}(i) \int \psi_{sp^3}(ii)\, d\tau$

$$=\frac{1}{4}\int(\psi_s+\psi_{px}+\psi_{py}+\psi_{pz})(\psi_s-\psi_{px}-\psi_{py}-\psi_{pz})\,d\tau$$

Cross terms do not contribute since atomic orbitals are orthogonal.

$$=\frac{1}{4}\int(\psi_s^2+\psi_{p_x}^2-\psi_{p_y}^2-\psi_{p_z}^2)\,d\tau$$

$$=\frac{1}{4}(1+1-1-1)=0$$

since orbitals are normalized.

11.8 Using the hydrogenlike orbitals of Table 10.1, write out the sp^2 orbitals of equations 11.54-11.56. For $\psi_{sp^2(i)}$, write the wavefunction as a function of r for $\theta = 0$ (along the positive z axis), $\theta = 90^\circ$ (in the xy plane), and $\theta = 180^\circ$ (along the negative z axis).

SOLUTION

e.g., $$\psi_{sp^2(i)}=\frac{1}{4\sqrt{2\pi}}\,[(2-\sigma)+\sqrt{2}\,\sigma\cos\theta]\frac{e^{-\sigma}}{\sqrt{3}}$$

$\theta = 0^\circ$ $$\psi=\frac{1}{4\sqrt{6\pi}}\,[2+0.414\,\sigma]\,e^{-\sigma}$$

$\theta = 90^\circ$ $$\psi=\frac{1}{4\sqrt{6\pi}}\,[2-\sigma]\,e^{-\sigma}$$

$\theta = 180^\circ$ $$\psi=\frac{1}{4\sqrt{6\pi}}\,[2-2.414\,\sigma]\,e^{-\sigma}$$

11.9 Show that the $\psi_{sp^2(i)}$ orbital has been normalized.

SOLUTION

$$\int\psi_{sp^2(i)}^2d\tau=\int\left[\frac{1}{3^{.5}}\,2s+\left(\frac{2}{3}\right)^{.5}2p_z\right]^2d\tau$$

$$=\int\left[\frac{1}{3}(2s)^2+\frac{2\,2^{.5}}{3}\,2s\,2p+\left(\frac{2}{3}\right)(2p_z)^2\right]d\tau$$

$$=\frac{1}{3}+\frac{2}{3}=1$$

since 2s and 2p are normalized and are orthogonal.

11.10 The solutions of equation 11.24 were given in Section 11.3. Actually solve the secular determinant to obtain these values. For H_2^+ the secular determinant can be written more simply as

$$\begin{vmatrix}\alpha-E & \beta-ES\\ \beta-ES & \alpha-E\end{vmatrix}=0$$

SOLUTION

The determinant yields

$(\alpha - E)^2 - (\beta - ES)^2 = 0$

Since $x^2 - y^2 = (x + y)(x - y)$, this can be written

$(\alpha - E + \beta - ES)(\alpha - E - \beta + ES) = 0$

The following two values of E are solutions:

$$E = \frac{\alpha + \beta}{1 + S} \text{ and } E = \frac{\alpha - \beta}{1 - S}$$

11.11 Bond order for a diatomic molecule can be defined by $\frac{1}{2}(N - N^*)$, where N is the number of electrons in bonding molecular orbitals and N* is the number of electrons in antibonding orbitals. Calculate the bond orders of H_2^+, N_2^+, N_2, and O_2.

SOLUTION

H_2^+: $\frac{1}{2}(1) = \frac{1}{2}$

N_2^+: $\frac{1}{2}(7 - 2) = 2.5$

N_2: $\frac{1}{2}(8 - 2) = 3$

O_2: $\frac{1}{2}(8 - 4) = 2$

11.12 Discuss the electronic structure of the methyl radical and the location of the unpaired electron.

SOLUTION

The CH_3 radical is planar with the 3 two-electron bonds due to the overlapping of the sp^2 orbitals of C with the s orbitals of H. The odd electron is in the remaining unhybridized p orbital, which is perpendicular to the plane.

11.13 How many electrons are involved in sigma and pi bonding orbitals in the following molecules: (a) ethylene, (b) ethane, (c) butadiene, (d) benzene ?

SOLUTION

(a) $10\sigma, 2\pi$ (b) $14\sigma, 0\pi$ (c) $18\sigma, 4\pi$ (d) $24\sigma, 6\pi$ (Not counting the 1s carbon orbitals.)

[σ electron pairs not labeled]

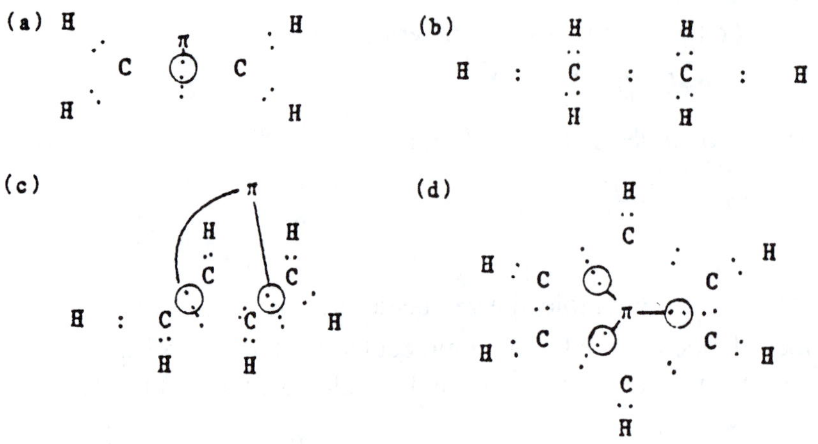

11.14 The heat of hydrogenation of cyclohexene is - 121 kJ mol^{-1}, and the heat of hydrogenation of benzene is - 209 kJ mol^{-1}. What is the reduction in the energy due to formation of the π bond system in benzene?

SOLUTION

If the molecule hexatriene existed, its heat of hydrogenation would be

expected to be 3(- 121 kJ mol^{-1}) = - 362 kJ mol^{-1}.

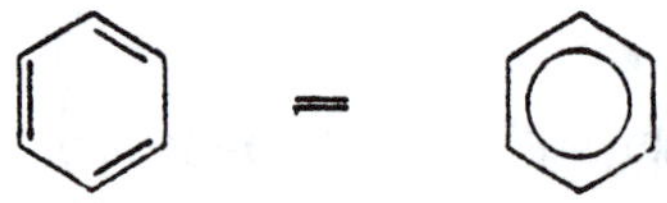

hexatriene benzene

The resonance stabilization is equal to the difference in heats of hydrogenation.

- 362 kJ mol^{-1} - (- 209 kJ mol^{-1}) = - 153 kJ mol^{-1}

11.15 Consider the Hückel molecular orbitals for butadiene given in equation 11.25. Each ϕ_i is an atomic p_z orbital on carbon atom i, so there is a nodal plane in the xy plane for each molecular orbital. There are other nodes in these orbitals as we move from atom 1 to atom 4 since the orbitals change sign. How many nodes are there for ψ_1 , ψ_2 , ψ_3 , and ψ_4? Where are they? Compare to Fig 11.21.

SOLUTION

ψ_1	no nodes	
ψ_2	1 node:	between atoms 2 and 3
ψ_3	2 nodes:	between 1 and 2 and between 3 and 4
ψ_4	3 nodes:	between 1 and 3, 2 and 3, and 3 and 4

11.16 In Hückel theory, the contribution to the electronic energy of a single π bond (see the discussion regarding ethylene in Section 11.7) is 2β. For butadiene, the contribution for two conjugated bonds is 4.472β, while for benzene (three conjugated bonds) it is 8β. What is the extra stabilization in butadiene and benzene due to the conjugation (in terms of β)? Using the data of problem 11.14, calculate the value of β for benzene and predict the value of the extra stabilization for butadiene.

SOLUTION

butadiene: $\Delta = 4.472\beta - 4\beta = 0.472\beta$
benzene: $\Delta = 8\beta - 6\beta = 2\beta$
$2\beta = -153$ kJ/mol from prob. 11.14
$\beta = -76.5$ kJ/mol
$\therefore \Delta_{\text{butadiene}} = 0.472(-76.5)$ kJ/mol $= -36.1$ kJ/mol

11.17 The delocalization energy of a conjugated molecule is the π electron energy minus the π electron energy for the corresponding amount of ethylene. Calculate the delocalization energies of 1,3-butadiene and benzene.

SOLUTION

The π electron energy of ethylene is $2\alpha + 2\beta$.
The π electron energy of 1,3-butadiene is $4\alpha + 4.472\beta$. Thus its delocalization energy is 0.472β.
The π electron energy of benzene is $6\alpha + 8\beta$. Thus its delocalization energy is 2β. This shows that the π electrons in benzene are more delocalized than in 1,3-butadiene.

*11.18 The Lennard-Jones parameters for nitrogen are $\varepsilon/k = 95.1$ K and $\sigma = 0.37$ pm. Plot the potential energy (expressed as V/k in K) for the interaction of two molecules of nitrogen.

SOLUTION

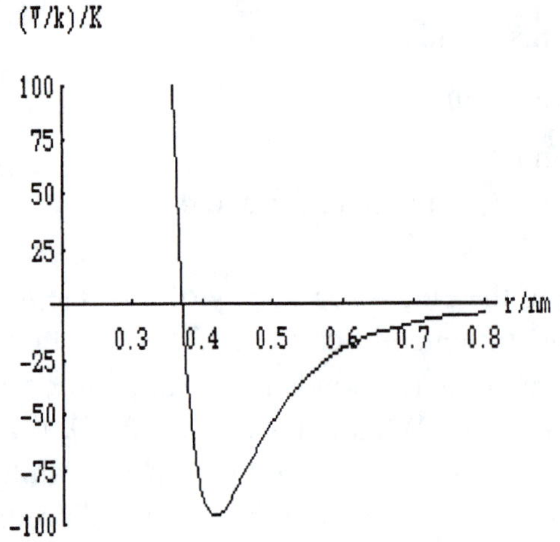

*11.19 For Ne the parameters of the Lennard-Jones 6-12 potential are $\varepsilon/k = 35.6$ K and $\sigma = 275$ pm. Plot V in J mol^{-1} versus R and calculate the distance R_m where dV/dr = 0.

SOLUTION

$r_m = 2^{1/6}\,\sigma = 2^{1/6}$ (0.275 nm) = 0.309 nm
ε = (8.314 J K^{-1} mol^{-1})(35.6 K) = 296.0 J mol^{-1}

$$V = 4\varepsilon\left[\left(\frac{\sigma}{R}\right)^{12} - \left(\frac{\sigma}{R}\right)^{6}\right]$$

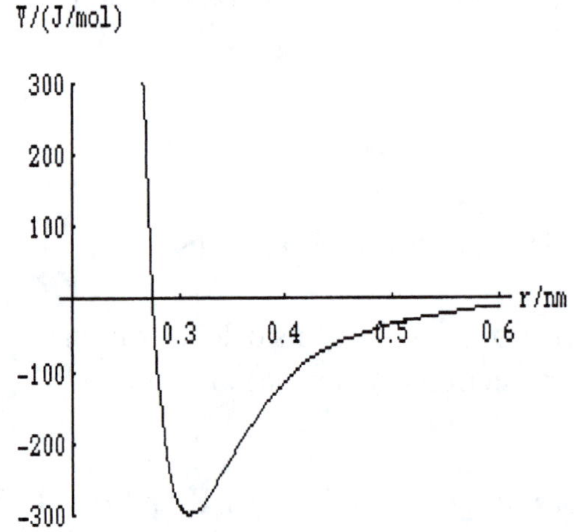

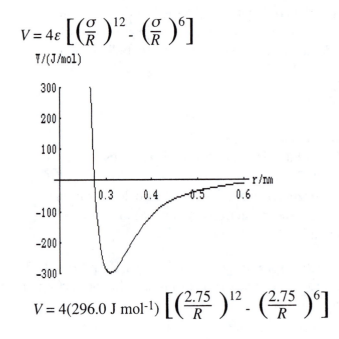

$$V = 4(296.0\text{ J mol}^{-1})\left[\left(\frac{2.75}{R}\right)^{12} - \left(\frac{2.75}{R}\right)^{6}\right]$$

11.20 For KF(g) the dissociation constant D_e is 5.18 eV and the dipole moment is 28.7 x 10^{-30} C m. Estimate these values assuming that the bonding is entirely ionic. The ionization potential K(g) is 4.34 eV, and the electron affinity of F(g) is 3.40 eV. The equilibrium internuclear distance in KF(g) is 0.217 nm.

SOLUTION

The work required to separate $K^+ F^-$ from the equilibrium internuclear distance to infinity is

$$\phi = \frac{Q_1 Q_2}{4\pi\varepsilon_0 R_e} = \frac{(1.602 \times 10^{-19}\ C)^2(0.8988 \times 10^{10}\ N\ m^2\ C^{-2})}{217 \times 10^{-12}\ m}$$

$$= 1.063 \times 10^{-18}\ J = \frac{1.063 \times 10^{-18}\ J}{1.602 \times 10^{-19}\ J\ eV^{-1}} = 6.64\ eV$$

Since KF actually dissociates into neutral atoms, the ionization potential of the metal has to be subtracted and the electron affinity of the nonmetal has to be added to calculate D_e:

$D_e = 6.64 - 4.34 + 3.40 = 5.70$ eV

This is higher than the experimental value (5.18 eV) because there is some repulsion of K^+ and F^- due to inner electron clouds. The dipole moment expected for the oversimplified structure $K^+ F^-$ is

$(1.602 \times 10^{-19}\ C)(217 \times 10^{-12}\ m) = 34.7 \times 10^{-30}$ C m

11.21 The equilibrium internuclear distance for NaCl(g) is 236.1 pm. What dipole moment is expected? The actual value is 3.003×10^{-29} C m. How do you explain the difference?

SOLUTION

$\mu = (1.602 \times 10^{-19}\ C)(0.2361 \times 10^{-9}\ m) = 3.783 \times 10^{-29}$ C m

The actual value is less because the ions polarize each other. The positive ion attracts the electrons of the negative ion toward itself, and the negative ion repels the electrons of the positive ion. The more polarizable the ions, the larger the effect is.

11.22 Calculate the dipole moment HCl would have if it consisted of a proton and a chloride ion (considered to be a point charge) separated by 127 pm (the internuclear distance obtained from the infrared spectrum). The experimental value is 3.44×10^{-30} C m. How do you explain the difference?

SOLUTION

$\mu = QR = (1.602 \times 10^{-19}\ C)(0.127 \times 10^{-9}\ m) = 20.4 \times 10^{-30}$ C m

$\mu_{expt} = 3.44 \times 10^{-30}$ C m

The charges are not completely separated in HCl.

11.23 The equilibrium distances in HCl, HBr, and HI are 127 pm, 141 pm, and 161 pm, respectively. Given the dipole moments in Table 22.2, find the effective fractional charges on the H and X ions. Is this in accord with the order of the electronegativities of the halogens?

SOLUTION

q = fractional charge

$\mu = qeR_e$

$$q_{HCl} = \frac{3.44 \times 10^{-30} \text{ C m}}{1.60 \times 10^{-19} \text{ C} \times 127 \times 10^{-12} \text{ m}} = 0.17$$

$q_{HBr} = 0.12$

$q_{HI} = 0.039$

This is in accord with the electronegativities.

11.24 Show that the dipole moment defined by equation 11.79 is independent of the location of the arbitrary origin if the net charge on the molecule is zero.

SOLUTION

If we select a Cartesian coordinate system with origin {0,0,0}, the dipole moment vector is given by

$$\boldsymbol{\mu} = \sum Q_a \boldsymbol{r}_a = \boldsymbol{i} \sum Q_a x_a + \boldsymbol{j} \sum Q_a y_a + \boldsymbol{k} \sum Q_a z_a = \boldsymbol{\mu}_x + \boldsymbol{\mu}_y + \boldsymbol{\mu}_z$$

If we move the origin to $\{x_1, y_1, z_1\}$, the x coordinate is changed to $x_{new} = x_a - x_1$. Substituting this in the expression for μ_x yields

$$\boldsymbol{\mu}_x = \boldsymbol{i} \sum Q_a(n_{new} + x_1) = \boldsymbol{i} \sum Q_a x_{new} + \boldsymbol{i} \sum Q_a x_1 = \mu_{x,new} + x_1\, \boldsymbol{i} \sum Q_a$$

If the net charge is zero, $\sum Q_a = 0$ so that $\mu_x = \mu_{x,new}$. The same applies to the other two coordinates, and so μ is independent of the location of the origin if the net charge is zero.

11.25 $1\sigma_g^2\ 1\sigma_u 2\sigma_g$

11.27

$$\begin{vmatrix} \alpha\text{-}E & \beta & 0 & \beta \\ \beta & \alpha-E & \beta & 0 \\ 0 & \beta & \alpha\text{-}E & \beta \\ \beta & 0 & \beta & \alpha\text{-}E \end{vmatrix} = 0$$

$E = \alpha + 2\beta,\ \alpha,\ \alpha,\ \alpha - 2\beta$

$E_\pi = \ 2(\alpha + 2\beta) + 2\alpha = 4\alpha + 4\beta$ No extra stabilization

11.28 (a) Ground state is $\psi_1^2\ \psi_2^2$.

First excited state is $\psi_1^2\ \psi_2\psi_3$.

$\Delta E = 1.49 \times 10^{-18}$ J, $\lambda = 133$ nm.

(b) Ground state is $\psi_1^2\ \psi_2^2\ \psi_3^2$.

First excited state is $\psi_1^2\ \psi_2^2\ \psi_3\psi_4$.

$\Delta E = 7.50 \times 10^{-19}$ J, $\lambda = 265$ nm.

11.29 Li_2^+	$1\sigma_g^2\ 1\sigma_{7u}^2\ 2\sigma_g$	B.O. = 1/2
Be_2^+	$1\sigma_g^2\ 1\sigma_{7u}^2\ 2\sigma_g^2 2\sigma_u$	B.O. = 1/2
B_2^+	$1\sigma_g^2\ 1\sigma_{7u}^2\ 2\sigma_g^2 2\sigma_u^2\ 3\sigma_g$	B.O. = 1/2
N_2^+	$1\sigma_g^2\ 1\sigma_{7u}^2\ 2\sigma_g^2 2\sigma_u^2\ 3\sigma_g^2\ 1p_u^3$	B.O. = 5/2

11.30 14.9%

11.31

$$\begin{vmatrix} \alpha-E & \beta_1 & 0 & \beta_2 \\ \beta_1 & \alpha-E & \beta_2 & 0 \\ 0 & \beta_2 & \alpha-E & \beta_1 \\ \beta_2 & 0 & \beta_1 & \alpha-E \end{vmatrix} = 0$$

$E_1 \approx \alpha + (1 + \gamma)\beta_1$, where $\gamma = \beta_2 / \beta_1$

$E_2 \approx \alpha + (1 - \gamma)\beta_1$

$E_3 \approx \alpha - (1 - \gamma)\beta_1$

$E_4 \approx \alpha - (1 + \gamma)\beta_1$

11.32 + ion: $\psi_1^2 \psi_2$

- ion: $\psi_1^2 \psi_2^2 \psi_3$

+ ion $|\psi_2|^2$: $(0.602)^2, (0.372)^2 < (-0.372)^2 < (-0.602)^2$

The unpaired electron densities are the same for the negative ion.

11.33 935 pm

11.34 31.4 eV

11.35 493.4 kJ mol^{-1} (Appendix C.3 gives 493.6 kJ mol^{-1})

11.38 787.38 kJ mol^{-1}

11.39

	Cl	Br	I
Electroneg.	3.0	2.8	2.5
μ	1.85	1.45	1.35

Because of the higher electronegativity of Cl, the Cl atom is more negative in HCl than the Br atom in HBr, causing the dipole moment of HCl to be larger than that of HBr.

11.41 429 pm

12

Symmetry

For problems 12.1-12.14, list the Schoenflies symbols and symmetry elements for each molecule.

12.1 H_2S

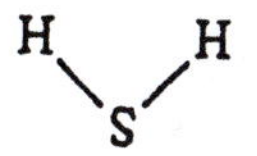

SOLUTION

C_{2v}: E, $C_2(z)$, $\sigma_v(xz)$, $\sigma_v'(yz)$

12.2 PCl_3

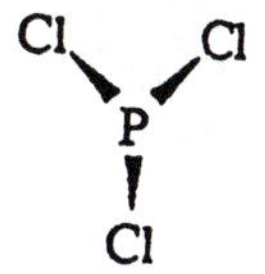

SOLUTION

C_{3v}: E, C_3, $3\sigma_v$

12.3 *trans*-$[CrBr_2(H_2O)_4]^+$ (ignoring the H's)

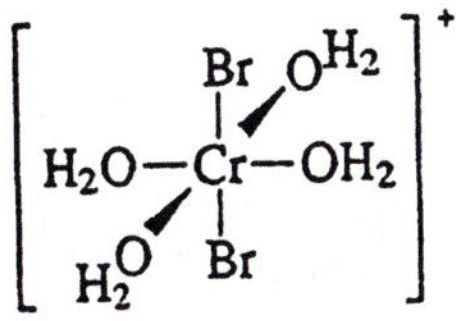

SOLUTION

D_{4h}: E, $C_4(z)$, $C_2(z)$, $S_4(z)$, $C_2'(x)$, $C_2'(y)$, $2C_2''$, i, σ_h, $\sigma_v(xz)$, $\sigma_v(yz)$, $2\sigma_d$

12.4 *gauche*-CH_2ClCH_2Cl

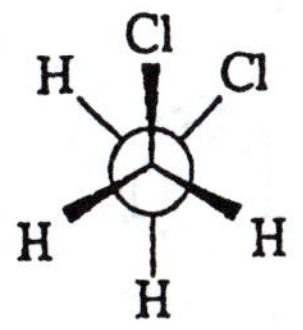

SOLUTION

C_2: E, C_2

12.5 1,3,5-tribromobenzene

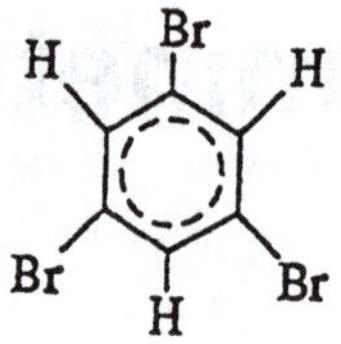

SOLUTION

C_{3h}: E, $C_3(z)$, $S_3(z)$, $3C_2$, σ_h, $3\sigma_v$

12.6 $CHClBr(CH_3)$

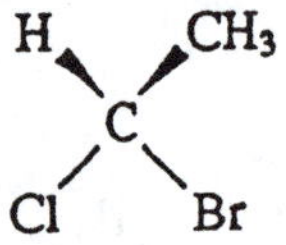

SOLUTION

C_1: E

12.7 IF_5

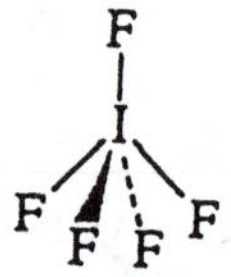

SOLUTION

C_{4v}: E, C_4, C_2, $2\sigma_v$, $2\sigma_d$

12.8 C_6H_{12}(cyclohexane)

SOLUTION

D_{3d}: E, C_3, $3C_2$, i, $3\sigma_d$, $2S_6$

12.9 B_2H_6

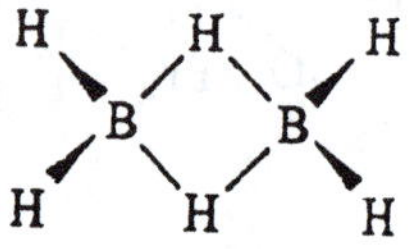

SOLUTION

D_{2h}: E, $C_2(z)$, $C_2(y)$, $C_2(x)$, i, $\sigma(xy)$, $\sigma(xz)$, $\sigma(yz)$

12.10 $C_{10}H_8$ (naphthalene)

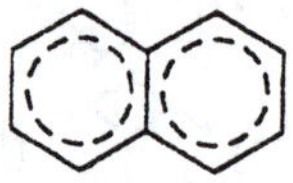

SOLUTION

D_{2h}: E, $C_2(z)$, $C_2(y)$, $C_2(x)$, i, $\sigma(xy)$, $\sigma(xz)$, $\sigma(yz)$

12.11 C_5H_8 (spiropentane)

(triangles ⊥ to each other)

SOLUTION

D_{2d}: E, $S_4(z)$, $C_2(z)$, $C_2'(x)$, $C_2'(y)$, $2\sigma_d$

12.12 C_4H_4S (thiophene)

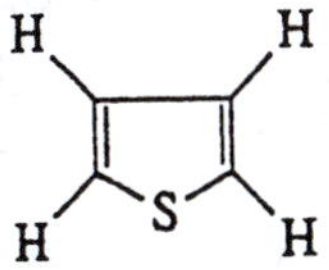

(planar)

SOLUTION

C_{2v}: E, $C_2(z)$, $\sigma_v(xz)$, $\sigma_v'(yz)$

12.13 *p*-dichlorobenzene $C_6H_4Cl_2$

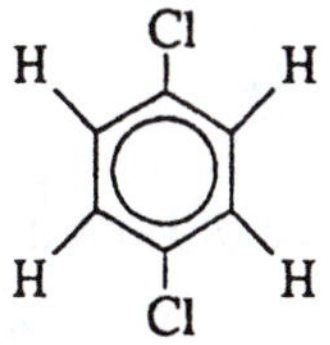

SOLUTION

D_{2h}: E, $C_2(z)$, $C_2(y)$, $C_2(x)$, i, $\sigma(xy)$, $\sigma(xz)$, $\sigma(yz)$

12.14 *trans*-CFClBrCFClBr

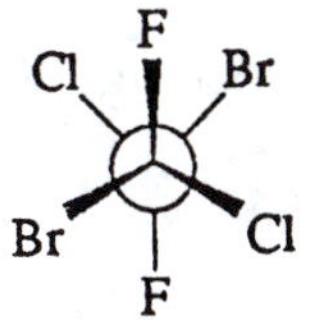

SOLUTION

C_i: E, i

12.15 The symmetry operations for the staggered form of ethane are given in Table 12.2, and it is in the D_{3d} point group. What are the operations for the eclipsed form of ethane (this is the sterically hindered form), and what is the point group?

SOLUTION

When the axis of the ethane molecule is vertical, it has a horizontal reflection plane σ_h. Figure 12.5 shows that this indicates the D_{3h} point group. Thus the point group changes when there is rotation about the C-C bond.

12.16 Construct the operator multiplication table for the point group C_{2h}.

SOLUTION

	E	C_2^1	σ_h	i
E	E	C_2^1	σ_h	i
C_2^1	C_2^1	E	i	σ_h
σ_h	σ_h	i	E	C_2^1
i	i	σ_h	C_2^1	E

12.17 List the operators associated with the S_6 elements and their equivalents, if any. How many distinct operations are produced?

SOLUTION

S_6^1	S_6^2	S_6^3	S_6^4	S_6^5	S_6^6
	C_3^1	i	C_3^2		E

Thus S_6 axis implies the existence of both a C_3 axis coincident with the S_6 axis and a center of symmetry (i). Only the S_6^1 and S_6^5 operations are distinct operations characteristic of only the S_6 axis.

12.18 Consider the three distinct isomers of dichloroethylene, $C_2H_2Cl_2$. To which symmetry group does each belong? Which can have a permanent dipole moment?

SOLUTION

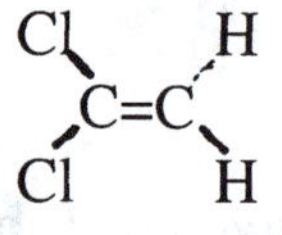

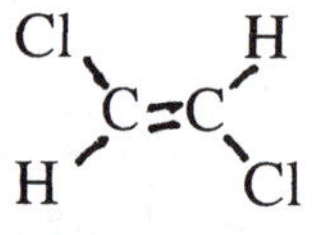

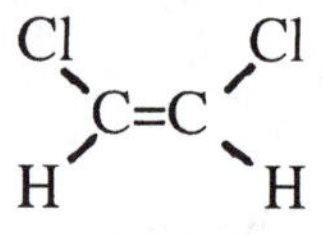

C_{2v} C_{2h} C_{2v}

dipole no dipole dipole

12.19 The first excited singlet state of ethylene is twisted so that the two hydrogens and carbon on one side are in a plane perpendicular to the plane containing the other three atoms. To which symmetry group does it belong? Does it have a dipole moment?

<u>SOLUTION</u>

D_{2d}, no dipole moment

12.20 Which of the molecules in problems 12.1 - 12.14 can have a permanent dipole moment?

<u>SOLUTION</u>

H_2S, PCl_3, $C_2H_4Cl_2$, CMeClBrH, IF_5, thiophene

12.21 Which of the molecules in problems 12.1 - 12.14 can be optically active?

<u>SOLUTION</u>

C(Me)ClBrH

12.22 Consider the three distinct isomers of dichlorobenzene. To which symmetry group does each belong? Which can have a dipole moment?

<u>SOLUTION</u>

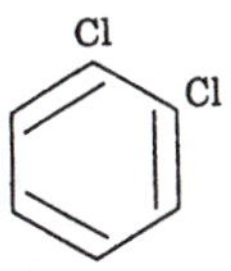

ortho
C_{2v}
dipole

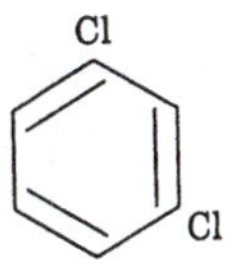

meta
C_{2v}
dipole

para
D_{2h}
no dipole

12.23 There are 10 distinct isomers of dichloronaphthalene, $C_{10}H_6Cl_2$. Two of them do not have a dipole moment. List these two and find the symmetry group to which each belongs.

<u>SOLUTION</u>

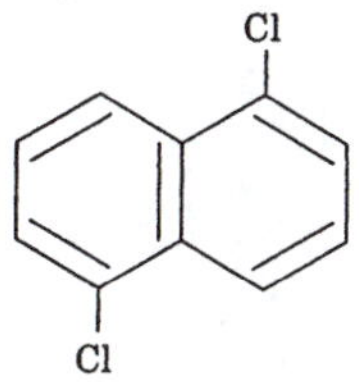

C_{2h}

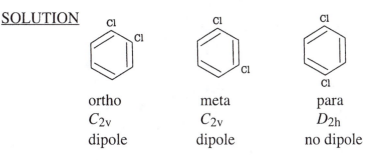

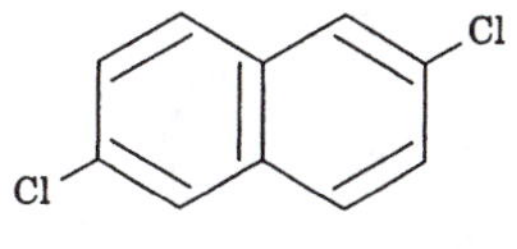

C_{2h}

12.24 Some of the excited electronic states of acetylene are cis-bent and some are trans-bent. What is the symmetry group of each of these structures? Cis-bent means that the hydrogens bend toward each other, while trans-bent means they bend away from one another.

SOLUTION

cis-bent
C_{2v}

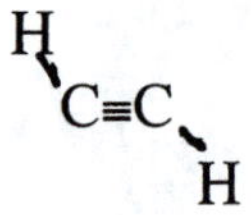

trans-bent
C_{2h}

12.25 D_{2d}: $E, 2S_4, C_2, 2\sigma_d, 2C_2^{1}$

12.26 C_s: E, σ

12.27 D_{5h}: $E, 5C_5, 5S_5, \sigma_h, 5\sigma_2, 5\sigma_v$

12.28 C_{2v}: $E, C_2, \sigma_v, \sigma_{v'}$

12.29 O_h: $E, 4C_3, 4S_6, 3C_4, 3S_4, 3C_2, 6C_2'$

12.30 D_{3h}: $E, 2C_3, 2S_3, \sigma_h, 3\sigma_v, 3C_2$

12.31 T_d: $E, 4C_3, 3S_4, 3C_2'$, $6\sigma_d$

12.32 C_{2v}: $E, C_2, \sigma_v, \sigma_{v'}$

12.33 C_s: E, σ

12.34 D_{3h}: $E, C_3, S_3, 3C_2, \sigma_h, 3\sigma_v$

12.35 C_3: E, C_3

12.36 C_{3v}: $E, C_3, 3\sigma_v$

12.37 T_d: $E, 4C_3, 3S_4, 3C_2, 6\sigma_d$

12.38 C_s: E, σ

12.39 D_{4h}: $E, C_4, C_2, S_4, 2C_2^{1}$, $2C_2''$, $i, \sigma_h, 2\sigma_{v'}, 2\sigma_d$

12.40

E	C_3	C_3^2	$\sigma_v^{(1)}$	$\sigma_v^{(2)}$	$\sigma_v^{(3)}$
C_3	C_3^2	E	$\sigma_v^{(3)}$	$\sigma_v^{(1)}$	$\sigma_v^{(2)}$
C_3^2	E	C_3	$\sigma_v^{(2)}$	$\sigma_v^{(3)}$	$\sigma_v^{(1)}$
$\sigma_v^{(1)}$	$\sigma_v^{(2)}$	$\sigma_v^{(3)}$	E	C_3	C_3^2
$\sigma_v^{(2)}$	$\sigma_v^{(3)}$	$\sigma_v^{(1)}$	C_3^2	E	C_3
$\sigma_v^{(3)}$	$\sigma_v^{(1)}$	$\sigma_v^{(2)}$	C_3	C_3^2	E

12.41 D_{2h}

12.42 $C_{\infty v}$, yes

13

Rotational and Vibrational Spectroscopy

13.1 Since the energy of a molecular quantum state is divided by kT in the Boltzmann distribution, it is of interest to calculate the temperature at which kT is equal to the energy of photons of different wavelength. Calculate the temperature at which kT is equal to the energy of photons of wavelength 10^3 cm, 10^{-1} cm, 10^{-3} cm, 10^{-5} cm.

SOLUTION

$$kT = \frac{hc}{\lambda}$$

$$T = \frac{hc}{k\lambda} = \frac{(6.626x10^{-34}\text{ J s})(2.998x10^{8}\text{m s}^{-1})}{(1.381x10^{-23}\text{J K}^{-1})(10\text{ m})}$$

$$= 1.44x10^{-4}\text{K for } \lambda = 10^{3}\text{ cm}$$

For $\lambda = 10^{-1}$cm, T = 14.4 K

For $\lambda = 10^{-3}$cm, T = 1440 K

For $\lambda = 10^{-5}$cm, T = 144,000 K

13.2 Most chemical reactions require activation energies ranging between 40 and 400 kJ mol^{-1}. What are the equivalents of 40 and 400 kJ mol^{-1} in terms of (a) nm, (b) wavenumbers, and (c) electron volts?

SOLUTION:

For 40 kJ mol^{-1}

(a) $\lambda = \frac{hc}{E}$

$$= \frac{(6.626 \text{ x } 10^{-34}\text{ J s})(2.998 \text{ x } 10^{8}\text{ m s}^{-1})(6.022 \text{ x } 10^{23}\text{ mol}^{-1})}{(4 \text{ x } 10^{4}\text{ J mol}^{-1})}$$

$$= 2.991 \text{ x } 10^{-6}\text{ m} = 2991\text{ nm}$$

(b) $\tilde{\nu} = \frac{1}{\lambda} = \frac{1}{2.991 \times 10^{-4}\ \text{cm}} = 3343\ \text{cm}^{-1}$

(c) $\frac{(4 \times 10^4\ \text{J mol}^{-1})}{(6.022 \times 10^{23}\ \text{mol}^{-1})(1.602 \times 10^{-19}\ \text{J eV}^{-1})} = 0.415\ \text{eV}$

For 400 kJ mol^{-1}

(a) $\lambda = (2991\ \text{nm}) \frac{4 \times 10^4\ \text{J mol}^{-1}}{4 \times 10^5\ \text{J mol}^{-1}} = 299.1\ \text{nm}$

(b) $\tilde{\nu} = \frac{1}{\lambda} = \frac{1}{2.991 \times 10^{-5}\ \text{cm}} = 33{,}430\ \text{cm}^{-1}$

(c) $(0.415\ \text{eV}) \frac{4 \times 10^5\ \text{J mol}^{-1}}{4 \times 10^4\ \text{J mol}^{-1}} = 4.15\ \text{eV}$

13.3 (a) What vibrational frequency in wavenumbers corresponds to a thermal energy of kT at 25 °C? (b) What is the wavelength of this radiation?

SOLUTION:

$hc\tilde{\nu} = kT$

(a) $\tilde{\nu} = \frac{kT}{hc} = \frac{(1.381 \times 10^{-23}\ \text{J K}^{-1})(298.15\ \text{K})(10^{-2}\ \text{m cm}^{-1})}{(6.626 \times 10^{-34}\ \text{J K}^{-1})(2.998 \times 10^8\ \text{m s}^{-1})} = 207\ \text{cm}^{-1}$

(b) $\lambda = \frac{1}{\tilde{\nu}} = 4.83 \times 10^{-3}\ \text{cm}$

13.4 Show that equation 13.17 is a solution of equation 13.9 by differentiating equation 13.17 and substituting it into equation 13.9.

SOLUTION

By use of $N_1 + N_2 = N_{\text{total}}$, equation 13.9 can be written as

$$-\frac{dN_2}{dt} = -B\rho_v(\nu_{12})N_{\text{total}} + [A + 2B\rho_v(\nu_{12})]N_2 \quad \text{(a)}$$

Differentiating equation 13.17 with respect to t yields

$$\frac{dN_2}{dt} = \frac{B\rho_v(\nu_{12})N_{\text{total}}}{A + 2B\rho_v(\nu_{12})}\left(A + 2B\rho_v(\nu_{12})\right) \exp\{-[A + 2B\rho_v(\nu_{12})]t\} \quad \text{(b)}$$

Substituting equation b into equation a and also equation 13.17 into equation a yields

$$-B\rho_v(\nu_{12})N_{\text{total}} \exp\{-[A + 2B\rho_v(\nu_{12})]t\}$$
$$= -B\rho_v(\nu_{12})N_{\text{total}} + B\rho_v(\nu_{12})N_{\text{total}}[1 - \exp\{-[A + 2B\rho_v(\nu_{12})]t\}] \quad \text{(c)}$$

Since the two sides of this equation are equal, equation 13.17 is a solution of equation 13.9.

13.5 Calculate the reduced mass and the moment of inertia of $D^{35}Cl$, given that R_e = 127.5 pm.

SOLUTION

$$\mu = \frac{m_D m_{Cl}}{m_D + m_{Cl}}$$

$$= \frac{(2.014\ 10 \times 10^{-3}\ \text{kg mol}^{-1})(34.968\ 85 \times 10^{-3}\ \text{kg mol}^{-1})}{[(2.014\ 10 + 34.968\ 85) \times 10^{-3}\ \text{kg mol}^{-1}](6.022 \times 10^{23}\ \text{mol}^{-1})}$$

$$= 3.162 \times 10^{-27}\ \text{kg}$$

$$I = \mu R_e^2 = (3.162 \times 10^{-27}\ \text{kg})(1.275 \times 10^{-10}\ \text{m})^2$$

$$= 5.141 \times 10^{-47}\ \text{kg m}^2$$

13.6 The H-O-H bond angle for ^1H_2O is 104.5°, and the H-O bond length is 95.72 pm. What is the moment of inertia of H_2O about its C_2 axis?

SOLUTION

The relative atomic mass of the hydrogen atom is 1.0078 x 10^{-3}, as shown on the back cover.

$$I = 2m_H r_H^2 = 2m_H R^2 \sin^2 \phi$$

where r_H is the distance from the axis, which is equal to $R \sin \phi$, and ϕ is half of the bond angle.

$$I = \frac{2(1.0078 \times 10^{-3}\ \text{kg mol}^{-1})(95.72 \times 10^{-12}\ \text{m})^2 \sin^2(52.25)}{6.022 \times 10^{23}\ \text{mol}^{-1}}$$

$$= 1.917 \times 10^{-47}\ \text{kg m}^2$$

13.7 Some of the following gas molecules have pure microwave absorption spectra and some do not: N_2, HBr, CCl_4, CH_3CH_3, CH_3CH_2OH, H_2O, CO_2, O_2. What is the gross selection rule for rotational spectra, and which molecules satisfy it?

SOLUTION

Molecules have a rotational spectrum only if they have a dipole moment that can interact with the electric vector of the electromagnetic radiation. The following molecules have permanent dipole moments: HBr, CH_3CH_2OH, H_2O.

13.8 Calculate the frequency in wavenumbers and the wavelength in cm of the first rotational transition ($J = 0 \rightarrow 1$) for $D^{35}Cl$.

SOLUTION

$$\tilde{\nu} = 2\tilde{B}\,J = \frac{h}{4\pi^2 cI} \quad \text{since } J = 1$$

The moment of inertia was calculated in problem 13.5.

$$\tilde{\nu} = \frac{(6.626 \text{ x } 10^{-34} \text{ J s})}{4\pi^2(2.998 \text{ x } 10^8 \text{ m s}^{-1})(5.141 \text{ x } 10^{-47} \text{ kg m}^2)} = 1089 \text{ m}^{-1}$$

$$= (1089 \text{ m}^{-1})(10^{-2} \text{ m cm}^{-1}) = 10.89 \text{ cm}^{-1}$$

$$\lambda = \frac{1}{\tilde{\nu}} = \frac{1}{(10.89 \text{ cm}^{-1})} = 0.09183 \text{ cm}$$

13.9 The pure rotational spectrum of $^{12}C^{16}O$ has transitions at 3.863 cm^{-1} and 7.725 cm^{-1}. Calculate the internuclear distance in $^{12}C^{16}O$. Predict the positions, in cm^{-1}, of the next two lines.

SOLUTION

$$\tilde{\nu} = \frac{h}{8\pi^2 Ic}\,[J(J+2) - J(J+1)] = \frac{2(J+1)h}{8\pi^2 Ic}$$

The only two transitions that are related 2:1 are 0→1 and 1→2.

$$J = 0 \quad \tilde{\nu} = 3.863 \text{ cm}^{-1} = \frac{h}{4\pi^2 Ic}$$

$$J = 1 \quad \tilde{\nu} = 7.725 \text{ cm}^{-1} = \frac{h}{2\pi^2 Ic}$$

$$\therefore \quad I = \frac{6.626 \text{ x } 10^{-34} \text{ J s}}{4(3.14159)^2(3.863 \text{ cm}^{-1})(10^2 \text{ cm m}^{-1})(2.998 \text{ x } 10^8 \text{ m s}^{-1})}$$

$$= 1.449 \text{ x } 10^{-46} \text{ kg m}^2$$

$$I = \mu R_e^2 \qquad \mu = \frac{m_O m_C}{m_O + m_C} = 1.139 \text{ x } 10^{-26} \text{ kg}$$

$$\therefore \quad R_e = 1.128 \text{ x } 10^{-10} \text{ m} = 112.8 \text{ pm}$$

Next two lines are $J = 2 \rightarrow 3$ and $J = 3 \rightarrow 4$

$$\tilde{\nu}_2 = \frac{6h}{8\pi^2 Ic} = 11.589 \text{ cm}^{-1}$$

$$\tilde{\nu}_3 = 15.452 \text{ cm}^{-1}$$

13.10 Assume the bond distances in $^{13}C^{16}O$, $^{13}C^{17}O$, and $^{12}C^{17}O$ are the same as in $^{12}C^{16}O$. Calculate the position, in cm^{-1}, of the first rotational transitions in these four molecules. (Use the information in problem 13.9.)

SOLUTION

$$I = \mu R_e^2 \qquad \text{and} \qquad \tilde{\nu} = \frac{h}{4\pi^2 Ic} = \frac{h}{4\pi^2 \mu R_e^2 c}$$

$\tilde{\nu}(^{12}C^{16}O) = 3.863$ cm^{-1} from problem 13.9.

$$\therefore \quad \frac{\tilde{\nu}(^{12}C^{16}O)}{\tilde{\nu}(^{13}C^{16}O)} = \frac{\mu(^{13}C^{16}O)}{\mu(^{12}C^{16}O)} \qquad \text{etc.}$$

$$\tilde{\nu}(^{13}C^{16}O) = \frac{\mu(^{12}C^{16}O)}{\mu(^{13}C^{16}O)}\,\tilde{\nu}(^{12}C^{16}O)$$

$$= \frac{m(^{12}C)\,[m(^{13}C) + m(^{16}O)]}{m(^{13}C)\,[m(^{12}C) + m(^{16}O)]}\,\tilde{\nu}(^{12}C^{16}O) \quad = 3.693 \text{ cm}^{-1}$$

Similarly

$\tilde{\nu}(^{13}C^{17}O) = 3.596 \text{ cm}^{-1}$

$\tilde{\nu}\ ^{12}C^{17}O) = 3.766 \text{ cm}^{-1}$

13.11 The far-infrared spectrum of HI consists of a series of equally spaced lines with $\Delta\tilde{\nu} = 12.8 \text{ cm}^{-1}$. What are (a) the moment of inertia and (b) the internuclear distance?

SOLUTION

(a) $$\Delta\tilde{\nu} = 2\tilde{B} = \frac{2h}{8\pi^2 cI}$$

$$I = \frac{h}{8\pi^2 c\tilde{B}} = \frac{(6.63 \text{ x } 10^{-34} \text{ J s})}{8\pi^2(2.998 \text{ x } 10^8 \text{ m s}^{-1})(6.4 \text{ cm}^{-1})(10 \text{ cm m}^{-1})}$$

$$= 4.37 \text{ x } 10^{-47} \text{ kg m}^2$$

(b) $$I = \frac{m_A m_B}{m_A + m_B} R_e^2$$

$$R_e = \left[\frac{(4.37 \text{ x } 10^{-47} \text{ kg m}^2)(128 \text{ x } 10^{-3} \text{ kg})(6.02 \text{ x } 10^{23} \text{ mol}^{-1})}{(127 \text{ x } 10^{-3} \text{ kg})(1 \text{ x } 10^{-3} \text{ kg})}\right]^{1/2}$$

$$= 163 \text{ pm}$$

13.12 For $H^{35}Cl$ calculate the relative populations of rotational levels, f_J/f_0, for the first three levels at 300 K and 1000 K.

SOLUTION

$$f_J/f_0 = (2J + 1)\exp[-hcJ(J + 1)\tilde{B}\ /kT]$$

$$\frac{hc\tilde{B}}{kT} = \frac{(6.626\text{x}10^{-34} \text{ J s})(2.998\text{x}10^8 \text{ m s}^{-1})(1059.34 \text{ m}^{-1})}{(1.3807\text{x}10^{-23} \text{ J K}^{-1})(300 \text{ K})}$$

$= 5.080\text{x}10^{-2}$ at 300 K

$= 1.524\text{x}10^{-2}$ at 1000 K

At 300 K,

$$f_J/f_0 = (2J + 1)\exp[-5.08\text{x}10^{-2}\ J(J + 1)]$$

For J = 0, 1, 2, 3,

$f_J/f_0 = 1, 2.710, 3.686, 3.805$

At 1000 K

$f_J/f_0 = (2J + 1)\exp[-1.524\text{x}10^{-2}\, J(J + 1)]$

For $J = 0, 1, 2, 3,$

$f_J/f_0 = 1, 2.910, 4.563, 5.830$

$J(0 \text{ -> } 1)$ occurs at $2\tilde{B} = 2118 \text{ m}^{-1}$

$J(1 \text{ -> } 2)$ occurs at $4\tilde{B} = 4236 \text{ m}^{-1}$

$J(2 \text{ -> } 3)$ occurs at $6\tilde{B} = 6354 \text{ m}^{-1}$

$J(3 \text{ -> } 4)$ occurs at $2\tilde{B} = 8472 \text{ m}^{-1}$

13.13 Using equation 13.44, show that J for the maximally populated level is given by

$$J_{\text{max}} = \sqrt{\frac{kT}{2hc\tilde{B}}} - \frac{1}{2}$$

SOLUTION

$$N_J = C(2J + 1)\, e^{-BhcJ(J+1)/kT}$$

$$\frac{\partial N_J}{\partial J} = 0 = 2e^{-[\,]} - (2J + 1)^2 \frac{\tilde{B}\, hc}{kT}\, e^{-[\,]}$$

$$\therefore \quad J_{\text{max}} = \left(\frac{kT}{2hc\tilde{B}}\right)^{1/2} - \frac{1}{2}$$

13.14 Using the result of problem 13.13, find the J nearest J_{max} at room temperature for $H^{35}Cl$ and $^{12}C^{16}O$. (a) What is the ratio of the population at that J to the population of $J = 0$? (b) What is the energy of that J relative to $J = 0$ in units of kT ?

SOLUTION

(a) For $H^{35}Cl$, $\tilde{B} = 10.59 \text{ cm}^{-1} = 10.59 \text{ x } 10^2 \text{ m}^{-1}$

$$J_{\text{max}} = \left[\frac{1.381 \text{ x } 10^{-23} \text{ x } 298}{2 \text{ x } 6.626 \text{ x } 10^{-34} \text{ x } 2.998 \text{ x } 10^8 \text{ x} 10.59 \text{ x } 10^2}\right]^{1/2} - \frac{1}{2}$$

$$= 2.64$$

$$\therefore \quad J = 3$$

$$\frac{E(3) - E(0)}{kT} = \frac{12\tilde{B}}{kT}$$

$$= \frac{12 \times 10.59 \times 10^2 \text{ m}^{-1} \times 6.626 \times 10^{-34} \times 2.998 \times 10^8}{1.381 \times 10^{-23} \times 298}$$

$$= 0.6134$$

$$\frac{N(3)}{N(0)} = 7\ e^{-0.6134} = 3.79$$

(b) For $^{12}C^{16}O$, $\tilde{B} = 1.931\ \text{cm}^{-1} = 193.1\ \text{m}^{-1}$

$J_{max} = 6.85 \quad \therefore J = 7$

$$\frac{E(7) - E(0)}{kT} = \frac{56\tilde{B}}{kT} = 0.5220$$

$$\frac{N(7)}{N(0)} = \frac{15}{1}\ e^{-0.5220} = 8.90$$

13.15 The moment of inertia of $^{16}O^{12}C^{16}O$ is 7.167×10^{-46} kg m^2. (a) Calculate the CO bond length, R_{CO}, in CO_2. (b) Assuming that isotopic substitution does not alter R_{CO}, calculate the moments of inertia of (1) $^{18}O^{12}C^{18}O$ and (2) $^{16}O^{13}C^{16}O$.

SOLUTION

(a) Since the CO_2 molecule is symmetrical, the carbon atom is on the axis of rotation and does not contribute to the moment of inertia.

$$I = 2mR_{CO}^2$$

$$R_{CO} = (I/2m)^{1/2}$$

$$= \left[\frac{(71.67 \times 10^{-47}\ \text{kg m}^2)(6.022 \times 10^{23}\ \text{mol}^{-1})}{2(15.994\,91 \times 10^{-3}\ \text{kg mol}^{-1})}\right]^{1/2}$$

$$= 0.1162\ \text{nm}$$

(b) For $^{18}O^{12}C^{18}O$

$$I = 2mR_{CO}^2$$

$$= \left[\frac{2(17.999\,159 \times 10^{-3}\ \text{kg mol}^{-1})(0.1162 \times 10^{-9}\ \text{m})^2}{6.022 \times 10^{23}\ \text{mol}^{-1}}\right]$$

$$= 8.071 \times 10^{-46}\ \text{kg m}^2$$

For $^{16}O^{13}C^{16}O$ the moment of inertia is the same as for $^{16}O^{12}C^{16}O$.

13.16 Derive the expression for the moment of inertia of a symmetrical tetrahedral molecule like CH_4 in terms of the bond length R and the masses of the four tetrahedral atoms. The easiest way to derive the expression is to consider an axis along one CH bond. Show that the same result is obtained if the axis is taken perpendicular to the plane defined by one group of three atoms HCH.

SOLUTION

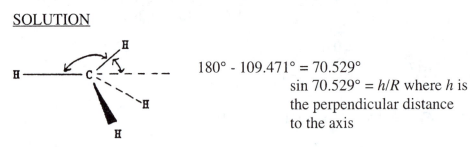

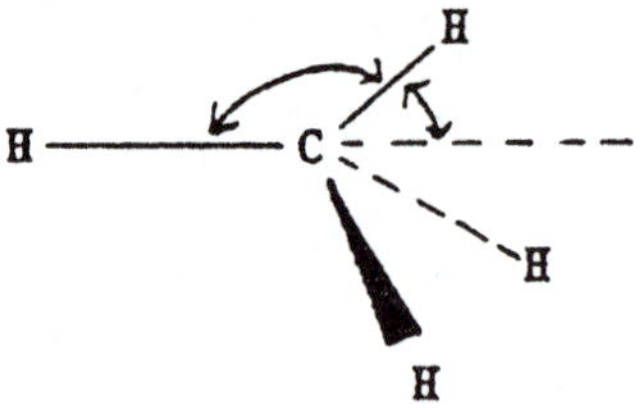

$180° - 109.471° = 70.529°$

$\sin 70.529° = h/R$ where h is the perpendicular distance to the axis

$$I = \sum m_i h_i^2 = 3mR^2 \sin^2(70.529°) = \frac{8}{3}\, mR^2$$

If the axis is taken perpendicular to the page through C

$$I = 2mR^2 + 2m\left[R^2 \sin^2\left(90° - \frac{109.471}{2}\right)\right] = \frac{8}{3}\, mR^2$$

13.17 What are the values of $\tilde{A}$ and $\tilde{B}$ (from equation 13.62) for the symmetric top NH_3 if $I_{\|} = 4.41 \times 10^{-47}$ kg m^2 and $I_{\perp} = 2.81 \times 10^{-47}$ kg m^2? What is the wavelength of the $J = 0$ to $J = 1$ transition? What are the wavelengths of the $J = 1$ to $J = 2$ transitions? (Remember the selection rules: $\Delta J = \pm 1$, $\Delta K = 0$, and find all allowed transitions.)

SOLUTION

$$\tilde{B} = \frac{\hbar}{4\pi c I_{\perp}} = \frac{1.054 \times 10^{-34}\text{ J s}}{4(3.14159)(2.998 \times 10^{8}\text{ m s}^{-1})(2.81 \times 10^{-47}\text{ kg m}^{2})}$$

$$= 9.95 \times 10^{2}\text{ m}^{-1} = 9.995\text{ cm}^{-1}$$

$$\tilde{A} = \frac{\hbar}{4\pi c I_{\|}} = 6.35\text{ cm}^{-1}$$

$$\tilde{\nu}_{0\to 1} = 2\tilde{B} = 19.90\text{ cm}^{-1}$$

$$\lambda = 1.01 \times 10^{-3}\text{ m}$$

$$\tilde{\nu}_{0\to 2}^{K=0} = 4\tilde{B} = 39.84\text{ cm}^{-1}$$

$$\lambda = 2.51 \times 10^{-4}\text{ m}$$

$$\tilde{\nu}_{1\to 2}^{K=+1} = 4\tilde{B} = 39.84\text{ cm}^{-1}$$

$$\lambda = 2.51 \times 10^{-4}\text{ m}$$

13.18 Consider a linear triatomic molecule, ABC. Find the center of mass (which by symmetry lies on the molecular axis). Show that the moment of inertia is given by

$$I = \frac{1}{M}\left[R_{AB}^2\, m_A m_B + R_{BC}^2\, m_B m_C + (R_{AB} + R_{BC})^2 m_A m_C\right]$$

where R_{AB} is the AB bond distance, R_{BC} is the BC bond distance, m_i are the masses of the atoms, and $M = m_A + m_B + m_C$. Show that if $R_{AB} = R_{BC}$ and $m_A = m_C$, $I = 2m_A R_{AB}^2$.

SOLUTION

(a) Center of mass: choose the origin at atom A, then

$$R_{CM} = \frac{m_B R_{AB} + m_C(R_{AB} + R_{BC})}{m_A + m_B + m_C} = \frac{m_B R_{AB} + m_C(R_{AB} + R_{BC})}{M}$$

(b) Moment of inertia

$$I = m_A(0 - R_{CM})^2 + m_B (R_{AB} - R_{CM})^2 + m_C(R_{AB} + R_{BC} - R_{CM})^2$$

$$= \frac{1}{M^2} \left\{m_A[m_B R_{AB} + m_C(R_{AB} + R_{BC})]^2 + m_B M^2(R_{AB} - R_{CM})^2 + m_C M^2(R_{AB} + R_{BC} - R_{CM})^2\right\}$$

$$= \frac{1}{M^2} \left\{R_{AB}^2 [m_A(m_B + m_C)M] + R_{BC}^2 M m_C(m_A + m_B) + 2R_{AB}R_{BC}M m_A m_C\right\}$$

$$= \left(\frac{1}{M}\right)\left[R_{AB}^2 m_A(m_B + m_C) + R_{BC}^2(m_A + m_B) m_C + 2R_{AB}R_{BC}m_A m_C\right]$$

$$I = \frac{1}{M} \left\{m_A m_B R_{AB}^2 + m_B m_C R_{BC}^2 + m_A m_C(R_{AB} + R_{BC})^2\right\}$$

If $R_{AB} = R_{BC}$ and $m_A = m_C$

$$I = \frac{1}{M}(2m_B m_A R_{AB}^2 + 4m_A^2 R_{AB}^2)$$

$$= \frac{(2m_A + m_B)}{M}(2m_A R_{AB}^2) = 2m_A R_{AB}^2$$

13.19 The fundamental vibration frequency of $H^{35}Cl$ is 8.967 x 10^{13} s^{-1} and of $D^{35}Cl$ is 6.428 x 10^{13} s^{-1}. What would be the separation between infrared absorption lines of $H^{35}Cl$ and $H^{37}Cl$ on one hand and those of $D^{35}Cl$ and $D^{37}Cl$ on the other, if the force constants are assumed to be the same in each pair?

SOLUTION

$$\mu = \frac{m_H m_{Cl}}{m_H + m_{Cl}}$$

For $H^{35}Cl$

$$\mu_{35} = \frac{(1.008 \times 10^{-3}\ \text{kg mol}^{-1})(34.97 \times 10^{-3}\ \text{kg mol}^{-1})}{(35.98 \times 10^{-3}\ \text{kg mol}^{-1})(6.022 \times 10^{23}\ \text{mol}^{-1})}$$

$= 1.627 \times 10^{-27}$ kg

For $H^{37}Cl$

$\mu_{37} = 1.629 \times 10^{-27}$ kg

$$\nu = \frac{1}{2\pi}\left(\frac{k}{\mu}\right)^{1/2}$$

$$\therefore \frac{\nu_{35}}{\nu_{37}} = \left(\frac{\mu_{37}}{\mu_{35}}\right)^{1/2} = \frac{\lambda_{37}}{\lambda_{35}}$$

$$\lambda_{37} = \lambda_{35}\left(\frac{\mu_{37}}{\mu_{35}}\right)^{1/2} \qquad \lambda_{37} - \lambda_{35} = \lambda_{35}\left[\left(\frac{\mu_{37}}{\mu_{35}}\right)^{1/2} - 1\right]$$

$$\lambda_{35} = \frac{2.998 \times 10^{8}\ \text{ms}^{-1}}{8.967 \times 10^{13}\text{s}^{-1}} = 3343 \times 10^{-9}\ \text{m}$$

$$\lambda_{37} - \lambda_{35} = 3343\left[\left(\frac{1.629}{1.627}\right)^{1/2} - 1\right] = 2.5\ \text{nm}$$

For $D^{35}Cl$

$$\frac{\nu_{35}}{\nu_{37}} = \left(\frac{\mu_{37}}{\mu_{35}}\right)^{1/2} = 1.0015$$

$$\nu_{35} = 6.428 \times 10^{13}\ \text{s}^{-1}$$

$$\lambda_{35} = 4664\ \text{nm}$$

$$\lambda_{37} - \lambda_{35} = 4664\ [(1.0015)^{1/2} - 1]) = 3.5\ \text{nm}$$

13.20 Find the force constants of the halogens $^{127}I_2$, $^{79}Br_2$, and $^{35}Cl_2$ using the data of Table 13.4. Is the order of these the same as the order of the bond energies?

SOLUTION

$$\tilde{\nu} = \frac{1}{2\pi c}\sqrt{\frac{k}{\mu}}$$

$$k = 4\pi^2 c^2 \tilde{\nu}^2 \mu$$

Note that the reduced mass μ of the iodine molecule is half of the mass of an iodine atom.

$$k_{I_2} = 4(3.14159)^2\ \frac{126.9 \times 10^{-3}}{2}\ \frac{(214.5 \times 10^2)^2\ \text{m}^{-2}}{6.022 \times 10^{23}}\ (2.998 \times 10^{8})^2\ \text{m}^2\ \text{s}^{-2}$$

$= 172.2$ N/m^2

$k_{Br_2} = 246.3$ N/m^2

$k_{Cl_2} = 323.0$ N/m^2

∴	I_2	Br_2	Cl_2	
D_o	1.54 eV	1.97 eV	2.48 eV	∴ same order
k	172.2 N/m^2	246.3 N/m^2	323.0 N/m^2	

13.21 Given the following fundamental frequencies of vibration, calculate ΔH° for the reaction
$H^{35}Cl\ (v = 0) + {}^2D_2\ (v = 0) = {}^2D^{35}Cl\ (v = 0) + H^2D\ (v = 0)$.

$H^{35}Cl$: 2989 cm^{-1} H^2D: 3817 cm^{-1}

${}^2D^{35}Cl$: 2144 cm^{-1} ${}^2D^2D$: 3119 cm^{-1}

SOLUTION

Since the electronic part of the bond energies D_o are the same on the left- and right-hand sides, only the zero point energy contributes to $\Delta H°$. N.B.

$D_o = D_e - \frac{1}{2}\ hc\tilde{v}$.

$$\Delta H° = E_{DCl} + E_{HD} - E_{D_2} - E_{HCl}$$

$$= +\frac{1}{2}\ hc(\tilde{v}_{DCl} + \tilde{v}_{HD} - \tilde{v}_{D_2} - \tilde{v}_{HCl})$$

$$= +\frac{1}{2}(6.626 \times 10^{-34} \times 2.998 \times 10^8)(2144 + 3817 - 2989 - 3119)$$

$$= (+147\ \text{cm}^{-1})\frac{1}{2}(6.626 \times 10^{-34} \times 2.998 \times 10^8)\ \times 10^2\ \text{m}^{-1}\ \text{cm}$$

$$= -\ 1.46 \times 10^{-21}\ \text{J}$$

$$= -\ 1.46 \times 10^{-21} \times 6.022 \times 10^{23} = -\ 879\ \text{J mol}^{-1}$$

*13.22 If the fundamental vibration frequency of ${}^1H_2$ is 4401.21 cm^{-1}, compute the fundamental vibration frequency of ${}^2D_2$ and ${}^1H^2D$ assuming the same force constants. If D_o for ${}^1H_2$ is 4.4781 eV, what is D_o for ${}^2D_2$ and ${}^1H^2D$? Neglect anharmonicities.

SOLUTION

$$\frac{\tilde{v}_{{}^2D_2}}{\tilde{v}_{{}^1H_2}} = \left(\frac{\mu_{{}^1H_2}}{\mu_{{}^2D_2}}\right)^{1/2} = \left(\frac{m_{{}^1H}}{m_{{}^2D}}\right)^{1/2} = 0.707$$

$$\frac{\tilde{v}_{{}^1H_2D}}{\tilde{v}_{{}^1H_2}} = \left(\frac{\mu_{{}^1H_2}}{\mu_{{}^1H{}^2D}}\right)^{1/2}$$

$$= \left[\frac{\frac{1}{2}m_{1_H}(m_{1_H} + m_{2_D})}{(m_{1_H}m_{2_D})}\right]^{1/2} = \left(\frac{m_{1_H} + m_{2_D}}{2m_{2_D}}\right)^{1/2} = 0.866$$

$\therefore \quad \tilde{\nu}_{^2D_2} = 3112.1\ \text{cm}^{-1} \quad \tilde{\nu}_{^1H^2D} = 3811.6\ \text{cm}^{-1}$

$D_o = D_e - \frac{1}{2}\, h\tilde{\nu} c$

$\therefore \quad D_o(D_2) = D_o(H_2) + \frac{hc}{2}(\tilde{\nu}_{H_2} - \tilde{\nu}_{D_2}) = 4.4781\ \text{eV} +$

$$\frac{6.626 \times 10^{-34} \times 2.998 \times 10^{8}}{2(1.602 \times 10^{-19})}(4401.2 - 3112.1) \times 10^{2}$$

$= 4.4781 + 0.080\ \text{eV} = 4.558\ \text{eV}$

$D_o(HD) = 4.4781 + 0.037\ \text{eV} = 4.515\ \text{eV}$

13.23 Using the values for $\tilde{\nu}_e$ and $\tilde{\nu}_e x_e$ in Table 13.4 for $^1H^{35}Cl$ estimate the dissociation energy assuming the Morse potential is applicable.

SOLUTION

$\tilde{\nu}_e = 2990.95\ \text{cm}^{-1} \quad \tilde{\nu}_e x_e = 52.819\ \text{cm}^{-1} \quad x_e = 0.01766$

The zero point energy lies at

$G(0) = \frac{1}{2}\tilde{\nu}_e - \frac{1}{4}\tilde{\nu}_e x_e = \frac{1}{2}(2991\ \text{cm}^{-1}) - \frac{1}{4}(52.8\ \text{cm}^{-1})$

$= 1482\ \text{cm}^{-1}$

$D_e = \frac{\tilde{\nu}_e}{4x_e} = \frac{2990.95\ \text{cm}^{-1}}{4(0.017\,66)} = 42{,}341\ \text{cm}^{-1}$

$D_0 = D_e - G(0) = 42{,}341 - 1482 = 40{,}859\ \text{cm}^{-1}$

$= (40{,}859\ \text{cm}^{-1})(1.2399 \times 10^{-4}\ \text{eV cm}) = 5.07\ \text{eV}$

The actual value is 4.434 eV.

13.24 Apply the Taylor expansion to the potential energy given by the Morse equation
$V(R) = D_e(1 - \exp[-a(R - R_0)])^2$
to show that the force constant k is given by $k = 2D_e a^2$.

SOLUTION

The Morse equation can be written
$V(x) = D_e(1 - e^{-ax})^2$
where $x = R - R_0$.

The first derivative of $V(x)$ is given by

$$\frac{dV}{dx} = 2D_e\{1 - e^{-ax}\}(-e^{-ax})(-a) = -2aD_e\{-e^{-ax} + e^{-2ax}\}$$

At $x = 0$, $dV/dx = 0$.

$$\frac{d^2V}{dx^2} = -2aD_e\{ae^{-ax} - 2ae^{-2ax}\}$$

At $x = 0$, $\frac{d^2V}{dx^2} = 2a^2D_e$

Thus according to the Taylor expansion

$V(x) = D_e a^2 x^2 + ...$

Thus in the neighborhood of R_0, the potential energy is given by a parabola.

Comparing this with $V(x) = (k/2)x^2$ for the harmonic oscillator yields

$k = 2D_e a^2$

13.25 (a) What fraction of $H_2(g)$ molecules are in the $v = 1$ state at room temperature? (b) What fractions of $Br_2(g)$ molecules are in the $v = 1$, 2, and 3 states at room temperature?

SOLUTION

(a) Using the equation $f_1 = (1 - e^{-h\nu/kT})\, e^{-h\nu/kT}$

$$\frac{h\nu}{kT} = \frac{hc\tilde{\nu}}{kT} = \frac{(6.626 \times 10^{-34}\text{ J s})(2.998 \times 10^{10}\text{ cm s}^{-1})(4401\text{ cm}^{-1})}{(1.381 \times 10^{-23}\text{ J K}^{-1})(298\text{ K})}$$

$= 21.2$

$f_1 = (1 - e^{-21.2})\, e^{-21.2} = 6.2 \times 10^{-10}$

(b) For Br_2, $\tilde{\nu}_e = 325.321\text{ cm}^{-1}$

$$\frac{h\nu}{kT} = \frac{hc\tilde{\nu}}{kT}$$

$$= \frac{(6.626 \times 10^{-34}\text{ J s})(2.998 \times 10^{10}\text{ cm s}^{-1})(325.3\text{ cm}^{-1})}{(1.381 \times 10^{-23}\text{ J K}^{-1})(298\text{ K})} = 1.570$$

$f_1 = (1 - e^{-1.570})\, e^{-1.570} = 0.165$

$f_2 = (1 - e^{-1.570})\, e^{-(2)(1.570)} = 0.034$

$f_3 = (1 - e^{-1.570})\, e^{-(3)(1.570)} = 0.007$

13.26 The first three lines in the R branch of the fundamental vibration-rotation band of $H^{35}Cl$ have the following frequencies in cm^{-1}: 2906.25(0), 2925.78(1), 2944.89(2), where the numbers in parentheses are the J values for the initial level. What are the values of $\tilde{\nu}_0$, $\tilde{B}_{v'}$, $\tilde{B}_{v''}$, $\tilde{B}_e$, and α?

SOLUTION

$\tilde{\nu} = \tilde{\nu}_0 + (\tilde{B}_{\upsilon}' + \tilde{B}_{\upsilon}'')m + (\tilde{B}_{\upsilon}' - \tilde{B}_{\upsilon}'')m^2$

$= a + bm + cm^2$ where $m = J'' + 1$

$2906.25 = a + b + c$

$2925.78 = a + 2b + 4c$

$2944.89 = a + 3b + 9c$

$a = 2886.30\ \text{cm}^{-1} = \omega_o$

$b = 20.16 = \tilde{B}_{\upsilon}' + \tilde{B}_{\upsilon}''$

$c = -0.21 = \tilde{B}_{\upsilon}' - \tilde{B}_{\upsilon}''$

$\tilde{B}_{\upsilon}' = 9.98\ \text{cm}^{-1} = \tilde{B}_e - \frac{3}{2}\alpha$

$\tilde{B}_{\upsilon}'' = 10.19\ \text{cm}^{-1} = \tilde{B}_e - \frac{1}{2}\alpha$

$\alpha = 0.21\ \text{cm}^{-1}$

13.27 In Table 11.1, D_e for H_2 is given as 4.7483 eV or 458.135 kJ mol^{-1}. Given the vibrational parameters for H_2 in Table 13.4 calculate the value you would expect for $\Delta_f H^o$ for H(g) at 0 K.

SOLUTION

$D_o = D_e - \frac{1}{2}\tilde{\nu}_e + \frac{1}{4}\tilde{\nu}_e x_e$

$= (4.7483\ \text{eV})(8065.478\ \text{cm}^{-1}\ \text{eV}^{-1}) - \frac{1}{2}(4400.39) + \frac{1}{4}(120.815)$

$= 36.127\ \text{cm}^{-1}$ or 432.175 kJ mol^{-1}

$H_2(g) = 2H(g)$

$\Delta H_0^o = 2\Delta_f H^o(\text{H}) = 432.175\ \text{kJ mol}^{-1}$

$\Delta_f H^o(\text{H}) = 216.088\ \text{kJ mol}^{-1}$

Table C.3 gives 216.037 kJ mol^{-1}

13.28 Calculate the wavelengths in (a) wavenumbers and (b) micrometers of the center two lines in the vibration-rotation spectrum of HBr for the fundamental vibration. The necessary data are to be found in Table 13.4.

SOLUTION

The reduced mass of $H^{80}Br$ is given in Table 13.4 as

$$\mu = \frac{0.995\,58 \times 10^{-3}\ \text{kg mol}^{-1}}{6.022\,137 \times 10^{-23}\ \text{mol}^{-1}} = 1.653\,20 \times 10^{-27}\ \text{kg}$$

$I = \mu R^2 = (1.653\,20 \times 10^{-27}\ \text{kg})(1.4138 \times 10^{-10}\ \text{m})^2$

$= 3.304\ 47 \times 10^{-47}$ kg m^2

$$\tilde{B} = \frac{h}{8\pi^2 cI} = \frac{(6.626 \times 10^{-34}\ \text{J s})(10^{-2}\ \text{m cm}^{-1})}{8\pi^2(2.998 \times 10^8\ \text{m s}^{-1})(3.3045 \times 10^{-47}\ \text{kg m}^2)} = 8.47\ \text{cm}^{-1}$$

(a) $\tilde{\nu}_P = \tilde{\nu}_0 - 2\tilde{B}J''$ where $J'' = 1, 2, 3, ...$

$= 2649.67\ \text{cm}^{-1} - 2(8.47\ \text{cm}^{-1}) = 2632.72\ \text{cm}^{-1}$

$\tilde{\nu}_R = \tilde{\nu}_0 + 2\tilde{B} + 2BJ''$ where $J'' = 0, 1, 2, ...$

$= 2649.67\ \text{cm}^{-1} + 2(8.47\ \text{cm}^{-1}) = 2666.61\ \text{cm}^{-1}$

(b) $\lambda_P = \dfrac{1}{\tilde{\nu}_P} = 3.798\ 34 \times 10^{-4}\ \text{cm} = 3.798\ 34\ \mu\text{m}$

$\lambda_R = \dfrac{1}{\tilde{\nu}_R} = 3.750\ 08 \times 10^{-4}\ \text{cm} = 3.750\ 08\ \mu\text{m}$

13.29 How many normal modes of vibration are there for (a) SO_2(bent), (b) H_2O_2(bent), (c) HC≡CH(linear), and (d) C_6H_6?

SOLUTION

The number of normal modes of vibration is $3N - 6$ for a nonlinear molecule and $3N - 5$ for a linear molecule.

(a) $3N - 6 = 9 - 6 = 3$ (c) $3N - 5 = 12 - 5 = 7$

(b) $3N - 6 = 12 - 6 = 6$ (d) $3N - 6 = 36 - 6 = 30$

13.30 List the numbers of translational, rotational, and vibrational degrees of freedom for (a) Ne, (b) N_2, (c) CO_2, and (d) CH_2O.

SOLUTION

	Molecule	Trans	Rot	Vib	Total = $3N$
(a)	Ne	3	0	0	3
(b)	N_2	3	2	1	6
(c)	CO_2	3	2	4*	9
(d)	CH_2O	3	3	6**	12

*For linear molecules $3N - 5$ **For nonlinear molecules $3N - 6$

13.31 Acetylene is a symmetrical linear molecule. It has seven normal modes of vibration, two of which are doubly degenerate. These normal modes may be represented as follows:

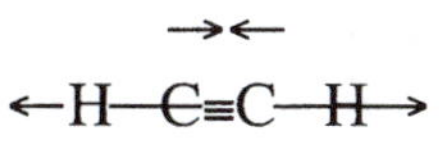

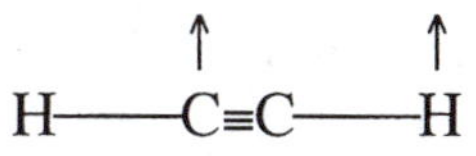

$\tilde{\nu}_1 = 3374\ \text{cm}^{-1}$

$\tilde{\nu}_4 = 612\ \text{cm}^{-1}$

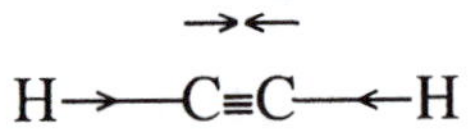

$\tilde{\nu}_2 = 1974\ \text{cm}^{-1}$

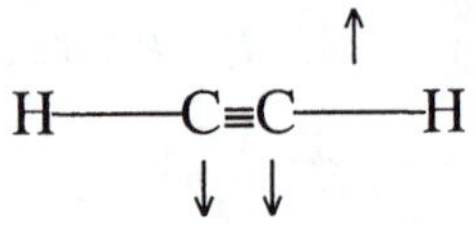

$\tilde{\nu}_5 = 729\ \text{cm}^{-1}$

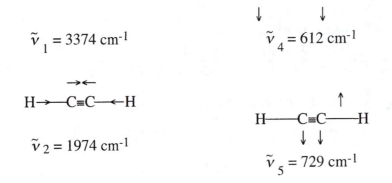

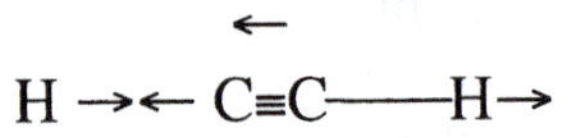

$\tilde{\nu}_3 = 3287\ \text{cm}^{-1}$

(a) Which are the doubly degenerate vibrations?
(b) Which vibrations are infrared active?
(c) Which vibrations are Raman active?

<u>SOLUTION</u>

(a) $\tilde{\nu}_4$ and $\tilde{\nu}_5$ since they can also take place in a plane perpendicular to the plane of the paper

(b) $\tilde{\nu}_3$ and $\tilde{\nu}_5$ because the dipole moment changes during the vibration

(c) $\tilde{\nu}_1$, $\tilde{\nu}_2$, and $\tilde{\nu}_4$ because the polarizability of the molecule changes during the vibrations.

13.32 Calculate the wavenumber and wavelength of the pure fundamental ($\upsilon = 0 \rightarrow 1$) vibrational transitions for (a) $^{12}C^{16}O$ and (b) $^{39}K^{35}Cl$ using data in Table 13.4.

<u>SOLUTION</u>

Equation 13.68 is

$\Delta G(\upsilon + \frac{1}{2}) = \tilde{\nu}_e - 2\tilde{\nu}_e x_e(\upsilon + 1)$, where υ is the vibrational quantum number for the lower of the two levels

(a) $\tilde{\nu} = 2169.814 - 2(13.288) = 2143.24\ \text{cm}^{-1}$

$\lambda = 1/\tilde{\nu} = 4.665\ 84\ \mu\text{m}$

(b) $\tilde{\nu} = 281 - 2(1.3) = 278.4\ \text{cm}^{-1}$
$\lambda = 35.92\ \mu\text{m}$

13.33 (a) Consider the four normal modes of vibration of a linear molecule AB_2 from the standpoint of changing dipole moment and changing polarizability. Which vibrational modes are infrared active, and which are Raman active? (Note the exclusion rule.) (b) Consider the three normal modes of a nonlinear molecule AB_2. Which vibrational modes are infrared active, and which are Raman active?

SOLUTION

(a)

Vib. mode	Changing μ	Changing α	IR	Raman
ν_1	No	Yes	No	Yes
ν_2	Yes	No*	Yes	No
ν_3	Yes	No*	Yes	No
ν_4	Yes	No*	Yes	No

*The exclusion rule is useful because it is difficult to judge qualitatively whether a vibrational mode involves a change in polarizability.

(b)

Vib. mode	Changing μ	Changing α	IR	Raman
ν_1	Yes	Yes	Yes	Yes
ν_2	Yes	Yes	Yes	Yes
ν_3	Yes	Yes	Yes	Yes

13.34 Calculate the fraction of Cl_2 molecules ($\tilde{\nu}$ = 559.7 cm^{-1}) in the i = 0, 1, 2, 3 vibrational states at 1000 K.

SOLUTION

$$f_i = (1 - e^{-h\nu/kT})\, e^{-ih\nu/kT}$$

$$\frac{h\nu}{kT} = \frac{hc\tilde{\nu}}{kT}$$

$$= \frac{(6.626\text{x}10^{-34}\text{ J s})(2.998\text{x}10^{8}\text{ m s}^{-1})(559.7\text{ cm}^{-1})(10^{2}\text{ cm m}^{-1})}{(1.381\text{x}10^{-23}\text{ J K}^{-1})(10^{3}\text{ K})}$$

$$= 0.8051$$

$f_0 = [1 - \exp(- 0.8051)] = 0.5530$

$f_1 = 0.5530 \exp(- 0.8051) = 0.2472$

$f_2 = 0.5530 \exp(- 2 \text{ x } 0.8051) = 0.1105$

$f_3 = 0.5530 \exp(- 3 \text{ x } 0.8051) = 0.0494$

13.35 When CCl_4 is irradiated with the 435.8 nm mercury line, Raman lines are obtained at 439.9, 441.8, 444.6, and 450.7 nm. Calculate the Raman frequencies of CCl_4 (expressed in wave numbers). Also calculate the wavelengths (expressed in μm) in the infrared at which absorption might be expected.

SOLUTION

$$\Delta\tilde{\nu}_R = \frac{1}{\lambda_{incid}} - \frac{1}{\lambda_{scatt}}$$

$$= \frac{1}{435.8 \text{ x } 10^{-7} \text{ cm}} - \frac{1}{439.9 \text{ x } 10^{-7} \text{ cm}} = 214 \text{ cm}^{-1}$$

$$\lambda = \frac{1}{\Delta\tilde{\nu}} = \frac{1}{214 \text{ cm}^{-1}} = 46.7 \text{ x } 10^{-6} \text{ m}$$

For the remaining lines

$\Delta\tilde{\nu}_R$/cm^{-1}	312	454	759
λ/μm	32.0	22.0	13.2

13.36 The first several Raman frequencies of $^{14}N_2$ are 19.908, 27.857, 35.812, 43.762, 51.721, and 59.662 cm^{-1}. These lines are due to pure rotational transitions with J = 1, 2, 3, 4, 5, and 6. The spacing between the lines is $4B_e$. What is the internuclear distance?

SOLUTION

$$\mu = \frac{(14.003\ 07 \text{ x } 10^{-3} \text{ kg})^2}{(28.006\ 14 \text{ x } 10^{-3} \text{ kg})(6.022\ 045 \text{ x } 10^{23} \text{ mol}^{-1})} = 1.162\ 651 \text{ x } 10^{-26}$$

kg

The average spacing between lines is 7.951 cm^{-1}, and so

$\tilde{B}_e = (7.951 \text{ cm}^{-1})(10^2 \text{ cm m}^{-1})/4 = 198.78 \text{ m}^{-1}$ since $\Delta J = 2$.

$$\tilde{B}_e = \frac{h}{8\pi^2 c\mu R_e^2}$$

$$R_e = \left(\frac{h}{8\pi^2 c\mu B_e^2}\right)^{1/2}$$

$$= \left[\frac{6.626\ 176 \text{ x } 10^{-34} \text{ J s}}{8\pi^2(2.997924\ 58 \text{ x } 10^8 \text{ m s}^{-1})(1.162651 \text{x } 10^{-26} \text{ kg})(198.78 \text{ m}^{-1})}\right]^{1/2}$$

$= 110$ pm

13.37 What Raman shifts are expected for the first four Stokes lines for CO_2?

SOLUTION

$I = 7.167 \times 10^{-46}$ kg m^2 (problem 13.15)

$$\tilde{B}_e = \frac{h}{8\pi^2 Ic} = \frac{(6.626 \times 10^{-34}\ \mathrm{J\ s})(10^{-2}\ \mathrm{m\ cm^{-1}})}{8\pi^2(7.167 \times 10^{-46}\ \mathrm{kg\ m^2})(2.9979 \times 10^8\ \mathrm{m\ s^{-1}})}$$

$= 0.3906$ cm^{-1} $\Delta\tilde{\nu}_R = 2B_e(2J + 3)$

J''	$\Delta\tilde{\nu}_R$
0	2.3436 cm^{-1}
1	3.9060
2	5.4684
3	7.0308

13.38 Some of the following gas molecules have a pure rotational Raman spectra and some do not: N_2, HBr, CCl_4, CH_3CH_3, CH_3CH_2OH, H_2O, CO_2, O_2. What is the gross selection rule for pure rotational Raman spectra, and which molecules satisfy it?

SOLUTION

Molecules have a pure rotational Raman spectrum only if the polarizability of the molecule is anisotropic. Thus all of these molecules have a pure rotational Raman spectrum except for CCl_4.

13.39 1 eV = 8065.541 cm^{-1} = 96.485 309 kJ mol^{-1}

13.40 11,600, 1.16×10^6, 1.16×10^7 K; 0.024 eV

13.41 (a) 1.1385×10^{-26} kg, (b) 1.4491×10^{-46} kg m^2

13.42 21.2, 42.36, 63.54, 190.6 cm^{-1}

472, 236, 157, 52.5 μm

13.43 (a) 4.607×10^{-48}, (b) 6.139×10^{-48}, (c) 6.904×10^{-48}, (d) 9.205×10^{-48} kg m^2

13.44 37.822 cm^{-1}, 0.452 kJ mol^{-1} for OH

20.02 cm^{-1}, 0.239 kJ mol^{-1} for OD

13.45 112.83 pm

13.46 1.938×10^{-47}, 1.950×10^{-47} kg m^2, $J = 2$, 86.64 cm^{-1} for ^{12}CO, 86.10 cm^{-1} for ^{13}CH

13.47 1.129×10^{-10} m

13.49 $6B$, $10B$, $14B$; $-6B$, $-10B$, $-14B$.

13.51 121.69×10^8, 243.38×10^8, 365.07×10^8 s^{-1}

13.52 1.188×10^{12}, 2.376×10^{12}, 3.564×10^{12} s^{-1}

13.53 9.10, 14.39 K

13.54 38,812 cm^{-1}; 42,358 cm^{-1}; 33,658 cm^{-1}

13.55 3.908, 4.618, 3.196 eV

13.56 516.4, 411.6, 314.1 N m^2

427.78, 362.56, 294.66 kJ mol^{-1}

13.57 (a) 5.94×10^{-2}, (b) 0.360

13.58 3.60, 3.03, 2.31 eV

13.59 5.29, 5.42, 5.49, 0.13 cm^{-1}; 127.0 pm ($^1H^{35}Cl$, R_e = 127.455 pm)

13.60 18 828.56 cm^{-1}

13.61 Molecules have an absorption only if vibration causes a change in dipole moment. Thus all of these molecules have absorption spectra except for N_2 and O_2.

13.62

	trans	rot	vib
Cl_2	3	2	1
H_2D	3	3	3
C_2H_2	3	2	7

13.63

	trans	rot	vib
NNO	3	2	4
NH_3	3	3	6

13.64 75.19 pm

13.65 112.6 pm

13.66 655 J mol^{-1}

14

Electronic Spectroscopy of Molecules

14.1 The spectroscopic dissociation energy of $H_2(g)$ into ground state hydrogen atoms is 4.4763 eV. What is the spectroscopic dissociation energy of $H_2(g)$ into one ground state H and one H atom in the 2p state? If H_2 is dissociated with photons of energy 15 eV, what is the velocity of the H atoms coming off in the 1s and 2p states?

SOLUTION

$H_2(g) = 2H(g;1s)$ $\qquad \Delta E = 4.4763$ eV

$H(g;1s) = H(g;2p)$ $\qquad \Delta E = 13.6(\frac{3}{4}) = 9.45$ eV

$H_2(g) = H(g;1s) + H(g;2p) \quad \Delta E = 13.93$ eV

The excess energy is 15 - 13.93 = 1.07 eV, all of which goes into kinetic energy of the two atoms:

$$2\left(\frac{1}{2}\ mv^2\right) = 1.07 \text{ eV}$$

$$v = \sqrt{\frac{1.07 \text{ eV x } 1.602 \text{ x } 10^{-19} \text{ J/eV}}{(1.008 \text{ x } 10^{-3}/6.022 \text{ x } 10^{23})\text{kg}}}$$

$$= 1.012 \text{ x } 10^4 \text{ m/s}$$

14.2 According to the hypothesis of Franck, the molecules of the halogens dissociate into one normal atom and one excited atom. The wavelength of the convergence limit in the spectrum of iodine is 499.5 nm. (a) What is the energy of dissociation in kJ mol^{-1} of iodine into one normal and one excited atom? (b) The thermochemical value of the heat of dissociation of I_2 into ground state atoms can be found in Appendix C.3. Calculate the energy of the excited state of I that is formed from the spectroscopic dissociation in kJ mol^{-1} and eV.

SOLUTION

(a) $E = hcN_A/\lambda$

$$= \frac{(6.626 \times 10^{-3}\ \text{J s})(2.998 \times 10^{8}\ \text{m s}^{-1})(6.022 \times 10^{23}\ \text{mol}^{-1})}{(499.5 \times 10^{-9}\ \text{m})(10^{3}\ \text{J kJ}^{-1})}$$

$= 239.5$ kJ mol^{-1}

(b) $I_2 = I + I^*$ $\quad \Delta E = 239.5$ kJ/mol

$I_2 = 2I$ $\Delta E = 2(107.25) - 65.52 = 148.98$ kJ/mol

$I = I^*$ $\quad \Delta E = 90.53$ kJ/mol $= 0.938$ eV

14.3 The ultraviolet absorption of O_2 includes a series of lines (the Schumann-Runge bands) due to transitions from the $^3\Sigma_g^-$ ground state to the excited electronic state $^3\Sigma_u^-$, which are shown in Fig. 14.1. These lines converge to 175.9 nm, which corresponds to dissociation to one O atom in its ground state ^3P and one O atom in an excited state ^1D. What is D_0 for O_2? How does this compare with the enthalpy of formation at 0 K? Given: The ^1D state of O is 1.970 eV above the ground state ^3P.

SOLUTION

The dissociation energy D_0 for $O_2(^3\Sigma_g^-) = O(^3P) + O(^1D)$ is

$$\Delta E = \frac{hcN_A}{\lambda}$$

$$= \frac{(6.626 \times 10^{-34}\ \text{J s})(2.998 \times 10^{8}\ \text{m s}^{-1})(6.022 \times 10^{23}\ \text{mol}^{-1})}{(175 \times 10^{-9}\ \text{m})}$$

$$= 680.2\ \text{kJ mol}^{-1} = \frac{(680.2\ \text{kJ mol}^{-1})}{(96.485\ \text{kJ eV}^{-1})} = 7.050\ \text{eV}$$

Since $O(^3P) = O(^1D)$ $\Delta E = 1.970$ eV

The dissociation energy of $O_2(^3\Sigma_g^-)$ into $2O(^3P)$ is

$D_0 = 7.050 - 1.970 = 5.080$ eV or 490.14 kJ mol^{-1}.

From Appendix C.3 $\Delta_f H^\circ$ (0 K) = 2(246.785) = 493.57 kJ mol^{-1}.

14.4 The spectroscopic dissociation energy of $^{127}I_2$ is 1.542 38 eV according to Table 13.4. What wavelength of light would you use to dissociate ground state molecules to ground state atoms if you wanted the atoms to fly away with velocity of
10^3 m s^{-1} ?

SOLUTION

$E = (1.542\ 38\ \text{eV})(96{,}485\ \text{J mol}^{-1}\ \text{eV}^{-1}) +$

$$2\left(\frac{1}{2}\right)\ 126.904 \times 10^{-3}\ \text{kg mol}^{-1})(10^3\ \text{m s}^{-1})^2$$

$= (1.488 \times 10^{-5} + 1.269 \times 10^5)\ \text{J mol}^{-1}$
$= 2.757 \times 10^5\ \text{J mol}^{-1} = N_A hc/\lambda$

$$\lambda = \frac{(6.022 \times 10^{23}\ \text{mol}^{-1})(6.626 \times 10^{-34}\ \text{J s})(2.998 \times 10^8\ \text{m s}^{-1})}{2.757 \times 10^5\ \text{J mol}^{-1}}$$

$= 4.339 \times 10^{-7}\ \text{m} = 433.9\ \text{nm}$

14.5 A solution of a dye containing 0.1 mol L^{-1} transmits 80% of the light at 435.6 nm in a glass cell 1 cm thick. (a) What percent of light will be absorbed by a solution containing 2 mol L^{-1} in a cell 1 cm thick? (b) What concentration will be required to absorb 50% of the light? (c) What percentage of the light will be transmitted by a solution of the dye containing 0.1 mol L^{-1} in a cell 5 cm thick? (d) What thickness should the cell be in order to absorb 90% of the light with a solution of this concentration?

SOLUTION

$\log (I_o/I) = \varepsilon cl$
$\log (100/80) = \varepsilon(0.1\ \text{mol L}^{-1})(1\ \text{cm})$
$\varepsilon = 0.969\ \text{L mol}^{-1}\ \text{cm}^{-1}$

(a) $I = I_o e^{-2.303\ \varepsilon cl}$

$= (100) \exp [-(2.303)(0.969\ \text{L mol}^{-1}\ \text{cm}^{-1})(2\ \text{mol L}^{-1})(1\ \text{cm})]$

$= 1.2\%$ (98.8% absorbed)

(b) $\log (100/50) = (0.969\ \text{L mol}^{-1}\ \text{cm}^{-1})(1\ \text{cm})c$

$c = 0.311\ \text{mol L}^{-1}$

(c) $I = (100) \exp[- (2.303)(0.969\ \text{L mol}^{-1}\ \text{cm}^{-1})(0.1\ \text{mol L}^{-1})(5\ \text{cm})]$

$= 32.8\%$

(d) $\log(100/10) = (0.969\ \text{L mol}^{-1}\ \text{cm}^{-1})(0.1\ \text{mol L}^{-1})l$

$l = 10.3\ \text{cm}$

14.6 Derive equation 14.19 for the integrated intensity of a Gaussian absorption line. A Gaussian line has the form $\varepsilon_m\, e^{-\sigma(\tilde{\nu} - \tilde{\nu}_m)^2}$, where $\tilde{\nu}_m$ is the frequency at the intensity maximum, ε_m. [Hint: Relate σ to the width of the line at half-maximum intensity and use the integral

$$\int_{-\infty}^{+\infty} e^{-\sigma x^2}\, dx = (\pi/\sigma)^{1/2}]$$

SOLUTION

$$\varepsilon(\tilde{\nu}) = \varepsilon_m e^{-\sigma(\tilde{\nu} - \tilde{\nu}_m)^2}$$

$$\int_{-\infty}^{+\infty} \varepsilon(\tilde{\nu})d\tilde{\nu} = \varepsilon_m \int_{-\infty}^{+\infty} d\tilde{\nu}\; e^{-\sigma(\tilde{\nu} - \tilde{\nu}_m)^2}$$

$$= \varepsilon_m \int_{-\infty}^{+\infty} dx\, e^{-\sigma x^2}$$

$$= \varepsilon_m \left(\frac{\pi}{\sigma}\right)^{1/2}$$

At $\tilde{\nu} = \tilde{\nu}_{1/2}$, $\varepsilon(\tilde{\nu}) = \varepsilon_m/2$

$$\therefore \tilde{\nu}_{1/2} = \tilde{\nu}_m \pm \frac{1}{\sigma^{1/2}}(\ln 2)^{1/2}$$

Full width at half maximum,

$$\therefore \Delta\tilde{\nu}_{1/2} = 2\sigma^{-1/2}(\ln 2)^{1/2}$$

$$\therefore \frac{1}{\sigma^{1/2}} = \frac{\Delta\tilde{\nu}_{1/2}}{2(\ln 2)^{1/2}}$$

$$\therefore \int_{-\infty}^{+\infty} \varepsilon(\tilde{\nu})d\tilde{\nu} = \varepsilon_m \bullet \frac{\pi^{1/2}}{2(\ln 2)^{1/2}} \bullet \Delta\tilde{\nu}_{1/2} = (1.064)\, \varepsilon_m \Delta\tilde{\nu}_{1/2}$$

14.7 According to equation 14.19, the integrated absorption coefficient $\int \kappa(\tilde{\nu})d\tilde{\nu}$ is equal to $1.06\kappa_{max}\Delta\tilde{\nu}_{1/2}$ when the absorption band is Gaussian in shape. The quantity $\Delta\tilde{\nu}_{1/2}$ is the width of the band when $\kappa(\tilde{\nu}) = \kappa_{max}/2$. For such a band, the Naperian absorption coefficient $\kappa(\tilde{\nu})$ is given by
$\kappa(\tilde{\nu}) = \kappa_{max} \exp[-\alpha(\tilde{\nu} - \tilde{\nu}_{max})^2]$
Derive equation 14.19.

SOLUTION

At the frequency $\tilde{\nu}_{1/2}$ that gives half of the maximum Naperian absorption coefficient,

$$\kappa(\tilde{\nu}_{1/2}) = \kappa_{max} \exp[-\alpha(\tilde{\nu}_{1/2} - \tilde{\nu}_{max})^2]$$

$$\frac{1}{2} = \exp[-\alpha(\tilde{\nu}_{1/2} - \tilde{\nu}_{max})^2]$$

$$\ln 2 = \alpha(\tilde{\nu}_{1/2} - \tilde{\nu}_{max})^2$$

$$\frac{0.833}{\alpha^{1/2}} = \tilde{\nu}_{1/2} - \tilde{\nu}_{max}$$

Since $\Delta\tilde{\nu}_{1/2}$ is the width of the band at half height,

$$\Delta\tilde{\nu}_{1/2} = 2(\tilde{\nu}_{1/2} - \tilde{\nu}_{max}) = \frac{2(0.833)}{\alpha^{1/2}}$$

Thus

$$\alpha = \left(\frac{2(0.833)}{\Delta\tilde{\nu}_{1/2}}\right)^2$$

Integrating the expression for the Naperian absorption coefficient over the entire frequency range yields

$$\int_{-\infty}^{\infty} \kappa(\tilde{\nu})d\tilde{\nu} = \kappa_{max}\int_{-\infty}^{\infty} \exp[-\alpha(\tilde{\nu} - \tilde{\nu}_{max})^2]d\tilde{\nu}$$

Table 17.1 shows that

$$\int_{-\infty}^{\infty} \exp[-ax^2)dx = \left(\frac{\pi}{a}\right)^{1/2}$$

Thus

$$\int_{-\infty}^{\infty} \kappa(\tilde{\nu})d\tilde{\nu} = \kappa_{max}\left(\frac{\pi}{\alpha}\right)^{1/2} = \frac{\pi^{1/2}}{2(0.833)}\kappa_{max}\Delta\tilde{\nu}_{1/2} = 1.06\,\kappa_{max}\Delta\tilde{\nu}_{1/2}$$

14.8 The following absorption data are obtained for solutions of oxyhemoglobin in pH 7 buffer at 575 nm in a 1-cm cell:

g/cm^3	Transmission, %
3 x 10^{-4}	53.5
5 x 10^{-4}	35.1
10 x 10^{-4}	12.3

The molar mass of hemoglobin in 64.0 kg mol^{-1}. (a) Is Beer's law obeyed? What is the molar absorption coefficient? (b) Calculate the percent transmission for a solution containing 10^{-4} g/cm^3.

SOLUTION

(a)

grams per cm^3	mol L^{-1}	I/I_o	$\log(I/I_o)$	ε/L mol^{-1} cm^{-1}
3 x 10^{-4}	4.69 x 10-6	0.535	- 0.272	5.80 x 10^4
5 x 10^{-4}	7.82 x 10^{-6}	0.351	- 0.455	5.82 x 10^4
10 x 10^{-4}	15.64 x 10^{-6}	0.123	- 0.910	5.82 x 10^4

Beer's law is obeyed and the molar absorption coefficient is

5.81 x 10^4 L mol^{-1} cm^{-1}.

(b) $\log(I/I_o) = -\varepsilon cl$

$= -(5.81 \times 10^4 \text{ L mol}^{-1} \text{ cm}^{-1})(1.564 \times 10^{-6} \text{ mol L}^{-1})(1 \text{ cm})$

$= -0.091$

$I/I_o = 0.81$ or 81% transmission

*14.9 The protein metmyoglobin and imidazole form a complex in solution. The molar absorption coefficients in L mol^{-1} cm^{-1} of the metmyoglobin (Mb) and the complex (C) are as follows:

$\frac{\lambda}{\text{nm}}$	$\frac{\varepsilon_{Mb}}{10^3 \text{ L mol}^{-1} \text{ cm}^{-1}}$	$\frac{\varepsilon_C}{10^3 \text{ L mol}^{-1} \text{ cm}^{-1}}$
500	9.42	6.88
630	3.58	1.30

An equilibrium mixture in a cell of 1 cm path length has an absorbance of 0.435 at 500 nm and 0.121 at 630 nm. What are the concentrations of metmyoglobin and complex?

SOLUTION

$\log(I_o/I) = A = (\varepsilon_1 c_1 + \varepsilon_2 c_2)1$

At 500 nm, $0.435 = 9.42 \times 10^3 c_1 + 6.88 \times 10^3 c_2$

At 630 nm, $0.121 = 3.58 \times 10^3 c_1 + 1.30 \times 10^3 c_2$

Solving these equations simultaneously gives

$c_1 = 2.17 \times 10^{-5}$ mol L^{-1} (metmyoglobin)

$c_2 = 3.37 \times 10^{-5}$ mol L^{-1} (complex)

14.10 The absorption spectrum for benzene in Fig. 14.18 shows maxima at about 180, 200, and 250 nm. Estimate the integrated absorption coefficients using ε_{max} and $\Delta\tilde{\nu}_{1/2}$ and assuming that the width at half-maximum is

5000 cm^{-1} in each case. What are the three oscillator strengths? (See Example 14.7.)

SOLUTION

At 180 nm

$$\varepsilon_{max}\Delta\tilde{\nu}_{1/2} = (50.1 \times 10^3 \text{ L mol}^{-1} \text{ cm}^{-1})(5000 \text{ cm}^{-1})$$

$$= 2.50 \times 10^8 \text{ L mol}^{-1} \text{ cm}^{-2}$$

$$f = (4.32 \times 10^{-9} \text{ L mol}^{-1} \text{ cm}^2)(2.50 \times 10^8 \text{ L mol}^{-1} \text{ cm}^{-2}) = 1.08$$

At 200 nm

$$\varepsilon_{max}\Delta\tilde{\nu}_{1/2} = (7000 \text{ L mol}^{-1} \text{ cm}^{-1})(5000 \text{ cm}^{-1})$$

$$= 3.5 \times 10^7 \text{ L mol}^{-1} \text{ cm}^{-2}$$

$$f = (4.32 \times 10^{-9} \text{ L mol}^{-1} \text{ cm}^2)(3.50 \times 10^7 \text{ L mol}^{-1} \text{ cm}^{-2}) = 0.152$$

At 250 nm

$$\varepsilon_{max}\Delta\tilde{\nu}_{1/2} = (100 \text{ L mol}^{-1} \text{ cm}^{-1})(5000 \text{ cm}^{-1})$$

$$= 5 \times 10^5 \text{ L mol}^{-1} \text{ cm}^{-2}$$

$$f = (4.32 \times 10^{-9} \text{ mol L}^{-1} \text{ cm}^2)(5 \times 10^5 \text{ L mol}^{-1} \text{ cm}^{-2}) = 0.0022$$

14.11 Relatively strong absorption bands have ε_{max} values of 10^4 to 10^5 L mol^{-1} cm^{-1} and $\Delta\tilde{\nu}_{1/2}$ of the order 1000 to 5000 cm^{-1}, while weak absorption bands have ε_{max} = 10 L mol^{-1} cm^{-1} and $\Delta\tilde{\nu}_{1/2}$ of the order 100 cm^{-1}. Assuming the absorption lines are Gaussian, compute the integrated absorption coefficient and the oscillator strengths for these bands.

SOLUTION

$$\int \varepsilon \, d\tilde{\nu} = 1.06\, \varepsilon_{max}\, \Delta\tilde{\nu}_{1/2}; \; 4.32 \times 10^{-9} \int \varepsilon \, d\tilde{\nu} = f$$

Strong

$$\int \varepsilon \, d\tilde{\nu} = 1.06 \times \begin{matrix} 10^4 \times 1000 \\ 10^5 \times 5000 \end{matrix} = \begin{matrix} 1.06 \times 10^7 \text{ L mol}^{-1} \text{ cm}^{-2} \\ 5.30 \times 10^8 \text{ L mol}^{-1} \text{ cm}^{-2} \end{matrix}$$

$f = 4.58 \times 10^{-2}$ to 2.29

Weak

$$\int \varepsilon \, d\tilde{\nu} = 1.06 \times 10 \times 100 = 1.06 \times 10^3 \text{ L mol}^{-1} \text{ cm}^{-2}$$

$f = 4.58 \times 10^{-6}$

14.12 The measured oscillator strength of a transition can be used to compute the transition moment, $|\boldsymbol{\mu}_{12}|^2$ by combining equations 14.15 and 14.18 to find

$$|\boldsymbol{\mu}_{12}|^2 = (3fhe^2/8\pi^2 m_e \nu)$$

For strong transitions (for which $f \approx 1$), moderately weak transitions ($f \approx 10^{-3}$) and weak transitions ($f \approx 10^{-6}$), calculate $|\boldsymbol{\mu}_{12}|$ and $|\boldsymbol{R}_{12}| = |\boldsymbol{\mu}_{12}|/e$, assuming a transition energy of 25 000 cm^{-1}.

SOLUTION

$$|\boldsymbol{\mu}_{12}|^2 = \left(\frac{3fhe^2}{8\pi 2m_e(25\ 000) \times 3 \times 10^{10}\ s^{-1}}\right)^{1/2}$$

$$= e\, f^{1/2}\ (60.7\ \text{pm})$$

	$\lvert\boldsymbol{\mu}_{12}\rvert$	$\lvert\boldsymbol{R}_{12}\rvert$
strong	9.72×10^{-30} C m	60.7 pm
moderate	3.08×10^{-31} C m	1.92 pm
weak	9.72×10^{-33} C m	0.061 pm

1 Debye = 3.34×10^{-30} C m

14.13 In Chapter 11, the Hückel molecular orbital model was introduced to describe the electronic states of conjugated molecules. In this chapter, the free electron model (FEMO) was introduced for the same systems. Consider the butadiene molecule in both descriptions. The Hückel model (equation 11.67) gives the energies and wavefunctions for four orbitals, while the FEMO model gives an infinite number of orbitals. Consider the lowest four in the FEMO model. Do they have the same number of nodes as the Hückel orbitals? Is there any way of choosing α and β in the Hückel model or a in the FEMO model to make the predictions for the energies of all four orbitals agree? Supposing we are content to make the lowest electronic absorption energy agree in both models, what is the formula for β in terms of a?

SOLUTION

FEMO $E = \dfrac{n^2h^2}{8ma^2}$ $\quad n = 1, 2, 3, 4$

$\phi_n = ()\sin\dfrac{n\pi x}{a}$ so n = # nodes - 1

$0 \le x \le a$ same as Hückel

Hückel: Eq 11.65

$E_1 = \alpha + 1.618\ \beta$

$E_2 = \alpha + 0.618\ \beta$

$E_3 = \alpha - 0.618\ \beta$

$E_4 = \alpha - 1.618\ \beta$

Since $(E_2 - E_1)_{FEMO} = \dfrac{3h^2}{8ma^2}$ $(E_2 - E_1)_{HU} = -\beta$

$(E_3 - E_2)_{FEMO} = \dfrac{5h^2}{8ma^2}$ $\qquad (E_3 - E_2)_{HU} = -1.236\beta$

$(E_4 - E_3)_{FEMO} = \dfrac{7h^2}{8ma^2}$ $\qquad (E_4 - E_3)_{HU} = -\beta$

OR: 3:5:7 1:1.236:1

There is no way to make the predictions agree for all four orbitals.

First excitation energy:

	FEMO	HÜCKEL
$(E_3 - E_2)$	$\dfrac{5h_2}{8ma^2}$	-1.236β

$$\therefore\ \beta = -\frac{5h^2}{8(1.236)ma^2}$$

14.14 The lifetimes of vibrationally excited states of molecules of a liquid are limited by the collision rates in the liquid. If one in ten collisions deactivates a vibrationally excited state, what is the broadening of vibrational lines if a molecule undergoes 10^{13} collisions per second?

SOLUTION

$$\Delta E = hc\Delta\tilde{\nu} \geq \frac{h}{2\pi\Delta t}$$

$$\Delta\tilde{\nu} \geq (2\pi c\Delta t)^{-1} = [2\pi(3 \times 10^{10}\ \text{cm s}^{-1})(10^{-12}\ \text{s})]^{-1}$$

$$\Delta\tilde{\nu} \geq 5.3\ \text{cm}^{-1}$$

14.15 Calculate the linewidth for (a) an electronic excited state with a lifetime of 10^{-8} s and (b) a rotational state with a lifetime of 10^3 s. In each case express the linewidth in cm^{-1} and MHz.

SOLUTION

(a) $$\Delta\tilde{\nu} = \frac{1}{2\pi c\tau} = \frac{1}{2\pi(3 \text{ x } 10^{10} \text{ cm s}^{-1})(10^{-8} \text{ s})} = 5.3 \text{ x } 10^{-4} \text{ cm}^{-1}$$

$$\Delta\nu = \frac{1}{2\pi\tau} = \frac{1}{2\pi(10^{-8} \text{ s})} = 16 \text{ MHz}$$

(b) $$\Delta\tilde{\nu} = \frac{1}{2\pi c\tau} = \frac{1}{2\pi(3 \text{ x } 10^{10} \text{ cm s}^{-1})(10^{3} \text{ s})} = 5.3 \text{ x } 10^{-15} \text{ cm}^{-1}$$

$$\Delta\nu = \frac{1}{2\pi\tau} = \frac{1}{2\pi(10^{3} \text{ s})} = 1.6\text{x}10^{-4} \text{ MHz}$$

14.16 A laser is powered by a 100 W flashlamp that produces pulses with a repetition rate of 100 Hz. If the efficiency of the laser is 1%, how many photons will there be in a laser pulse with a wavelength of 500 nm?

SOLUTION

The energy of a pulse from the flashlamp is 100 J s^{-1}/100 s^{-1} = 1 J. If 1% of this energy is converted into a laser pulse of 500 nm, the number of photons in a pulse is

$$N = \frac{E\lambda}{hc} = \frac{(0.01 \text{ J})(500\text{x}10^{-9} \text{ m})}{(6.625\text{x}10^{-34} \text{ J s})(2.998\text{x}10^{8} \text{ m s}^{-1})} = 2.517\text{x}10^{16}$$

14.17 What is the width of the frequency distribution for a 10 fs pulse and for a 100 fs pulse from a laser?

SOLUTION

$$\Delta\nu = \frac{1}{2\pi\Delta t}$$

For a 10 fs pulse, $\Delta\nu = \dfrac{1}{2\pi(10\text{x}10^{-15} \text{ s})} = 1.59\text{x}10^{13} \text{ s}^{-1}$ or $\Delta\nu/c = 531 \text{ cm}^{-1}$

For a 100 fs pulse, $\Delta\nu = \dfrac{1}{2\pi(100\text{x}10^{-15} \text{ s})} = 1.59\text{x}10^{12} \text{ s}^{-1}$ or $\Delta\nu/c = 53.1 \text{ cm}^{-1}$

14.18 A laser operating at 700 nm produces 20 fs pulses with a repetition rate of 100 MHz. The average radiant power of the laser is 1 W. (a) What is the radiant power in each pulse? (b) How many photons are there in a pulse? (c) How many photons are emitted by the laser in one second?

SOLUTION

(a) $\frac{1 \text{ J s}^{-1}}{10^8 \text{ s}^{-1}} = 10^{-8} \text{ J}$

(b) $h\nu = \frac{hc}{\lambda} = \frac{(6.626\text{x}10^{-34} \text{ J s})(2.998\text{x}10^8 \text{ m s}^{-1})}{800\text{x}10^{-9} \text{ m}} = 2.48\text{x}10^{-19} \text{ J}$

$\frac{10^{-8} \text{ J}}{2.48\text{x}10^{-19} \text{ J}} = 4.03\text{x}10^{10}$ photons

(c) $(4.03\text{x}10^{10}$ photons$)(10^8 \text{ s}^{-1}) = 4.03\text{x}10^{18}$ photons s^{-1}

14.19 A laser powered by a 100 W light source produces photons with 1000 nm wavelength at a repetition rate of 10 Hz. The actual number of photons emitted per pulse is 10^{17}. What is the efficiency of converting energy from the light source to laser output?

SOLUTION

The energy available for each pulse is 10 J. The energy of an emitted photon is

$\frac{hc}{\lambda} = \frac{(6.626\text{x}10^{-34} \text{ J s})(2.998\text{x}10^8 \text{ m s}^{-1})}{(1000\text{x}10^{-9} \text{ m})} = 1.99\text{x}10^{-19} \text{ J}$

The maximum number of these photons that could be produced in a pulse corrresponding to 10 J is

$\frac{10 \text{ J}}{1.99\text{x}10^{-19} \text{ J}} = = 5.03\text{x}10^{19}$

$\text{Efficiency} = \frac{10^{17}}{5.03\text{x}10^{19}} = 0.00199 \text{ or } 0.199 \%$

14.20 The first ionization potentials of Ar, Kr, and Xe are 15.755 eV, 13.966 eV, and 12.130 eV, respectively. Calculate the velocity of the emitted electrons when photons from a He discharge lamp with $\lambda = 58.43$ nm are used to record the photoelectron spectra of these gases.

SOLUTION

$\lambda = 58.43 \text{ nm}; E_{\text{photon}} = \frac{hc}{\lambda} = \frac{6.626 \text{ x } 10^{34} \text{ x } 2.998 \text{ x } 10^8}{58.43 \text{ x } 10^{-9}} \text{ J}$

$= 3.3998 \text{ x } 10^{-18} \text{ J}$

$= 21.22 \text{ eV}$

(a) Ar

$\Delta E = 21.22 - 15.755 \text{ eV} = 5.47 \text{ eV}$

$$v = \left\{\frac{2\Delta E}{m_{Ar}}\right\}^{1/2} = \left\{\frac{2x5.47x1.602x10^{-19}\text{J}}{(39.948/6.022\text{x}10^{23})x10^{-3}\text{kg}}\right\}^{1/2} = 5.14 \text{ km/s}$$

(b) Kr

$\Delta E = 21.22 - 13.966 = 7.26 \text{ eV}$

$$\upsilon = \left\{\frac{2 \text{ x } 7.26 \text{ x } 1.602 \text{ x } 10^{-19}}{[83.80/(6.022 \text{ x } 10^{23})] \text{ x } 10^{-3}}\right\}^{1/2} = 4.09 \text{ km/s}$$

(c) Xe

$\Delta E = 21.22 - 12.13 = 9.09 \text{ eV}$

$$\upsilon = \left\{\frac{2 \text{ x } 9.09 \text{ x } 1.602 \text{ x } 10^{-19}}{[131.29/(6.022 \text{ x } 10^{23})] \text{ x } 10^{-3}}\right\}^{1/2} = 3.66 \text{ km/s}$$

14.21 A sample of oxygen gas is irradiated with $MgK\alpha_1\alpha_2$ radiation of 0.99 nm (1253.6 eV). A strong emission of electrons with velocities of $1.57 \text{ x } 10^7$ m s^{-1} is found. What is the binding energy of these electrons?

SOLUTION

$$\frac{1}{2} m\upsilon^2 = h\nu - I$$

$$I = h\nu - \frac{1}{2} m\upsilon^2$$

$$= 1253.6 \text{ eV} - \frac{1}{2}\frac{(9.109 \text{ x } 10^{-31} \text{ kg})(1.57 \text{ x } 10^7 \text{ m s}^{-1})^2}{(1.602 \text{ x } 10^{-19} \text{ J eV}^{-1})} = 552.8 \text{ eV}$$

14.22 The photoelectron spectrum of molecules shows that similar atoms in different chemical environments have slightly different core orbital binding energies. For example, the 1s binding energy of carbon in CH_4 is 290 eV, while it is 293 eV in CH_3F. (a) Explain this shift based on the electronegativity difference between carbon and fluorine. (b) In the molecule $F_3CCOOCH_2CH_3$, predict the order of the carbon 1s binding energies in the four carbon atoms.

SOLUTION

(a) Since F is more electronegative than C and H, it pulls electrons toward itself, thereby decreasing the shielding at the carbon nucleus, and increasing the binding energy of the 1s electrons

(b) The binding energy should be greatest for the CF_3 carbon, next largest for the COO carbon, next for the OCH_2 carbon, and smallest for the CH_3 carbon.

14.23 When α-D-mannose ($[\alpha]^{20}$ = +29.3°) is dissolved in water, the optical rotation decreases as β-D-mannose is formed until at equilibrium $[\alpha]^{20}$ = +14.2°. This process is referred to as mutarotation. As expected, when β-D-mannose ($[\alpha]^{20}$ = -17.0°) is dissolved in water, the optical rotation increases until $[\alpha]^{20}$ = +14.2° is obtained. Calculate the percentage of the α form in the equilibrium mixture.

SOLUTION

$[\alpha]_\alpha = 29.3°$ $\qquad [\alpha]_\beta = -17.0°$

$$[\alpha]_{mixture} = +14.2° = 29.3°\, f_\alpha - 17.0°\, f_\beta$$
$$= 29.3°\, f_\alpha - 17.0°\,(1 - f_\alpha)$$
$$= -17.0° + 36.3\, f_\alpha$$

$$f_\alpha = \frac{31.2°}{46.3°} = 0.67$$

The percentage of α form in the equilibrium mixture is 67%.

14.24 - 92.2 kJ mol^{-1}

14.25 (a) 16,065, (b) 1.992 eV

14.26 (a)

	$E/10^{-19}$ J	E/kJ mol^{-1}
$n = 1$	2.410	145.14
$n = 2$	9.640	580.57

(b) 275 nm

14.27 210 nm

14.28 196 L mol^{-1} cm^{-1}, 24.7 %

14.29 8.48 x 10^7 L mol^{-1} cm^{-2}, 0.366

14.30 0.091, 88.2%

14.31 7.7 x 10^{-5} g cm^{-3}

14.32 2.6 μg/μL

14.33 (a) 0.324×10^{-4}, 0.512×10^{-4} mol L^{-1}

(b) 2.95×10^{4} L mol^{-1}

14.34 $\varepsilon_{app} = \varepsilon_{base}/[1 + 10^{-(pH - pK)}]$

14.35 7.3×10^{-4}

14.36 14.04 eV

14.37 5.20 eV

14.38 63.6%

15

Magnetic Resonance Spectroscopy

15.1 NMR spectrometers usually have fixed frequencies between 60 and 750 MHz. Calculate the magnetic flux densities needed to give an NMR transition frequency for hydrogen equal to these frequencies.

SOLUTION

For 60 MHz

$$B = \frac{h\nu}{|g_N|\mu_N} = \frac{(6.626 \times 10^{-34}\ \text{J s})(6 \times 10^{7}\ \text{s}^{-1})}{(5.585)(5.0508 \times 10^{-27}\ \text{J T}^{-1})} = 1.409\ \text{T}$$

For 750 MHz

$$B = 17.61\ \text{T}$$

15.2 For the frequencies of problem 15.1, calculate the corresponding energies in kJ mol^{-1} and compare these with RT at 300 K.

SOLUTION

(a) 60 MHz

$$E = h\nu = \frac{6.626 \times 10^{-34} \times 6 \times 10^{7}\ \text{s}^{-1} \times 6.022 \times 10^{23}\ \text{mol}^{-1}}{10^{3}\ \text{J kJ}^{-1}}$$

$$= 2.394 \times 10^{-5}\ \text{kJ mol}^{-1}$$

$$RT = 8.314 \times 10^{-3} \times 300\ \text{kJ mol}^{-1} = 2.494\ \text{kJ mol}^{-1}$$

$$\therefore \frac{E(60\ \text{MHz})}{RT} = 0.960 \times 10^{-5}$$

(b) 750 MHz

$$E = 2.993 \times 10^{-2}\ \text{kJ mol}^{-1}$$

$$\frac{E(750\ \text{mHz})}{RT} = 1.200 \times 10^{-4}$$

15.3 (a) What are the energy levels for a ^{23}Na nucleus in a magnetic field of 2 T? (b) What is the absorption frequency?

SOLUTION

(a) $E = |g_N|\mu_N m_I B = (1.478)(5.0508 \times 10^{-27}\ \mathrm{J\ T^{-1}})(2\ \mathrm{T})\, m_I$

$= (1.493 \times 10^{-26}\ \mathrm{J}) m_I$

Since $I = \frac{3}{2}$, $m_I = -\frac{3}{2}, -\frac{1}{2}, \frac{1}{2}, \frac{3}{2}$ and

$E = -2.240 \times 10^{-26}, -0.7465 \times 10^{-26}, 0.7465 \times 10^{-26}$, and 2.240×10^{-26} J

(b) $$\nu = \frac{\Delta E}{h} = \frac{1.493 \times 10^{-26}\ \mathrm{J}}{6.6262 \times 10^{-34}\ \mathrm{J\ s}} = 22.53\ \mathrm{MHz}$$

15.4 The magnetogyric ratio γ for a nucleus is defined by $\mu = \gamma |I|$. What is the value of γ for H?

SOLUTION

$\gamma = \mu / |I|$ and equation 15.10 shows that $\mu / |I| = g_N \mu_N / \hbar$

Therefore,

$$\gamma_N = g_N \mu_N / \hbar$$

$$= \frac{2\pi(5.585)(5.0508 \times 10^{-17}\ \mathrm{J\ T^{-1}})}{(6.6262 \times 10^{-34}\ \mathrm{J\ s})} = 2.675 \times 10^{8}\ \mathrm{s^{-1}\ T^{-1}}$$

15.5 What is the difference in fractional populations of ^{13}C spins between the upper and lower states in a magnetic field of 2 T at room temperature?

SOLUTION

The fractional population of spins in the lower state is given by

$$\frac{N_l - N_u}{N_l} = \frac{\frac{N_l}{N_u} - 1}{\frac{N_l}{N_u}}$$

Example 15.3 shows that

$$\frac{N_l}{N_u} = 1 + \frac{g_N \mu_N B}{kT}$$

Therefore,

$$\frac{N_l - N_u}{N_l} = \frac{g_N \mu_N B/kT}{1 + g_N \mu_N B/kT}$$

$$= \frac{(1.405)(5.052 \times 10^{-27})(2)/(1.3807 \times 10^{-23} \times 298.15)}{1 + (3.5 \times 10^{-6})}$$

$$= 3.4485 \times 10^{-6}$$

15.6 In a magnetic field of 2 T, what fraction of the protons have their spin lined up with the field at room temperature?

SOLUTION

$$\frac{N_l}{N_h} = 1 + \frac{g_N \mu_N B}{kT}$$

$$= 1 + \frac{(5.585)(5.05 \times 10^{-27}\ \mathrm{J\ T^{-1}})(2\ \mathrm{T})}{(1.38 \times 10^{-23}\ \mathrm{J\ K^{-1}})(298\ \mathrm{K})} = 1.000\ 013\ 72$$

$$\frac{N_l}{N_l + N_u} = \frac{1}{1 + \frac{N_h}{N_l}} = \frac{1}{1 + \frac{1}{(N_h/N_l)}} = \frac{1}{1 + \frac{1}{1.000\ 013\ 72}} = 0.500\ 003\ 43$$

15.7 It is now possible to do NMR experiments at very low temperatures. Calculate the ratio of the number of protons in the upper spin state to that in the lower spin state in a magnetic field of 2 T at 1 mK and 10 mK.

SOLUTION

$$\frac{N_u}{N_l} = e^{-\Delta E/kT} = e^{-g_N \mu_N B/kT}$$

(a) At 1 mK

$$\frac{g_N \mu_N B}{kT} = \frac{(5.585)(5.05 \times 10^{-27}\ \mathrm{J\ T^{-1}})(2\ \mathrm{T})}{(1.38 \times 10^{-23}\ \mathrm{J\ K^{-1}})(0.001\ \mathrm{K})} = 4.08$$

$$\frac{N_u}{N_l} = e^{-4.08} = 1.69 \times 10^{-2}$$

(b) At 10 mK

$$\frac{N_u}{N_l} = e^{-0.408} = 0.665$$

15.8 (a) What is the value of the magnetogyric ratio for the proton? (b) What is the Larmor frequency for the proton at 10 T?

SOLUTION

(a) $\gamma = \dfrac{g_I\mu_N}{\hbar} = \dfrac{(5.586)(5.0508 \text{ x } 10^{-27} \text{ J T}^{-1})}{1.054 \text{ x } 10^{-34} \text{ J s}}$

$= 2.676 \text{ x } 10^{8} \text{ s}^{-1} \text{ T}^{-1}$

(b) $\nu_L = \dfrac{\gamma B}{2\pi} = \dfrac{2.676 \text{ x } 10^{8} \text{ s}^{-1} \text{ T}^{-1} (10 \text{ T})}{2\pi} = 425.8 \text{ MHz}$

15.9 Calculate the magnetic fields required for resonance at 300 MHz for (a) ^{31}P and (b) ^{33}S.

SOLUTION

(a) $B = \dfrac{h\nu}{g_N\mu_N} = \dfrac{(6.626 \text{ x } 10^{-34} \text{ J s})(300 \text{ x } 10^{6} \text{ s}^{-1})}{(2.2634)(5.0508 \text{ x } 10^{-27} \text{ J T}^{-1})} = 17.39 \text{ T}$

(b) $B = \dfrac{h\nu}{g_N\mu_N} = \dfrac{(6.626 \text{ x } 10^{-34} \text{ J s})(300 \text{ x } 10^{6} \text{ s}^{-1})}{(0.4289)(5.0508 \text{ x } 10^{-27} \text{ J T}^{-1})} = 9.176 \text{ T}$

15.10 Using information from Tables 15.2 and 15.3, sketch the spectrum you would expect for ethyl acetate ($CH_3CO_2CH_3$).

SOLUTION

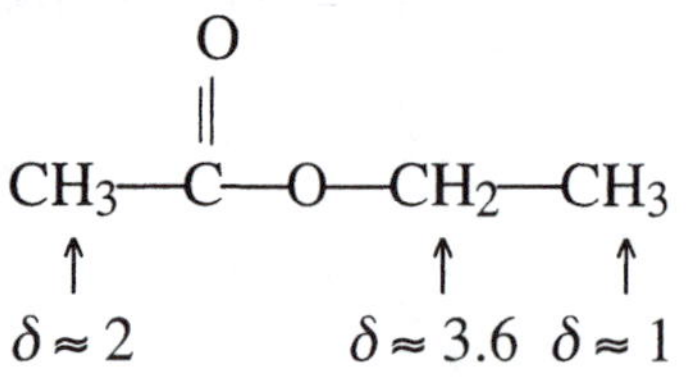

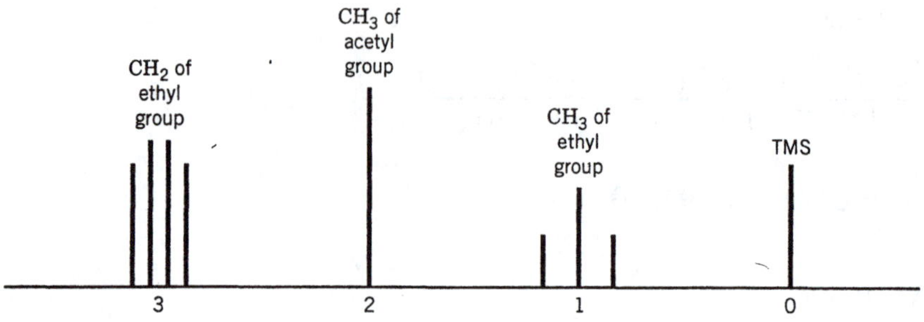

15.11 Chemical shifts δ are expressed in ppm, but they can also be expressed in Hz. What magnetic fields are necessary to produce frequency shifts of 100 Hz and 500 Hz for protons with a $\delta = 1$?

SOLUTION

$$\nu_i = |g_i| \mu_N (1 - \sigma_i) \frac{B}{h} \qquad \Delta\nu = |g_i| \mu_N \sigma_i \frac{B}{h}$$

For 100 Hz $\quad B = \dfrac{h\Delta\nu}{|g_i| \mu_N \sigma_i} = \dfrac{(6.626 \times 10^{-34}\ \text{J s})(100\ \text{s}^{-1})}{(5.585)(5.051 \times 10^{-27}\ \text{J T}^{-1})(10^{-6})}$

$= 2.35$ T

For 500 Hz $\quad B = 5 \times 2.35$ T $= 11.7$ T

15.12 What is the separation of the methyl and methylene proton resonances in ethanol at (a) 60 MHz? (b) At 500 MHz? (See Table 15.2.)

SOLUTION

$\delta_{CH_3} = 1.17 \times 10^{-6} \qquad \delta_{CH_2} = 3.59 \times 10^{-6}$

(a) $\nu_0 \delta = (60 \times 10^6\ \text{Hz})(2.42 \times 10^{-6}) = 145.2$ Hz

(b) $\nu_0 \delta = (500 \times 10^6\ \text{Hz})(2.42 \times 10^{-6}) = 1210$ Hz

15.13 At a magnetic field strength of 1.41 T, the frequency separation between protons in benzene and protons in tetramethylsilane is 436.2 Hz. What is the chemical shift?

SOLUTION

$$\nu_0 = \frac{B g_N \mu_N}{h} = \frac{(1.41\ \text{T})(5.585)(5.0508 \times 10^{-27}\ \text{J T}^{-1})}{6.6262 \times 10^{-34}\ \text{J s}} = 60.0\ \text{MHz}$$

$\delta = 436.2$ Hz/60 MHz $= 7.270 \times 10^{-6}$

which is usually given as 7.270 ppm

15.14 Equation 15.26 indicates that the chemical shift δ measured with respect to a reference is a million times greater than the difference in shielding constants σ for the reference and the group of interest. Since the reference is arbitrary, we can also apply this equation to the difference between two groups. In the ethanol molecule, the chemical shift is 1.17 ppm for the protons in CH_3 and 3.59 ppm for the protons in CH_2. (a) What is the difference in shielding constants for these two types of protons? (b) What is the difference in the magnetic field at the protons in CH_3 and CH_2 when the applied field B_0 is 1 T and (c) 2 T?

SOLUTION

(a) $\Delta\delta = 10^6\Delta\sigma = 1.17 - 3.59 = -2.42$
$\Delta\sigma = -2.42 \times 10^{-6}$

(b) $\Delta B = B_0\Delta\sigma = (1\ \text{T})(-2.42 \times 10^{-6}) = -2.42\ \mu\text{T}$

(c) $\Delta B = B_0\Delta\sigma = (2\ \text{T})(-2.42 \times 10^{-6}) = -4.82\ \mu\text{T}$
The field is weaker at the protons in CH_3 than the protons in CH_2. The protons are more shielded in CH_3.

15.15 (a) Using equation 15.26, show that

$$\nu_1 - \nu_2 = \nu_{ref} \times 10^{-6}(\delta_1 - \delta_2)$$

where ν_1 and ν_2 are the resonance frequencies for protons in groups 1 and 2. (b) What is the difference in resonance frequencies for protons in CH_3 and CH_2 at 60 MHz?

SOLUTION

(a) According to equation 15.27, $10^{-6}\ \nu_{ref}\ \delta = \nu_S - \nu_R$
$\nu_1 = \nu_{ref}(1 + 10^{-6}\delta_1)$
$\nu_2 = \nu_{ref}(1 + 10^{-6}\delta_2)$
$\nu_1 - \nu_2 = \nu_{ref} \times 10^{-6}(\delta_1 - \delta_2)$

(b) $\nu_1 - \nu_2 = (60\ \text{Hz})(1.17 - 3.59) = -145.2\ \text{Hz}$
Thus CH_3 comes into resonance at a lower frequency, and the difference in chemical shifts can be stated as - 145.2 Hz at 60 MHz.

15.16 Sketch the proton resonance spectrum of $D_2CHCOCD_3$ (deuteroacetone containing a little hydrogen). Indicate the relative intensities of the lines.

SOLUTION

The deuteron has a spin of $I = 1$. Some of the methyl groups will be CD_2H-groups, so that the proton sees two equivalent deuterons. The first deuteron will split the proton resonance into a triplet, and the second deuteron will further split these lines with the same spin-spin coupling constant.

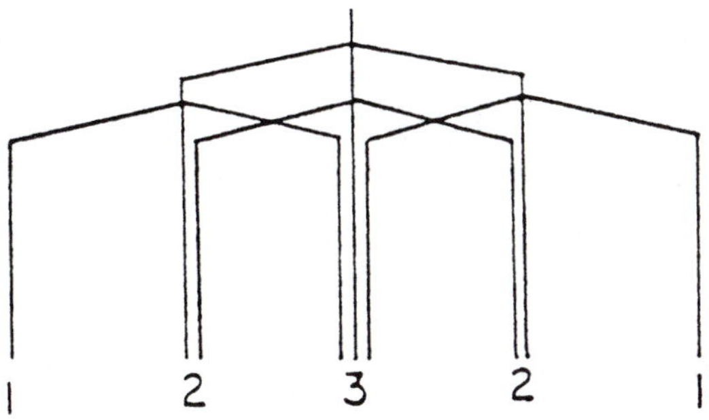

*15.17 The proton resonance pattern of 2,3-dibromothiophene shows an AB-type spectrum with lines at 405.22, 410.85, 425.07, and 430.84 Hz measured from tetramethylsilane at 1.41 T [K.F. Kuhlmann and C. L. Braun, *J. Chem. Ed.* **56**, 750 (1969)]. (a) What is the coupling constant *J*? (b) What is the difference in the chemical shifts of the A and B hydrogens? (c) At what frequencies would the lines be found at 2 T?

SOLUTION

(a) The average spacing of the two doublets is $J = 5.70 \pm 0.07$ Hz.

(b) $\nu_0\delta = [(a - d)(b - c)]^{1/2} = [(25.62)(14.22)]^{1/2} = 19.09$ Hz

$\delta = (19.09 \text{ Hz})/(60 \times 10^6 \text{ Hz}) = 0.318 \times 10^{-6} = 0.318$ ppm

(c) At 2 T $\nu_0' = \frac{2}{1.41}\ 60 \times 10^6 \text{ Hz} = 85.1 \times 10^6$ Hz

The center of the spectrum shifts by ν_0'/ν_0

$$418.00\frac{85.1 \times 10^6}{60 \times 10^6} = 592.86$$

$$\text{Distance of } b \text{ and } c \text{ from center} = \frac{[(\nu_0\delta)^2 + J^2]^{1/2} - J}{2}$$

$$= \frac{1}{2}\left\{[(85.1 \times 0.318)^2 + 5.70^2]^{1/2} - 5.70\right\} = 10.98$$

$a = 592.86$	$- 10.98$	$- 5.70$	$= 576.18$ Hz
$b = 592.86$	$- 10.98$		$= 581.88$ Hz
$c = 592.86$	$+ 10.98$		$= 603.84$ Hz
$d = 592.86$	$+ 10.98$	$+ 5.70$	$= 609.54$ Hz

*15.18 Derive the equations for the frequencies for an AX system from the equations for the energy levels for $\nu_0(\sigma_1 - \sigma_2) = 50$ s^{-1} and $J_{12} = 5$ s^{-1}.

SOLUTION

Equations 15.42 to 15.45 with $J_{12}^2 << \nu_0^2(\sigma_1 - \sigma_2)^2$ are

$$E_1 = -h\nu_0\left(1 - \frac{\sigma_1 + \sigma_2}{2}\right) + \frac{hJ_{12}}{4}$$

$$E_2 = -\frac{h\nu_0}{2}(\sigma_1 - \sigma_2) - \frac{hJ_{12}}{4}$$

$$E_3 = \frac{h\nu_0}{2}(\sigma_1 - \sigma_2) - \frac{hJ_{12}}{4}$$

$$E_4 = h\nu_0\left(1 - \frac{\sigma_1 + \sigma_2}{2}\right) + \frac{hJ_{12}}{4}$$

Since the selection rule requires that only one type of nucleus at a time can undergo a transition, the four transitions are given by

$$\nu(1\text{->}2) = \frac{1}{h}(E_2 - E_1) = \nu_0(1 - \sigma_1) - \frac{J_{12}}{2}$$

$$\nu(1\text{->}3) = \frac{1}{h}(E_3 - E_1) = \nu_0(1 - \sigma_2) - \frac{J_{12}}{2}$$

$$\nu(2\text{->}4) = \frac{1}{h}(E_4 - E_2) = \nu_0(1 - \sigma_2) + \frac{J_{12}}{2}$$

$$\nu(3\text{->}4) = \frac{1}{h}(E_4 - E_3) = \nu_0(1 - \sigma_1) + \frac{J_{12}}{2}$$

15.19 Calculate the relative frequencies and relative intensities for an AB system in which $\nu_0(\sigma_1 - \sigma_2) = 10\ \text{s}^{-1}$ and $J_{12} = 5\ \text{s}^{-1}$.

SOLUTION

$$\nu(1\text{->}2) = -\frac{1}{2}[\nu_0^2(\sigma_1 - \sigma_2)^2 + J_{12}^2]^{1/2} - \frac{J_{12}}{2} = -\frac{1}{2}(125)^{1/2} - 2.5 = -8.090$$

$$\nu(1\text{->}3) = \frac{1}{2}[\nu_0^2(\sigma_1 - \sigma_2)^2 + J_{12}^2]^{1/2} - \frac{J_{12}}{2} = \frac{1}{2}(125)^{1/2} - 2.5 = 3.090$$

$$\nu(2\text{->}4) = \frac{1}{2}[\nu_0^2(\sigma_1 - \sigma_2)^2 + J_{12}^2]^{1/2} + \frac{J_{12}}{2} = \frac{1}{2}(125)^{1/2} + 2.5 = 8.090$$

$$\nu(3\text{->}4) = -\frac{1}{2}[\nu_0^2(\sigma_1 - \sigma_2)^2 + J_{12}^2]^{1/2} + \frac{J_{12}}{2} = -\frac{1}{2}(125)^{1/2} + 2.5 = -3.090$$

$$r = \left[\frac{(\Delta^2 + J_{12}^2)^{1/2} + \Delta}{(\Delta^2 + J_{12}^2)^{1/2} - \Delta}\right]^{1/2} = \left[\frac{(125)^{1/2} + 10}{(125)^{1/2} - 10}\right]^{1/2} = 4.236$$

The relative intensities I are given by

$$I(1\text{->}3) = I(3\text{->}4) = 1$$

$$I(1\text{->}2) = I(2\text{->}4) = \left(\frac{r - 1}{r + 1}\right)^2 = \left(\frac{4.236 - 1}{4.236 + 1}\right)^2 = 0.382$$

Transition	Relative freq/s^{-1}	Relative intensity I
1->2	-8.090	0.382
3->4	-3.090	1
1->3	3.090	1
2->4	8.090	0.382

15.20 Verify the equation

$$E_2^{(0)} = - \hbar\, \gamma B_0\left(\frac{\sigma_1 - \sigma_2}{2}\right)$$

for the energy corresponding to $\psi_2 = \beta(1)\alpha(2)$ for a molecule containg two hydrogen atoms that do not interact.

SOLUTION

The Schrö dinger equation for this system is

$$\hat{H}^{(0)}\, \beta(1)\alpha(2) = - \gamma B_0(1 - \sigma_1)\hat{I}_{z1}\, \beta(1)\alpha(2) - \gamma B_0(1 - \sigma_2)\hat{I}_{z2}\, \beta(1)\alpha(2)$$

Since

$$\hat{I}_{zj}\, \beta(j) = - \frac{\hbar}{2}\, \beta(j) \text{ and } \hat{I}_{zj}\, \alpha(j) = - \frac{\hbar}{2}\, \alpha(j)$$

then

$$\hat{H}^{(0)}\, \beta(1)\alpha(2) = \frac{\hbar}{2}\, \gamma B_0(1 - \sigma_1)\beta(1)\alpha(2) + \frac{\hbar}{2}\, \gamma B_0(1 - \sigma_2)\beta(1)\alpha(2)$$

$$= - \frac{1}{2}\hbar\, \gamma B_0(\sigma_1 - \sigma_2)\psi_2$$

The eigenvalue for the energy is given by the coefficient of ψ_2.

15.21 Show that for a Lorenzian absorption line, the width at half maximum intensity $\Delta\nu_{1/2}$ is given by $1/\pi T_2$.

SOLUTION

$$I = \frac{T_2}{1 + T_2^2(2\pi)^2(\nu - \nu_0)^2}$$

The maximum intensity is at $\nu = \nu_0$. At $I(\nu)/I(\nu_0) = 1/2$,

$T_2^2(2\pi)^2(\nu - \nu_0)^2 = 1$

Therefore, when $\nu = \nu_0 \pm \Delta\nu_{1/2}/2$,

$$T_2^2(2\pi)^2\left(\frac{\Delta\nu_{1/2}}{2}\right)^2 = 1$$

or

$$T_2(2\pi)\left(\frac{\Delta\nu_{1/2}}{2}\right) = 1$$

Therefore, $\Delta\nu_{1/2} = 1/\pi T_2$

15.22 At room temperature the chemical shift of cyclohexane protons is an average of the chemical shifts of the axial and equatorial protons. Explain.

SOLUTION

At room temperature the rate of conversion of cyclohexane from boat to chair forms is so fast that the protons are at the average local magnetic field.

15.23 The two lines in the proton magnetic resonance spectrum for the two methyl groups connected to nitrogen in *N,N*-dimethylacetamide coalesce when the temperature is raised. What is the rate constant for the cis-trans isomerization when the multiplet structure is just lost at 331 K? The difference in chemical shifts between the two peaks is 10.85 Hz.

SOLUTION

$$k = \frac{\pi\Delta\nu}{\sqrt{2}} = \frac{\pi(10.85\text{ Hz})}{\sqrt{2}} = 24.1\text{ s}^{-1}$$

15.24 Calculate the transition (Larmor) frequency of a free electron in a 3 T field. What energy in cm^{-1} does this correspond to?

SOLUTION

$$\nu = \frac{Bg_e\mu_B}{h} = \frac{(3\text{ T})(2.0023)(9.2742\text{ x }10^{-24}\text{ J T}^{-1})}{6.6262\text{ x }10^{-34}\text{ J s}}$$

$$= 8.4074\text{ x }10^{10}\text{ s}^{-1} = 84\,074\text{ MHz}$$

$$\frac{1}{\lambda} = \frac{\nu}{c} = \frac{8.4074\text{ x }10^{10}\text{ s}^{-1}}{3\text{ x }10^{10}\text{ cm s}^{-1}} = 2.80\text{ cm}^{-1}$$

15.25 Line separations in ESR may be expressed in G or MHz. Show how the conversion factor 1 T = 2.80 x 10^4 MHz is obtained.

SOLUTION

The resonance frequency for electrons in a 1 tesla field is given by

$$\nu = \frac{g_e B \mu_B}{h} = \frac{(2.00)(9.274 \times 10^{-24}\ \mathrm{J\ T^{-1}})(1\ \mathrm{T})}{6.626 \times 10^{-34}\ \mathrm{J\ s}}$$

$$= 2.80 \times 10^{10}\ \mathrm{s^{-1}} = 2.80 \times 10^{4}\ \mathrm{MHz}$$

15.26 1.410×10^{-28} J T^{-1}

15.27 (a) 0.4669, (b) 0.1248 T

15.28 511 MHz, 0.017 cm^{-1}

15.29 425.8 MHz

15.30 4.99, 18.67 T

15.31 For protons, 6.856×10^{-6}

For fluorines, 6.453×10^{-6}

15.32

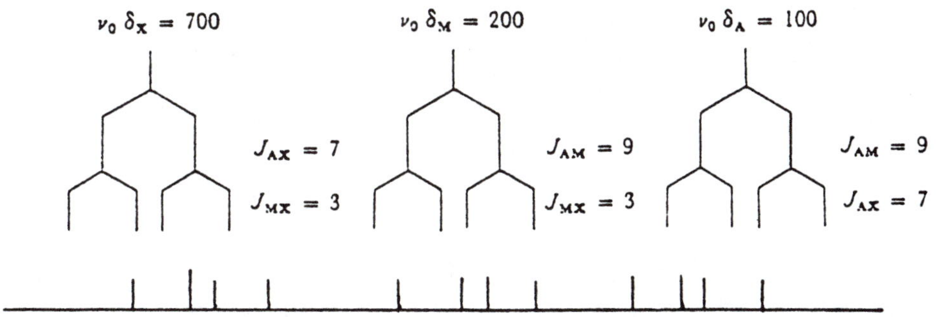

15.33 The proton resonance would be split into three lines by the deuteron, and the deuteron resonance would be split into two lines by the proton.

15.34 $J_{24} \sim 0.5$-4, $J_{25} \sim 0$, $J_{45} \sim 6$-9

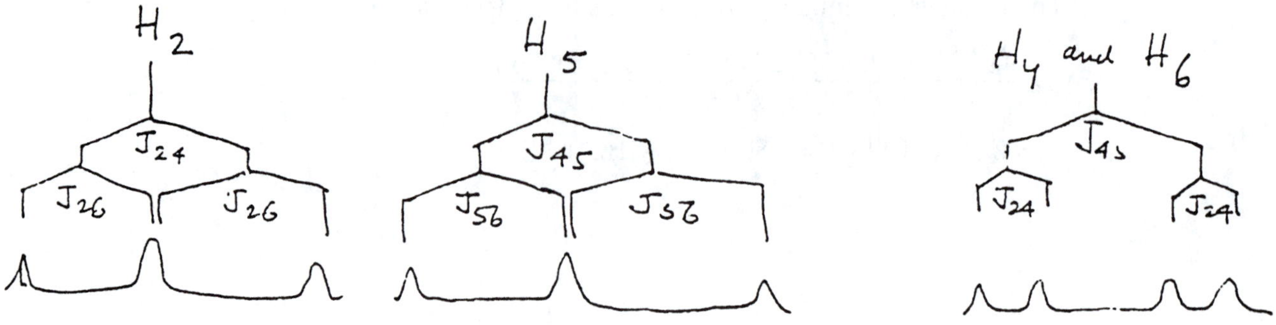

15.35 The protons in 1,2-difluoroethylene are not magnetically equivalent because each proton is coupled differently to the two fluorine nuclei; therefore, they are merely chemically equivalent. This molecule can be represented by AA'XX'. The two protons in 1,2-difluoroallene are magnetically equivalent because they are each coupled in the same way to the two fluorine nuclei. This molecule can be represented by A_2X_2.

15.36 0.31 cm^{-1}

16

Statistical Mechanics

16.1 Using the Boltzmann distribution calculate the ratio of populations at 25 °C of energy levels separated by (a) 100 cm^{-1} and (b) 10 kJ mol^{-1}.

SOLUTION

(a) $$\exp\left[-\frac{hc\tilde{\nu}}{kT}\right] = \exp\left[-\frac{(6.626 \times 10^{-34}\ \text{J s})(2.998 \times 10^{8}\ \text{m s}^{-1})(10^{4}\ \text{m}^{-1})}{(1.380 \times 10^{-23}\ \text{J K}^{-1})(298.15\ \text{K})}\right]$$

$$= 0.617$$

(b) $$\exp\left[-\frac{10^{4}\ \text{J mol}^{-1}}{(8.314\ \text{J K}^{-1}\ \text{mol}^{-1})(298\ \text{K})}\right] = 0.0177$$

16.2 Calculate the ratio of populations at 25 °C of energy levels separated by (a) 1 eV and (b) 10 eV. (c) Calculate the ratios at 1000 °C.

SOLUTION

(a) $$\frac{N_A}{N_B} = \frac{e^{-E_A/RT}}{e^{-E_B/RT}} = e^{-(E_A - E_B)/RT}$$

1 eV = (1.602 x 10^{-19} J)(6.022 x 10^{23} mol^{-1}) = 9.647 x 10^{4} J mol^{-1}

$$\frac{N_A}{N_B} = \exp\left(\frac{-9.647 \times 10^{4}\ \text{J mol}^{-1}}{8.314\ \text{J K}^{-1}\ \text{mol}^{-1} \times 298.15\ \text{K}}\right)$$

$$= 1.253 \times 10^{-17}$$

(b) $$\frac{N_A}{N_B} = \exp\left(\frac{-9.647 \times 10^{5}\ \text{J mol}^{-1}}{8.315\ \text{J K}^{-1}\ \text{mol}^{-1} \times 298.15\ \text{K}}\right)$$

$$= 9.60 \times 10^{-170}$$

(c) At 1000 K: $(N_A/N_B)_{1\ \text{eV}} = 1.100 \times 10^{-4}$

$$(N_A/N_B)_{10\ \text{eV}} = 2.596 \times 10^{-40}$$

16.3 Starting with the definition of the molecular partition function (equation 16.32) and $U = \Sigma_i N_i\varepsilon_i$, derive equation 16.25 for U.

SOLUTION

$$U = \left(\frac{N}{q}\right) \Sigma_i\, g_i\, \varepsilon_i \exp(-\varepsilon_i/kT)$$

$$q = \Sigma_i\, g_i \exp(-\varepsilon_i/kT)$$

$$\left(\frac{\partial q}{\partial T}\right)_V = \frac{1}{kT^2} \Sigma_i\, g_i \varepsilon_i \exp(-\varepsilon_i/kT) = q\left(\frac{\partial \ln q}{\partial T}\right)_V$$

Eliminating the summation between the first and third equations,

$$\frac{1}{kT^2}\left(\frac{qU}{N}\right) = q\left(\frac{\partial \ln q}{\partial T}\right)_V$$

$$U = NkT^2\left(\frac{\partial \ln q}{\partial T}\right)_V$$

16.4 Show that the energy E of a system of N independent protons in a magnetic field B is $-N\hbar\gamma B/2$ in the limit as $T \to 0$ and is equal to 0 in the limit as $T \to \infty$. (See equation 15.18.) How do you interpret these results?

SOLUTION

The molecular partition function is

$$q = \exp(\hbar\gamma B/2kT) + \exp(-\hbar\gamma B/2kT)$$

The energy is given by

$$U = NkT^2\left(\frac{\partial \ln q}{\partial T}\right)_V$$

$$= \frac{N\hbar\gamma B}{2}\frac{[-\exp(\hbar\gamma B/2kT) + \exp(-\hbar\gamma B/2kT)]}{\exp(\hbar\gamma B/2kT) + \exp(-\hbar\gamma B/2kT)}$$

In the limit as $T \to 0$, $U = -\dfrac{N\hbar\gamma B}{2}$.

In the limit as $T \to \infty$, $U = 0$.

This can be understood as follows: As $T \to 0$, thermal energy goes to zero and the protons become oriented in the direction of the field. As $T \to \infty$, The thermal energy becomes so great that a particular proton is equally likely to be oriented in the direction of the field or opposed to it. This is another example of a fact that we encountered in discussing lasers: It is impossible to get a population inversion by irradiating a two-level system.

16.5 Use the relation between pressure P and the molecular partition function q_t to derive the equation of state of an ideal gas starting with equation 16.39 for q_t.

SOLUTION

q_t = (function not involving V)V

$\ln q_t = \ln$ (function not involving V) + $\ln V$

$$P = NkT\left(\frac{\partial \ln q_t}{\partial V}\right)_T$$

$$\left(\frac{\partial \ln q_t}{\partial V}\right)_T = \frac{1}{V}$$

$$P = \frac{NkT}{V} = \frac{nRT}{V}$$ since $N = nN_A$ and $R = N_A k$.

16.6 Derive the expression for the translational partition function for a molecule that is moving along a line, rather than in three-dimensional space. This is of interest in connection with transition state theory (Section 19.4).

SOLUTION

Converting the summation into an integration,

$$q_t = \int_0^\infty \exp(-h^2n^2/8ma^2)dn = \left(\frac{2\pi mkT}{\hbar^2}\right)^{1/2} a$$

where a is the length of the line.

16.7 Show that the same expression is obtained for the chemical potential from A_{indis} (equation 16.21) as from G_{indis} (equation 16.27) for an ideal gas.

SOLUTION

$$A_{indis} = -NkT\left[\ln\left(\frac{q}{N}\right) + 1\right]$$

$$= -kT(N\ln q - N\ln N + N)$$

$$\mu = (\partial A_{indis}/\partial N)_{T,V}$$

$$\mu = -kT(\ln q - \ln N)$$

$$= -kT\ln\left(\frac{q}{N}\right)$$

$$G_{indis} = -NkT\ln\left(\frac{q}{N}\right)$$

$= - kT(N\ln q - N\ln N)$

$\mu = \left(\frac{\partial G_{indis}}{\partial N}\right)_{T,P}$

Since we have to hold P constant for the differentiation, G_{indis} has to be written as a function of P. The molecular partition function for our ideal gas is directly proportional to the volume; $q = q'\,V$. For an ideal gas, $V = \frac{NkT}{P}$ so that

$G_{indis} = - NkT\ln\left(q'\frac{kT}{P}\right)$

$= - kT\left[N\ln(q'kT) - N\ln P\right]$

$\mu = \left(\frac{\partial G_{indis}}{\partial N}\right)_{T,P}$

$= - kT[\ln(q'kT) - \ln P]$

$= - kT\ln\left(\frac{q'kT}{P}\right)$

$= - kT\ln\left(\frac{qkT}{PV}\right)$

$= - kT\ln\left(\frac{q}{N}\right)$

16.8 What is the ratio of the thermal wavelength to the length of one side of the container for (a) a hydrogen atom in a cube 1 nm on a side at 2 K and (b) an oxygen molecule in 0.25 m^3 at 300 K?

SOLUTION

(a) $\Lambda = \left(\frac{h^2}{2\pi mkT}\right)^{1/2}$

For H(g) at 2 K

$$\Lambda = \frac{6.626x10^{-34}\ \text{J s}}{\left[2\pi\left(\frac{1.008x10^{-3}\text{kg mol}^{-1}}{6.022\text{x}10^{23}\text{mol}^{-1}}\right)\left(1.381x10^{-23}\text{J K}^{-1}\right)\left(2\ \text{K}\right)\right]^{1/2}}$$

$= 1.23x10^{-9}\text{m}$

$\frac{\Lambda}{10^{-9}\text{m}} = 1.23$

(b) For O_2(g) at 300 K

$$\Lambda = \frac{6.626 \text{ x } 10^{-34} \text{ J s}}{\left[2\pi\left(\frac{32 \text{ x } 10^{-3}}{6.022 \text{ x } 10^{23}}\right)(1.381 \text{ x } 10^{-23} \text{ J K}^{-1})(300 \text{ K})\right]^{1/2}}$$

$= 1.782 \text{ x } 10^{-11}$ m

$$\frac{\Lambda}{(0.25 \text{ m}^3)^{1/3}} = \frac{1.782 \text{ x } 10^{-11} \text{ m}}{(0.25 \text{ m}^3)^{1/3}} = 2.83 \text{ x } 10^{-11}$$

16.9 Calculate the translational partition function for $H_2(g)$ at 1000 K and 1 bar.

SOLUTION

$$V = \frac{nRT}{P} = \frac{(1 \text{ mol})(8.314 \text{ J K}^{-1})(1000 \text{ K})}{10^5 \text{ N m}^{-2}} = 0.08314 \text{ m}^3$$

$$m = \frac{2(1.0078 \text{ x } 10^{-3} \text{ kg mol}^{-1})}{(6.022 \text{ x } 10^{23} \text{ mol}^{-1})}$$

$= 3.347 \text{ x } 10^{-27}$ kg

$$q = \left(\frac{2\pi mkT}{h^2}\right)^{3/2} V$$

$$= \left[\frac{2\pi(3.347 \text{ x } 10^{-27} \text{ kg})(1.3806 \text{ x } 10^{-23} \text{ J K}^{-1})(1000 \text{ K})}{(6.626 \text{ x } 10^{-34} \text{ J s})^2}\right]^{3/2} \text{x } (0.083\,14 \text{ m}^3)$$

$= 1.414 \text{ x } 10^{30}$

16.10 What are the translational partition functions of hydrogen atoms and hydrogen molecules at 500 K in a volume of $4.157 \text{ x } 10^{-2} \text{ m}^3$? (This is the molar volume of an ideal gas at this temperature and a pressure of 1 bar.)

SOLUTION

$$q_t = \left(\frac{2\pi mkT}{h^2}\right)^{3/2} V$$

For H

$$q_t = \left[\frac{2\pi(1.008 \text{ x } 10^{-3} \text{ kg mol}^{-1})(1.3806 \text{ x } 10^{-23} \text{ J K}^{-1})(500 \text{ K})}{(6.022 \text{ x } 10^{23} \text{ mol}^{-1})(6.626 \text{ x } 10^{-34} \text{ J s})^2}\right]^{3/2}$$

$$\text{x } (4.157 \text{ x } 10^{-2} \text{ m}^3)$$

$= 8.84 \text{ x } 10^{28}$

For H_2, $m_{H_2} = 2m_H$

$q_{tH_2} = 2^{3/2}\, q_{tH} = 2.50 \text{ x } 10^{29}$

16.11 Calculate the molar entropy of one mole of H-atom gas at 1000 K and (a) 1 bar and (b) 1000 bar.

SOLUTION

(a) Since there are only translational contributions and electronic contributions,

$$\bar{S}^\circ = R\left\{\ln\left[\left(\frac{2\pi mkT}{h^2}\right)^{3/2}\left(\frac{kT}{P}\right)\right] + 5/2\right\} + R\ln 2$$

$$\left(\frac{2\pi mkT}{h^2}\right)^{3/2} = \left[\frac{2\pi(1.673 \times 10^{-27}\text{ kg})(1.381 \times 10^{-23}\text{ J K}^{-1})(1000\text{ K})}{(6.626 \times 10^{-34}\text{ J s})^2}\right]$$

$$= 6.012 \times 10^{30}\text{ m}^{-3}$$

$$\bar{S}^\circ = R\ln\left[\frac{(6.012 \times 10^{30}\text{ m}^{-3})(1.381 \times 10^{-23}\text{ J K}^{-1})(1000\text{ K})}{(10^5\text{ Pa})}\right]$$

$$+ R\ln 2 + R\ln\frac{5}{2}$$

$$= 139.86\text{ J K}^{-1}\text{ mol}^{-1}$$

(b) At 1000 K and 1000 bar,

$$\bar{S}^\circ = 82.4\text{ J K}^{-1}\text{ mol}^{-1}$$

16.12 Calculate the molar entropy of neon at 25 °C and 1 bar.

SOLUTION

$$m = \frac{20.179 \times 10^{-3}\text{ kg mol}^{-1}}{6.022 \times 10^{23}\text{ mol}^{-1}} = 3.350 \times 10^{-26}\text{ kg}$$

$$\bar{V} = \frac{RT}{P} = \frac{(8.314\text{ J K}^{-1}\text{ mol}^{-1})(298.15\text{ K})}{10^5\text{ N m}^{-2}}$$

$$= 2.479 \times 10^{-2}\text{ m}^3\text{ mol}^{-1}$$

$$\bar{S}^\circ = R\left\{\ln\left[\left(\frac{2\pi mkT}{h^2}\right)^{3/2}\frac{\bar{V}}{N_A}\right] + \frac{5}{2}\right\} = 146.3\text{ J K}^{-1}\text{ mol}^{-1}$$

16.13 Write out the summation

$$S = \sum_{i=1}^{3}\sum_{j=0}^{1} x^i y^j$$

and show that the summation can also be written as

$$S = \sum_{i=1}^{3} x^i \sum_{j=0}^{1} y^j$$

SOLUTION

$$S = \sum_{i=1}^{3} x^i(1 + y) = (x + x^2 + x^3)(1 + y)$$

$$S = (x + x^2 + x^3)(1 + y)$$

16.14 Write out the summation

$$S = \sum_{i=1}^{2} \sum_{j=1}^{2} x^{i+j}$$

and show that the summation can also be written as

$$S = \sum_{i=1}^{2} x^i \sum_{j=1}^{2} x^j$$

SOLUTION

$$S = \sum_{i=1}^{2} (x^{i+1} + x^{i+2}) = (x^2 + x^3) + (x^3 + x^4) = x^2 + 2x^3 + x^4$$

$$S = (x + x^2)(x + x^2) = x^2 + 2x^3 + x^4$$

16.15 Derive the expression for the vibrational contribution to the internal energy

$$U_v = \frac{RTx}{e^x - 1}$$

where $x = \frac{h\nu}{kT}$. What is the limit of the vibrational contribution to the internal energy at high temperatures?

SOLUTION

$$q_v = \frac{1}{1 - e^{-h\nu/kT}}$$

$$U_v = RT^2 \frac{\partial \ln q_v}{\partial T}$$

$$\ln q_v = -\ln(1 - e^{-h\nu/kT})$$

$$\frac{\partial \ln q_v}{\partial T} = \frac{e^{-h\nu/kT}}{1 - e^{-h\nu/kT}}\left(\frac{h\nu}{kT^2}\right) = \frac{x}{T(e^x - 1)} \quad \text{where } x = h\nu/kT.$$

$$U_v = \frac{RTx}{e^x - 1}$$

You can show that $x/(e^x - 1) \rightarrow 1$ as $x \rightarrow 0$, so that $U_v(T \rightarrow \infty) = RT$.

16.16 By use of series expansions show the vibrational contribution to $\bar{C}_V^o$ for a diatomic molecule approaches R as $T \rightarrow \infty$.

SOLUTION

$$(\bar{C}_V^o)_v = R\left(\frac{\Theta_v}{T}\right)^2 \frac{e^{\Theta_v/T}}{(e^{\Theta_v/T} - 1)^2}$$

$$e^x = 1 + x + \frac{x^2}{2!} + \cdots$$

$$e^{\Theta_v/T} = 1 + \frac{\Theta_v}{T} + \cdots$$

$$(\bar{C}_V^o)_v = R\left(\frac{\Theta_v}{T}\right)^2 \frac{(1 + e^{\Theta_v/T} + \cdots)}{\left(\frac{\Theta_v}{T}\right)^2}$$

As $T \rightarrow \infty$, $(\bar{C}_V^o)_v \rightarrow R$.

16.17 According to Fig. 13.14, the normal mode vibrational frequencies of H_2O are 3657, 1595, and 3756 cm^{-1}. What is the value of the vibrational partition function of H_2O at 2000 K?

SOLUTION

$$q_{vib} = \frac{1}{1 - \exp\left(-\frac{hc\tilde{\nu}}{kT}\right)}$$

$\tilde{\nu}$ /cm^{-1}	3657	1595	3756
$\dfrac{hc\tilde{\nu}}{kT}$	2.631	1.1474 2.702	
q_{vib}	1.072	1.465	1.072

The vibrational partition function for the H_2O molecule is the product of the vibrational partition functions for the three normal modes.
$q_{\text{vib}} = 1.684$

16.18 What are the rotational contributions to $\bar{C}_P^{\text{o}}$, $\bar{S}^{\text{o}}$, $\bar{H}^{\text{o}}$, and $\bar{G}^{\text{o}}$ for NH_3(g) at 25°C?

SOLUTION

$$(\bar{C}_P^{\text{o}})_{\text{r}} = \frac{3}{2} R = 12.471 \text{ J K}^{-1}\text{mol}^{-1}$$

$$\bar{H}_{\text{r}}^{\text{o}} = \frac{3}{2} R = 3.718 \text{ kJ mol}^{-1}$$

$$\bar{S}_{\text{r}}^{\text{o}} = R\left\{ \ln\left[\frac{\left(\frac{T^3\pi}{\Theta_{\text{a}}\Theta_{\text{b}}\Theta_{\text{c}}}\right)^{1/2}}{\sigma}\right] + \frac{3}{2} \right\}$$

$$= 47.822 \text{ J K}^{-1} \text{ mol}^{-1}$$

since $\Theta_{\text{a}}\Theta_{\text{b}}\Theta_{\text{c}} = 1876.0$ K and $\sigma = 3$,

$$\bar{G}_{\text{r}}^{\text{o}} = - RT \ln\left[\frac{\left(\frac{T^3\pi}{q_{\text{a}}q_{\text{b}}q_{\text{c}}}\right)^{1/2}}{\sigma}\right]$$

$$= - 10.540 \text{ kJ mol}^{-1}$$

16.19 What fraction of HCl molecules is in the state $\upsilon = 2$, $J = 7$ at 500 °C? The characteristic vibrational and rotational temperatures are given in Table 16.2.

SOLUTION

$$\frac{N_i}{N} = \frac{g_i\exp(-\varepsilon_i/kT)}{q} = \frac{g_{\text{v}}\exp(-\varepsilon_{\text{v}}/kT)g_{\text{r}}\exp(-\varepsilon_{\text{r}}/kT)}{q_{\text{v}}q_{\text{r}}}$$

$g_{\text{v}} = 1 \qquad g_{\text{r}} = 2J + 1 = 15$

$$\frac{\varepsilon_{\text{v}}}{kT} = \frac{\upsilon h\nu}{kT} = \frac{\upsilon\Theta_{\text{v}}}{T} = \frac{2(4301 \text{ K})}{773 \text{ K}}$$

$$\frac{\varepsilon_{\text{r}}}{kT} = \frac{J(J+1)h^2}{8\pi^2 IkT} = \frac{J(J+1)\Theta_{\text{r}}}{T} = \frac{(7)(8)(15.2 \text{ K})}{773 \text{ K}}$$

$q_v = (1 - e^{-\Theta_v/T})^{-1}$ (equation 16.65)

$q_r = \dfrac{T}{\sigma\Theta_r}$ (equation 16.73)

$q_v = (1 - e^{-4301/773})^{-1} = 1.0039$

$q_r = \dfrac{773}{15.2} = 50.85$

$$\frac{N_i}{N} = \frac{\exp\left[-\frac{(2)(4301)}{773}\right] \times 15 \times \exp\left[-\frac{(56)(15.2)}{773}\right]}{(1.0039)(50.85)} = 1.43 \times 10^{-6}$$

16.20 Calculate the translational partition functions for H, H_2, and H_3 at 1000 K and 1 bar. What are the rotational partition functions of H_2 and H_3 (linear) at 1000 K? The internuclear distances in H_3 are 94 pm.

SOLUTION

$$V = \frac{nRT}{P} = \frac{(1\text{ mol})(8.314\ 51\text{ J K}^{-1}\text{ mol}^{-1})(1000\text{ K})}{10^5\text{ N m}^{-2}}$$

$= 0.083\ 1451\text{ m}^3$

$$q_t = \frac{(2\pi mkT)^{3/2}\,V}{h^3}$$

$$= \frac{\left[2\pi\left(\frac{1.0079 \times 10^{-3}\text{ kg}}{6.022 \times 10^{23}\text{ mol}^{-1}}\right)(1.38 \times 10^{-23}\text{ J K}^{-1})(10^3\text{ K})\right]^{3/2} V}{(6.62 \times 10^{-34}\text{ J s})^3}$$

$= (6.026 \times 10^{30}\text{ m}^{-3})\ V = 5.00 \times 10^{29}$

For H_2 $q_t = 6.026 \times 10^{30}\ V\ 2^{3/2} = 1.42 \times 10^{30}$

For H_3 $q_t = 6.026 \times 10^{30}\ V\ 3^{3/2} = 2.60 \times 10^{30}$

$$q_r = \frac{8\pi^2 IkT}{2h^2}$$

For H_2

$$q_r = \frac{8\pi^2(4.6054 \times 10^{-48}\text{ kg m}^2)(1.3806 \times 10^{-23}\text{ J K}^{-1})(1000\text{ K})}{2(6.626 \times 10^{-34}\text{ J s})^2} = 5.72$$

For H_3

$$I = \frac{m_1 m_3}{m_1 + m_3} R^2$$

$$= \frac{(1.0079 \times 10^{-3}\text{ kg mol}^{-1})^2(1.88 \times 10^{-10}\text{ m})^2}{2(1.00789 \times 10^{-3}\text{ kg mol}^{-1})(6.022 \times 10^{23}\text{ mol}^{-1})}$$

$= 2.96 \times 10^{-47}\text{ kg m}^2$

(Note that m_2 is on the axis of rotation and does not contribute to the moment of inertia.)

$$q_t = 5.72\,\frac{29.6 \times 10^{-48}\ \text{kg m}^2}{4.60 \times 10^{-48}\ \text{kg m}^2} = 36.8$$

16.21 What are the symmetry numbers of the following organic molecules, assuming free rotation of methyl groups? (a) ethane, (b) propane, (c) 2-methylpropane, (d) 2,2-dimethylpropane.

SOLUTION

(a) CH_3CH_3 $\sigma = 3^2 \times 2 = 18$

(b) $CH_3CH_2CH_3$ $\sigma = 3^2 = 9$

```
            H
            |
(c)   CH3—C—CH3
            |
            CH3
```

$\sigma = 3^3 \times 3 = 81$

```
            CH3
            |
(d)   CH3—C—CH3
            |
            CH3
```

$\sigma = 12 \times 3^4 = 972$

16.22 Calculate the symmetry numbers of methane (CH_4) and ethylene (C_2H_4) by adding up the number of distinct proper rotational operations in Table 12.3 plus the identity operation.

SOLUTION

CH_4 is T_d and has $8C_3$, $3C_2$, and E. Thus its symmetry number is 12.
C_2H_4 is D_{2h} and has $C_2(x)$, $C_2(y)$, $C_2(z)$, and E. Thus its symmetry number is 4.

16.23 (a) Calculate the symmetry number for the ethane structures shown in Table 12.3. This is the symmetry number of the rigid structure. (b) Since there is essentially free rotation about the C—C bond, there are three equivalent positions of the second CH_3 group with respect to the first. What is the symmetry number of a freely rotating ethane molecule?

SOLUTION

(a) C_2H_6 is D_{3d} and has $3C_3$, $3C_2$, and E. Thus the symmetry number is 6.

(b) $\sigma = 3 \times 6 = 18$.

16.24 What are the electronic contributions to $\overline{S}^{\text{o}}$ and $\overline{G}^{\text{o}}$ for I(g) at 298.15 K and
3000 K?

SOLUTION

For I(g) $g_{o} = 4$, $g_{1} = 2$, $\Theta_{e}/\text{K} = 10939.3(2)$

$$\overline{S}_{e}^{o} = R\ln\left(g_{o} + g_{1}\,e^{-\varepsilon_i/kT} + \cdots\right)$$

$$\overline{S}_{e}^{o} = 8.314\ \ln(4 + 2\,e^{-10939.3/298})$$

$$= 11.53\ \text{J K}^{-1}\ \text{mol}^{-1}$$

$$\overline{G}_{e}^{o} = -\,RT\ln\left(g_{o} + g_{1}\,e^{-\varepsilon_i/kT} + \cdots\right)$$

$$\overline{G}_{e}^{o} = -\,8.314(298.15)\ln(4 + 2e^{-0939.3/248})$$

$$= -\,3.44\ \text{kJ mol}^{-1}$$

16.25 What is the electronic partition function for C(g) at 1000 K, according to the data of Table 16.2? What are the relative populations of these levels?

SOLUTION

$\Theta_{el} = 0(1), 23.6(3), 62.6(5)$

$q_e = 1 + 3\exp(-\,23.6/1000) + 5\exp(-\,62.6/1000) = 8.627$

$f_1 = 1/8.627 = 0.1159$

$f_2 = 2.930/8.627 = 0.3396$

$f_3 = 4.697/8.627 = 0.5445$

16.26 Calculate the electronic contribution to the standard molar entropy and standard molar Gibbs energy of C(g) at 1000 K, to the degree of completeness we have used here?

SOLUTION

$$\overline{S}_{e}^{o} = R\ln q_e = (8.314\ \text{J K}^{-1}\ \text{mol}^{-1})\ln 8.627 = 17.92\ \text{J K}^{-1}\ \text{mol}^{-1}$$

$$\overline{G}_{e}^{o} = -\,T\overline{S}_{e}^{o} = -\,(1000\ \text{K})(17.92\ \text{J K}^{-1}\ \text{mol}^{-1}) = -\,17.92\ \text{kJ mol}^{-1}$$

16.27 Calculate the molar entropies of H(g) and N(g) at 25 °C and 1 bar. The degeneracies of the ground states are 2 and 4, respectively. Compare these values with those in Appendix C.2.

SOLUTION:

$$\overline{S} = R\left\{\ln\left[(2\pi mkT/h^2)^{3/2}(kT/P)\right] + 5/2 + \ln g_i\right\}$$

$$= R\left[- 1.151\,693 + \frac{3}{2}\ln A_r - \ln(P/P^o) + \frac{5}{2}\ln(T/K) + \ln g_i\right\}$$

For H(g),

$$\overline{S} = 8.31451\left(- 1.151\,693 + \frac{3}{2}\ln 1.00794 + \frac{5}{2}\ln 298.15 + \ln 2\right)$$

$= 114.718$ J K^{-1} mol^{-1} (JANAF, 114.716)

For N(g),

$$\overline{S} = 8.31451\left(- 1.151\,693 + \frac{3}{2}\ln 4.0067 + \frac{5}{2}\ln 298.15 + \ln 4\right)$$

$= 153.301$ J K^{-1} mol^{-1} (JANAF, 153.300)

16.28 A molecule has a ground state and two excited electronic energy levels, all of which are nondegenerate; $\varepsilon_0 = 0$, $\varepsilon_1 = 1 \times 10^{-20}$ J, $\varepsilon_2 = 3 \times 10^{-20}$ J. What fraction of each level is occupied at 298 K and 1000 K?

SOLUTION

$N_i/N = (1/q)\, e^{-\varepsilon_i/kT}$

$q = 1 + \exp(- 1 \times 10^{-20}/kT) + \exp(- 3 \times 10^{-20}/kT)$

At 298 K

$q = 1 + 0.0879 + 0.00068 = 1.0886$

$$N_0 = \frac{1}{1.0886} = 0.919$$

$N_1/N = 0.0879/1.0886 = 0.081$

$N_2/N = 0.00068/1.0886 = 0.006$

At 1000 K

$q = 1 + 0.485 + 0.114 = 1.599$

$$N_0 = \frac{1}{1.599} = 0.625$$

$N_1/N = 0.485/1.599 = 0.303$

$N_2/N = 0.114/1.599 = 0.072$

16.29 The ground state of Cl(g) is fourfold degenerate. The first excited state is 875.4 cm^{-1} higher in energy and is twofold degenerate. What is the value of the electronic partition function at 25 °C? At 1000 K?

SOLUTION

$$q = g_0 e^{-0} + g_1 e^{-\varepsilon_1/kT} = 4 + 2e^{-hc\tilde{\nu}/kT}$$

$$= 4 + 2 \exp\left[\frac{-(6.626 \times 10^{-34}\text{ J s})(2.9979 \times 10^{8}\text{ m s}^{-1})(8.754 \times 10^{4}\text{ m}^{-1})}{(1.3806 \times 10^{-23}\text{ J K}^{-1})(298.15\text{ K})}\right]$$

$$= 4 + 2e^{-4.224} = 4.029$$

At 1000 K $\quad q = 4 + 2e^{-1.259} = 4.568$

16.30 What is the partition function for oxygen atoms at 1000 K according to the data in Table 16.2? What are the relative populations of these levels at equilibrium?

SOLUTION

$$q_e = 5 + 3e^{-228.1/1000} + e^{-325.9/1000} = 8.11$$

$$P_D = 5/q_e = 0.617$$

$$P_1 = \frac{3e^{-228.1/1000}}{q_e} = 0.294$$

$$P_2 = \frac{e^{-325.9/1000}}{q_e} = 0.089$$

16.31 Derive the expression for the electronic internal energy of an atom or molecule. What is the electronic energy per mole for a chlorine atom at 298 K and 1000 K? (See problem 16.34).

SOLUTION

$$U = N\frac{\sum_i \varepsilon_i e^{-\varepsilon_i/kT}}{\sum_i e^{-\varepsilon_i/kT}} = N\frac{2\varepsilon_1 e^{-\varepsilon_1/kT}}{4 + 2e^{-\varepsilon_1/kT}}$$

where ε_0 is taken as zero and $\varepsilon_1 = 875.4\text{ cm}^{-1}$

$$\Theta_1 = (875.4\text{ cm}^{-1})(1.439\text{ K cm}^{-1}) = 1{,}260\text{ K}$$

$$N\varepsilon_1 = Nhc\tilde{\nu}_1 = (6.022 \times 10^{23}\text{ mol}^{-1})(6.626 \times 10^{-34}\text{ J s})$$

$$\times(2.998 \times 10^{8}\text{ m s}^{-1})(875.4\text{ cm}^{-1})(100\text{ cm m}^{-1})$$

$$= 10.47\text{ kJ mol}^{-1}$$

$$U_{298} = \frac{(10.47 \text{ kJ mol}^{-1})e^{-1260/298}}{2 + e^{-1260/298}} = 0.076 \text{ kJ mol}^{-1}$$

$$U_{1000} = \frac{(10.47 \text{ kJ mol}^{-1})e^{-1260/1000}}{2 + e^{-1260/1000}} = 1.301 \text{ kJ mol}^{-1}$$

16.32 Calculate the fraction of hydrogen atoms that at equilibrium at 1000 °C would have $n = 2$.

SOLUTION

Since the fraction will be very small, it is given by the ratio of the number with $n = 2$ to the number with $n = 1$.

$$\text{Fraction} = \frac{e^{-E_2/kT}}{e^{-E_1/kT}} = e^{-(E_2 - E_1)/kT}$$

From Example 10.1 $E = \dfrac{-2.179 \times 10^{-18} \text{ J}}{n^2}$

$$\text{Fraction} = \exp\left[\frac{(-2.179 \times 10^{-18} \text{ J})(0.75)}{(1.380 \times 10^{-23} \text{ J K}^{-1})(1273 \text{ K})}\right]$$

$$= 4 \times 10^{-41}$$

16.33 A quantum mechanical system has two energy levels ε_1 and ε_2. Derive equations for the probability p_1 that the system will be in state 1 and the probability that the system will be in state 2. What are the probabilities at $T/\text{K} = 0$ and ∞? What are the values of p_1 and p_2 at $\Delta\varepsilon = kT$?

SOLUTION

$$\frac{p_2}{p_1} = \exp(-\Delta\varepsilon/kT)$$

$$p_1 + p_2 = 1$$

$$p_1 + p_1 \exp(-\Delta\varepsilon/kT) = 1$$

$$p_1 = \frac{1}{1 + \exp(-\Delta\varepsilon/kT)}$$

$$p_2 = \frac{\exp(-\Delta\varepsilon/kT)}{1 + \exp(-\Delta\varepsilon/kT)}$$

T	p_1	p_2
0	1	0
$\Delta\varepsilon/k$	0.73	0.27
∞	1/2	1/2

16.34 Calculate $\bar{C}_P^{\circ}$ for NH_3(g) at 1000 K. The characteristic vibrational temperatures for the six normal modes are given in Table 16.2.

SOLUTION

The vibrational contribution of a normal mode is calculated from

$$\frac{\bar{C}_P}{R} = \frac{(\Theta_{vi}/T)^2\exp(\Theta_{vi}/T)}{[\exp(\Theta_{vi}/T) - 1]^2}$$

For Θ_{vi} = 1367, 2341, 4800, and 4955 K, these contributions are 0.858, 0.646, 0.193, and 0.176, respectively. The degeneracies of these normal modes have to be taken into account in adding up the contributions.

$$\frac{\bar{C}_P}{R} = 5/2 + 0.858 + 2(0.646) + 0.193 + 2(0.176) + 3/2 = 6.695$$

$$\bar{C}_P^{\circ} = 55.67\ \text{J K}^{-1}\ \text{mol}^{-1} \qquad \text{(Appendix C.3 gives 56.491 J K}^{-1}\text{mol}^{-1}\text{.)}$$

16.35 Calculate the entropy of nitrogen gas at 25 °C and 1 bar pressure. The equilibrium separation of atoms is 109.5 pm and the vibrational wavenumber is 2330.7 cm^{-1}.

SOLUTION

$$m = \frac{1(14.0067 \times 10^{-3}\ \text{kg mol}^{-1})}{6.022 \times 10^{23}\ \text{mol}^{-1}} = 4.651 \times 10^{-26}\ \text{kg}$$

$$\frac{kT}{P^{\circ}} = \frac{(1.380\,6 \times 10^{-23}\ \text{J K}^{-1})(298.15\ \text{K})}{10^5\ \text{N m}^{-2}} = 4.1164 \times 10^{-26}\ \text{m}^3$$

$$\bar{S}_t^{\circ} = R\left\{\ln\left[\left(\frac{2\pi mkT}{h^2}\right)^{3/2}\frac{kT}{P^{\circ}}\right] + \frac{5}{2}\right\}$$

$$\ln\left[\left(\frac{2\pi(4.651 \times 10^{-26}\ \text{kg})(1.3806 \times 10^{-23}\ \text{J K}^{-1})(298.15\ \text{K})}{(6.626 \times 10^{-34}\ \text{J s})^2}\right)^{3/2} \times (4.116\,4 \times 10^{-26}\ \text{m}^3)\right] = 15.591$$

$\overline{S}_t^o = (8.314 \text{ J K}^{-1} \text{ mol}^{-1})(15.591 + 5/2) = 150.42 \text{ J K}^{-1} \text{ mol}^{-1}$

$$\Theta_r = \frac{h^2}{8\pi^2 IK}$$

$$\mu = \frac{m_N}{2} = \frac{14.0067 \times 10^{-13} \text{ kg mol}^{-1}}{2(6.022 \times 10^{23} \text{ mol}^{-1})} = 1.163 \times 10^{-26} \text{ kg}$$

$$I = \mu R^2 = (1.163 \times 10^{-26} \text{ kg})(1.095 \times 10^{-10} \text{ m})^2$$

$$= 1.394 \times 10^{-46} \text{ kg m}^2$$

$$\Theta_r = \frac{(6.626 \times 10^{-34} \text{ J s})^2}{8\pi^2(1.394 \times 10^{-46} \text{ kg m}^2)(1.380 \times 10^{-23} \text{ J K}^{-1})}$$

$$= 2.888 \text{ K}$$

$$\overline{S}_r^\theta = R\left[\ln\left(\frac{T}{\sigma\Theta_r}\right) + 1\right]$$

$$= (8.314 \text{ J K}^{-1} \text{ mol}^{-1}) \left\{\ln\left[\frac{(298.15 \text{ K})}{(2)(2.888 \text{ K})}\right] + 1\right\}$$

$$= 41.10 \text{ J K}^{-1} \text{ mol}^{-1}$$

$$x = \frac{h\tilde{\nu}}{kT} = \frac{hc\tilde{\nu}}{kT}$$

$$= \frac{(6.626 \times 10^{-34} \text{ J s})(2.9979 \times 10^8 \text{ m s}^{-1})(2.3307 \times 10^5 \text{ m}^{-1})}{(1.380 \times 10^{-23} \text{ J K}^{-1})(298.15 \text{ K})}$$

$$= 11.25$$

$$\overline{S}_v^\theta = R\left[\frac{x}{e^x - 1} - \ln(1 - e^{-x})\right]$$

$$= (8.314 \text{ J K}^{-1} \text{ mol}^{-1})\left[\frac{11.25}{e^{11.25} - 1} - \ln(1 - e^{-11.25})\right]$$

$$= 1.21 \times 10^{-3} \text{ J K}^{-1} \text{ mol}^{-1}$$

$$\overline{S}^o = \overline{S}_t^\theta + \overline{S}_r^\theta + \overline{S}_v^\theta = 150.42 + 41.10 + 0 = 191.52 \text{ J K}^{-1} \text{ mol}^{-1}$$

16.36 Calculate $\overline{C}_P^o$ for CO_2 at 1000 K. Compare the actual contributions to $\overline{C}_P^o$ from the various normal modes with the classical expectations.

SOLUTION

$$\overline{C}_P^o = \frac{5}{2} R + R + \sum_{i=1}^{4} \frac{R x_i^2 e^{x_i}}{\left(e^{x_i} - 1\right)^2}$$

$$x_i = \frac{hc\tilde{\nu}}{kT} = (1.438 \times 10^{-5} \text{ m})\tilde{\nu}$$

$\tilde{\nu}_i$ / m^{-1}	1.3512×10^5	6.722×10^4	6.722×10^4	2.3964×10^5
x_i	1.943	0.967	0.967	3.446

$$\bar{C}_P^{\circ} = \frac{7}{2}(8.314) + 6.127 + 2(7.695) + 3.357 = 53.97 \text{ J K}^{-1}\text{mol}^{-1}$$

Classically $\bar{C}_P^{\circ} = \bar{C}_{Pt}^{\circ} + \bar{C}_{Pr\mathrm{v}}^{\circ} + \bar{C}_{P\mathrm{v}}^{\circ} = \frac{5}{2}R + R + 4R = \frac{15}{2}R = 62.36 \text{ J K}^{-1}\text{ mol}^{-1}$

16.37 Calculate the equilibrium constant for the isotope exchange reaction D + H_2 = H + DH at 25 °C. Assume that the equilibrium distance and force constants of H_2 and DH are the same.

SOLUTION

Since the force constants are the same

$$\nu_{HD} = \nu_{H_2}\left(\frac{\mu_{H_2}}{\mu_{HD}}\right)^{1/2} = \nu_{H_2}\left[\frac{m_H + m_D}{2m_D}\right]^{1/2}$$

The change in energy for the reaction is due to the difference in zero point energies for H_2 and HD.

$$\Delta\varepsilon_0 = \frac{h}{2}(\nu_{HD} - \nu_{H_2}) = \frac{h}{2}\nu_{H_2}\left[\left(\frac{m_H + m_D}{2m_D}\right)^{1/2} - 1\right]$$

$$= \frac{h}{2}(1.319 \times 10^{-14}\text{ s}^{-1})\left[\left(\frac{3}{4}\right)^{1/2} - 1\right] = -5.85 \times 10^{-21}\text{ J}$$

$$K = \frac{\left(\frac{f_{HD}}{P^{\circ}}\right)\left(\frac{f_H}{P^{\circ}}\right)}{\left(\frac{f_{H_2}}{P^{\circ}}\right)\left(\frac{f_H}{P^{\circ}}\right)} = \frac{q_H q_{HD}}{q_D q_{H_2}} e^{-\Delta\varepsilon_0/kT}$$

$$\frac{q_{tH}q_{tHD}}{q_{tD}q_{tH_2}} = \left(\frac{m_H m_{HD}}{m_D m_{H_2}}\right)^{3/2} = \left(\frac{2}{4}\right)^{3/2} = 0.353$$

$$\frac{q_{rHD}}{q_{rH_2}} = \frac{2\left(\frac{1}{m_H} + \frac{1}{m_H}\right)}{\left(\frac{1}{m_H} + \frac{1}{m_D}\right)} = \frac{2(2/m_H)}{(3/2m_H)} = \frac{8}{3} = 2.66$$

$$\frac{q_{vHD}}{q_{vH_2}} = 1$$

$$K = (0.353)(2.66)\exp\left[(5.85 \times 10^{-21}\text{J})/(1.38 \times 10^{-23}\text{ J K}^{-1})(298\text{ K})\right]$$

$$= 3.90$$

16.38 Calculate the equilibrium constant at 25 °C for the reaction $H_2 + D_2 = 2HD$. It may be assumed that the equilibrium distance and force constant k are the same for all three molecular species, so that the additional vibrational frequencies required may be calculated from $2\pi\nu = (k/\mu)^{1/2}$. Because of the zero point vibration, $\Delta\varepsilon_0$ for this reaction is given by $\Delta\varepsilon_0 = \frac{1}{2} N_A h(2\nu_{HD} - \nu_{H_2} - \nu_{D_2})$

SOLUTION

Since the force constants are the same

$$\nu_H = \nu_{H_2}\left(\frac{\mu_{H_2}}{\mu_{HD}}\right)^{1/2} = \nu_{H_2}\left[\frac{(m_H + m_D)}{2m_D}\right]^{1/2}$$

$$\nu_{D_2} = \nu_{H_2}\left(\frac{m_H}{m_D}\right)^{1/2}$$

$$\Delta\varepsilon_0 = \frac{1}{2} h\nu_{H_2}\left[\sqrt{2}\left(\frac{m_H + m_D}{m_D}\right)^{1/2} - \left(\frac{m_H}{m_D}\right)^{1/2} - 1\right]$$

$$= \frac{h\nu_{H_2}}{2}\left[\sqrt{2}\left(\frac{3}{2}\right)^{1/2} - \left(\frac{1}{2}\right)^{1/2} - 1\right]$$

$$= \frac{(6.626 \times 10^{-34}\ \text{J s})(1.319 \times 10^{14}\ \text{s}^{-1})(0.0249)}{2}$$

$$= 1.09 \times 10^{-21}\ \text{J}$$

$$K = \frac{\left(\frac{f_{HD}}{P^o}\right)^2}{\left(\frac{f_{H_2}}{P^o}\right)\left(\frac{f_{D_2}}{P^o}\right)} = \frac{q_{HD}^2}{q_{H_2}q_{D_2}}\, e^{-\Delta\varepsilon_0/RT}$$

$$= \frac{q_{t\,HD}^2}{q_{tH}q_{tD_2}}\,\frac{q_{rHD}^2}{q_{rH_2}q_{rD_2}}\, e^{-\Delta\varepsilon_0/RT}$$

The vibrational partition functions are essentially equal to unity.

$$\frac{q_{t\,HD}^2}{q_{tH_2}q_{tD_2}} = \frac{m_{HD}^3}{m_{H_2}^{3/2} m_{D_2}^{3/2}} = \frac{3^3}{2^{3/2}4^{3/2}} = 1.19$$

$$\frac{q^r{}_{HD}{}^2}{q_{rH_2}q_{rD_2}} = \frac{2^2\left(\frac{1}{m_H} + \frac{1}{m_H}\right)}{\left(\frac{1}{m_H} + \frac{1}{m_D}\right)^2} = \frac{16\, m_H\, m_D}{(m_D + m_D)^2}$$

$$= \frac{16(1)(2)}{3^2} = 3.55$$

$$K = (1.19)(3.55)\, e^{-(1.09 \times 10^{-21}\ \mathrm{J})/kT} = 4.22\, e^{-79/T}$$
$$= 4.22\, e^{-79/298} = 3.23$$

16.39 Express the equilibrium constant for the reaction $H_2 + I_2 = 2HI$ in terms of molecular properties.

SOLUTION

$$K = \frac{(q_{HI}/V)^2}{(q_{H_2}/V)(q_{I_2}/V)}$$

$$= \left(\frac{m_{HI}^2}{m_{H_2} m_{I_2}}\right)^{3/2} \frac{4\Theta_{rH_2}\Theta_{rI_2}}{\Theta_{rHI}^2} \times \frac{(1 - e^{-\Theta_{vH_2}/T})(1 - e^{-\Theta_{vI_2}/T})}{(1 - e^{-\Theta_{vHI}/T})}$$
$$\times \exp\left[\frac{-(2D_{0HI} - D_{0H_2} - D_{0I_2})}{RT}\right]$$

16.40 The classical limits of heat capacities of molecules of ideal gases are readily calculated using the principle of equipartition. Calculate $\bar{C}_P^o/R$ and $\bar{C}_V^o/R$ for Ar, O_2, CO_2, and CH_4 and compare $\bar{C}_P^o/R$ with values in Appendix C.3 at 3000 K.

SOLUTION

	trans	rot	vib	$\bar{C}_V^o/R$	$\bar{C}_P^o/R$	Table C.3
Ar	1.5	0	0	1.5	2.5	2.500
O_2	1.5	1	1	3.5	4.5	4.806
CO_2	1.5	1	4	6.5	7.5	7.404
CH_4	1.5	1.5	9	12	13	12.195

The $\bar{C}_P^o$ of oxygen at 3000 K is high because of electronic excitation.

16.41 Considering H_2O to be a rigid nonlinear molecule, what value of $\bar{C}_P^o$ for the gas would be expected classically? If vibration is taken into account, what value is expected? Compare these values of $\bar{C}_P^o$ with the actual values of 298 and 3000 K in Appendix C.2.

SOLUTION

A rigid molecule has translational and rotational energy. The translational contribution to $\bar{C}_V$ is $\frac{3}{2}R = 12.47$ J K^{-1} mol^{-1}. Since H_2O is a nonlinear molecule, it has three rotational degrees of freedom, and so the rotational contribution to $\bar{C}_V$ is $\frac{3}{2}R = 12.47$ J K^{-1} mol^{-1}. Thus $\bar{C}_P$ for the rigid molecule is $\bar{C}_P = \bar{C}_V + R = 33.26$ J K^{-1} mol^{-1}. Since H_2O is a nonlinear molecule, the number of vibrational degrees of freedom is $3N - 6 = 3$. Since each vibrational degree of freedom contributes R to the heat capacity, classical theory predicts

$$\bar{C}_P = 33.26 \text{ J K}^{-1}\text{ mol}^{-1} + 3R$$
$$= 58.20 \text{ J K}^{-1}\text{ mol}^{-1}$$

The experimental value of $\bar{C}_P$ at 298 K is 33.58 J K^{-1} mol^{-1}, which is only slightly higher than the value expected for a rigid molecule. The experimental value of $\bar{C}_P$ at 3000 K is 55.66 J K^{-1} mol^{-1}, which is only slightly less than the classical expectations for a vibrating water molecule.

16.42 Show how $P = kT(\partial \ln Q/\partial V)_T$ leads to $PV = nRT$.

SOLUTION

$$Q = \frac{q^N}{N!} = \frac{q^N}{N^N e^{-N}}$$

$$\ln Q = N\ln q - N\ln N + N$$

The only contribution to the molecular partition function that depends on the volume is the translational contribution. $q_t = V/\Lambda^3$, where Λ is the thermal wavelength.

$$\frac{\partial \ln Q}{\partial V} = N\frac{\partial \ln Q}{\partial V} = \frac{N}{V}$$

$$P = \frac{NkT}{V} = \frac{NRT}{N_A V}$$

$PV = nRT$, where $n = N/N_A$

16.43 Show that the statistical mechanical expressions for $\bar{H}$, $\bar{S}$, and $\bar{G}$ of a monatomic gas without electronic excitation are consistent with

$$\bar{G} = \bar{H} - T\bar{S}.$$

SOLUTION

$$\bar{H} = \frac{5}{2} RT$$

$$\bar{S} = R \ln F + \frac{5}{2} R$$

$$\bar{G} = - RT \ln F$$

where F is a complicated function. Substituting the expressions for the molar enthalpy and the molar entropy in

$$\bar{G} = \bar{H} - T\bar{S}$$

yields the expected expression for the molar Gibbs energy.

16.44 The average energy and average square of the energy of a macroscopic system are given by

$$< E > = \frac{\sum E_j e^{-\beta E_j}}{Q}$$

and

$$< E^2 > = \frac{\sum E_j^2 e^{-\beta E_j}}{Q}$$

Show that the square of the standard deviation of the energy is given by

$\sigma_E^2 = < E^2 > - < E >^2 = k_B T^2 C_V$

Calculate $\sigma_E / < E >$ for a monatomic ideal gas for which $< E > = (3/2)Nk_B T$ and $C_V = (3/2)Nk_B$. What do you think of the chances of observing a fluctuation in the energy of a macroscopic system?

SOLUTION

The square of the standard deviation in the energy is given by

$$\sigma_E^2 = \frac{\sum E_j^2 e^{-\beta E_j}}{Q} - \left(\frac{\sum E_j e^{-\beta E_j}}{Q} \right)^2$$

The value of $- \sigma_E^2$ can be obtained by taking the derivative of $< E >$ with respect to β. In order to show this we substitute $Q = \sum e^{-\beta E_j}$ into the expression for $< E >$:

$$< E > = \frac{\sum E_j e^{-\beta E_j}}{\sum e^{-\beta E_j}}$$

$$\left(\frac{\partial <E>}{\partial \beta} \right) = - \frac{\sum E_j^2 e^{-\beta E_j}}{\sum e^{-\beta E_j}} + \frac{(\sum E_j e^{-\beta E_j})(\sum E_j e^{-\beta E_j})}{\sum e^{-2\beta E_j}}$$

$$= -\frac{\sum E_j^2 e^{-\beta E_j}}{Q} + \left(\frac{\sum E_j e^{-\beta E_j}}{Q}\right)^2 = -\sigma_E^2$$

Since $C_V = (\partial\langle E\rangle/\partial T)_{N,V}$,

$$C_V = \left(\frac{\partial \langle E\rangle}{\partial T}\right)_{N,V} = -\frac{1}{k_B T^2}\left(\frac{\partial \langle E\rangle}{\partial \beta}\right)_{N,V} = \frac{1}{k_B T^2}\,\sigma_E^2$$

The standard deviation of the energy divided by the average energy is given by

$$\frac{\sigma_E}{\langle E \rangle} = \frac{(k_B T^2 C_V)^{1/2}}{\langle E \rangle}$$

Subsituting the values of C_V and $\langle E \rangle$ for an ideal monatomic gas yields

$$\frac{\sigma_E}{\langle E \rangle} = \frac{1}{(\frac{3}{2})^{1/2} N^{1/2}}$$

If the system contains a mole, a fluctuation is of the order of 10^{-12} of the average energy. If the system contains a micromole, a fluctuation is of the order of 10^{-9} of the average energy, and so fluctuations are not often seen - but there are examples. Problem 17.4 shows that the fluctuations of individual molecules are of the order of the energy of the molecule.

16.45 The canonical ensemble partition function Q for a mixture of two monatomic ideal gases is given by

$$Q = \frac{q_1^{N_1}}{N_1!}\,\frac{q_2^{N_2}}{N_2!}$$

Show that

$$U = \frac{3}{2}(n_1 + n_2)\,RT$$

and

$$PV = (n_1 + n_2)RT$$

SOLUTION

$$\ln Q = N_1 \ln q_1 + N_2 \ln q_2 - \ln(N_1!) - \ln(N_2!)$$
$$= N_1 \ln q_1 + N_2 \ln q_2 - N_1 \ln N_1 + N_1 - N_2 \ln N_2 + N_2$$

The partition function for monatomic gas i not involving electronic excitation is given by

$$q_i = \left(\frac{2\pi m_i kT}{h^2}\right)^{3/2} V$$

We will be interested in two special Cases: (a) If the volume is held constant, the dependence of $\ln q_i$ on T is given by $\ln q_i = (3/2) \ln T +$ terms not dependent on T.

(b) If the temperature is held constant, the dependence of $\ln q_i$ on V is given by $\ln q_i = \ln V +$ terms not dependent on V.

In Case a,

$\ln Q = N_1(3/2)\ln T + N_2(3/2)\ln T +$ terms not dependent on T

The energy of the mixture is given by equation 16.7:

$$U = kT^2\left(\frac{\partial \ln Q}{\partial T}\right)_{N_1,N_2,V}$$

Since

$$\left(\frac{\partial \ln Q}{\partial T}\right)_{N_1,N_2,V} = \frac{3}{2}\frac{N_1 + N_2}{T}$$

$$U = \frac{3}{2}(N_1 + N_2)\ kT = \frac{3}{2}(n_1 + n_2)\ RT$$

In Case b,

$\ln Q = N_1 \ln V + N_2 \ln V +$ terms not dependent on V

The pressure of the mixture is given by:

$$P = kT\left(\frac{\partial \ln Q}{\partial V}\right)_{N_1,N_2,T}$$

Since

$$\left(\frac{\partial \ln Q}{\partial V}\right)_{N_1,N_2,T} = \frac{N_1 + N_2}{V}$$

$$P = (N_1 + N_2)kT/V = (n_1 + n_2)RT/V$$

16.46 Since the Helmholtz energy A of an ideal gas is given by $-NkT\ln(qe/N)$, all the other thermodynamic properties can be calculated if $q = q_t q_v q_r q_e$ can be expressed as a function of T, V, and N. Derive these relations, which are given in equations 16.22 to 16.27.

SOLUTION

$$A = -NkT\ln q + NkT\ln N - NkT\ln \mathrm{e}$$

$$S = -\left(\frac{\partial A}{\partial T}\right)_{V,N} = Nk\left[\ln\frac{q}{N} + T\left(\frac{\partial \ln q}{\partial T}\right)_V + \ln \mathrm{e}\right] = Nk\left[\ln\frac{q\mathrm{e}}{N} + T\left(\frac{\partial \ln q}{\partial T}\right)_V\right]$$

$$P = -\left(\frac{\partial A}{\partial V}\right)_{T,N} = NkT\left(\frac{\partial \ln q}{\partial V}\right)_{T,N}$$

$$\mu = \left(\frac{\partial A}{\partial N}\right)_{T,V} = -kT\ln\left(\frac{q}{N}\right)$$

The expressions for *U, H*, and *G* can be obtained from their definitions.

$$U = A + TS = NkT^2\left(\frac{\partial \ln q}{\partial T}\right)_{V,N}$$

$$H = U + PV = NkT\left[T\left(\frac{\partial \ln q}{\partial T}\right)_{V,N} + 1\right]$$

$$G = A + PV = -NkT\ln\left(\frac{qe}{T}\right) + nkT\ln e = -NkT\ln\left(\frac{q}{N}\right)$$

16.48 (a) 3.85×10^{-173}, (b) 2.47×10^{-23}

16.49 7.74×10^{21}, 4.76×10^{22}, 5.32×10^{23}

16.50 (a) 4.596×10^{-11} m (b) The de Broglie wavelength is a little larger than the thermal wavelength. (c) Both the de Broglie wavelength and the thermal wavelength are small compared with the mean distance between atoms.

16.51 154.8, 20.8 J K^{-1} mol^{-1}

16.52 126.154 J K^{-1} mol^{-1}

16.53 $q_t(I)/q(H) = (126.90/1.0079)^{3/2}$

16.54 (a) 3.08×10^{-10} , 1.79×10^{-11} m, (b) 5.18×10^{-10}, 3.45×10^{-11} m

(c) Applicable at 298 K, but not at 1 K.

16.55 0.8971, 0.0923, 0.0095, 0.00098

16.56 0.048, 48, 48 000, 48 x 10^6 K

16.57 12.47, 42.37 J K^{-1} mol^{-1}

16.58 4, 3, 18, 27, 32

16.59 (a) 2, (b) 12

16.60 0.0407

16.62 22,000 K

16.63 (a) 0.0828 x 10^{-20} J, 0.519 x 10^{-20} J, (b) same as (a)

16.64 (a) 3.369, (b) 8.144

16.65 < 10^{-99}, 1.49 x 10^{-23}

16.66 No substance remains a gas all the way down to 0 K. The Debye theory shows that $S \rightarrow 0$ for a solid as $T \rightarrow 0$.

16.67 - 1.151 693

16.68 29.10, 32.93 J K^{-1} mol^{-1}

16.69 222.86 J K^{-1} mol^{-1}

16.70 20.79 J K^{-1} mol^{-1}, 162.71 J K^{-1} mol^{-1}, 62.36 kJ^{-1} mol^{-1}, - 425.76 kJ mol^{-1}

16.71 34.91 J K^{-1} mol^{-1}, -337.52 kJ mol^{-1}, 201.01 J K^{-1} mol^{-1}, - 940.56 kJ mol^{-1}

16.72 6.67 J K^{-1} mol^{-1}, 462.24 kJ mol^{-1}, 124.39 J K^{-1} mol^{-1}, 89.03 kJ mol^{-1}, 2.82 x 10^{-2} ,0.084

16.73 432.070, 493.580, 239.242, 427.783, 1071.780 kJ mol^{-1}

16.74 2.5, 4.5, 7, 10

17

Kinetic Theory of Gases

17.1 If the diameter of a gas molecule is 0.4 nm and each is imagined to be in a separate cube, what is the length of the side of the cube in molecular diameters at 0 °C and pressure of (a) 1 bar and (b) 1 Pa?

SOLUTION

(a) $$\left[\frac{(22.7\ \text{L mol}^{-1})(1000\ \text{cm}^3\ \text{L}^{-1})(10^7\ \text{nm cm}^{-1})^3}{6.02 \times 10^{23}\ \text{mol}^{-1}}\right]^{1/3}$$

$$= 3.43\ \text{nm} = \frac{3.43\ \text{nm}}{0.4\ \text{nm}} = 8.6 \text{ molecular diameters}$$

(b) $$\left[\frac{(2271\ \text{m}^3\ \text{mol}^{-1})(10^9\ \text{nm m}^{-1})^3}{6.022 \times 10^{23}\ \text{mol}^{-1}}\right]^{1/3}$$

$$= 156\ \text{nm} = \frac{156\ \text{nm}}{0.4\ \text{nm}} = 390 \text{ molecular diameters}$$

*17.2 Plot the probability density $f(v)$ of molecular speeds versus speed for oxygen at 25 °C.

SOLUTION

$$F(v) = 4\pi v^2 \left(\frac{M}{2\pi RT}\right)^{3/2} e^{-Mv^2/2RT}$$

$$= 4\pi v^2 \left[\frac{32 \times 10^{-3}\ \text{kg mol}^{-1}}{2\pi(8.314\ \text{J K}^{-1}\ \text{mol}^{-1})(298\ \text{K})}\right]^{3/2}$$

$$\times \exp\left[-\,(32 \times 10^{-3}\ \text{kg mol}^{-1})v^2/2(8.324\ \text{J K}^{-1}\ \text{mol}^{-1})(298\ \text{K})\right]$$

$$= v^2\,(3.704 \times 10^{-8}\ \text{s m}^{-1})\, e^{-6.458 \times 10^{-6} v^2}$$

where v is in m s^{-1}

v/ m s^{-1}	$f(v)/10^{-4}$ s m^{-1}
100	3.47
300	18.64
500	18.42
700	7.67

1000	0.58

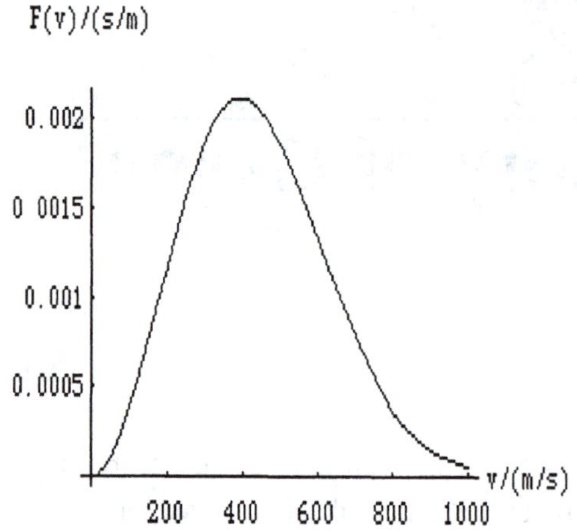

17.3 What is the ratio of the probability that gas molecules have two times the mean speed to the probability that they have the mean speed?

SOLUTION

$$F(<\upsilon>) \propto <\upsilon>^2 e^{-<\upsilon>^2 m/2kT}$$

$$F(2<\upsilon>) \propto 4<\upsilon>^2 e^{-4<\upsilon>^2 m/2kT}$$

$$\frac{F(2<\upsilon>)}{F(<\upsilon>)} = 4\, e^{-3<\upsilon>^2 m/2kT}$$

Since $<\upsilon>^2 = \dfrac{8kT}{\pi m}$

$$\frac{F(2<\upsilon>)}{F(<\upsilon>)} = 4e^{-12/\pi} = 0.0877$$

17.4 (a) Use equation 17.21 to calculate the average value of ε^2 for a molecule of an ideal gas. (b) Calculate the standard deviation σ_ε of the molecular energy from (a) and equation 9.59. (c) Calculate the ratio of the standard deviation of the translational energy to the average translational energy.

SOLUTION

(a) $$< \varepsilon^2 > = \frac{2\pi}{(\pi kT)^{3/2}} \int_0^\infty \varepsilon^{5/2} e^{-\varepsilon/kT}\, d\varepsilon = \frac{2}{(kT)^{3/2}\pi^{1/2}} \left(\frac{15}{(2/kT)^3}\right) (\pi kT)^{1/2}$$

$$= \frac{15}{4} (kT)^2$$

(b) $$\sigma_\varepsilon^2 = < \varepsilon^2 > - < \varepsilon>^2 = \frac{15}{4} (kT)^2 - \frac{9}{4} (kT)^2 = \frac{3}{2} (kT)^2$$

$$\sigma_\varepsilon = \left(\frac{3}{2}\right)^{1/2} kT$$

(c) Since $< \varepsilon > = (3/2)kT$,

$$\frac{\sigma_\varepsilon}{< \varepsilon >} = \left(\frac{2}{3}\right)^{1/2}$$

Thus the standard deviation of the translational energy of a molecule is of the same order of magnitude as the average molecular energy. This can be compared with the fact that the standard deviation of the translational energy of a mole of gas molecules is tiny compared with the translational energy (see problem 16.44).

17.5 Calculate the mean speed and the root-mean-square speed for the following set of molecules: 10 molecules moving 5 x 10^2 m s^{-1}, 20 molecules moving 10 x 10^2 m s^{-1}, and 5 molecules moving 15 x 10^2 m s^{-1}.

SOLUTION

$$<v> = \frac{\Sigma N_i v_i}{\Sigma N_i} = \frac{10(500) + 20(1000) + 5(1500)}{35} = 928 \text{ m s}^{-1}$$

$$<v^2>^{1/2} = \left(\frac{\Sigma N_i v_i^2}{\Sigma N_i}\right)^{1/2} = \left[\frac{10(500)^2 + 20(1000)^2 + 5(1500)^2}{35}\right]^{1/2} = 982 \text{ m s}^{-1}$$

17.6 Calculate the most probable, mean, and root-mean-square speeds for oxygen molecules at 25 °C.

SOLUTION

$$v_p = \left(\frac{2RT}{M}\right)^{1/2} = \left[\frac{(2)(8.314 \text{ J K}^{-1} \text{ mol}^{-1})(298 \text{ K})}{32 \text{ x}10^{-3} \text{ kg mol}^{-1}}\right]^{1/2} = 393 \text{ m s}^{-1}$$

$$<v> = \left(\frac{8RT}{\pi M}\right)^{1/2} = \left[\frac{(8)(8.314 \text{ J K}^{-1} \text{ mol}^{-1})(298 \text{ K})}{(3.1416)(32 \text{ x } 10^{-3} \text{ kg mol}^{-1})}\right]^{1/2} = 444 \text{ m s}^{-1}$$

$$<v^2>^{1/2} = \left[\frac{(3)(8.314 \text{ J K}^{-1} \text{ mol}^{-1})(298 \text{ K})}{32 \text{ x } 10^{-3} \text{ kg mol}^{-1}}\right]^{1/2} = 481 \text{ m s}^{-1}$$

17.7 The mean speed of H_2 at 298 K is 1769 m s^{-1}. What is the mean speed of a hydrogen molecule relative to another hydrogen molecule? How do you rationalize your calculation?

SOLUTION

If two H_2 molecules are moving parallel to each other, their relative speed is zero. If they are moving directly at each other their relative speed is 3538 m s^{-1}. We have seen earlier that in the typical encounter their paths are at right angles. When their paths are at right angles, one molecule is approaching the other's path at 1769 m s^{-1}, but the other molecule is moving. Using the construction of Fig. 17.11,

$<v_{11}> = (1769^2 + 1769^2)^{1/2} = 2^{1/2}(1769 \text{ m s}^{-1}) = 2502 \text{ m s}^{-1}$
or
$<v_{11}> = 2^{1/2} < v_1 >$

17.8 What fraction of oxygen molecules at 300 K have velocities (a) between 400 and 410 m s^{-1} and (b) between 800 and 810 m s^{-1}. You can assume that $F(v)$ is independent of v in each of these intervals.

SOLUTION

$$F(v) = 4\pi v^2\left(\frac{m}{2\pi kT}\right)^{3/2}\exp\left(-\frac{mv^2}{2kT}\right)$$

$$\frac{m}{2kT} = \frac{M}{2RT} = \frac{2(15.9994 \text{ x } 10^{-3} \text{ kg mol}^{-1})}{2(8.3145 \text{ J K}^{-1} \text{ mol}^{-1})(300 \text{ K})} = 6.4142 \text{ x } 10^{-6} \text{ m}^{-2} \text{ s}^2$$

(a) $F(405 \text{ m s}^{-1}) = 4\pi(405 \text{ m s}^{-1})^2(6.4142 \text{ x } 10^{-6} \text{ m}^{-2} \text{ s}^2/\pi)^{3/2}$
$\text{x } \exp(-6.4142 \text{ x } 10^{-6} \text{ x } 405^2) = 2.099 \text{ x } 10^{-3} \text{ s m}^{-1}$

The probability that a molecule will have a velocity between 400 and 410 m s^{-1} is given by $F(v)\text{d}v$, which we are approximating with $F(v)\Delta v$:

$F(405 \text{ m s}^{-1})(10 \text{ m s}^{-1}) = (2.099 \text{ x } 10^{-3} \text{ s m}^{-1})(10 \text{ m s}^{-1})$

$= 2.099 \text{ x } 10^{-2}$

Thus the probability is 2.099 %.

(b) $F(805 \text{ m s}^{-1}) = 4\pi(805 \text{ m s}^{-1})^2(6.4142 \text{ x } 10^{-6} \text{ m}^{-2} \text{ s}^2/\pi)^{3/2}$
$\text{x } \exp(-6.4142 \text{ x } 10^{-6} \text{ x } 805^2) = 3.721 \text{ x } 10^{-6} \text{ s m}^{-1}$

The probability that a molecule will have a velocity between 800 and 810 m s^{-1} is given by

$F(805 \text{ m s}^{-1})(10 \text{ m s}^{-1}) = (3.721 \text{ x } 10^{-6} \text{ s m}^{-1})(10 \text{ m s}^{-1})$

$= 3.721 \text{ x } 10^{-5}$

Thus the probability is 0.003721 %.

17.9 The standard deviation σ of a distribution is given by
$\sigma = [<x^2> - <x>^2]^{1/2}$
What is the standard deviation of the distribution of speeds v of hydrogen molecules at 298.15 K?

SOLUTION

Equation 17.26 shows that the mean speed squared is

$$<v>^2 = \frac{8RT}{\pi M}$$

Equation 17.28 shows that the mean square speed is

$$<v^2> = \frac{3RT}{M}$$

The standard deviation σ_v of the speed distribution is given by

$$\sigma_v = \left[\left(3 - \frac{8}{\pi}\right)\frac{RT}{M}\right]^{1/2}$$

For molecular hydrogen at 298.15 K,

$$\sigma_v = \left[\left(3 - \frac{8}{\pi}\right)\frac{(8.3145 \text{ J K}^{-1} \text{ mol}^{-1})(298.15 \text{ K})}{0.002014 \text{ kg mol}^{-1}}\right]^{1/2} = 746 \text{ m s}^{-1}$$

This can be compared with the mean speed of 1769 m s^{-1}.

17.10 Derive equations for $\bar{U}$ and $\bar{C}_V$ for any monatomic gas from kinetic theory.

SOLUTION

$$\bar{U} = \frac{N_A}{2} m \int_0^\infty v^2 \, F(v)dv$$

$$= \frac{N_A}{2} m <v^2> = \frac{N_A}{2} m\left(\frac{3kT}{m}\right) = \frac{3}{2} RT$$

$$\bar{C}_V = \left(\frac{\partial \bar{U}}{\partial T}\right)_V = \frac{3}{2} R$$

17.11 Calculate the velocity of sound in nitrogen gas at 25 °C. (See Section 17.4.)

SOLUTION

$$v_s = \left(\frac{\bar{C}_P RT}{\bar{C}_V M}\right)^{1/2}$$

$$= \left[\frac{(29.125 \text{ J K}^{-1} \text{ mol}^{-1})(8.3144 \text{ J K}^{-1} \text{ mol}^{-1})(298.15 \text{ K})}{(20.811 \text{ J K}^{-1} \text{ mol}^{-1})(2)(14.0067 \times 10^{-3} \text{ kg mol}^{-1})}\right]^{1/2} = 352 \text{ m s}^{-1}$$

$$= \frac{(352 \text{ m s}^{-1})(10^2 \text{ cm m}^{-1})(60 \text{ s min}^{-1})(60 \text{ min hr}^{-1})}{(2.54 \text{ cm in}^{-1})(12 \text{ in ft}^{-1})(5280 \text{ ft mile}^{-1})} = 787 \text{ miles hr}^{-1}$$

17.12 Calculate the speed of sound at 25 °C in (a) H_2O(g) and (b) CO_2(g). The molar heat capacities at constant pressure are given in Appendix C.2 and $\bar{C}_P - \bar{C}_V = R$.

SOLUTION

	$\bar{C}_P$/J K^{-1} mol^{-1}	$\bar{C}_V$/J K^{-1} mol^{-1}	M/kg mol^{-1}
H_2O(g)	33.57	25.26	0.018
CO_2(g)	37.11	28.80	0.044

$$v_s = \left(\frac{\bar{C}_P RT}{\bar{C}_V M}\right)^{1/2}$$

(a) $$v_s(H_2O) = \left(\frac{(33.58\ \text{J K}^{-1}\ \text{mol}^{-1})(8.3145\ \text{J K}^{-1}\ \text{mol}^{-1})(298.15\ \text{K})}{(25.26\ \text{J K}^{-1}\ \text{mol}^{-1})(0.018\ \text{kg mol}^{-1})}\right)^{1/2}$$

$$= 428\ \text{m s}^{-1}$$

(b) $$v_s(CO_2) = \left(\frac{(37.11\ \text{J K}^{-1}\ \text{mol}^{-1})(8.3145\ \text{J K}^{-1}\ \text{mol}^{-1})(298.15\ \text{K})}{(28.80\ \text{J K}^{-1}\ \text{mol}^{-1})(0.044\ \text{kg mol}^{-1})}\right)^{1/2}$$

$$= 269\ \text{m s}^{-1}$$

17.13 (a) Calculate the collision frequency for a nitrogen molecule in nitrogen at 1 bar pressure and 25 °C. (b) What is the collision density? What is the effect on the collision density of (c) doubling the absolute temperature at constant pressure and (d) doubling the pressure at constant temperature?

SOLUTION

(a) $$z_{11} = 2^{1/2}\ \rho\pi d^2 \langle v\rangle$$

$$\rho = \frac{PN_A}{RT} = \frac{(10^5\ \text{N m}^{-2})(6.022 \times 10^{23}\ \text{mol}^{-1})}{(8.314\ \text{J K}^{-1}\ \text{mol}^{-1})(298.15\ \text{K})} = 2.429 \times 10^{25}\ \text{m}^{-3}$$

$$\langle v\rangle = \left(\frac{8RT}{\pi M}\right)^{1/2} = \left[\frac{8(8.314)(298.15)}{\pi(28 \times 10^{-3})}\right]^{1/2} = 475\ \text{m s}^{-1}$$

$$z_{11} = 2^{1/2}\,(2.429 \times 10^{25})\pi(0.375 \times 10^{-9})^2(475) = 7.21 \times 10^9\ \text{s}^{-1}$$

(b) $$Z_{11} = \frac{1}{2^{1/2}}\ \rho^2\pi d^2\langle v\rangle = 2^{-1/2}\,(2.429 \times 10^{25})^2\pi(0.375 \times 10^{-9})^2(475)$$

$$= 8.75 \times 10^{34}\ \text{m}^{-3}\ \text{s}^{-1} = (8.75 \times 10^{34}\ \text{m}^{-3}\ \text{s}^{-1})(10^{-2}\ \text{m cm}^{-1})^3$$

$$= 8.75 \times 10^{28}\ \text{cm}^{-3}\ \text{s}^{-1}$$

(c) $\rho \propto T^{-1}$ and $\langle v\rangle \propto T^{1/2}$

$Z_{11} \propto \rho^2 \langle v\rangle$, which is $T^{-2}\,T^{1/2} \propto T^{-1.5}$

$$\frac{Z_{11}(2T)}{Z_{11}(T)} = \frac{T^{1.5}}{(2T)^{1.5}} = \frac{1}{2^{1.5}} = 0.354$$

(d) $Z_{11} \propto P^2$

$$\frac{Z_{11}(2P)}{Z_{11}(P)} = 4$$

17.14 (a) Calculate the mean free path for hydrogen gas ($d = 0.247$ nm) at 1 bar and 0.1 Pa at 25 °C. (b) Repeat the calculation for chlorine gas ($d = 0.496$ nm).

SOLUTION

At 1 bar

$$\rho = \frac{N_A P}{RT} = \frac{(6.022 \times 10^{23}\ \text{mol}^{-1})(10^5\ \text{N m}^{-2})}{(8.314\ \text{J K}^{-1}\ \text{mol}^{-1})(298\ \text{K})}$$

$$= 2.44 \times 10^{25}\ \text{m}^{-3}$$

At 0.1 Pa

$$\rho = \frac{N_A P}{RT} = \frac{(6.022 \times 10^{23}\ \text{mol}^{-1})(0.1\ \text{N m}^{-2})}{(8.314\ \text{J K}^{-1}\ \text{mol}^{-1})(298\ \text{K})}$$

$$= 2.44 \times 10^{19}\ \text{m}^{-3}$$

(a) $$\lambda = \frac{1}{2^{1/2}\pi d^2 \rho}$$

$$= \frac{1}{2^{1/2}\,\pi(0.247 \times 10^{-9}\ \text{m})^2(2.44 \times 10^{25}\ \text{m}^{-3})}$$

$$= 1.52 \times 10^{-7}\ \text{m at 1 bar}$$

$$\lambda = \frac{1}{2^{1/2}\,\pi(0.247 \times 10^{-9}\ \text{m})^2(2.44 \times 10^{19}\ \text{m}^{-3})}$$

$$= 0.152\ \text{m at 0.1 Pa}$$

(b) $$\lambda = \frac{1}{2^{1/2}\,\pi(0.496 \times 10^{-9}\ \text{m})^2(2.44 \times 10^{25}\ \text{m}^{-3})}$$

$$= 3.77 \times 10^{-8}\ \text{m at 1 bar}$$

$$\lambda = \frac{1}{2^{1/2}\,\pi(0.496 \times 10^{-9}\ \text{m})^2(2.44 \times 10^{19}\ \text{m}^{-3})}$$

$$= 0.037\ \text{m at 0.1 Pa}$$

17.15 The pressure in interplanetary space is estimated to be of the order of 10^{-14} Pa. Calculate (a) the average number of molecules per cubic centimeter, (b) the collision frequency, and (c) the mean free path in miles. Assume that only hydrogen atoms are present and that the temperature is 1000 K. Assume $d = 0.2$ nm.

SOLUTION

(a) $$\rho = \frac{N_A P}{RT} = \frac{(6.022 \times 10^{23}\ \text{mol}^{-1})(10^{-14}\ \text{N m}^{-2})}{(8.314\ \text{J K}^{-1}\ \text{mol}^{-1})(10^3\ \text{K})}$$

$$= 0.724 \times 10^6\ \text{m}^{-3} = 0.724\ \text{cm}^{-3}$$

(b) $$\langle v \rangle = \left(\frac{8RT}{\pi M}\right)^{1/2} = \left[\frac{8(8.314\ \text{J K}^{-1}\ \text{mol}^{-1})(10^3\ \text{K})}{\pi(1 \times 10^{-3}\ \text{kg mol}^{-1})}\right]^{1/2} = 4600\ \text{m s}^{-1}$$

$$z_{11} = 2^{1/2}\,\rho\pi d^2 \langle v \rangle$$

$$= 2^{1/2}\,(0.724 \times 10^6\ \text{m}^{-3})\pi(0.2 \times 10^{-9}\ \text{m})^2(4600\ \text{m s}^{-1}) = 5.92 \times 10^{-10}\ \text{s}^{-1}$$

(c) $\lambda = \dfrac{1}{2^{1/2}\pi d^2\rho} = \dfrac{1}{2^{1/2}\pi(0.2 \times 10^{-9}\ \text{m})^2(0.724 \times 10^6\ \text{m}^{-3})}$

$= 7.77 \times 10^{12}$ m

$= \dfrac{(7.77 \times 10^{12}\ \text{m})(10^2\ \text{cm m}^{-1})}{(2.54\ \text{cm in}^{-1})(12\ \text{in ft}^{-1})(5280\ \text{ft mile}^{-1})} = 4.83 \times 10^9$ miles

17.16 Calculate the collision frequency z_{11} and the collision density Z_{11} for molecular chlorine at 25 °C and 1 bar. The collision diameter is 0.544×10^{-9} m.

SOLUTION

$$\rho = \frac{N_A P}{RT} = \frac{(6.022 \times 10^{23}\ \text{mol}^{-1})(10^5\ \text{Pa})}{(8.314\ \text{J K}^{-1}\ \text{mol}^{-1})(298\ \text{K})} = 2.43 \times 10^{25}\ \text{m}^{-3}$$

$$<v> = \left(\frac{8RT}{\pi M}\right)^{1/2} = \left[\frac{8\ (8.314\ \text{J K}^{-1}\ \text{mol}^{-1})(298\ \text{K})}{\pi\ (70.9 \times 10^{-3}\ \text{kg mol}^{-1})}\right]^{1/2} = 298\ \text{m s}^{-1}$$

$$z_{11} = 2^{1/2}\rho\pi d^2 <v> = 2^{1/2}(2.43 \times 10^{25}\ \text{m}^{-3})\pi(0.544 \times 10^{-9}\ \text{m})^2(298\ \text{m s}^{-1})$$

$$= 9.52 \times 10^9\ \text{s}^{-1}$$

$$Z_{11} = 2^{-1/2}\rho^2\pi d^2 <v>$$

$$= 2\ (2.43 \times 10^{25}\ \text{m}^{-3})^2\ \pi\ (0.544 \times 10^{-9}\ \text{m})^2\ (298\ \text{m s}^{-1}) = 1.157 \times 10^{25}\ \text{s}^{-1}\ \text{m}^{-3}$$

$$= (1.157 \times 10^{25}\ \text{s}^{-1}\ \text{m}^{-3})(10^{-3}\ \text{m}^{-3}\ \text{L}^{-1})/(6.022 \times 10^{23}\ \text{mol}^{-1})$$

$$= 1.921 \times 10^8\ \text{mol L}^{-1}\ \text{s}^{-1}$$

17.17 A gas mixture contains H_2 at 0.666 bar and O_2 at 0.333 bar at 25 °C. (a) What is the collision frequency z_{12} of a hydrogen molecule with an oxygen molecule? (b) What is the collision frequency z_{21} of an oxygen molecule with a hydrogen molecule? (c) What is the collision density Z_{12} between hydrogen molecules and oxygen molecules in mol L^{-1} s^{-1}? The collision diameters of H_2 and O_2 are 0.272 nm and 0.360 nm, respectively.

SOLUTION

$$<v_{12}> = \left(\frac{8kT}{\pi\mu}\right)^{1/2} = 1824\ \text{m s}^{-1}\ \text{from Example 17.6}$$

$$\rho_1 = \frac{N_A P_1}{RT} = \frac{P_1}{kT} = \frac{0.666 \times 10^5\ \text{Pa}}{(1.381 \times 10^{-23}\ \text{J K}^{-1})(298\text{K})} = 16.18 \times 10^{24}\ \text{m}^{-3}$$

$$\rho_2 = \frac{N_A P_2}{RT} = \frac{P_2}{kT} = \frac{0.333 \times 10^5\ \text{Pa}}{(1.381 \times 10^{-23}\ \text{J K}^{-1})(298\text{K})} = 8.092 \times 10^{24}\ \text{m}^{-3}$$

$$d_{12} = (1/2)(0.272 + 0.360) \times 10^{-9}\ \text{m} = 0.316 \times 10^{-9}\ \text{m}$$

(a) $z_{12} = \rho_2\, \pi\, d_{12}^2 <v_{12}>$

$= (8.092 \times 10^{24}\ \text{m}^{-3})\ \pi\ (0.316 \times 10^{-9}\ \text{m})^2\ (1824\ \text{m s}^{-1}) = 4.630 \times 10^9\ \text{s}^{-1}$

(b) $z_{21} = \rho_1 \pi d_{12}^2 <v_{12}>$
$= (16.18 \times 10^{24} \text{ m}^{-3}) \pi (0.316 \times 10^{-9} \text{ m})^2 (1824 \text{ m s}^{-1}) = 9.258 \times 10^9 \text{ s}^{-1}$

(c) $Z_{12} = \rho_1 \rho_2 \pi d_{12}^2 <v_{12}> = \rho_1 z_{12} = (16.18 \times 10^{24} \text{ m}^{-3})(4.630 \times 10^9 \text{ s}^{-1})$
$= 7.491 \times 10^{34} \text{ s}^{-1} \text{ m}^{-3}$
$= (7.491 \times 10^{34} \text{ s}^{-1} \text{ m}^{-3})(10^{-3} \text{ m}^3 \text{ L}^{-1})/(6.022 \times 10^{23} \text{ mol}^{-1})$
$= 1.244 \times 10^8 \text{ mol L}^{-1} \text{ s}^{-1}$

17.18 For O_2(g), $d = 0.361$ nm. $\Theta_r = 2.079$ K. $\Theta_v = 2273.64$ K, and $M = 31.9988$ g mol^{-1}. At 1 bar and 25 °C what is the average time between collisions? How many vibrational oscillations will have occurred during this time?

SOLUTION

From Table 17.2, $<v> = 444 \text{ m s}^{-1}$
From Example 17.8, $\lambda = 7.11 \times 10^{-8}$ m
$\Delta t = \lambda/<v> = (7.11 \times 10^{-8} \text{ m})/(444 \text{ m s}^{-1})$
$= 1.58 \times 10^{-10}$ s

$$\nu = \frac{k}{h}\Theta_v = \frac{(1.381 \times 10^{-23} \text{ J K}^{-1})(2274 \text{ K})}{6.626 \times 10^{-34} \text{ J s}}$$

$= 4.74 \times 10^{13} \text{ s}^{-1}$

The average number of oscillations between collisions is
$(1.58 \times 10^{-10} \text{ s})(4.74 \times 10^{13} \text{ s}^{-1}) = 7590$

17.19 (a) How many molecules of H_2 strike the wall per unit area per unit time at one bar at 298 K? 1000 K? (b) How many molecules of O_2 strike the wall per unit area per unit time at one bar at 298 K? 1000 K?

SOLUTION

(a) For H_2 at 1 bar and 298 K.

$$m_{H_2} = \frac{2.016 \times 10^{-3} \text{ kg mol}^{-1}}{6.022 \times 10^{23} \text{ mol}^{-1}} = 3.348 \times 10^{-27} \text{ kg}$$

$$J_N = \frac{P}{(2\pi mkT)^{1/2}}$$

$$= \frac{10^5 \text{ N m}^{-2}}{\left[(2\pi)(3.348 \times 10^{-27} \text{ kg})(1.381 \times 10^{-23} \text{ J K}^{-1})(298 \text{ K})\right]^{1/2}}$$

$= 1.075 \times 10^{28} \text{ m}^{-2} \text{ s}^{-1}$

At 1000 K,
$J_N = (1.075 \times 10^{28} \text{ m}^{-2} \text{ s}^{-1})(298/1000)^{1/2}$
$= 5.866 \times 10^{27} \text{ m}^{-2} \text{ s}^{-1}$

(b) For O_2 at 1 bar and 298 K

$$m_{O_2} = \frac{32 \times 10^{-3}\ \text{kg mol}^{-1}}{6.022 \times 10^{23}\ \text{mol}^{-1}} = 5.314 \times 10^{-26}\ \text{kg}$$

$$J_N = \frac{10^5\ \text{N m}^{-2}}{\left[(2\pi)(5.314 \times 10^{-26}\ \text{kg})(1.381 \times 10^{-23}\ \text{J K}^{-1})(298\ \text{K})\right]^{1/2}}$$

$$= 2.698 \times 10^{27}\ \text{m}^{-2}\ \text{s}^{-1}$$

At 1000 K, $J_N = (2.698 \times 10^{27}\ \text{m}^{-2}\ \text{s}^{-1})(298/1000)^{1/2}$

$$= 1.473 \times 10^{27}\ \text{m}^{-2}\ \text{s}^{-1}$$

17.20 In Section 17.6, we derived the expression for the flux J_N of gas molecules through a surface in terms of velocities v_x of molecules in the direction perpendicular to the surface. The derivation can be made in a more general way by considering that the molecules can approach the surface with velocity v at angles θ and ϕ in the system of spherical coordinates. In this case, the differential of the flux is given by

$$dJ_N = \frac{\rho}{4\pi}\, vF(v)dv \cos\theta \sin\theta\, d\theta d\phi$$

Integrate this equation to obtain the flux J_N.

SOLUTION

$$J_N = \frac{\rho}{4\pi} \int_0^{\infty} vF(v)dv \int_0^{\pi/2} \cos\theta \sin\theta\, d\theta \int_0^{2\pi} d\phi$$

The integral of $\cos\theta \sin\theta$ is equal to 1/2, and so the integrations yield

$$J_N = \frac{\rho}{4\pi} < v > (\frac{1}{2})\ 2\pi$$

$$= \frac{\rho < v >}{4}$$

17.21 A Knudsen cell containing crystalline benzoic acid (M = 122 g mol^{-1}) is carefully weighed and placed in an evacuated chamber thermostated at 70 °C for 1 hr. The circular hole through which effusion occurs is 0.60 mm in diameter. Calculate the sublimation pressure of benzoic acid at 70 °C in Pa from the fact that the weight loss is 56.7 mg.

SOLUTION

$$P = \frac{w}{tA}\left(\frac{2\pi RT}{M}\right)^{1/2}$$

$$= \frac{56.7 \times 10^{-6}\ \text{kg}}{(60 \times 60\ \text{s})\pi(0.3 \times 10^{-3}\ \text{m})^2}\left[\frac{2\pi(8.314\ \text{J K}^{-1}\ \text{mol}^{-1})(343\ \text{K})}{122 \times 10^{-3}\ \text{kg mol}^{-1}}\right]^{1/2}$$

$$= 21.3\ \text{N m}^{-2} = 21.3\ \text{Pa}$$

17.22 R. B. Holden, R. Speiser, and H. L. Johnston [*J. Am. Chem. Soc.* **70**, 3897 (1948)] found the rate of loss of weight of a Knudsen effusion cell containing finely divided beryllium to be 19.8×10^{-7} g cm^{-2} s^{-1} at 1320 K and 1210×10^{-7} g cm^{-2} s^{-1} at 1537 K. Calculate ΔH_{sub} for this temperature range.

SOLUTION

According to equations 17.38 and 17.39 the two vapor pressures are proportional to $\Delta w T^{1/2}$.

$$\Delta_{sub}H = \frac{RT_1T_2}{(T_2 - T_1)} \ln \frac{P_2}{P_1}$$

$$= \frac{(8.314)(1320)(1537)}{(217)} \ln \frac{1210(1537)^{1/2}}{19.8(1320)^{1/2}} = 326 \text{ kJ mol}^{-1}$$

17.23 A 5 mL container with a hole 10 μm in diameter is filled with hydrogen. This container is placed in an evacuated chamber at 0 °C. How long will it take for 90% of the hydrogen to effuse out?

SOLUTION

$$J_N = \frac{\rho\langle v\rangle}{4} \text{ may be written } -\frac{1}{A}\frac{dN}{dt} = \frac{N}{V}\frac{\langle v\rangle}{4}$$

$$\langle v\rangle = 1.69 \times 10^3 \text{ m s}^{-1}$$

$$-\frac{dN}{dt} = \frac{\pi(5 \times 10^{-6} \text{ m})^2(1.69 \times 10^3 \text{ m s}^{-1})N}{(5 \times 10^{-6} \text{ m}^3)(4)}$$

$$= (6.64 \times 10^{-3} \text{ s}^{-1})N$$

$$\int_{N_0}^{N_t} \frac{dN}{N} = -(6.64 \times 10^{-3} \text{ s}^{-1}) \int_0^t dt$$

$$\ln \frac{N_t}{N_0} = -(6.64 \times 10^{-3} \text{ s}^{-1})t$$

$$t = \frac{\ln 0.1}{-6.64 \times 10^{-3} \text{ s}^{-1}} = 347 \text{ s}$$

17.24 The vapor pressure of naphthalene ($M = 128.16$ g mol^{-1}) is 17.7 Pa at 30 °C. Calculate the weight loss in a period of 2 h of a Knudsen cell filled with naphthalene and having a round hole 0.50 mm in diameter.

SOLUTION

$$J_N = \frac{P}{(2\pi mkT)^{1/2}} = \frac{\Delta w}{mtA}$$

$$A = \pi r^2 = \pi(0.025 \text{ cm})^2(0.01 \text{ m cm}^{-1})^2 = 1.96 \times 10^{-7} \text{ m}^2$$

$$m = \frac{128.16 \text{ g mol}^{-1}}{6.022 \times 10^{23} \text{ mol}^{-1}} = 2.128 \times 10^{-22} \text{ g}$$

$$= 2.128 \times 10^{-25} \text{ kg}$$

$$\Delta w = \frac{mtAP}{(2\pi mkT)^{1/2}}$$

$$= \frac{(2.128 \text{ x } 10^{-25} \text{ kg})(2 \text{ x } 60 \text{ x } 60 \text{ s})(1.96 \text{ x } 10^{-7} \text{ m}^2)(17.7 \text{ Pa})}{[2\pi(2.128 \text{ x } 10^{-25} \text{ kg})(1.381 \text{ x } 10^{-23} \text{ J K}^{-1})(303 \text{ K})]^{1/2}}$$

$$= 0.0711 \text{ g}$$

17.25 Atoms and molecules can escape from the uppermost layer of the earth's atmosphere only if they have the escape velocity. The minimum escape velocity is the velocity in the direction perpendicular to the surface of the earth. (a) Show that the minimum escape velocity v_e is given by

$$v_e = \left(\frac{2GM_{earth}}{R_{earth}}\right)^{1/2}$$

where M_{earth} is the mass of the earth, R_{earth} is the radius of the earth (radius of the uppermost layer of the atmosphere), and G is the gravitational constant, which is defined by

$$V(r) = -\frac{Gm_1m_2}{r}$$

Given: $G = 6.67\text{x}10^{-11}$ J m kg^{-1}, $M_{earth} = 5.98\text{x}10^{24}$ kg, and $R_{earth} = 6.36\text{x}10^6$ m. An atom or molecule with mass m can escape if its kinetic energy is equal and opposite to the potential energy between the atom or molecule and the earth. (b) Calculate the escape velocity of an atom or molecule.

SOLUTION

(a) The minimum escape velocity is given by

$$\frac{1}{2} mv_e^2 = \frac{GmM_{earth}}{R_{earth}}$$

$$v_e = \left(\frac{2GM_{earth}}{R_{earth}}\right)^{1/2}$$

Note that the escape velocity is indpendent of the molar mass of the atom or molecule.

(b) $$v_{escape} = \left[\frac{2(6.67\text{x}10^{-11} \text{ J m kg}^{-1})(5.98\text{x}10^{24} \text{ kg})}{6.36\text{x}10^6 \text{ m}}\right]^{1/2} = 11\ 200 \text{ m s}^{-1}$$

*17.26 The viscosity of helium is 1.88 x 10^{-5} Pa s at 0 °C. Calculate (a) the collision diameter, and (b) the diffusion coefficient at 1 bar.

SOLUTION

(a) $$d = \left[\frac{5}{16}\frac{(\pi mkT)^{1/2}}{\pi\eta}\right]^{1/2}$$

$$m = \frac{4.0026 \text{ x } 10^{-3} \text{ kg mol}^{-1}}{6.022 \text{ x } 10^{23} \text{ mol}^{-1}} = 6.647 \text{ x } 10^{-27} \text{ kg}$$

$$d = \left[\frac{5}{16}\frac{(\pi 6.647 \text{ x } 10^{-27} \text{ x } 1.38 \text{ x } 10^{-23} \text{ x } 273)^{1/2}}{\pi(1.88 \text{ x } 10^{-5})}\right]^{1/2}$$

$$= 0.217 \text{ nm}$$

(b) $$\rho = \frac{PN_A}{RT} = \frac{(10^5 \text{ Pa m}^{-2})(6.022 \times 10^{23} \text{ mol}^{-1})}{(8.314 \text{ J K}^{-1} \text{ mol}^{-1})(273 \text{ K})}$$

$$= 2.65 \times 10^{25} \text{ m}^{-3}$$

$$D = \frac{3}{8}\frac{(\pi mkT)^{1/2}}{\pi d^2 \rho m}$$

$$= 1.28 \times 10^{-4} \text{ m}^2 \text{ s}^{-1}$$

17.27 What is the self-diffusion coefficient of radioactive CO_2 in ordinary CO_2 at 1 bar and 25 °C? The collision diameter is 0.40 nm.

SOLUTION

$$m = \frac{44 \times 10^{-3} \text{ kg mol}^{-1}}{6.022 \times 10^{23} \text{ mol}^{-1}} = 7.307 \times 10^{-26} \text{ kg}$$

$$\rho = \frac{N}{V} = \frac{PN_A}{RT} = \frac{(1 \text{ bar})(6.022 \times 10^{23} \text{ mol}^{-1})(10^3 \text{ L mol}^{-1})}{(0.083\ 14 \text{ L bar K}^{-1} \text{ mol}^{-1})(298 \text{ K})}$$

$$= 2.43 \times 10^{25} \text{ m}^{-3}$$

$$D = \frac{3}{8}\frac{(\pi mkT)^{1/2}}{\pi d^2 \rho m}$$

$$= \frac{3\left[\pi(7.307 \times 10^{-26} \text{ kg})(1.38 \times 10^{-23} \text{ J K}^{-1})(298 \text{ K})\right]^{1/2}}{8\pi(0.4 \times 10^{-9} \text{ m})^2(2.43 \times 10^{25} \text{ m}^{-3})(7.307 \times 10^{-26} \text{ kg})}$$

$$= 1.26 \times 10^{-5} \text{ m}^2 \text{ s}^{-1}$$

17.28 The probability F(ε) that the molecular translational energy is in the range ε + dε is given by equation 17.20. Use this equation to calculate the most probable translational energy.

SOLUTION

$$F(\varepsilon) = \frac{2\pi}{(\pi kT)^{3/2}}\varepsilon^{1/2}e^{-\varepsilon/kT}$$

$$\frac{\partial F}{\partial \varepsilon} = \frac{2\pi}{(\pi kT)^{1/2}}\left[\varepsilon^{1/2}e^{-\varepsilon/kT}\left(\frac{-1}{kT}\right) + e^{-\varepsilon/kT}\frac{1}{2}\varepsilon^{-1/2}\right] = 0$$

$$\varepsilon = \frac{kT}{2}$$

17.29 374 m s^{-1}

17.31 0.199

17.32 4.18 x 10^2, 4 x 10^2, 4.4 x 10^2 m s^{-1}

17.33 7.9 x 10^2 m s^{-1}, 802 K

17.34 281 m s^{-1}

17.35 (a) 1016, (b) 351.9 m s^{-1}

17.36 8.314 51 J K^{-1} mol^{-1}

17.37 (a) 1.6 x 10^{-10}, (b) 1.60 x 10^{-4}, (c)160 s

17.38 1.4 x 10^{-10} s

17.39 (a) 4.75 x 10^{29} m^{-3} s^{-1}, (b) 2.86 x 10^{-5} m

17.40 (a) 1.908 x 10^9 s^{-1}, 1.908 x 10^9 s^{-1}, (b) 3.87 x 10^7 mol L^{-1} s^{-1}

17.41 (a) 6.24 x 10^6 s^{-1}, (b) 1.26 x 10^2 mol L^{-1} s^{-1}, (c) 160 x 10^{-7} s

17.42 7.1 m, 62.6 s^{-1}

17.43 2.697 x 10^{23} cm^{-2} s^{-1}

17.44 (a) 6590 m, (b) 2.88 x 10^{16} m^{-2} s^{-1}

17.45 (a) 1.136 x 10^{26} m^{-2} s^{-1}, (b) 20.39 g cm^{-2} min^{-1}

17.46 2.0 x 10^{-10} kg

18

Experimental Kinetics and Gas Reactions

18.1 Nitrogen pentoxide (N_2O_5) gas decomposes according to the reaction
$2N_2O_5 = 4NO_2 + O_2$
At 328 K, the rate of reaction υ under certain conditions is 0.75×10^{-4} mol L^{-1} s^{-1}. Assuming that none of the intermediates have appreciable concentrations, what are the values of $d[N_2O_5]/dt$, $d[NO_2]/dt$, and $d[O_2]/dt$?

SOLUTION

$$\upsilon = 0.75 \times 10^{-4} \text{ mol L}^{-1} \text{ s}^{-1}$$
$$= \frac{1}{2}\frac{d[N_2O_5]}{dt} = \frac{1}{4}\frac{d[NO_2]}{dt} = \frac{d[O_2]}{dt}$$
$$\frac{d[N_2O_5]}{dt} = -2\upsilon = -1.5 \times 10^{-4} \text{ mol L}^{-1} \text{ s}^{-1}$$
$$\frac{d[NO_2]}{dt} = 4\upsilon = 3.0 \times 10^{-4} \text{ mol L}^{-1} \text{ s}^{-1}$$
$$\frac{d[O_2]}{dt} = \upsilon = 0.75 \times 10^{-4} \text{ mol L}^{-1} \text{ s}^{-1}$$

18.2 In studying the decomposition of ozone
$2O_3(g) = 3O_2(g)$
in a 2 L reaction vessel, it is found that $d[O_3]/dt = -1.5 \times 10^{-2}$ mol L^{-1} s^{-1}.
(a) What is the rate of reaction υ? (b) What is the rate of conversion $d\xi/dt$?
(c) What is the value of $d[O_2]/dt$?

SOLUTION

(a) $$\upsilon = -\frac{1}{2}\frac{d[O_3]}{dt} = 0.75 \times 10^{-2} \text{ mol L}^{-1} \text{ s}^{-1}$$

(b) $$\frac{d\xi}{dt} = V\upsilon = (2 \text{ L})(0.75 \times 10^{-3} \text{ mol L}^{-1} \text{ s}^{-1}) = 1.50 \times 10^{-3} \text{ mol s}^{-1}$$

(c) $$\frac{d[O_2]}{dt} = 3\upsilon = 3(0.75 \times 10^{-2} \text{ mol L}^{-1} \text{ s}^{-1}) = 2.25 \times 10^{-2} \text{ mol L}^{-1} \text{ s}^{-1}$$

18.3 The decomposition of N_2O_5
$2N_2O_5 = 4NO_2 + O_2$

is studied by measuring the concentration of oxygen as a function of time, and it is found that

$d[O_2]/dt = (1.5 \times 10^{-4}\ s^{-1})[N_2O_5]$

at constant temperature and pressure. Under these conditions the reaction goes to completion to the right. What is the half-life of the reaction under these conditions?

SOLUTION

$$-\frac{1}{2}\frac{d[N_2O_5]}{dt} = \frac{d[O_2]}{dt} = (1.5 \times 10^{-4}\ s^{-1})[N_2O_5]$$

Thus k in

$$-\frac{d[A]}{dt} = k[A]$$

is $3.0 \times 10^{-4}\ s^{-1}$ and

$$t_{1/2} = \frac{0.693}{3.0 \times 10^{-4}\ s^{-1}} = 2310\ s$$

*18.4 The following data were obtained on the rate of hydrolysis of 17% sucrose in 0.099 mol L^{-1} HCl aqueous solution at 35 °C.

t /min	9.82	59.60	93.18	142.9	294.8	589.4
Sucrose remaining,%	96.5	80.3	71.0	59.1	32.8	11.1

What is the order of the reaction with respect to sucrose, and what is the value of the rate constant k?

SOLUTION

$$\ln\frac{[A]}{[A]_o} = -kt \qquad \text{slope} = -k = -3.76 \times 10^{-3}\ min^{-1}$$

$k = 3.76 \times 10^{-3}\ min^{-1}$

$= (3.76 \times 10^{-3}\ min^{-1})(\frac{1}{60}\ min\ s^{-1})$

$= 6.27 \times 10^{-5}\ s^{-1}$

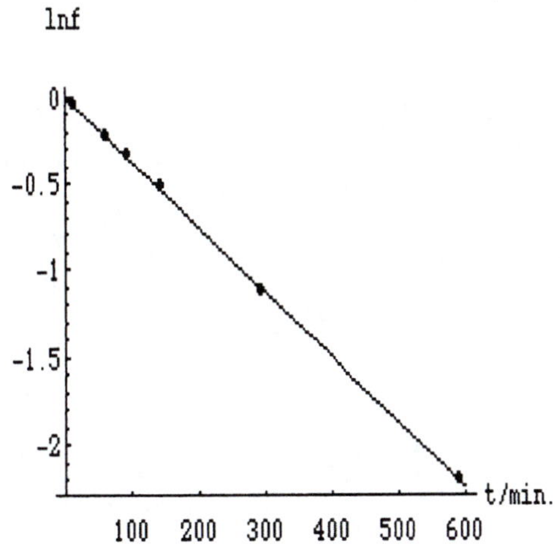

*18.5 Methyl acetate is hydrolyzed in approximately 1 mol L^{-1} HCl at 25 °C. Aliquots of equal volume are removed at intervals and titrated with a solution of NaOH. Calculate the first-order rate constant form the following experimental data:

t /s	339	1242	2745	4546	∞
V/cm^3	26.34	27.80	29.70	31.81	39.81

SOLUTION

Any quantity proportional to the concentration of reactant A that remains may be used in the equation

$\ln[A] = -kt + \ln[A]_o$

In this case the concentration of A remaining is proportional to 39.81 cm^3 - V. Therefore ln(39.81 cm^3 - V) is plotted versus t.

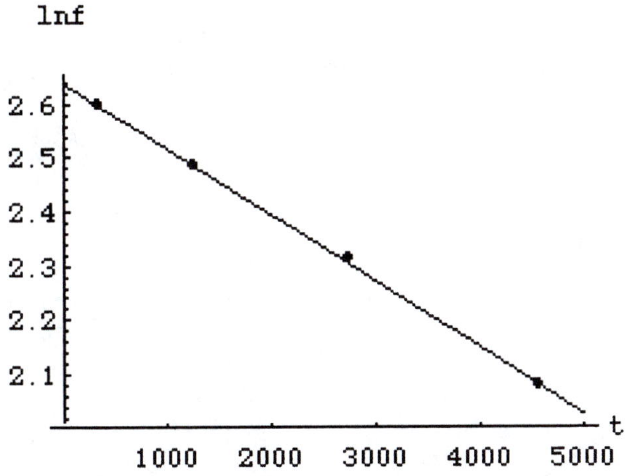

$$\text{slope} = -k = -\frac{0.66}{5000} = -1.23 \times 10^{-4}\ s^{-1}$$

$k = 1.23 \times 10^{-4}\ s^{-1}$

18.6 Prove that in a first-order reaction, where $dn/dt = -kn$, the average life, that is, the average life expectancy of the molecules, is equal to $1/k$.

SOLUTION

$$\text{Average life} = \frac{\int_0^\infty n dt}{n_0}$$

$$\text{Average life} = \frac{-\frac{1}{k}\int_0^\infty \frac{dn}{dt}dt}{n_0}$$

$$= \frac{-\frac{1}{k}(0 - n_0)}{n_0}$$

$$= \frac{1}{k}$$

The average life is also referred to as the relaxation time.

*18.7 The hydrolysis of 1-chloro-1-methylcycloundecane in 80% ethanol has been studied at 25 $^{\circ}$C. The extent of hydrolysis was measured by titrating the acid formed after measured intervals of time with a solution of NaOH. The data are as follows:

t/h	0	1.0	3.0	5.0	9.0	12	∞
x/cm^3	0.035	0.295	0.715	1 .055	1.505	1.725	2.197

(a) What is the order of the reaction? (b) What is the value of the rate constant?
(c) What fraction of the 1-chloro-1-methylcycloundecane will be left unhydrolyzed after 8 h?

SOLUTION
(a)

t/h	0	1.0	3.0	5.0	9.0	12
$\ln(x_\infty - x)$	0.771	0.643	0.393	0.133	- 0.368	- 0.751

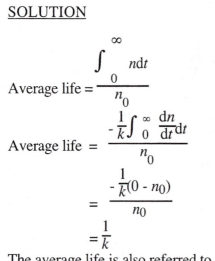

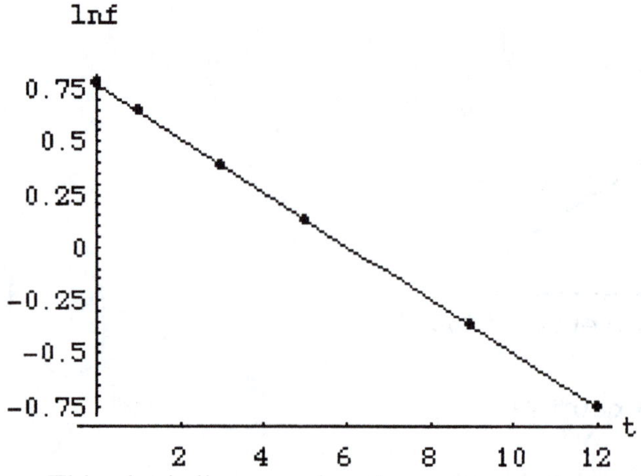

This plot is linear, and so the reaction is first order.

(b) $\text{slope} = \frac{-0.51 - 0.77}{10} = -0.128\ \text{h}^{-1} = -k$

$k = 0.128\ \text{h}^{-1}$

(c) At 8 hours, $\ln c = \ln(x_\infty - x) = -(0.128\ \text{h}^{-1})(8\ \text{h}) + 0.771 = -0.253$

$x_\infty - x = 0.776$

$\text{Fraction unhydrolyzed} = \frac{2.197 - 0.776}{2.197 - 0.035} = 0.325$

18.8 The following values of percent transmission are obtained with a spectrophotometer at a series of times during the decomposition of a substance absorbing light at a particular wavelength. Calculate k, $t_{1/2}$, and τ assuming the reaction is first order.

t /min	Percent transmission
5	14.1
10	57.1
∞	100

Beer's law: $\log(100/T) = abc$, where T = percent transmission, a = absorbancy index, b = cell thickness, and c = concentration.

SOLUTION

The concentration of the reactant is proportional to $\log(100/T)$

t /min	$\log(100/T)$
5	0.851
10	0.243
∞	0

$$\ln\frac{c_1}{c_2} = k(t_2 - t_1)$$

$$\ln\frac{0.851}{0.243} = k5$$

$k = 0.251\ \text{min}^{-1}$

$$t_{1/2} = \frac{0.693}{0.251} = 2.76\ \text{min}$$

$$\tau = \frac{1}{0.251} = 3.98\ \text{min}$$

18.9 Since radioactive decay is a first-order process, the decay rate for a particular nuclide is commonly given as the half-life. Given that potassium contains 0.0118% ^{40}K, which has a half-life of 1.27×10^9 years, how many disintegrations per second are there in a gram of KCl?

SOLUTION

$t_{1/2} = (1.27 \times 10^9\ \text{y})(365\ \text{d y}^{-1})(24\ \text{h d}^{-1})(60\ \text{min h}^{-1})(60\ \text{s min}^{-1})$

$= 4.01 \times 10^{16}\ \text{s}$

The number N of molecules of ^{40}K is 1.18 x 10^{-4} times the number of molecules of KCl, which is [(1 g)/(74.55 g mol^{-1})]N_A.

$$\frac{dN}{dt} = \frac{0.693}{t_{1/2}} N$$

$$= \frac{(0.693)(1\ g)(1.18 \times 10^{-4})(6.02 \times 10^{23}\ mol^{-1})}{(74.55\ g\ mol^{-1})(4.01 \times 10^{16}\ s)} = 15.7\ s^{-1}$$

18.10 The decomposition of HI to $H_2 + I_2$ at 508 °C has a half-life of 135 min when the initial pressure of HI is 0.1 atm and 13.5 min when the pressure is 1 atm.
(a) Show that this proves that the reaction is second order. (b) What is the value of the rate constant in L mol^{-1} s^{-1}? (c) What is the value of the rate constant in bar^{-1} s^{-1}? (d) What is the value of the rate constant in cm^3 s^{-1}?

SOLUTION

(a) The reaction studied is
$2HI = H_2 + I_2$
If the reaction is second order,

$$v = k[HI]^2 = -\frac{1}{2}\frac{d[HI]}{dt}$$

According to eq. 18.23

$$2kt = \frac{1}{[HI]} - \frac{1}{[HI]_0}$$

$$2kt_{1/2} = \frac{2}{[HI]_0} - \frac{1}{[HI]_0} = \frac{1}{[HI]_0}$$

$$t_{1/2} = \frac{1}{2k[HI]_0}$$

This is in agreement with the fact that the half-life is reduced by a factor of 10 when the pressure is increased by a factor of 10.

(b)

$$k = \frac{1}{2t_{1/2}[HI]_0}$$

$$P_{HI} = [HI]RT$$

$$[HI]_0 = P_{HI}/RT = (1\ bar)/(0.08314\ L\ bar\ K^{-1}\ mol^{-1})(781\ K)$$

$$= 1.54 \times 10^{-2}\ mol\ L^{-1}$$

$$k = \frac{1}{2(13.5 \times 60\ s)(1.54 \times 10^{-2}\ mol\ L^{-1})} = 3.96 \times 10^{-2}\ L\ mol^{-1}\ s^{-1}$$

(c)

$$k = \frac{1}{2t_{1/2}P_0} = \frac{1}{2(13.5 \times 60\ s)(1.013\ bar)}$$

$$= 6.17 \times 10^{-4}\ bar^{-1}\ s^{-1}$$

(d)

$$k = \frac{(3.96 \times 10^{-2}\ L\ mol^{-1}\ s^{-1})(10^3\ cm^3\ L^{-1})}{(6.022 \times 10^{-23}\ mol^{-1})}$$

$$= 6.58 \times 10^{-22}\ cm^3\ s^{-1}$$

18.11* The reaction between propionaldehyde and hydrocyanic acid has been studied at 25 °C. In a certain aqueous solution at 25 °C the concentrations at various times were as follows:

t /min	2.78	5.33	8.17	15.23	19.80	∞
$\frac{[HCN]}{mol\ L^{-1}}$	0.0990	0.0906	0.0830	0.0706	0.0653	0.0424
$\frac{[C_3H_7CHO]}{mol\ L^{-1}}$	0.0566	0.0482	0.0406	0.0282	0.0229	0.0000

What is the order of the reaction and the value of the rate constant k?

SOLUTION

The data do not give a linear ln concentration versus t plot, and so the data are tested in the integrated equation for a second order reaction. Equation 18.27 is

$$\frac{1}{([A]_0 - [B]_0)} \ln \frac{[A][B]_0}{[A]_0[B]} = kt$$

In order to get $[A]_0$ and $[B]_0$ for $t = 0$, the clock is started at the first experimental point. This yields the following data to be plotted:

t /min	0	2.55	5.39	12.45	17.02
$\frac{1}{([A]_0-[B]_0)} \ln \frac{[A][B]_0}{[A]_0[B]}$	0	1.695	3.677	8.458	11.523

The slope of this plot, 0.675 L mol^{-1} min^{-1}, is the second order rate constant.

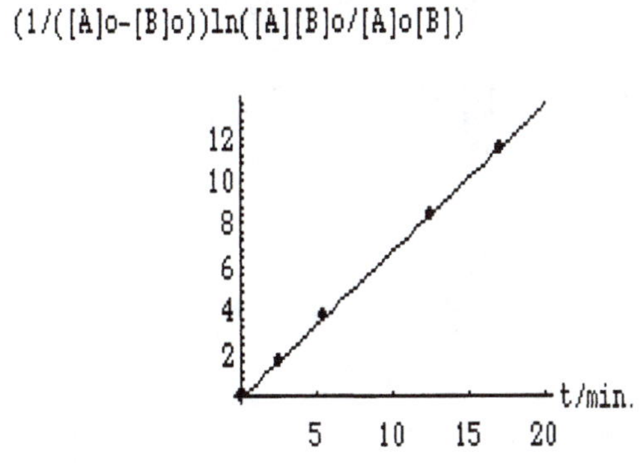

*18.12 Hydrogen peroxide reacts with thiosulfate ion in slightly acidic solution as follows:

$H_2O_2 + 2S_2O_3^{2-} + 2H^+ \rightarrow 2H_2O + S_4O_6^{2-}$

This reaction rate is independent of the hydrogen-ion concentration in the pH range 4.0 to 6.0. The following data were obtained at 25 °C and pH 5.0. Initial concentrations: $[H_2O_2] = 0.036\ 80$ mol L^{-1}; $[S_2O_3^{2-}] = 0.020\ 40$ mol L^{-1}.

t /min	16	36	43	52
$\frac{[S_2O_3^{2-}]}{10^{-3}\ \text{mol L}^{-1}}$	10.30	5.18	4.16	3.13

(a) What is the order of reaction? (b) What is the rate constant?

SOLUTION

(a) In the first 16 minutes, $[S_2O_3^{2-}]$ is approximately halved. In the next 20 minutes, $[S_2O_3^{2-}]$ is approximately halved. In the next 16 minutes, $[S_2O_3^{2-}]$ is considerably less than halved. Therefore, the order is higher than first. The next section shows that the reaction is first order in H_2O_2, first order in $S_2O_3^{2-}$, and second order overall.

(b) Let A = H_2O_2 and B = $S_2O_3^{2-}$. Equation 18.26 becomes

$$kt = \frac{1}{(2[A]_0-[B]_0)} \ln \frac{[A][B]_0}{[A]_0[B]}$$

$$\ln \frac{[A]}{[B]} = \ln\frac{[A]_0}{[B]_0} + (2[A]_0 - [B]_0)\, kt$$

t /min	0	16	36	43	52
$\frac{[B]}{10^{-3}\ \text{mol L}^{-1}}$	20.40	10.30	5.18	4.16	3.13
$\frac{[A]}{10^{-3}\ \text{mol L}^{-1}}$	36.80	31.75	29.19	28.68	28.17
$\ln \frac{[A]}{[B]}$	0.590	1.126	1.729	1.931	2.197

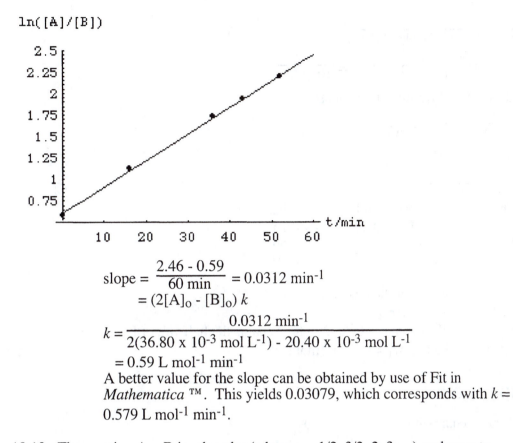

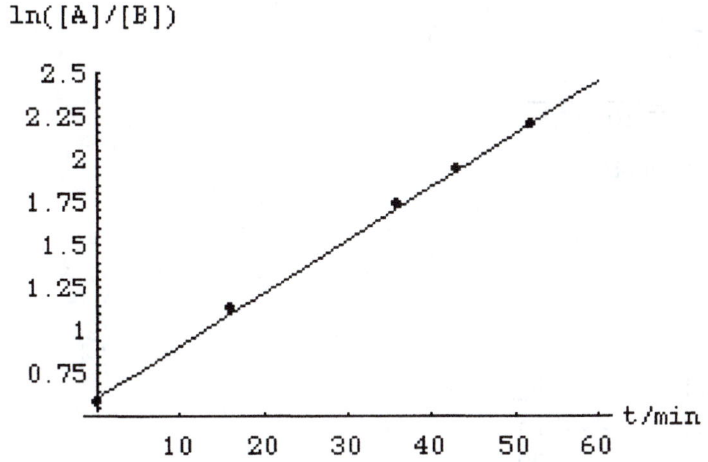

$$\text{slope} = \frac{2.46 - 0.59}{60\ \text{min}} = 0.0312\ \text{min}^{-1}$$
$$= (2[A]_o - [B]_o)\, k$$
$$k = \frac{0.0312\ \text{min}^{-1}}{2(36.80 \times 10^{-3}\ \text{mol L}^{-1}) - 20.40 \times 10^{-3}\ \text{mol L}^{-1}}$$
$$= 0.59\ \text{L mol}^{-1}\ \text{min}^{-1}$$

A better value for the slope can be obtained by use of Fit in *Mathematica* ™. This yields 0.03079, which corresponds with $k = 0.579\ \text{L mol}^{-1}\ \text{min}^{-1}$.

18.13 The reaction A = B is nth order (where $n = 1/2, 3/2, 2, 3, \cdots$) and goes to completion to the right. Derive the expression for the half life in terms of k, n, and $[A]_0$.

SOLUTION

$$\frac{1}{[A]^{n-1}} - \frac{1}{[A]_0^{n-1}} = (n - 1)kt.$$

If $[A] = \frac{[A]_0}{2}$, $t = t_{1/2}$.

$$\frac{2^{n-1}}{[A]_0^{n-1}} - \frac{1}{[A]^{n-1}} = (n - 1)kt_{1/2}$$

$$t_{1/2} = \frac{2^{n-1} - 1}{(n - 1)k[A]_0^{n-1}}$$

18.14 A gas reaction 2A = B is second order in A and goes to completion in a reaction vessel of constant volume and temperature with a half-life of 1 hour. If the initial pressure of A is 1 bar, what are the partial pressures of A and B, and the total pressure at 1 hour, 2 hours, and at equilibrium?

SOLUTION

$$2kt = \frac{1}{[A]} - \frac{1}{[A]_0}$$

$$2kt_{1/2} = \frac{2}{[A]} - \frac{1}{[A]_0} = \frac{1}{[A]_0}$$

$$t_{1/2} = \frac{1}{2k[A]_0} = 1 \text{ h} \quad \text{or} \quad k = \frac{1}{2t_{1/2}[A]_0}$$

Eliminating k from the first equation,

$$\frac{t}{t_{1/2}[A]_0} = \frac{1}{[A]} - \frac{1}{[A]_0}$$

or

$$[A] = \frac{[A]_0}{(t/t_{1/2}) + 1}$$

t/h	P_A/bar	P_B/bar	P/bar
1	0.50	0.25	0.75
2	0.33	0.33	0.66
∞	0	0.50	0.50

18.15 The rate constant for the reaction

$$I + I + Ar \rightarrow I_2 + Ar$$

is 0.59×10^{16} cm^6 mol^{-2} s^{-1} at 293 K. What is the half-life of I if $[I]_0 = 2 \times 10^{-5}$ mol L^{-1} and $[Ar] = 5 \times 10^{-3}$ mol L^{-1}?

SOLUTION

The rate equation is

$$v = k[I]^2[Ar] = -\frac{1}{2}\frac{d[I]}{dt} = \frac{d[I_2]}{dt}$$

so that

$$-\frac{d[I]}{dt} = (2)(0.59 \times 10^{16} \text{ cm}^6 \text{ mol}^{-2} \text{ s}^{-1})[I]^2[Ar]$$

$$[Ar] = (5 \times 10^{-3} \text{ mol L}^{-1})(10^{-3} \text{ L cm}^{-3})$$
$$= 5 \times 10^{-6} \text{ mol cm}^{-3}$$

$$-\frac{d[I]}{dt} = (5.9 \times 10^{10} \text{ cm}^3 \text{ mol}^{-1} \text{ s}^{-1})[I]^2$$

$$t_{1/2} = \frac{1}{k[I]_0} = \frac{1}{(5.9 \times 10^{10} \text{ cm}^3 \text{ mol}^{-1} \text{ s}^{-1})(10^{-3} \text{ L cm}^3)(2 \times 10^{-5} \text{ mol L}^{-1})}$$
$$= 0.85 \times 10^{-3} \text{ s}$$

18.16 A solution of A is mixed with an equal volume of a solution of B containing the same number of moles, and the reaction A + B = C occurs. At the end of 1 h A is 75% reacted. How much of A will be left unreacted at the end of 2 h if the reaction is (a) first order in A and zero order in B; (b) first order in both A and B; and (c) zero order in both A and B?

SOLUTION

(a) If the reaction is first order in A and zero order in B, the half time is 1/2 hour, and the following table may be constructed doing calculations in one's head.

t/h	0	0.5	1	1.5	2
% Unreacted	100	50	25	12.5	6.25

% Reacted	0	50	75	8.75	93.75

(b) If the reaction is first order in both A and B, the initial concentrations are equal, and the stoichiometry is 1:1, the concentration of A will follow

$$k = \frac{1}{t}\left(\frac{1}{[A]} - \frac{1}{[A]_0}\right) = \frac{1}{t}\frac{([A]_0 - [A])}{[A]_0\,[A]}$$

$$k[A]_0 = \frac{1}{t}\left(\frac{[A]_0}{[A]} - 1\right) = \frac{1}{1\text{ h}}\left(\frac{100}{25} - 1\right) = 3\text{ h}^{-1}$$

After 2 h

$$3\text{ h}^{-1} = \frac{1}{2\text{ h}}\left(\frac{100}{[A]} - 1\right)$$

$$(A) = \frac{100}{7} = 14.3\%$$

(c) If the reaction is zero order in both A and B, both will be completely gone in $1\frac{1}{3}$ h.

18.17 Show that for a first order reaction R → P the concentration of product can be represented as a function of time by $[P] = a + bt + ct^2 \cdots$ and express a, b, and c in terms of $[R]_0$ and k.

SOLUTION

$[P] = [R]_0 - [R] = [R]_0(1 - e^{-kt})$

Expanding e^{-kt} as a power series gives

$$[P] = [R]_0\left(kt - \frac{(kt)^2}{2!} + \frac{(kt)^3}{3!} - \cdots\right)$$

Thus $a = 0$, $b = [R]_0k$, and $c = \frac{[R]_0k^2}{2}$.

*18.18 For a reaction A → X, the following concentrations of A were found in a single kinetics experiment

$[A]/\text{mol L}^{-1}$	1.000	0.952	0.909	0.870	0.833	0.800
t/h	0	0.05	0.10	0.15	0.20	0.25

What is the rate v of this reaction at $[A] = 1.000\text{ mol L}^{-1}$?

SOLUTION

$$v = -\frac{d[A]}{dt}$$

As a first approximation $v = \frac{0.048}{0.05} = 0.960\text{ mol L}^{-1}\text{ h}^{-1}$

A better value may be obtained by using

$$\frac{[P]}{t} = b + ct$$

for the first 10% of the reaction.

$[X]/\text{mol L}^{-1}$	0.048	0.091	0.130	0.167	0.200
t/h	0.05	0.10	0.15	0.10	0.25

$$\frac{0.048}{0.05} = b + c\,(0.05) = 0.960$$

$$\frac{0.091}{0.10} = b + c\,(0.10) = 0.910$$

Taking the difference between these equations

$0.050 \quad = -\ 0.05\ c$

$c \qquad = -\ 1.00, \qquad b = 1.010 = v$

More experimental points can be included by linear fitting.

Linear fitting through 13% reaction yields

$v = 1.005\ \text{mol L}^{-1}\ \text{h}^{-1}$

Linear fitting through 16.7% reaction yields

$v = 0.998\ \text{mol L}^{-1}\ \text{h}^{-1}$

Linear fitting through 20% reaction yields

$v = 0.993\ \text{mol L}^{-1}\ \text{h}^{-1}$

As the extent of reaction increases, this truncated series expansion becomes less adequate, and so the initial velocity is $1.00 \pm 0.005\ \text{mol L}^{-1}\ \text{h}^{-1}$, rather than $0.96\ \text{mol L}^{-1}\ \text{h}^{-1}$, obtained as a first approximation.

18.19 The following table gives kinetic data for the following reaction at 25 °C.
$OCl^- + I^- = OI^- + Cl^-$

$[OCl^-]/\text{mol L}^{-1}$	$[I^-]/\text{mol L}^{-1}$	$[OH^-]/\text{mol L}^{-1}$	$\dfrac{d[IO^-]/dt}{10^{-4}\ \text{mol L}^{-1}\ \text{s}^{-1}}$
0.0017	0.0017	1.00	1.75
0.0034	0.0017	1.00	3.50
0.0017	0.0034	1.00	3.50
0.0017	0.0017	0.5	3.50

What is the rate law for the reaction, and what is the value of the rate constant?

SOLUTION

When other concentrations are held constant, doubling $[OCl^-]$ doubles the rate, doubling

$[I^-]$ doubles the rate, and halving $[OH^-]$ doubles the rate. Therefore the rate law is

$$\frac{d[OI^-]}{dt} = \frac{k[OCl^-][I^-]}{[OH^-]}$$

Substituting the values for the first experiment

$$1.75 \times 10^{-4}\ \text{mol L}^{-1}\ \text{s}^{-1} = \frac{k(0.0017\ \text{mol L}^{-1})(0.0017\ \text{mol L}^{-1})}{(1\ \text{mol L}^{-1})}$$

$k = 61\ \text{s}^{-1}$

18.20 For a reversible first-order reaction

$$A \underset{k_2}{\overset{k_1}{\rightleftarrows}} B$$

$k_1 = 10^{-2}\ s^{-1}$ and $[B]_{eq}/[A]_{eq} = 4$. If $[A]_o = 0.01$ mol L^{-1} and $[B]_0 = 0$, what will be the concentration of B at 30 s?

SOLUTION

$$[B] = \frac{k_1[A]_0}{k_1 + k_2}\left[1 - e^{-(k_1+k_2)t}\right]$$

$$\frac{[B]_{eq}}{[A]_{eq}} = 4 = \frac{k_1}{k_2}$$

$$k_2 = 2.5 \times 10^{-3}\ s^{-1}$$

$$[B] = \frac{(10^{-2})(10^{-2})}{1.25 \times 10^{-2}}\left[1 - e^{-(1.25 \times 10^{-2})(30)}\right]$$

$$= 2.50 \times 10^{-3}\ \text{mol L}^{-1}$$

18.21 The first three steps in the decay of ^{238}U are

$$^{238}U \xrightarrow[4.5\times10^9\ y]{\alpha} {}^{234}Th \xrightarrow[24.1\ d]{\beta} {}^{234}Pa \xrightarrow[1.14\ min]{\beta} {}^{234}U$$

If we start with pure ^{238}U, what fraction will be ^{234}Th after 10, 20, 40, and 80 days?

SOLUTION

$$k_1 = \frac{0.693}{(4.5 \times 10^9\ y)(365\ d\ y^{-1})} = 4.2 \times 10^{-13}\ d^{-1}$$

$$k_2 = \frac{0.693}{24.1\ d} = 0.0288\ d^{-1}$$

$$F = \frac{[^{234}Th]}{[^{238}U]_0} = \frac{k_1}{k_2 - k_1}\left(e^{-k_1t} - e^{-k_2t}\right)$$

t/d	10	20	40	80
F	3.6×10^{-12}	6.39×10^{-12}	9.98×10^{-12}	13.14×10^{-12}

18.22 Equation 18.27 for a second order reaction becomes indeterminant when $[A]_0 = [B]_0$, but the text states that when the initial concentrations of A and B in a reaction A + B = X are equal,

$$\frac{1}{[A]} = \frac{1}{[A]_0} + kt \text{ and } \frac{1}{[B]} = \frac{1}{[B]_0} + kt$$

Show that this is correct by using L'Hopital's rule. According to L'Hopital's rule, if a function of a variable x is indeterminant as $x \to 0$ because the function becomes 0/0, then the limit can be found by taking the limit of the derivative of the numerator divided by the derivative of the denominator.

SOLUTION

If the initial concentrations of A and B are nearly equal, we can write $[A]_0 = [B]_0 + x$ and $[A] = [B] + x$, where x is a small quantity that we will allow to approach zero. With these relations equation 18.27 can be written

$$kt = \frac{1}{x} \ln \frac{([B] + x)[B]_0}{[B]([B]_0 + x)}$$

Taking the limit as $x \to 0$ yields

$$\lim_{x \to 0} kt = \lim_{x \to 0} \frac{\ln([B]+ x) - \ln[B] + \ln[B]_0 - \ln([B]_0 + x)}{x}$$

This ratio can be replaced by the ratio of derivatives:

$$\lim_{x \to 0} kt = \lim_{x \to 0} \left(\frac{1}{[B] + x} - \frac{1}{[B]_0 + x} \right) = \frac{1}{[B]} - \frac{1}{[B]_0}$$

18.23 Suppose the transformation of A to B occurs by both a reversible first-order reaction and a reversible second-order reaction involving hydrogen ion.

$$A \underset{k_2}{\overset{k_1}{\rightleftarrows}} B \qquad A + H^+ \underset{k_4}{\overset{k_3}{\rightleftarrows}} B + H^+$$

What is the relationship between these four rate constants?

SOLUTION

$$\frac{[B]_{eq}}{[A]_{eq}} = \frac{k_1}{k_2}$$

$$\frac{[B]_{eq}[H^+]_{eq}}{[A]_{eq}[H^+]_{eq}} = \frac{k_3}{k_4} \quad \text{or} \quad \frac{[B]_{eq}}{[A]_{eq}} = \frac{k_3}{k_4}$$

$$\text{Therefore } \frac{k_1}{k_2} = \frac{k_3}{k_4} \quad \text{or} \quad k_1 k_4 = k_2 k_3$$

18.24 Use the rapid equilibrium approximation to derive the rate law for the mechanism

$$\begin{array}{ccc} A^- & \xrightarrow{k_A} & B^- \\ \uparrow\downarrow K_{HA} & & \uparrow\downarrow K_{HB} \\ HA & \xrightarrow[k_B]{} & HB \end{array}$$

The acid dissociation reactions are rapid in comparison with the isomerization reactions.

SOLUTION

The total concentrations of A and B are represented by

$[A] = [A^-] + [HA]$

$[B] = [B^-] + [HB]$

Since the acid dissociations are assumed to be at equilibrium,

$$[A^-] = \frac{[A]}{1 + [H^+]/K_{HA}}$$

$$[HA] = \frac{[A]}{1 + K_{HA}/[H^+]}$$

The rate law is

$$v = -\frac{d([A^-] + [HA])}{dt} = k_A[A^-] + k_{HA}[HA]$$

$$= \left(\frac{k_A}{1 + [H^+]/K_{HA}} + \frac{k_{HA}}{1 + K_{HA}/[H^+]}\right)[A]$$

$$= k_A'[A]$$

where k_A' is a pH-dependent first order rate constant.

If $k_A = 0$,

$$v = \frac{k_{HA}[A]}{1 + K_{HA}/[H^+]}$$

so that the pH dependent rate constant decreases as pH increases.

If $k_{HA} = 0$,

$$v = \frac{k_A[A]}{1 + [H^+]/K_{HA}}$$

so that the pH dependent rate constant increases as pH decreases.

18.25 Suppose that

$$A \xrightarrow{k_1} B \xrightarrow{k_2} C \longrightarrow \ldots$$

and you are interested in isolating the largest possible amount of B. Given the values of k_1 and k_2, derive an equation for the time that the concentration of B goes through a maximum. Now consider two cases: (a) A reacts more rapidly than B and (b) A reacts less rapidly than B. For a given value of k_2, in which case would you wait the longer time for B to go through its maximum?

SOLUTION

Equation 18.49 can be used since it applies whether C decomposes or not. When B goes through a maximum, d[B]/dt = 0, and so we take the derivative of equation 18.49 to obtain

$$\frac{d[B]}{dt} = \frac{-k_1[A]_0}{k_2 - k_1}\left[k_1 \exp(-k_1 t) - k_2 \exp(-k_2 t)\right]$$

Thus B goes through its maximum when

$k_1 \exp(-k_1 t) = k_2 \exp(-k_2 t)$

or $t = \dfrac{1}{k_1 - k_2} \ln \dfrac{k_1}{k_2}$

For a given value of k_2, you would wait the longer time in case (b) for B to go through its maximum concentration.

18.26 The hydrolysis of $(CH_2)_6C(Cl)CH_3$ in 80% ethanol follows the first-order rate equation. The values of the specific reaction rate constants are as follows:

t /°C	0	25	35	45
k/s^{-1}	1.06×10^{-5}	3.19×10^{-4}	9.86×10^{-4}	2.92×10^{-3}

(a) Plot ln k against 1/T. (b) Calculate the activation energy. (c) Calculate the pre-exponential factor.

SOLUTION

(a)

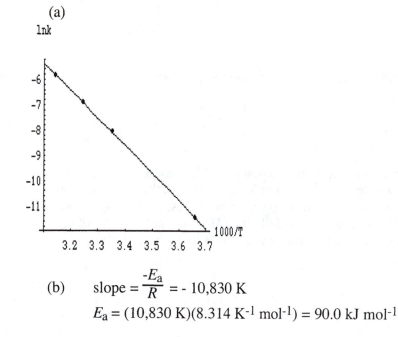

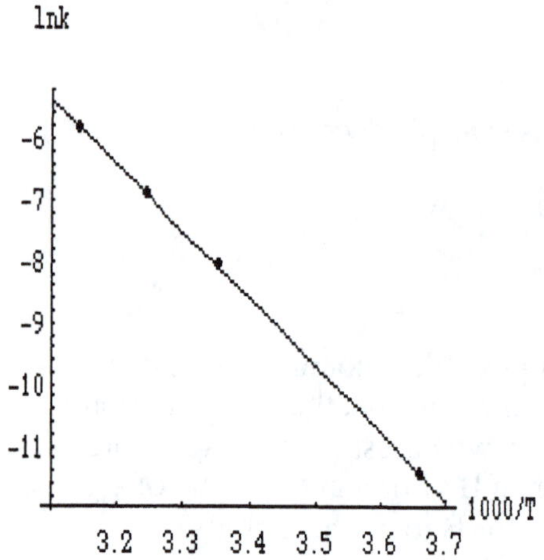

(b) slope = $\dfrac{-E_a}{R}$ = - 10,830 K

E_a = (10,830 K)(8.314 K^{-1} mol^{-1}) = 90.0 kJ mol^{-1}

(c) Taking the 35 °C point and $E_a = 90.0$ kJ mol^{-1}

$$\ln k = \frac{-E_a}{RT} + \ln A$$

$$-6.922 = \frac{-90\,000 \text{ J mol}^{-1}}{(8.314 \text{ K}^{-1} \text{ mol}^{-1})(308.15 \text{ K})} + \ln A$$

$$A = \exp(\ln A) = 1.77 \times 10^{12} \text{ s}^{-1}$$

18.27 If a first-order reaction has an activation energy of 104,600 J mol^{-1} and a pre-exponential factor A of 5×10^{13} s^{-1}, at what temperature will the reaction have a half-life of (a) 1 min and (b) 30 days?

SOLUTION

$$k = \frac{0.693}{t_{1/2}} = Ae^{-E_a/RT}$$

$$T = \frac{-E_a}{R \ln\left(\frac{0.693}{At_{1/2}}\right)}$$

(a) $$T = \frac{-104\,600 \text{ J mol}^{-1}}{(8.314 \text{ J K}^{-1} \text{ mol}^{-1}) \ln\left[\frac{0.693}{(5 \times 10^{13} \text{ s}^{-1})(60 \text{ s})}\right]} = 349 \text{ K or } 76\ ^{o}\text{C}$$

(b) $$T = \frac{-104\,600 \text{ J mol}^{-1}}{(8.314 \text{ J K}^{-1} \text{ mol}^{-1}) \ln\left[\frac{0.693}{(5 \times 10^{13})(60)^2(24)(30)}\right]}$$

$$= 270 \text{ K or } -3\ °\text{C}$$

18.28 Isopropenyl allyl ether in the vapor state isomerizes to allyl acetone according to a first-order rate equation. The following equation gives the influence of temperature on the rate constant (in s^{-1}):
$k = 5.4 \times 10^{11} \exp(-123\,000/RT)$
where the activation energy is expressed in J mol^{-1}. At 150 °C, how long will it take to build up a partial pressure of 0.395 bar of allyl acetone, starting with 1 bar of isopropenyl allyl ether?

SOLUTION

$$k = (5.4 \times 10^{11} \text{ s}^{-1})\, e^{-123{,}000/(8.314)(423.15)}$$

$$= 3.53 \times 10^{-4} \text{ s}^{-1}$$

$$t = \frac{1}{k} \ln \frac{P_0}{P} = \frac{1}{3.53 \times 10^{-4} \text{ s}^{-1}} \ln \frac{1 \text{ bar}}{0.605 \text{ bar}}$$

$$= 1420 \text{ s}$$

18.29 The pre-exponential factor for the trimolecular reaction $2NO + O_2 \rightarrow 2NO_2$ is 10^9 cm^6 mol^{-2} s^{-1}. What is the value in L^2 mol^{-2} s^{-1} and cm^6 s^{-1}?

SOLUTION

$$(10^9 \text{ cm}^6 \text{ mol}^{-2} \text{ s}^{-1})(10^{-1} \text{ dm cm}^{-1})^6 = 10^3 \text{ dm}^6 \text{ mol}^{-2} \text{ s}^{-1}$$
$$= 10^3 \text{ L}^2 \text{ mol}^{-2} \text{ s}^{-1}$$
$$= \frac{(10^3 \text{ L}^2 \text{ mol}^{-2} \text{ s}^{-1})(10^3 \text{ cm}^3 \text{ L}^{-1})^2}{(6.02 \times 10^{23} \text{ mol}^{-1})^2}$$
$$= 2.76 \times 10^{-39} \text{ cm}^6 \text{ s}^{-1}$$

18.30 A reaction A + B + C → D follows the mechanism

$A + B \rightleftarrows AB \qquad\qquad AB + C \rightarrow D$

in which the first step remains essentially in equilibrium. Show that the dependence of rate on temperature is given by

$k = Ae^{-(E_a + \Delta H)/RT}$

where ΔH is the enthalpy change for the first reaction.

SOLUTION

For A + B = AB

$$K = \frac{[AB]}{[A][B]} = e^{\Delta S^o/R} e^{-\Delta H^o/RT}$$

$$[AB] = [A][B] e^{\Delta S^o/R} e^{-\Delta H^o/RT}$$

$$\frac{d[D]}{dt} = k'[AB][C] = k'[A][B][C] e^{\Delta S^o/R} e^{-\Delta H^o/RT}$$

$$= k[A][B][C]$$

where

$$k = k' e^{\Delta S^o/R} e^{-\Delta H^o/RT}$$

Assuming

$$k' = A' e^{-E_a/RT}$$

$$k = A' e^{\Delta S^o/R} e^{-(E_a + \Delta H^o)/RT}$$

$$= Ae^{-(E_a + \Delta H^o)/RT}$$

Note that the activation energy for this kind of reaction can b negative (see Computer Problem 18.J).

18.31 Consider the following mechanism

$$A + B \underset{k_2}{\overset{k_1}{\rightleftarrows}} C \qquad\qquad C \xrightarrow{k_3} D$$

(a) Derive the rate law using the steady state approximation to eliminate the concentration of C. (b) Assuming that $k_3 << k_2$, express the pre-exponential factor A and E_a for the apparent second-order rate constant in terms of A_1, A_2, and A_3 and E_{a1}, E_{a2}, and E_{a3} for the three steps.

SOLUTION

(a) $$\frac{d[C]}{dt} = k_1[A][B] - (k_2 + k_3)[C] = 0$$

$$\frac{d[D]}{dt} = k_3[C] = \frac{k_1k_3[A][B]}{k_2 + k_3}$$

(b) For $k_2 >> k_3$,

$$k_{app} = \frac{k_1k_3}{k_2} = \frac{A_1^{-E_{a1}/RT} A_3^{-E_{a3}/RT}}{A_2 e^{-E_{a2}/RT}}$$

$$= \frac{A_1A_3}{A_2} e^{-(E_{a1} + E_{a3} - E_{a2})/RT}$$

$$A_{app} = \frac{A_1A_3}{A_2} \qquad E_{app} = E_{a1} + E_{a3} - E_{a2}$$

18.32 For the two parallel reactions $A \xrightarrow{k_1} B$ and $A \xrightarrow{k_2} C$, show that the activation energy E' for the disappearance of A is given in terms of the activation energies E_1 and E_2 for the two paths by

$$E' = \frac{k_1E_1 + k_2E_2}{k_1 + k_2}$$

SOLUTION

The rate equation for A is

$$\frac{d[A]}{dt} = k_1[A] + k_2[A] = (k_1 + k_2)[A] = k'[A]$$

where $k' = k_1 + k_2 = Ae^{-E'/RT}$

$$\frac{d\ln k'}{dt} = \frac{E}{RT^2} = \frac{d\ln(k_1 + k_2)}{dT} = \frac{d(k_1 + k_2)}{(k_1 + k_2)dT} = \frac{1}{(k_1 + k_2)}\left(\frac{dk_1}{dT} + \frac{dk_2}{dT}\right)$$

$$= \frac{1}{(k_1 + k_2)}\left(k_1\frac{d\ln k_1}{dT} + k_2\frac{d\ln k_2}{dT}\right)$$

$$= \frac{1}{(k_1 + k_2)}\left(\frac{k_1E_1}{RT^2} + \frac{k_2E_2}{RT^2}\right)$$

$$E' = \frac{k_1E_1 + k_2E_2}{k_1 + k_2}$$

18.33 Set up the rate expressions for the following mechanism

$$A \underset{k_2}{\overset{k_1}{\rightleftarrows}} B \qquad B + C \xrightarrow{k_3} D$$

If the concentration of B is small compared with the concentrations of A, C, and D, the steady-state approximation may be used to derive the rate law.

Show that this reaction may follow the first-order equation at high pressures and the second-order equation at low pressure.

SOLUTION

Since the concentration of B is small, it is assumed to be in a steady state.

$$\frac{d[B]}{dt} = k_1[A] - (k_2 + k_3[C])[B] = 0$$

$$[B] = \frac{k_1[A]}{k_2 + k_3[C]}$$

$$\frac{d[D]}{dt} = k_3[B][C] = \frac{k_1k_3[A][C]}{k_2 + k_3[C]}$$

At high pressure, $k_3[C] >> k_2, \frac{d[D]}{dt} = k_1[A]$

At low pressure, $k_2 >> k_3[C], \frac{d[D]}{dt} = \frac{k_1k_3}{k_2}[A][C]$

18.34 A dimerization $2A \longrightarrow A_2$ is found to be first order, with a half life of 666 s. This somewhat surprising result is explained by postulating the following mechanism:

$$A \xrightarrow{k_1} A^*$$

$$A^* + A \xrightarrow{k_2} A_2$$

where $k_2 >> k_1$. (a) What is the value for the rate constant k_1? (b) If the initial concentration of A is 0.05 M, how much time is required to reach [A] = 0.0125 M?

SOLUTION

(a) The reaction will behave like a first order reaction with rate constant $2k_1$ because

$$-d[A]/dt = k_1[A] + k_2[A^*][A] = 2k_1[A]$$

using the steady state assumption for [A*]

$$t_{1/2} = 0.693/2k_1$$

$$k_1 = 0.693/2(666\text{ s}) = 5.20 \times 10^{-4}\text{ s}^{-1}$$

(b) $$\ln\frac{[A]_o}{[A]} = 2k_1t$$

$$t = (\ln 4)/2(5.20 \times 10^{-4}\text{ s}^{-1}) = 1333\text{ s}$$

*18.35 The reaction $NO_2Cl = NO_2 + \frac{1}{2}Cl_2$ is first order and appears to follow the mechanism

$NO_2Cl \xrightarrow{k_1} NO_2 + Cl$ $\qquad$ $NO_2Cl + Cl \xrightarrow{k_2} NO_2 + Cl_2$

(a) Assuming a steady state for the chlorine atom concentration, show that the empirical first-order rate constant can be identified with $2k_1$. (b) The following data were obtained at 180 °C. In a single experiment the reaction is first order, and the empirical rate constant is represented by k. Show that the reaction is second order at these low gas pressures and calculate the second-order rate constant.

$c/10^{-8}$ mol cm^{-3}	5	10	15	20
$k/10^{-4}$ s^{-1}	1.7	3.4	5.2	6.9

SOLUTION

(a) $\frac{d[Cl]}{dt} = k_1[NO_2Cl] - k_2[NO_2Cl][Cl] = 0$

Therefore $[Cl] = \frac{k_1}{k_2}$

$$\frac{-d[NO_2Cl]}{dt} = k_1[NO_2Cl] + k_2[NO_2Cl][Cl]$$
$$= k_1[NO_2Cl] + k_2[NO_2Cl]k_1/k_2$$
$$= 2k_1[NO_2Cl]$$

(b) At low pressures the rate-determining step is the first step, and the reaction becomes second order.

$$NO_2Cl + NO_2Cl \xrightarrow{k_1'} NO_2Cl + NO_2 + Cl$$

$$\frac{-d[NO_2Cl]}{dt} = k_1'\,[NO_2Cl]^2$$

assuming $[NO_2Cl]$ is approximately constant

$k = k_1'[NO_2Cl]$

$1.7 \times 10^{-4} = k_1'(5 \times 10^{-8})$

$k_1' = 3.4 \times 10^3$ cm^3 mol^{-1} s^{-1}

18.36 The reaction

$2SO_2 + O_2 = 2\ SO_4$

is catalyzed by the mechanism

$$2\,NO + O_2 \underset{k_{-1}}{\overset{k_1}{\rightleftarrows}} 2NO_2$$

$$NO_2 + SO_2 \underset{k_{-2}}{\overset{k_2}{\rightleftarrows}} NO + SO_3$$

In order to obtain the overall reaction from this mechanism, the second step has to be taken twice, and so the stoichiometric number s_2 of the second step is said to be 2. The equilibrium constant K_c for an overall reaction is related to the rate constants for the individual steps, k_i and k_{-i}, by

$$K_c = \prod_{i=1}^{S}\left(\frac{k_i}{k_{-i}}\right)^{s_i}$$

where s_i is the stoichiometric coefficient of the ith step and S is the number of steps. Verify this relation for the above mechanism.

SOLUTION

The standard Gibbs energy for the overall reaction is given by

$$\Delta G^o = \Delta G_1^o + 2\Delta G_2^o$$

$$-RT\ln K_c = -RT\ln K_1 - RT\ln K_2^2$$

$$= -RT\ln\left(\frac{k_1}{k_{-1}}\right) - RT\ln\left(\frac{k_2}{k_{-2}}\right)^2$$

where the equilibrium constants of the elementary reactions have been expressed in terms of the rate constants. The last form of the equation may be written

$$K_c = \left(\frac{k_1}{k_{-1}}\right)\left(\frac{k_2}{k_{-2}}\right)$$

which may be generalized to

$$K_c = \prod_{i=1}^{S}\left(\frac{k_i}{k_{-i}}\right)^{s_i}$$

18.37 What is the rate constant for the following reaction at 500 K?

$$H + HCl \rightarrow Cl + H_2$$

The data required are to be found in Tables 18.3 and Appendix C.3.

SOLUTION

The rate constant for the backward reaction is given by

$$k_b = (10^{10.9}\ \text{L mol}^{-1}\ \text{s}^{-1})\ e^{-23{,}000/8.314 \times 500}$$

$$= 3.1 \times 10^8\ \text{L mol}^{-1}\ \text{s}^{-1}$$

$$\Delta G^o = \Delta_f G^o(\text{Cl}) - \Delta_f G^o(\text{H}) - \Delta_f G^o(\text{HCl})$$

$$= 94.203 - 192.957 + 97.166 = -1.588\ \text{kJ mol}^{-1}$$

$K_P = e^{1588/8.324 \times 500} = 1.47$

$K_P = K_c = \dfrac{k_f}{k_b}$

$k_f = (1.47)(3.1 \times 10^8 \text{ L mol}^{-1} \text{ s}^{-1}) = 4.5 \times 10^8 \text{ L mol}^{-1} \text{ s}^{-1}$

18.38 For the gas reaction

$$O + O_2 + M \underset{k'}{\overset{k}{\rightleftarrows}} O_3 + M$$

When M = O_2, the rate constant is given by $k = (6.0 \times 10^7 \text{ L}^2 \text{ mol}^{-2} \text{ s}^{-1})$ $e^{2.5/RT}$ where the activation energy is in kJ mol^{-1}. Calculate the values of the parameters in the Arrhenius equation for the reverse reaction assuming ΔH^o and ΔS^o are independent of temperature.

SOLUTION

$\Delta H^o = 142.7 - 249.170 = -106.5 \text{ kJ mol}^{-1}$

$\Delta S^o = 238.82 - 160.946 - 205.029 = -127.16 \text{ J K}^{-1} \text{ mol}^{-1}$

$K_P = e^{-\Delta H^o/RT} e^{\Delta S^o/R}$

$= e^{106\ 500/RT} e^{-127.16/R}$

$K_c = K_P \left(\dfrac{P^o}{c^o RT}\right)^{\Sigma \nu_i} = K_P \left(\dfrac{1}{24.46}\right)^{-1} \text{L mol}^{-1} = \dfrac{k}{k'}$

$k' = \dfrac{k}{24.46\, K_P}$

$k' = \dfrac{6 \times 10^7\, e^{2500/RT}}{24.46\, e^{(106\ 500)/RT}\, e^{-127.16/R}}$

$= \left(\dfrac{6 \times 10^7}{24.46}\, e^{127.16R}\right) e^{(2500 - 106\ 500)/RT}$

$= (1.1 \times 10^{13} \text{ L mol}^{-1} \text{ s}^{-1})\, e^{-(104\ 000)/RT}$

18.39 (a) Write the steady state rate equations for A and B in reaction 18.61. (b) Use these rate equations to derive the equilibrium expressions for [B]/[A] and [C]/[A] by use of the principle of detailed balance that requires that $k_1k_3k_6 = k_2k_4k_6$.

SOLUTION

[a] The steady-state rate expression for A is

$d[A]/dt = -(k_1+k_6)[A] + k_2[B] + k_5[C] = 0$ [I]

The steady-state rate expression for B is

$d[B]/dt = -(k_2+k_3)[B] + k_1[A] + k_4[C] = 0$ [II]

[b] Dividing equations I and II by [A] yields

$k_2[B]/[A] + k_5[C]/[A] = k_1 + k_6$ [III]

$-(k_2 + k_3)[B]/[A] + k_4[C]/[A] = -k_1$ [IV]

These simultaneous linear equations can be solved for [B]/[A] and [C]/[A].

$$\frac{[B]}{[A]} = \frac{k_1k_4 + k_1k_5 + k_4k_6}{k_2k_4 + k_2k_5 + k_3k_5} \quad \text{(V)}$$

$$\frac{[C]}{[A]} = \frac{k_1k_3 + k_2k_6 + k_3k_6}{k_2k_4 + k_2k_5 + k_3k_5} \quad \text{(VI)}$$

Now you can use equation 18.65 to eliminate one of the rate constants in each of these equations and obtain

$$\frac{[B]}{[A]} = \frac{k_1}{k_2}$$

$$\frac{[C]}{[A]} = \frac{k_6}{k_5}$$

18.40 The mechanism of the pyrolysis of acetaldehyde at 520 °C and 0.2 bar is

$$CH_3CHO \xrightarrow{k_1} CH_3 + CHO$$

$$CH_3 + CH_3CHO \xrightarrow{k_2} CH_4 + CH_3CO$$

$$CH_3CO \xrightarrow{k_3} CO + CH_3$$

$$CH_3 + CH_3 \xrightarrow{k_4} C_2H_6$$

What is the rate law for the reaction of acetaldehyde, using the usual assumptions? (As a simplification further reactions of the radical CHO have been omitted and its rate equation may be ignored.)

SOLUTION

(1) $$\frac{d[CH_3CHO]}{dt} = -(k_1 + k_2[CH_3])[CH_3CHO]$$

(2) $$\frac{d[CH_3]}{dt} = k_1[CH_3CHO] - k_2[CH_3][CH_3CHO] + k_3[CH_3CO] - 2k_4[CH_3]^2 = 0$$

(3) $$\frac{d[CH_3CO]}{dt} = k_2[CH_3][CH_3CHO] - k_3[CH_3CO] = 0$$

Equation 3 yields $[CH_3CO] = k_2[CH_3][CH_3CHO]/k_3$

Substituting this in equation 2 yields

$$[CH_3] = \left(\frac{k_1[CH_3CHO]}{2k_4}\right)^{1/2}$$

Substituting this in equation 1 yields

$$\frac{d[CH_3CHO]}{dt} = -\left\{k_1 + k_2\left(\frac{k_1}{2k_4}\right)^{1/2}[CH_3CHO]^{1/2}\right\}[CH_3CHO]$$

If k_1 is small,

$$\frac{d[CH_3CHO]}{dt} = -k_2(k_1/2k_4)^{1/2}[CH_3CHO]^{3/2}$$

18.41 (a) 1.2×10^{-3} mol L^{-1} s^{-1}, (b) 2.4×10^{-3} mol L^{-1} s^{-1}, (c) 3.6×10^{-3} mol s^{-1}, (d) 3.2×10^{-3} mol

18.42 (a) 0.05 mol L^{-1} s^{-1}, (b) -0.05 mol L^{-1} s^{-1}, (c) -0.15 mol L^{-1} s^{-1}

18.43 $v = k[H_2][Br_2]^{1/2}$

18.44 1.56%

18.45 0.0396, 0.0421 min^{-1}

18.46 14.6% reacted

18.47 (b) 20.0 min

18.48

t/min	P_A/bar	P_B/bar
10	0.50	1.00
20	0.25	1.50
∞	0	2.00

18.49 (a) 1.38×10^{-10} %, (b) 6950 g

18.50 7160 y

18.51 3989 s

18.52 (a) 221 s, (b) 82.4 s

18.53 26.3 min

18.54 3.73×10^9 L mol^{-1} s^{-1}

18.55 (a) 0.107 L mol^{-1} s^{-1}, (b) 2.85×10^3 s

18.56 666 s

18.57 2×10^3 s

18.58 $10^{-12.6}$ cm^3 s^{-1}

18.59 $[A]_0^{1/2} - [A]^{1/2} = (k/2)t$

$t_{1/2} = (\sqrt{2}/k)(\sqrt{2} - 1)[A]_0^{1/2}$

18.60 $[A]^{-1/2} - [A]_0^{-1/2} = kt/2t_{1/2} = 2(\sqrt{2} - 1)/k[A]_0^{1/2}$

18.61 (a) 80%, (b) 67.1%, (c) 61.5%

18.62 $\ln([B]/[B]_0) = -k[B]_0 t$

18.63 5.78 x 10^{-4} s^{-1}

18.64 $[B] = k_1[A]_0 t\, e^{-kt}$

$[C] = [A]_0[1 - e^{-k_1 t}(1 + k_1 t)]$

18.65 $k[B][C]/[A]$, $k[B]^{1/2}[C]^{1/2}$

18.66 (a) 97.0 kJ mol^{-1}, (b) 8.99 x 10^{13} s^{-1}, (c) 1.71 s

18.67 (a) 721 K, (b) 663 K

18.68 (a) 4.6 x 10^6 s, (b) 2430 s

18.69 (a) 1105 K, (b) 736 K, (c) The rate of reaction with the higher activation energy increases more rapidly with increasing temperature than the rate of the reaction with the lower activation energy.

18.70 $k = (5.98 \times 10^{10}\ \text{L}^2\ \text{mol}^{-2}\ \text{s}^{-1}) \exp[5730\ \text{J mol}^{-1}/(8.314\ \text{J K}^{-1}\ \text{mol}^{-1})(T/\text{K})]$

18.71 (a) 15.1 kJ mol^{-1}, (b) 2.27

18.72 $d[NO_2]/dt = k'[NO]^2[O_2]$

18.73 $d[I_2]/dt = k_1K[I]^2[M] - k_{-1}[I_2][M]$

$[I_2]/[I]^2 = k_1K/k_{-1}$

18.74 $d[A_2]/dt = k_1k_2[M][A]^2/(k_{-1} + k_2[M])$

18.76 E_a = 18,931 J mol^{-1}, A = 4.74 x 10^{10} L mol^{-1} s^{-1}

18.77 3.0 x 10^{16} L mol^{-1} s^{-1}

18.78 $d[O_2]/dt = 2k_1[NO][O_3]$

18.79 $$\frac{d[COCl_2]}{dt} = \frac{k_2k_3(2k_1/k_{-1})^{1/2}[Cl_2]^{3/2}[CO]}{k_{-2} + k_3[Cl_2]}$$

19

Chemical Dynamics and Photochemistry

19.1 Use the pre-exponential factor $A = 3 \times 10^{13}$ cm^3 mol^{-1} s^{-1} for the reaction Br + H_2 → HBr + H to calculate the reaction cross section and collision diameter for this reaction at 400 K.

SOLUTION

$$\mu N_A = \frac{1}{\dfrac{1}{79.904 \times 10^{-3}\ \text{kg}} + \dfrac{1}{2.0158 \times 10^{-3}\ \text{kg}}}$$

$$= 1.966 \times 10^{-3}\ \text{kg mol}^{-1}$$

$$\left(\frac{8RT}{\pi \mu N_A}\right)^{1/2} = \left[\frac{8(8.314\ \text{J K}^{-1}\ \text{mol}^{-1})(400\ \text{K})}{\pi(1.966 \times 10^{-3}\ \text{kg mol}^{-1})}\right]^{1/2}$$

$$= 2075\ \text{m s}^{-1}$$

$$A = N_A \pi d_{12}^2 \left(\frac{8RT}{\pi \mu N_A}\right)^{1/2}$$

$$d_{12} = \left[\frac{A}{N_A \pi \left(\dfrac{8RT}{\pi \mu N_A}\right)^{1/2}}\right]^{1/2}$$

$$= \left[\frac{(3 \times 10^{13}\ \text{cm}^3\ \text{mol}^{-1}\ \text{s}^{-1})(10^{-2}\ \text{m cm}^{-1})^3}{\pi(6.02 \times 10^{23}\ \text{mol}^{-1})(2075\ \text{m s}^{-1})}\right]^{1/2} = 87.4\ \text{pm}$$

The cross section is $\pi d_{12}^2 = 2.40 \times 10^{-2}$ nm^2.

19.2 (a) Calculate the second-order rate constant for collisions of dimethyl ether molecules with each other at 777 K. It is assumed that the molecules are spherical and have a radius of 0.25 nm. If every collision was effective in producing decomposition what would be the half-life of the reaction (b) at 1 bar pressure, and (c) at a pressure of 0.13 Pa?

SOLUTION

(a) $k = 2N_A\pi d_{12}^2 \left(\frac{8RT}{\pi\mu N_A}\right)^{1/2}$ since A_1 and A_2 are identical

$\mu N_A = \frac{M}{2} = \frac{46 \times 10^{-3}\ \text{kg mol}^{-1}}{2} = 23 \times 10^{-3}\ \text{kg mol}^{-1}$

$d_{12} = 0.5 \times 10^{-9}$ m

$$k = 2(6.02 \times 10^{23}\ \text{mol}^{-1})\pi(0.5 \times 10^{-9}\ \text{m})^2 \left[\frac{8(8.314\ \text{J K}^{-1}\ \text{mol}^{-1})(777\ \text{K})}{(3.1416)(23 \times 10^{-3}\ \text{kg mol}^{-1})}\right]^{1/2}$$

$= 8.00 \times 10^8\ \text{m}^3\ \text{mol}^{-1}\ \text{s}^{-1}$

$= (8.00 \times 10^8\ \text{m}^3\ \text{mol}^{-1}\ \text{s}^{-1})(10^3\ \text{L m}^{-3})$

$= 8.00 \times 10^{11}\ \text{L mol}^{-1}\ \text{s}^{-1}$

(b) $t_{1/2} = \frac{1}{k[\text{A}]_0}$

$[\text{A}]_0 = \frac{P}{RT} = \frac{1\ \text{bar}}{(0.08314\ \text{L bar k}^{-1}\ \text{mol}^{-1})(777\ \text{K})}$

$= 0.0155\ \text{mol L}^{-1}$

$t_{1/2} = \frac{1}{(8.00 \times 10^{11}\ \text{L mol}^{-1}\ \text{s}^{-1})(0.0155\ \text{mol L}^{-1})}$

$= 8.1 \times 10^{-11}$ s

(c) $P = 0.13 \times 10^{-5}$ bar

$[\text{A}]_0 = \frac{P}{RT} = \frac{0.13 \times 10^{-5}}{0.08314 \times 777}$

$= 2.0 \times 10^{-8}\ \text{mol L}^{-1}$

$t_{1/2} = \frac{1}{(8.00 \times 10^{11}\ \text{L mol}^{-1}\ \text{s}^{-1})(2.0 \times 10^{-8}\ \text{mol L}^{-1})}$

$= 6.2 \times 10^{-5}$ s

19.3 Show that the transition-state theory yields the simple collision theory result when it is applied to the reaction of two rigid spherical molecules.

SOLUTION

Assuming that the molecules react on their first collision (that is, the activation energy is zero), transition state theory yields

$$k = \frac{RT}{h}\frac{q''_{AB^\ddagger}}{q'_A q'_B} \tag{1}$$

where q'_A and q'_B are molecular partition fucntions without the volume factor, and

$q''_{AB}{}^{\ddagger}$ is the molecular partition function for the transition state without the volume factor and without the vibrational factor.

$$q'_A = \left(\frac{2\pi m_A kT}{h^2}\right)^{3/2} \qquad (2)$$

$$q'_B = \left(\frac{2\pi m_B kT}{h^2}\right)^{3/2} \qquad (3)$$

$$q''_{AB} = \left[\frac{2\pi(m_A + m_B)kT}{h^2}\right]^{3/2} \frac{8\pi^2\mu(R_A + R_B)^2 kT}{h^2} \qquad (4)$$

$$\mu = \frac{m_A m_B}{m_a + m_B} \qquad (5)$$

Substituting equations 2, 3, 4 and 5 in equation 1 yields

$$k = N_A\left(\frac{8\pi kT}{\mu}\right)^{1/2}(R_A + R_B)^2$$

which may be compared with equation 19.4

$$k = \pi d_{12}^2 \left(\frac{8RT}{\pi\mu N_A}\right)^{1/2} = \left(\frac{8\pi kT}{\mu}\right)^{1/2}(R_A + R_B)^2$$

The difference between these expressions of a factor of N_A is simply a matter of units.

19.4 The rate constant for the elementary reaction

$K + Br_2 \longrightarrow KBr + Br$

is 1.0 x 10^{12} L mol^{-1} s^{-1} independent of temperature. Calculate the rate constant expected from collision theory at 298 K. The fact that the rate constant is greater than would be expected from collision theory is explained by the harpoon mechanism. According to this mechanism an electron jumps from K to Br_2 when these two molecules come within a certain distance that is greater than the collision diameter d_{12}, which is 400 pm.

SOLUTION

The second order rate constant obtained from collision theory is given by equation 19.4 in m^3 mol^{-1} s^{-1}. To obtain a value of the rate constant to be compared with 1.0 x 10^{12} L mol^{-1} s^{-1}, it is necessary to multiply by 10^3 L m^{-3}. The reduced mass μ_{12} for this collision is given by

$$\mu_{12} = \frac{(39.1)(2 \text{ x } 79.9)(10^{-3} \text{ kg g}^{-1})}{(39.1 + 2 \text{ x } 79.9)(6.02 \text{ x } 10^{23} \text{ mol}^{-1})}$$

$$= 5.21 \text{ x } 10^{-26} \text{ kg}$$

The mean relative speed of these two reactants is given by equation 17.46.

$$< v_{12} > = \left[\frac{8(1.381 \text{ x } 10^{-23} \text{ J K}^{-1})(298 \text{ K})}{\pi(5.21 \text{ x } 10^{-26} \text{ kg})}\right]^{1/2}$$

$= 449 \text{ m s}^{-1}$

$k = \pi(400 \times 10^{-12} \text{ m})^2(449 \text{ m s}^{-1})(6.02 \times 10^{23} \text{ mol}^{-1})(10^3 \text{ L m}^{-3})$

$= 1.4 \times 10^{11} \text{ L mol}^{-1} \text{ s}^{-1}$

The experimental value is about 7 times greater than that expected from collision theory because of the harpoon mechanism.

19.5 A certain photochemical reaction requires an excitation energy of 126 kJ mol^{-1}. To what value does this correspond in the following units: (a) frequency of light, (b) wave number, (c) wavelength in nanometers, and (d) electron volts?

SOLUTION

(a) $$\nu = \frac{E}{h} = \frac{1.26 \times 10^5 \text{ J mol}^{-1}}{(6.63 \times 10^{-34} \text{ J s})(6.02 \times 10^{23} \text{ mol}^{-1})}$$
$$= 3.16 \times 10^{14} \text{ s}^{-1}$$

(b) $$\tilde{\nu} = \frac{\nu}{c} = \frac{(3.16 \times 10^{14} \text{ s}^{-1})(10^{-2} \text{ m cm}^{-1})}{2.998 \times 10^8 \text{ m s}^{-1}}$$
$$= 10{,}500 \text{ cm}^{-1}$$

(c) $$\lambda = \frac{1}{\tilde{\nu}} = \frac{1}{10\,500 \text{ cm}^{-1}} = 9.52 \times 10^{-5} \text{ cm} = 925 \text{ nm}$$

(d) $$\frac{126 \text{ kJ mol}^{-1}}{96\,485 \text{ C mol}^{-1}} = 1.31 \text{ eV}$$

19.6 How may moles of photons does a laser with an intensity of 0.1 watt at 560 nm produce in one hour?

SOLUTION

$$\frac{E\lambda}{N_A hc} = \frac{(0.1 \text{ J s}^{-1})(60 \times 60 \text{ s})(560 \times 10^{-9} \text{ m})}{(6.022 \times 10^{23} \text{ mol}^{-1})(6.626 \times 10^{-34} \text{ J s})(2.998 \times 10^8 \text{ m s}^{-1})}$$

$$= 1.7 \times 10^{-3} \text{ mol}$$

19.7 A sample of gaseous acetone is irradiated with monochromatic light having a wavelength of 313 nm. Light of this wavelength decomposed the acetone according to the equation

$(CH_3)_2CO \rightarrow C_2H_6 + CO$

The reaction cell used has a volume of 59 cm^3. The acetone vapor absorbs 91.5% of the incident energy. During the experiment the following data are obtained:

Temperature of reaction	= 56.7 °C
Initial pressure	= 102.16 kPa
Final pressure	= 104.42 kPa
Time of radiation	= 7 hr
Incident energy	= 48.1×10^{-4} J s^{-1}

What is the quantum yield?

SOLUTION

Since $PV = nRT$, $V\Delta P = RT\Delta n$

$$\Delta n = \frac{V\Delta P}{RT} = \frac{(59 \times 10^{-6}\ \text{m}^{-3})[(104.42 - 102.16) \times 10^3\ \text{Pa}]}{(8.314\ \text{J K}^{-1}\ \text{mol}^{-1})(329.85\ \text{K})}$$

$= 4.86 \times 10^{-5}$ mol

$= (4.86 \times 10^{-5}\ \text{mol})(6.02 \times 10^{23}\ \text{mol}^{-1})$

$= 2.93 \times 10^{19}$ molecules reacting

$$E = \frac{hc}{\lambda} = \frac{(6.626 \times 10^{-34}\ \text{J s})(2.988 \times 10^{8}\ \text{m s}^{-1})}{313 \times 10^{-9}\ \text{m}})$$

$= 6.36 \times 10^{-19}$ J

Amount reacting per unit time per unit volume = v

$$= \frac{4.86 \times 10^{-5}\ \text{mol}}{(7\ \text{hr})(60\ \text{min hr}^{-1})(60\ \text{s min}^{-1})(0.059\ \text{L})} = 3.27 \times 10^{-8}\ \text{mol L}^{-1}\ \text{s}^{-1}$$

$$I_a = \frac{(4.81 \times 10^{-3}\ \text{J s}^{-1})(0.915)}{(6.36 \times 10^{-19}\ \text{J})(6.022 \times 10^{23}\ \text{mol}^{-1})(0.059\ \text{L})}$$

$= 1.95 \times 10^{-7}$ mol L^{-1} s^{-1}

$\phi = v/I_a = (3.7 \times 10^{-8}\ \text{mol L}^{-1}\ \text{s}^{-1})/(1.95 \times 10^{-7}\ \text{mol L}^{-1}\ \text{s}^{-1}$

$= 0.167$

19.8 A 100-cm^3 vessel containing hydrogen and chlorine was irradiated with light of 400 nm. Measurements with a thermopile showed that 11×10^{-7} J of light energy was absorbed by the chlorine per second. During an irradiation of 1 min the partial pressure of chlorine, as determined by the absorption of light and the application of Beer's law, decreased from 27.3 to 20.8 kPa (corrected to 0 °C). What is the quantum yield?

SOLUTION

$$E = \frac{hc}{\lambda} = \frac{(6.626 \times 10^{-34}\ \text{J s})(2.998 \times 10^{8}\ \text{m s}^{-1})}{400 \times 10^{-9}\ \text{m}}$$

$= 4.97 \times 10^{-19}$ J

$$I_a = \frac{(11 \times 10^{-7}\ \mathrm{J\ s^{-1}})}{(4.97 \times 10^{-19}\ \mathrm{J})\ (6.02 \times 10^{23}\ \mathrm{mol^{-1}})(0.100\ \mathrm{L})}$$

$$= 3.68 \times 10^{-11}\ \mathrm{mol\ L^{-1}\ s^{-1}}$$

Since $PV = nRT$, $V\Delta P = RT\Delta n$

$$\Delta N = \frac{N_A V\Delta P}{RT} = \frac{(6.02 \times 10^{23}\ \mathrm{mol^{-1}})(0.1 \times 10^{-3}\ \mathrm{m^3})[(27.3 - 20.8) \times 10^3\ \mathrm{Pa}]}{(8.314\ \mathrm{J\ K^{-1}mol^{-1}})(273\ \mathrm{K})}$$

$= 1.72 \times 10^{20}$ molecules of Cl_2 react.

Since $H_2 + Cl_2 = 2\ HCl$

$$\Delta n = V\Delta P/RT$$

$$= \frac{(0.1 \times 10^{-3}\ \mathrm{m^3})[(27.3 - 20.8) \times 10^3\ \mathrm{Pa}]}{(8.314\ \mathrm{J\ K^{-1}\ mol^{-1}})(273\ \mathrm{K})}$$

$$= 2.86 \times 10^{-4}\ \mathrm{mol}$$

$$\upsilon = \frac{(2.86 \times 10^{-4}\ \mathrm{mol})}{(60\ \mathrm{s})(0.100\ \mathrm{L})} = 4.77 \times 10^{-5}\ \mathrm{mol\ L^{-1}\ s^{-1}}$$

$$\phi = \frac{\upsilon}{I_a} = \frac{4.77 \times 10^{-5}\ \mathrm{mol\ L^{-1}\ s^{-1}}}{3.68 \times 10^{-11}\ \mathrm{mol\ L^{-1}\ s^{-1}}} = 1.3 \times 10^6$$

$$= -\frac{1}{I_a}\frac{\mathrm{d}[Cl_2]}{\mathrm{d}t} = \frac{1}{2I_a}\frac{\mathrm{d[HCl]}}{\mathrm{d}t}$$

19.9 Show that if a solute follows the Beer-Lambert law, the intensity of absorbed radiation I_a in moles of photons per unit volume per second is given by

$$I_a = \frac{I_0}{lN_A h\nu}(1 - e^{-\kappa cl})$$

where l is the length of the cell in the direction of the incident monochromatic radiation and I_0 is in energy per unit area per unit time.

SOLUTION

Consider that monochromatic radiation of intensity I_0 is perpendicular to the face of a reaction cell of area A and optical path length l. The rate that energy enters the cell is I_0A, and the rate that energy leaves the cell is IA, where $I = I_0 \exp(-\kappa cl)$. The rate I_a of absorption of electromagnetic radiation in moles of photons per unit volume per second is given by

$$I_a = \frac{I_0A - IA}{VN_A h\nu} = \frac{I_0}{lN_A h\nu}(1 - e^{-\kappa cl})$$

If the cell is thick enough that the radiation is essentially all absorbed, $I_a = I_0/lN_A h\nu$.

19.10 When CH_3I molecules in the vapor state absorb 253.7 nm light, they dissociate into methyl radicals and iodine atoms. The energy required to rupture the C-I bond is

209 kJ mol^{-1}. What are the velocities of the iodine atom and the methyl radical, assuming all of the excess energy goes into translational motion?

SOLUTION

$$E = \frac{hc}{\lambda} = \frac{(6.626 \times 10^{-34}\ \text{J s})(2.998 \times 10^{8}\ \text{m s}^{-1})}{(253.7 \times 10^{-9}\ \text{m})}$$

$= 7.83 \times 10^{-19}$ J

$$\text{Excess energy} = 7.83 \times 10^{-19} - \frac{209\,000}{6.022 \times 10^{23}}$$

$$= 4.36 \times 10^{-19}\ \text{J}$$

$$= \frac{1}{2} m_I v_I^2 + \frac{1}{2} m_{CH_3} v_{CH_3}^2$$

But the momenta of the fragments have to be equal and opposite.

$m_I v_I = m_{CH_3} v_{CH_3}$

$$4.36 \times 10^{-19} = \frac{1}{2} m_I v_I^2 + \frac{1}{2} m_{CH_3} \left(\frac{m_I}{m_{CH_3}}\right)^2 v_I^2$$

$$= \frac{1}{2} m_I v_I^2 \left(1 + \frac{m_I}{m_{CH_3}}\right)$$

$$m_I = \frac{126.9 \times 10^{-3}\ \text{kg mol}^{-1}}{6.022 \times 10^{23}\ \text{mol}^{-1}} = 2.107 \times 10^{-25}\ \text{kg}$$

$$m_{CH_3} = \frac{15.0 \times 10^{-3}\ \text{kg mol}^{-1}}{6.022 \times 10^{23}\ \text{mol}^{-1}} = 2.491 \times 10^{-26}\ \text{kg}$$

$$1 + \frac{m_I}{m_{CH_3}} = 9.459$$

$$v_I = \left[\frac{2(4.36 \times 10^{-19}\ \text{J})}{(2.107 \times 10^{-25}\ \text{kg})(9.459)}\right]^{1/2} = 662\ \text{m s}^{-1}$$

$v_{CH_3} = 8.459\ v_I = 5595$ m s^{-1}

19.11 The phosphorescence of butyrophenone in acetonitrile is quenched by 1,3-pentadiene (P). The following quantum yields were measured at 25 °C.

[P]/10^{-3} mol L^{-1}	0	1.0	2.0
ϕ / ϕ_0	1	0.61	0.43

Assuming that the quenching reaction is diffusion controlled and the rate constant has a value of 10^{10} L mol^{-1} s^{-1}, what is the lifetime of the triplet state?

SOLUTION

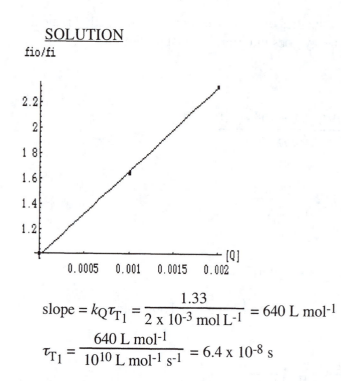

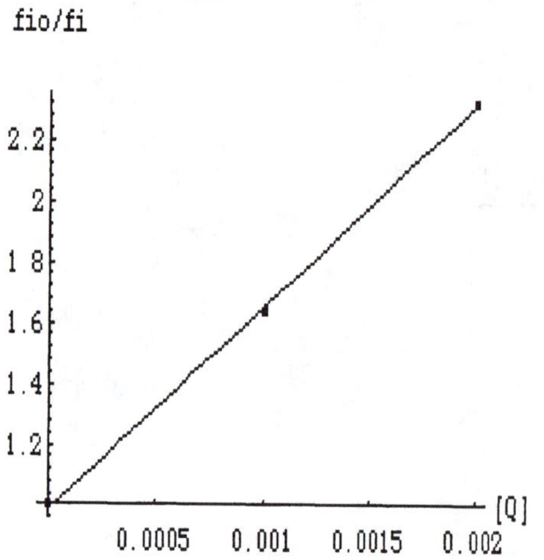

$$\text{slope} = k_Q \tau_{T1} = \frac{1.33}{2 \times 10^{-3} \text{ mol L}^{-1}} = 640 \text{ L mol}^{-1}$$

$$\tau_{T1} = \frac{640 \text{ L mol}^{-1}}{10^{10} \text{ L mol}^{-1} \text{ s}^{-1}} = 6.4 \times 10^{-8} \text{ s}$$

19.12 Biacetyl triplets have a quantum yield of 0.25 for phosphorescence and a measured lifetime of the triplet state of 10^{-3} s. If its phosphorescence is quenched by a compound Q with a diffusion-controlled rate (10^{10} L mol^{-1} s^{-1}), what concentration of Q is required to cut the phosphorescence yield by half? (See Section 20.5.)

SOLUTION

$$\frac{I_P^o}{I_P} = 1 + k_Q \tau_{T1}[Q]$$

$$2 = 1 + (10^{10} \text{ L mol}^{-1} \text{ s}^{-1})(10^{-3} \text{ s}) [Q]$$

$$[Q] = 10^{-7} \text{ mol L}^{-1}$$

19.13 For 900 s, light of 436 nm was passed into a carbon tetrachloride solution containing bromine and cinnamic acid. The average power absorbed was 19.2×10^{-4} J s^{-1}. Some of the bromine reacted to give cinnamic acid dibromide, and in this experiment the total bromine content decrease by 3.83×10^{19} molecules. (a) What was the quantum yield? (b) State whether or not a chain reaction was involved.

SOLUTION

(a) $$E = \frac{h}{\lambda} = \frac{(6.63 \times 10^{-34} \text{ J s})(3 \times 10^{8} \text{ m s}^{-1})}{(436 \times 10^{-9} \text{ m})}$$

$$= 4.56 \times 10^{-19} \text{ J}$$

$$I_a = \frac{19.2 \times 10^4 \text{ J s}^{-1}}{(4.56 \times 10^{-19} \text{ J})(6.02 \times 10^{13} \text{ mol}^{-1})(0.500 \text{ L})}$$

$$= 1.40 \times 10^{-8} \text{ mol L}^{-1} \text{ s}^{-1}$$

$$v = \frac{6.36 \times 10^{-5} \text{ mol}}{(900 \text{ s})(0.500 \text{ L})}$$

$$\phi = \frac{v}{I_a} = \frac{1.41 \times 10^{-7} \text{ mol L}^{-1} \text{ s}^{-1}}{1.40 \times 10^{-8} \text{ mol L}^{-1} \text{ s}^{-1}} = 10.1$$

(b) Since $\phi > 1$, a chain reaction is involved.

19.14 The following calculations are made on a uranyl oxalate actinometer on the assumption that the energy of all wavelengths between 254 and 435 nm is completely absorbed. The actinometer contains 20 cm^3 of 0.05 mol L^{-1} oxalic acid, which also is 0.01 mol L^{-1} with respect to uranyl sulfate. After 2 hr of exposure to ultraviolet light, the solution required 34 cm^3 of potassium permanganate, $KMnO_4$, solution to titrate the undecomposed oxalic acid. The same volume, 20 cm^3, of unilluminated solution required 40 cm^3 of the $KMnO_4$ solution. If the average energy of the quanta in this range may be taken as corresponding to a wavelength of 350 nm, how many joules were absorbed per second in this experiment? ($\phi = 0.57$.)

SOLUTION

Moles of oxalic acid decomposed $= (0.02 \text{ L})(0.05 \text{ mol L}^{-1})\left(\frac{40 - 34}{40}\right)$

$= 1.5 \times 10^{-4}$ mol

Moles of photons required per second $= \frac{1.5 \times 10^{-4} \text{ mol}}{(2 \times 60 \times 60 \text{ s})(0.57)}$

$= 3.65 \times 10^{-8}$ mol s^{-1}

$$E = \frac{hc}{\lambda} = \frac{(6.266 \times 10^{-34} \text{ J s})(2.998 \times 10^8 \text{ m s}^{-1})}{3.5 \times 10^{-7} \text{ m}}$$

$= 5.68 \times 10^{-19}$ J

Energy flux $= (3.65 \times 10^{-8} \text{ mol}^{-1} \text{ s}^{-1})(6.02 \times 10^{23} \text{ mol}^{-1})(5.68 \times 10^{-19} \text{ J}) = 1.25$

$\times 10^{-2}$ J s^{-1}

19.15 A solution of a dye is irradiated with 400 nm light to produce a steady concentration of triplet state molecules. If the triplet state quantum yield is 0.9, and the triplet state lifetime is 20 x 10^{-6} s, what light intensity, expressed in watts per liter, is required to maintain a steady triplet concentration of 5 x 10^{-6} mol L^{-1} in a liter of solution. Assume that all of the light is absorbed.

SOLUTION

$$\frac{d[T_1]}{dt} = 0.9\, I - \frac{1}{20 \times 10^{-6}} [T_1] = 0$$

$$I = \frac{(5 \times 10^{-6} \text{ mol L}^{-1})}{(20 \times 10^{-6} \text{ s})(0.9)} = 0.278 \text{ mol L}^{-1} \text{ s}^{-1}$$

$$E = \frac{N_A hc}{\lambda} = \frac{(6.022 \times 10^{23} \text{ mol}^{-1})(6.62 \times 10^{-34} \text{ J s})(3 \times 20^{8} \text{ m s}^{-1})}{400 \times 10^{-9} \text{ m}}$$

$$= 2.99 \times 10^{5} \text{ J mol}^{-1}$$

$$I = (0.278 \text{ mol L}^{-1} \text{ s}^{-1})(2.99 \times 10^{5} \text{ J mol}^{-1}) = 83 \text{ kW L}^{-1}$$

19.16 The photochemical chlorination of chloroform

$CHCl_3 + Cl_2 = CCl_4 + HCl$

is believed to proceed by the following mechanism.

$$Cl_2 + h\nu \xrightarrow{I_a} 2Cl$$

$$Cl + CHCl_3 \xrightarrow{k_1} CCl_3 + HCl$$

$$CCl_3 + Cl_2 \xrightarrow{k_2} CCl_4 + Cl$$

$$2CCL_3 + Cl_2 \xrightarrow{k_3} 2CCl_4$$

Derive the steady state rate law for the production of carbon tetrachloride.

SOLUTION

$$\frac{d[CCl_4]}{dt} = k_2[CCl_3][Cl_2] + 2k_3[CCl_3]^2[Cl_2]$$

$$\frac{d[CCl_3]}{dt} = k_1[Cl][CHCl_3] - k_2[CCl_3][Cl_2] - 2k_3[CCl_3]^2[Cl_2] = 0$$

$$\frac{d[Cl]}{dt} = 2I_a - k_1[Cl][CHCl_3] + k_2[CCl_3][Cl_2] = 0$$

Adding the two steady state expressions yields

$$I_a = k_3[CCl_3]^2[Cl_2]$$

This makes it possible to eliminate $[CCl_3]$ from the first equation to obtain

$$\frac{d[CCl_4]}{dt} = k_2 I_a^{1/2} [Cl_2]^{1/2}/k_3^{1/2} + 2I_a$$

19.17 The mechanism for quenching fluorescence is

$A + h\nu \longrightarrow A^* \; I_a$

$A^* + Q \longrightarrow A + Q \quad k_q$

$A^* \longrightarrow A + h\nu_f \quad k_f = I_f/[A^*]$

where I_a is the amount of exciting radiation absorbed per liter of solution per second, k_q is the rate constant for quenching, k_f is the rate constant for fluorescence, and I_f is the amount of fluorescence radiation per liter per second. Assuming a steady state is reached, derive the equation for the intensity of fluorescence radiation I_f as a function of [Q]. Describe how the data should be plotted to determine the rate constant for quenching k_q.

SOLUTION

In the steady state

$d[A^*]/dt = I_a - k_f[A^*] - k_q[Q][A^*] = 0$

Thus

$$[A^*] = \frac{I_a}{k_f + k_q[Q]}$$

Multiplying by k_f yields the fluorescence intensity

$$I_f = \frac{k_f I_a}{k_f + k_q[Q]}$$

which can be rearranged to

$I_a/I_f = 1 + (k_q/k_f)[Q]$

so that k_q/k_f is the slope of the plot of I_a/I_f versus [Q]. The rate constant for fluorescence k_f can be calculated from the half life of the fluorescence, and so k_q can be calculated from the slope.

19.18 When a solution of anthracene in benzene is exposed to ultraviolet light, anthracene molecules are excited and form dimers with unexcited anthracene molecules. If the excited anthracene molecules fluoresce before they react with unexcited anthracene molecules to form dimers, they do not form dimers. In concentrated solutions of anthracene, the quantum yield ϕ for the formation of dimers is high, but in dilute solutions it is low because the excitation is lost in fluorescence. Formulate a mechanism to represent these facts, and derive the quantum yield as a function of the concentration of anthracene. It is useful to assume that the excited anthracene molecules are in a steady state.

SOLUTION

$A + h\nu \rightarrow A^*$	I_a
$A^* + A \rightarrow A_2$	k
$A^* \rightarrow A + h\nu_f$	k_f

The rate equations for A and A* are

$$\frac{d[A]}{dt} = - I_a - k[A^*][A] + k_f[A^*]$$

$$\frac{d[A^*]}{dt} = I_a - k[A^*][A] - k_f[A^*] = 0$$

In the steady state the rate of change of the concentration of A* is zero. Eliminating

[A*] between these equations yields

$$\frac{d[A]}{dt} = \frac{-2kI_a[A]}{k_f + k[A]}$$

The quantum yield ϕ is defined by v/I_a, where $v = \frac{d[A_2]}{dt} = -\frac{1}{2}\frac{d[A]}{dt}$

$$\phi = \left(-\frac{1}{2}\frac{d[A]}{dt}\right)/I_a$$

Therefore,

$$\phi = \frac{k[A]}{k_f + k[A]}$$

As [A] is decreased, the quantum yield approaches zero; and as [A] is increased, the quantum yield approaches 1.

19.19 Prof. Mario Molina (MIT) made an important contribution to the understanding of the role of chlorine atoms in the stratosphere by suggesting that the decomposition of ozone is catalyzed by the reactions

$2ClO + M \longrightarrow ClOOCl + M$

$ClOOCl + h\nu \longrightarrow Cl + ClOO$

$ClOO + M \longrightarrow Cl + O_2 + M$

$Cl + O_3 \longrightarrow ClO + O_2$

The steps in this mechanism add up to $2O_3 + h\nu \longrightarrow 3O_2$, but what is the stoichiometric number of the last step, if the stoichiometric numbers of the first three steps are taken as unity?

SOLUTION

The first three steps add up to

$2ClO + h\nu \longrightarrow 2Cl + O_2$

In order to obtain

$2O_3 + h\nu \longrightarrow 3O_2$

we have to add

$2Cl + 2O_3 \longrightarrow 2ClO + 2O_2$

Therefore, the stoichiometric number for step 4 is 2; $s_4 = 2$.

19.20 Sunlight between 290 and 313 nm can produce sunburn (erythema) in 30 min. The intensity of radiation between these wavelengths in summer and at 45° latitude is about 50 μW cm^{-2}. Assuming that 1 photon produces chemical change in 1 molecule, how many molecules in a square centimeter of human skin must be photochemically affected to produce evidence of sunburn?

SOLUTION

$$E = \frac{hc}{\lambda} = \frac{(6.63 \times 10^{-34})(3 \times 10^{8})}{3 \times 10^{-7}} = 6.63 \times 10^{-19}\ \text{J}$$

$$(50 \times 10^{-6}\ \text{J s}^{-1}\ \text{cm}^{-2})(30 \times 60\ \text{s}) = 0.09\ \text{J cm}^{-2}$$

$$\frac{0.09}{6.63 \times 10^{-19}} = 1.36 \times 10^{17}\ \text{molecules cm}^{-2}$$

19.21 4.87×10^{11} L mol^{-1} s^{-1}, 0.21

19.22 5.69×10^{11} L mol^{-1} s^{-1}

19.24 $E_a = RT/2 + E_c$

19.25 $K^* = \prod \left(\frac{P_i}{P^*} \right)^{\nu_i}$

Thus

$$K^* = K \left(\frac{P^*}{P^o} \right)^{-\Sigma \nu_i} = 100(10^5)^{-1} = 10^{-3}$$

19.26 2.51×10^{-6} mol s^{-1}

19.27 (a) 0.1709 J s^{-1}, (b) 0.3988 J s^{-1}, (c) 0.1709 watt, (d) 0.3988 watt

19.28 (a) \$1.55, (b) \$0.02

19.29 1.27×10^{-3} mol

19.30 26.9 m

19.31 5×10^{-9} mol s^{-1}

19.32 370 ns

19.33 9×10^{-9} mol L^{-1}

19.34 21.8 hr

19.35 47.1 s

19.36 0.050 J s^{-1}

19.37 (a) 10^{14} quanta, (b) 4.5 x 10^{-9} g day^{-1}

19.38 617 g m^{-2}

19.39 700 W m^{-2}

19.40 10^{-3}

20

Kinetics in the Liquid Phase

20.1 A steel ball ($\rho = 7.86$ g cm^{-3}) 0.2 cm in diameter falls 10 cm through a viscous liquid ($\rho_0 = 1.50$ g cm^{-3}) in 25 s. What is the viscosity at this temperature?

SOLUTION

$$\eta = \frac{2r^2(\rho - \rho_0)g}{9\frac{dx}{dt}}$$

$$= \frac{2(1 \times 10^{-3}\ \text{m})^2[(7.86 - 1.50) \times 10^3\ \text{kg m}^{-3}](9.8\ \text{m s}^{-2})}{9\left(\frac{0.10\ \text{m}}{25\ \text{s}}\right)}$$

$$= 3.46\ \text{Pa s}$$

20.2 Estimate the rate of sedimentation of water droplets of 1 μm diameter in air at 20 °C. The viscosity of air at this temperature is 1.808×10^{-5} Pa s.

SOLUTION

$$\frac{dx}{dt} = \frac{2r^2(\rho - \rho_o)g}{9\eta}$$

$$= \frac{2(0.5 \times 10^{-6}\ \text{m})^2(0.998 \times 10^3\ \text{kg m}^{-3})(9.8\ \text{m s}^{-2})}{9(1.808 \times 10^{-5}\ \text{Pa s})}$$

$$= 3.01 \times 10^{-5}\ \text{m s}^{-1}$$

20.3 The viscosity of mercury is 1.661×10^{-3} Pa s at 0 °C and 1.476×10^{-3} Pa s at 35 °C. What is the activation energy, and what viscosity is expected at 50 °C?

SOLUTION

$$\ln\frac{\eta_1}{\eta_2} = \frac{E_a(T_2 - T_1)}{RT_1T_2}$$

$$E_a = \frac{RT_1T_2}{(T_2 - T_1)} \ln\frac{\eta_1}{\eta_2}$$

$$= \frac{(8.314 \text{ J K}^{-1} \text{ mol}^{-1})(273.15 \text{ K})(308.15\text{K})}{35 \text{ K}} \ln \frac{1.661}{1.476}$$

$$= 2.36 \text{ kJ mol}^{-1}$$

$$A = \frac{1}{\eta} e^{E_a/RT}$$

$$= \frac{e^{2360/(8.314)(273.15)}}{1.661 \times 10^{-3} \text{ Pa s}}$$

$$= 1.70 \times 10^{3} \text{ Pa}^{-1} \text{ s}^{-1}$$

At 50 °C

$$\eta = \frac{e^{2360/(8.314)(323.15)}}{1.70 \times 10^{3} \text{ Pa}^{-1} \text{ s}^{-1}}$$

$$= 1.41 \times 10^{-3} \text{ Pa s}$$

20.4 The viscosity of a liquid can be determined by measuring the falling velocity of a sphere of known density in the liquid. The force of gravity on the sphere is given by the apparent mass of the sphere (its mass minus the mass of liquid displaced) times the acceleration of gravity, $g = 9.807 \text{ m s}^{-2}$. The retarding force is given by the frictional coefficient of the sphere times the velocity of fall. Derive the equation for the velocity of fall.

SOLUTION

$$\frac{4}{3}\pi r^3 (\rho - \rho_0) g = 6\pi\eta r\left(\frac{dx}{dt}\right)$$

The density of the sphere is given by ρ and the density of the liquid is given by ρ_0.

$$\frac{dx}{dt} = \frac{2r^2(\rho - \rho_0)g}{9\eta}$$

20.5 How fast does a bubble of air rise in water at 25 °C if its diameter is 1 mm?

SOLUTION

See problem 20.4 and Table 20.1. The density of air is negligible compared with the density of liquid water, and so

$$\frac{dx}{dt} = \frac{-2r^2 \rho_0 g}{9\eta} = \frac{-2(5 \times 10^{-4} \text{m})^2 (10^3 \text{kg m}^{-3})(9.8 \text{ m s}^{-2})}{9(8.91 \times 10^{-4} \text{kg m}^{-1} \text{ s}^{-1})} = -0.61 \text{ m s}^{-1}$$

20.6 A sharp boundary is formed between a dilute aqueous solution of sucrose and water at 25 °C. After 5 h the standard deviation of the concentration gradient is 0.434 cm. (a) What is the diffusion coefficient for sucrose under these conditions? (b) What will be the standard deviation after 10 h?

SOLUTION

(a) $$D = \frac{\sigma^2}{2t} = \frac{(4.34 \times 10^{-3}\ \text{m})^2}{2(5 \times 60 \times 60\ \text{s})} = 5.23 \times 10^{-10}\ \text{m}^2\ \text{s}^{-1}$$

(b) $$\sigma = \sqrt{2Dt} = \sqrt{2(5.23 \times 10^{-10}\ \text{m}^2\ \text{s}^{-1})(10 \times 60 \times 60\ \text{s})}$$
$$= 6.14 \times 10^{-3}\ \text{m}$$

20.7 (a) Calculate the time required for the half-width of a freely diffusing boundary of dilute potassium chloride in water to become 0.5 cm at 25 °C ($D = 1.99 \times 10^{-9}\ \text{m}^2\ \text{s}^{-1}$). (b) Calculate the corresponding time for serum albumin ($D = 6.15 \times 10^{-11}\ \text{m}^2\ \text{s}^{-1}$).

SOLUTION

(a) $$D = \sigma^2/2t$$
$$t = \frac{\sigma^2}{2D} = \frac{(0.005\ \text{m})^2}{2(1.99 \times 10^{-9}\ \text{m}^2\ \text{s}^{-1})}$$
$$= 6.28 \times 10^3\ \text{s} = 1.75\ \text{h}$$

(b) $$t = \frac{\sigma^2}{2D} = \frac{(0.005\ \text{m})^2}{2(6.15 \times 10^{-11}\ \text{m}^2\ \text{s}^{-1})}$$
$$= 2.03 \times 10^5\ \text{s} = 56.5\ \text{h}$$

20.8 The standard deviation σ of a freely diffusing boundary between dilute salt solution and water at 25 °C is 3.8 mm after 1 h. What is the diffusion coefficient of the salt in water? What will the standard deviation be after 2 h?

SOLUTION

$$D = \frac{\sigma^2}{2t} = \frac{(0.38\ \text{cm})^2}{2(3600\ \text{s})} = 2.00 \times 10^{-5}\ \text{cm}^2\ \text{s}^{-1}$$
$$\sigma = \sqrt{2Dt} = \sqrt{2(2.00 \times 10^{-5}\ \text{cm}^2\ \text{s}^{-1})(7200\ \text{s})}$$
$$= 0.54\ \text{cm}$$

20.9 Using a table of the probability integral, calculate enough points on a plot of c versus x (like Fig. 20.2) to draw in the smooth curve for diffusion of 0.1 mol L^{-1} sucrose into water at 25 °C after 4 h and 29.83 min

($D = 5.23 \times 10^{-10}\ \text{m}^2\ \text{s}^{-1}$).

SOLUTION

Equation 20.18 is

$$c = \frac{c_0}{2}\left[1 + \frac{2}{\sqrt{\pi}} \int_0^{x/2\sqrt{Dt}} e^{-\beta^2}\, d\beta\right]$$

where $\beta = x/2\sqrt{Dt}$. Thus

$$c = \frac{c_0}{2}\left[\,1 \pm \mathrm{erf}(x/2\sqrt{Dt}\,)\right]$$

where the positive sign gives the concentrations at positive values of x and the negative sign gives the values at negative values of x. The following table shows the calculation of concentrations for positive values of x.

x/m	$x/2\sqrt{Dt}$	$\mathrm{erf}(x/2\sqrt{Dt}\,)$	c/mol L^{-1}
0	0	0	0.0500
0.001	0.171	0.191	0.600
0.002	0.342	0.371	0.0686
0.003	0.513	0.531	0.0766
0.004	0.717	0.689	0.0845
0.006	1.03	0.855	0.0928
0.008	1.37	0.947	0.974

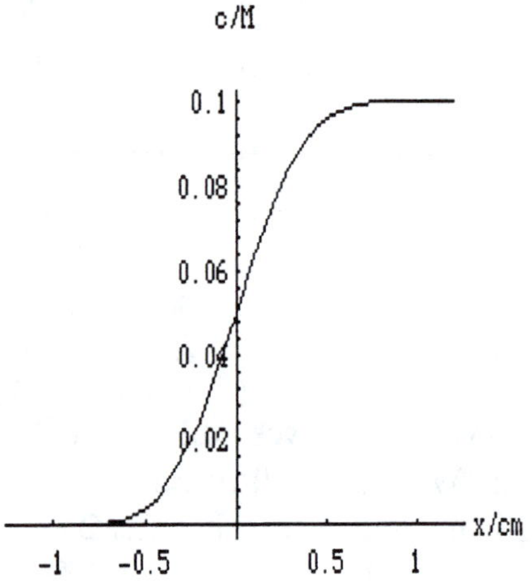

20.10 Calculate the conductivity of 0.001 mol L^{-1} HCl at 25 °C. The limiting ion mobilities may be used for this problem.

SOLUTION

$$\kappa = Fc(u_{H+} + u_{Cl-})$$
$$= (96{,}485\ \mathrm{C\ mol^{-1}})(1\ \mathrm{mol\ m^{-3}})[(36.25 + 7.91) \times 10^{-8}\ \mathrm{m^2\ V^{-1}\ s^{-1}}]$$
$$= 0.0426\ \Omega^{-1}\ \mathrm{m^{-1}}$$

20.11 One hundred grams of sodium chloride is dissolved in 10,000 L of water at 25 °C, giving a solution that may be regarded in these calculations as infinitely dilute. (a) What is the conductivity of the solution? (b) This dilute solution is placed in a glass tube of 4-cm diameter provided with electrodes filling the tube and placed 20 cm apart. How much current will flow if the potential drop between the electrodes is 80 V?

SOLUTION

(a) $$c = \frac{(100\ \text{g})/(58.5\ \text{g mol}^{-1})}{(10^7\ \text{cm}^3)(10^{-2}\ \text{m cm}^{-1})^3}$$
$$= 1.71 \times 10^{-1}\ \text{mol m}^{-3}$$
$$\kappa = (96{,}485\ \text{C mol}^{-1})(1.71 \times 10^{-1}\ \text{mol m}^{-3})(13.105 \times 10^{-8}\ \text{m}^2\ \text{V}^{-1}\ \text{s}^{-1})$$
$$= 2.16 \times 10^{-3}\ \Omega^{-1}\ \text{m}^{-1}$$

(b) $$R = \frac{1}{\kappa A} = \frac{0.2\ \text{m}}{(2.16 \times 10^{-3}\ \Omega^{-1}\ \text{m}^{-1})(2 \times 10^{-2}\ \text{m})^2 \pi}$$
$$= 7.36 \times 10^4\ \Omega$$
$$I = \frac{E}{R} = \frac{80\ \text{V}}{7.36 \times 10^4\ \Omega} = 1.09 \times 10^{-3}\ \text{A}$$

20.12 It is desired to use a conductance apparatus to measure the concentration of dilute solutions of sodium chloride. If the electrodes in the cell are each 1 cm^2 in area and are 0.2 cm apart, calculate the resistance that will be obtained for 1, 10, and 100 ppm NaCl at 25 °C.

SOLUTION

$$1\ \text{ppm} = 1\ \text{g NaCl in } 10^6\ \text{g H}_2\text{O} = \frac{(1\ \text{g})/(58.45\ \text{g mol}^{-1})}{1\ \text{m}^3}$$

Using electric mobilities at infinite dilution,

$$\kappa = Fc(u_{\text{Na}^+} + u_{\text{Cl}^-})$$
$$= (96{,}485\ \text{C mol}^{-1})(1.71 \times 10^{-2}\ \text{mol m}^{-3})[(5.192 + 7.913) \times 10^{-8}\ \text{m V}^{-1}\ \text{s}^{-1}]$$
$$= 2.16 \times 10^{-4}\ \Omega^{-1}\ \text{m}^{-1}$$
$$R = \frac{1}{\kappa A} = \frac{0.2 \times 10^{-2}\ \text{m}}{(2.16 \times 10^{-4}\ \Omega^{-1}\ \text{m}^{-1})(0.01\ \text{m})^2} = 9.26 \times 10^4\ \Omega$$

For 10 ppm, $R = 9.26 \times 10^3\ \Omega$

For 100 ppm, $R = 926\ \Omega$

20.13 Derive the expression for log k for a reaction

$A^{zA} + B^{zB} + C^{zC}$ -> Products

As a function of ionic strength.

SOLUTION

$$\log k = \log k^\circ + \log\gamma_A + \log\gamma_B + \log\gamma_C - \log\gamma_{\text{complex}}$$

$$= \log k^{o} + AI^{1/2}[z_A^{\ 2} + z_B^{\ 2} + z_c^{\ 2} - (z_A + z_B + z_C)^2]$$
$$= \log k^{o} + \ 2AI^{1/2}[z_A z_B + z_A z_C + z_B z_C)$$

20.14 Derive the expressions for the relaxation times τ for the following two reactions:

(a) $A \underset{k_2}{\overset{k_1}{=}} B$

(a) $A + B \underset{k_2}{\overset{k_1}{=}} C + D$

<u>SOLUTION</u>

(a) $\left(\dfrac{d[B]}{dt}\right) = k_1[A] - k_2[B]$

Substitute

$[A] = [A]_{eq} - \Delta[B]$

$[B] = [B]_{eq} + \Delta[B]$

Substitute these in the rate equation, and use the equilibrium expression to simplify it. This yields

$$\left(\frac{d[B]}{dt}\right) = -(k_1 + k_2)\Delta[B] = -\frac{\Delta[B]}{\tau}$$

$\tau = (k_1 + k_2)^{-1}$

(b) $\left(\dfrac{d[B]}{dt}\right) = k_1[A][B] - k_2[C][D]$

Substitute equations like equations 20.36 to 20.38 in the rate equation, and use the equilibrium expression to simplify it. This yields

$\tau = \{k_1\ ([A]_{eq} + [B]_{eq}) + k_2([C]_{eq} + [D]_{eq})\}^{-1}$

20.15 Show that if A and B can be represented by spheres of the same radius that react when they touch, the second-order rate constant is given by

$$k_a = \frac{8 \times 10^3\ RT}{3\eta}\ \text{L mol}^{-1}\ \text{s}^{-1}$$

where R is in J K^{-1} mol^{-1}. To obtain this result the diffusion coefficient is expressed in terms of the radius of a spherical particle by use of equation 20.12. For water at 25 °C, $\eta = 8.95 \times 20^{-4}$ kg m^{-1} s^{-1}. Calculate k at 25 °C.

SOLUTION

The diffusion coefficient for a spherical particle of radius r is given by

$$D = \frac{RT}{N_A 6\pi\eta r}$$

Substituting in equation 20.30

$$k_a = 4\pi(10^3\ \text{L m}^{-3})N_A(D_1 + D_2)R_{12}$$
$$= 4\pi(10^3\ \text{L m}^{-3})\left(\frac{2RT}{6\pi\eta r}\right)(2r)$$
$$= \frac{8 \times 10^3\ RT}{3\eta}\ \text{L mol}^{-1}\ \text{s}^{-1}$$

where RT is expressed in J mol^{-1} and η is expressed in kg m^{-1} s^{-1}.

For water at 25°C

$$k_a = \frac{8(10^3\ \text{L m}^{-3})(8.314\ \text{J K}^{-1}\ \text{mol}^{-1})(298\ \text{K})}{3(8.95 \times 10^{-4}\ \text{kg m}^{-1}\ \text{s}^{-1})}$$
$$= 7.4 \times 10^9\ \text{L mol}^{-1}\ \text{s}^{-1}$$

20.16 What is the reaction radius for the reaction

$$H^+ + OH^- \xrightarrow{1.4 \times 10^{11}\ \text{L mol}^{-1}\ \text{s}^{-1}} H_2O$$

at 25 °C given that the diffusion coefficients of H^+ and OH^- at this temperature are 9.1×10^{-9} m^2 s^{-1} and 5.2×10^{-9} m^2 s^{-1}.

SOLUTION

$$k_a = 4\pi N_A(D_1 + D_2)R_{12}f$$
$$R_{12}f = \frac{k_a}{4\pi N_A(D_1 + D_2)}$$
$$= \frac{(1.4 \times 10^{11}\ \text{L mol}^{-1}\ \text{s}^{-1})(10^{-3}\ \text{m}^3\ \text{L}^{-1})}{4\pi(6.02 \times 10^{23}\ \text{mol}^{-1})(14.3 \times 10^{-9}\ \text{m}^2\ \text{s}^{-1})}$$
$$= 1.29 \times 10^{-9}\ \text{m}$$

To calculate the electrostatic factor f, we have to know R_{12}. Successive approximations have to be made. Let us begin by estimating $R_{12} = 0.9$ nm.

$$x = \frac{z_1 z_2 e^2}{4\pi\varepsilon_o \kappa kTR_{12}}$$
$$= \frac{-(1.602 \times 10^{-19}\ \text{C})^2\ (8.988 \times 10^9\ \text{N C}^{-2}\ \text{m}^2)}{(78.3)(1.38 \times 10^{-23}\ \text{J K}^{-1})(298\ \text{K})(0.9 \times 10^{-9}\ \text{m})} = -0.796$$

$$f = x(e^x - 1)^{-1} = -0.796(e^{-0.796} - 1)^{-1} = 1.45$$

Thus $$R_{12} = \frac{1.29 \times 10^{-9}\ \text{m}}{1.45} = 0.89\ \text{nm}$$

20.17 For acetic acid in dilute aqueous solution at 25°C, $K = 1.73 \times 10^{-5}$, and the relaxation time is 8.5×10^{-9} s for a 0.1 M solution. Calculate k_a and k_d in

$$CH_3CO_2H \underset{k_a}{\overset{k_d}{\rightleftarrows}} CH_3CO_2^- + H^+$$

SOLUTION

$$K = \frac{[H^+][CH_3CO_2^-]}{[CH_3CO_2H]} = \frac{x^2}{0.10} = 1.73 \times 10^{-5} = \frac{k_d}{k_a}$$

$$x = (1.73 \times 10^{-6})^{1/2} = 1.32 \times 10^{-3} = [H^+] = [CH_3CO_2^-]$$

$$\tau = \frac{1}{k_d + k_a([H^+] + [CH_3CO_2^-])}$$

$$8.5 \times 10^{-9}\ s = \frac{1}{1.73 \times 10^{-5} k_a + k_a(2.64 \times 10^{-3})}$$

$$k_a = 4.42 \times 10^{10}\ L\ mol^{-1}\ s^{-1}$$

$$k_d = (1.73 \times 10^{-5})k_a = 7.65 \times 10^5\ s^{-1}$$

20.18 Derive the relation between the relaxation time τ and the rate constants for the reaction A + B ⇔ C + D, which is subjected to a small displacement from equilibrium.

SOLUTION

$d[C]/dt = k_1[A][B] - k_2[C][D] = 0$
At equilibrium, $0 = k_1[A]_{eq}[B]_{eq} - k_2[C]_{eq}[D]_{eq}$
$[A] = [A]_{eq} - \Delta[C]$
$[B] = [B]_{eq} - \Delta[C]$
$[C] = [C]_{eq} - \Delta[C]$
$[D] = [D]_{eq} - \Delta[C]$
Thus the equation in the first line can be written

$$\frac{d\Delta[C]}{dt} = k_1([A]_{eq} - \Delta[C])([B]_{eq} - \Delta[C]) - k_2([C]_{eq} + \Delta[C])([D]_{eq} + \Delta[C])$$

Multiplying this out, using the equation in the second line, and dropping the $(\Delta[C])^2$ term yields

$$\frac{d\Delta[C]}{dt} = -\frac{\Delta[C]}{\tau}$$

where

$$\frac{1}{\tau} = k_1([A]_{eq} + [B]_{eq}) + k_2([C]_{eq} + [D]_{eq})$$

20.19 Derive the relation between the relaxation time τ and the rate constants for the mechanism.

k_1

$$\begin{array}{ccccc} & \mathrm{A} & \rightleftarrows & \mathrm{B} & \\ & & k_{-1} & & \\ K_{\mathrm{A}} & \uparrow\downarrow & & \uparrow\downarrow & K_{\mathrm{B}} \\ & & k_1' & & \\ & \mathrm{A'} & \rightleftarrows & \mathrm{B'} & \\ & & k_{-1}' & & \end{array}$$

which is subjected to a small displacement from equilibrium. It is assumed that the equilibria, A ⇔ A', K_A = [A']/[A], and B ⇔ B', K_B = [**B**']/[**B**], are adjusted very rapidly so that these steps remain in equilibrium.

SOLUTION

$[\mathrm{A'}] = K_{\mathrm{A}}[\mathrm{A}]$, $[\mathrm{B'}] = K_{\mathrm{B}}[\mathrm{B}]$

$d[\mathrm{A}]/dt = -k_1[\mathrm{A}] + k_{-1}[\mathrm{B}] - k_1' K_{\mathrm{A}}[\mathrm{A}] + k'_{-1} K_{\mathrm{B}}[\mathrm{B}]$

$\quad = -(k_1 + k_1' K_{\mathrm{A}})[\mathrm{A}] + (k_{-1} + k'_{-1} K_{\mathrm{B}})[\mathrm{B}]$

At equilibrium, $d[\mathrm{A}]/dt = 0 = (-k_1 + k_1' K_{\mathrm{A}})[\mathrm{A}]_{\mathrm{eq}} + (k_{-1} + k'_{-1}K_{\mathrm{B}})[\mathrm{B}]_{\mathrm{eq}}$

Set $[\mathrm{A}] = [\mathrm{A}]_{\mathrm{eq}} + \Delta[\mathrm{A}]$

$\quad [\mathrm{B}] = [\mathrm{B}]_{\mathrm{eq}} - \Delta[\mathrm{A}]$

$$\frac{d\Delta[\mathrm{A}]}{dt} = -(k_1 + k_1' K_{\mathrm{A}})\Delta[\mathrm{A}] - (k_{-1} + k'_{-1}K_{\mathrm{B}})\Delta[\mathrm{A}] = -\frac{\Delta[\mathrm{A}]}{\tau}$$

$$\tau = \frac{1}{k_1 + k_1'K_{\mathrm{A}} + k_{-1} + k'_{-1}K_{\mathrm{B}}}$$

20.20 Calculate the first-order rate constants for the dissociation of the following weak acids: acetic acid, acid form of imidazole ($C_3N_2H_5^+$), NH_4^+. The corresponding acid dissociation constants are 1.75 x 10^{-5}, 1.2 x 10^{-7}, and 5.71 x 10^{-10}, respectively. The second-order rate constants for the formation of the acid forms from a proton plus the base are 4.5 x 10^{10}, 1.8 x 10^{10}, and 4.3 x 10^{10} L mol^{-1} s^{-1}, respectively.

SOLUTION

For HA = H^+ + A^-

$$K = \frac{[\mathrm{H^+}][\mathrm{A^-}]}{[\mathrm{HA}]} = \frac{k_1}{k_2}$$

For acetic acid, $k_1 = k_2K = (4.5\text{x}10^{10}\ \mathrm{L\ mol^{-1}\ s^{-1}})(1.75\text{x}10^{-5}\ \mathrm{mol\ L^{-1}}) = 7.9\text{x}10^{-5}\ \mathrm{s^{-1}}$

For imidazole, $k_1 = k_2K = (1.8\text{x}10^{10}\ \mathrm{L\ mol^{-1}\ s^{-1}})(1.2\text{x}10^{-7}\ \mathrm{mol\ L^{-1}})$

$\quad = 2.1\text{x}10^{-5}\ \mathrm{s^{-1}}$ WRONG

For NH_4^+, $k_1 = k_2K = (4.3\text{x}10^{10}\ \text{L mol}^{-1}\ \text{s}^{-1})(5.71\text{x}10^{-10}\ \text{mol L}^{-1}) = 25\ \text{s}^{-1}$

20.21 The hydrolysis of pyrophosphate ($P_2O_7^{4-}$) at pH = 7 at 25 °C by the enzyme pyrophosphatase occurs with an apparent first order rate constant of $k' = 0.001\ \text{s}^{-1}$. The reaction is first order because the concentration of pyrophosphate is much lower than the Michaelis constant. Calculate the apparent first order rate constant at pH = 6 and pH = 8 assuming that the mechanism is

$$P_2O_7^{4-} + H_2O \xrightarrow{k} 2HPO_4^{2-}$$

$\parallel$ $K_{HA} = 10^{-8.95}$

$HP_2O_7^{3-}$

$\parallel$ $K_{H_2A} = 10^{-6.12}$

$H_2P_2O_7^{2-}$

and that the acid dissociations are fast compared with the hydrolysis. The reaction goes so far to the right that we do not have to be concerned with the reverse reaction.

SOLUTION

$$k' = \frac{k}{1 + [H^+]/K_{HA} + [H^+]^2/K_{HA}K_{H_2A}}$$

At pH = 7, $k' = k/101.9 = 0.001\ \text{s}^{-1}$ and so $k = 0.102\ \text{s}^{-1}$

At pH = 8, $k' = 0.102\ \text{s}^{-1}/10.03 = 0.0102\ \text{s}^{-1}$

At pH = 6, $k' = 0.102\ \text{s}^{-1}/2071 = 4.93 \times 10^{-5}\ \text{s}^{-1}$

The reaction goes more slowly at the lower pH because a smaller fraction of the pyrophosphate is in the form $P_2O_7^{4-}$.

20.22 The solution reaction

$$I^- + OCl^- = OI^- + Cl^-$$

is believed to go by the mechanism

$$OCl^- + H_2O \underset{}{\overset{K_1}{\rightleftarrows}} HOCl + OH^- \quad \text{(fast)}$$

$$I^- + HOCl \xrightarrow{k} HOI + Cl^- \qquad \text{(slow}$$

$$HOI + OH^- \overset{K_2}{\rightleftarrows} H_2O + OI^- \quad \text{(fast)}$$

Derive the rate equation for the forward rate of this reaction that shows the effect of the concentration of OH^-.

SOLUTION

For the slow step,

$$\frac{d[Cl^-]}{dt} = k[I^-][HOCl]$$

Since $K_1 = [HOCl][OH^-]/[OCl^-]$,

$[HOCl] = K_1[OCl^-]/[OH]$

Substituting this in the rate equation for the slow step yields

$$\frac{d[Cl^-]}{dt} = \frac{kK_1[I^-][OCl^-]}{[OH^-]}$$

20.23 The mutarotation of glucose is first order in glucose concentration and is catalyzed by acids (A) and bases (B). The first-order rate constant may be expressed by an equation of the type that is encountered in reactions with parallel paths.

$$k = k_0 + k_{H^+}[H^+] + k_A[A] + k_B[B]$$

where k_0 is the first-order rate constant in the absence of acids and bases other than water. The following data were obtained by J. H. Brönsted and E. A. Guggenheim [*J. Am. Chem. Soc.* **49**, 2554 (1927)] at 18 °C in a medium containing 0.02 mol L^{-1} sodium acetate and various concentrations of acetic acid.

$[CH_3CO_2H]$/mol L^{-1}	0.020	0.105	0.199
$k/10^{-4}$ min^{-1}	1.36	1.40	1.46

Calculate k_0 and k_A. The term involving k_{H^+} is negligible under these conditions.

SOLUTION

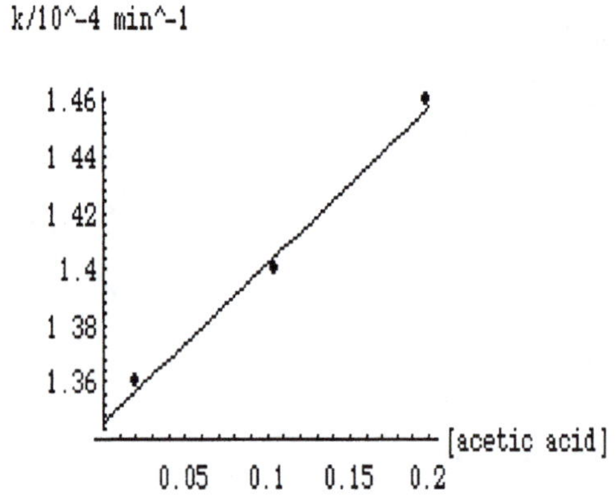

Intercept = k_0 = 1.35 x 10^{-4} min^{-1}

$$\text{Slope} = \frac{(1.46 - 1.35) \times 10^{-4}}{0.2 \text{ mol L}^{-1}} = 5.5 \times 10^{-5} \text{ L mol}^{-1} \text{ min}^{-1} = k_A$$

20.24 The rate of a reaction between oppositely charged ions is measured at an ionic strength of 0.01 mol L^{-1}. How will the rate be affected if the ionic strength is raised to 0.05 mol L^{-1} if the reaction is (a) $A^+ + B^-$ or (b) $A^{2+} + B^{2+}$?

SOLUTION

(a)

$$\log k_1 = \log k^o + 2z_Az_B A I_1^{1/2}$$

$$\log k_2 = \log k^o + 2z_Az_B A I_2^{1/2}$$

$$\log \frac{k_2}{k_1} = 2z_Az_B A(I_2^{1/2} - I_1^{1/2})$$

$$\log \frac{k_2}{k_1} = -2[0.509 \text{ (mol L}^{-1})^{-1/2}][(0.05 \text{ mol L}^{-1})^{1/2} - (0.01 \text{ mol L}^{-1})^{1/2}]$$

$$\frac{k_2}{k_1} = 0.748$$

Raising the ionic strength reduces k.

(b)

$$\log \frac{k_2}{k_1} = (2)(2)(2)(0.509)(0.05^{1/2} - 0.01^{1/2})$$

$$\frac{k_2}{k_1} = 3.20$$

Raising the ionic strength increases k.

20.25 Suppose an enzyme has a turnover number of 10^4 min^{-1} and a molar mass of 60,000 g mol^{-1}. How many moles of substrate can be turned over per hour per gram of enzyme if the substrate concentration is twice the Michaelis constant? It is assumed that the substrate concentration is maintained constant by a preceding enzymatic reaction and that products do not accumulate and inhibit the reaction.

SOLUTION

$$v = \frac{V}{1 + K/[S]} = \frac{(10^4\ \text{min}^{-1})[E]_0}{1 + 0.5}$$

$$= \frac{(10^4\ \text{min}^{-1})(1\ \text{g}/60{,}000\ \text{g mol}^{-1})(60\ \text{min hr}^{-1})}{1.5} = 6.7\ \text{mol hr}^{-1}$$

20.26 The kinetics of the fumarase reaction
fumarate + H_2O = L-malate is
studied at 25 °C using an 0.01 ionic strength buffer of pH 7. The rate of reaction is obtained using a recording ultraviolet spectrometer to measure the fumarate concentration. The following rates of the forward reaction are obtained using a fumarase concentration of 5 x 10^{-10} mol L^{-1}:

$[F]/10^{-6}$ mol L^{-1}	$v_F/10^{-7}$ mol L^{-1} s^{-1}
2	2.2
40	5.9

The following rates of the reverse reaction are obtained using a fumarase concentration of 5 x 10^{-10} mol L^{-1}:

$[M]/10^{-6}$ mol L^{-1}	$v_M/10^{-7}$ mol L^{-1} s^{-1}
5	1.3
100	3.6

(a) Calculate the Michaelis constants and turnover numbers for the two substrates. In practice many more concentrations would be studied. (b) Calculate the four rate constants in the mechanism

$$\text{E} + \text{F} \underset{k_{-1}}{\overset{k_1}{\rightleftarrows}} \text{X} \underset{k_{-2}}{\overset{k_2}{\rightleftarrows}} \text{E} + \text{M}$$

where E represents the catalytic site. There are four catalytic sites per fumarase molecule. (c) Calculate K_{eq} for the reaction catalyzed. The concentration of H_2O is omitted in the expression for the equilibrium constant because its concentration cannot be varied in dilute aqueous solutions.

SOLUTION

(a) $$v_F = \frac{V_F}{1 + K_F/[F]}$$

$$v_F + \frac{v_F}{[F]} K_F = V_F$$

$$2.2 \times 10^{-7} + 0.110\ K_F = V_F$$

$$5.9 \times 10^{-7} + 0.0148\,K_F = V_F$$

$$-3.7 \times 10^{-7} + 0.0952\,K_F = 0$$

$$K_F = \frac{3.7 \times 10^{-7}}{0.0952} = 3.9 \times 10^{-6}\ \text{mol L}^{-1}$$

$$V_F = 2.2 \times 10^{-7} + (0.110)(3.9 \times 10^{-6}) = 6.5 \times 10^{-7}\ \text{mol L}^{-1}\ \text{s}^{-1}$$

$$\upsilon_M = \frac{V_M}{1 + K_M/[\text{M}]}$$

$$\upsilon_M + \frac{\upsilon_M}{[\text{M}]} K_M = V_M$$

$$1.3 \times 10^{-7} + 0.260\,K_M = V_M$$

$$3.6 \times 10^{-7} + 0.0036\,K_M = V_M$$

$$-2.3 \times 10^{-7} + 0.0224\,K_M = 0$$

$$K_M = \frac{2.3 \times 10^{-7}}{0.0224} = 1.03 \times 10^{-5}\ \text{mol L}^{-1}$$

$$V_M = 1.3 \times 10^{-7} + 0.026(1.03 \times 10^{-5}) = 4.0 \times 10^{-7}\ \text{mol L}^{-1}\ \text{s}^{-1}$$

(b) The rate constants should be expressed in terms of the concentration of enzymatic sites and so

$$k_2 = \frac{V_F}{[\text{E}]_0} = \frac{6.5 \times 10^{-7}\ \text{mol L}^{-1}\ \text{s}^{-1}}{4(5 \times 10^{-10}\ \text{mol L}^{-1})} = 3.3 \times 10^{2}\ \text{s}^{-1}$$

$$k_{-1} = \frac{V_M}{[\text{E}]_0} = \frac{4.0 \times 10^{-7}\ \text{mol L}^{-1}\ \text{s}^{-1}}{4(5 \times 10^{-10}\ \text{mol L}^{-1})} = 2.0 \times 10^{2}\ \text{s}^{-1}$$

$$K_F = \frac{k_2 + k_{-1}}{k_1}$$

$$k_1 = \frac{k_2 + k_{-1}}{K_F} = \frac{(3.3 + 2.0) \times 10^{2}\ \text{s}^{-1}}{3.9 \times 10^{-6}\ \text{mol L}^{-1}} = 1.4 \times 10^{8}\ \text{L mol}^{-1}\ \text{s}^{-1}$$

$$K_M = \frac{k_2 + k_{-1}}{k_{-2}}$$

$$k_{-2} = \frac{k_2 + k_{-1}}{K_M} = \frac{(3.3 + 2.0) \times 10^{2}\ \text{s}^{-1}}{1.03 \times 10^{-5}\ \text{mol L}^{-1}} = 5.1 \times 10^{7}\ \text{L mol}^{-1}\ \text{s}^{-1}$$

(c) $$K = \frac{[M]_{eq}}{[F]_{eq}} = \frac{V_F K_M}{V_M K_F} = \frac{(6.5 \times 10^{-7})(1.03 \times 10^{-5})}{(4.0 \times 10^{-7})(3.9 \times 10^{-6})} = 4.3$$

$$K = \frac{k_1 k_2}{k_{-1} k_{-2}} = \frac{(1.4 \times 10^{8})(3.3 \times 10^{2})}{(2.0 \times 10^{2})(5.1 \times 10^{7})} = 4.5$$

20.27 Derive the steady-state rate equation for the mechanism

$$E + S \underset{k_2}{\overset{k_1}{\rightleftarrows}} X \xrightarrow{k_3} E + P \qquad E + I \underset{k_5}{\overset{k_4}{\rightleftarrows}} EI$$

for the case that $[S] >> [E]_0$ and $[I] >> [E]_0$.

SOLUTION

$$[E]_0 = [E] + [EI] + [X]$$

$$= [E](1 + [I]/K_I) + [X] \qquad (1)$$

Since $K_I = \frac{[E][I]}{[EI]} = \frac{k_5}{k_4}$

Assuming that X is in a steady state

$$\frac{d[X]}{dt} = k_1[E][S] - (k_2 + k_3)[X] = 0 \qquad (2)$$

Solving equation 1 for [E] and substituting this expression in equation 2 yields

$$\frac{k_1[S][E]_0}{1 + [I]/K_I} = [X]\left[\frac{k_1[S]}{1 + [I]/K_I} + k_2 + k_3\right]$$

$$\frac{d[P]}{dt} = k_3[X] = \frac{k_3[E]_0}{1 + \frac{k_2 + k_3}{k_1[S]}\left(1 + \frac{[I]}{K_I}\right)}$$

$$= \frac{V_S}{1 + \frac{K_S}{[S]}\left(1 + \frac{[I]}{K_I}\right)}$$

20.28 The following initial velocities were determined spectrophotometrically for solutions of sodium succinate to which a constant amount of succinoxidase was added. The velocities are given as the change in absorbancy at 250 nm in 10 s. Calculate V, K_M and K_I for malonate.

$\frac{[\text{Succinate}]}{10^{-3}\ \text{mol L}^{-1}}$	$\frac{\Delta A \times 10^3}{10\ \text{s}}$ No Inhibitor	15×10^{-6} mol L^{-1} malonate
10	16.7	14.9
2	14.2	10.0
1	11.3	7.7
0.5	8.8	4.9
0.33	7.1	--

SOLUTION

The first step is to calculate the Michaelis constant and maximum velocity for succinate. The following type of plot can be used:

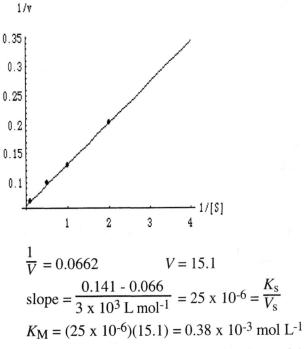

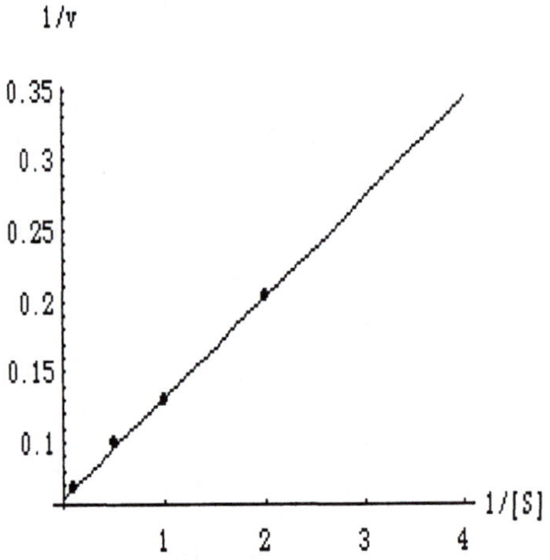

$$\frac{1}{V} = 0.0662 \qquad V = 15.1$$

$$\text{slope} = \frac{0.141 - 0.066}{3 \times 10^3\ \text{L mol}^{-1}} = 25 \times 10^{-6} = \frac{K_s}{V_s}$$

$$K_M = (25 \times 10^{-6})(15.1) = 0.38 \times 10^{-3}\ \text{mol L}^{-1}$$

The second step is to determine the slope of this type of plot when the inhibitor (malonate) is present.

$$v = \frac{d[P]}{dt} = \frac{V}{1 + \frac{K_M}{[S]}\left(1 + \frac{[I]}{K_I}\right)}$$

$$\frac{1}{v} = \frac{1}{V} + \frac{K_M}{V[S]}\left(1 + \frac{[I]}{K_I}\right)$$

$$\text{slope} = \frac{0.205 - 0.066}{2 \times 10^3\ \text{L mol}^{-1}} = 0.069 = \frac{K_M}{V}\left(1 + \frac{[I]}{K_I}\right)$$

$$0.069 = \frac{0.38}{15.1}\left(1 + \frac{15 \times 10^{-6} \text{ mol L}^{-1}}{K_I}\right)$$

$$K_I = \frac{15 \times 10^{-6} \text{ mol L}^{-1}}{\frac{(15.1)(0.069)}{0.38} - 1} = 8.6 \times 10^{-6} \text{ mol L}^{-1}$$

20.29 In the Eadie-Hostee method for determining k_{cat} and K_M for an enzymatic reaction, $v/[E]_0[S]$ is plotted versus $v/[E]_0$. How are the kinetic parameters obtained from this plot?

SOLUTION

The Michaelis-Menten equation can be arranged in the form

$$\frac{v}{[E]_0[S]} = \frac{k_{cat}}{K_M} - \frac{v}{K_M[E]_0}$$

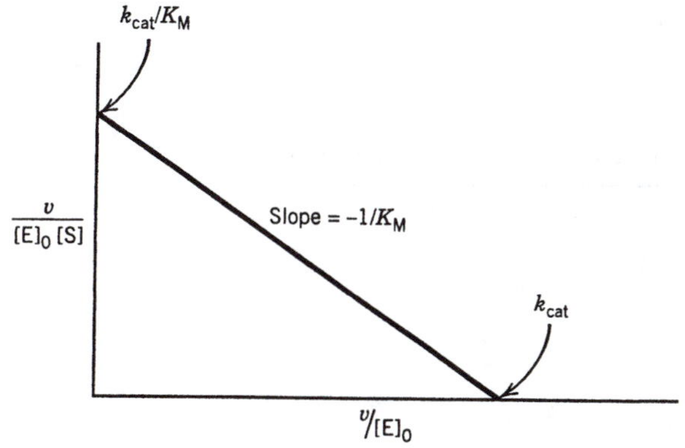

20.30 The maximum initial velocities for an enzymatic reaction are determined at a series of pH values:

pH	6.0	7.0	7.5	8.0	8.5	9.0
V	11	74	129	147	108	53

Calculate the values of the parameters V', K_a, K_b in

$$V = \frac{V'}{1 + [H^+]/K_a + K_b/[H^+]}$$

See problem 20.27.

SOLUTION

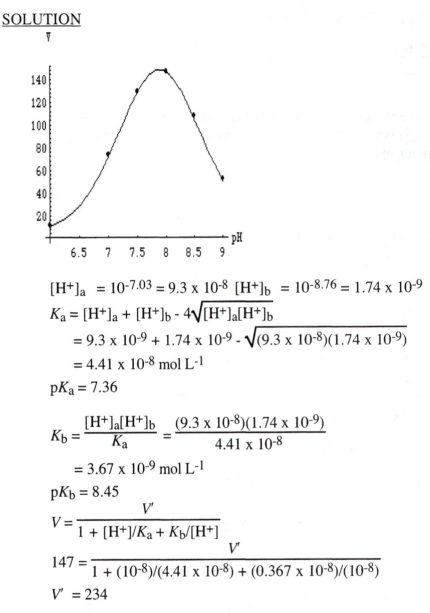

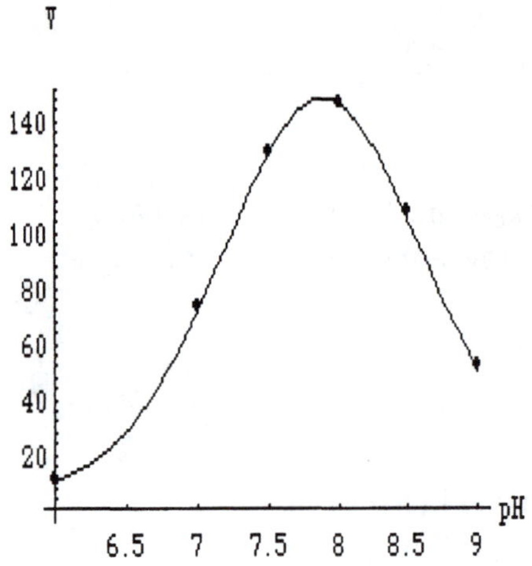

$[H^+]_a = 10^{-7.03} = 9.3 \times 10^{-8}$ $[H^+]_b = 10^{-8.76} = 1.74 \times 10^{-9}$

$K_a = [H^+]_a + [H^+]_b - 4\sqrt{[H^+]_a[H^+]_b}$

$= 9.3 \times 10^{-9} + 1.74 \times 10^{-9} - \sqrt{(9.3 \times 10^{-8})(1.74 \times 10^{-9})}$

$= 4.41 \times 10^{-8} \text{ mol L}^{-1}$

$pK_a = 7.36$

$$K_b = \frac{[H^+]_a[H^+]_b}{K_a} = \frac{(9.3 \times 10^{-8})(1.74 \times 10^{-9})}{4.41 \times 10^{-8}}$$

$= 3.67 \times 10^{-9} \text{ mol L}^{-1}$

$pK_b = 8.45$

$$V = \frac{V'}{1 + [H^+]/K_a + K_b/[H^+]}$$

$$147 = \frac{V'}{1 + (10^{-8})/(4.41 \times 10^{-8}) + (0.367 \times 10^{-8})/(10^{-8})}$$

$V' = 234$

20.31 Use equation 20.87 to show that

$$V_S = \frac{V_{max}\,(1 + 2(K_b/K_a)^{1/2})}{1 + [H^+]/K_a + K_b/[H^+]}$$

where V_{max} is the maximum initial velocity in the plot of V_S versus pH. Further show that $K_aK_b = [H^+]_a[H^+]_b$

where $[H^+]_a$ is the hydrogen ion concentration at which $v/V_{max} = 1/2$ on the acidic side of the plot and $[H^+]_b$ is the hydrogen ion concentration at which $v/V_{max} = 1/2$ on the basic side of the plot. Further show that the value of K_a can be calculated using $K_a = [H^+]_a + [H^+]_b - 4([H^+]_a[H^+]_b)^{1/2}$

SOLUTION

At $[H^+]_a$,

$$\frac{V_S}{V_{max}} = \frac{1}{2} = \frac{1 + 2(K_b/K_a)^{1/2}}{1 + [H^+]_a/K_a + K_b/[H^+]_a} \quad \text{(a)}$$

At $[H^+]_b$,

$$\frac{V_S}{V_{max}} = \frac{1}{2} = \frac{1 + 2(K_b/K_a)^{1/2}}{1 + [H^+]_b/K_a + K_b/[H^+]_b} \quad \text{(b)}$$

Therefore,

$$\frac{[H^+]_a}{K_a} + \frac{K_b}{[H^+]_a} = \frac{[H^+]_b}{K_a} + \frac{K_b}{[H^+]_b} \text{ or } K_aK_b = [H^+]_a[H^+]_b \quad \text{(c)}$$

Use this to eliminate K_b from equation (a) to obtain

$$K_a = [H^+]_a + [H^+]_b - 4([H^+]_a[H^+]_b)^{1/2} \quad \text{(d)}$$

20.32 For the reaction

$$CO_2 \ (+ H_2O) \underset{k_2}{\overset{k_1}{\rightleftarrows}} H_2CO_3$$

where the parentheses indicate that H_2O is not included in the equilibrium constant expression or in the rate equation, the following data were obtained: $\Delta_rH^o = 4730$ J mol^{-1} and $\Delta_rS^o = -33.5$ J K^{-1} mol^{-1}. At 25 °C, $k_1 = 0.0375$ s^{-1} and at 0 °C, $k_1 = 0.0021$ s^{-1}. Assuming that Δ_rH^o and Δ_rS^o are independent of temperature in this range, (a) calculate the equilibrium constant and k_2 values at 25 °C and and 0 °C, and (b) calculate the activation energies for the forward and backward reactions.

SOLUTION

(a) Since $\Delta_rG^o = 14\ 718$ J mol^{-1} at 25 °C, the equilibrium constant is given by

$$K = \frac{[H_2CO_3]}{[CO_2]} = 0.00264 = \frac{k_1}{k_2}$$

Therefore, k_2 at 25 °C is 14.21 s^{-1}.

Since $\Delta_rG^o = 13\ 880$ J mol^{-1} at 0 °C, the equilibrium constant is equal to 0.002217. Therefore, k_2 at 0 °C is 0.947 s^{-1}.

(b) The activation energy is given by

$$E_a = \frac{RT_1T_2}{T_2 - T_1} \ln \frac{k(T_2)}{k(T_1)}$$

For the forward reaction,

$$E_{a1} = \frac{8.3145 x 298.15 x 273.15}{25} \ln \frac{0.0375}{0.0021} = 78\ 070 \text{ J mol}^{-1}$$

For the backward reaction,

$$E_{a2} = \frac{8.3145 x 298.15 x 273.15}{25} \ln \frac{14.2}{0.9471} = 73\ 340 \text{ J mol}^{-1}$$

The value of $\Delta_r H^o$ is equal to the dfference between these activation energies:

$\Delta_r H^o = 78\ 070 - 73\ 340 = 4730\ \text{J mol}^{-1}$

The activation energy for the backward reaction is smaller, and, therefore, the backward reaction is faster and the equilibrium constant is less than 1 at both temperatures.

20.33	8.57 minutes
20.34	0.66 s
20.35	17.4 kJ mol^{-1}
20.36	7.09×10^{-11} m^2 s^{-1}
20.37	(a) 0.0705 cm (b) 0.141 cm
20.38	1.83, 3.66, 5.49 mm
20.39	3.34×10^{-9} m^2 s^{-1}
20.41	0.1058 Ω^{-1} m^{-1}
20.42	1.55×10^{-2} Ω^{-1} m^{-1}
20.43	1.99×10^{-9} m^2 s^{-1}
20.44	520 nm
20.45	1.4×10^{11} L mol^{-1} s^{-1}
20.46	2.7×10^{-4} s
20.47	$\tau^{-1} = k_1 + k_2$
20.48	$k_1 = 1.8 \times 10^{10}$ L mol^{-1} s^{-1}, $k_{-1} = 1.11 \times 10^3$ s^{-1}
20.49	$k_o = 1.21 \times 10^{-4}$ min^{-1}
	$k_{H^+} = 3.4 \times 10^{-3}$ L mol^{-1} s^{-1}
20.50	$V = 1.2 \times 10^{-6}$ M s^{-1}, $K_M = 0.48 \times 10^{-3}$ mol L^{-1}
20.51	(a) $k_1 = 1.3 \times 10^8$ M^{-1} s^{-1}
	$k_2 = 0.33 \times 10^3$ s^{-1}
	$k_{-1} = 0.20 \times 10^3$ s^{-1}
	$k_{-2} = 5.3 \times 10^7$ M^{-1} s^{-1}
	(b) - 3.4 kJ mol^{-1}
20.52	$k_1 = 4.9 \times 10^8$ M^{-1} s^{-1}, $k_2 = 0.8 \times 10^3$ s^{-1}
	$k_{-1} = 2.6 \times 10^3$ s^{-1}, $k_{-2} = 3.4 \times 10^7$ M^{-1} s^{-1}
20.54	33% inhibition

20.55 $$\frac{v}{[S]} = \frac{V}{K_M(1 + [I]/K_I)} = \frac{v}{K_M(1 + [I]/K_I)}$$

20.56 $$V = k_2[E]_0/(1 + [H^+]/K_{EHS})$$

$$K_M = \frac{k_{-1} + k_2}{k_1} \frac{[1 + [H^+]/K_{EH}]}{[1 + [H^+]/K_{EHS}\}}$$

21

Macromolecules

21.1 A polymer solution contains 250 molecules of molar mass 75,000 g mol^{-1}, 500 molecules of molar mass 100,000 g mol^{-1}, and 250 molecules of molar mass 125,000 g mol^{-1}. Calculate $\bar{M}_n$, $\bar{M}_m$, and the ratio $\bar{M}_m/\bar{M}_n$ (the polydispersity).

SOLUTION

$$\bar{M}_n = \frac{250(75{,}000)}{1000} + \frac{500(100{,}000)}{1000} + \frac{250(125{,}000)}{1000}$$

$$= 100{,}000 \text{ g mol}^{-1}$$

$$\bar{M}_m = \frac{250(75{,}000)^2 + 500(100{,}000)^2 + 250(125{,}000)^2}{250(75{,}000) + 500(100{,}000) + 250(125{,}000)}$$

$$= 103{,}000 \text{ g mol}^{-1}$$

$$\frac{\bar{M}_m}{\bar{M}_n} = 1.03$$

*21.2 Plot the probability density $W(r)$ for random walk in three dimensions after 1000 steps with a step length of unity. Indicate the root-mean-square end-to-end distance on this plot.

SOLUTION

$$W(r) = \left(\frac{3}{2\pi n l^2}\right)^{3/2} \exp\left(-\frac{3r^2}{2nl^2}\right)$$

$$= \left(\frac{3}{2\pi 1000}\right)^{3/2} \exp\left(-\frac{3r^2}{2000}\right)$$

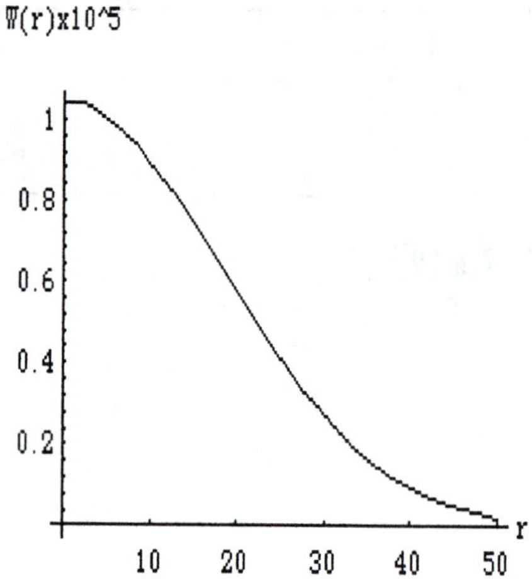

21.3 In polyethylene $H(CH_2{-}CH_2)_nH$ the bond length l is 0.15 nm. What is the root-mean-square end-to-end distance for a molecule with universal joints for a molar mass of 10^5 g mol^{-1}? Taking into account the fact that carbon forms tetrahedral bonds, what is $(\overline{r^2})^{1/2}$?

SOLUTION

why divide by 14

$$N = \frac{10^5 \text{ g mol}^{-1}}{14 \text{ g mol}^{-1}}$$

$$(\overline{r^2})^{1/2} = n^{1/2}\, l = \left(\frac{10^5}{14}\right)^{1/2} 0.15 \text{ nm} = 12.7 \text{ nm}$$

$$(\overline{r^2})^{1/2} = n^{1/2} l \times \left(\frac{1 + \cos\theta}{1 - \cos\theta}\right)^{1/2} = (12.7 \text{ nm})\left(\frac{1 + \cos 71°}{1 - \cos 71°}\right)^{1/2} = 17.8 \text{ nm}$$

21.4 Derive the expression for the mean separation $< r >$ for the ends of a freely jointed chain of N bonds of length l. See the definite integrals in Table 17.1 or Appendix D.3.

SOLUTION

$$< r > = \int_0^\infty r W(r) 4\pi r^2 \mathrm{d}r$$

$$= 4\pi\left(\frac{3}{2\pi N l^2}\right)^{3/2} \int_0^\infty \mathrm{e}^{-3r^2/2Nl^2} r^3 \mathrm{d}r$$

The needed integral from Table 17.1 is

$$\int_0^\infty e^{-ax^2} x^3 dx = \frac{1}{2a^2}$$

$$< r > = \left(\frac{8}{3\pi}\right)^{1/2} N^{1/2}\, l$$

21.5 Derive the expression for the root-mean-square separation $< r^2 >^{1/2}$ for the ends of a freely jointed chain of N bonds of length l. See the definite integrals in Table 17.1 or Appendix D.3.

SOLUTION

$$< r^2 > = \int_0^\infty r^2 W(r) 4\pi r^2 dr$$

$$= 4\pi\left(\frac{3}{2\pi N l^2}\right)^{3/2} \int_0^\infty e^{-3r^2/2Nl^2} r^4 dr$$

The needed integral from Table 17.1 is

$$\int_0^\infty e^{-ax^2} x^4 dx = \frac{3}{8}\left(\frac{\pi}{a^5}\right)^{1/2}$$

$$< r^2 >^{1/2} = N^{1/2}\, l$$

21.6 For a condensation polymerization of a hydroxy acid in which 99% of the acid groups are used up, calculate (a) the average number of monomer units in the polymer molecules, (b) the probability that a given molecule will have the number of residues given by this value, and (c) the weight fraction having this particular number of monomer units.

SOLUTION

(a) The number-average degree of polymerization is given by

$$\bar{X}_n = \frac{1}{1 - p} = \frac{1}{1 - 0.99} = 100$$

(b) The probability that a given molecule will have 100 residues is given by

$$\pi_i = p^{i-1}(1 - p)$$

$$\pi_{100} = 0.99^{100-1}(1 - 0.99) = 3.70 \times 10^{-3}$$

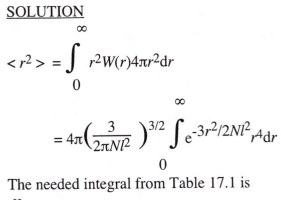

(c) The weight fraction having this number of residues is given by

$W_i = ip^{i-1}(1 - p)^2$

$W_{100} = (100)(0.99)^{100-1}(1 - 0.99)^2$

$= 3.70 \times 10^{-3}$

21.7 A hydroxy acid $HO\text{-}(CH_2)_5\text{-}CO_2H$ is polymerized, and it is found that the product has a number average molar mass of 20,000 g mol^{-1}. (a) What is the extent of reaction p? (b) What is the degree of polymerization $\overline{X}_n$? (c) What is the mass-average molar mass?

SOLUTION

The molar mass of a monomer unit is 114 g mol^{-1} = M_o

(a) $\overline{M}_n = M_o/(1 - p)$

$p = 1 - M_o/\overline{M}_n = 1 - 114/20{,}000 = 0.9943$

(b) $\overline{X}_n = \dfrac{1}{1 - p} = \dfrac{1}{1 - 0.9943} = 175$

(c) $\overline{M}_m = M_o \dfrac{1 + p}{1 - p} = 114 \dfrac{1.9943}{1 - 0.9943}$

$= 39{,}900$ g mol^{-1}

21.8 A general polymerization reaction in the liquid phase can be written

$M = \frac{1}{n} P_n$

The values of $\Delta_r H^o$ and $\Delta_r G^o$ have been determined for some polymerization reactions. This makes it possible to calculate $\Delta_r S^o$, and some values at 25 ^{o}C are shown in the following table:

Monomer	$-\Delta_r H^o$/k J mol^{-1}	$-\Delta_r S^o$/J K^{-1} mol^{-1}	$-\Delta_r G^o$/k J mol^{-1}
Styrene	69.9	104	38.5
α-Methylstyrene	35.2	104	4.2
Tetrafluoro-ethylene	154.8	112	121

If we assume that $\Delta_r H^o$ and $\Delta_r S^o$ are independent of temperature, we can calculate the temperature at which the equilibrium constant for the polymerization reaction is unity. At this temperature, depolymerization occurs, and that temperature is called the ceiling temperature. Calculate the ceiling temperatures of these three polymers, and interpret these temperatures.

SOLUTION

$$T_c = \frac{\Delta_r H^o}{\Delta_r S^o}$$

The ceiling temperature for polystyrene is 69,900/104 = 672 K.
The ceiling temperature for poly-α-methylstyrene is 35,200/104 = 338 K. The bonding is not as strong as for styrene, as indicated by $\Delta_r H^o$, and so depolymerization occurs at a lower temperature. The methyl group apparently prevents as close packing as in polystyrene. The ceiling temperature for tetrafluoroethylene is 154,800/112 = 1382 K, and so Teflon is used on cooking utensils. The bonding is very strong, as indicated by $\Delta_r H^o$.

21.9 Show that the intrinsic viscosity can also be defined by

$$[\eta] = \lim_{c \to 0} \left(\frac{1}{c}\right) \ln\left(\frac{\eta}{\eta_0}\right)$$

(Hint: $\ln(1 + x) \approx x$ if $x \ll 1$.)

SOLUTION

$$\ln\left(\frac{\eta}{\eta_0}\right) = \ln\left(1 + \frac{\eta - \eta_0}{\eta_0}\right) \approx \frac{\eta - \eta_0}{\eta_0} = \frac{\eta}{\eta_0} - 1$$

$$[\eta] = \lim_{c \to 0} \left(\frac{1}{c}\right) \ln\left(\frac{\eta}{\eta_0}\right) = \lim_{c \to 0} \frac{(\eta/\eta_0) - 1}{c}$$

In treating experimental data, it is advantageous to plot $\frac{\eta_r - 1}{c}$ and $\ln \frac{\eta_r}{c}$ versus c because both plots extrapolate to $[\eta]$ at $c = 0$.

*21.10 The relative viscosities of a series of solutions of a sample of polystyrene in toluene were determined with an Ostwald viscometer at 25 °C.

$c/10^{-2}$ g cm^{-3}	0.249	0.499	0.999	1.998
η/η_o	1.355	1.782	2.879	6.090

The ratio η_{sp}/c is plotted against c and extrapolated to zero concentration to obtain the intrinsic viscosity. If the constants in equation 21.39 are $K = 3.7 \times 10^{-2}$

and $a = 0.62$ for this polymer, when concentrations are expressed in g/cm^3, calculate the molar mass.

SOLUTION

$c/10^{-2}$ g cm^{-3}	0.249	0.499	0.999	1.998
$\dfrac{\eta/\eta_o - 1}{c}$	142.6	156.7	188.1	254.8

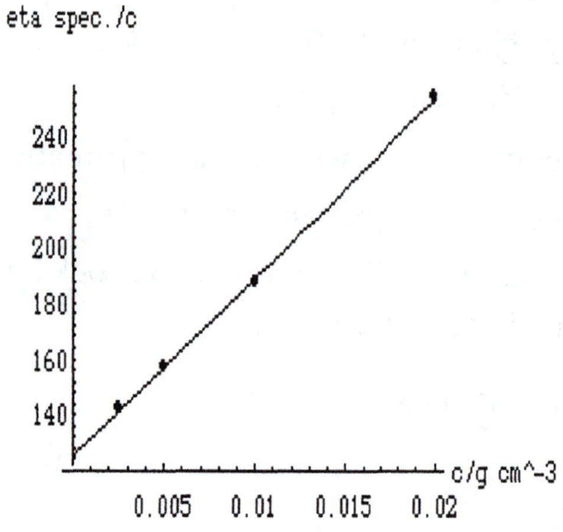

$$\ln \frac{[\eta]}{3.7 \times 10^{-2}} = 0.62 \ln M$$

$$M = \exp\left[\frac{1}{0.62} \ln \frac{126}{3.7 \times 10^{-2}}\right] = 500{,}000 \text{ g mol}^{-1}$$

21.11 At 34 °C the intrinsic viscosity of a sample of polystyrene in toluene is 84 cm^3 g^{-1}. The empirical relation between the intrinsic viscosity of polystyrene in toluene and molar mass is $[\eta] = 1.15 \times 10^{-2} M^{0.72}$. What is the molar mass of this sample?

SOLUTION

$$\ln \frac{[\eta]}{1.15 \times 10^{-2}} = 0.72 \ln M$$

$$M = \exp\left[\frac{1}{0.72} \ln \frac{84}{1.15 \times 10^{-2}}\right] = 230{,}000 \text{ g mol}^{-1}$$

21.12 Given that the intrinsic viscosity of myosin is 217 cm^3 g^{-1}, approximately what concentration of myosin in water would have a relative viscosity of 1.5?

SOLUTION

$$\frac{\frac{\eta}{\eta_o} - 1}{c} = 217 \text{ cm}^3 \text{ g}^{-1}$$

$$\frac{0.5}{c} = 217 \text{ cm}^3 \text{ g}^{-1}$$

$$c = \frac{0.5}{217 \text{ cm}^3 \text{ g}^{-1}} = 2.30 \times 10^{-3} \text{ g cm}^{-3}$$

21.13 The sedimentation coefficient of myoglobin at 20 °C is 2.06×10^{-13} s. What molar mass would it have if the molecules were spherical? Given: $\upsilon = 0.749 \times 10^{-3}$ $\text{m}^3 \text{ kg}^{-1}$, $\rho = 0.9982 \times 10^3 \text{ kg m}^{-3}$, and $\eta = 0.001\,005$ Pa s.

SOLUTION

$$S = \frac{M(1 - \upsilon\rho)}{N_A f} = \frac{M(1 - \upsilon\rho)}{N_A 6\pi\eta}\left(\frac{4\pi N_A}{3M\upsilon}\right)^{1/3}$$

$$M = \left(\frac{N_A 6\pi\eta S}{1 - \upsilon\rho}\right)^{3/2}\left(\frac{3\upsilon}{4\pi N_A}\right)^{1/2}$$

$$M = \left[\frac{(6.022 \times 10^{23} \text{ mol}^{-1})(6\pi)(0.001\,005)(2.06 \times 10^{-13})}{1 - (0.749 \times 10^{-3})(0.9982 \times 10^3)}\right]^{3/2}$$

$$\times \left(\frac{3 \times 0.749 \times 10^{-3}}{4\pi 6.022 \times 10^{23}}\right)^{1/2}$$

$$= 15.5 \text{ kg mol}^{-1} = 15{,}500 \text{ g mol}^{-1}$$

21.14 The sedimentation and diffusion coefficients for hemoglobin corrected to 20 °C in water are 4.41×10^{-13} s and $6.3 \times 10^{-11} \text{ m}^2 \text{ s}^{-1}$, respectively. If $\upsilon = 0.749 \text{ cm}^3 \text{ g}^{-1}$ and $\rho_{H_2O} = 0.998 \text{ g cm}^{-3}$ at this temperature, calculate the molar mass of the protein. If there is 1 mol of iron per 17,000 g of protein, how many atoms of iron are there per hemoglobin molecule?

SOLUTION

$$M = \frac{RTS}{D(1 - \upsilon\rho)}$$

$$= \frac{(8.31 \text{ J K}^{-1} \text{ mol}^{-1})(203 \text{ K})(4.41 \times 10^{-13} \text{ s})}{(6.3 \times 10^{-11} \text{ m}^2 \text{ s}^{-1})[1 - (0.749)(0.998)]}$$

$$= 67.6 \text{ kg mol}^{-1} = 67{,}600 \text{ g mol}^{-1}$$

$$\frac{67\,000 \text{ g mol}^{-1}}{17\,000 \text{ g (g-atom Fe)}^{-1}} = 4 \text{ g-atom Fe mol}^{-1} = 4 \text{ Fe per molecule}$$

21.15 Given the diffusion coefficient for sucrose at 20 °C in water ($D = 45.5 \times 10^{-11}$ $m^2\ s^{-1}$), calculate its sedimentation coefficient. The partial specific volume v is 0.630 $cm^3\ g^{-1}$.

SOLUTION

$$S = \frac{MD(1 - \upsilon\rho)}{RT}$$

$$= \frac{(342 \times 10^{-3}\ \text{kg mol}^{-1})(45.5 \times 10^{-11}\ \text{m}^2\ \text{s}^{-1})[1 - (0.630\ \text{cm}^3\ \text{g}^{-1})(1\ \text{g cm}^{-3})]}{(8.31\ \text{J K}^{-1}\ \text{mol}^{-1})(293\ \text{K})}$$

$= 0.236 \times 10^{-13}$ s

21.16 A beam of sodium D light (589 nm) is passed through 100 cm of an aqueous solution of sucrose containing 10 g sucrose per 100 cm^3. Calculate I/I_o, where I_o is the intensity that would have been obtained with pure water, given that M = 342.30 g mol^{-1} and $dn/dc = 0.15\ g^{-1}\ cm^3$ for sucrose. The refractive index of water at 20 °C is 1.333 for the sodium D line.

SOLUTION

$$\tau = \frac{32\pi^3 n_o^2 (dn/dc)^2 Mc}{3N_A\lambda^4}$$

$$= \frac{32\pi^3(1.333)^2(0.15\ \text{g}^{-1}\ \text{cm}^3)^2(342.30\ \text{g mol}^{-1})(0.1\ \text{g cm}^{-3})}{3(6.02 \times 10^{23}\ \text{mol}^{-1})(5.89 \times 10^{-5}\ \text{cm}^4)}$$

$= 6.25 \times 10^{-5}\ cm^{-1}$

$$\frac{I}{I_o} = e^{-\tau x} = e^{-(6.25 \times 10^{-5}\ \text{cm}^{-1})(100\ \text{cm})} = 0.9938$$

21.17 (a) Integrate equation 21.41 to obtain S as a function of t_1, t_2, r_1, and r_2. (b) In an ultracentrifuge experiment, the sedimenting boundary of a protein is 5.949 cm from the axis at t_1 and 6.731 cm from the axis at t_1 + 70 minutes. If the speed of the rotor is 50 400 rpm, what is the sedimentation coefficient of the protein?

SOLUTION

(a) $$\int_{t_1}^{t_2} S dt = \frac{1}{\omega^2}\int_{r_1}^{r_2} d\ln r$$

$$S = \frac{1}{\omega^2(t_2 - t_1)}\ln\left(\frac{r_2}{r_1}\right)$$

(b) $\omega^2 = (2\pi(50,400/60))^2 = 2.79 \times 10^7 s^{-2}$

$$S = \frac{1}{(2.79 \times 10^7 s^{-2})(60 \times 70s)} \ln\left(\frac{6.731}{5.949}\right) = 10.5 \times 10^{-13} s$$

21.18 (a) 150 000 g mol^{-1}, (b) 167 000 g mol^{-1}

21.19 (a) 36.8 kg mol^{-1}, (b) 48.5 kg mol^{-1}

21.20 0.0182 bar

21.21 (a) 20, (b) 0.0189, (c) 0.0189

21.22 20 M_o , 39.0 M_o

21.23 (a) 20,000 g mol^{-1}, (b) 39,800 g mol^{-1}

21.24 8.36 x 10^6 g mol^{-1}

21.25 638,000 g mol^{-1}

21.26 1.9 cm^3 g^{-1}

21.27 $c_2/c_1 = 2.05$

21.28 177 x 10^{-13} s

21.29 0.33 cm

21.30 64,000 g mol^{-1}

21.31 511,000 g mol^{-1}

21.32 100,000 g mol^{-1}

21.33 27.7 x 10^6 g mol^{-1}

21.34 1.72 x 10^4 g mol^{-1}

21.35 19,500 g mol^{-1}, 0.9980

22

Electric and Magnetic Properties of Molecules

22.1 If a molecule has two groups with dipole moments μ_1 and μ_2, the square of the dipole moment of the molecule is given by

$\mu^2 = \mu_1{}^2 + \mu_2{}^2 + 2\mu_1\mu_2 \cos\theta$

where θ is the angle between the vectors. Show that when the dipole moments of the groups are equal

$\mu = 2\mu_1 \cos(\theta/2)$

Given: There is a trigonometric identity $\cos 2x = 2\cos^2 x - 1$.

SOLUTION

If the dipole moments of the groups have the same magnitude

$\mu^2 = 2\mu_1{}^2(1 + \cos\theta)$

The trignometric identity shows that

$1 + \cos\theta = 2\cos^2(\theta/2)$

Thus

$\mu = 2\mu_1 \cos(\theta/2)$

22.2 Calculate the SI units of the electric susceptibility χ from its definition in equation 22.6.

SOLUTION

$$\chi = \frac{P}{\varepsilon_0 E}$$

The polarizability P has the units of dipole moment per unit volume or C m m^{-3} = C m^{-2}. The permittivity has the units of C^2 N^{-1} m^{-2}, and the electric field has the units of V m^{-1} or J C^{-1} m^{-1}. Substituting these units in the above equation yields unity, and so the electric susceptibility is dimensionless.

22.3 The relative permittivity of HI(g) at 1 atm and 273 K is 1.002 34. Given that its dipole moment is 1.40×10^{-30} C m, calculate its polarizability.

SOLUTION

$$\frac{\varepsilon_r - 1}{\varepsilon_r + 2} = \frac{0.002\,34}{3.002\,34} = 7.79 \times 10^{-4}$$

The factor on the right is given by Example 22.1:

$$\frac{N_A P}{RT}\frac{1}{3\varepsilon_0} = 1.0119 \times 10^{36}\ \mathrm{C^{-2}\ m^{-2}\ J}$$

The contribution from orientation polarization is given by

$$\frac{\mu^2}{3kT} = \frac{(1.40 \times 10^{-30}\ \mathrm{C\ m})^2}{3(1.381 \times 10^{-23}\ \mathrm{J\ K^{-1}})(273\ \mathrm{K})}$$

$$= 1.73 \times 10^{-40}\ \mathrm{C^2\ m^2\ J^{-1}}$$

$$\alpha = \frac{7.79 \times 10^{-4}}{1.0119 \times 10^{36}\ \mathrm{C^{-2}\ m^{-2}\ J}} - 1.73 \times 10^{-40}\ \mathrm{C^2\ m^2\ J^{-1}}$$

$$= 5.97 \times 10^{-40}\ \mathrm{C^2\ m^2\ J^{-1}}$$

Table 22.2 gives $6.06 \times 10^{-40}\ \mathrm{C^2\ m^2\ J^{-1}}$.

22.4 Given that the mean electric polarizability α of CH_4 is $2.90 \times 10^{-40}\ \mathrm{J^{-1}\ C^2\ m^2}$, express α' in units of a_0^3, where a_0 is the Bohr radius. Compare the volume α' with the volume corresponding with the molecular diameter of CH_4 obtained from kinetic theory (0.414 nm in Table 17.4).

SOLUTION

$$\alpha' = \frac{\alpha}{4\pi\varepsilon_0} = \frac{2.90 \times 10^{-40}\ \mathrm{C^2\ m^2\ J^{-1}}}{4\pi(8.854 \times 10^{-12}\ \mathrm{J^{-1}\ C^2\ m^{-1}})} = 2.61 \times 10^{-30}\ \mathrm{m^3}$$

$$a_0^3 = (0.529 \times 10^{-10}\ \mathrm{m})^3 = 1.48 \times 10^{-31}\ \mathrm{m^3}$$

The volume α' expressed in units of a_0^3 is

$$\frac{\alpha'}{a_0^3} = \frac{2.61 \times 10^{-30}\ \mathrm{m^3}}{1.48 \times 10^{-31}\ \mathrm{m^3}} = 17.6$$

The volume of a methane molecule calculated from the collision diameter is

$$\frac{4}{3}\pi\left(\frac{d}{2}\right)^3 = \frac{4}{3}\pi\left(\frac{0.414 \times 10^{-9}\ \mathrm{m}}{2}\right)^3 = 3.72 \times 10^{-29}\ \mathrm{m^3}$$

22.5 The magnetic susceptibility of molecular oxygen at 1 bar and 300 K is 1.9×10^{-6}. What does this tell us about the spin of an oxygen molecule?

SOLUTION

Replacing N/V in equation 22.39 with its value, N_AP/RT, for an ideal gas yields

$$\chi_{mag} = \frac{N_A P}{RT}\frac{\mu_0}{3kT}\ S(S+1)g_e^2\mu_B^2$$

$$S(S+1) = \chi_{mag} \frac{RT}{N_A P} \frac{1}{g_e{}^2 \mu_B{}^2} \frac{3kT}{\mu_0}$$

$$= 1.9 \times 10^{-6} \frac{(8.314 \text{ J K}^{-1} \text{ mol}^{-1})(300 \text{ K})}{(6.022 \times 10^{23} \text{ mol}^{-1})(10^5 \text{ N m}^{-2})} \frac{(9.89 \times 10^{-15} \text{ m A}^2)}{(3.45 \times 10^{-46} \text{ J}^2 \text{ T}^{-2})}$$

$$= 2.2$$

Since $O_2(g)$ has two unpaired electrons, each with spin 1/2, the spin S is expected to be 1, and $S(S + 1)$ is expected to be 2. The value of the magnetic susceptibility indicates a small contribution from orbital angular momentum.

*22.6 Given that the molar magnetic susceptibility of NO(g) is 2.0×10^{-8} m^3 mol^{-1} at 293 K and 1 bar, calculate the total spin quantum number S.

SOLUTION

The molar magnetic susceptibility for an ideal gas is given by

$$\chi_{mag,m} = \frac{N_A \mu_0 S(S+1) g_e{}^2 \mu_B{}^2}{3kT}$$

Solving for $S(S + 1)$ yields

$$S(S+1) = (2.0 \times 10^{-8} \text{ m}^3 \text{ mol}^{-1}) \left(\frac{3kT}{\mu_0}\right)\left(\frac{1}{N_A g_e{}^2 \mu_B{}^2}\right)$$

Since $3kT/\mu_0 = 9.65 \times 10^{-15}$ m A^2 and $N_A g_e{}^2 \mu_B{}^2 = 2.08 \times 10^{-22}$ J^2 T^{-2} mol^{-1},

$$S(S+1) = (2.0 \times 10^{-8} \text{ m}^3 \text{ mol}^{-1}) \frac{(9.65 \times 10^{-15} \text{ m A}^2)}{(2.08 \times 10^{-22} \text{ J}^2 \text{ T}^{-2} \text{ mol}^{-1})} = 0.93$$

Thus $S^2 + S - 0.93 = 0$, and

$$S = \frac{-1 + [1 + 4(0.63)]^{1/2}}{2} = 0.59$$

As shown by Table 14.2, NO(g) has two closely spaced levels ($\Omega = 1/2$ and $\Omega = 3/2$, where Ω is the total angular momentum along the internuclear axis for a diatomic molecule) at the ground state.

*22.7 Use the molar magnetic susceptibility of $MnSO_4 \cdot 4H_2O$ to calculate the total spin quantum number at 293 K.

SOLUTION

$$\chi_{mag,m} = \frac{N_A \mu_0 S(S+1) g_e{}^2 \mu_B{}^2}{3kT} = 18.1 \times 10^{-8} \text{ m}^3 \text{ mol}^{-1}$$

$$S(S+1) = (18.1 \times 10^{-8} \text{ m}^3 \text{ mol}^{-1}) \frac{(9.82 \times 10^{-15} \text{ m A}^2)}{(6.02 \times 10^{23} \text{ mol}^{-1})(3.45 \times 10^{-46} \text{ J}^2 \text{ T}^{-2})}$$

$$= 8.56$$

$S^2 + S - 8.56 = 0$

Using the quadratic formula yields

$$S = \frac{-1 + [1 + 4(8.56)]^{1/2}}{2} = 2.47$$

This corresponds with a spin of 5/2.

22.8 0.896×10^{-40} C^2 m^2 J^{-1}

22.9 5.29×10^{-30} C m

22.11 4.48 K

22.12 3.8×10^{-6} at 200 K and 0.15×10^{-6} at 1000 K

23

Solid-State Chemistry

23.1 What is the equation for the distances between planes 110 for a crystal with mutually perpendicular axes?

SOLUTION

$$\frac{1}{d} = \left(\frac{1}{a^2} + \frac{1}{b^2}\right)^{1/2}$$

$$d = \left(\frac{1}{a^2} + \frac{1}{b^2}\right)^{-1/2} = \left(\frac{b^2 + a^2}{a^2b^2}\right)^{-1/2} = \frac{ab}{(a^2 + b^2)^{1/2}}$$

23.2 Calculate the angles at which the first-, second-, and third-order reflections are obtained from planes 500 pm apart, using X rays with a wavelength of 100 pm.

SOLUTION

$$\sin\theta = \frac{\lambda}{2d_{hkl}} = \frac{\lambda}{2(d/n)}$$

$$\sin\theta_1 = \frac{100\text{ pm}}{2(500\text{ pm}/1)} \qquad \theta_1 = 5.74°$$

$$\sin\theta_2 = \frac{100\text{ pm}}{2(500\text{ pm}/2)} \qquad \theta_2 = 11.54°$$

$$\sin\theta_3 = \frac{100\text{ pm}}{2(500\text{ pm}/3)} \qquad \theta_3 = 17.46°$$

23.3 Calculate the structure factor for a cubic unit cell of AB in which the B atoms occupy the body centered position. Which reflections will be strong and which weak?

SOLUTION

$$F_{hkl} = f_A + f_B\, e^{2\pi i(h/2 + k/2 + l/2)}$$

$$= f_A + f_B\, e^{\pi i(h + k + l)}$$

$$= f_A + f_B(-1)^{h + k + l}$$

If $h + k + l$ is even, $F_{hkl} = f_A + f_B$, strong reflections.

If $h + k + l$ is odd, $F_{hkl} = f_A - f_B$, weak reflections.

23.4 The crystal unit cell of magnesium oxide is a cube 420 pm on an edge. The structure is interpenetrating face centered. What is the density of crystalline MgO?

SOLUTION

$$\frac{4(40.32 \times 10^{-3} \text{ kg mol}^{-1})}{(6.022 \times 10^{23} \text{ mol}^{-1})(4.20 \times 10^{-10} \text{ m})^3} = 3.615 \times 10^3 \text{ kg m}^{-3} = 3.615 \text{ g cm}^{-3}$$

23.5 Platinum forms face-centered cubic crystals. If the radius of a platinum atom is 139 pm, what is the length of the side of the unit cell? What is the density of the crystal?

SOLUTION

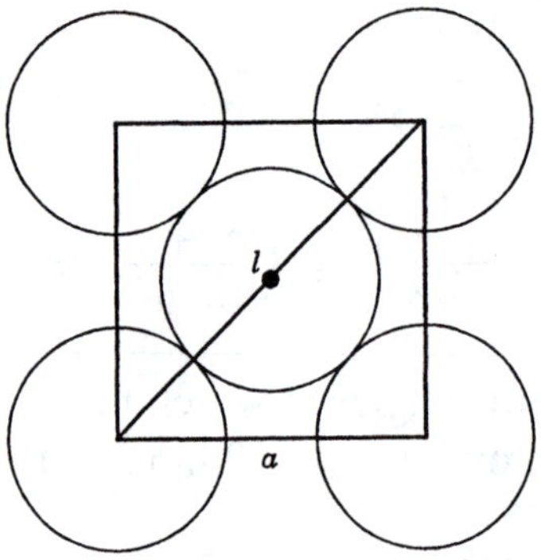

$$l^2 = 2a^2 = [4(139 \text{ pm})]^2$$

$$a = \frac{1}{2^{1/2}} [4(139 \text{ pm})] = 393 \text{ pm}$$

$$d = \frac{4(195.08 \times 10^{-3} \text{ kg mol}^{-1})}{(6.022 \times 10^{23} \text{ mol}^{-1})(393 \times 10^{-12} \text{ m})^3}$$

$$= 21.32 \times 10^3 \text{ kg m}^{-3}$$

23.6 Tungsten forms body-centered cubic crystals. From the fact that the density of tungsten is 19.3 g cm^{-3}, calculate (a) the length of the side of this unit cell and (b) d_{200}, d_{110}, and d_{222}.

SOLUTION

(a) $$19.3 \times 10^3 \text{ kg m}^{-3} = \frac{2(183.85 \times 10^{-3} \text{ kg mol}^{-1})}{(6.022 \times 10^{23} \text{ mol}^{-1})a^3}$$

$$a = 316 \times 10^{-12} \text{ m} = 316 \text{ pm}$$

(b) $$d_{200} = \frac{a}{\sqrt{2^2 + 0 + 0}} = \frac{316 \text{ pm}}{2} = 158 \text{ pm}$$

$$d_{110} = \frac{a}{\sqrt{1 + 1 + 0}} = \frac{316 \text{ pm}}{\sqrt{2}} = 223 \text{ pm}$$

$$d_{222} = \frac{a}{\sqrt{4 + 4 + 4}} = \frac{316 \text{ pm}}{\sqrt{12}} = 91.2 \text{ pm}$$

23.7 (a) Metallic iron at 20 °C is studied by the Bragg method, in which the crystal is oriented so that a reflection is obtained from the planes parallel to the sides of the cubic crystal, then from planes cutting diagonally through opposite edges, and finally from planes cutting diagonally through opposite corners. Reflections are first obtained at $\theta = 11°\ 36'$, $8°\ 3'$, and $20°\ 26'$, respectively. What type of cubic lattice does iron have at 20 °C? (b) Metallic iron also forms cubic crystals at 1100 °C, but the reflections determined as described in (a) occur at $\theta = 9°\ 8'$, $12°\ 57'$, and $7°\ 55'$, respectively. What type of cubic lattice does iron have at 1100 °C? (c) The density of iron at 20 °C is 7.86 g cm^{-3}. What is the length of a side of the unit cell at 20 °C? (d) What is the wavelength of the X rays used? (e) What is the density of iron at 1100 °C?

SOLUTION

(a)(b) The three orientations will give reflections from planes $a00$, $bb0$, and ccc, respectively, where a, b and c are small integers. The smallest suitable integers are as follows: primitive lattice; $a = b = c = 1$; body-centered lattice; $b = 1$, $a = c = 2$ (100 and 111 reflections are missing); face-centered lattice; $c = 1$, $a = b = 2$.

$$\sin \theta = \frac{\lambda}{2d_{hkl}} = \frac{\lambda\sqrt{h^2 + k^2 + l^2}}{2a}$$

Since $\lambda/2a$ is constant for a particular experiment, the three lattice types can be distinguished simply on the basis of the relative magnitudes of the three θ values. Thus the crystal at 20 °C is body-centered since the second angle is smaller than the first and the third is the largest of all (ratio of sin θ is 2: $\sqrt{2}$: $\sqrt{12}$). At 1100 °C the face-centered form is found (ratio of sin θ is 2:$\sqrt{8}$:$\sqrt{3}$). Note that a primitive lattice would give ratios of sin θ of 1:$\sqrt{2}$:$\sqrt{3}$.

(c) $$d = 7.86 \times 10^3 \text{ kg m}^{-3} = \frac{2(55.847 \times 10^{-3} \text{ kg mol}^{-1})}{(6.022 \times 10^{23} \text{ mol}^{-1})a^3}$$

$$a = \left[\frac{(2)(55.847 \times 10^{-3} \text{ kg mol}^{-1})}{(6.022 \times 10^{23} \text{ mol}^{-1})(7.86 \times 10^3 \text{ kg m}^{-3})}\right]^{1/3}$$

$$= 286.8 \times 10^{-12} \text{ m} = 286.8 \text{ pm}$$

(d) $\lambda = 2d_{hkl} \sin\theta = \dfrac{2a \sin\theta}{\sqrt{h^2 + k^2 + l^2}}$

$\dfrac{573.6 \text{ pm}}{2} \sin 11° \; 36' = 57.7 \text{ pm}$

$\dfrac{573.6 \text{ pm}}{\sqrt{2}} \sin 8° \; 3' = 56.8 \text{ pm}$

$\dfrac{573.6 \text{ pm}}{\sqrt{12}} \sin 20° \; 26' = 57.8 \text{ pm}$ Average = 57.4 pm

(e) $d_{hkl} = \dfrac{\lambda}{2 \sin\theta}$

$d_{200} = \dfrac{57.4 \text{ pm}}{2 \sin 9° \; 8'} = 180.8 \text{ pm} \quad a = d_{200}\sqrt{4} = 361.6 \text{ pm}$

$d_{220} = \dfrac{57.4 \text{ pm}}{2 \sin 12° \; 57'} = 128.1 \text{ pm} \quad a = d_{220}\sqrt{8} = 362.3 \text{ pm}$

$d_{111} = \dfrac{57.4 \text{ pm}}{2 \sin 7° \; 55'} = 208.4 \text{ pm} \quad a = d_{111}\sqrt{3} = 360.9 \text{ pm}$

Average = 361.6 pm

$$d = \frac{4(55.847 \times 10^{-3} \text{ kg mol}^{-1})}{(6.022 \times 10^{23} \text{ mol}^{-1})(361.6 \times 10^{-12} \text{ m})^3}$$
$= 7.846 \times 10^3 \text{ kg m}^{-3}$

23.8 Cesium chloride, bromide, and iodide form interpenetrating simple cubic crystals instead of interpenetrating face-centered cubic crystals like the other alkali halides. The length of the side of the unit cell of CsCl is 412.1 pm. (a) What is the density? (b) Calculate the ionic radius of Cs^+, assuming that the ions touch along a diagonal through the unit cell and that the ion radius of Cl^- is 181 pm.

SOLUTION

(a) $$d = \frac{(168.36 \times 10^{-3} \text{ kg mol}^{-1})}{(6.022 \times 10^{23} \text{ mol}^{-1})(412.1 \times 10^{-12} \text{ m})^3}$$
$= 3.995 \times 10^3 \text{ kg m}^{-3}$

(b) The diagonal plane through the cubit unit cell is as follows:

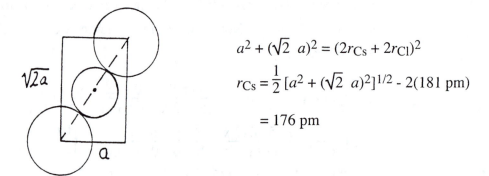

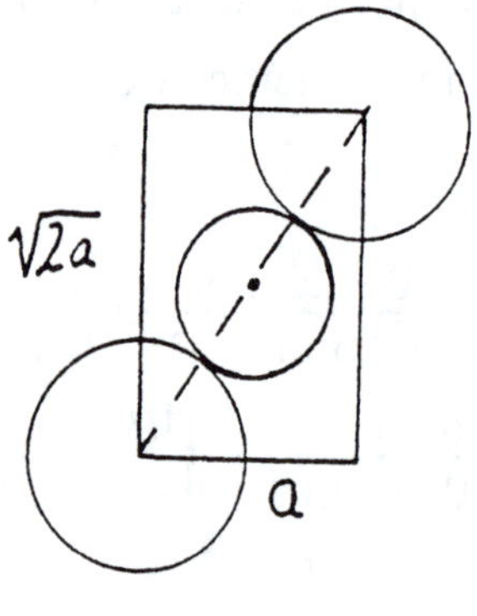

$a^2 + (\sqrt{2}\; a)^2 = (2r_{Cs} + 2r_{Cl})^2$

$r_{Cs} = \frac{1}{2}[a^2 + (\sqrt{2}\; a)^2]^{1/2} - 2(181 \text{ pm})$

$= 176 \text{ pm}$

23.9 Deslattes et al. [*Phys. Rev. Lett.* **33**: 463 (1974)] found the following values for a single crystal of very pure silicon at 25 °C: $\rho = 2.328\,992$ g cm^{-3}, $a = 543.1066$ pm. Silicon has a face-centered cubic lattice like diamond. The atomic mass is 28.085 41 g mol^{-1}. What value of Avogadro's constant is obtained from these values?

SOLUTION

$$\rho = \frac{nM}{N_A a^3}$$

$$N_A = \frac{nM}{\rho a^3}$$

$$= \frac{(8)(28.085\,41 \times 10^{-3} \text{ kg mol}^{-1})}{(2.328\,992 \times 10^3 \text{ kg m}^{-3})(543.106\,6 \times 10^{-12} \text{ m})^3}$$

$$= 6.022\,093 \times 10^{23} \text{ mol}^{-1}$$

23.10 Insulin forms crystals of the orthorhombic type with unit-cell dimensions of 13.0 x 7.48 x 3.09 nm. If the density of the crystal is 1.315 g cm^{-3} and there are six insulin molecules per unit cell, what is the molar mass of the protein insulin?

SOLUTION

$$d = 1.315 \times 10^3 \text{ kg m}^{-3}$$

$$= \frac{6M}{(6.022 \times 10^{23} \text{ mol}^{-1})(13.0 \times 7.48 \times 3.09 \times 10^{-27} \text{ m}^3)}$$

$$M = \frac{1}{6}(1.315 \times 10^3 \text{ kg m}^{-3})(6.022 \times 10^{23} \text{ mol}^{-1})(13.0 \times 7.48 \times 3.09 \times 10^{-27} \text{ m}^3))$$

$$= 39.7 \text{ kg mol}^{-1} = 39{,}700 \text{ g mol}^{-1}$$

23.11 Molybdenum forms body-centered cubic crystals and, at 20 °C, the density is 10.3 g cm^{-3}. Calculate the distance between the centers of the nearest molybdenum atoms.

SOLUTION

$$10.3 \times 10^3 \text{ kg m}^3 = \frac{2(95.94 \times 10^{-3} \text{ kg mol}^{-1})}{(6.022\,045 \times 10^{23} \text{ mol}^{-1})a^3}$$

$$a = 314 \times 10^{-12} \text{ m} = 314 \text{ pm}$$

Consider a plane bisecting the cube and cutting diagonally through opposite faces.

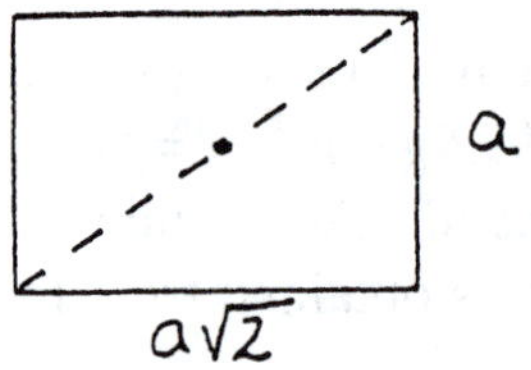

length of diagonal = $\sqrt{a^2 + 2b^2} = a\sqrt{3}$

distance between atoms = $\frac{a\sqrt{3}}{2}$ = 273 pm

23.12 Silicon has a face-centered cubic structure with two atoms per lattice point, just like diamond. At 25 °C, a = 543.1 pm. What is the density of silicon?

SOLUTION

$$n = 8 \text{ x } \frac{1}{8} + 6 \text{ x } \frac{1}{2} + 4 = 8$$

= corners + faces + interstitial

$$d = \frac{8(28.0855 \text{ x } 10^{-3} \text{ kg mol}^{-1})}{(6.022 \text{ x } 10^{23} \text{ mol}^{-1})(5.431 \text{ x } 10^{-10} \text{ m})^3} = 2.33 \text{ x } 10^3 \text{ kg m}^{-3}$$

23.13 The common form of ice has a tetrahedral structure with protons located on the lines between oxygen atoms. A given proton is closer to one oxygen atom than the other and is said to belong to the closer oxygen atom. How many different orientations of a water molecule in space are possible in this lattice?

SOLUTION

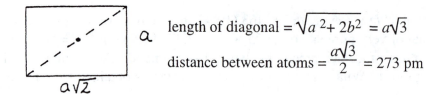

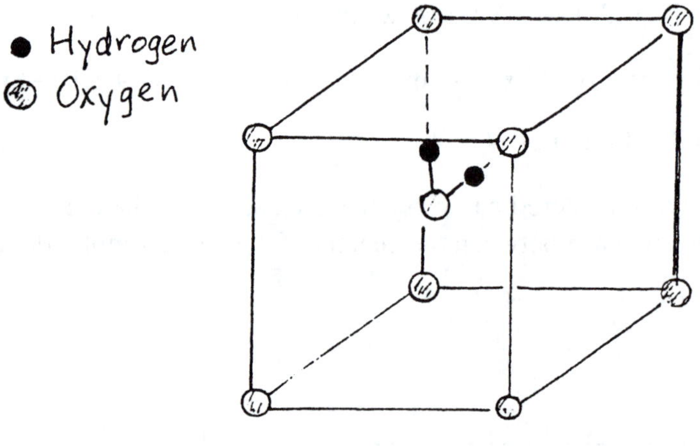

On each of the 6 faces there are 2 pairs of oxygens so that the H's can point to 12 possible orientations

23.14 Diamond has a face-centered cubic crystal lattice, and there are eight atoms in a unit cell. Its density is 3.51 g cm^{-3}. Calculate the first six angles at which reflections would be obtained using an X-ray beam of wavelength 71.2 pm.

SOLUTION

$$d = 3.51 \times 10^3 \text{ kg m}^{-3} = \frac{8(12.011 \times 10^{-3} \text{ kg mol}^{-1})}{(6.022 \times 10^{23} \text{ mol}^{-1})a^3}$$

$a = 357 \times 10^{-12}$ m = 357 pm

$$\theta = \sin^{-1} \frac{\lambda}{2a}\sqrt{h^2 + k^2 + l^2} = \sin^{-1} \frac{71.2}{2(357)}\sqrt{h^2 + k^2 + l^2}$$

For face-centered cubic the first six reflections are:

hkl	θ
111	9.95
200	11.50°
220	16.38°
311	19.31°
222	20.21°
400	23.51°

23.15 Derive the structure factor for a body-centered cubic unit cell of identical atoms. The relative coordinates of the lattice points are given by (0,0,0), (1,0,0), (0,1,0), (0,0,1), (1,1,0), (1,0,1), (0,1,1), (1,1,1), and (1/2,1/2,1/2).

SOLUTION

The atoms in the corners are each shared with eight other unit cells, and so the contributions of each corner atom in the sum in equation 23.22 should be multiplied by (1/8).

$$F(h,k,l) = \frac{1}{8} f[e^{2\pi i(0)} + e^{2\pi i(h)} + e^{2\pi i(k)} + e^{2\pi i(l)} + e^{2\pi i(h+k)} + e^{2\pi i(k+l)} + e^{2\pi i(h+l)} + e^{2\pi i(h+k+l)}]$$
$$+ f[e^{2\pi i(h/2+k/2+l/2)}]$$

Since $e^{2\pi i} = \cos 2\pi + i \sin 2\pi = 1$ and $e^{\pi i} = -1$,

$$F(h,k,l) = \frac{1}{8} f[1^0 + 1^h + 1^k + 1^l + 1^{h+k} + 1^{k+l} + 1^{h+l} + 1^{h+k+l}] + f(-1)^{h+k+l}$$
$$= f[1 + (-1)^{h+k+l}]$$

When $h+k+l$ is an even number, $F(h,k,l) = 2f$. When $h+k+l$ is odd, $F(h,k,l) = 0$.

This shows that there will be reflections for a body-centered unit cell only when h + k + l is even.

23.16 Calculate the ratio of the radii of small and large spheres for which the small spheres will just fit into octahedral sites in a close-packed structure of the large spheres.

SOLUTION

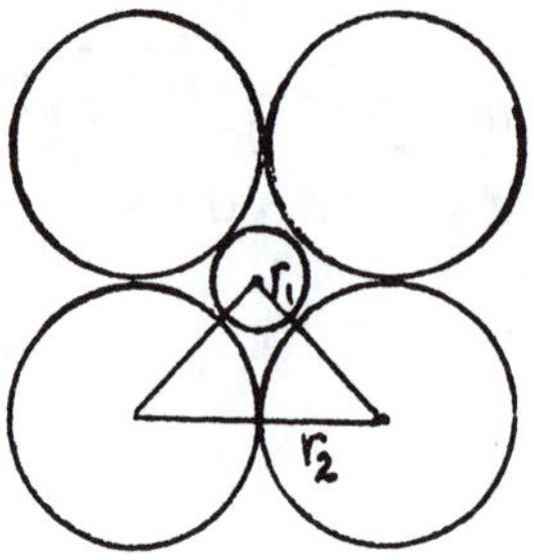

$$2(r_1 + r_2)^2 = (2r_2)^2$$

$$r_1 + r_2 = \frac{2r_2}{\sqrt{2}} = \sqrt{2}\ r_2$$

$$\frac{r_1}{r_2} + 1 = \sqrt{2}$$

$$\frac{r_1}{r_2} = 0.414$$

23.17 A close-packed structure of uniform spheres has a cubic unit cell with a side of 800 pm. What is the radius of the spherical molecule?

SOLUTION

Since the cubic close-packed structure is face-centered cubic, a face diagonal is equal to four radii.

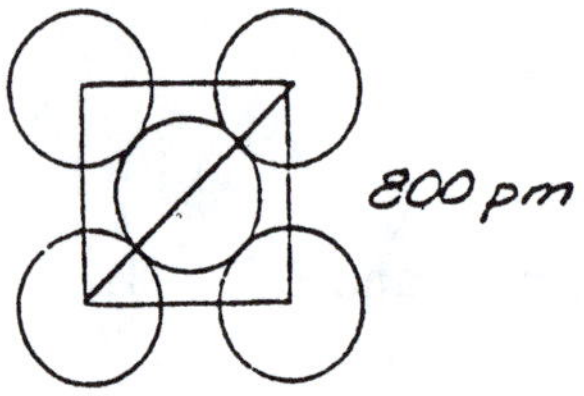

$$\text{Length of face diagonal} = \sqrt{128 \times 10^4}$$

$$\text{Radius of sphere} = \frac{\sqrt{128 \times 10^4}}{4} = 282.8 \text{ pm}$$

23.18 Titanium forms hexagonal close-packed crystals. Given the atomic radius of 146 pm, what are the unit cell dimensions, and what is the density of crystals?

SOLUTION

$a = b = 2(146 \text{ pm}) = 292 \text{ pm}$ $\qquad c = \sqrt{2}\ 2\, a/\sqrt{3} = 477 \text{ pm}$

$V = a^2 c(1 - \cos^2 \gamma)^{1/2} = a^2 c \sin \gamma$

$= (292 \text{ pm})^2 (477 \text{ pm}) \sin 120° = 35.22 \times 10^{-30} \text{ m}^3$

$$d = \frac{2(47.90 \times 10^{-3} \text{ kg mol}^{-1})}{(6.022 \times 10^{23} \text{ mol}^{-1})(35.22 \times 10^{-30} \text{ m}^3)}$$

$= 4.517 \times 10^3 \text{ kg m}^{-3}$

23.19 What neutron energy in electron volts is required for a wavelength of 100 pm?

SOLUTION

$$\text{Kinetic energy} = \frac{p^2}{2m_n} = \frac{h^2}{\lambda^2 2m_n} \qquad \text{since } \lambda = h/p$$

$$Ee = \frac{h^2}{2\lambda^2 m_n}$$

$$E = \frac{h^2}{2\lambda^2 m_n e}$$

$$= \frac{(6.626 \times 10^{-34} \text{ J s})^2}{2(10^{-10} \text{ m})^2(1.675 \times 10^{-27} \text{ kg})(1.602 \times 10^{-19} \text{ C})}$$

$= 0.0818 \text{ V}$ or 0.0818 eV

23.20 236.5, 193.1, 149.6, 136.6 pm

23.21 7

23.22 When $h + k$ is odd, the structure factor $F(hkl)$ is zero and the reflection is extinguished.

23.23 392.4 pm

23.24 74.69 g mol^{-1}

23.25 2826 kg m^{-3}

23.26 3.516 x 10^3 kg m^{-3}

23.27 (a) 2, (b) 328, (c) 164, (d) 232, (e) 94.7 pm

23.28 287, 124 pm

23.29 8.84 g cm^{-3}

23.30 (a) 2.698 x 10^3 kg m^{-3}, (b) 202.5, 143.2, 233.8 pm

23.31 152 pm

23.32 1.295 x 10^{19}, 9.157 x 10^{18}, 1.495 x 10^{19} m^{-2}

23.33 1414.2, 1154.7 pm

23.34 200.3 pm

23.35 184 pm

24

Surface Dynamics

24.1 In an ultrahigh vacuum chamber ($P = 5 \times 10^{-10}$ torr), how many molecules strike 1 cm^2 of surface in one second at 298 K, if the gas is (a) helium and (b) mercury vapor?

SOLUTION

$J_n = PN_A/(2\pi MRT)^{1/2}$

$P = (5 \times 10^{-10} \text{ torr})(133.3 \text{ Pa torr}^{-1}) = 6.66 \times 10^{-8} \text{ Pa}$

(a) $$J_n = \frac{(6.66 \times 10^{-8} \text{ Pa})(6.02 \times 10^{23} \text{ mol}^{-1})}{[2\pi(4.00 \times 10^{-3} \text{ kg mol}^{-1})(8.314 \text{ J K}^{-1} \text{ mol}^{-1})(298 \text{ K})]^{1/2}}$$

$$= (5.08 \times 10^{15} \text{ m}^{-2} \text{ s}^{-1})(0.01 \text{ m cm}^{-1})^2$$

$$= 5.08 \times 10^{11} \text{ cm}^{-2} \text{ s}^{-1}$$

(b) $$J_n = (5.08 \times 10^{11} \text{ m}^{-2} \text{ s}^{-1})(4/200.6)^{1/2}$$

$$= 7.17 \times 10^{10} \text{ cm}^{-2} \text{ s}^{-1}$$

24.2 A readily oxidized metal surface with 10^{15} metal atoms per square centimeter is exposed to molecular oxygen at 10^{-5} Pa at 298 K. How long will it take to completely oxidize the surface if the oxide formed is MO?

SOLUTION

$$J_n = \frac{(10^{-5} \text{ Pa})(6.02 \times 10^{23} \text{ mol}^{-1})}{[2\pi(32 \times 10^{-3} \text{ kg mol}^{-1})(8.314 \text{ J K}^{-1})(298 \text{ K})]^{1/2}}$$

$$= (2.7 \times 10^{17} \text{ m}^{-2} \text{ s}^{-1})(0.01 \text{ m cm}^{-1})^2$$

$$= 2.7 \times 10^{13} \text{ cm}^{-2} \text{ s}^{-1}$$

$$t = \frac{10^{15} \text{ cm}^{-2}}{2(2.7 \times 10^{13} \text{ cm}^{-2} \text{ s}^{-1})} = 18.5 \text{ s}$$

This is assuming every collision causes a reaction.

24.3 In problem 17.19, we found that at 1 bar and 298 K, 1.075 x 10^{28} molecules of molecular hydrogen strike a surface per square meter per second. When the 100 plane of metallic copper is exposed to molecular hydrogen under these conditions, what is the rate of collisions with atoms of copper? Copper forms face-centered cubic crystals with the length of the side of the unit cell equal to 361 pm.

SOLUTION

The area of copper atoms in the face of the unit cell is
$(1/2)(361 \times 10^{-12}\ \text{m})^2$ since there are two atoms in the area of a face, and so the rate of collisions is
$(1.074 \times 10^{28}\ \text{m}^{-2}\ \text{s}^{-1})(1/2)(361 \times 10^{-12}\ \text{m})^2 = 7.00 \times 10^8\ \text{s}^{-1}$

24.4 For the adsorption of nitrogen molecules on a certain sample of carbon, the pressure required to half-saturate the surface at 298 K is 2 x 10^{-5} Pa. If the enthalpy of adsorption is -10 kJ mol^{-1} and the sticking coefficient s^* is unity, what is the rate constant of desorption k_d?

SOLUTION

$$k_d = \frac{PN_A}{(2\pi MRT)^{1/2}\exp(\Delta_{ads}H/RT)}$$

$$= \frac{(2 \times 10^{-5}\ \text{Pa})(6.02 \times 10^{23}\ \text{mol}^{-1})\exp(10\,000/8.314 \times 298\ \text{K})}{[2\pi(28 \times 10^{-3}\ \text{kg mol}^{-1})(8.314\ \text{J K}^{-1}\ \text{mol}^{-1})(298\ \text{K})]^{1/2}}$$

$$= 3.26 \times 10^{19}\ \text{m}^{-2}\ \text{s}^{-1}$$

24.5 The pressure of nitrogen required for adsorption of 1.0 $\text{cm}^3\ \text{g}^{-1}$ (25 °C, 1.013 bar) of gas on graphitized carbon black are 24 Pa at 77.5 K and 290 Pa at 90.1 K. Calculate the enthalpy of adsorption at this fraction of surface coverage.

SOLUTION

$$\Delta_{ads}H = \frac{RT_1T_2 \ln (P_1/P_2)}{T_2 - T_1}$$

$$= \frac{(8.314\ \text{J K}^{-1}\ \text{mol}^{-1})(77.5\ \text{K})(90.1\ \text{K}) \ln (24/290)}{(90.1\ \text{K} - 77.5\ \text{K})}$$

$$= -\ 11.6\ \text{kJ mol}^{-1}$$

24.6 A mixture of A and B is adsorbed on a solid for which the adsorption isotherm follows the Langmuir equation. If the mole fractions in the gas at equilibrium are y_A and y_B, what is the equation for the adsorption isotherm in terms of total pressure? What is the expression for the mole fraction of A in the adsorbed gas in terms of K_A, K_B, y_A, and y_B?

SOLUTION

See Example 24.2

$$\upsilon = \upsilon_m(\Theta_A + \Theta_B) = \frac{\upsilon_m(y_A K_A P + y_B K_B P)}{1 + y_A K_A P + y_B K_B P}$$

$$= \frac{\upsilon_m(y_A K_A + y_B K_B)P}{1 + (y_A K_A + y_B K_B)P} = \frac{\upsilon_m K' P}{1 + k' P}$$

The mole fraction of adsorbed A is given by

$$x_A = \frac{\Theta_A}{\Theta_A + \Theta_B} = \frac{y_A K_A P}{y_A K_A P + y_B K_B P} = \frac{y_A K_A}{y_A K_A + y_B K_B}$$

*24.7 The following table gives the volume of nitrogen (reduced to 0 °C and 1 bar) adsorbed per gram of active carbon at 0 °C at a series of pressures.

P/Pa	524	1731	3058	4534	7497
υ/cm^3 g^{-1}	0.987	3.04	5.08	7.04	10.31

Plot the data according to the Langmuir isotherm, and determine the constants.

SOLUTION

$$\frac{1}{\upsilon} = \frac{1}{\upsilon_m} + \frac{1}{\upsilon_m K P}$$

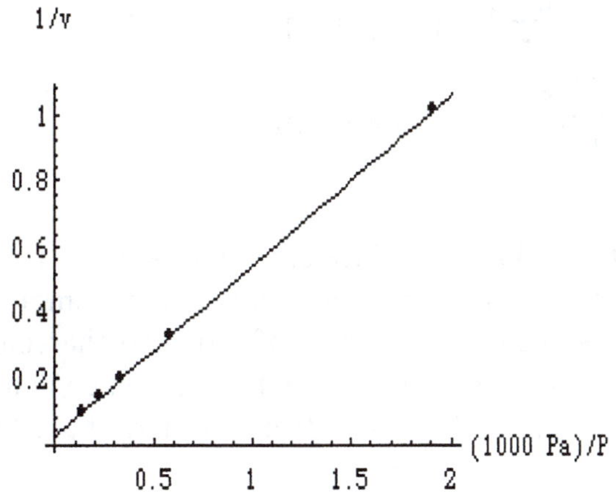

Intercept = 0.025 g cm^{-3}

υ_m = 1/(0.025 g cm^{-3}) = 40 cm^3 g^{-1}

$$\text{Slope} = \frac{(1.00 - 0.025)\text{ g cm}^{-3}}{1.9 \times 10^{-3}\text{ Pa}^{-1}} = 513 \text{ g cm}^{-3}\text{ Pa}$$

$$= \frac{1}{v_m K} = \frac{1}{(40\ \text{cm}^3\ \text{g}^{-1})K}$$

$K = [(513\ \text{g cm}^{-3}\ \text{Pa})(40\ \text{cm}^3\ \text{g}^{-1})]^{-1} = 4.8 \times 10^{-5}\ \text{Pa}^{-1}$

24.8 According to problem 24.7, the Langmuir constant for the adsorption of molecular nitrogen on active carbon at 0 °C is $K = 4.8 \times 10^{-5}\ \text{Pa}^{-1}$. What pressures of molecular nitrogen are required to cover 10%, 50%, and 90% of the surface at 0 °C?

SOLUTION

$P = \Theta/K(1 - \Theta)$

$P = 0.1/(4.8 \times 10^{-5}\ \text{Pa})(0.9) = 2.31 \times 10^3\ \text{Pa}$

$P = 0.5/(4.8 \times 10^{-5}\ \text{Pa})(0.5) = 20.8 \times 10^3\ \text{Pa}$

$P = 0.9/(4.8 \times 10^{-5}\ \text{Pa})(0.1) = 188 \times 10^3\ \text{Pa}$

24.9 According to problem 17.43, the rate with which oxygen molecules strike a surface at 1 bar and 25 °C is $2.69 \times 10^{23}\ \text{cm}^{-2}\ \text{s}^{-1}$. If the oxygen molecules are striking a platinum surface, what are the frequencies of collisions per atom on (100), (110), and (111) planes? (See problem 23.33.)

SOLUTION

100 planes $\dfrac{(2.69 \times 10^{23}\ \text{cm}^{-2}\ \text{s}^{-1})(10^2\ \text{cm m}^{-1})^2}{(1.295 \times 10^{19}\ \text{m}^{-2})} = 2.08 \times 10^8\ \text{s}^{-1}$

110 planes $\dfrac{(2.69 \times 10^{27}\ \text{m}^{-2}\ \text{s}^{-1})}{(9.157 \times 10^{18}\ \text{m}^{-2})}) = 2.94 \times 10^8\ \text{s}^{-1}$

111 planes $\dfrac{(2.69 \times 10^{27}\ \text{m}^{-2}\ \text{s}^{-1})}{(1.495 \times 10^{19}\ \text{m}^{-2})} = 1.80 \times 10^8\ \text{s}^{-1}$

24.10 One gram of activated charcoal has a surface area of 1000 m^2. If complete surface coverage is assumed, as a limiting case, how much ammonia, at 25 °C and 1 bar, could be adsorbed on the surface of 45 g of activated charcoal? The diameter of the NH_3 molecule is 3×10^{-10} m, and it is assumed that the molecules just touch each other in a plane so that four adjacent spheres have their centers at the corners of a square.

SOLUTION

Area per NH_3 molecule $= (3 \times 10^{-10}\ \text{m})^2$

Number of NH_3 molecules required $= \dfrac{45 \times 10^3\ \text{m}^2}{(3 \times 10^{-10}\ \text{m})^2} = 5 \times 10^{23}$

$$V = nRT/P = \left(\frac{5 \times 10^{23}}{6.022 \times 10^{23}\ \text{mol}^{-1}}\right)\frac{(0.08314\ \text{L bar K}^{-1}\ \text{mol}^{-1})(298\ \text{K})}{1\ \text{bar}}$$

$= 20.6$ L

24.11 Calculate the surface area of a catalyst that adsorbs 103 cm^3 of nitrogen (calculated at 1.013 bar and 0 °C) per gram in order to form a monolayer. The adsorption is measured at -195 °C, and the effective area occupied by a nitrogen molecule on the surface is 16.2 x 10^{-20} m^2 at this temperature.

SOLUTION

The surface area per gram is equal to the number of molecules adsorbed per gram times the effective area per molecule. The number of molecules is PVN_A/RT.

$$A_s = \frac{(1.013\ \text{bar})(0.103\ \text{L})(6.022 \times 10^{23}\ \text{mol}^{-1})(16.2 \times 10^{-20}\ \text{m}^2)}{(0.083\ \text{L bar K}^{-1}\ \text{mol}^{-1})(273\ \text{K})}$$

$= 449$ m^2

24.12 The de Broglie wavelength of an electron that has been accelerated through a potential difference of ϕ is given by

$$\lambda = \left(\frac{h^2}{2me\phi}\right)^{1/2}$$

Derive this equation and verify that it is correct to write equation 24.12.

SOLUTION

$(1/2)mv^2 = e\phi$

$(1/2)p^2/m = e\phi$

$p = (2me\phi)^{1/2}$

$$\lambda = h/p = \left(\frac{h^2}{2me\phi}\right)^{1/2}$$

$$= \left(\frac{(6.6261\text{x}10^{-34}\ \text{J s})^2}{2(9.1094\text{x}10^{-31}\ \text{kg})(1.6022\text{x}10^{-19}\ \text{C})\phi}\right)^{1/2}$$

$$= \left(\frac{1.504\text{x}10^{-18}\ \text{V m}^2}{\phi}\right)^{1/2}$$

Note that J = kg m^2 s^{-2} and V = J C^{-1}.

24.13 In LEED experiments, acceleration voltages of 10 to 200 V are generally used. (a) Calculate the energies of electrons accelerated by these voltages in kJ mol^{-1}. (b) Calculate the wavelengths of electrons accelerated by these voltages in nm.

SOLUTION

(a) $\text{Ø} \, e \, N_A/1000 = (10 \text{ V})(1.602\text{x}10^{-19} \text{ C})(6.022\text{x}10^{23} \text{ mol}^{-1})/(1000 \text{ J kJ}^{-1})$
$= 965 \text{ kJ mol}^{-1}$
$20\text{x}965 \text{ kJ mol}^{-1} = 1930 \text{ kJ mol}^{-1}$

(b) wavelength at 10 V = $(1.504/10)^{1/2} = 0.387$ nm
wavelength at 200 V = $(1.504/200)^{1/2} = 0.0867$ nm

24.14 2×10^{-3} L

24.15 - 36.8 kJ mol^{-1}

24.16 - 29.6 kJ mol^{-1}

24.17 (a) 90 Pa, (b) 30 Pa

24.18 $2.70 \times 10^{20} \text{ m}^{-2} \text{ s}^{-1}$

24.20 560 $\text{m}^2 \text{ g}^{-1}$

24.21 48.0 L

SECOND SECTION: PROBLEMS THAT REQUIRE A PERSONAL COMPUTER WITH A MATHEMATICAL APPLICATION

Chapter 1 Zeroth Law of Thermodynamics

1.A Problem 1.7 yields $B = 0.135$ L mol^{-1} and $C = 4.3 \times 10^{-4}$ L^2 mol^{-2} for H_2 (g) at 0 °C. Calculate the molar volumes of molecular hydrogen at 75 and 150 bar and compare these molar volumes with the molar volume of an ideal gas.

SOLUTION

The virial equation in terms of molar volumes is

$P\overline{V}/RT = 1 + B/\overline{V} + C/\overline{V}^2$

Multiplying the virial equation through by $\overline{V}^2$, substituting P = 75 bar, and using Solve yields

```
Off[General::spell1];
Off[General::spell];

Solve[((75 / (.0831451 * 273.15)) * v^3) - (v^2) - (.135 * v) - (4.3 * 10^-4) == 0, v]

{{v → -0.0985587}, {v → -0.003265}, {v → 0.404638}}
```

The first two values are impossible molar volumes because they are negative, and so the answer is 0.405 L mol^{-1}. The molar volume for an ideal gas is given by

```
(.0831451*273.15)/75

0.302814
```

Use of Solve at 150 bar yields

```
Solve[((150 / (.0831451 * 273.15)) * v^3) - (v^2) - (.135 * v) - (4.3 * 10^-4) == 0, v]

{{v → -0.0836472}, {v → -0.0032659}, {v → 0.23832}}
```

The molar volume for the real gas is 0.238 L mol^{-1}, and the molar volume for an ideal gas is

```
(0.0831451 * 273.15) / 150

0.151407
```

Further calculations can be made on other gases for which virial coefficients are given in Table 1.1.

1.B (*a*) Plot the pressure of ethane versus its molar volume in the range $0 < P < 200$ bar and molar volumes up to 0.5 mol L^{-1} using the van der Waals equation at 265, 280, 310.671, 350, and 400 K, where 310.671 K is the critical temperature calculated with the van der Waals constants. (*b*) Discuss the significance of the plots and the extent to which they represent reality. (*c*) Calculate the molar volumes at 400 K and $P = 150$ bar and at 265 K and 20 bar.

SOLUTION

```
ClearAll["Global`*"]
```

(a) The pressure function is

```
pf = ((8.3145 * 10^-2) * t / (vm - .0638)) - (5.562 / (vm^2))
```

$$\frac{0.083145\,t}{-0.0638 + vm} - \frac{5.562}{vm^2}$$

The functions we want to plot are given by

```
pp = pf /. {t -> {265, 280, 310.671, 350, 400}}
```

$$\left\{\frac{22.0334}{-0.0638 + vm} - \frac{5.562}{vm^2}, \frac{23.2806}{-0.0638 + vm} - \frac{5.562}{vm^2}, \frac{25.8307}{-0.0638 + vm} - \frac{5.562}{vm^2}, \frac{29.1008}{-0.0638 + vm} - \frac{5.562}{vm^2}, \frac{33.258}{-0.0638 + vm} - \frac{5.562}{vm^2}\right\}$$

```
Plot[Evaluate[pp], {vm, .07, .5},
  AxesLabel -> {"V̄/L mol^-1", "P/bar"}, PlotRange -> {0, 200}];
```

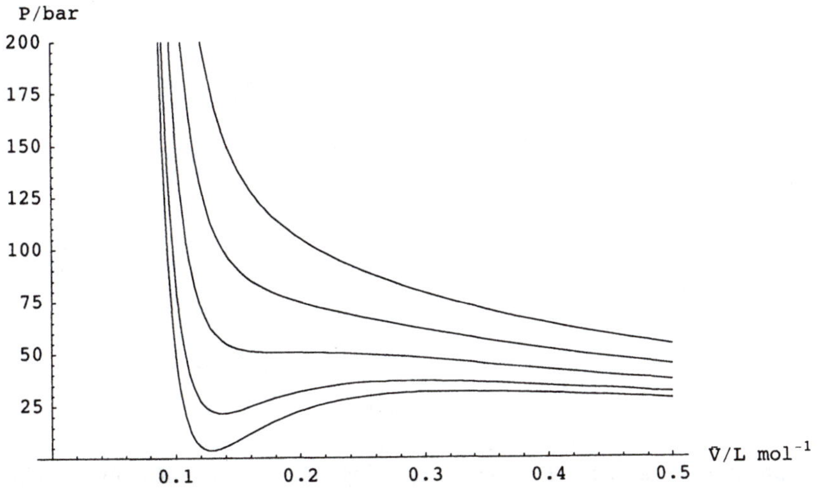

(b) At the critical point 310.671 K, there is an inflection at the critical point. At lower temperatures, the van der Waals equation gives three different molar volumes at a specified pressure. This is a physically impossible result, and the Maxwell construction has to be used to calculate the vapor pressure of the liquid.

(c) At the highest temperature there are two imaginary solutions and one real solution.

```
NSolve[150 == ((8.3145 * 10^-2) * 400 / (vm - .0638)) - (5.562 / (vm^2)), vm]

{{vm → 0.141652}, {vm → 0.071934 + 0.107361 i}, {vm → 0.071934 - 0.107361 i}}
```

At the lowest temperature there are three real solutions to the van der Waals equation. At 265 K between the smaller molar volume and the largest molar volume, there are two phases .

```
NSolve[20 == ((8.3145 * 10^-2) * 265 / (vm - .0638)) - (5.562 / (vm^2)), vm]

{{vm → 0.868918}, {vm → 0.187857}, {vm → 0.108697}}
```

1.C This is a follow-up on Computer Problem 1.B on the van der Waals equation. (*a*) Plot the derivative of the pressure with respect to the molar volume for ethane at 265 K. (*b*) Plot the derivative at the critical temperature. (*c*) Plot the second derivative of the pressure with respect to the molar volume at the critical temperature. In each case, what is the significance of the maxima and minima?

SOLUTION

(a) Plot the first derivative at 265 K.

```
dvt = D[((8.3145 * 10^-2) * 265 / (vm - .0638)) - (5.562 / (vm^2)), vm]
```

$$-\frac{22.0334}{(-0.0638 + \text{vm})^2} + \frac{11.124}{\text{vm}^3}$$

```
Plot[dvt, {vm, .1, .5}, AxesLabel → {"V̄/ mol⁻¹", "dvt"}];
```

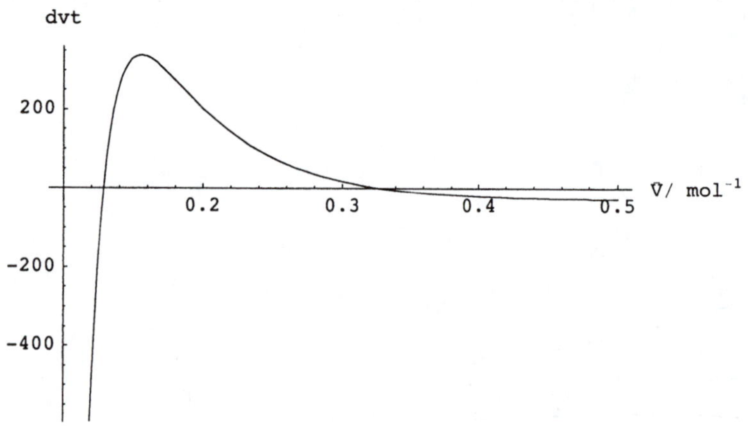

The slope is negative until the minimum pressure is reached. Then the positive slope increases to a maximum and reaches zero at the weak maximum.

(b) Plot the first derivative at the critical temperature.

```
dvtatTc = D[((8.3145 * 10^-2) * 310.671 / (vm - .0638)) - (5.562 / (vm^2)), vm]
```

$$-\frac{25.8307}{(-0.0638 + vm)^2} + \frac{11.124}{vm^3}$$

```
Plot[dvtatTc, {vm, .1, .5}, PlotRange -> {-70, 70}, AxesLabel → {"V̄/ mol⁻¹", "dvtatTc"}];
```

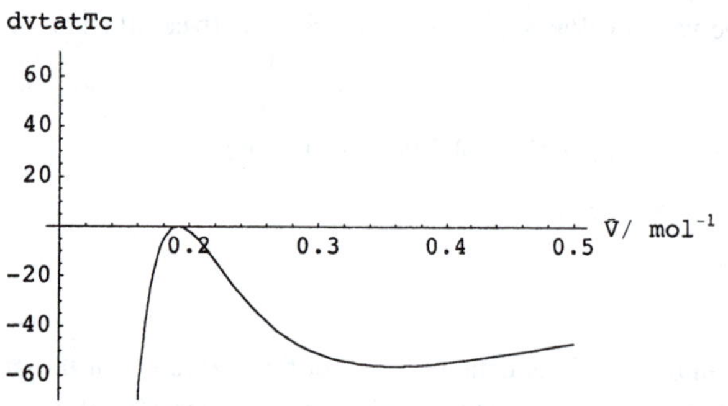

The slope is negative until the critical volume is reached, where the slope is zero. As the molar volume is increased, the slope decreases and eventually reaches that for an ideal gas.

(c) Plot the second derivative at the critical point.

```
secdvtatTc = D[dvtatTc, vm]
```

$$\frac{51.6615}{(-0.0638 + vm)^3} - \frac{33.372}{vm^4}$$

```
Plot[secdvtatTc, {vm, .1, .5},
  PlotRange -> {-750, 750}, AxesLabel → {"V̄/ mol^-1", "secdvtatTc"}];
```

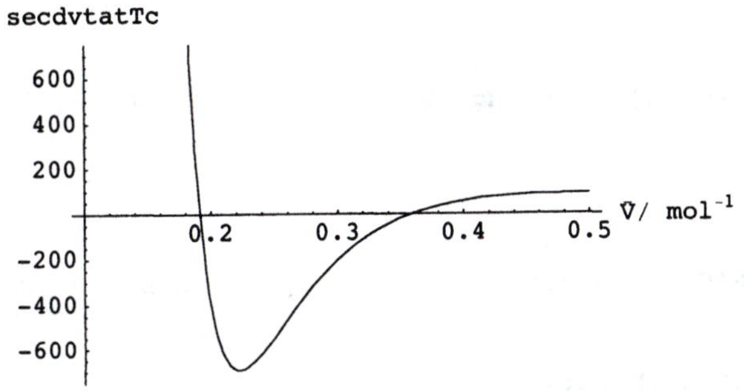

The second derivative is positive below the critical point and zero at the critical point. Then the second derivative becomes more negative. At higher molar volumes, the second derivative becomes positive.

1.D (*a*) Express the compressibility factors for N_2 and O_2 at 298.15 K as a function of pressure using the virial coefficients in Table 1.1. (*b*) Plot these compressibility factors versus P from 0 to 1000 bar.

SOLUTION

(a) For N_2

```
ClearAll["Global`*"]

bpprimeN2=-(4.5*10^-6*10^5)/(8.315*298.15)

-0.000181516

cpprimeN2=(1100-4.5^2)*(10^-12)*10^10/(8.315^2*298.15^2)

1.75683 × 10^-6

zN2=1-.0001815*p+(1.757*10^-6)*p^2

1 - 0.0001815 p + 1.757 × 10^-6 p^2
```

The factors 10^5 and 10^{10} are required because the other quantities are expressed in SI units, but here the pressure is expressed in bars, rather than pascals.

For O_2

```
bprimeO2=-(16.1*10^-6*10^5)/(8.315*298.15)

-0.000649425

cprimeO2=((1200-16.1^2)*10^-12*10^10)/(8.315^2*298.15^2)

1.53073 × 10^-6
```

```
zO2=1-.0006494*p+1.531*10^-6*p^2
```

$1 - 0.0006494\, p + 1.531 \times 10^{-6}\, p^2$

(b) Plot for N_2

```
plotN2=Plot[zN2,{p,0,1000},AxesOrigin->{0,.5},PlotRange->{.5,2.6},AxesLabel->{"P"/"bar"
,"Z"},DisplayFunction->Identity];
```

Plot for O_2

```
plotO2=Plot[zO2,{p,0,1000},AxesOrigin->{0,.5},PlotRange->{.5,2.6},AxesLabel->{"P/bar","
Z"},DisplayFunction->Identity];
```

```
Show[plotN2,plotO2,DisplayFunction->$DisplayFunction];
```

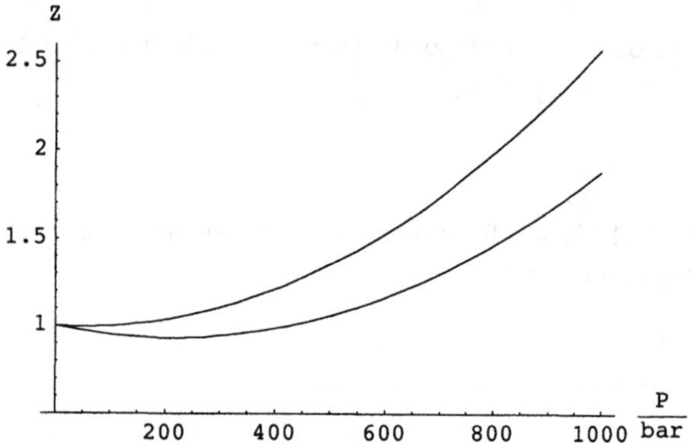

The upper plot is for N_2 and the lower plot is for O_2.

1.E The second virial coefficients of N_2 at a series of temperatures are given by

T/K	75	100	125	150	200	250	300	400	500	600	700
$B'/\mathrm{cm^3\ mol^{-1}}$	-274	-160	-104	-71.5	-35.2	-16.2	-4.2	9	16.9	21.3	24

(*a*) Fit these data to the function
$B' = \alpha + \beta T + \gamma T^2$
(*b*) Plot this function versus temperature. (*c*) Calculate the Boyle temperature of molecular nitrogen..

SOLUTION
(a)

```
t={75,100,125,150,200,250,300,400,500,600,700};

bprime={-274,-160,-104,-71.5,-35.2,-16.2,-4.2,9,16.9,21.3,24};

data=Transpose[{t,bprime}];
```

```
fn=Fit[data,{1,1/tt,1/tt^2},tt]
```

$$40.2596 - \frac{1.02386\times 10^{6}}{tt^{2}} - \frac{9883.51}{tt}$$

(b)

```
Plot[fn, {tt, 0, 1000}, PlotRange → {-200, 25}, AxesLabel → {"T/K", "B'/cm³ mol⁻¹"}];
```

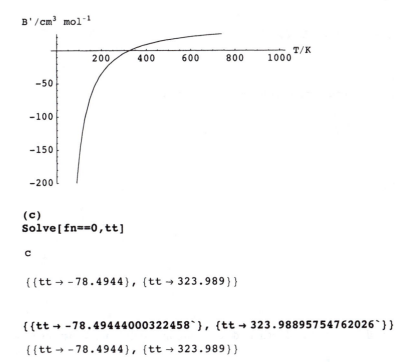

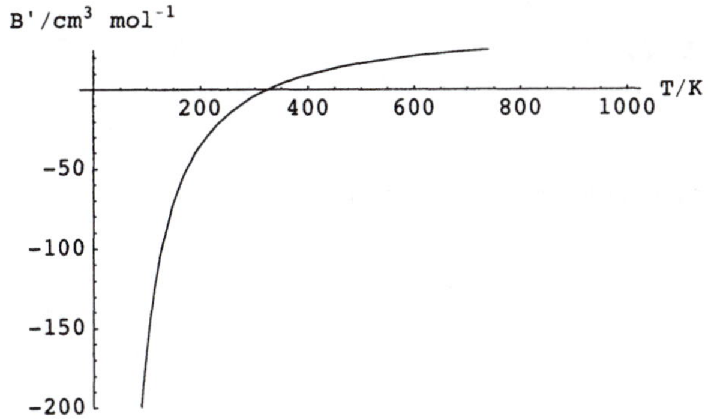


```
(c)
Solve[fn==0,tt]
```

c

{{tt → -78.4944}, {tt → 323.989}}

```
{{tt → -78.49444000322458`}, {tt → 323.98895754762026`}}
```

{{tt → -78.4944}, {tt → 323.989}}

Thus the Boyle temperature is 324 K. This is the temperature at which nitrogen behaves like an ideal gas over a range of low pressures.

1.F Nitrogen tetroxide (N_2O_4) gas is placed in a 500 – cm^3 glass vessel, and the reaction $N_2O_4 = 2\,NO_2$ goes to equilibrium at 25 °C. The density of the gas at equilibrium at 1.0133 bar is 3.176 g L^{-1}. Assuming that the gas mixture is ideal, what are the partial pressures of the two gases at equilibrium?

SOLUTION

The density of the gas is m/V = 3.176 g L^{-1} = PM/RT where M is the average molar mass. Solving for the average molar mass yields

```
3.176 * .083145 * 298.15 / 1.0133
```

77.6986

The molar mass of N_2O_4 is

```
2 * 14.0067 + 4 * 15.9994
```

```
92.011
```

The molar mass of NO_2 is half this. The equilibrium molar mass is given by

```
77.6986 == y * 92.011 + (1 - y) * 92.011 / 2
```

```
77.6986 == 46.0055 (1 - y) + 92.011 y
```

```
Solve[77.6986 == y * 92.011 + (1 - y) * 92.011 / 2, y]
```

```
{{y → 0.688898}}
```

which is the equilibrium mole fraction of N_2O_4. The equilibrium mole fraction of NO_2 is

```
1 - .6889
```

```
0.3111
```

The equilibrium partial pressures of N_2O_4 and NO_2 are

```
.6889 * 1.0133
```

```
0.698062
```

```
.3111 * 1.0133
```

```
0.315238
```

This method of determining partial pressures will be very useful later in determining the equilibrium constants of gas reactions.

1.G Calculate the molar volume of ethane at 350 K and 70 bar using the van der Waals constants in Table 1.3.

SOLUTION

The van der Waals equation for the pressure can be solved for the molar volume v as follows:

```
Solve[70 == .08315 * 350 / (v - .0638) - 5.562 / (v^2), v]
```

```
{{v → 0.124906 - 0.0804012 i}, {v → 0.124906 + 0.0804012 i}, {v → 0.229738}}
```

Two of the roots of this cubic equation are complex, and are therefore eliminated. The real root indicates that the molar volume of ethane under these conditions is 0.2297 L mol^{-1}. The ideal gas law yields 0.416 L mol^{-1}.

1.H Plot the partial pressures of oxygen, nitrogen, and the total pressure in bars versus height above the surface of the earth from zero to 50,000 feet assuming that the temperature is constant at 273 K.

SOLUTION

For O_2,

```
po2=1.01325*.2*Exp[-9.8*32*10^-3*h*12*2.54*10^-2/(8.314*273.15)]
```

$0.20265\, e^{-0.0000420901\, h}$

For N_2,

```
pn2=1.01325*.8*Exp[-9.8*28*10^-3*h*12*2.54*10^-2/(8.314*273.15)]
```

$0.8106\, e^{-0.0000368288\, h}$

```
Plot[{po2,pn2,po2+pn2},{h,0,50000},AxesLabel->{"h/feet","P/bar"}];
```

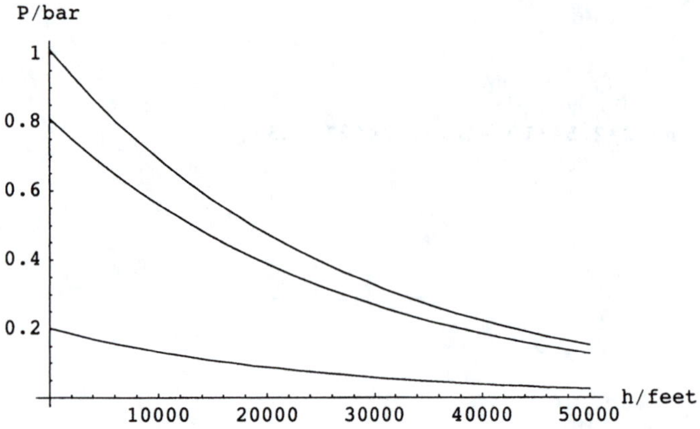

Chapter 2 First Law of Thermodynamics

2.A In Section 2.11 we have seen that knowledge of the temperature dependence of C_P°, which can be represented by a polynomial in T (see Table 2.2), makes it possible to calculate the change in molar enthalpy $\bar{H}^\circ$ with temperature. In Section 2.15 we have seen that the standard enthalpy of reaction at other temperatures can be calculated from $\Delta_r H^\circ$ (298.15 K) by integration of $\Delta_r\ C_P^\circ\ dT$. It is difficult to make these calculations with a hand-held calculator, but they can be conveniently made using a mathematical application that can integrate. By putting in empirical equations for C_P°, we can solve the following problems:

(*a*) Print out the values of C_P° at 298.15 K and 1000 K for the following gases: N_2, O_2, H_2, CO, CO_2, H_2O, NH_3, and CH_4. Compare these values with those in Table C.3 of the Appendix.

(*b*) Plot C_P° for CO_2, H_2O, NH_3, and CH_4 versus temperature from 298.15 K to 1500 K.

(*c*) Calculate $\bar{H}^\circ$ (1000 K) - $\bar{H}^\circ$ (298.15 K) for N_2 and compare it with the value in Table C.3.

(*d*) Calculate $\Delta_r\ H^\circ$(1000 K) for the following four gas reactions:

$H_2 + (1/2)O_2 = H_2O$

$(1/2)N_2 + (3/2)H_2 = NH_3$

$CO_2 + 4H_2 = CH_4 + 2H_2O$

$CO + 3H_2 = CH_4 + H_2O$

SOLUTION

(a) The equations for the molar heat capacities from Table 2.2 are put in explicitely. Since we want SetDelayed, rather than an assignment, the := operator is used rather than the immediate assignment operator =.

```
Clear[t]
Off[General::spell1];
Off[General::spell];

cpN2 := 28.883 - .157 10^-2 t + .808 10^-5 t^2 - 2.871 10^-9 t^3;

cpO2 := 25.46 + 1.519 10^-2 t - .715 10^-5 t^2 + 1.311 10^-9 t^3;

cpH2 := 29.088 - .192 10^-2 t + .4 10^-5 t^2 - .87 10^-9 t^3;

cpCO := 28.142 + .167 10^-2 t + .537 10^-5 t^2 - 2.221 10^-9 t^3;

cpCO2 := 22.243 + 5.977 10^-2 t - 3.499 10^-5 t^2 + 7.464 10^-9 t^3;

cpH2O := 32.218 + .192 10^-2 t + 1.055 10^-5 t^2 - 3.593 10^-9 t^3;

cpNH3 := 24.619 + 3.75 10^-2 t - .138 10^-5 t^2;

cpCH4 := 19.875 + 5.021 10^-2 t + 1.268 10^-5 t^2 - 11.004 10^-9 t^3;

values298 = {cpN2, cpO2, cpH2, cpCO, cpCO2, cpH2O, cpNH3, cpCH4} /. t -> 298.15

{29.0571, 29.3881, 28.8481, 29.0584, 37.1509, 33.633, 35.677, 35.6806}

values1000 = {cpN2, cpO2, cpH2, cpCO, cpCO2, cpH2O, cpNH3, cpCH4} /. t -> 1000

{32.522, 34.811, 30.298, 32.961, 54.487, 41.095, 60.739, 71.761}
```

```
TableForm[Transpose[{values298, values1000}],
 TableHeadings -> {{"cpN2", "cpO2", "cpH2", "cpCO",
    "cpCO2", "cpH2O", "cpNH3", "cpCH4"}, {298.15 K, 1000 K}}]
```

	298.15 K	1000 K
cpN2	29.0571	32.522
cpO2	29.3881	34.811
cpH2	28.8481	30.298
cpCO	29.0584	32.961
cpCO2	37.1509	54.487
cpH2O	33.633	41.095
cpNH3	35.677	60.739
cpCH4	35.6806	71.761

These values are in quite good agreement with those in Table C.3, but the values in Table C.3 are better for reasons we will see later.

(b)

```
Clear[t]

Plot[{cpCO2, cpH2O, cpNH3, cpCH4}, {t, 300, 1500},
 AxesOrigin -> {300, 30}, AxesLabel -> {"T/K", "C/J K^-1 mol^-1"}];
```

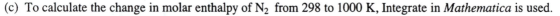

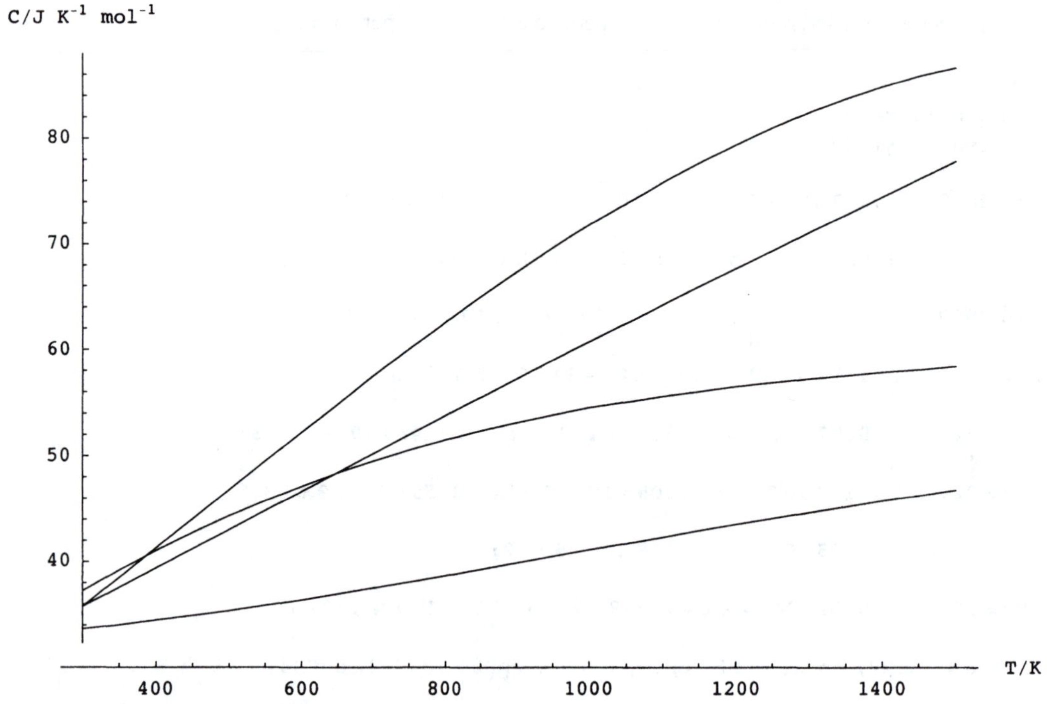

(c) To calculate the change in molar enthalpy of N_2 from 298 to 1000 K, Integrate in *Mathematica* is used.

```
Integrate[cpN2, {t, 298.15, 1000}]

21466.2
```

The value given in Table C.3 is 21 463 J mol^-1.

(d) The standard enthalpies of formation in kJ mol^{-1} of the eight species are entered as follows:

```
enthN2 = 0;
enthO2 = 0;
enthH2 = 0;
enthCO = -110.525;
enthCO2 = -393.509;
enthH2O = -241.818;
enthNH3 = -46.11;
enthCH4 = -74.81;

hrx1 = (enthH2O - enthH2 - .5 * enthO2) +
  (10^-3) * Integrate[cpH2O - cpH2 - .5 * cpO2, {t, 298.15, 1000}]

-247.819

hrx2 = (enthNH3 - .5 * enthN2 - 1.5 * enthH2) +
  (10^-3) * Integrate[cpNH3 - .5 * cpN2 - 1.5 * cpH2, {t, 298.15, 1000}]

-53.8633

hrx3 = (enthCH4 + 2 * enthH2O - enthCO2 - 4 * enthH2) +
  (10^-3) * Integrate[cpCH4 + 2 * cpH2O - cpCO2 - 4 * cpH2, {t, 298.15, 1000}]

-190.519

hrx4 = (enthCH4 + enthH2O - enthCO - 3 * enthH2) +
  (10^-3) * Integrate[cpCH4 + cpH2O - cpCO - 3 * cpH2, {t, 298.15, 1000}]

-225.449

TableForm[{{hrx1}, {hrx2}, {hrx3}, {hrx4}},
 TableHeadings -> {{"H2+(1/2)O2=H2O", "(1/2)N2+(3/2)H2=NH3",
   "CO2+4H2=CH4+2H2O", "CO+3H2=CH4+H2O"}, {DfHo / ("kJ mol^-1")}}]
```

	$\frac{\text{DfHo}}{\text{kJ mol}^{-1}}$
H2+(1/2)O2=H2O	-247.819
(1/2)N2+(3/2)H2=NH3	-53.8633
CO2+4H2=CH4+2H2O	-190.519
CO+3H2=CH4+H2O	-225.449

Heats of other reactions can be calculated in this way if there are empirical equations for the heat capacities.

2.B Plot the molar heat capacities in J K^{-1} mol^{-1} at constant pressure for gaseous He, N_2, H_2O, CO_2, NH_3, and CH_4 from 300 to 1800 K using the parameters in Table 2.2.

SOLUTION

```
cpHe = (5 / 2) * 8.3145;

cpN2 := 28.883 - .157 10^-2 t + .808 10^-5 t^2 - 2.871 10^-9 t^3;

cpH2 := 29.088 - .192 10^-2 t + .4 10^-5 t^2 - .87 10^-9 t^3;

cpCO2 := 22.243 + 5.977 10^-2 t - 3.499 10^-5 t^2 + 7.464 10^-9 t^3;

cpH2O := 32.218 + .192 10^-2 t + 1.055 10^-5 t^2 - 3.593 10^-9 t^3;

cpNH3 := 24.619 + 3.75 10^-2 t - .138 10^-5 t^2;

cpCH4 := 19.875 + 5.021 10^-2 t + 1.268 10^-5 t^2 - 11.004 10^-9 t^3;

Plot[{cpHe, cpN2, cpCO2, cpH2O, cpNH3, cpCH4}, {t, 300, 1800},
 AxesOrigin → {300, 0}, PlotRange → {0, 100}, AxesLabel → {"T/K", "C/J K^-1 mol^-1"}];
```

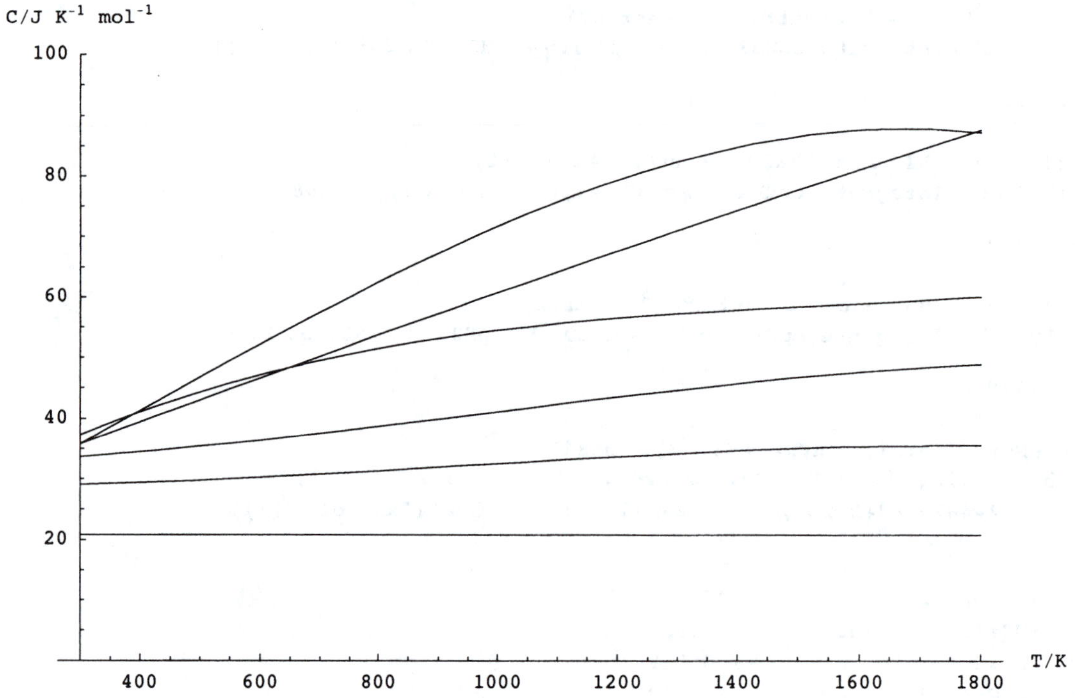

The plots can be identified by their ordinates at 1000 K.

```
{cpHe, cpN2, cpCO2, cpH2O, cpNH3, cpCH4} /. t -> 1000

{20.7863, 32.522, 54.487, 41.095, 60.739, 71.761}
```

2.C Calculate the standard enthalpy of reaction at 298 K and 1000 K for the gas reaction $CO_2 + 4\,H_2 = CH_4 + 2\,H_2O$.

SOLUTION

```
ClearAll["Global`*"]

enthchg298=-74.81+2*(-241.82)-(-393.51)

-164.94

cpH2 := 29.088 - .192 10^-2 t + .4 10^-5 t^2 - .87 10^-9 t^3;

cpCO2 := 22.243 + 5.977 10^-2 t - 3.499 10^-5 t^2 + 7.464 10^-9 t^3;

cpH2O := 32.218 + .192 10^-2 t + 1.055 10^-5 t^2 - 3.593 10^-9 t^3;

cpCH4 := 19.875 + 5.021 10^-2 t + 1.268 10^-5 t^2 - 11.004 10^-9 t^3;

enthchg1000=enthchg298+Integrate[cpCH4+2*cpH2O-cpCO2-4*cpH2,{t,298,1000}]/1000

-190.53
```

The answer is in kJ mol^{-1}. The factor of 1000 is required because the molar heat capacities are in joules, whereas the standard heat capacities are in J K^{-1} mol^{-1}.

2.D Calculate the standard enthalpy of formation of CO_2 at 1000 K from the standard enthalpy of formation at 298.15 K and empirical equations for the heat capacities. The standard molar heat capacities of graphite are given in Table C.3 as a function of temperature. It is convenient to fit these data to a function of temperature.

SOLUTION

```
ClearAll["Global`*"]

cpCexpt={8.517,14.623,21.610,24.094};

texpt={298.15,500,1000,2000};

data=Transpose[{texpt,cpCexpt}];

cpC=Fit[Transpose[{texpt,cpCexpt}],{1,t,t^2},t];
```

Paste in equations for cpO2 and cpCO2 from Computer Problem 2.A.

```
cpO2 := 25.46 + 1.519 10^-2 t - .715 10^-5 t^2 + 1.311 10^-9 t^3;

cpCO2 := 22.243 + 5.977 10^-2 t - 3.499 10^-5 t^2 + 7.464 10^-9 t^3;
```

```
deltaH2000=-393.52+(Integrate[cpCO2,{t,298.15,1000}]-Integrate[cpC+cpO2,{t,298.14,1000}
])/1000

-394.428
```

The value given in Table C.3, -394.623 kJ mol^{-1}, is more accurate.

2.E (a) Calculate the work of reversible isothermal expansion of a mole of carbon dioxide at 298.15 K from an initial volume of 5 L to a final volume of 20 L on the assumption that carbon dioxide is an ideal gas. (b) Calculate the reversible work using the van der Waals constants for carbon dioxide. (c) Explain the difference.

SOLUTION

(a)

```
wideal = -8.31451 * 298.15 * Log[20 / 5]

-3436.58
```

This is the work that is done on the gas in joules.

(b) The van der Waals constants in Table 1.3 are a = 3.640 L^2 bar mol^{-2} and b = 0.04267 L mol^{-1}.

```
wvdW = -8.31451 * 298.15 * Log[(20 - .04267) / (5 - .04267)] +
  3.640 * ((1 / 5) - (1 / 20)) * 8.31451 / .0831451

-3397.94
```

Note that it is essential to express the two terms in the same units. The first term is in J mol^{-1}, but a/V is in L bar mol^{-1}. Therefore, the second term is multiplied by the ratio of the gas constant in J K^{-1} mol^{-1} to the gas constant in L bar K^{-1} mol^{-1}. This calculation shows that less work on the surroundings by the van der Waals gas.

(c) The pressures in bars of the ideal gas at the initial and final volumes are

```
.08315 * 298.15 / {5, 20}

{4.95823, 1.23956}
```

The pressures of the van der Waals gas at the initial and final volumes are

```
.08315 * 298.15 / {5 - .04267, 20 - .04267} - 3.640 / {5^2, 20^2}

{4.85531, 1.23311}
```

The van der Waals gas exerts a lower pressure and does less work because of intermolecular attractions.

Chapter 3 Second and Third Laws of Thermodynamics

3.A Calculate the standard molar entropies at 1000 K for the eight gases in Computer Problem 2.A using the empirical equations given there for the molar heat capacities as a function of temperatureand the values of the standard molar entropies at 298.15 K in Table C.2. Compare these calculated values with values in Table C.3 in the Appendix.

SOLUTION

(a) Copy and paste polynomials fro computer Problem 2.A.

```
Clear[t]
Off[General::spell];
Off[General::spell];

cpN2 := 28.883 - .157 10^-2 t + .808 10^-5 t^2 - 2.871 10^-9 t^3;

cpO2 := 25.46 + 1.519 10^-2 t - .715 10^-5 t^2 + 1.311 10^-9 t^3;

cpH2 := 29.088 - .192 10^-2 t + .4 10^-5 t^2 - .87 10^-9 t^3;

cpCO := 28.142 + .167 10^-2 t + .537 10^-5 t^2 - 2.221 10^-9 t^3;

cpCO2 = 22.243 + 5.977 10^-2 t - 3.499 10^-5 t^2 + 7.464 10^-9 t^3;

cpH2O := 32.218 + .192 10^-2 t + 1.055 10^-5 t^2 - 3.593 10^-9 t^3;

cpNH3 := 24.619 + 3.75 10^-2 t - .138 10^-5 t^2;

cpCH4 := 19.875 + 5.021 10^-2 t + 1.268 10^-5 t^2 - 11.004 10^-9 t^3;
```

Enter data for the standard molar entropies of the gases at 298.15 K and integrate the standard molar heat capacities divided by temperature.

```
sbar298 = {191.61, 205.138, 130.684, 197.684, 213.74, 188.825, 192.45, 186.264};

sbar1000 = sbar298 + Integrate[{cpN2 / t, cpO2 / t, cpH2 / t,
    cpCO / t, cpCO2 / t, cpH2O / t, cpNH3 / t, cpCH4 / t}, {t, 298.15, 1000}]

{228.21, 243.778, 166.077, 234.638, 269.089, 232.802, 247.934, 247.761}
```

The values for the molar entropies at 1000 K are given in J K^{-1} mol^{-1}. The corresponding values from Table C.3 are given by

```
sbar1000C3 = {228.17, 243.87, 166.22, 234.54, 209.30, 232.74, 246.49, 247.66};
```

Comparison of standard molar entropies at 298.15 K, the values calculated here for 1000 K, and the values in Appendix C.3:

```
TableForm[Transpose[{sbar298, sbar1000, sbar1000C3}], TableHeadings ->
  {{"N2", "O2", "H2", "CO", "CO2", "H2O", "NH3", "CH4"}, {298.15 K, 1000 K, 1000 KC3}}]
```

	298.15 K	1000 K	1000 KC3
N2	191.61	228.21	228.17
O2	205.138	243.778	243.87
H2	130.684	166.077	166.22
CO	197.684	234.638	234.54
CO2	213.74	269.089	209.3
H2O	188.825	232.802	232.74
NH3	192.45	247.934	246.49
CH4	186.264	247.761	247.66

The agreement with Table C.3 is good, but the values in Table C.3 are better for reasons we will see later.

3.B Plot the standard molar entropy of water vapor versus temperature from 300 to 1000 K by use of the empirical equation for the molar heat capacity of water vapor.

SOLUTION

```
cpH2O := 32.218 + .192 10^-2 t + 1.055 10^-5 t^2 - 3.593 10^-9 t^3;

sbar=188.83+Integrate[{cpH2O/t},t]
```

$\{188.83 + 0.00192\,t + 5.275\times10^{-6}\,t^2 - 1.19767\times10^{-9}\,t^3 + 32.218\,\text{Log}[t]\}$

```
Plot[sbar, {t, 300, 1000}, AxesOrigin → {300, 370},
 AxesLabel → {"T/K", "S/J K^-1 mol^-1"}, PlotRange → {370, 430}];
```

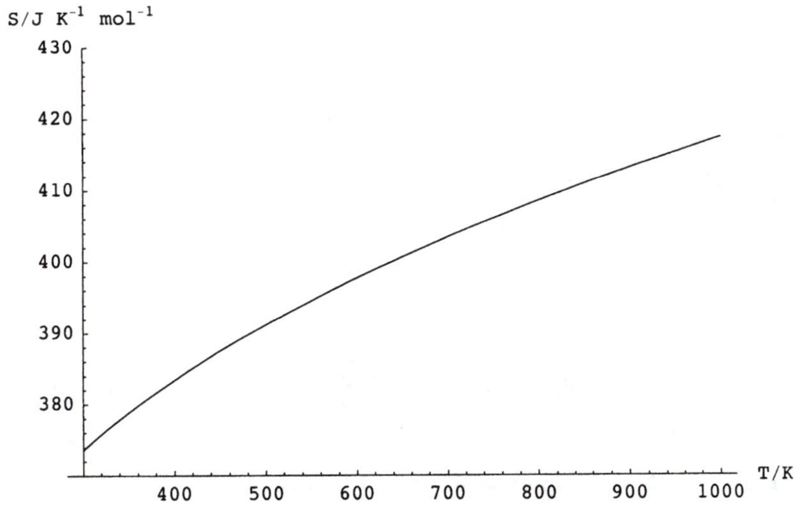

3.C Plot the entropy of mixing of two ideal gases to form a mole of mixture versus the mole fraction of one of the species. Investigate the slope of this plot as y_2 ->1 and 0.

SOLUTION

It is convenient to plot the ratio of the molar entropy of mixing to the gas constant R.

```
smix=-(1-y2)*Log[1-y2]-y2*Log[y2];
```

```
Plot[smix, {y2, 0, 1}, AxesLabel → {"y2", "(Δ\ S)mix/R"}];
```

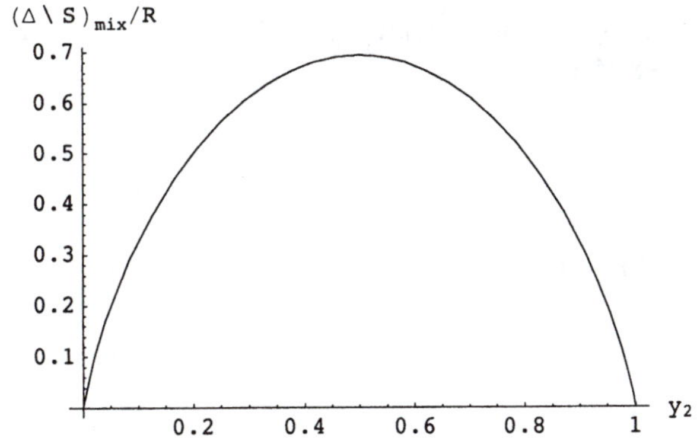


```
dvt=D[smix,y2]
```

```
Log[1 - y2] - Log[y2]
```

```
Plot[dvt,{y2,0,1},AxesLabel->{"y2","dvt"}];
```

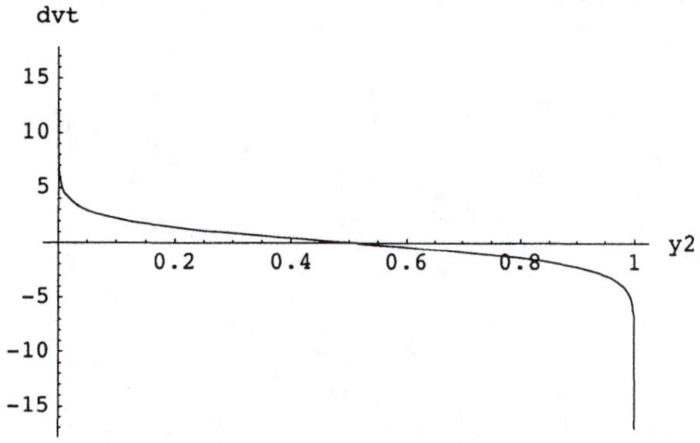

The slope appears to approach infinity as y_2 -> 0 and negative infinity as y_2 ->1. We can pursue this by expanding the scale near y_2 ->1.

```
Plot[dvt,{y2,.99,1},AxesOrigin->{.99,-16},AxesLabel->{"y2","dvt"}];
```

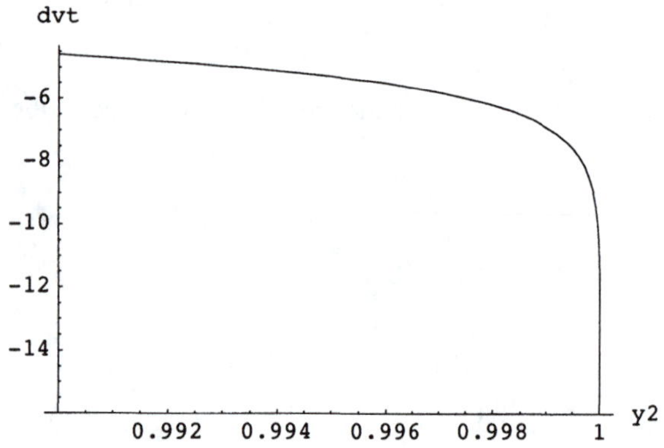

3.D Plot the molar entropy in J K^{-1} mol^{-1} of a monatomic ideal gas versus pressure P from 1 bar to 100 bar and T from 298.15 K to 500 K , assuming that its molar entropy is zero at 298.15 K and 1 bar.

SOLUTION

The Sakur-Tetrode equation is used.

```
smolar[t_,p_]=8.31451*(Log[(t/298.15)^(5/2)]-Log[p])
```

$8.31451\ (-\text{Log}[p] + \text{Log}[6.51498 \times 10^{-7}\ t^{5/2}])$

```
Plot3D[smolar[t,p],{p,1,100},{t,298.15,500},AxesLabel->{"P/bar","T/K","S"}];
```

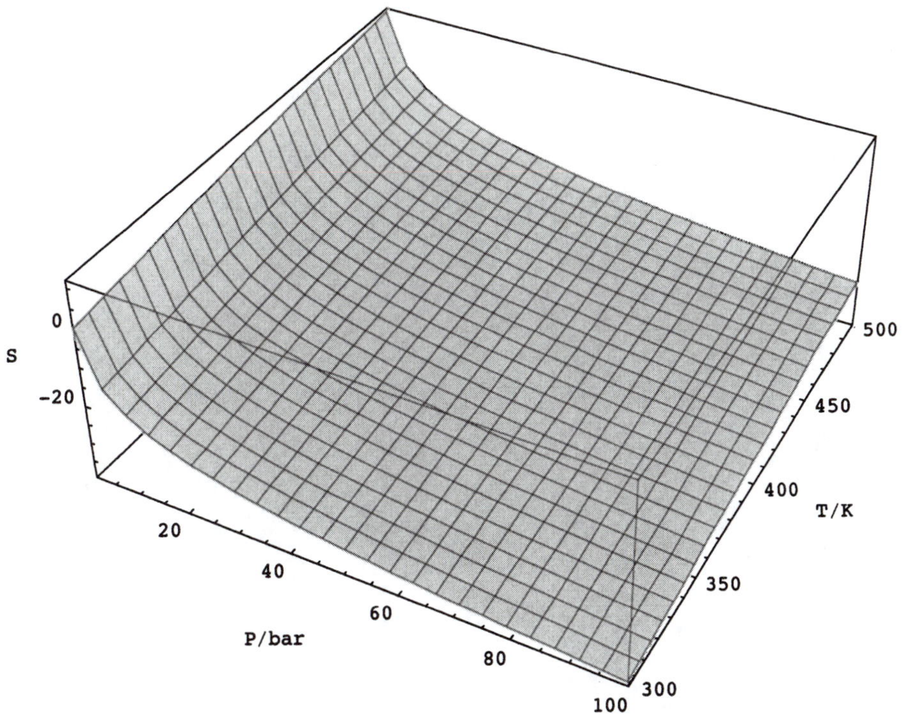

Chapter 4 Fundamental Equations

4.A We know a good deal about the thermodynamics of monatomic ideal gases, as we have already seen. Plot (*a*) μ, (*b*) $\overline{S}$, and (*c*) $\overline{V}$ versus temperature from 100 to 500 K and pressure from 0.1 to 20 bar and discuss the slopes of each of the plot of μ in the two directions. The first term in the expression for the chemical potential is neglected here. The necessary equations are

(*a*) $\mu = -T \ln (T^{5/2}/P)$

(*b*) $\overline{S} = \ln (T^{5/2}/P)$

(*c*) $\overline{V} = T/P$

SOLUTION

(a)

```
mu[t_, p_] = -1 * t * Log[t^(5/2) p^(-1)]
```

$$-t\, \text{Log}\left[\frac{t^{5/2}}{p}\right]$$

```
Plot3D[mu[t, p], {p, .1, 20}, {t, 100, 500}, Shading -> False,
  ViewPoint -> {-1, -2.4, 2}, AxesLabel -> {"P/bar", " T/K", "    μ/J mol^-1"}];
```

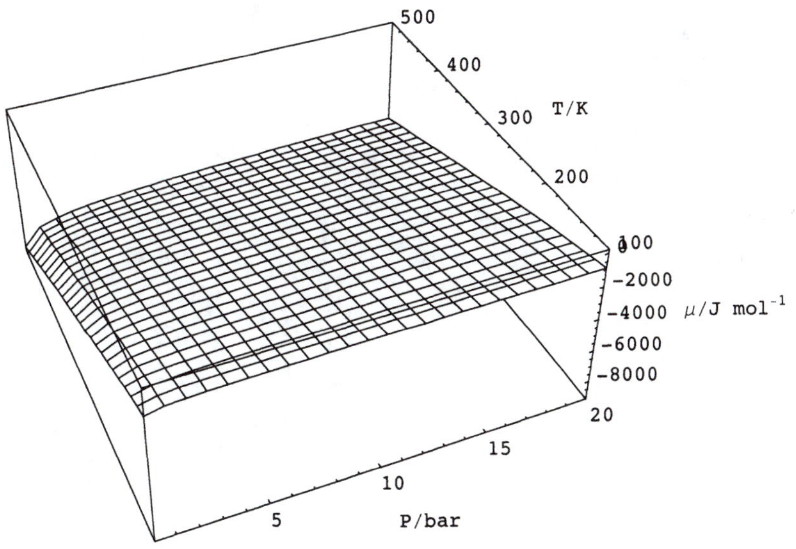

$d\mu = -\overline{S}\,dT + \overline{V}\,dP$

Therefore, the slopes in the two directions of the μ surface give use information on the molar entropy and the molar volume of the monatomic ideal gas. The slope of the μ surface in the direction of increasing temperature is always negative, and so the molar entropy is always positive. The slope of the mu surface in the direction of increasing pressure is always positive, and so the molar volume is always positive.

(b)

```
ent[t_, p_] = Log[(t^(5/2)) (p^-1)]
```

$\text{Log}\left[\frac{t^{5/2}}{p}\right]$

```
Plot3D[ent[t, p], {p, .1, 20}, {t, 100, 500}, Shading -> False,
  ViewPoint -> {-1, -2.4, 2}, AxesLabel -> {"P/bar", " T/K", " S̄/J K^-1 mol^-1"}];
```

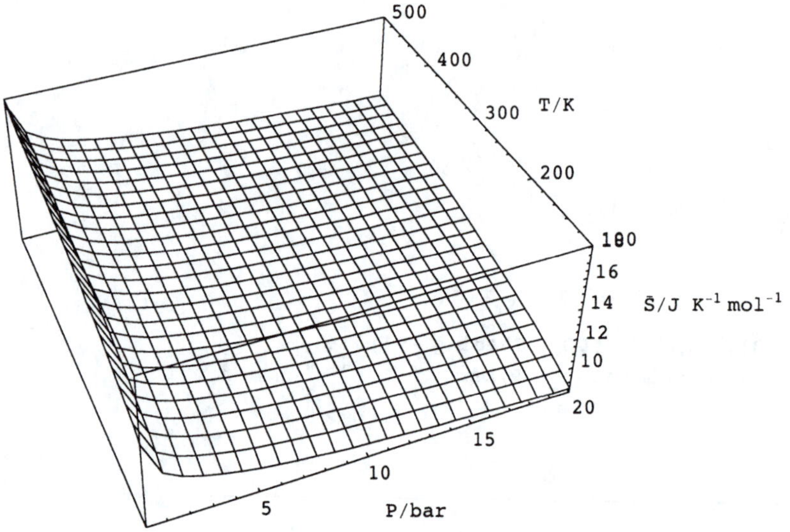

(c)

```
v[t_, p_] = t/p
```

$\frac{t}{p}$

```
Plot3D[v[t, p], {p, .1, 20}, {t, 100, 500}, Shading -> False,
  ViewPoint -> {-1, -2.4, 2}, AxesLabel -> {"P/bar", " T/K", "  V̄/L mol^-1"}];
```

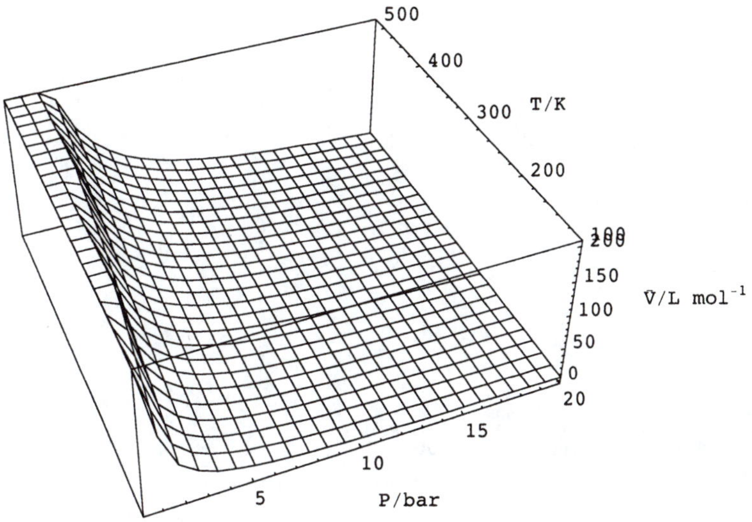

The left side of this plot is truncated because the volume continues to increase as the pressure is lowered, but it is evident that the front and back faces are rectangular hyperbolas and that the molar volume always increases with temperature.

4.B Given the virial equation for N_2 in terms of pressure at 298.15 K, plot the compressibility factor and the fugacity from 0 to 1000 bar versus the pressure.

SOLUTION

```
zN2=1-.0001815*p+(1.757*10^-6)*p^2;
```

```
Plot[zN2, {p, 0, 1000}, AxesLabel → {"\!\(\*
StyleBox[\"P\",\nFontSlant->\"Italic\"]\)/bar", "\!\(\*
StyleBox[\"Z\",\nFontSlant->\"Italic\"]\)"}];
```

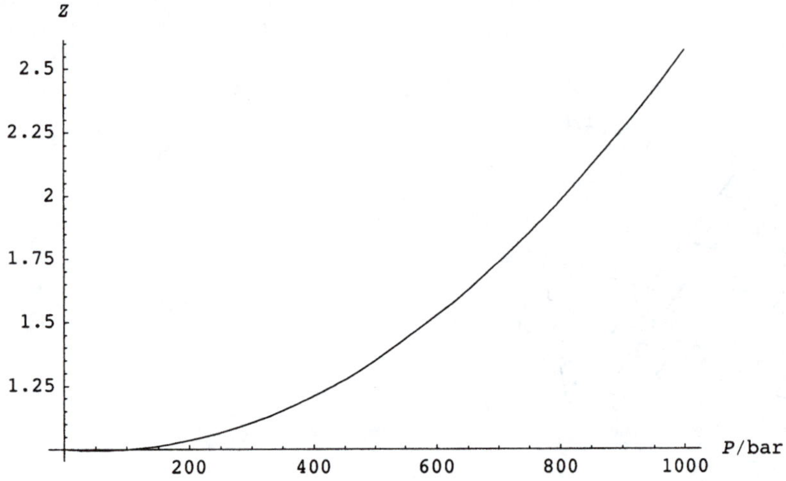


```
fugacity[pp_]:=pp*
Exp[Integrate[(zN2-1)/p,{p,0,pp}]];

Plot[fugacity[p],{p,0,1000},AxesLabel->{"P/bar","f/bar"}];
```

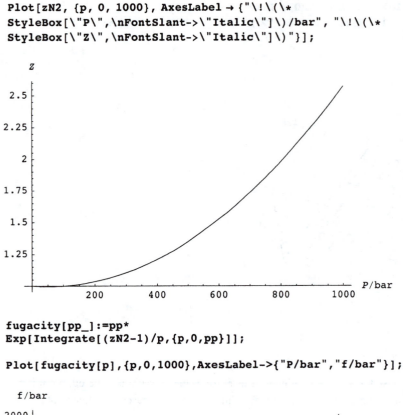

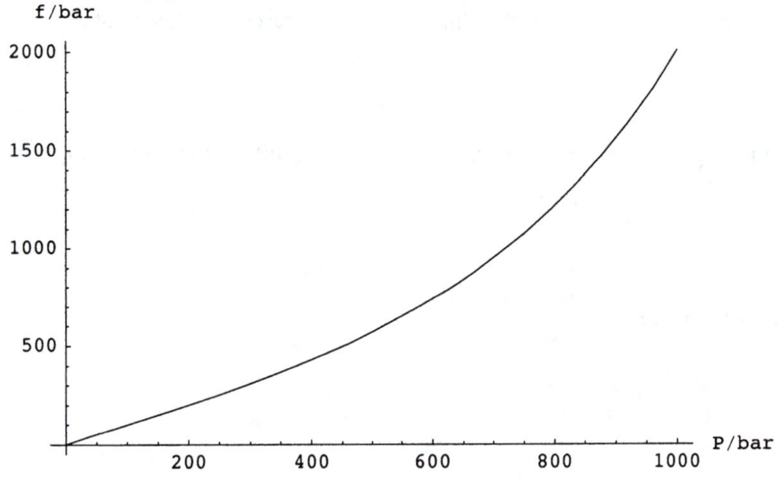

4.C Given the virial coefficients for O_2 at 298.15 K in Table 1.1, plot the compressibility factor and the fugacity of the gas versus its pressure up to 1000 bar.

SOLUTION

First we have to calculate the value of B' in bar^{-1}:

Since $B' = B/RT$ and $RT = (8.3145 \times 10^{-2}\ \mathrm{L\ bar\ K^{-1}\ mol^{-1}})(10^{-3}\ \mathrm{m^3\ L^{-1}})(298.15\ \mathrm{K})$,

```
bprime = (-16.1 * 10^-6) / ((8.3145 * 10^-2) * 10^-3 * 298.15)

-0.000649464
```

Second we have to calculate the value of C' in bar^{-2}:
Since $C' = (C - B^2)/(RT)^2$,

```
cprime = (1200 * 10^-12 - (16.1 * 10^-6)^2) / (((8.3145 * 10^-2) * 10^-3 * 298.15)^2)
```

1.53091×10^{-6}

```
zO2 = 1 - .0006494 * p + 1.5309 * 10^-6 * p^2
```

$1 - 0.0006494\,p + 1.5309 \times 10^{-6}\,p^2$

```
Plot[zO2, {p, 0, 1000}, AxesLabel → {"P/bar", "Z"}];
```

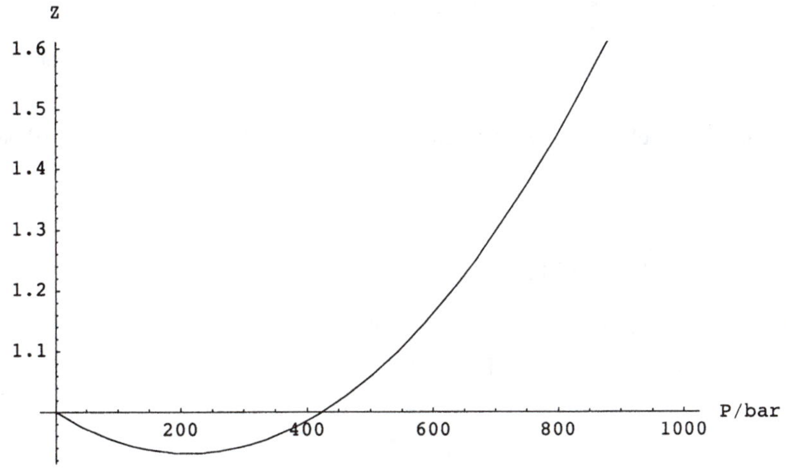


```
fugacity[pp_] := pp * Exp[Integrate[(zO2 - 1) / p, {p, 0, pp}]]
```

```
Plot[fugacity[p], {p, 0, 1000}, AxesLabel → {"P/bar", "f/bar"}];
```

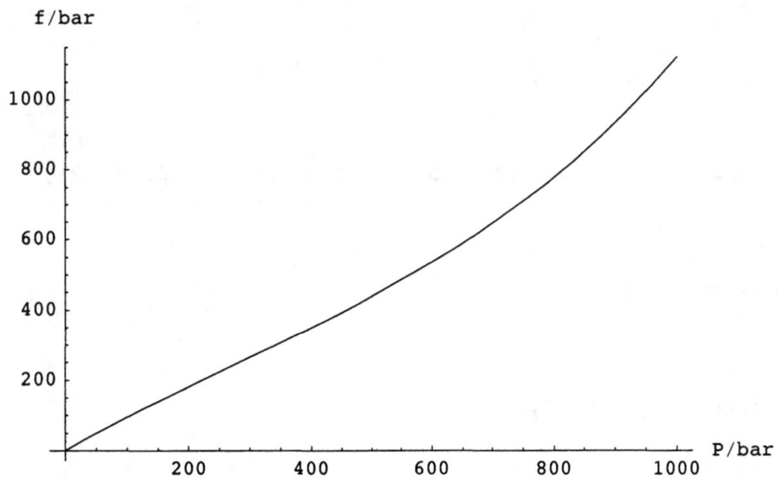

4.D Plot the Gibbs energy of formation of an ideal gas relative to its standard Gibbs energy at 298.15 K from 0 to 10 bar.

SOLUTION

```
gfminusstdgf = (8.3145 * 10^-3) * (298.15) * Log[p]

2.47897 Log[p]
```

```
Plot[gfminusstdgf, {p, 0, 10}, AxesLabel → {"P/bar", "(ΔfG\!\(\*
StyleBox[\"-\",\nFontSize->11]\)ΔfG°)/kJ mol^-1"},
  PlotRange -> {-7.5, 7.5}, AspectRatio -> 1.5];
```

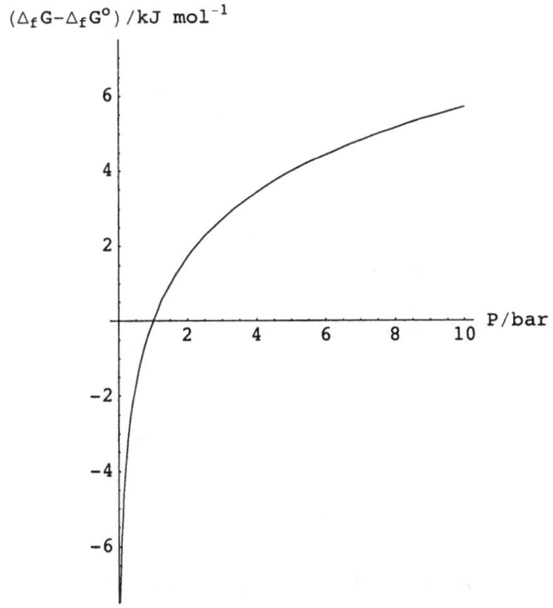

4.E The 3D plots of molar entropy and molar volume for a monatomic ideal gas made in Computer Problem 4.A can be made in another way by using

$S_m = -(\partial\mu/\partial T)_P$

$V_m = (\partial\mu/\partial P)_T$

(a) Plot S_m, (b) Plot V_m, (c) Plot $-(\partial\mu/\partial T)_P$, (d) Plot $(\partial\mu/\partial P)_T$.

SOLUTION

(a) S_m

```
mu[y_,p_]=-1*t*Log[t^(5/2)*p^(-1)]
```

$-t\,\mathrm{Log}\left[\frac{t^{5/2}}{p}\right]$

```
dvtt=-D[mu[t,p],t]
```

$\frac{5}{2}+\mathrm{Log}\left[\frac{t^{5/2}}{p}\right]$

```
Plot3D[Evaluate[dvtt],{p,.1,20},{t,100,500},ViewPoint->{-1,-2.4,2},AxesLabel->{"P/bar",
"T/K"," "}];
```

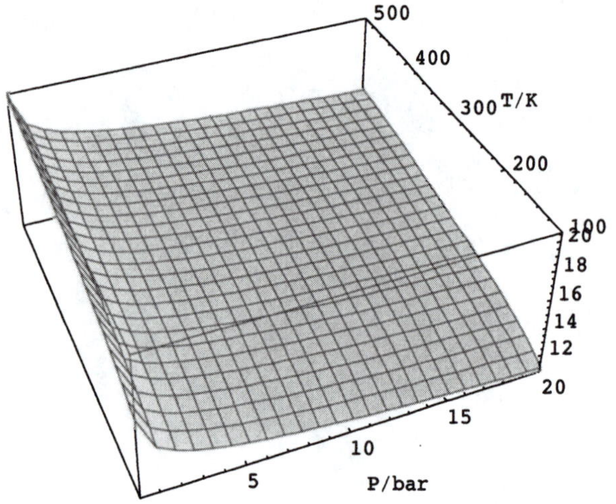

(b) V_m

```
dvtp=D[mu[t,p],p]
```

$\frac{t}{p}$

```
Plot3D[Evaluate[dvtp],{p,.1,20},{t,100,500},ViewPoint->{-1,-2.4,2},AxesLabel->{"P/bar",
"T/K"," "}];
```

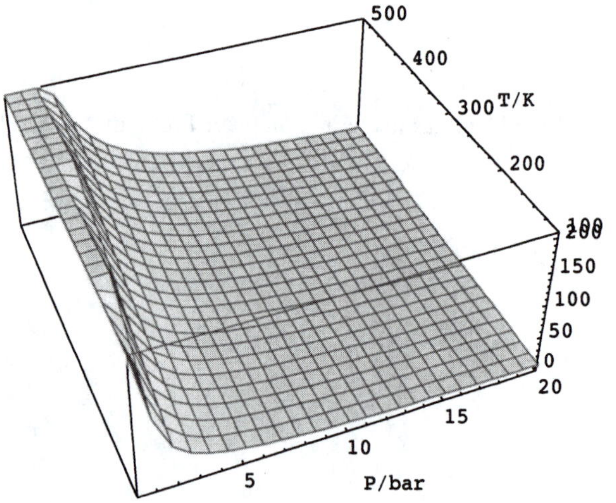

(c) $-(\partial\mu / \partial T)_P$

```
dvttp=-D[dvtt,p]
```

$\frac{1}{p}$

```
Plot3D[Evaluate[dvttp],{p,.1,20},{t,100,500},ViewPoint->{-1,-2.4,2},AxesLabel->{"P/bar"
,"T/K"," "}];
```

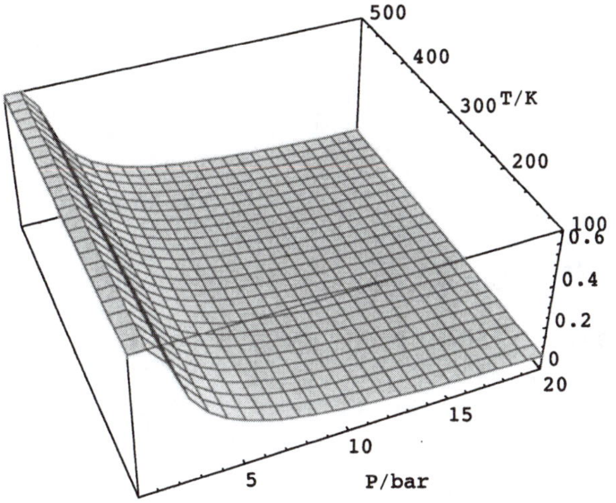

(d) $(\partial\mu / \partial P)_T$

```
dvtpt=D[dvtp,t]
```

$\frac{1}{p}$

This will obviously give the same plot as (c).

4.F (a) Calculate the fugacity of molecular hydrogen at 100 bar and 25 ^oC using the virial coefficients in Table 1.1. (b) Plot the fugacity of molecular hydrogen from p = 0 to 1000 bar at 25 $^o C$.

SOLUTION
(a)Use the equations in Example 1.5 to calculate the virial coefficients in the virial expansion in terms of pressure (equation 1.12).

```
bprime = 1.4 * 10^-6 / (.08315 * 298.15 * 10^-3)
```

0.0000564717

The 10^{-3} m^3 L^{-1} is required to convert RT to m^3 mol^{-1}.

```
cprime = (350 * 10^-12 - (14.1 * 10^-6)^2) / ((.08315 * 298.15 * 10^-3)^2)
```

2.45997×10^{-7}

lnf = lnP + bprimeP + cprimeP^2 /2

```
lnf = Log[100] + 5.647 * 10^-5 * 100 + (2.4599 * 10^-7) * 100^2 / 2
```

4.61205

```
fugacity = Exp[lnf]

100.69
```

The fugacity is expressed in bars.

(b) ln(f/p) is given by

```
5.64717 * p + (2.45997 * 10^-7 / 2) * p^2
```

$5.64717\,p + 1.22999 \times 10^{-7}\,p^2$

The ratio of fugacity to pressure is given by Exp[Log[f/p]] and the fugacity is given by p*Exp[Log[f/p]].

```
plot1 = Plot[p * Exp[5.647 * 10^-5 * p + (2.4599 * 10^-7) * p^2 / 2], {p, 0, 1000}];
```

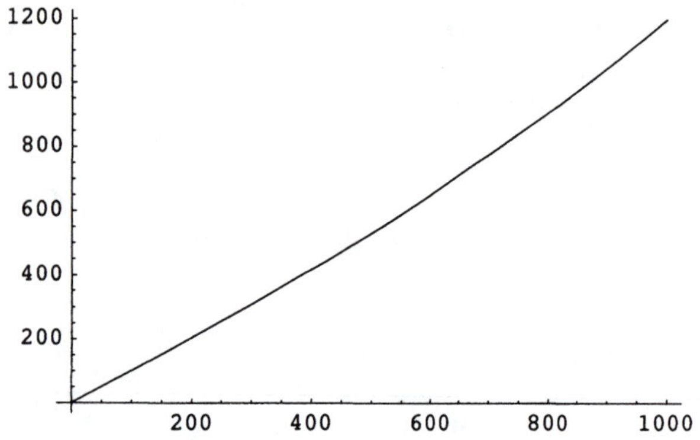


```
plot2 = Plot[p, {p, 0, 1000}];
```

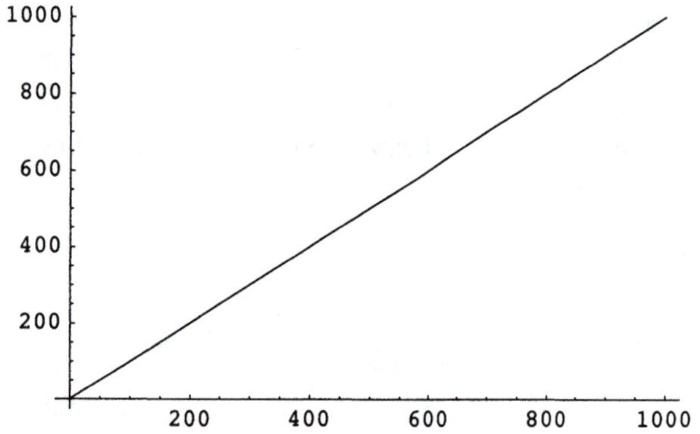

```
Show[plot1, plot2, AxesLabel -> {"p/bar", "f/bar"}];
```

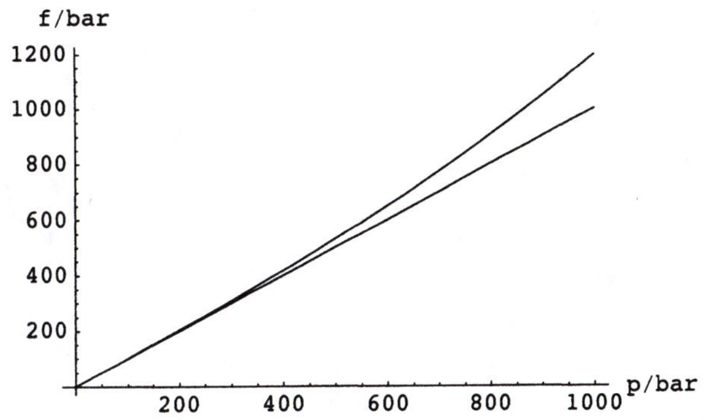

Chapter 5 Chemical Equilibrium

5.A For the reaction A(g) = B(g), assume that the standard Gibbs energy of formation of A is 50 kJ mol^{-1} and that of B is 45 kJ mol^{-1} at 298.15 K. (*a*) For a reaction starting with a mole of A at a pressure of 1 bar, plot the Gibbs energy of the mixture versus extent of reaction from zero to unity and identify the approximate extent of reaction at equilibrium. (*b*) Identify the equilibrium extent of reaction more precisely by plotting the derivative of the Gibbs energy of the mixture with respect to extent of reaction. Note that the slope of this plot goes to minus infinity at $\xi = 0$ and positive infinity at $\xi = 1$. (*c*) Calculate the equilibrium constant to verify the equilibrium extent of reaction.

SOLUTION

(a) The Gibbs energy of the reaction mixture is given by

```
Off[General::spelll];
Off[General::spell];

Clear[x]

g = 50 - 5 * x + (8.31451 * 10^-3) * 298.15 * ((1 - x) * Log[1 - x] + x * Log[x])

50 - 5 x + 2.47897 ((1 - x) Log[1 - x] + x Log[x])

Plot[g, {x, 0, 1}, AxesOrigin -> {0, 44},
  AxesLabel → {"ξ/mol", "G/kJ mol^-1"}, PlotRange → {44, 51}];
```

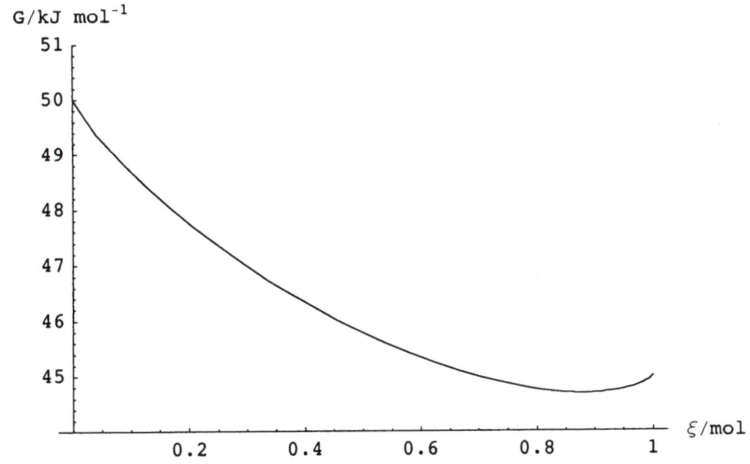

The approximate extent of reaction at equilibrium 298 K and 1 bar is 0.9. It looks like the limiting slope at $\xi = 1$ is about 10, but we can look at this more closely by changing the scale.

```
Plot[g, {x, .999, 1}, AxesOrigin -> {.999, 44.985}, AxesLabel → {"ξ/mol", "G/kJ mol⁻¹"}];
```

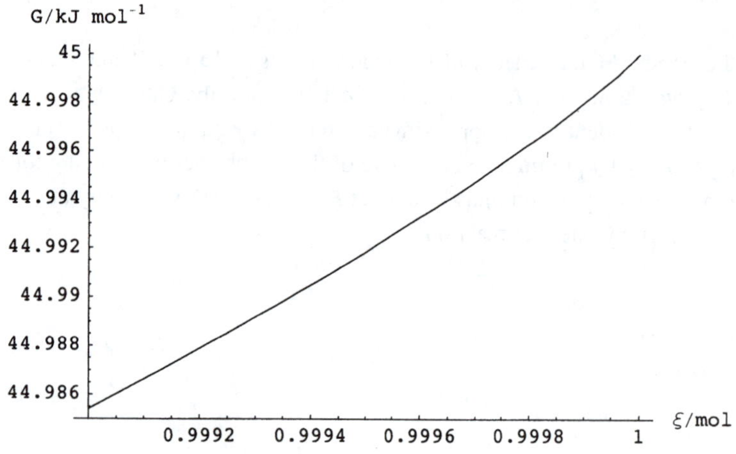

(b) The derivative of G with respect to ξ is given by

```
dvt = D[g, x]

-5 + 2.47897 (-Log[1 - x] + Log[x])

Plot[dvt, {x, 0, 1}, AxesLabel → {"ξ/mol", "dG/dξ"}];
```

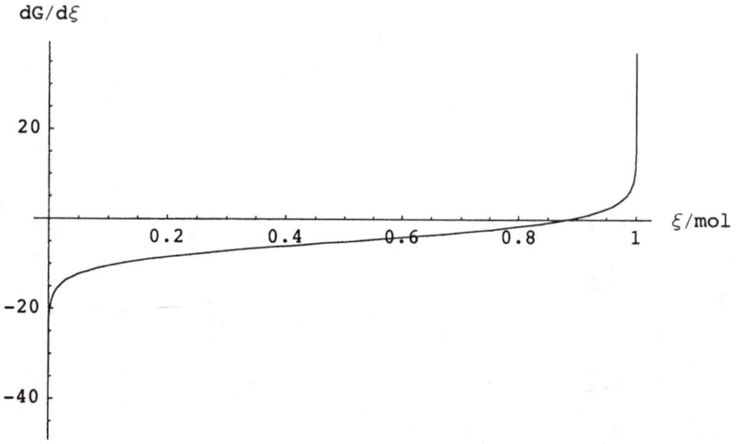

(c) The equilibrium constant is given by $K = \xi/(1-\xi)$, and so the equilibrium extent of reaction is given by ξ=k/(1+k).

```
k = Exp[5 / ((8.3145 * 10^-3) * 298.15)]

7.5155
```

```
k / (1 + k)

0.882567
```

which agrees with the preceding plot. Further calculations: If $\Delta_r G^o$ is more negative, the equilibrium will be further to the right. Try $\Delta_f G^o$ (B) = 30 kJ mol^{-1} and make sure that there is a minimum by plotting $\xi = 0.999$ to 1.

5.B Calculate $\Delta_r H^o$ and $\Delta_r S^o$ for the reaction
$N_2(g) + O_2(g) = 2NO(g)$
from the following values of the equilibrium constant:

T/K	1900	2000	2100	2200	2300	2400	2500	2600
$K/10^{-4}$	2.31	4.08	6.86	11.0	16.9	25.1	36.0	50.3

SOLUTION

```
t = {1900, 2000, 2100, 2200, 2300, 2400, 2500, 2600};

k = 10^-4 * {2.31, 4.08, 6.86, 11.0, 16.9, 25.1, 36, 50.3};

data = Transpose[{1 / t, Log[k]}] // N;

plot1 = ListPlot[data, Prolog → AbsolutePointSize[4],
   AxesOrigin → {.00038, -9}, DisplayFunction → Identity];

eq = Fit[data, {1, x}, x]

3.07613 - 21756.3 x
```

The intercept is equal to $\Delta_r S^o/R$ and so $\Delta_r S^o$ in J K^{-1} mol^{-1} is equal to

```
8.3145 * 3.076

25.5754
```

The intercept is equal to $-\Delta_r H^o/R$ and so $\Delta_r H^o$ in kJ mol^{-1} is equal to

```
8.3145 * 10^-3 * 21756

180.89

plot2 =
  Plot[eq, {x, 1 / 2600, 1 / 1900}, AxesOrigin → {.00038, -9}, DisplayFunction → Identity];
```

```
Show[plot1, plot2, AxesOrigin → {.00038, -9},
  AxesLabel → {"K/T", "ln K"}, DisplayFunction -> $DisplayFunction];
```

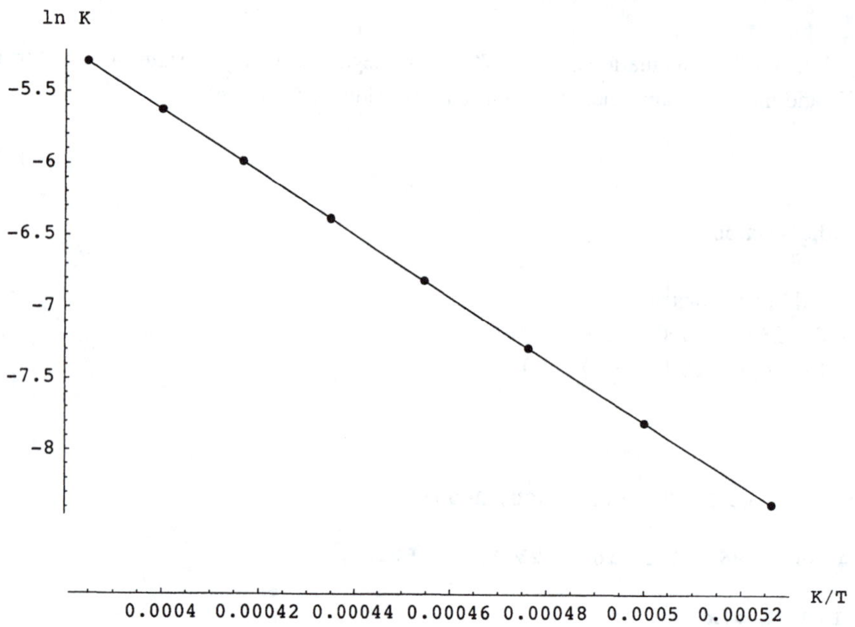

5.C Calculate the equilibrium composition for the reaction
N_2 (g) + 3 H_2 (g) = 2 NH_3 (g)
at 600 K and 15 bar, starting with 2 moles of N_2 and 1 mole of H_2. The equilibrium constant at 600 K is 0.001715.

SOLUTION

Note that kf is the expression for the equilibrium constant ranther than its numerical value.

```
ClearAll["Global`*"]

kf = pNH3^2 / pN2 / pH2^3;
pT = 15.;
names = {"N2", "H2", "NH3"};
initialmoles = {2, 1, 0};
moles = {2 - x, 1 - 3 x, 2 x};
molT = Apply[Plus, moles];
molX = moles / molT;
pp = pT * molX;
{pN2, pH2, pNH3} = pp;
kf
```

$$\frac{0.0177778\,(3 - 2\,x)^2\,x^2}{(1 - 3\,x)^3\,(2 - x)}$$

```
ans = NSolve[kf == .001715, x]

{{x → 1.67673 + 0.714119 i}, {x → 1.67673 - 0.714119 i}, {x → -0.446664}, {x → 0.0932039}}
```

The fourth power polynomial has four solutions, but only one of them, the last, is a solution to the physical problem, since it is real and positive. We can check this solution by substituting it into the expression for the equilibrium constant.

```
kf /. ans[[4]]
```

```
0.001715
```

Let us make the familiar table to summarize the calculation.

```
TableForm[Join[{names, initialmoles, moles, molX}]]
```

N2	H2	NH3
2	1	0
2 - x	1 - 3 x	2 x
$\frac{2-x}{3-2x}$	$\frac{1-3x}{3-2x}$	$\frac{2x}{3-2x}$

This table can be compared with the one in Section 5.4. The equilibrium mole fractions and partial pressures are given by the following table:

```
TableForm[{names, molX /. ans[[4]], pp /. ans[[4]]}]
```

N2	H2	NH3
0.677709	0.256039	0.0662526
10.1656	3.84058	0.993789

This approach can be applied to many other gas reactions.

5.D Calculate the equilibrium constants for the following four reactions at 500, 1000, and 2000 K using Table C.3 and make a table.

$H_2 (g) = 2H(g)$

$2HI(g) = H_2 (g) + I_2 (g)$

$I_2 (g) = 2I(g)$

$HI(g) = H(g) + I(g)$

SOLUTION

```
krx1 = Exp[- (2 / 8.31451) * {192.957 / .5, 165.484 / 1, 106.76 / 2}]
```

$\{4.84015 \times 10^{-41}, 5.15753 \times 10^{-18}, 2.65198 \times 10^{-6}\}$

```
krx2 = Exp[- (2 / 8.31451) * {10.088 / .5, 14.006 / 1, 21.009 / 2}]
```

```
{0.00780335, 0.0344223, 0.0799153}
```

```
krx3 = Exp[- (2 / 8.31451) * {50.203 / .5, 24.039 / 1, -29.42 / 2}]
```

$\{3.24281 \times 10^{-11}, 0.00308125, 34.4116\}$

```
krx4 = Exp[- (1 / 8.31451) * {(192.957 + 50.203 + 10.088) / .5,
    (165.485 + 24.039 + 14.006) / 1, (16.76 - 29.41 + 21.009) / 2}]
```

$\{3.4997 \times 10^{-27}, 2.33858 \times 10^{-11}, 0.60491\}$

```
TableForm[{krx1, krx2, krx3, krx4},
 TableHeadings -> {{"H2(g) = 2H(g)\\n", "2HI(g) = H2(g) + I2(g)",
   "\\nI2(g) = 2I(g)", "\\nHI(g) = H(g) + I(g)"}, {"500 K", "1000 K", "2000 K"}}]
```

	500 K	1000 K	2000 K
H2(g) = 2H(g)\n	4.84015×10^{-41}	5.15753×10^{-18}	2.65198×10^{-6}
2HI(g) = H2(g) + I2(g)	0.00780335	0.0344223	0.0799153
\nI2(g) = 2I(g)	3.24281×10^{-11}	0.00308125	34.4116
\nHI(g) = H(g) + I(g)	3.4997×10^{-27}	2.33858×10^{-11}	0.60491

5.E Sufficient data to calculate $\Delta_r G^o$ and $\Delta_r H^o$ for the reaction 2HI(g) = H_2 (g) + I_2 are given in Table C.3. calculate the standard reaction Gibbs energy at 500, 1000, 2000, and 3000 K. Use these data to calculate the standard reaction enthalpy at these temperatures by use of the Gibbs-Helmholtz equation, and compare the values calculated in this way with the values calculated from Table C.3

SOLUTION

The five values of the standard reaction Gibbs energy are given by

```
gerx = {19.325, 0, 0, 0, 0} - 2 * {1.56, -10.088, -14.006, -21.009, -27.114}

{16.205, 20.176, 28.012, 42.018, 54.228}
```

Plot gerx/T versus 1/T.

```
t = {298.15, 500, 1000, 2000, 3000}

{298.15, 500, 1000, 2000, 3000}

gerx / t

{0.0543518, 0.040352, 0.028012, 0.021009, 0.018076}

Transpose[{1 / t, gerx / t}] // N

{{0.00335402, 0.0543518}, {0.002, 0.040352},
 {0.001, 0.028012}, {0.0005, 0.021009}, {0.000333333, 0.018076}}

g1 = ListPlot[Transpose[{1 / t, gerx / t}], AxesOrigin -> {0, .01},
   Prolog -> AbsolutePointSize[4], AxesLabel → {"K/T", "ΔrG°/T"},
   PlotRange → {.01, .055}, DisplayFunction -> Identity];
```

Fit to a cubic equation.

```
Clear[x]

fn = Fit[Transpose[{1 / t, gerx / t}] // N, {1, x, x^2, x^3}, x]

0.0125465 + 17.7105 x - 2417.65 x^2 + 254532. x^3

g2 = Plot[fn, {x, .0003, .004}, AxesOrigin -> {0, .01}, DisplayFunction → Identity];
```

```
Show[g1, g2, DisplayFunction -> $DisplayFunction];
```

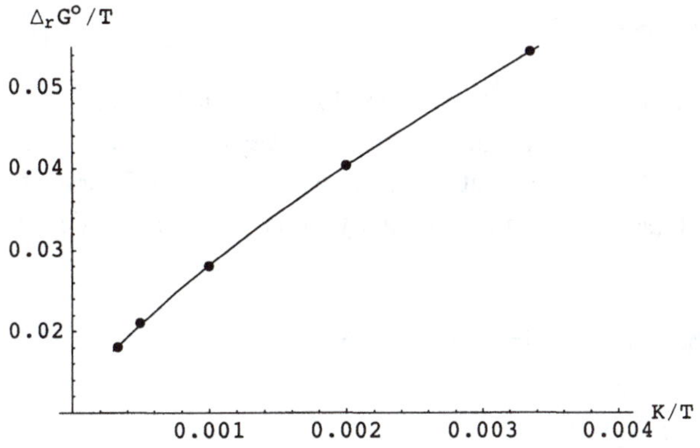

According to the Gibbs-Helmholtz equation, the derivative of $\Delta_r G^o/T$ with respect to 1/T is equal to the standard enthalpy of reaction

```
dvt = D[fn, x]
```

$17.7105 - 4835.29\,x + 763597.\,x^2$

where x = 1/T.

The values of the derivative at the five temperatures are

```
dHcalc = dvt /. x → 1 / t

{10.0828, 11.0943, 13.6388, 15.4837, 16.1835}
```

The JANAF Tables give the following values for the standard enthalpy of reaction:

```
dH = {62.421, 0, 0, 0, 0} - 2 * {26.359, -5.622, -6.754, -7.589, -10.489}

{9.703, 11.244, 13.508, 15.178, 20.978}
```

The errors in the standard enthalpies of reaction in kJ mol^{-1} due to fitting with a cubic equation are

```
dH - dHcalc

{-0.379826, 0.149744, -0.13076, -0.30571, 4.79446}
```

This indicates that a polynomial equation is not very good in representing the contributions of translational, vibrational, rotational, and electronic contributions to the Gibbs energies and enthalpies of the reactants over a wide range of temperatures. Further calculations: Show that equation 5.60 is better at fitting these data. Equation 5.60 requires 5 parameters, but it is limited in the sense that it is based on the assumption that the change in heat capacity in the reaction is indepedent of temperature.

5.F Calculate $\Delta_r H^\circ$ (1000 K), $\Delta_r S^\circ$ (1000 K), and K for the four following gas reactions:

$H_2 + (1/2)O_2 = H_2O$
$(1/2)N_2 + (3/2)H_2 = NH_3$
$CO_2 + 4H_2 = CH_4 + 2H_2O$
$CO + 3H_2 = CH_4 + H_2O$

This problem continues the use of the empirical dependencies of C_P° on temperature used in problems in Chapters 2 and 3.

(a) Calculate $\Delta_r H^\circ$ (1000 K) for these reactions.
(b) Calculate $\Delta_r S^\circ$ (1000 K) for these reactions.
(c) Use these values of $\Delta_r H^\circ$ (1000 K) and $\Delta_r S^\circ$ (1000 K) for these reactions to calculate K(1000 K) for each reactions.
(d) Make a table of the values calculated in (a) - (c) and discuss the relative importance of $\Delta_r H^\circ$ and $\Delta_r S^\circ$ in determining whether a reaction goes foward or backward as written.

SOLUTION

The empirical equations for the molar heat capacities used in Computer Problems 2.A and 3.A are used here again.

```
ClearAll["Global`*"]

cpN2 := 28.883 - .157 10^-2 t + .808 10^-5 t^2 - 2.871 10^-9 t^3;

cpO2 := 25.46 + 1.519 10^-2 t - .715 10^-5 t^2 + 1.311 10^-9 t^3;

cpH2 := 29.088 - .192 10^-2 t + .4 10^-5 t^2 - .87 10^-9 t^3;

cpCO := 28.142 + .167 10^-2 t + .537 10^-5 t^2 - 2.221 10^-9 t^3;

cpCO2 := 22.243 + 5.977 10^-2 t - 3.499 10^-5 t^2 + 7.464 10^-9 t^3;

cpH2O := 32.218 + .192 10^-2 t + 1.055 10^-5 t^2 - 3.593 10^-9 t^3;

cpNH3 := 24.619 + 3.75 10^-2 t - .138 10^-5 t^2;

cpCH4 := 19.875 + 5.021 10^-2 t + 1.268 10^-5 t^2 - 11.004 10^-9 t^3;
```

The following lists give $\Delta_r H^\circ$, $\Delta_r G^\circ$, and S° at 298.15 K in kJ mol^{-1} for the first two properties and in J K^{-1} mol^{-1} for the last property.

```
stdthpropN2 = {0, 0, 191.61};

stdthpropO2 = {0, 0, 205.138};
```

```
stdthpropH2 = {0, 0, 130.684};

stdthpropCO = {-110.525, -137.168, 197.674};

stdthpropCO2 = {-393.509, -394.359, 213.74};

stdthpropH2O = {-241.818, -228.572, 188.825};

stdthpropNH3 = {-46.11, -16.45, 192.45};

stdthpropCH4 = {-74.81, -50.72, 186.264};

TableForm[{stdthpropN2, stdthpropO2, stdthpropH2,
  stdthpropCO, stdthpropCO2, stdthpropH2O, stdthpropNH3, stdthpropCH4},
 TableHeadings -> {{"N2", "O2", "H2", "CO", "CO2", "H2O", "NH3", "CH4"},
   {DfHo / ("kJ mol^-1"), DfGo / ("kJ mol^-1"), Smo / ("J K^-1 mol-1")}}]
```

	DfHo / kJ mol^-1	DfGo / kJ mol^-1	Smo / J K^-1 mol-1
N2	0	0	191.61
O2	0	0	205.138
H2	0	0	130.684
CO	-110.525	-137.168	197.674
CO2	-393.509	-394.359	213.74
H2O	-241.818	-228.572	188.825
NH3	-46.11	-16.45	192.45
CH4	-74.81	-50.72	186.264

The calculations of the standard reaction enthalpies of four gas reactions at 1000 K in kJ mol^{-1} are repeated here.
(b)

```
hrx1 = (stdthpropH2O - stdthpropH2 - .5 * stdthpropO2) [[1]] +
  (10^-3) * Integrate[cpH2O - cpH2 - .5 * cpO2, {t, 298.15, 1000}]

-247.819

hrx2 = (stdthpropNH3 - .5 * stdthpropN2 - 1.5 * stdthpropH2) [[1]] +
  (10^-3) * Integrate[cpNH3 - .5 * cpN2 - 1.5 * cpH2, {t, 298.15, 1000}]

-53.8633

hrx3 = (stdthpropCH4 + 2 * stdthpropH2O - stdthpropCO2 - 4 * stdthpropH2) [[1]] +
  (10^-3) * Integrate[cpCH4 + 2 * cpH2O - cpCO2 - 4 * cpH2, {t, 298.15, 1000}]

-190.519

hrx4 = (stdthpropCH4 + stdthpropH2O - stdthpropCO - 3 * stdthpropH2) [[1]] +
  (10^-3) * Integrate[cpCH4 + cpH2O - cpCO - 3 * cpH2, {t, 298.15, 1000}]

-225.449

srx1=(stdthpropH2O-stdthpropH2-.5*stdthpropO2)[[3]]+Integrate[(cpH2O-cpH2-.5*cpO2)/t,{t
,298.15,1000}]

-55.1648
```

```
srx2=(stdthpropNH3-.5*stdthpropN2-1.5*stdthpropH2)[[3]]+Integrate[(cpNH3-.5*cpN2-1.5*cp
H2)/t,{t,298.15,1000}]
```

-115.288

```
srx3=(stdthpropCH4+2*stdthpropH2O-stdthpropCO2-4*stdthpropH2)[[3]]+Integrate[(cpCH4+2*c
pH2O-cpCO2-4*cpH2)/t,{t,298.15,1000}]
```

-220.035

```
srx4=(stdthpropCH4+stdthpropH2O-stdthpropCO-3*stdthpropH2)[[3]]+Integrate[(cpCH4+cpH2O-
cpCO-3*cpH2)/t,{t,298.15,1000}]
```

-252.297

(c) Calculate equilibrium constants of four reactions at 1000 K using $\Delta_r H^\circ$ (1000 K) and $\Delta_r S^\circ$ (1000 K).

```
krx1=Exp[-(1000*hrx1-1000*srx1)/(8.31451*1000)]
```

1.156×10^{10}

```
krx2=Exp[-(1000*hrx2-1000*srx2)/(8.31451*1000)]
```

0.000618865

```
krx3=Exp[-(1000*hrx3-1000*srx3)/(8.31451*1000)]
```

0.0287289

```
krx4=Exp[-(1000*hrx4-1000*srx4)/(8.31451*1000)]
```

0.0395921

(d) Make a table.

```
TableForm[Transpose[{{hrx1,hrx2,hrx3,hrx4},1000*(10^-3)*{srx1,srx2,srx3,srx4},{krx1,krx
2,krx3,krx4}}],TableHeadings->{{"H2+(1/2)O2=H2O","(1/2)N2+(3/2)H2=NH3","CO2+4H2=CH4+2H2
O","CO+3H2=CH4+H2O"},{DrHo/("kJ mol^-1"),"T*DrSo"/("kJ mol^-1"),"K"}}]
```

	DrHo / kJ mol^-1	T*DrSo / kJ mol^-1	K
H2+(1/2)O2=H2O	-247.819	-55.1648	1.156×10^{10}
(1/2)N2+(3/2)H2=NH3	-53.8633	-115.288	0.000618865
CO2+4H2=CH4+2H2O	-190.519	-220.035	0.0287289
CO+3H2=CH4+H2O	-225.449	-252.297	0.0395921

Note that the entropies of reaction are all negative because there is a decrease in the number of moles of gas in each of the reactions. Thus the entropy changes favor the reactants rather than the products. In the first reaction there is such a negative enthalpy change that the reaction goes strongly to the right. Although the enthalpy changes are favorable for the other three reactions, the entropy change dominates and so that these reactions are not spontaneous in the direction written.

To calculate equilibrium constants for other reactions, you can type in the expressions for the heat capacities, put in the lists of properties, and use equations of the form used here. From these printouts, you can see how to change the temperature.

5.G Plot the equilibrium extent of the reaction A(g) = 2B(g) as a function of pressure for $K = 0.1$, 1, and 10. Equation 5.31 can be used to calculate the equilibrium extent of reaction.

SOLUTION

```
extent[k_, p_] := 1/ (1 + (4 * p) / k) ^.5
```

```
Plot[{extent[.1, p], extent[1, p], extent[10, p]}, {p, 0, 5}, AxesLabel → {"\!\(\*
StyleBox[\"P\",\nFontSlant->\"Italic\"]\)/\!\(\*
StyleBox[\(\*
StyleBox[\"P\",\nFontSlant->\"Italic\"]°\)]\)", "ξ"}];
```

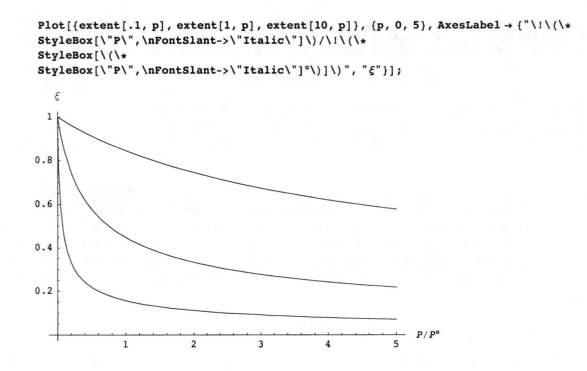

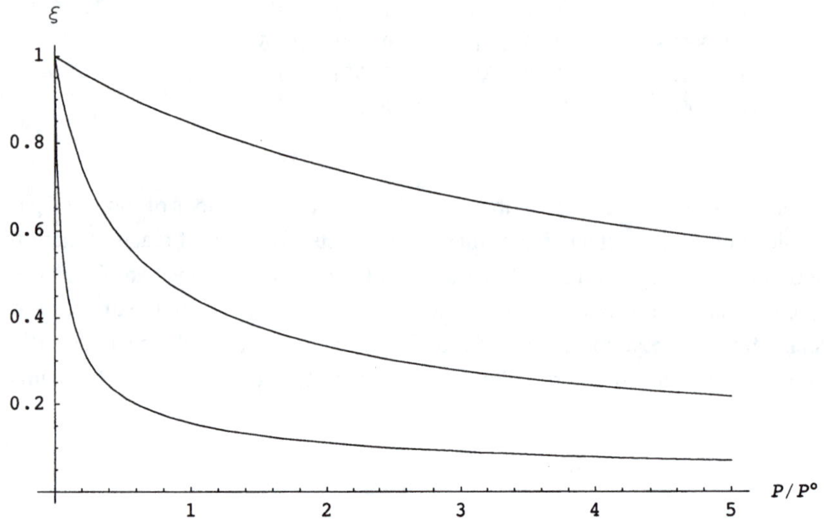

5.H Plot the equilibrium extent of reaction for the reaction A(g) = 2B(g) versus pressure for reactions with equilibrium constants of 0.1, 1, and 10 by solving for the extent of reaction, rather than using the analytic expression for the extent of reaction. This method can be used for more complicated reactions.

SOLUTION

The relation between the equilibrium constant and the equilibrium extent of reaction for this reaction is given by equation 5.32.

```
k=4.*x^2*p/(1-x^2)
```

$$\frac{4.\,p\,x^2}{1 - x^2}$$

```
extentrx=Solve[.1==k,x]
```

$$\left\{\left\{x \to -\frac{1.}{\sqrt{1. + 40.\,p}}\right\}, \left\{x \to \frac{1.}{\sqrt{1. + 40.\,p}}\right\}\right\}$$

The extent of reaction must be positive, and so we will plot the second solution.

```
plot1=Plot[1/Sqrt[1+40*p],{p,0,5},AxesLabel->{"P/bar","ξ"},DisplayFunction->Identity];

extentrx=Solve[1==k,x]
```

$$\left\{\left\{x \to -\frac{1.}{\sqrt{1. + 4.\,p}}\right\}, \left\{x \to \frac{1.}{\sqrt{1. + 4.\,p}}\right\}\right\}$$

```
plot2=Plot[1/Sqrt[1+4*p],{p,0,5},AxesLabel->{"P/bar","ξ"},DisplayFunction->Identity];
```

```
extentrx=Solve[10==k,x]
```

$$\left\{\left\{x \to -\frac{2.23607}{\sqrt{5.+2.\,p}}\right\},\ \left\{x \to \frac{2.23607}{\sqrt{5.+2.\,p}}\right\}\right\}$$

```
plot3=Plot[Sqrt[5]/Sqrt[5+2*p],{p,0,5},AxesLabel->{"P/bar","ξ"},DisplayFunction->Identi
ty];

Show[plot1,plot2,plot3,DisplayFunction->$DisplayFunction];
```

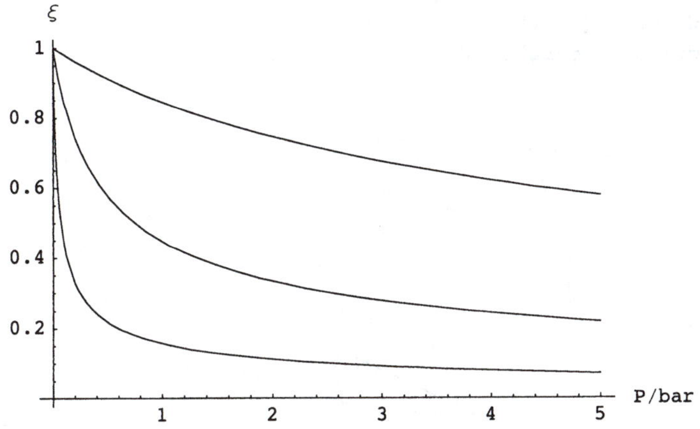

5.I Plot the equilibrium extent of the reaction A(g) = 2B(g) versus temperature at pressures of 0.1, 1, and 10 bar for a reaction having an equilibrium constant of 0.1 at 298 K and a standard enthalpy of reaction of 50 kJ mol^{-1} that is independent of temperature.

SOLUTION

The equilibrium extent of reaction is given by

$\xi = 1/[1 + (4P/K]^{1/2}$

```
ClearAll["Global`*"]

extent = 1 / (1 + (4 * p / k)) ^ .5
```

$$\frac{1}{(1+\frac{4\,p}{k})^{0.5}}$$

The equilibrium constant is given as a function of temperature by

```
k = c * Exp[-50000 / (8.3145 * t)]
```

$c\, e^{-6013.59/t}$

```
c = .1 / Exp[-50000 / (8.3145 * 298)]
```

```
5.80752 × 10^7
```

```
plota = Plot[extent /. p → .1, {t, 300, 400}, DisplayFunction → Identity];
```

```
plotb = Plot[extent /. p → 1, {t, 300, 400}, DisplayFunction -> Identity];
```

```
plotc = Plot[extent /. p → 10, {t, 300, 400}, DisplayFunction → Identity];
```

```
Show[{plota, plotb, plotc}, PlotRange → {0, 1},
  DisplayFunction → $DisplayFunction, AxesLabel → {"T/K", "ξ"}];
```

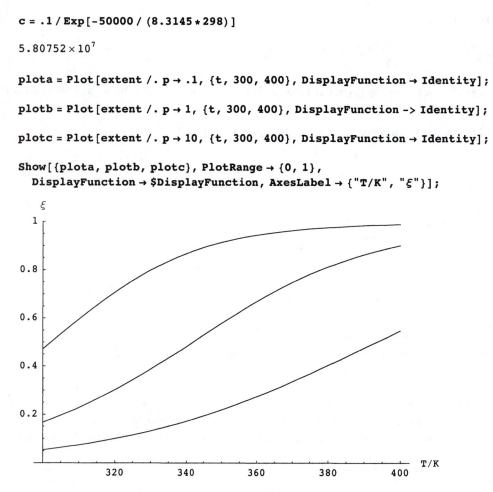

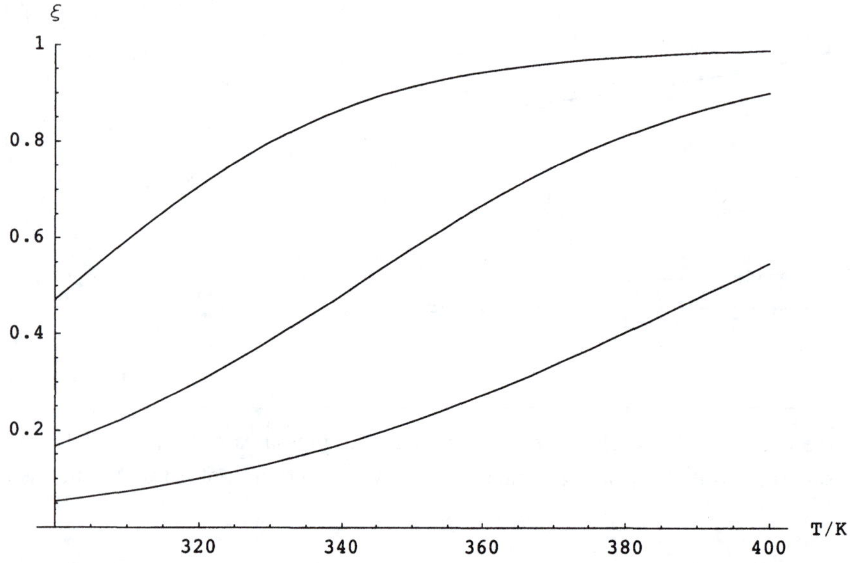

The top curve is for the lowest pressure (0.1 bar), the middle curve is for the intermediate pressure (1 bar), and the bottom curve is for the highest pressure (10 bar).

5.J Calculate the equilibrium composition of a reaction mixture containing CO, CO_2, H_2, H_2O, and CO_2 at 1000 K and a total presssure of 1 bar. In doing that also calculate the volume. The mixture initially consists of 2 mol CH_4 and 3 mol H_2O. There are two independent reactions, which can be written as

$CH_4 + H_2O = CO + 3H_2$

$CO + H_2O = CO_2 + H_2$

SOLUTION

First we need the values of the two equilibrium constants at 1000 K.

```
ClearAll["Global`*"]
```

```
dG1 = -200.275 - 19.492 + 192.59
```

```
-27.177
```

```
k1 = Exp[-dG1 / ((8.3145 * 10^-3) (1000))]
```

```
26.2752
```

```
dG2 = -395.886 + 200.275 + 192.59

-3.021

k2 = Exp[-dG2 / ((8.3145 * 10^-3) (1000))]

1.43813
```

As a simplification, V/RT is replaced with in the conversions of amounts to partial pressure.

```
soln = Solve[{k1 == pco * ph2^3 / (pch4 * ph2o), k2 == pco2 * ph2 / (pco * ph2o),
   pco == 83.145 * nco / v, ph2o == 83.145 * nh2o / v, pco2 == 83.145 * nco2 / v, ph2 == 83.145 * nh2 / v,
   pch4 == 83.145 * nch4 / v, 1 == 83.145 * (nco + nh2o + nco2 + nh2 + nch4) / v,
   14 == 2 * nh2 + 4 * nch4 + 2 * nh2o, 2 == nco + nco2 + nch4, 3 == 2 * nco2 + nco + nh2o},
  {nch4, nh2o, nco, nco2, nh2, pch4, ph2o, pco, pco2, ph2, v}]

{{nh2 → -8.9239 - 6.03163 i, nch4 → 3.49501 + 0.677928 i,
  nh2o → 8.93388 + 4.67577 i, nco → 2.94386 + 3.31992 i, nco2 → -4.43887 - 3.99784 i,
  pco2 → -0.595664 - 2.39081 i, pch4 → 1.03868 + 1.03794 i, ph2o → 1.97627 + 3.6594 i,
  pco → 0.240842 + 1.81418 i, v → 167.12 - 112.733 i, ph2 → -1.66013 - 4.12071 i},
 {nh2 → -8.9239 + 6.03163 i, nch4 → 3.49501 - 0.677928 i, nh2o → 8.93388 - 4.67577 i,
  nco → 2.94386 - 3.31992 i, nco2 → -4.43887 + 3.99784 i, pco2 → -0.595664 + 2.39081 i,
  pch4 → 1.03868 - 1.03794 i, ph2o → 1.97627 - 3.6594 i, pco → 0.240842 - 1.81418 i,
  v → 167.12 + 112.733 i, ph2 → -1.66013 + 4.12071 i}, {nh2 → -7.68767,
  nch4 → 5.96715, nh2o → 2.75338, nco → -8.18091, nco2 → 4.21376, pco2 → -1.43604,
  pch4 → -2.03359, ph2o → -0.938346, pco → 2.78804, v → -243.972, ph2 → 2.61994},
 {nh2 → -2.23495, nch4 → 4.1049, nh2o → 1.02515, nco → -6.18464, nco2 → 4.07974,
  pco2 → 5.16289, pch4 → 5.19473, ph2o → 1.29733, pco → -7.82663, v → 65.7016,
  ph2 → -2.82832}, {nh2 → 5.79617, nch4 → 0.174437, nh2o → 0.854954,
  nco → 1.50608, nco2 → 0.319482, pco2 → 0.0369296, pch4 → 0.0201634,
  ph2o → 0.0988258, pco → 0.174091, v → 719.298, ph2 → 0.66999},
 {nh2 → 9.64103, nch4 → 2.02338, nh2o → -6.68778, nco → -9.73454, nco2 → 9.71116,
  pco2 → 1.96056, pch4 → 0.408495, ph2o → -1.35018, pco → -1.96528, v → 411.838,
  ph2 → 1.94641}, {nh2 → 12.3332, nch4 → -2.25988, nh2o → -0.813465,
  nco → 4.7063, nco2 → -0.446415, pco2 → -0.0330194, pch4 → -0.167154,
  ph2o → -0.0601686, pco → 0.348105, v → 1124.1, ph2 → 0.912237}}

TableForm[soln]
```

nh2 → -8.9239 - 6.03163 i	nch4 → 3.49501 + 0.677928 i	nh2o → 8.93388 + 4.67577 i	r
nh2 → -8.9239 + 6.03163 i	nch4 → 3.49501 - 0.677928 i	nh2o → 8.93388 - 4.67577 i	r
nh2 → -7.68767	nch4 → 5.96715	nh2o → 2.75338	r
nh2 → -2.23495	nch4 → 4.1049	nh2o → 1.02515	r
nh2 → 5.79617	nch4 → 0.174437	nh2o → 0.854954	r
nh2 → 9.64103	nch4 → 2.02338	nh2o → -6.68778	r
nh2 → 12.3332	nch4 → -2.25988	nh2o → -0.813465	r

Only the fifth solution is realistic physically because all of the partial pressures are positive real numbers.

```
soln[[5]]

{nh2 → 5.79617, nch4 → 0.174437, nh2o → 0.854954,
 nco → 1.50608, nco2 → 0.319482, pco2 → 0.0369296, pch4 → 0.0201634,
 ph2o → 0.0988258, pco → 0.174091, v → 719.298, ph2 → 0.66999}
```

Note that 8.65113 is the total number of moles of species.

```
{nh2, nch4, nh2o, nco, nco2} /. soln[[5]]
```

```
{5.79617, 0.174437, 0.854954, 1.50608, 0.319482}
```

```
Apply[Plus, {nh2, nch4, nh2o, nco, nco2} /. soln[[5]]]
```

```
8.65113
```

You can make more calculations to test the effect of initial composition and pressure of the equilibrium composition. This procedure can also be applied to other multireaction systems.

5.K There is another way that the equilibrium composition of a multi-reaction system can be calculated, and that is by use of the Newton-Raphson method. This method starts with an estimate of the equilibrium composition and improves it in an iterative process. A very efficient program for carrying this out, called equcalc in *Mathematica*, has been written by F. Krambeck of Mobil Research and Development (A. M. Sapre and F. J. Krambeck (Eds.), Chemical Reactions in Complex Systems, Van Nostrand Reinhold, New York, 1991). The input consists of a conservation matrix as, a vector of Gibbs energies of formation at the desired pressure multiplied by (-1/RT), and a vector of initial amounts of species no. Calculate the equilibrium amounts for the preceding problem using equcalc.

SOLUTION

```
equcalc[as_,lnk_,no_]:=Module[{s,l,x,b,ac,m,n,e,k},
(*  as=conservation matrix
lnk=-(1/RT)(Gibbs energy of formation vector at T,P)
no=initial composition vector *)
(*Setup*)
{m,n}=Dimensions[as];
s=as.no;
ac=as-Outer[ Times,s/s[[1]],as[[1]] ];
ac[[1]]=Table[1.,{n}];
b=Table[0.,{m}];b[[1]]=1.;
(*Initialize*)
l=LinearSolve[ as.Transpose[as],-as.(lnk+Log[n]) ];
(*Solve*)
Do[ e=b-ac.(x=E^(lnk+l.as) );
If[(10^-10)>Max[ Abs[e] ], Break[] ];
l=l+LinearSolve[ac.Transpose[as*Table[x,{m}]],e],
{k,100}];
If[ k=100,Return["Algorithm Failed"] ];
(*Normalize*)
Return[x*s[[1]]/(as[[1]].x)]
]
```

```
as={{0,1,0,1,1},{2,0,2,0,4},{1,1,0,2,0}};
```

```
lnk=-(1/(8.31451*1.000))*{-192.590,-200.275,0,-395.886,19.492}
```

```
{23.1631, 24.0874, 0, 47.6139, -2.34434}
```

```
TableForm[as]
```

0	1	0	1	1
2	0	2	0	4
1	1	0	2	0

```
no={3,0,0,0,2}
```

```
{3, 0, 0, 0, 2}
```

```
amounts=equcalc[as,lnk,no]

{0.854955, 1.50608, 5.79617, 0.319482, 0.174437}

TableForm[{amounts},TableHeadings->{None,{"H2O","CO","H2","CO2","CH4"}}]
```

H2O	CO	H2	CO2	CH4
0.854955	1.50608	5.79617	0.319482	0.174437

This calculated composition is the same as in the previous problem.

```
as.amounts

{2., 14., 3.}

as.no

{2, 14, 3}

Apply[Plus, amounts]

8.65113
```

5.L (a) Calculate the standard enthalpy of dissociation of calcite from the data in Table 5.1 by fitting the data to lnP = A + B/T. Compare $\Delta_r H°$ with the NBS value at 298.15 K. (b) Calculate the standard Gibbs Energy of the reaction at 298.15 K using the parameters obtained in (a).

SOLUTION

(a)

```
p = {9.2 * 10^-5, 2.39 * 10^-3, 2.88 * 10^-2, .2217, .987, 3.82, 11.35, 28.31}

{0.000092, 0.00239, 0.0288, 0.2217, 0.987, 3.82, 11.35, 28.31}

t = 273.15 + {500, 600, 700, 800, 897, 1000, 1100, 1200}

{773.15, 873.15, 973.15, 1073.15, 1170.15, 1273.15, 1373.15, 1473.15}

lnp = Log[p];

input = Transpose[{1 / t, lnp}];

plot1 = ListPlot[input, AxesOrigin → {6 * 10^-4, -10},
   Prolog → AbsolutePointSize[4], DisplayFunction → Identity];

fn = Fit[input, {1, x}, x]

17.4601 - 20545.4 x

plot2 =
  Plot[fn, {x, 6 * 10^-4, .0013}, AxesOrigin → {6 * 10^-4, -10}, DisplayFunction → Identity];
```

```
Show[plot1, plot2, AxesOrigin → {6 * 10 ^ -4, -10},
  DisplayFunction → $DisplayFunction, AxesLabel → {"K/T", "ln K"}];
```

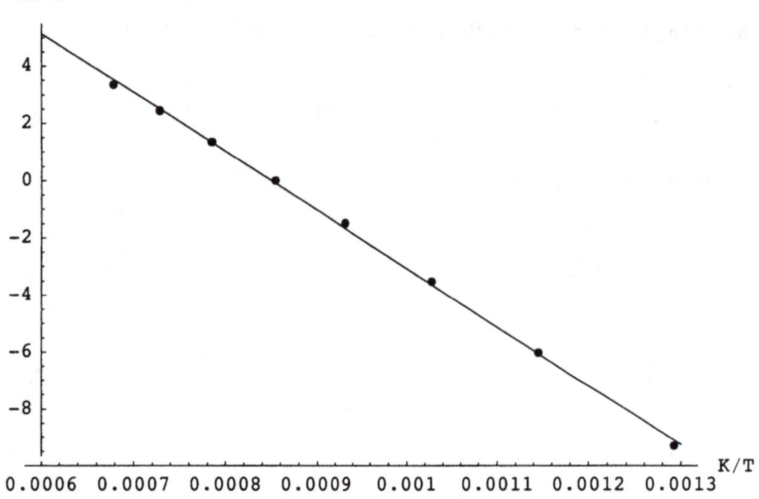

The intercept is equal to -$\Delta_r H°$ /R and so $\Delta_r H°$ in kJ mol^{-1} is equal to

```
8.31451 * 10 ^ -3 * 20545

170.822
```

The NBS Tables yield $\Delta_r H°$ (298.15 K) in kJ mol^{-1}

```
-635.09 - 393.51 + 1206.92

178.32
```

(b) The value of $\Delta_r G°$ (298.15 K) in kJ mol^{-1} calculated from Fig. 5.6 is

```
-8.31451 * 10 ^ -3 * 298.15 * fn /. x → 1 / 298.15

127.542
```

The NBS Tables yield

```
-604.03 - 394.36 + 1128.79

130.4
```

Note K(298.15K) is extremely small.

```
Exp[-130.4 / (8.31451 * .29815)]
```

```
1.42902 × 10^-23
```

These data require a more complete treatment because $\Delta_r C_P{}^\circ$ varies with temperature.

5.M The standard Gibbs energies of the isomers of the pentanes at 700 K are as follows:

n-pentane	193.26 kJ mol^-1
2-mtheylbutane	190.75
2,2-dimethylpropane	202.05

What are the equilibrium mole fractions of the three isomers? If the gases are ideal, will changing the pressure change this distribution.

SOLUTION

```
dGiso=-8.3145*.700*Log[Exp[-193.26/(8.3145*.700)]+Exp[-190.75/(8.3145*.700)]+Exp[-202.0
5/(8.3145*.700)]]
```

```
187.351
```

```
Exp[(187.35-{193.26,190.75,202.05})/(8.3145*.700)]
```

```
{0.362244, 0.557564, 0.0800016}
```

Changing the pressure will not change the distribution.

5.N The reaction A = B, where A and B are ideal gases, is studied at 298.15 K. The standard chemical potential of A at this temperature is 40 kJ mol^{-1} and for B is 37 kJ mol^{-1}. (a) Plot $G_{avg} = (1 - \xi)\mu_A{}^\circ + \xi\mu_B{}^\circ$ versus ξ from 0 to 1. (b) Plot the Gibbs energy of mixing versus ξ. (c) Plot the sum of (a) and (b) versus ξ. (d) Plot $dG/d\xi$ versus ξ. (e) Plot $d^2 G/d\xi^2$ versus ξ.

SOLUTION

(a)

```
Plot[(1-ξ)*40+ξ*37,{ξ,0,1},AxesLabel->{"ξ","\!\(G\_avg\)"}];
```

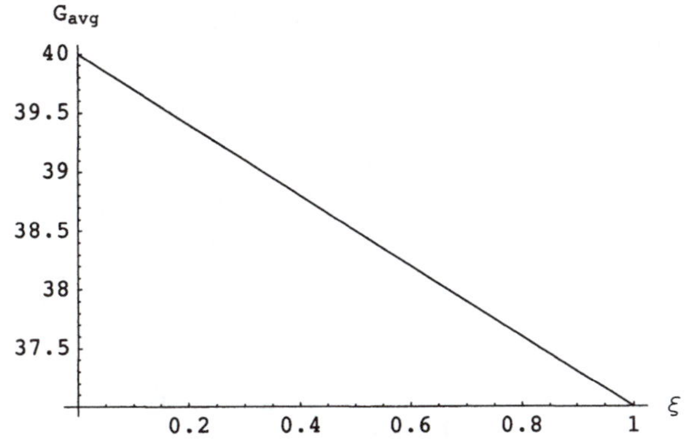

(b)

```
Plot[8.31451*.29815*((1-ξ)*Log[1-ξ]+ξ*Log[ξ]),{ξ,0,1},AxesOrigin->{0,-1.8},AxesLabel->{
"ξ","\!\(G\_mix\)"}];
```

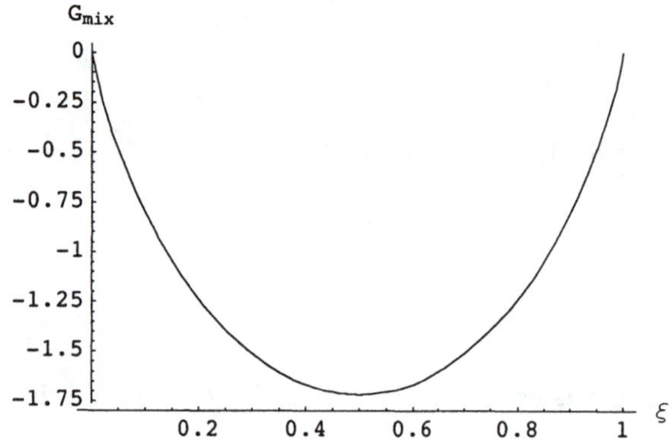

(c)

```
plot1=Plot[(1-ξ)*40+ξ*37+8.31451*.29815*((1-ξ)*Log[1-ξ]+ξ*Log[ξ]),{ξ,0,1},AxesLabel->{"
ξ","G"},PlotRange->{36,40}];
```

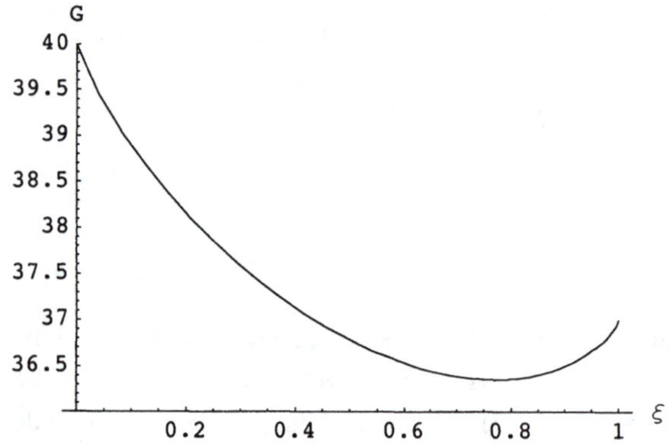

(d)

```
plot2=Plot[Evaluate[D[(1-ξ)*40+ξ*37+8.31451*.29815*((1-ξ)*Log[1-ξ]+ξ*Log[ξ]),ξ]],{ξ,0,1
},AxesLabel->{"ξ","dG/dξ"},PlotRange->{-15,15}];
```

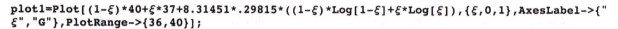

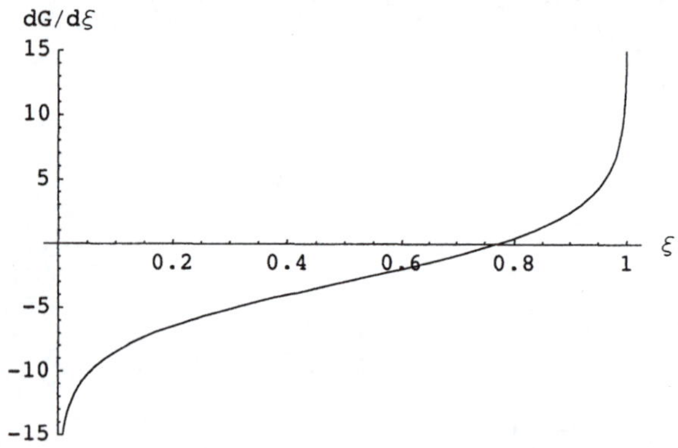

(e)

```
plot3=Plot[Evaluate[D[(1-ξ)*40+ξ*37+8.31451*.29815*((1-ξ)*Log[1-ξ]+ξ*Log[ξ]),{ξ,2}]],{ξ
,0,1},AxesLabel->{"ξ","\!\(d\^2\)G/d\!\(ξ\^2\)"},PlotRange->{0,100}];
```

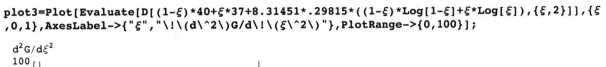

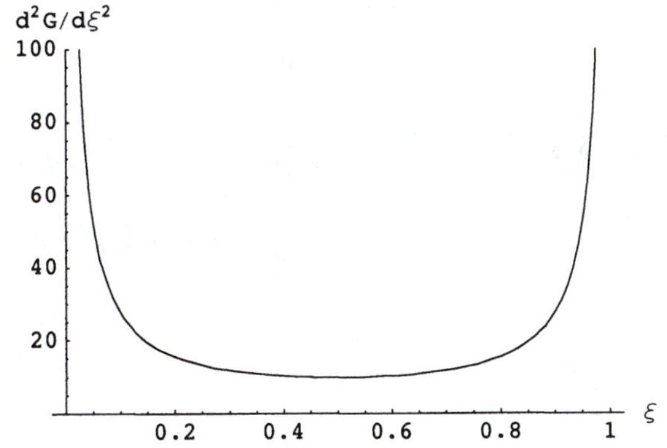

```
Show[GraphicsArray[{plot1,plot2,plot3}]];
```

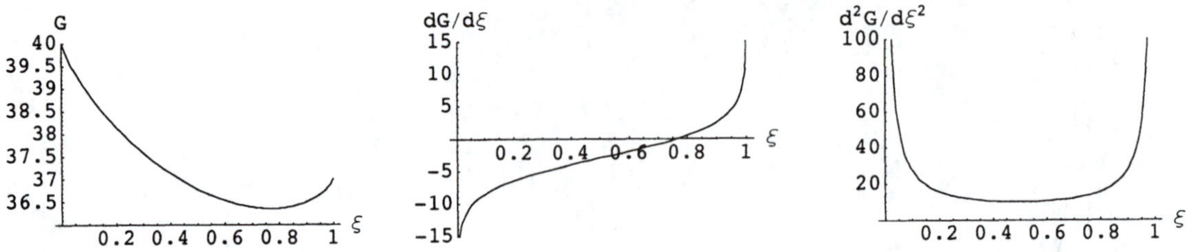

Chapter 6 Phase Equilibrium

6.A The vapor pressure of ice from -40 ^{o} C to 0 ^{o} C is given in Table 6.1. Fit these data to the equation
$\ln P = -a/T + b \ln T + cT + d$
The form of this equation is sugested by equation 5.60. Plot this equation as ln P versus $1/T$. Calculate $\Delta_{sub} H^o$ as a function of temperature, and calculate its value at each of the temperatures in Table 6.1.

SOLUTION
The temperatures are in K, and the pressures are in kPa.

```
Off[General::spell1];
Off[General::spell];

t = 273.15 + {-40, -30, -20, -10, 0};

p = {.013, .038, .103, .260, .611};

lnp = Log[{.013, .038, .103, .260, .611}];

input = Transpose[{1 / t, lnp}]

{{0.00428908, -4.34281}, {0.00411269, -3.27017},
 {0.00395023, -2.27303}, {0.00380011, -1.34707}, {0.00366099, -0.492658}}

g1 = ListPlot[Transpose[{1 / t, lnp}],
  Prolog -> AbsolutePointSize[4], DisplayFunction → Identity];
```

We want to fit lnp to fn=-a/T+blnT+cT+d. But since we want to take the derivative with respect to 1/T, we will define a new variable x = 1/T and write the function representing the vapor pressure by fn=-ax-blnx+c/x+d.

```
Clear[x]

fn = Fit[input, {-x, -Log[x], 1 / x, 1}, x]
```

$$-568.856 - \frac{0.205901}{x} + 7578.19\,x - 106.392\,\text{Log}[x]$$

```
g2 = Plot[fn, {x, .0036, .0043}, DisplayFunction → Identity];
```

```
Show[g1, g2, AxesOrigin → {.0036, -5}, PlotRange → {-5, 0},
  AxesLabel → {"K/T", "ln P"}, DisplayFunction → $DisplayFunction];
```

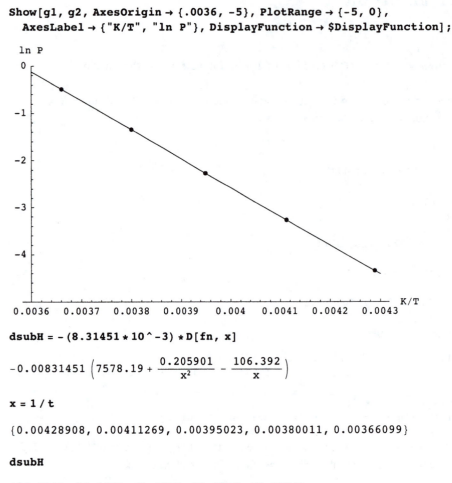

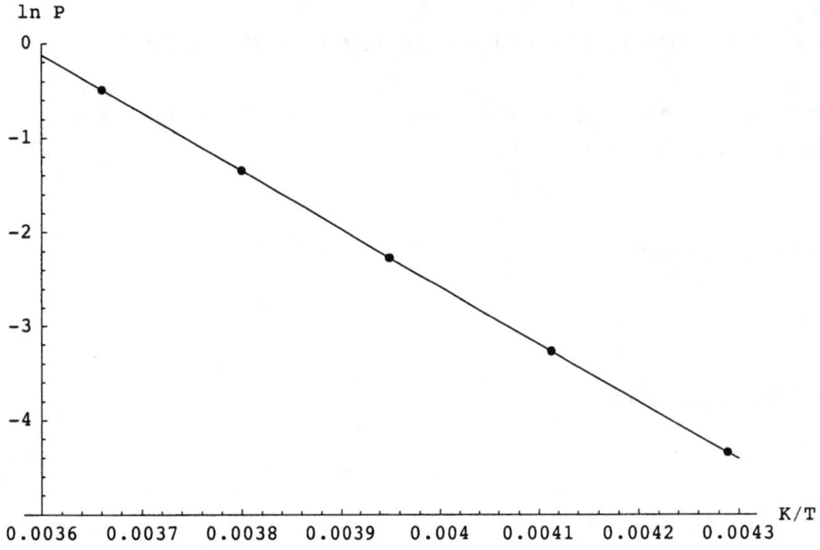


```
dsubH = - (8.31451 * 10^-3) * D[fn, x]
```

$$-0.00831451\left(7578.19+\frac{0.205901}{x^2}-\frac{106.392}{x}\right)$$

```
x = 1 / t
```

```
{0.00428908, 0.00411269, 0.00395023, 0.00380011, 0.00366099}
```

```
dsubH
```

```
{50.1743, 50.8661, 51.2156, 51.2227, 50.8874}
```

These are the standard enthalpies of sublimation of ice in kJ mol^{-1} at the experimental temperatures.

6.B The vapor pressure of water from 0 ^oC to 100 ^oC is given in Table 6.1. Fit this data to the equation
$\ln P = -a/T + b \ln T + cT + d$
Plot the data on ln P versus $1/T$. Determine the parameters a, b, c, and d, and test this equation to see how well it represents the vapor pressure data. Calculate $\Delta_{vap} H^o$ as a function of temperature, and calculate its value at each of the temperatures in Table 6.1. Calculate the standard enthalpy of vaporization at 298.15 K using Table C.3 and compare it with the value calculated with this empirical equation.

SOLUTION

```
ClearAll["Global`*"]
```

The temperatures are in K, and the pressures are in kPa.

```
t = 273.15 + {0, 10, 20, 30, 40, 60, 100};

1 / t;

p = {.611, 1.228, 2.338, 4.245, 7.381, 19.933, 101.325};

lnp = Log[{.611, 1.228, 2.338, 4.245, 7.381, 19.933, 101.325}];

input = Transpose[{1 / t, lnp}];

g1 = ListPlot[Transpose[{1 / t, lnp}],
  Prolog -> AbsolutePointSize[4], DisplayFunction → Identity];
```

We want to fit lnp to the function fn=-a/T+blnT+cT+d. But later since we want to take the derivative with respect to 1/T, we will define a new variable x = 1/T and write the function representing the vapor pressure by fn=-ax-blnx+c/x+d.

```
fn = Fit[input, {-x, -Log[x], 1 / x, 1}, x]
```

$$68.8367 + \frac{0.00519681}{x} - 7196.38\,x + 7.91493\,\mathrm{Log}[x]$$

```
g2 = Plot[fn, {x, .0026, .0037}, DisplayFunction → Identity];

Show[g1, g2, AxesOrigin → {.0026, -1}, PlotRange → {-1, 5},
  AxesLabel → {"K/T", "ln P"}, DisplayFunction → $DisplayFunction];
```

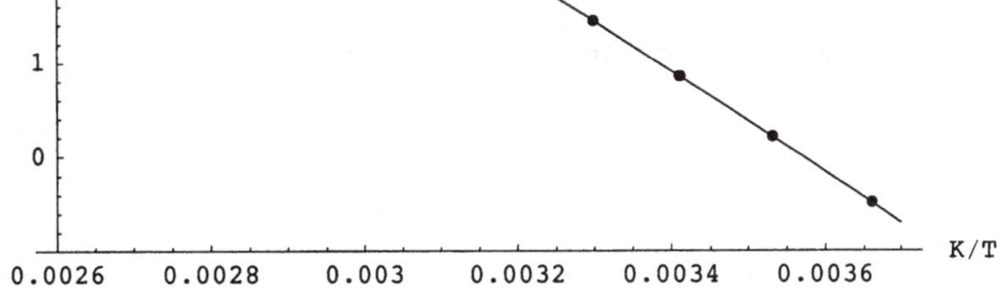

```
dvapH = - (8.31451 * 10^-3) * D[fn, x]
```

$$-0.00831451\left(-7196.38-\frac{0.00519681}{x^2}+\frac{7.91493}{x}\right)$$

```
x = 1 / t
```

```
{0.00366099, 0.0035317, 0.00341122, 0.0032987, 0.00319336, 0.00300165, 0.00267989}
```

```
dvapH
```

```
{45.0826, 44.6648, 44.2558, 43.8553, 43.4635, 42.7059, 41.2943}
```

```
x = 1 / 298.15
```

```
0.00335402
```

```
dvapH
```

```
44.0545
```

The value calculated using Table C.2 is 44.01 kJ mol^{-1}.

6.C The vapor pressure of methanol in bars is given in the following table as a function of temperature. Fit these vapor pressures to equation 6.17 using a computer and calculate the heat of vaporization at each of these temperatures using equation 6.18. Calculate the standard enthalpy of vaporization of methanol at 298.15 K using Table C.2 and compare it with the value calculated with the empirical equation.

$t/^oC$	-44.0	-16.2	5.0	21.2	49.9	64.7
P_{vap}/bar	0.001333	0.01333	0.0533	0.1333	0.533	1.0132

SOLUTION

```
ClearAll["Global`*"]

p = {0.001333, 0.01333, 0.0533, 0.1333, 0.533, 1.0132};

lnp = Log[p];

t = 273.15 + {-44.0, -16.2, 5.0, 21.2, 49.9, 64.7};

input = Transpose[{1 / t, lnp}];

plot1 = ListPlot[input, Prolog -> AbsolutePointSize[4], DisplayFunction → Identity];

fn = Fit[input, {-x, -Log[x], 1 / x, 1}, x]
```

$$335.483+\frac{0.0953766}{x}-12976.6\,x+56.5529\,\mathrm{Log}[x]$$

```
plot2 = Plot[fn, {x, .0029, .0044}, DisplayFunction → Identity];
```

```
Show[plot1, plot2, AxesOrigin → {.0029, -7}, PlotRange → {-7, 1},
  AxesLabel → {"K/T", "ln P"}, DisplayFunction → $DisplayFunction];
```

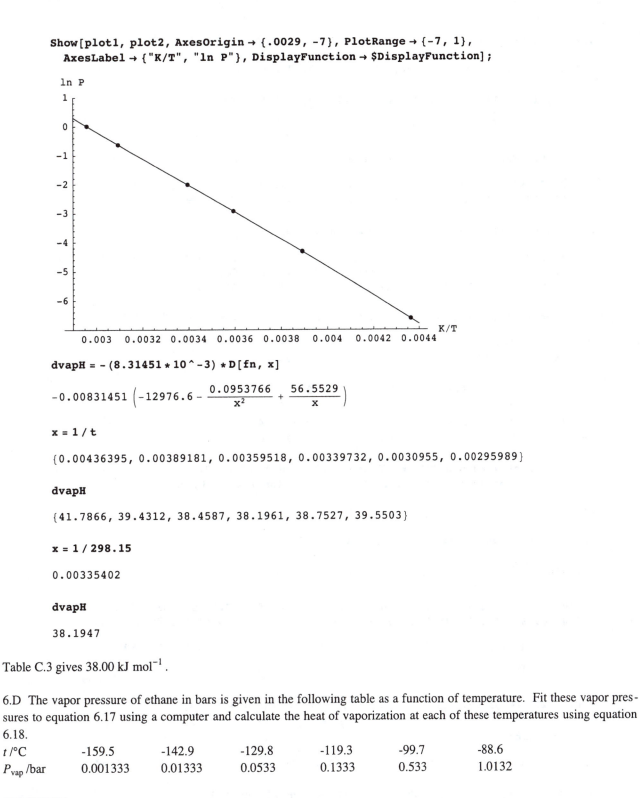

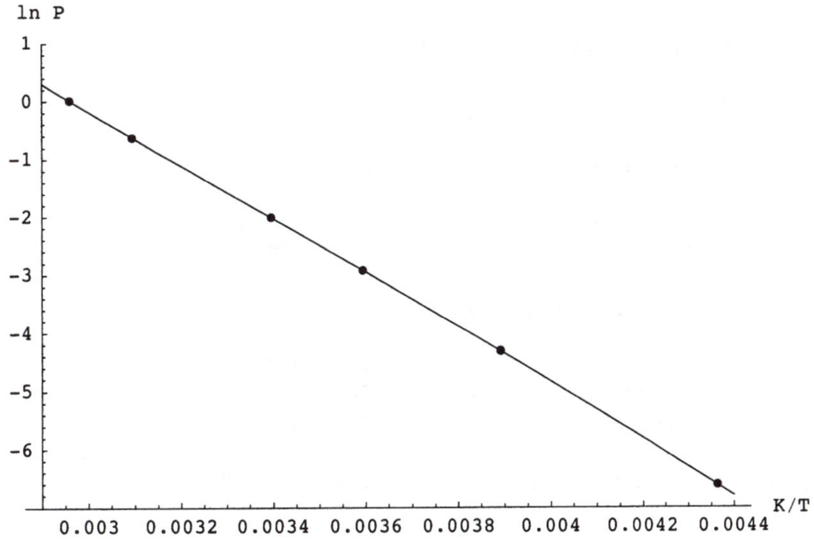


```
dvapH = -(8.31451 * 10^-3) * D[fn, x]
```

$$-0.00831451\left(-12976.6-\frac{0.0953766}{x^2}+\frac{56.5529}{x}\right)$$

```
x = 1 / t

{0.00436395, 0.00389181, 0.00359518, 0.00339732, 0.0030955, 0.00295989}

dvapH

{41.7866, 39.4312, 38.4587, 38.1961, 38.7527, 39.5503}

x = 1 / 298.15

0.00335402

dvapH

38.1947
```

Table C.3 gives 38.00 kJ mol^{-1}.

6.D The vapor pressure of ethane in bars is given in the following table as a function of temperature. Fit these vapor pressures to equation 6.17 using a computer and calculate the heat of vaporization at each of these temperatures using equation 6.18.

t /°C	-159.5	-142.9	-129.8	-119.3	-99.7	-88.6
P_{vap} /bar	0.001333	0.01333	0.0533	0.1333	0.533	1.0132

SOLUTION

```
pz = {.001333, .01333, .0533, .1333, .533, 1.0132}

{0.001333, 0.01333, 0.0533, 0.1333, 0.533, 1.0132}
```

```
lnpzdata = Log[pz]

{-6.62032, -4.31774, -2.93182, -2.01515, -0.629234, 0.0131136}

tz = 273.15 + {-159.5, -142.9, -129.8, -119.3, -99.7, -88.6}

{113.65, 130.25, 143.35, 153.85, 173.45, 184.55}

newinputz = Transpose[{1 / tz, lnpzdata}]

{{0.00879894, -6.62032}, {0.00767754, -4.31774}, {0.00697593, -2.93182},
 {0.00649984, -2.01515}, {0.00576535, -0.629234}, {0.00541859, 0.0131136}}

datatablez = Transpose[{tz, 1 / tz, pz, lnpzdata}]

{{113.65, 0.00879894, 0.001333, -6.62032}, {130.25, 0.00767754, 0.01333, -4.31774},
 {143.35, 0.00697593, 0.0533, -2.93182}, {153.85, 0.00649984, 0.1333, -2.01515},
 {173.45, 0.00576535, 0.533, -0.629234}, {184.55, 0.00541859, 1.0132, 0.0131136}}

TableForm[datatablez, TableHeadings → {None, {"T/K", "1/T", "P", "ln P"}}]
```

T/K	1/T	P	ln P
113.65	0.00879894	0.001333	-6.62032
130.25	0.00767754	0.01333	-4.31774
143.35	0.00697593	0.0533	-2.93182
153.85	0.00649984	0.1333	-2.01515
173.45	0.00576535	0.533	-0.629234
184.55	0.00541859	1.0132	0.0131136

```
g1z = ListPlot[Transpose[{1 / tz, lnpzdata}], PlotRange → {-8, 1},
   Prolog -> AbsolutePointSize[4], DisplayFunction → Identity];

Clear[x]

fnz = Fit[newinputz, {-x, -Log[x], 1 / x, 1}, x]
```

$$91.7358 + \frac{0.0403068}{x} - 3344.42\,x + 15.5309\,\mathrm{Log}[x]$$

```
x = {0.00879894412670479653`, 0.00767754318618042397`, 0.00697593303104290374`,
  0.00649983750406240012`, 0.00576535024502738657`, 0.00541858574911947954`}

{0.00879894, 0.00767754, 0.00697593, 0.00649984, 0.00576535, 0.00541859}

fnz

{-6.62023, -4.31806, -2.93193, -2.01431, -0.630182, 0.0135549}

fnz - lnpzdata

{0.0000908392, -0.000317593, -0.000113543, 0.000847124, -0.0009481, 0.000441273}

g2z = Plot[fnz, {x, .005, .009}, DisplayFunction → Identity];
```

```
Show[g1z, g2z, AxesLabel → {"K/T", "ln P"}, PlotRange → {-8, 1},
  AxesOrigin → {.005, -8}, DisplayFunction -> $DisplayFunction];
```

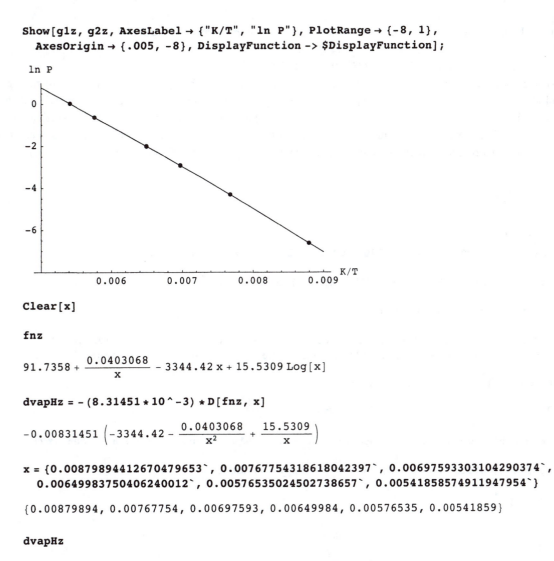

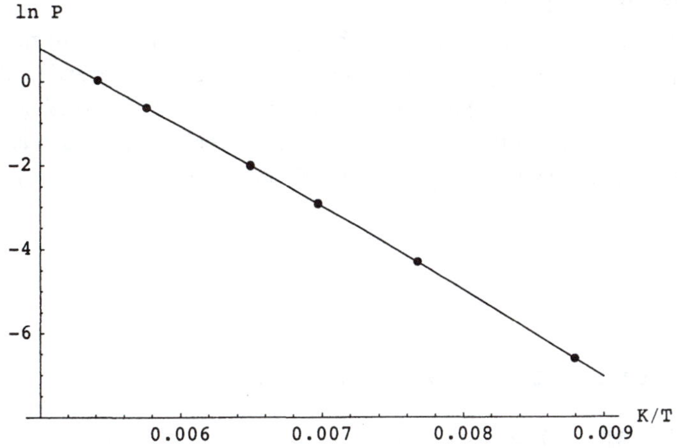


```
Clear[x]
```

```
fnz
```

$$91.7358 + \frac{0.0403068}{x} - 3344.42\,x + 15.5309\,\text{Log}[x]$$

```
dvapHz = - (8.31451 * 10^-3) * D[fnz, x]
```

$$-0.00831451\left(-3344.42 - \frac{0.0403068}{x^2} + \frac{15.5309}{x}\right)$$

```
x = {0.00879894412670479653`, 0.00767754318618042397`, 0.00697593303104290374`,
  0.00649983750406240012`, 0.00576535024502738657`, 0.00541858574911947954`}
```

{0.00879894, 0.00767754, 0.00697593, 0.00649984, 0.00576535, 0.00541859}

```
dvapHz
```

{17.4601, 16.6734, 16.1829, 15.8729, 15.4918, 15.3902}

These are the enthalpies of vaporization of ethane in kJ mol^-1.

6.E The molar volumes of the following solutions of water and methanol at 0 ^oC and 1 bar are given by

x(methanol)	0	.114	.197	.249	.495	.692	.785	.892	1
Molar vol.	.0181	.0203	.0219	.0230	.0283	.0329	.0352	.0379	.0407

where the molar volumes are in m^3 $kmol^{-1}$. (*a*) Plot the molar volumes versus the mole fraction of methanol and fit these data to a quadratic polynomial. (*b*) Calculate the polynomials that give the partial molar volumes of water and methanol and calculate the partial molar volumes in each of the nine solutions. (*c*) Show that the molar volumes of the solutions are given by the sum of the contributions of the two liquids. (*d*) Show that the equation for the partial molar volume of water can be calculated using the equation for the partial molar volume of methanol by use of the Gibbs-Duhem equation for the molar volume.

SOLUTION

(a)

```
mv = {.0181, .0203, .0219, .0230, .0283, .0329, .0352, .0379, .0407};

xmeth = {0, .114, .197, .249, .495, .692, .785, .892, 1};

data = Transpose[{xmeth, mv}];

plot1 = ListPlot[data, AxesOrigin -> {0, .015}, AxesLabel -> {"xM", "molar vol"},
   Prolog -> AbsolutePointSize[3], DisplayFunction → Identity, PlotRange → {.015, .05}];

vbar = Fit[data, {1, xm, xm^2}, xm]

0.0181007 + 0.018658 xm + 0.00395452 xm^2

plot2 = Plot[vbar, {xm, 0, 1}, AxesOrigin -> {0, .015}, DisplayFunction → Identity];

Show[plot1, plot2, DisplayFunction -> $DisplayFunction];
```

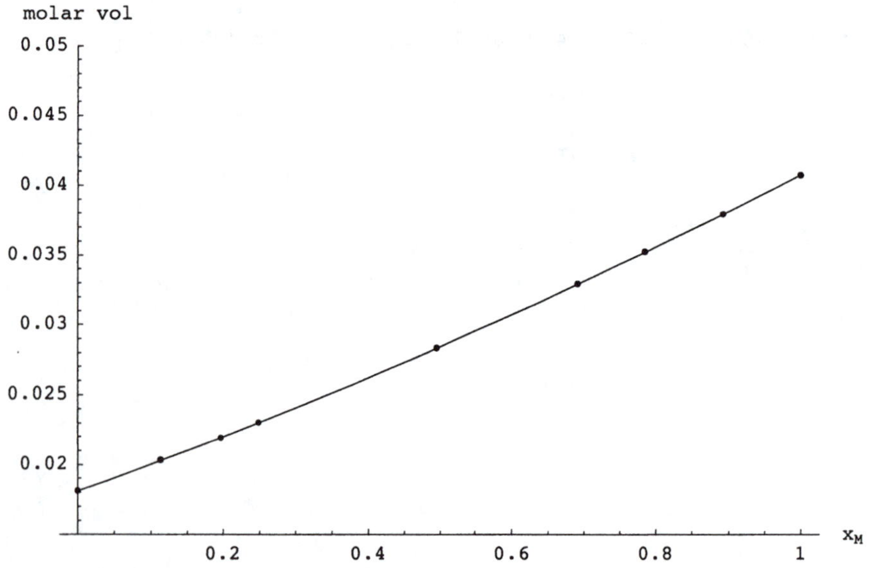

(b) The two partial molar volumes are given by

vbarH2O=vbar-(xm*dvt)

vbarCH3OH=vbar+(1-xm)*dvt

where dvt is the partial derivative of the molar volume with respect to the mole fraction of methanol.

```
dvt = D[vbar, xm]

0.018658 + 0.00790905 xm

vbarH2O = Simplify[vbar - (xm * dvt)] // Chop

0.0181007 - 0.00395452 xm^2

xm = xmeth

{0, 0.114, 0.197, 0.249, 0.495, 0.692, 0.785, 0.892, 1}

vbarH2O

{0.0181007, 0.0180493, 0.0179473, 0.0178555,
 0.0171318, 0.0162071, 0.0156639, 0.0149543, 0.0141462}

Clear[xm]

vbarCH3OH = Simplify[vbar + (1 - xm) * dvt]

-0.00395452 (-4.20864 + xm) (2.20864 + xm)

xm = xmeth

{0, 0.114, 0.197, 0.249, 0.495, 0.692, 0.785, 0.892, 1}

vbarCH3OH

{0.0367587, 0.0376089, 0.0381633, 0.0384829,
 0.0397047, 0.0403381, 0.0405304, 0.0406671, 0.0407132}
```

(c) The calculated molar volumes are given by

```
mvcalc = (1 - xm) * vbarH2O + xm * vbarCH3OH

{0.0181007, 0.0202791, 0.0219298, 0.0229918,
 0.0283054, 0.0329057, 0.0351841, 0.0378901, 0.0407132}

mvcalc - mv

{7.33883 × 10^-7, -0.0000208651, 0.0000298241, -8.24837 × 10^-6,
 5.38382 × 10^-6, 5.72484 × 10^-6, -0.0000158871, -9.88898 × 10^-6, 0.0000132228}
```

(d) This calculation is based on $x_M \, \mathrm{d}\overline{V}_M + x_w \, \mathrm{d}\overline{V}_w = 0$.

```
Clear[xm]

dvtvbarCH3OH = Simplify[D[vbarCH3OH, xm]]

0.00790905 - 0.00790905 xm
```

```
0.0181 - Integrate[(xm / (1 - xm)) * dvtvbarCH3OH, xm] // Chop
```

$0.0181 - 0.00395452\ xm^2$

This result agrees with that obtained above.

6.F Calculate the standard and normal boiling points of water using the expression calculated in Computer Problem 6.B for $\ln P_{vap}$ for water. The standard boiling point is at the standard pressure of 1 bar, and the normal boiling point is at a pressure of 1 atmosphere (1.013 25 bar).

SOLUTION

```
ClearAll["Global`*"]
```

The expression for ln P for liquid water is

```
fn2=68.836708+.0051968111/x-7196.379214*x+7.914932*Log[x]
```

$68.8367 + \frac{0.00519681}{x} - 7196.38\ x + 7.91493\ \text{Log}[x]$

```
FindRoot[fn2==Log[100],{x,0.002}]
```

{x → 0.00268254}

```
1/x/.x->.00268254
```

372.781

This is the standard boiling point expected at 1 bar (100 kPa).

```
FindRoot[fn2==Log[101.325],{x,0.002}]
```

{x → 0.00267989}

```
1/x/.x->.00267989
```

373.15

This is in agreement with the normal boiling point at 1 atmosphere (101.325 kPa).

6.G When pressure is applied to the surface of a liquid, the vapor pressure of the liquid is increased because molecules are squeezed out of the liquid into the vapor phase. This pressure can be applied by a gas. The following equation for the ratio of the vapor pressure P at the higher pressure to the vapor pressure P^* in the absence of the pressurizing gas can be calculated using

$\ln(P/P^*) = \bar{V}\Delta P/RT$

This equation is based on the assumption that the pressurizing gas is insoluble. Calculate P/P^* at 298.15 K and $\Delta P = 100$ bar for n-octane, which has a density of 0.7036 g mL at this temperature.

SOLUTION

```
m = 8 * 12.01 + 18 * 1.008

114.224

v = 114.224 / .7036

162.342
```

The molar volume is in mL.
ln(P/P*) is given by

```
0.1623 * 100 / (.083145 * 298.15)

0.654708
```

P/P* is given by

```
Exp[0.6547]

1.92457
```

Thus the vapor pressure of octane is increased by about a factor of 2.

6.H The ratio of the vapor pressure P of a liquid droplet is greater than the vapor pressure P^* of the liquid over a flat surface. Conversely, the ratio of the vapor pressure P inside a bubble is smaller than the vapor pressure P^* of the liquid over a flat surface. (a) Plot P/P^* of water at 25 °C versus log r of the droplet, where r is the radius in nm. (b) Plot P/P^* of water at 25 °C versus log r of the bubble, where r is the radius in nm. The surface tension of water at 25 C is 0.07197 N m^{-1}.

SOLUTION

The Kelvin equation is $RT\ln(P/P^*) = 2\gamma\bar{V}/r$

where r is the radius of the droplet. This equation applies to the vapor pressure in a bubble when the sign of the right hand side is reversed.

(a)

```
pratio = Exp[2 * 18 * 10^-6 * .07197 / (r * 10^-9 * 8.3145 * 298.15)]

e^(1.04516/r)

<< Graphics`Graphics`

? LogLinearPlot

LogLinearPlot[f, {x, xmin, xmax}] generates a plot of f as a function of Log[x]. More...
```

```
plot1 =
  LogLinearPlot[pratio, {r, 1, 10^3}, PlotRange → {0, 3}, AxesLabel → {"r/nm", "P/P*"}];
```

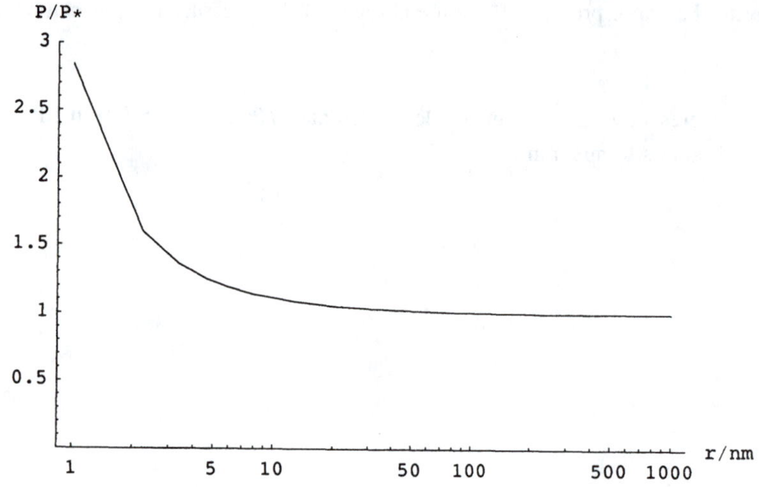

(b)

```
pratiobub = Exp[-2 * 18 * 10^-6 * .07197 / (r * 10^-9 * 8.3145 * 298.15)]
```

$e^{-1.04516/r}$

```
plot2 =
  LogLinearPlot[pratiobub, {r, 1, 10^3}, PlotRange → {0, 3}, AxesLabel → {"r/nm", "P/P*"}];
```

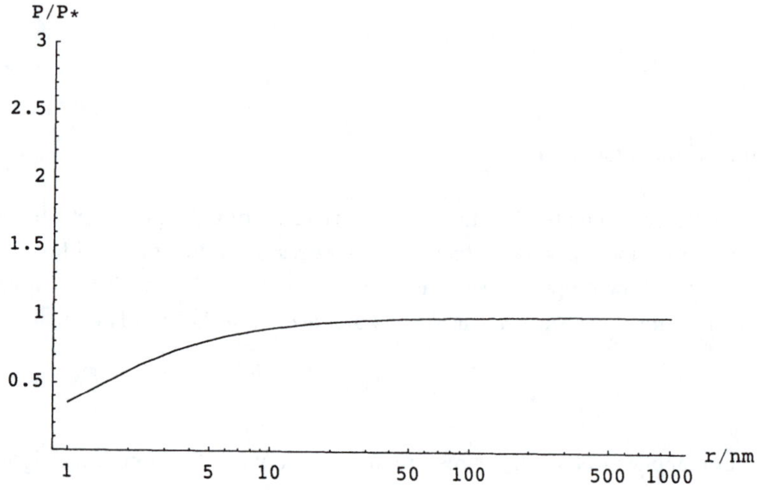

```
Show[plot1, plot2, AxesLabel → {"r/nm", "P/P*"}];
```

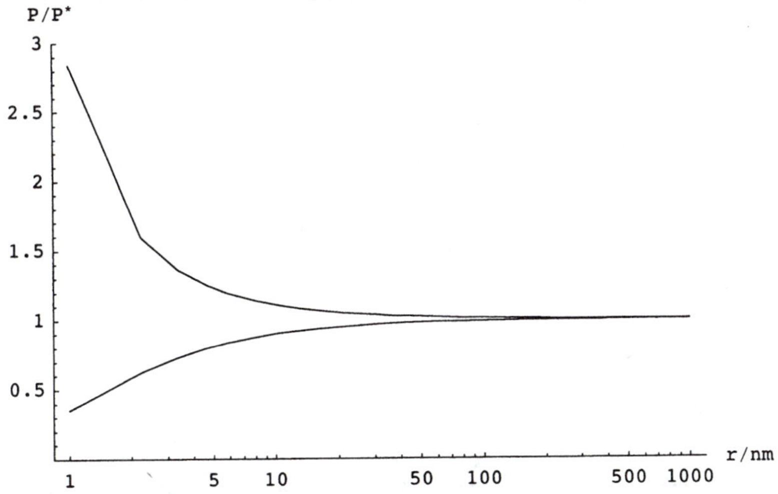

6.I (a) Fit the data in Table 6.1 on the vapor pressure of liquid water to equation 6.17, and plot lnP_{vap} calculated with this equation between -40 and 180 °C. (b) Fit the data in Table 6.1 on the vapor pressure of ice to equation 6.17 without the C term, and plot lnP_{ice} calculated with this equation between -40 and 180 °C. (c) Superimpose these two plots to see the relation between the vapor pressure of supercooled water and ice below 0 °C.

SOLUTION

(a)

```
t = 273.15 + {0, 10, 20, 30, 40, 60, 100, 140, 180};

p = {.611, 1.228, 2.338, 4.245, 7.381, 19.933, 101.325, 361.21, 1001.9};

lnp = Log[p];

input = Transpose[{1/t, lnp}];

fn = Fit[input, {-x, -Log[x], 1/x, 1}, x]
```

$$77.5278 + \frac{0.00753318}{x} - 7437.14\,x + 9.42081\,\text{Log}[x]$$

```
plot1 = Plot[fn, {x, 1 / (180 + 273.15), 1 / (273.15 - 40)},
   AxesLabel → {"K/T", "ln P"}, AxesOrigin → {1 / (180 + 273.15), -4}, PlotRange → {-4, 7}];
```

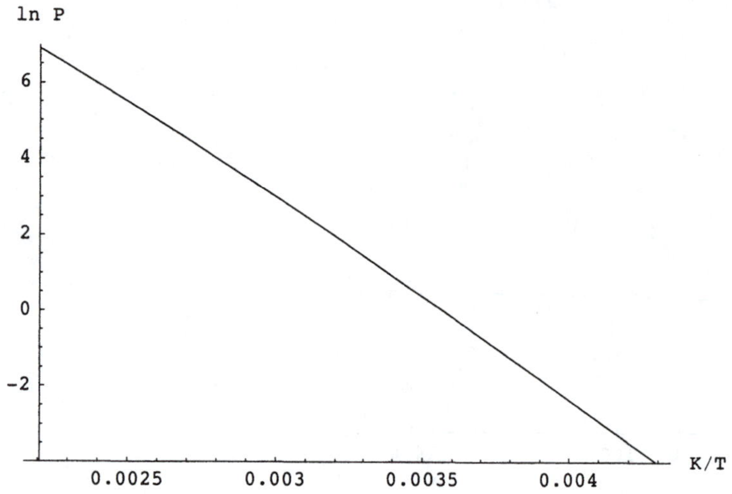

(b) The C term in equation 6.17 is omitted because there are only five data points.

```
tt = 273.15 + {-40, -30, -20, -10, 0};

pp = {.013, .038, .103, .260, .611};

lnpp = Log[pp];

inputice = Transpose[{1 / tt, lnpp}];

fnice = Fit[inputice, {-x, -Log[x], 1}, x]

6.32175 - 5530.37 x - 2.39458 Log[x]

plot2 = Plot[fnice, {x, 1 / (180 + 273.15), 1 / (273.15 - 40)},
   AxesLabel → {"K/T", "ln P"}, AxesOrigin → {1 / (180 + 273.15), -4}, PlotRange → {-4, 7}];
```

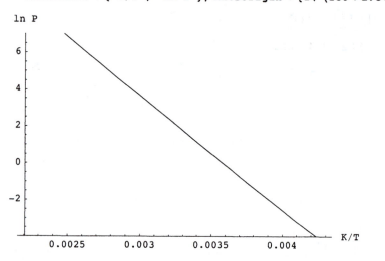

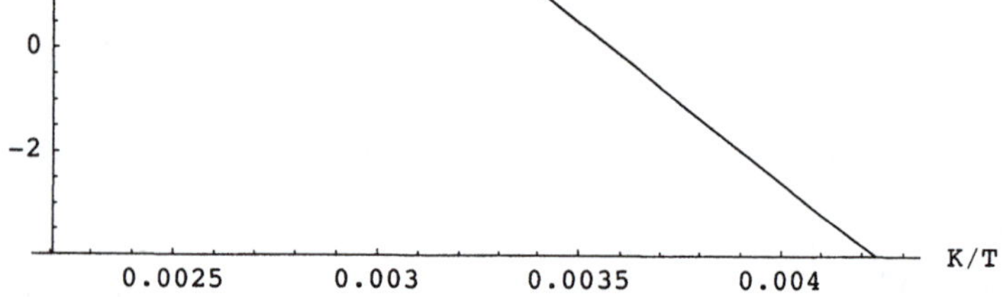

```
Show[plot1, plot2];
```

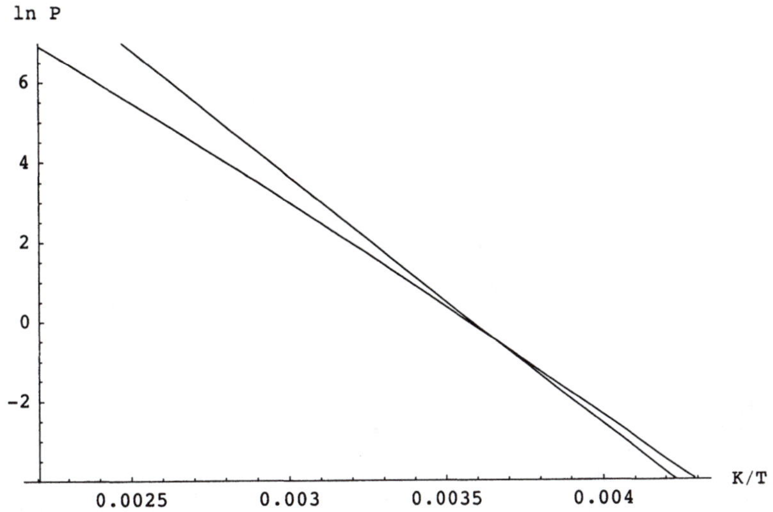

These two vapor pressure plots cross at 1/273.15 K = 0.00366 K^{-1}. At lower temperatures (that is to the right of the intersection), liquid water has a higher vapor pressure than ice and is therefore unstable with respect to ice.

Chapter 7 Electrochemical Equilibrium

7.A Calculate the values of B in equation 7.65 that fit each data point for the mean ionic activity coefficient of (*a*) HCl, (*b*) NaCl, (*c*) CsCl up to 0.40 molal in Table 7.1. Also plot the data as the base 10 logarithms of the mean ionic activity coefficients versus the square root of the ionic strength. (*d*) Calculate the percent error in the mean ionic activity coefficients calculated for HCl, NaCl, and CsCl using equation 7.65 using 1.6 $kg^{1/2}$ $mol^{-1/2}$ at each ionic strength.

SOLUTION

(a) The values of the mean ionic activity coefficients for HCl are

```
Off[General::spelll];
Off[General::spell];

achcl = {.905, .875, .830, .796, .767, .755};
```

The molal concentrations are

```
mhcl = {.01, .02, .05, .1, .2, .4};

b1 = -.509 / Log[10, achcl] - mhcl^(-.5)

{1.74125, 1.70601, 1.81788, 1.97463, 2.18216, 2.58918}

data = Transpose[{mhcl^.5, Log[10, achcl]}];

phcl = ListPlot[data, AxesOrigin -> {0, 0}, AxesLabel -> {"I^1/2", "log γ"},
   Prolog -> AbsolutePointSize[4], DisplayFunction → Identity];
```

(b) NaCl

```
acnacl = {.902, .870, .825, .790, .757, .74};

b2 = -.509 / Log[10, acnacl] - mhcl^(-.5)

{1.36327, 1.34483, 1.62031, 1.80974, 1.97388, 2.31124}

datanacl = Transpose[{mhcl^.5, Log[10, acnacl]}];

pnacl = ListPlot[datanacl, AxesOrigin -> {0, 0}, AxesLabel -> {"I^1/2", "log γ"},
   Prolog -> AbsolutePointSize[5], DisplayFunction → Identity];
```

(c) CsCl

```
accscl = {.899, .865, .807, .756, .718, .628};
```

```
b3 = -.509 / Log[10, accscl] - mhcl^(-.5)

{1.00771, 1.01036, 0.99355, 1.02777, 1.30171, 0.93816}

datacscl = Transpose[{mhcl^.5, Log[10, accscl]}];

pcscl = ListPlot[datacscl, AxesOrigin -> {0, 0}, AxesLabel -> {"I^1/2", "log γ"},
   Prolog -> AbsolutePointSize[6], DisplayFunction → Identity];
```

This is the function in equation 7.65 that involves the squareroot of the ionic strength (sri):

```
fn2 = -(.509 * sri) / (1 + 1.6 * sri)
```

$$-\frac{0.509\ \text{sri}}{1 + 1.6\ \text{sri}}$$

Plot of equation 7.65 with B = 1.6

```
edht = Plot[fn2, {sri, 0, .63}, DisplayFunction → Identity];

Show[phcl, pnacl, pcscl, edht, DisplayFunction → $DisplayFunction];
```

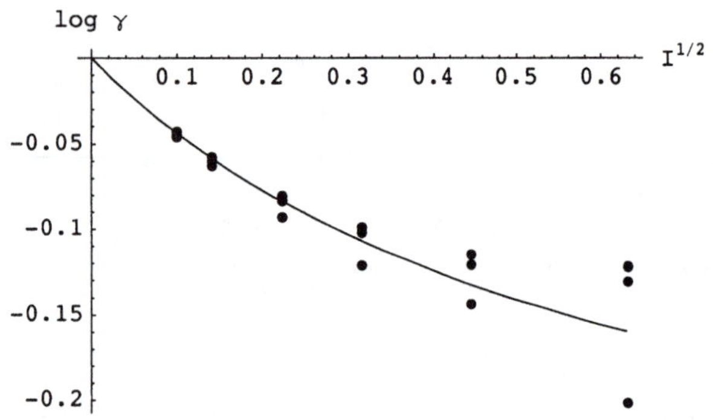

Note that the upper set of points is for HCl and the lower set of points is for CsCl.

(d) Calculate the percent error in the mean ionic activity coefficient calculated for HCl, NaCl, and CsCl using equation 7.65 using 1.6 kg^.5 mol^-.5 at each ionic strength.

```
calcerror[ac_] := Module[{gdht, sri},
  (*Calculates percent error in the mean ionic activity coefficients at each ionic
    strength predicted by equation 7.65 using B = 1.6 for 1-1 electrolytes.*)
      sri = {.01, .02, .05, .1, .2, .4}^.5;
      gdht = 10^(-(.509 * sri) / (1 + 1.6 * sri));
      ((gdht - ac) / ac) * 100]

calcerror[achcl]

{-0.121477, -0.163123, -0.666353, -1.77879, -3.94573, -8.36794}

calcerror[acnacl]

{0.210713, 0.410652, -0.0643311, -1.0328, -2.67685, -6.51053}
```

```
calcerror[accscl]

{0.54512, 0.991061, 2.16472, 3.4181, 2.60951, 10.1627}

calcgdht[mconc_] :=
 Module[{sri}, (*Calculates the mean ionic activity coefficients predicted by equation
     7.65 using B = 1.6 for 1-1 electrolytes at the specified ionic strengths.*)
    sri = mconc^.5;
        10^(-(.509 * sri) / (1 + 1.6 * sri))]

calcgdht[mhcl]

{0.903901, 0.873573, 0.824469, 0.781841, 0.736736, 0.691822}
```

Make a table of percent error:

```
names = {"m", "HCl", "NaCl", "CsCl"}

{m, HCl, NaCl, CsCl}
```

```
TableForm[Transpose[{mhcl, calcerror[achcl], calcerror[acnacl], calcerror[accscl]}],
 TableHeadings → {{}, names}]
```

m	HCl	NaCl	CsCl
0.01	-0.121477	0.210713	0.54512
0.02	-0.163123	0.410652	0.991061
0.05	-0.666353	-0.0643311	2.16472
0.1	-1.77879	-1.0328	3.4181
0.2	-3.94573	-2.67685	2.60951
0.4	-8.36794	-6.51053	10.1627

Note that in the ionic strength range 0.1-0.2 the errors are from 1% to 4%.

7.B Calculate the values of B in equation 7.65 that give the best fit of the data on the mean ionic activity coefficients of (*a*) $CaCl_2$ and (*b*) $LaCl_3$ up to 0.05 molal. The values of the mean ionic activity coefficients are given by

m/m^o	0.001	0.005	0.01	0.05
$\gamma_\pm$ ($CaCl_2$)	0.888	0.789	0.732	0.584
$\gamma_\pm$ ($LaCl_3$)	0.790	0.636	0.560	0.388

In addition do two things for these two electrolytes: (i) Plot the experimental data as base 10 logarithms of the mean ionic activity coefficients versus the square root of the ionic strength and compare the data with the extended Debye-Huckel equation with $B = 1.6$. (ii) Calculate the percent errors in the mean ionic activity coefficients calculated for $CaCl_2$ and $LaCl_3$ using equation 7.65 with $B = 1.6$ kg$^{1/2}$ mol$^{-1/2}$ at each ionic strength.

SOLUTION

(a) The values of the mean ionic activity coefficients for $CaCl_2$ are

```
accacl2 = {.888, .789, .732, .584};
```

The molal concentrations are

```
mcacl2 = {.001, .005, .01, .05};
```

The ionic strengths are

```
icacl2 = 3 * mcacl2;

b4 = -2 * .509 / Log[10, accacl2] - icacl2^(-.5)

{1.47622, 1.72592, 1.74003, 1.77613}
```

(i)

```
datacacl2 = Transpose[{icacl2^.5, Log[10, accacl2]}];

pcacl2 = ListPlot[datacacl2, AxesOrigin -> {0, 0}, AxesLabel -> {"I^(1/2)", "log γ"},
   Prolog -> AbsolutePointSize[4], DisplayFunction → Identity];

fn4 = -2 * (.509 * sri) / (1 + 1.6 * sri);
```

Plot equation 7.65 with B = 1.6.

```
eqdhtcacl2 = Plot[fn4, {sri, 0, .4}, DisplayFunction → Identity];
```

```
Show[pcacl2, eqdhtcacl2, AxesOrigin → {0, -.25}, DisplayFunction → $DisplayFunction];
```

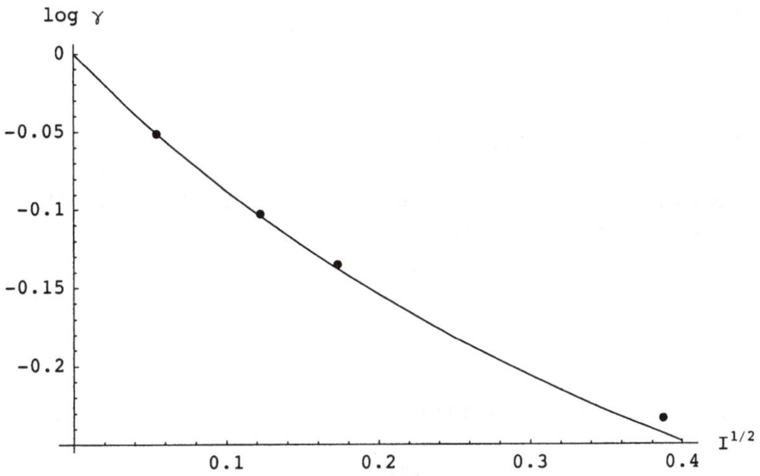

(ii) Calculate the percent errors in the mean ionic activity coefficients calculated for CaCl2 using equation 7.65 using 1.6 kg^.5 mol^-.5 at each ionic strength.

```
calcerror2[ac_] := Module[{gdht, sri},
  (*Calculates percent error in the mean ionic activity coefficients at each ionic
    strength predicted by equation 7.65 using B = 1.6 for 2-1 electrolytes.*)
      sri = {0.003`, 0.015`, 0.03`, 0.15`}^.5;
      gdht = 10^(-(2 * .509 * sri) / (1 + 1.6 * sri));
      ((gdht - ac) / ac) * 100]

calcerror2[accacl2]

{0.0740691, -0.305142, -0.590706, -2.23975}
```

Note that at I = 0.15, the error is 2%.

(b) The values of mean ionic activity coefficients of $LaCl_3$ are

```
aclacl3 = {.790, .636, .560, .388};
```

The ionic strengths of these solutions are

```
ilacl3 = 6 * mcacl2;
```

```
b5 = -3 * .509 / Log[10, aclacl3] - ilacl3 ^ (-.5)
```

```
{2.00611, 1.99579, 1.98157, 1.88807}
```

(i)

```
datalacl3 = Transpose[{ilacl3^.5, Log[10, aclacl3]}];
```

```
placl3 = ListPlot[datalacl3, AxesOrigin -> {0, -.5}, AxesLabel -> {"I^(1/2)", "log γ"},
   Prolog -> AbsolutePointSize[4], DisplayFunction → Identity];
```

```
fn3 = -3 * (.509 * sri) / (1 + 1.6 * sri);
```

Plot of equation 7.65 with B = 1.6

```
eqdht = Plot[fn3, {sri, 0, .6}, DisplayFunction → Identity];
```

```
Show[placl3, eqdht, DisplayFunction -> $DisplayFunction];
```

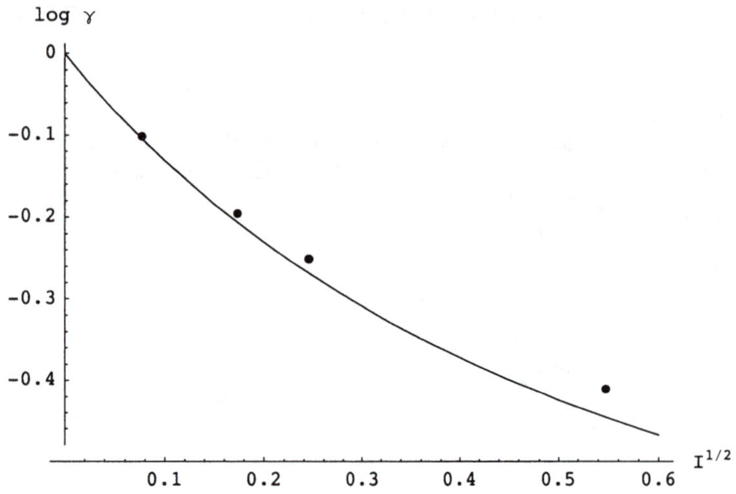

(ii) Calculate the percent error in the mean ionic activity coefficients calculated for $LaCl_3$ using equation 7.65 using 1.6 at each ionic strength.

```
calcerror3[ac_] := Module[{gdht, sri},
  (*Calculates percent error in the mean ionic activity coefficients at each ionic
    strength predicted by equation 7.65 using B = 1.6 for 3-1 electrolytes.*)
      sri = {0.006`, 0.03`, 0.06`, 0.3`}^.5;
      gdht = 10^(-(3 * .509 * sri) / (1 + 1.6 * sri));
      ((gdht - ac) / ac) * 100]
```

```
calcerror3[aclacl3]

{-0.657579, -2.39996, -3.81853, -7.65244}
```

Note that the error in the 0.06-0.3 ionic strength range is 4-8%.

7.C Calculate the standard electrode potential of (*a*) Cd^{2+} | Cd, (*b*) Cl^- | Cl_2 (g) | Pt, and (*c*) Cl^- | AgCl(s) | Ag(s) at 0.25 m ionic strength using the extended Debye-Huckel equation. See the values at I = 0 in Example 7.14.

SOLUTION

```
electrodepot[dfGo_, nue_, snuz2_, i_] := Module[{},
  (*Calculates the standard electrode potential at the specified ionic strength i,
   given the standard Gibbs energy of reaction for the electrode reaction dfGo,
   the number nue of formal electrons involved in the half reaction,
   and the sum of the stoichiometric numbers times charges of ions squared for the
     electrode reaction.  The standard Gibbs energies of reaction are in kJ mol^-1.*)
   -(1 / (96.485 * nue)) (dfGo - 2.91482 * snuz2 * i^.5 / (1 + 1.6 * i^.5))]
```

(a)

```
electrodepot[77.612, 2, -4, .25]

-0.418981
```

(b)

```
electrodepot[-262.456, 2, 2, .25]

1.36848
```

(c)

```
electrodepot[-21.439, 1, 1, .25]

0.230592
```

7.D Plot E^o (*I*), $\Delta_r G^o$ (*I*), and *K*(*I*) for the reaction
(1/2)H_2 (g) + AgCl(s) = H^+ + Cl^- + Ag(s)
versus the ionic strength from 0 to 0.25 m using the extended Debye-Huckel equation.

SOLUTION

```
dGH = -2.91482 * i^.5 / (1 + 1.6 * i^.5)
```

$$-\frac{2.91482\, i^{0.5}}{1 + 1.6\, i^{0.5}}$$

```
dGCl = -131.228 - 2.91482 * i^.5 / (1 + 1.6 * i^.5)
```

$$-131.228 - \frac{2.91482\, i^{0.5}}{1 + 1.6\, i^{0.5}}$$

```
drGo = dGH + dGCl + 109.789
```

$$-21.439 - \frac{5.82964\, i^{0.5}}{1 + 1.6\, i^{0.5}}$$

```
Plot[drGo, {i, 0, .25}, AxesLabel → {"I^1/2", "ΔrG°"}];
```

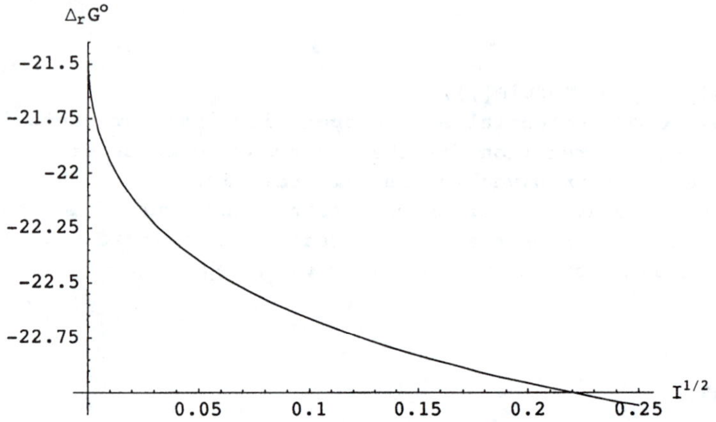


```
Plot[-drGo / 96.485, {i, 0, .25}, AxesLabel → {"I^1/2", "E°"}];
```

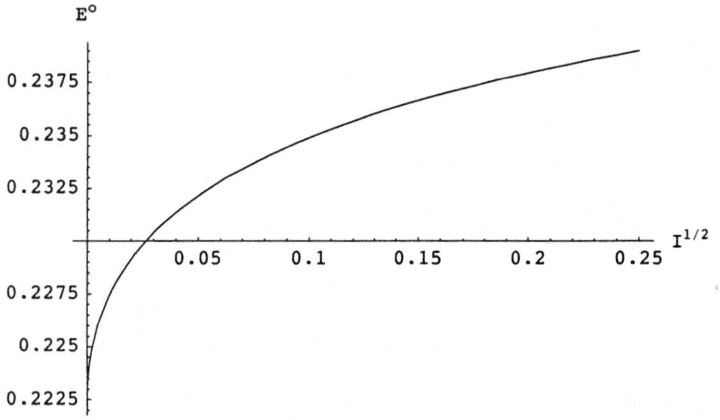

```
Plot[Exp[-drGo / (8.3145 * .29815)], {i, 0, .25}, AxesLabel → {"I^1/2", "K"}];
```

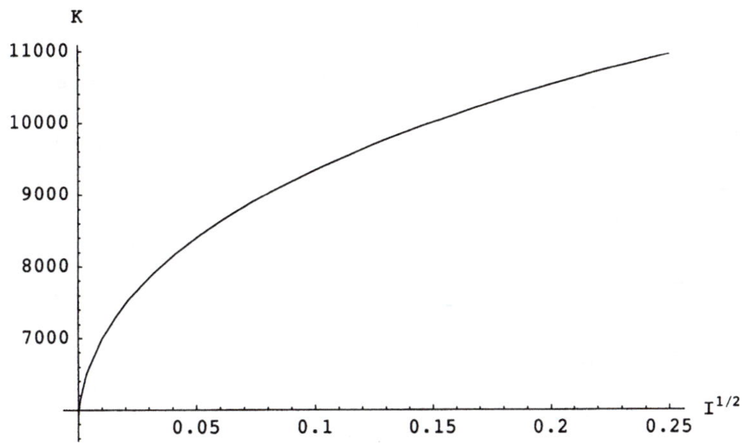

7.E (*a*) Plot the log (base 10) of the mean ionic activity coefficient versus the square root of the ionic strength, up to 0.5, at 25 ^{o}C for $z_+ z_- = -1, -2, -3$, and -4 using the extended Debye-Huckel equation. (*b*) Plot the mean ionic activity coefficients versus the ionic strength up to 0.25 mol kg^{-1} for these cases.

SOLUTION

(a) Represent the square root of the ionic strength by x.

```
loggamma[n_] := .509 * n * x / (1 + 1.6 * x)

Plot[{loggamma[-1], loggamma[-2], loggamma[-3], loggamma[-4]},
  {x, 0, .5}, AxesLabel → {"√I", "log10γ±"}, AxesOrigin → {0, -.6}];
```

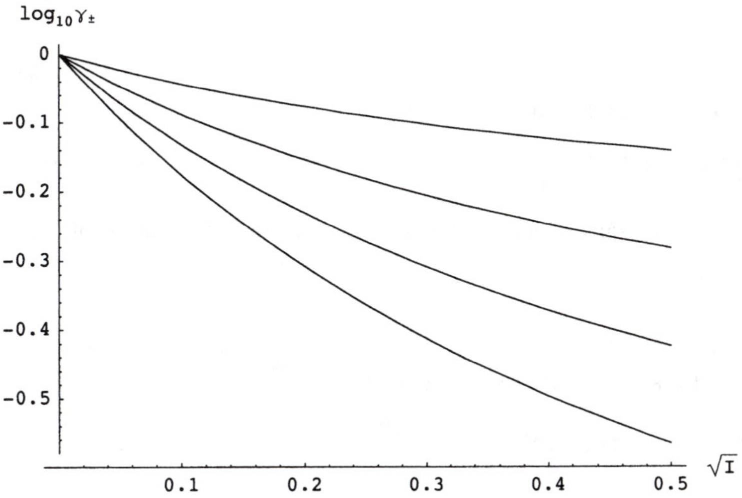

(b)

```
gamma[n_] := 10^(-.509 * n * i^.5 / (1 + 1.6 * i^.5))
```

```
Plot[{gamma[1], gamma[2], gamma[3], gamma[4]}, {i, 0, .25},
  AxesOrigin → {0, 0}, AxesLabel → {"I/mol kg^-1", "γ±"}, PlotRange → {0, 1}];
```

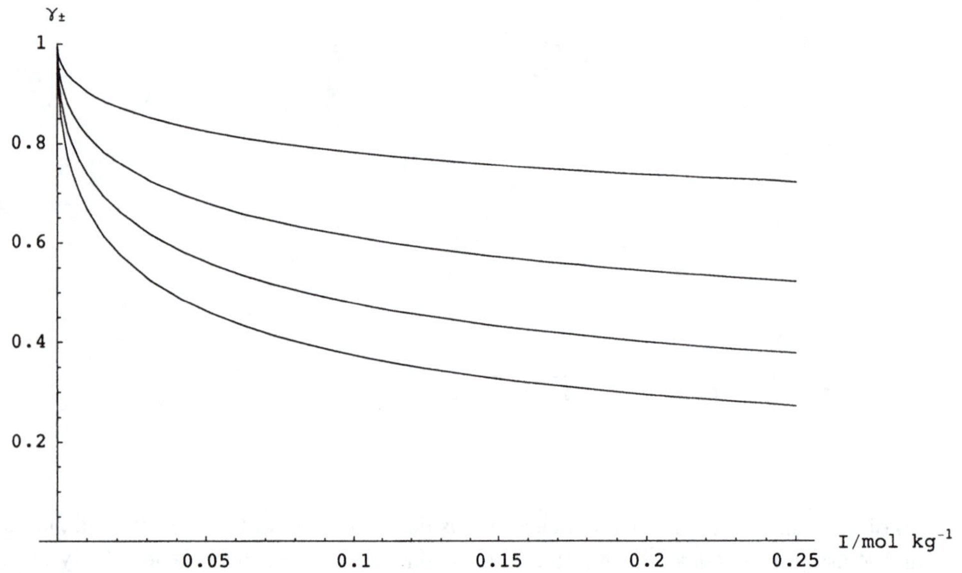

7.F In the range 0-100 ^oC the Debye-Huckel coefficient varies with temperature according to

$\alpha = 1.10708 - 1.54508 \times 10^{-3}\,T + 5.95584 \times 10^{-6}\,T^2$

(a) Calculate the temperature dependencies of the ionic strength coefficients in the equations for $\Delta_f G°$, $\Delta_f H°$, and $\Delta_f S°$.

(b) Calculate the values of these coefficients at 0 ^oC, 25 ^oC, and 40 ^oC and make a table.

SOLUTION

```
Clear[t]
```

(a) Alpha is given by

```
a = 1.10708 - 1.54508 * 10^-3 * t + 5.95584 * 10^-6 * t^2

1.10708 - 0.00154508 t + 5.95584 × 10^-6 t^2
```

The temperature dependence of the coefficient in the equation for the standard Gibbs energy of formation of ion i is given by RTα.

```
coeffG = (8.31451 * 10^-3) * t * a

0.00831451 t (1.10708 - 0.00154508 t + 5.95584 × 10^-6 t^2)

gvalues = coeffG /. t → {273.15, 298.15, 313.25}

{2.56502, 2.9149, 3.14497}
```

The temperature dependence of the coefficient for the standard enthalpy of formation of ion i can be obtained by use of the Gibbs-Helmholtz equation.

```
coeffH = -t^2*D[(coeffG/t), t]

-0.00831451 (-0.00154508 + 0.0000119117 t) t^2

hvalues = coeffH /. t → {273.15, 298.15, 313.25}

{-1.05993, -1.48293, -1.78369}
```

The coefficient for the standard entropy of formation of ion i can be obtained using S = -(dG/dT).

```
coeffS = -D[coeffG, t]

-0.00831451 (-0.00154508 + 0.0000119117 t) t -
 0.00831451 (1.10708 - 0.00154508 t + 5.95584 × 10^-6 t^2)

svalues = coeffS /. t → {273.15, 298.15, 313.25}

{-0.0132709, -0.0147504, -0.0157339}
```

(b)

```
TableForm[{gvalues, hvalues, svalues},
 TableHeadings → {{"G", "H", "S"}, {"0 °C", "25 °C", "40 °C"}}]
```

	0 °C	25 °C	40 °C
G	2.56502	2.9149	3.14497
H	-1.05993	-1.48293	-1.78369
S	-0.0132709	-0.0147504	-0.0157339

The coefficient for $\Delta_f G°$ is in kJ $mol^{-2/3}\ kg^{1/2}$. The coefficient for $\Delta_f H°$ is in kJ $mol^{-2/3}\ kg^{1/2}$. The coefficient for $\Delta_f S°$ is in kJ $K^{-1}\ mol^{-2/3}\ kg^{1/2}$. At 25 °C slightly different values are given in the text because those values were obtained using a slightly different empirical equation.

7.G Calculate and plot versus ionic strength the ionic strength contributions to $\Delta_f G°$, $\Delta_f H°$, and $\Delta_f S°$ for ions with charges of 1, 2, 3, and 4 up to $I = 0.25$ M at 298.15 K.

SOLUTION

```
g = -2.91482*z^2*i^.5 / (1 + 1.6*i^.5);

h = 1.4775*z^2*i^.5 / (1 + 1.6*i^.5);

s = (h - g) / .29815;
```

```
Plot[Evaluate[g /. z → {1, 2, 3, 4}], {i, 0, .25},
  AxesOrigin → {0, -15}, PlotRange → {-15, 0}, AxesLabel → {"I/M", "G/kJ mol⁻¹"}];
```

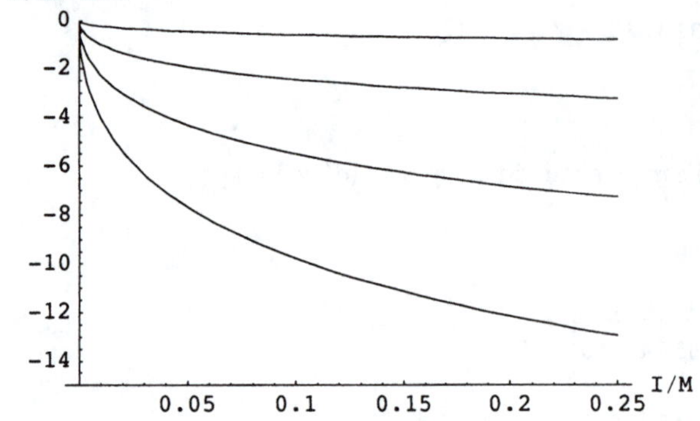


```
Plot[Evaluate[h /. z → {1, 2, 3, 4}], {i, 0, .25},
  AxesOrigin → {0, 0}, PlotRange → {8, 0}, AxesLabel → {"I/M", "H/kJ mol⁻¹"}];
```

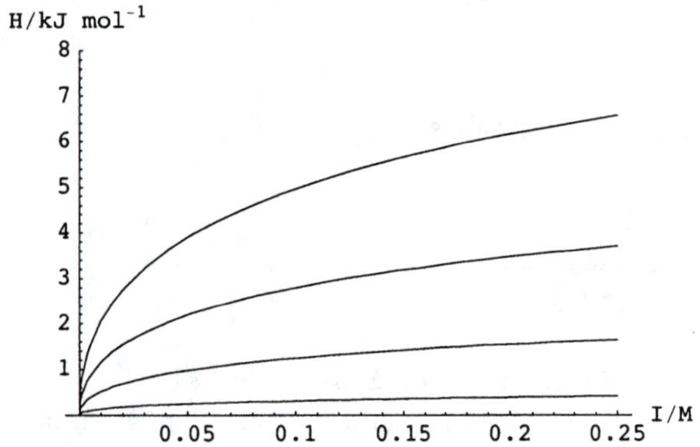

```
Plot[Evaluate[s /. z → {1, 2, 3, 4}], {i, 0, .25}, AxesOrigin → {0, 1},
  PlotRange → {0, 70}, AxesLabel → {"I/M", "S/kJ K^-1 mol^-1"}];
```

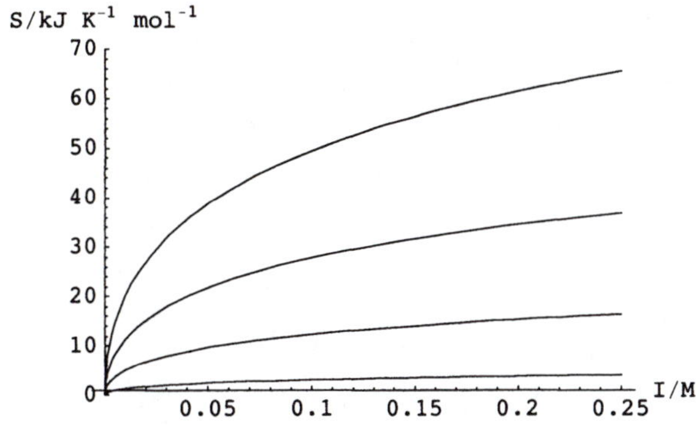

The plot of the Gibbs energy of formation indicates that raising the ionic strength at constant T and P increases the thermodynamic stability of the ion. The plot of the enthalpy of formation indicates that the enthalpy of the ion increases with ionic strength. The plot of the entropy indicates that the effect on the entropy opposes the effect on the enthalpy and is responsible for the increasing stability of the ion.

Chapter 8 Thermodynamics of Biochemical Reactions

8.A Write a program to calculate $\Delta_f G'^o$ and $\Delta_f H'^o$ for inorganic phosphate at 298.15 K and pH 7 using data in Example 8.10. The program should be writtten to handle any number of pseudoisomers so that it can be used for other reactants.

SOLUTION

```
calcthermoprops[dGi_, dHi_] := Module[{rt},
  (*Calculates dGiso and dHiso in a pseudoisomer group from input of standard
    transformed Gibbs energies of formation of species (dGi) and standard transformed
    enthalpies of formation of species (dHi) in kJ mol^-1 at 298.15 K.  dGi and
    dHi are vectors.  The program will work for any number of pseudoisomers.*)
    rt = (8.31451 * 10^-3) * (298.15);
    dGiso = -rt * Log[Apply[Plus, Exp[-dGi / rt]]];
         ri = Exp[(Table[dGiso, {Length[dGi]}] - dGi) / rt];
                 dHiso = ri.dHi;
         {ri, {dGiso, dHiso}}]
```

```
dGi = {-1058.57, -1056.58}
```

```
{-1058.57, -1056.58}
```

```
dHi = {-1297.77, -1303.01}
```

```
{-1297.77, -1303.01}
```

```
calcthermoprops[dGi, dHi]
```

```
{{0.690563, 0.309437}, {-1059.49, -1299.39}}
```

The first two numbers are equilibrium mole fractions, and the second two are $\Delta_f G'^o$ and $\Delta_f H'^o$.

8.B (*a*) Write a program to calculate the apparent equilibrium constant *K'* for ATP + H_2O = ADP + P_i at 298.15 K, pH 7, and ionic strength 0.25 M. The acid dissociation constants needed are given in Example 8.7. (*b*) Plot *K'* versus pH. (*c*) Plot $\Delta_r G^o$ versus pH.

SOLUTION
(a)

```
ClearAll["Global`*"]
```

```
k' := .15 * (1 + (10^-pH) / (4.7 * 10^-7)) *
   (1 + (10^-pH) / (2.23 * 10^-7)) / ((10^-pH) * (1 + (10^-pH) / (3.42 * 10^-7)))
```

```
k'
```

$$\frac{0.15\ 10^{pH}\ (1 + 2.12766 \times 10^6\ 10^{-pH})\ (1 + 4.4843 \times 10^6\ 10^{-pH})}{1 + 2.92398 \times 10^6\ 10^{-pH}}$$

```
pH = 7

7

k'

2.03878 × 10^6
```

(b)

```
Plot[k', {pH, 5, 9}, AxesLabel -> {"pH", "K'"}];
```

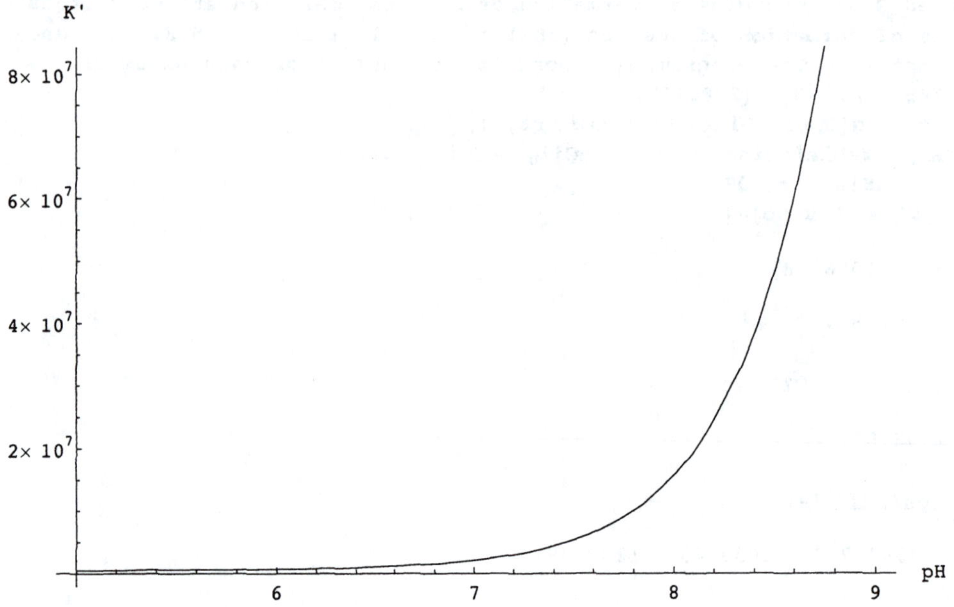

(c)

```
Plot[-8.3145 * .29815 * Log[k'], {pH, 5, 9}, AxesLabel -> {"pH", "ΔrG'°/kJ mol^-1"}];
```

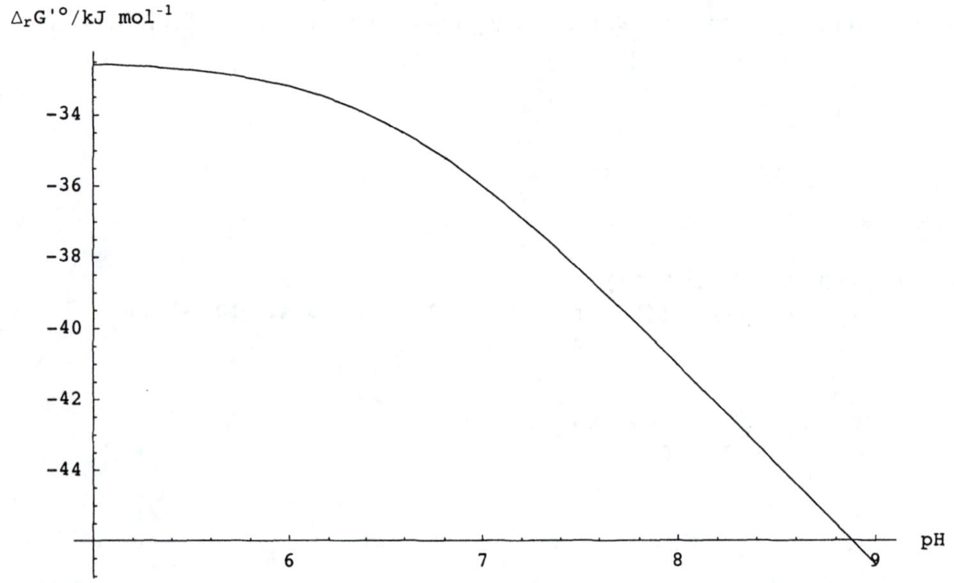

8.C It can be shown that the change in binding of hydrogen ions in a biochemical reaction at a specified pH is given by the negative of the derivative of log *K'* with respect to pH at constant *T* and *P*. (*a*) Use a mathematical program to take the derivative of log *K'* for the hydrolysis of ATP to ADP and inorganic phosphate with respect to pH from the previous problem. (*b*) Plot the production of H^+ versus pH in the range pH 5-9. (*c*) Interpret this plot in terms of the predominant chemical reactions at pH 5 and at pH 9.

SOLUTION

(a)

```
ClearAll["Global`*"]

k' = .15 * (1 + (10^-pH) / (4.7 * 10^-7)) *
   (1 + (10^-pH) / (2.23 * 10^-7)) / ((10^-pH) * (1 + (10^-pH) / (3.42 * 10^-7)))
```

$$\frac{0.15\ 10^{pH}\ (1+2.12766\times10^{6}\ 10^{-pH})\ (1+4.4843\times10^{6}\ 10^{-pH})}{1+2.92398\times10^{6}\ 10^{-pH}}$$

```
dvt = D[Log[10, k'], pH]
```

$$\Bigg(2.8953\ 10^{-pH}\ (1+2.92398\times10^{6}\ 10^{-pH})\ \Bigg(-\frac{1.54882\times10^{6}\ (1+2.12766\times10^{6}\ 10^{-pH})}{1+2.92398\times10^{6}\ 10^{-pH}}+\frac{1.00991\times10^{6}\ (1+2.12766\times10^{6}\ 10^{-pH})\ (1+4.4843\times10^{6}\ 10^{-pH})}{(1+2.92398\times10^{6}\ 10^{-pH})^{2}}-\frac{734868.\ (1+4.4843\times10^{6}\ 10^{-pH})}{1+2.92398\times10^{6}\ 10^{-pH}}+\frac{0.345388\ 10^{pH}\ (1+2.12766\times10^{6}\ 10^{-pH})\ (1+4.4843\times10^{6}\ 10^{-pH})}{1+2.92398\times10^{6}\ 10^{-pH}}\Bigg)\Bigg)\Big/\left((1+2.12766\times10^{6}\ 10^{-pH})\ (1+4.4843\times10^{6}\ 10^{-pH})\right)$$

Note that in this calculation the minus sign has been omitted so that the amount of H^+ produced is calculated, rather than the change in binding of H^+ .

(b)

```
Plot[dvt, {pH, 5, 9}, AxesLabel -> {"pH", "H+ produced"}];
```

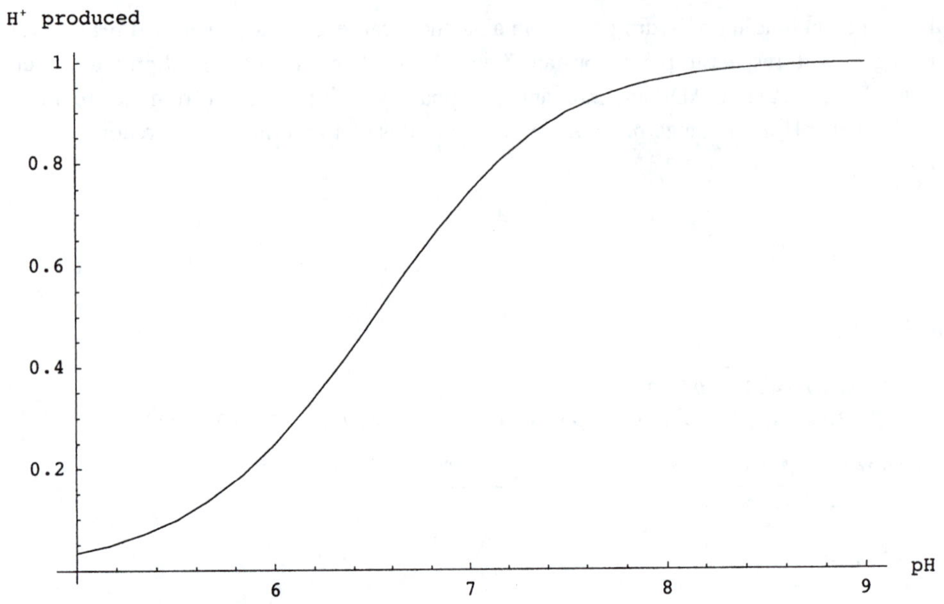

(c) At pH 5, the predominant chemical reaction is
HATP3- + H2O = HADP2- + H2PO4-
which shows that there is no change in binding of hydrogen ions in the reaction.
At pH 7, the predominant chemical reaction is
ATP4- + H2O = ADP3- + HPO42- + H+
which shows that one mole of H+ is produced per mole of ATP hydrolyzed.

8.D (*a*) Write a program to calculate the apparent equilibrium constant *K*' for glucose 6-phosphate + H_2O = glucose + P_i at 298.15 K, pH 7, and ionic strength 0.25 M. The standard Gibbs energies of formation of the species involved at 298.15 K and ionic strength 0.25 M are given in the following table:

	$\Delta_f G^o$ /kJ mol^{-1}
glucose 6 – phosphate^{2-}	-1767.18
Hglucose 6 – phosphate^{-}	-1801.4
H_2O	-237.19
glucose	-915.9
$HPO_4{}^{2-}$	-1099.34
$H_2PO_4{}^{-}$	-1138.11
H^+	-0.81

(*b*) Plot *K*' versus pH. (*c*) Plot $\Delta_r G'^o$ versus pH.

SOLUTION
(a)

First calculate the acid dissociation constants and the equilibrium constant for the reference chemical reaction glucose 6 – phosphate^{2-} + H_2O = glucose + $HPO_4{}^{2-}$

```
Clear[pH]

kG6P = Exp[- (-0.81 - 1767.18 + 1801.4) / (8.31451 * .29815)]

1.40234 × 10^-6

kphos = Exp[- (-.81 - 1099.34 + 1138.11) / (8.31451 * .29815)]

2.23735 × 10^-7

kref = Exp[- (-915.9 - 1099.34 + 1767.18 + 237.19) / (8.3145 * .29815)]

80.2293

k' := 80.229 * (1 + (10^-pH) / kphos) / (1 + (10^-pH) / kG6P)

k'
```

$$\frac{80.229\ (1 + 4.46958 \times 10^{6}\ 10^{-pH})}{1 + 713092.\ 10^{-pH}}$$

```
pH = 7

7

k'

108.361
```

(b)

```
Plot[k', {pH, 5, 9}, PlotRange → {0, 500}, AxesLabel -> {"pH", "K'"}];
```

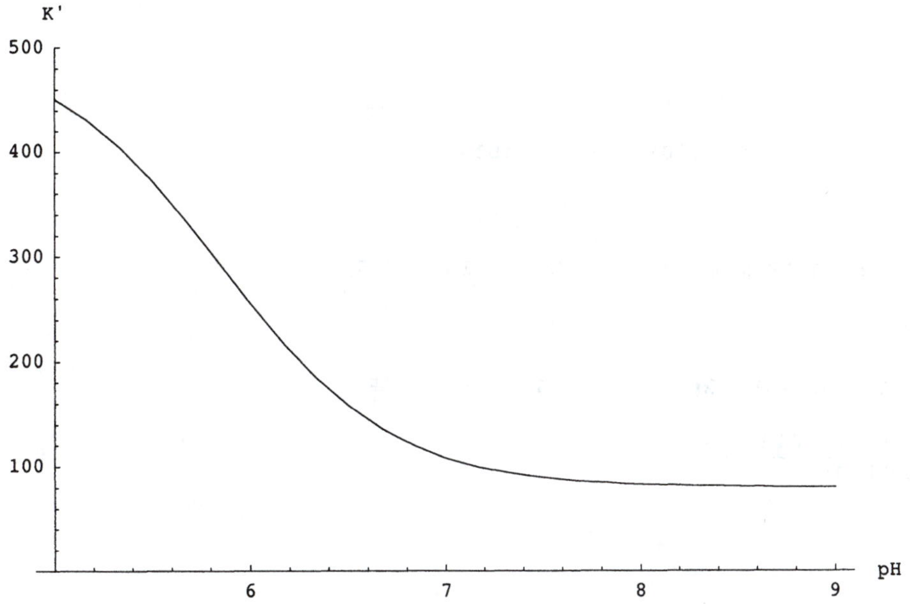

(c)

```
Plot[-8.3145 * .29815 * Log[k'], {pH, 5, 9},
  PlotRange → {-16, -10}, AxesLabel -> {"pH", "ΔrG'°/kJ mol-1"}];
```

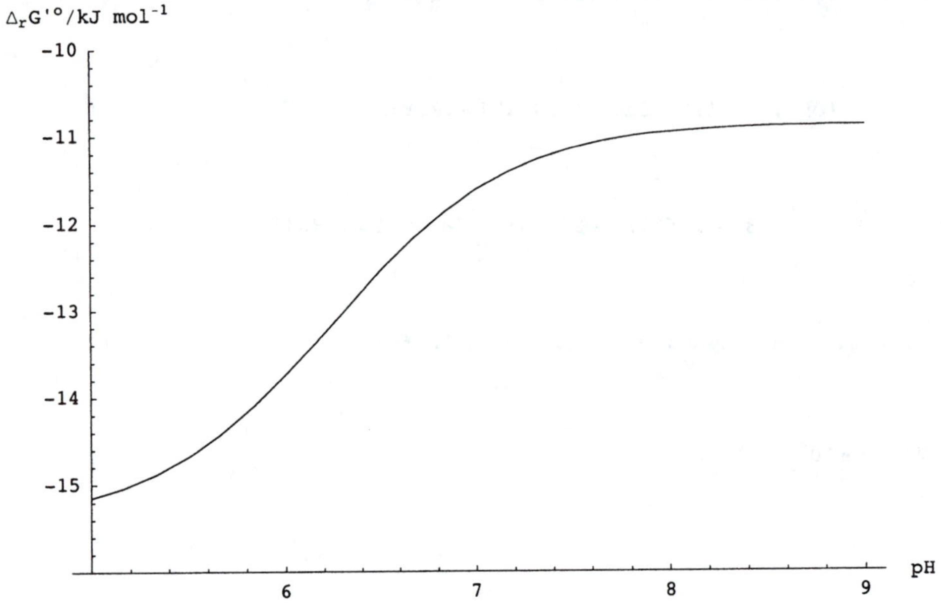

8.E It can be shown that the change in binding of hydrogen ions in a biochemical reaction at a specified pH is given by the negative of the derivative of log K' with respect to pH at constant T and P. (*a*) Use a mathematical program to take the derivative of log K' for glucose 6-phosphate + H_2O = glucose + P_i with respect to pH from the previous problem. (*b*) Plot the production of H^+ versus pH in the range pH 5-9. (*c*) Interpret this plot in terms of the predominant chemical reactions.

SOLUTION

(a)

```
Clear[pH]

kG6P = Exp[- (-0.81 - 1767.18 + 1801.4) / (8.31451 * .29815)]

1.40234 × 10^-6

kphos = Exp[- (-.81 - 1099.34 + 1138.11) / (8.31451 * .29815)]

2.23735 × 10^-7

k2 ' = 80.229 * (1 + (10^-pH) / kphos) / (1 + (10^-pH) / kG6P)
```

$$\frac{80.229\,(1 + 4.46958 \times 10^{6}\ 10^{-pH})}{1 + 713092.\ 10^{-pH}}$$

```
dvt2 = D[Log[10, k2'], pH]
```

$$\frac{0.00541319\,(1+713092.\ 10^{-pH})\left(-\frac{8.25683\times10^{8}\ 10^{-pH}}{1+713092.\ 10^{-pH}}+\frac{1.31732\times10^{8}\ 10^{-pH}\,(1+4.46958\times10^{6}\ 10^{-pH})}{(1+713092.\ 10^{-pH})^{2}}\right)}{1+4.46958\times10^{6}\ 10^{-pH}}$$

Note that in this calculation the minus sign has been omitted so that the amount of H^+ produced is calculated, rather than the change in binding of H^+.

(b)

```
Plot[dvt2, {pH, 5, 9}, AxesLabel -> {"pH", "H+ produced"},
  AxesOrigin -> {5, -.5}, PlotRange -> {-.5, 0}];
```

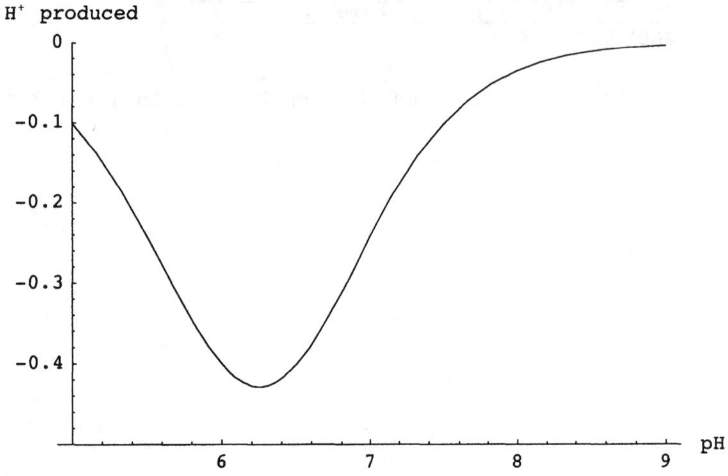

(c) At pH 5, the predominant chemical reaction is
Hglucose 6-phosphate^- + H2O = glucose + H2PO4^-
At pH 9, the predominant chemical reaction is
glucose 6-phosphate^2- + H2O = glucose + HPO4^2-
Thus at very low and very high pH, we would not expect any acid production. However, glucose 6-phosphate is a stronger acid, and so at pH 6 the following reaction is significant:
glucose 6-phosphate^2- + H2O + H^+ = glucose + H2PO4^-
and so under these intermediate conditions H^+ is consumed.

8.F The reaction of two single strands of DNA to form a double helix AA' is represented by
A + A' = AA'
The standard reaction Gibbs energy for the reaction in problem 8.xx is -20.5 kJ mol^{-1} , and the standard reaction enthalpy is -128.4 kJ mol^{-1} at 298 K and pH 7 in 1 M NaCl (G. G. Hammes, Thermodynamics and Kinetics for the Biological Sciences, Wiley, New York, 2000). (a) Calculate the equilibrium constants at a series of temperatures between 273 K and 313 K on the assumption that the standard reaction enthalpy is independent of temperature. (b) Calculate the equilibrium concentration of AA' at each of these temperatures for initial concentrations of A and A' of 10^{-4} M. (c) Calculate the fraction f_S of the DNA in the single strand form at each temperature and plot f_S versus temperature. (d) According to this plot, what is the melting temperature of the double strand. Verify that the equilibrium constant at the melting temperature is given by $K = x/(c-x)^2$.

SOLUTION

(a) Calculate equilibrium constants at 273, 283, 293, 303, and 313 K. Here the standard properties are stated in J mol^{-1} .

```
Clear[t]

kfT = Exp[-((t / 298) * (-20.5 * 10^3) + (1 - t / 298) * (-128.4 * 10^3)) / (8.31451 * t)]
```

$$e^{\frac{0.120272\,(128400.\,(1-\frac{t}{298})+68.7919\,t)}{t}}$$

```
kfT /. t → {273, 283, 293, 303, 313}

{451037., 61111.7, 9490.56, 1666.66, 327.095}
```

(b) Calculate equilibrium concentrations of AA'

A + A' = AA'

init c c 0

equil c-x c-x x

$K = x/(c-x)^2$

The fraction in the single-strand form is given by f_S = 2(c-x)/2c = 1 - x/c.

At 273 K,

```
Solve[4.51 * 10^5 == x / ((10^-4 - x)^2)][[1]]

{x → 0.0000861768}

1 - (8.63 * 10^-5) / (10^-4)

0.137
```

At 283 K

```
Solve[6.11 * 10^4 == x / ((10^-4 - x)^2)][[1]]

{x → 0.0000669083}

1 - (6.72 * 10^-5) / (10^-4)

0.328
```

At 293 K

```
Solve[9.49 * 10^3 == x / ((10^-4 - x)^2)][[1]]

{x → 0.0000373036}

1 - (3.77 * 10^-5) / (10^-4)

0.623
```

At 303 K

```
Solve[1.67 * 10^3 == x / ((10^-4 - x)^2)][[1]]

{x → 0.0000127213}

1 - (1.29 * 10^-5) / (10^-4)

0.871
```

At 313 K

```
Solve[327 == x / ((10^-4 - x)^2)][[1]] // N

{x → 3.07217 × 10^-6}
```

```
1 - (3.13 * 10^-6) / (10^-4)

0.9687

data = Transpose[{{273, 283, 293, 303, 313}, {.137, .328, .623, .871, .969}}]

{{273, 0.137}, {283, 0.328}, {293, 0.623}, {303, 0.871}, {313, 0.969}}

ListPlot[data, Prolog → AbsolutePointSize[4],
  PlotRange → {0, 1}, AxesOrigin → {273, 0}, AxesLabel → {"T/K", "fs"}];
```

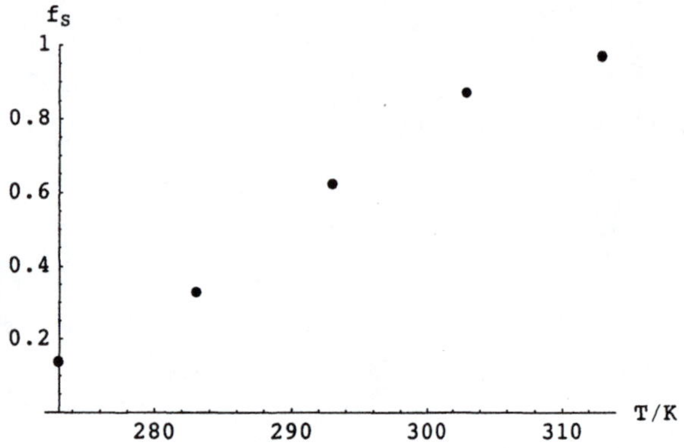

(d) The melting temperature of this double strand is 289 K.

```
kfT /. t → 289

19683.7
```

The concentration of double helix is given by

```
Solve[1.97 * 10^4 == x / ((10^-4 - x)^2)][[1]] // N

{x → 0.0000497479}
```

The fraction single strands is given by

```
1 - (4.97 * 10^-5) / (10^-4)

0.503
```

The concentration of each single strand is given by.

```
(10^-4 - 4.97 * 10^-5)

0.0000503
```

The expression for the equilibrium constant yields

```
(4.97 * 10^-5) / ((10^-4 - 4.97 * 10^-5) ^2)

19643.6
```

8.G The fraction of a repressor protein that is denatured at equilibrium in 3 M urea at pH 8 in 0.1 M NaCl is given by the following table (G. S. Huang and T. G. Oas, Biochemistry 35, 6173 (1996)).

$t/^o$ C	-5	0	10	21	28	30	32	36	40	42	46
f_D	0.30	0.20	0.17	0.20	0.30	0.40	0.50	0.60	0.70	0.80	0.90

These data are especially interesting because they show that when the temperature is reduced below 10 ^{o} C, the denaturation increases. (a) Calculate K_D at each temperature and plot it versus absolute temperature. (b) Calculate ΔG_D^o and plot it versus T. (c) Fit the data in (b) with equation 8.85, and use this function to plot ΔG_D^o versus temperature. (d) Calculate ΔH_D^o using the Gibbs-Duhem equation and plot it versus T. (e) Calculate ΔS_D^o using -dΔG_D^o/dT and plot it versus T. (f) Calculate ΔC_D^o using -dΔH_D^o/dT and plot it versus T.

SOLUTION

(a)

```
t = {-5, 0, 10, 21, 28, 30, 32, 36, 40, 42, 46};

fd = {.3, .2, .17, .2, .3, .4, .5, .6, .7, .8, .9};

kd = fd / (1 - fd)

{0.428571, 0.25, 0.204819, 0.25, 0.428571, 0.666667, 1., 1.5, 2.33333, 4., 9.}

data = Transpose[{t + 273, kd}]

{{268, 0.428571}, {273, 0.25}, {283, 0.204819}, {294, 0.25}, {301, 0.428571},
 {303, 0.666667}, {305, 1.}, {309, 1.5}, {313, 2.33333}, {315, 4.}, {319, 9.}}

kdplot = ListPlot[data, AxesOrigin → {263, 0},
   AxesLabel → {"T/K", "K_D"}, Prolog → AbsolutePointSize[4]];
```

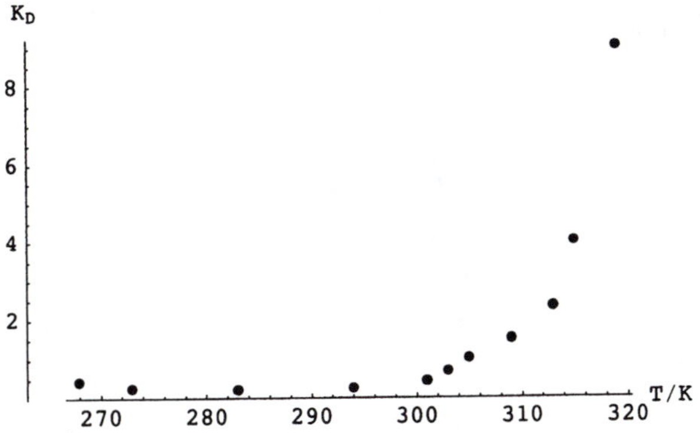

(b)

```
data2 = Transpose[{t + 273, -8.31451 * 10^-3 * (t + 273) * Log[kd]}]

{{268, 1.88802}, {273, 3.1467}, {283, 3.73099},
 {294, 3.38875}, {301, 2.1205}, {303, 1.02149}, {305, 0.},
 {309, -1.04171}, {313, -2.20504}, {315, -3.6308}, {319, -5.82776}}
```

```
plot2 = ListPlot[data2, AxesOrigin -> {263, 0},
   AxesLabel → {"T/K", "ΔG°D/kJ mol⁻¹"}, Prolog → AbsolutePointSize[4]];
```

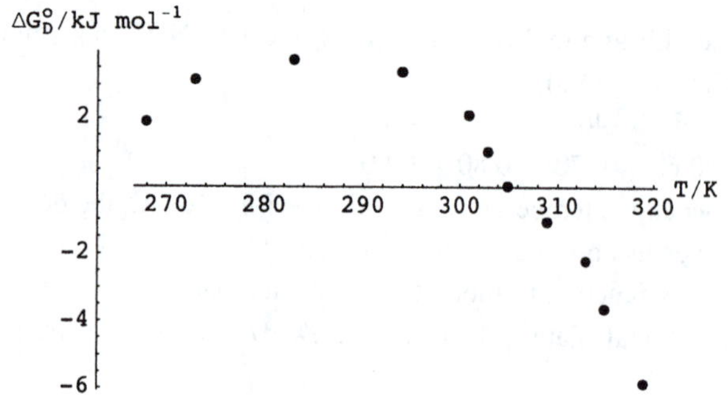

(c) Note that x is used as the symbol for absolute temperature in the following sections. Note that T_g is 305 K..

```
gibbsfx = Fit[data2, {1 - x / 305, (305 - x + x * Log[x / 305])}, x]
```

$$97.3515 \left(1 - \frac{x}{305}\right) - 4.13495 \left(305 - x + x \,\mathrm{Log}\left[\frac{x}{305}\right]\right)$$

This indicates that $\Delta H_D^o = 97.35$ kJ mol^{-1} and $\Delta C_P^o = 4.13$ kJ K^{-1} mol^{-1} at 305 K

```
plot3 =
  Plot[gibbsfx, {x, 263, 320}, AxesOrigin -> {263, 0}, AxesLabel → {"T/K", "ΔG°D/kJ mol⁻¹"}];
```

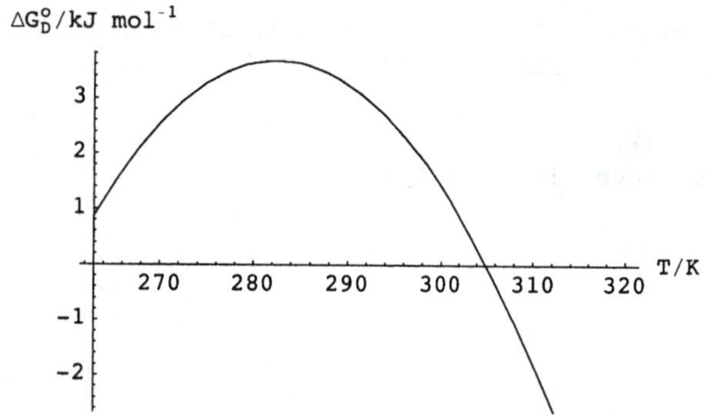

```
Show[plot2, plot3];
```

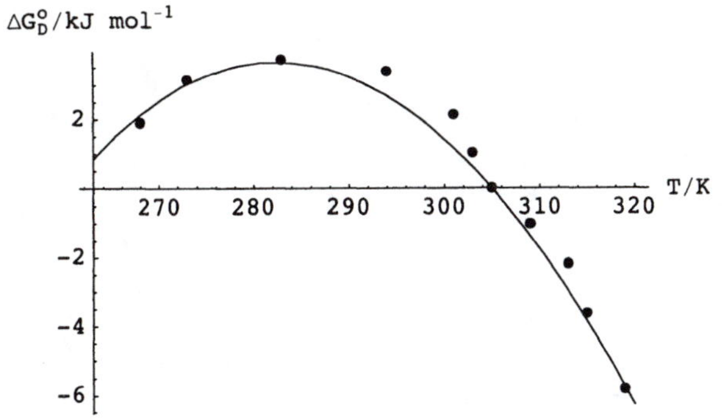

(d)

```
enthalpyfx = -x^2 * D[gibbsfx / x, x]
```

$$-x^2 \left(\frac{-0.319185 - 4.13495\,\mathrm{Log}[\frac{x}{305}]}{x} - \frac{97.3515\,(1 - \frac{x}{305}) - 4.13495\,(305 - x + x\,\mathrm{Log}[\frac{x}{305}])}{x^2}\right)$$

```
hdplot = Plot[enthalpyfx, {x, 263, 320},
   AxesOrigin -> {263, 0}, AxesLabel → {"T/K", "ΔH°D/kJ mol-1"}];
```

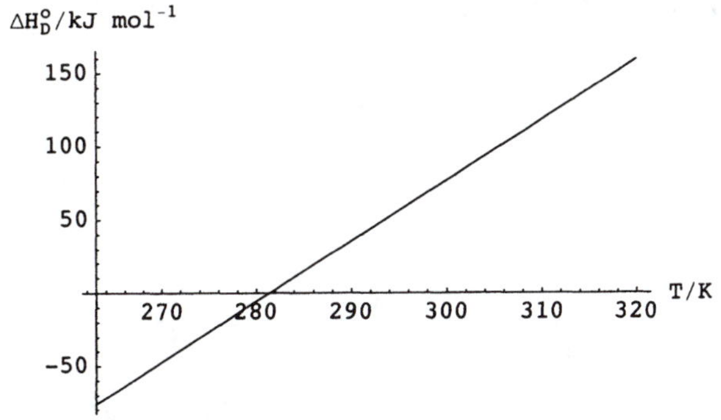

Note that the enthalpy changes rapidly with temperature, and that it changes sign at 283 K (10 °C). The value of the enthalpy change at 305 K is in agreement with the value of the parameter $\Delta H_D^o(305\,K) = 97.35$ kJ mol^{-1} determined in the fitting process.

(e)

```
entropyfx = -D[gibbsfx, x]
```

$$0.319185 + 4.13495\,\mathrm{Log}\left[\frac{x}{305}\right]$$

```
sdplot = Plot[entropyfx, {x, 263, 320},
   AxesOrigin -> {263, 0}, AxesLabel → {"T/K", "ΔS°D/kJ mol-1"}];
```

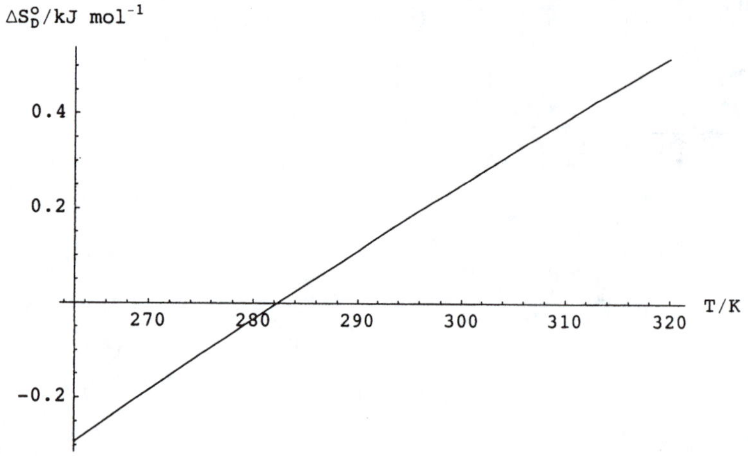

(f)

```
heatcapfx = D[enthalpyfx, x];

Plot[heatcapfx, {x, 263, 320}, AxesOrigin -> {263, 3},
  PlotRange → {3, 6}, AxesLabel → {"T/K", "ΔC°D/kJ mol-1"}];
```

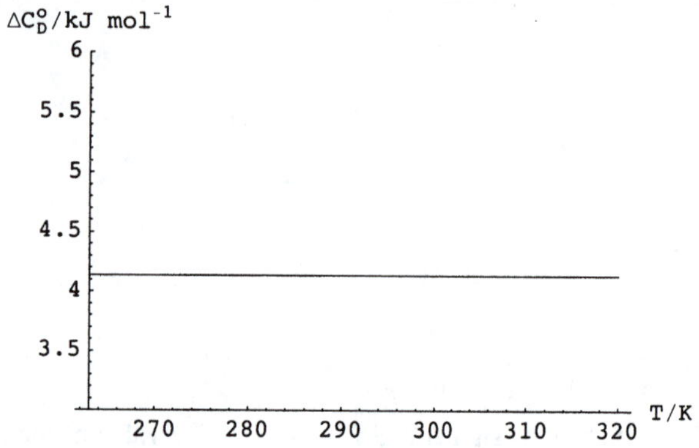

The change in heat capacity in the denaturation reaction had to turn out to be constant, because the fitting of the data was made on that assumption.

```
Show[GraphicsArray[{{kdplot, plot3}, {hdplot, sdplot}}, GraphicsSpacing → 0]];
```

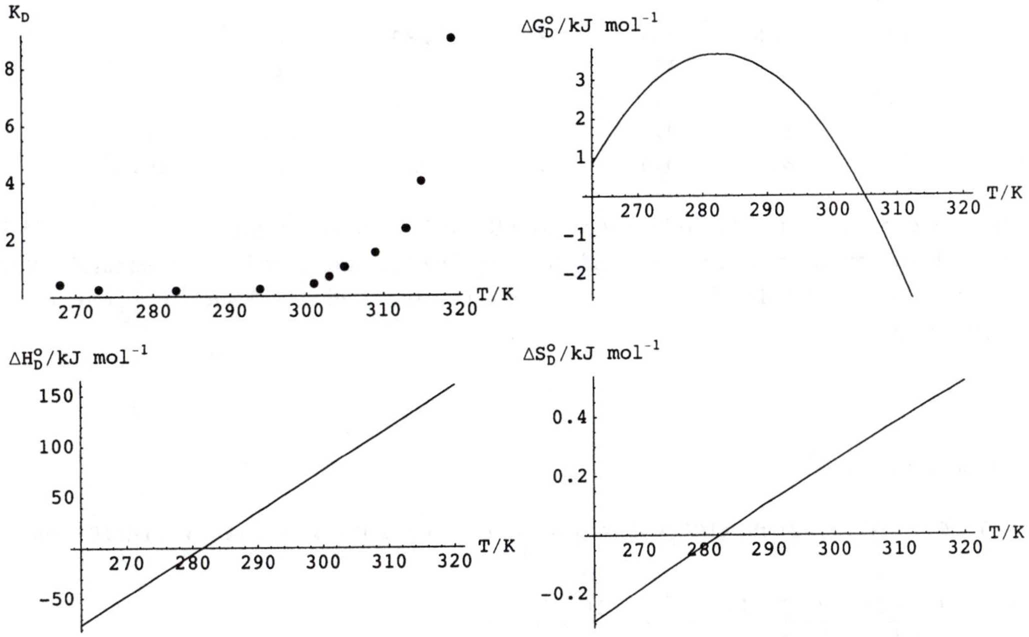

8.H Since the hydrolysis of a sodium salt of a monoprotic weak acid produces equal concentrations of HA and OH^-, K_h = $[OH^-]^2/c$, where c is is the molar concentration of the salt. We know that $K_h = K_w/K_a$ and $[OH^-] = K_h/10^{-pH}$. Therefore it can be shown that

pH = (1/2)(14.00 + pK_a + log$c$)

at 298.15 K. Make a table of the pHs of sodium salts of weak acids with pKs of 5 to 11 and concentrations of 0.10, 0.01, 0.001, and 0.0001 M at 298.15 K.

SOLUTION

```
calcpHofNaA[pK_,c_]:=Module[{},(*Calculates the pH of a sodium salt of a weak acid HA
with a specified pK at 25 oC and a specified concentration c in moles per liter.*)
.5*(14.0+pK+Log[c]/Log[10])]

funct=calcpHofNaA[pK,c]
```

$$0.5\left(14. + pK + \frac{Log[c]}{Log[10]}\right)$$

```
funct/.pK->{5,6,7,8,9,10,11}/.c->{.1,.01,.001,.0001}
```

```
{{9., 8.5, 8., 7.5}, {9.5, 9., 8.5, 8.}, {10., 9.5, 9., 8.5}, {10.5, 10., 9.5, 9.},
 {11., 10.5, 10., 9.5}, {11.5, 11., 10.5, 10.}, {12., 11.5, 11., 10.5}}
```

```
TableForm[Transpose[calcpHofNaA[pK,c]/.pK->{5,6,7,8,9,10,11}/.c->{.1,.01,.001,.0001}],T
ableHeadings->{{"0.1M","0.01M","0.001M","0.0001M"},{"pK5","pK6","pK7","pK8","pK9","pK10
","pK11"}}]
```

	pK5	pK6	pK7	pK8	pK9	pK10	pK11
0.1M	9.	9.5	10.	10.5	11.	11.5	12.
0.01M	8.5	9.	9.5	10.	10.5	11.	11.5
0.001M	8.	8.5	9.	9.5	10.	10.5	11.
0.0001M	7.5	8.	8.5	9.	9.5	10.	10.5

8.I (a) Plot the titration curve for a liter of 0.10 M acetic acid (pK = 4.756) with concentrated NaOH at 298.15 K on the assumption that the ionic strength can be taken as zero and that a the NaOH solution used is so concentrated that there is not a significant change in volume (see Problem 8.2). (b) Plot the corresponding titration curves for monoprotic weak acids with pKs of 4, 5, 6, 7, 8, 9, and 10.

SOLUTION

(a) Titration curve for acetic acid

```
n=(.1*10^-4.756-10^(-2*pH)-10^(-pH-4.756)-10^(-14)+10^(-14+pH-4.756))/(10^-pH+10^-4.756
)
```

$$\frac{1.75388\times 10^{-6} - 10^{-4.756-pH} + 10^{-18.756+pH} - 10^{-2\,pH}}{0.0000175388 + 10^{-pH}}$$

```
Plot[n,{pH,2,13},PlotRange->{0,.2},AxesLabel->{"pH","n(NaOH)"}];
```

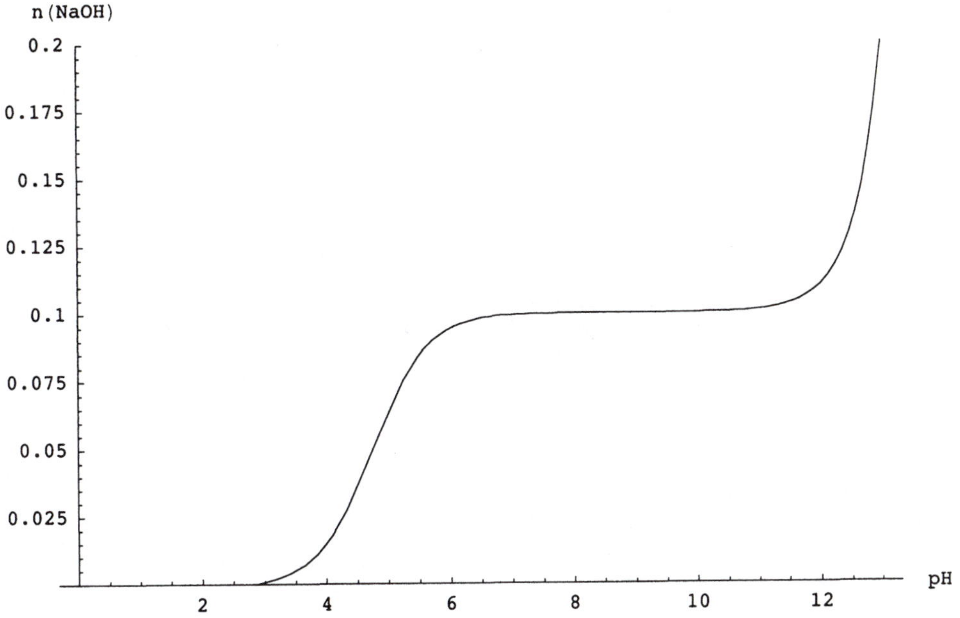

(b) Titration curves for seven different weak acids

```
n4=(.1*10^-4-10^(-2*pH)-10^(-pH-4)-10^(-14)+10^(-14+pH-4))/(10^-pH+10^-4);

p4=Plot[n4,{pH,2,13},PlotRange->{0,.2},AxesLabel->{"pH","n(NaOH)"},DisplayFunction->Ide
ntity];

n5=(.1*10^-5-10^(-2*pH)-10^(-pH-5)-10^(-14)+10^(-14+pH-5))/(10^-pH+10^-5);

p5=Plot[n5,{pH,2,13},PlotRange->{0,.2},AxesLabel->{"pH","n(NaOH)"},DisplayFunction->Ide
ntity];

n6=(.1*10^-6-10^(-2*pH)-10^(-pH-6)-10^(-14)+10^(-14+pH-6))/(10^-pH+10^-6);

p6=Plot[n6,{pH,2,13},PlotRange->{0,.2},AxesLabel->{"pH","n(NaOH)"},DisplayFunction->Ide
ntity];

n7=(.1*10^-7-10^(-2*pH)-10^(-pH-7)-10^(-14)+10^(-14+pH-7))/(10^-pH+10^-7);

p7=Plot[n7,{pH,2,13},PlotRange->{0,.2},AxesLabel->{"pH","n(NaOH)"},DisplayFunction->Ide
ntity];

n8=(.1*10^-8-10^(-2*pH)-10^(-pH-8)-10^(-14)+10^(-14+pH-8))/(10^-pH+10^-8);

p8=Plot[n8,{pH,2,13},PlotRange->{0,.2},AxesLabel->{"pH","n(NaOH)"},DisplayFunction->Ide
ntity];

n9=(.1*10^-9-10^(-2*pH)-10^(-pH-9)-10^(-14)+10^(-14+pH-9))/(10^-pH+10^-9);

p9=Plot[n9,{pH,2,13},PlotRange->{0,.2},AxesLabel->{"pH","n(NaOH)"},DisplayFunction->Ide
ntity];

n10=(.1*10^-10-10^(-2*pH)-10^(-pH-10)-10^(-14)+10^(-14+pH-10))/(10^-pH+10^-10);

p10=Plot[n10,{pH,2,13},PlotRange->{0,.2},AxesLabel->{"pH","n(NaOH)"},DisplayFunction->I
dentity];
```

```
Show[p4,p5,p6,p7,p8,p9,p10,DisplayFunction->$DisplayFunction];
```

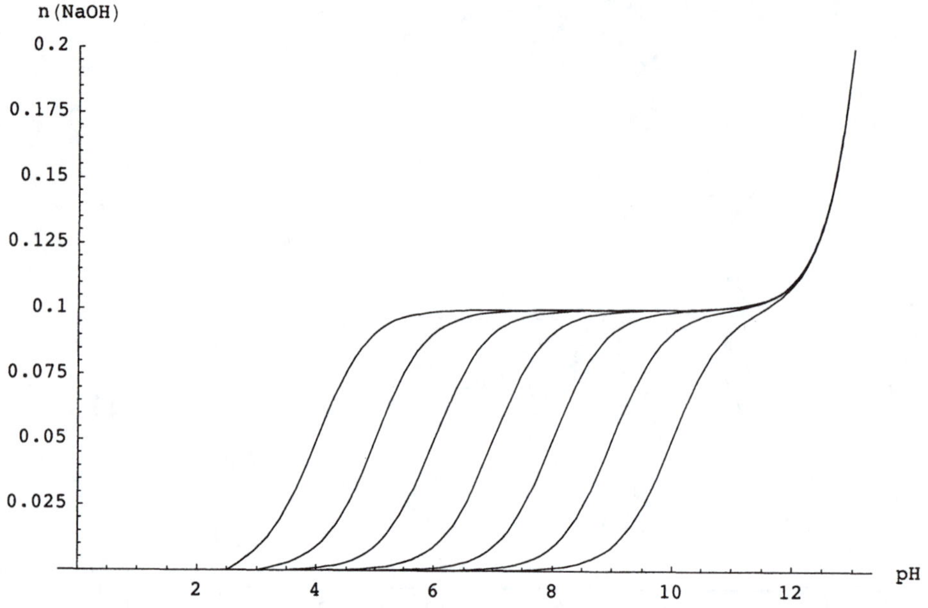

Note that the initial solutions have pHs determined by the pK of the weak acid. The pHs at the end points of these titrations are all different, and each corresponds with the pH of the corresponding pure salt.

8.J Calculate the plot of $\bar{N}_H$ versus pH for phosphoric acid at 298.15 K using its binding polynomial.

SOLUTION

```
p=1+(10^-pH)/10^-12+(10^-pH)^2/(10^-12*6.34*10^-8)+(10^-pH)^3/(10^-12*6.34*10^-8*7.11*1
0^-3)
```

$$1 + 10^{12-pH} + 2.21841 \times 10^{21}\ 10^{-3\,pH} + 1.57729 \times 10^{19}\ 10^{-2\,pH}$$

```
nH=-D[Log[10,p],pH]
```

$$-\frac{-3.63184 \times 10^{19}\ 2^{1-2\,pH}\ 5^{-2\,pH} - 1.53242 \times 10^{22}\ 10^{-3\,pH} - 10^{12-pH}\ \mathrm{Log}[10]}{(1 + 10^{12-pH} + 2.21841 \times 10^{21}\ 10^{-3\,pH} + 1.57729 \times 10^{19}\ 10^{-2\,pH})\ \mathrm{Log}[10]}$$

```
Plot[nH,{pH,0,14},AxesLabel->{"pH","\!\(N\_H\)"}];
```

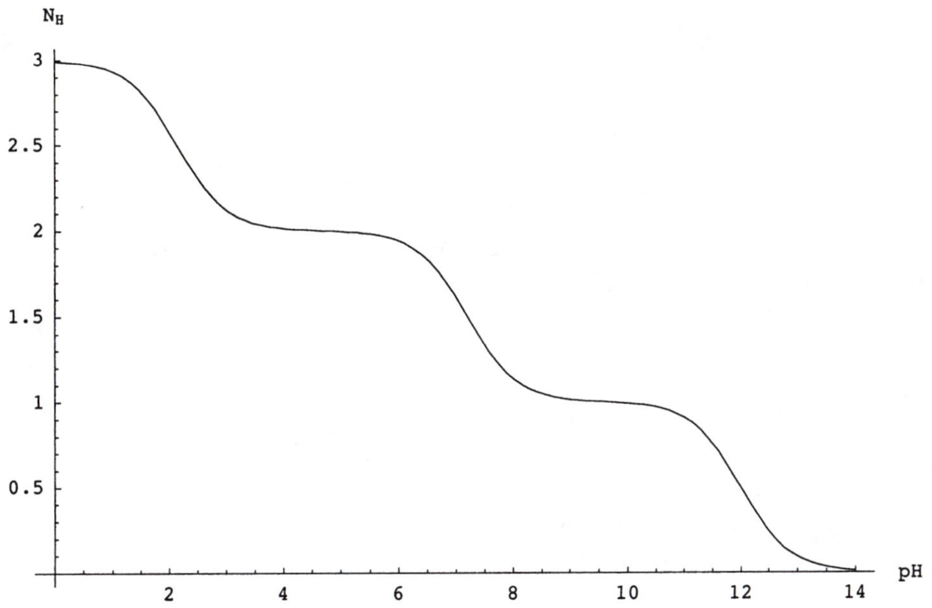

Chapter 9 Quantum Theory

9.A (*a*) Plot the wavefunctions and probability densities for the first three levels of an electron in a box 0.1 nm in length. (*b*) Test the normalization and orthogonality of these wavefunctions. (*c*) Calculate the average value of x and the average value of x^2 for electrons in these energy levels. (*d*) Calculate the average value of p and the average value of p^2. (e) Check these values against the Heisenberg uncertainty principle.

SOLUTION

(a) The wave function for a particle in a box with quantum number n is given by

```
ClearAll["Global`*"]

Off[General::spell1];
Off[General::spell];

psiPB[n_] = Sqrt[2 / .1] * Sin[n * Pi * x / .1];

psi1 = Plot[psiPB[1], {x, 0, .1}, AxesLabel -> {"\!\(\*
StyleBox[\"x\",\nFontSlant->\"Italic\"]\)/m", "ψ1"}, DisplayFunction → Identity];

psi2 = Plot[psiPB[2], {x, 0, .1}, AxesLabel -> {"\!\(\*
StyleBox[\"x\",\nFontSlant->\"Italic\"]\)/m", "ψ2"}, DisplayFunction → Identity];

psi3 = Plot[psiPB[3], {x, 0, .1}, AxesLabel -> {"\!\(\*
StyleBox[\"x\",\nFontSlant->\"Italic\"]\)/m", "ψ3"}, DisplayFunction → Identity];

Show[GraphicsArray[{psi1, psi2, psi3}]];
```

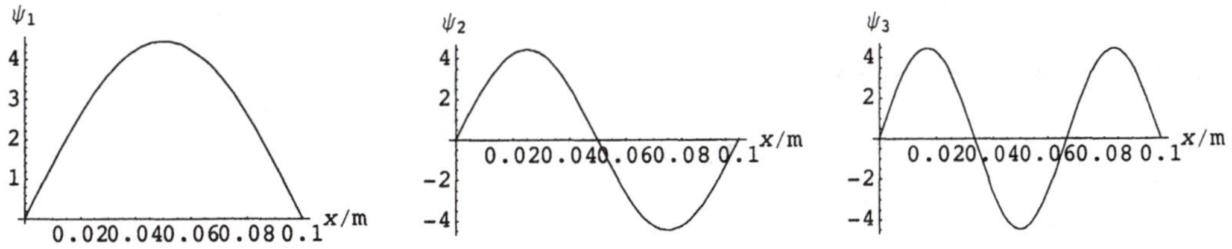


```
psi12 = Plot[psiPB[1]^2, {x, 0, .1}, AxesLabel -> {"\!\(\*
StyleBox[\"x\",\nFontSlant->\"Italic\"]\)/m", "ψ1^2"}, DisplayFunction → Identity];

psi22 = Plot[psiPB[2]^2, {x, 0, .1}, AxesLabel -> {"\!\(\*
StyleBox[\"x\",\nFontSlant->\"Italic\"]\)/m", "ψ2^2"}, DisplayFunction → Identity];

psi32 = Plot[psiPB[3]^2, {x, 0, .1}, AxesLabel -> {"\!\(\*
StyleBox[\"x\",\nFontSlant->\"Italic\"]\)/m", "ψ3^2"}, DisplayFunction → Identity];
```

```
Show[GraphicsArray[{psi12, psi22, psi32}]];
```

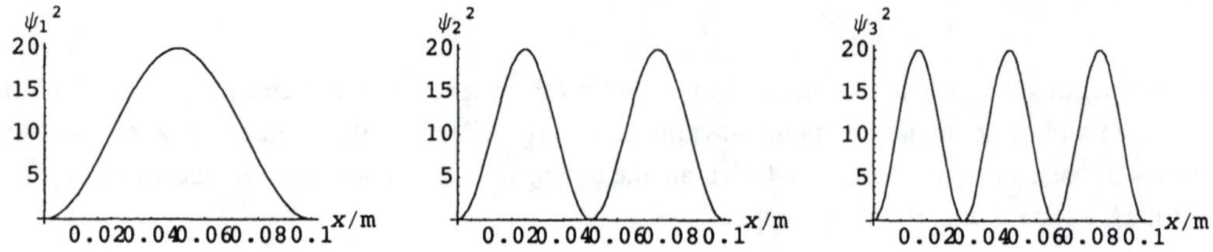

(b) The normalization and orthogonality are checked by integration .

```
Off[FactorSquareFree::lrgexp]

Integrate[psiPB[1]^2, {x, 0, .1}]

1.

Integrate[psiPB[2]^2, {x, 0, .1}]

1.

Integrate[psiPB[3]^2, {x, 0, .1}]

1.

Integrate[psiPB[1]*psiPB[2], {x, 0, .1}]

0.
```

This shows that psi1 and psi2 are orthogonal.

(c) The average values of x and x^2 are given by

```
Integrate[x*psiPB[1]^2, {x, 0, .1}] // N

0.05

Integrate[x*psiPB[3]^2, {x, 0, .1}] // N

0.05

avx21 = N[Integrate[x^2*psiPB[1]^2, {x, 0, .1}]]

0.00282673

avx22 = N[Integrate[x^2*psiPB[2]^2, {x, 0, .1}]]

0.00320668

avx23 = N[Integrate[x^2*psiPB[3]^2, {x, 0, .1}]]

0.00327704
```

(d) It can be shown that the average momentum of a particle in a box is zero, independent of the quantum number.

```
avp = N[Integrate[psiPB[2] * D[psiPB[2], {x, 1}], {x, 0, .1}]]

0.
```

The operator for p^2 is -(h bar^2)(d^2/dx^2). In the following three integrals the factor hbar^2 has been taken out. as a simplification. The average value of the momentum squared is given by the integral of the complex conjugate of the wave function times the operator times the wave function.

```
avp21 = N[Integrate[psiPB[1] * -1 * D[psiPB[1], {x, 2}], {x, 0, .1}]]

986.96

avp22 = N[Integrate[psiPB[2] * -1 * D[psiPB[2], {x, 2}], {x, 0, .1}]]

3947.84

avp23 = N[Integrate[psiPB[3] * -1 * D[psiPB[3], {x, 2}], {x, 0, .1}]]

8882.64
```

The standard deviations for distance and momentum are given by the square root of the difference between the average value of the square of the property minus the square of the average value.

```
dx1 = (avx21 - .05^2)^.5

0.0180756

dx2 = (avx22 - .05^2)^.5

0.0265835

dx3 = (avx23 - .05^2)^.5

0.0278755
```

The factor h bar has been omitted in the following three expressions:

```
dp1 = avp21^.5

31.4159

dp2 = avp22^.5

62.8319

dp3 = avp23^.5

94.2478
```

(e) The Heisenberg products without the hbar are

```
dx1 * dp1

0.567862
```

```
dx2 * dp2

1.67029

dx3 * dp3

2.6272
```

According to the Heisenberg uncertainty principle, these products should be greater than 0.5, and they are. The text gives the following equation for ΔxΔp for a particle in a one dimensional box:
$\Delta x \Delta p = (\text{h bar}/2)(\frac{n^2 \pi^2}{3} - 2)^{1/2}$
These last three results agree with this equation for n = 1, 2, and 3.

9.B (*a*) Plot the first four wavefunctions for a harmonic oscillator to see the shapes of the wavefunctions. (*b*) Plot the corresponding probability densities for these four levels.

SOLUTION
(a) The wave functions are represented in *Mathematica* as HermiteH[v,x]*Exp[-x^2/2].

```
psi0 = Plot[HermiteH[0, x] * Exp[-x^2 / 2],
   {x, -4, 4}, AxesLabel -> {"x", "ψ0"}, DisplayFunction -> Identity];

psi1 = Plot[HermiteH[1, x] * Exp[-x^2 / 2],
   {x, -4, 4}, AxesLabel -> {"x", "ψ1"}, DisplayFunction -> Identity];

psi2 = Plot[HermiteH[2, x] * Exp[-x^2 / 2],
   {x, -4, 4}, AxesLabel -> {"x", "ψ2"}, DisplayFunction -> Identity];

psi3 = Plot[HermiteH[3, x] * Exp[-x^2 / 2],
   {x, -4, 4}, AxesLabel -> {"x", "ψ3"}, DisplayFunction -> Identity];
```

```
Show[GraphicsArray[{{psi0,psi1},{psi2,psi3}},GraphicsSpacing->.01]];
```

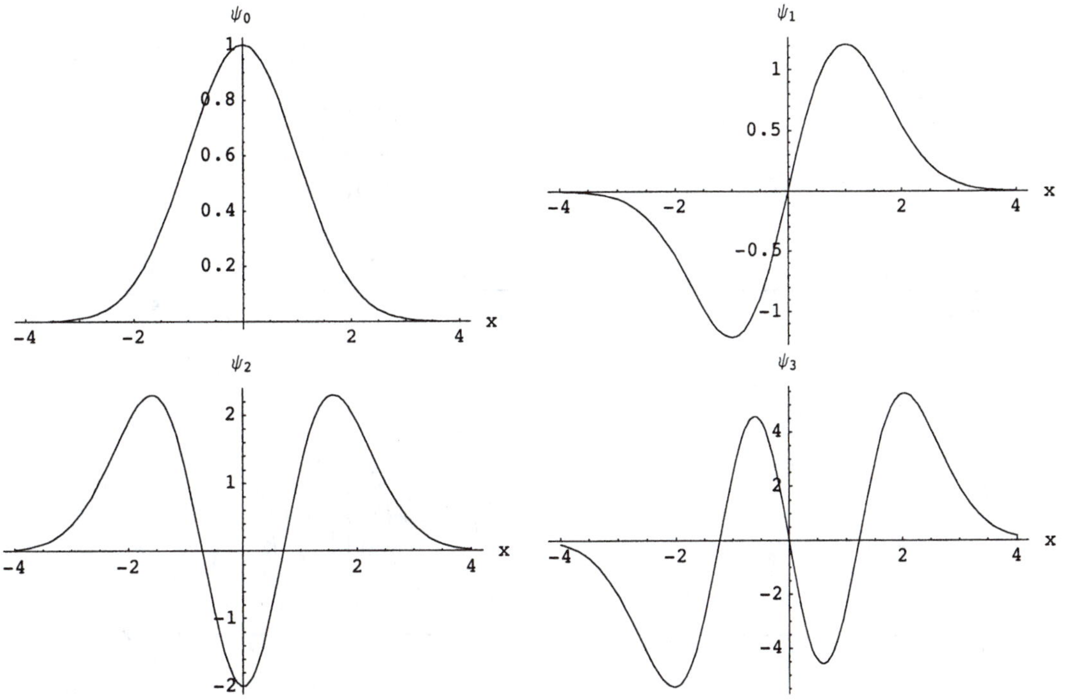

The option DisplayFunction->Identity is used to avoid immediate printing of plots.

(b) The probability densities are proportional to the squares of the wavefunctions.

```
psi02 = Plot[(HermiteH[0, x] * Exp[-x^2 / 2])^2, {x, -4, 4},
   AxesLabel -> {"x", "ψ0^2"}, PlotRange -> All, DisplayFunction -> Identity];

psi12 = Plot[(HermiteH[1, x] * Exp[-x^2 / 2])^2,
   {x, -4, 4}, AxesLabel -> {"x", "ψ1^2"}, DisplayFunction -> Identity];

psi22 = Plot[(HermiteH[2, x] * Exp[-x^2 / 2])^2,
   {x, -4, 4}, AxesLabel -> {"x", "ψ2^2"}, DisplayFunction -> Identity];

psi32 = Plot[(HermiteH[3, x] * Exp[-x^2 / 2])^2,
   {x, -4, 4}, AxesLabel -> {"x", "ψ3^2"}, DisplayFunction -> Identity];
```

```
Show[GraphicsArray[{{psi02,psi12},{psi22,psi32}},GraphicsSpacing->.1]];
```

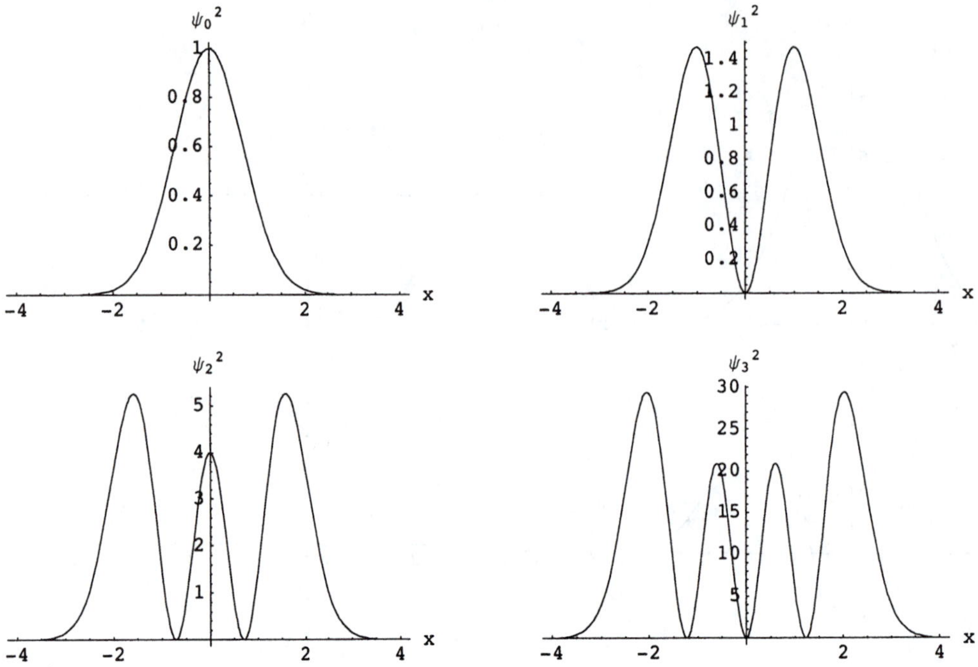

Plots for higher quantum numbers can be calculated in the same way.

9.C In the wavefunction for the harmonic oscillator, α is taken as unity here because it is only needed to provide units in calculations. (*a*) Plot normalized wavefunctions and probability densities for vibrational quantum numbers of 0 and 1. (*b*) Check the normalization and orthogonality of these two wavefunctions. (*c*) Plot the wavefunction and probability density for $v = 30$.

SOLUTION

(a) Normalized wavefunctions for the harmonic oscillator can be calculated using
psiHO[v_]:=HermiteH[v,x]*Exp[-x^2/2]/Sqrt[2^v*v!*Sqrt[Pi]]

```
psiHO[v_] := HermiteH[v, x] * Exp[-x^2 / 2] / Sqrt[2^v * v! * Sqrt[Pi]]

psi0 = Plot[psiHO[0], {x, -6, 6}, AxesLabel -> {"x", "ψ0"}, DisplayFunction → Identity];

psi1 = Plot[psiHO[1], {x, -6, 6}, PlotPoints -> 100,
   AxesLabel -> {"x", "ψ1"}, DisplayFunction → Identity];
```

```
Show[GraphicsArray[{psi0, psi1}]];
```

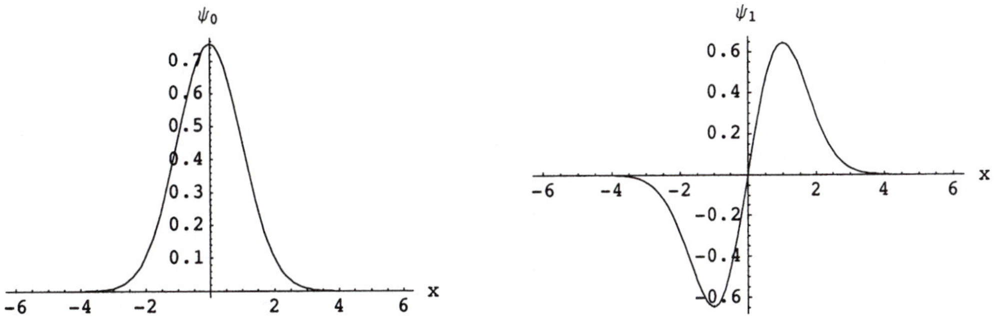

(b) Check the normalization and orthogonality.

```
Integrate[psiHO[0]^2, {x, -Infinity, Infinity}]
```

1

```
Integrate[psiHO[1]^2, {x, -Infinity, Infinity}]
```

1

```
Integrate[psiHO[0]*psiHO[1], {x, -Infinity, Infinity}]
```

0

(c) Make plots for v = 30.

```
Plot[psiHO[30], {x, -10, 10}, PlotPoints -> 100, AxesLabel -> {"x", "ψ30"}];
```

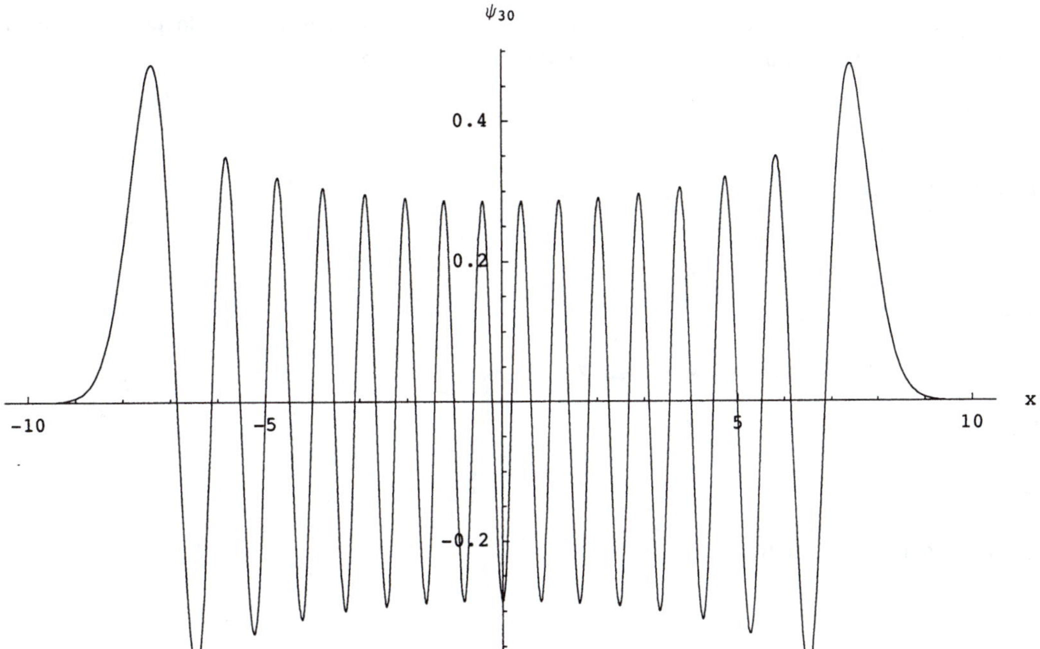

```
Plot[psiHO[30]^2, {x, -10, 10}, PlotPoints -> 100, AxesLabel -> {"x", "ψ30²"}];
```

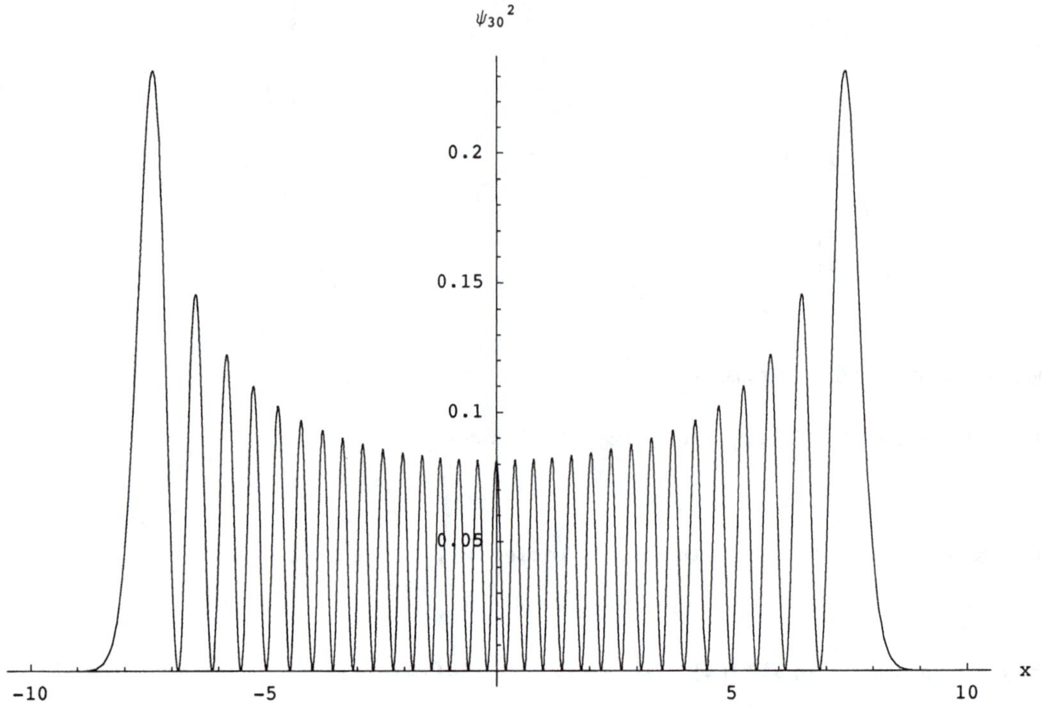

The default value of PlotPoints is 25, and so 100 is used to obtain smoother plots. You can make plots using other quantum numbers using these functions.

9.D (a) Assuming that the $^{12}C^{16}O$ molecule is a harmonic oscillator with $\mu = 1.1385 \times 10^{-26}$ kg and $k = 1886$ N m^{-1} write the expression for the vibrational wavefunction. (b) Plot the wavefunction versus the distance x for $v = 0$ and $v = 1$. (c) Calculate the value of the average value of x, the average of the square of x, and the standard deviation of the internuclear distance for $v = 0$ and $v = 1$. Note that the mean internuclear distance is 113 pm.

SOLUTION

```
ClearAll["Global`*"]
```

(a)

```
alpha=(k*mu)^.5/hbar;

k=1886;mu=1.1385*10^-26;hbar=1.054572*10^-34;

alpha

4.39401×10^22
```

The normalization constant N_v is given by

```
norm=((2^v*v!)^-.5)*(alpha/Pi)^.25
```

$$\frac{343897.}{(2^v\ v!)^{0.5}}$$

The wavefunction for the harmonic oscillator is given by

```
((2^v*v!)^-.5)*(alpha/Pi)^.25*HermiteH[v,alpha^.5*x]*Exp[-alpha*x^2/2]
```

$$\frac{343897.\ e^{-2.19701\times10^{22}\ x^2}\ \text{HermiteH}[v,\ 2.09619\times10^{11}\ x]}{(2^v\ v!)^{0.5}}$$

To use this function in calculations we write

```
psihosc[v_]:=((2^v*v!)^-.5)*(alpha/Pi)^.25*HermiteH[v,alpha^.5*x]*Exp[-alpha*x^2/2]
```

```
Plot[psihosc[0],{x,-2*10^-11,2*10^-11},AxesLabel->{"x/m","ψ0"}];
```

```
Plot[psihosc[1],{x,-2*10^-11,2*10^-11},AxesLabel->{"x/m","ψ1"}];
```

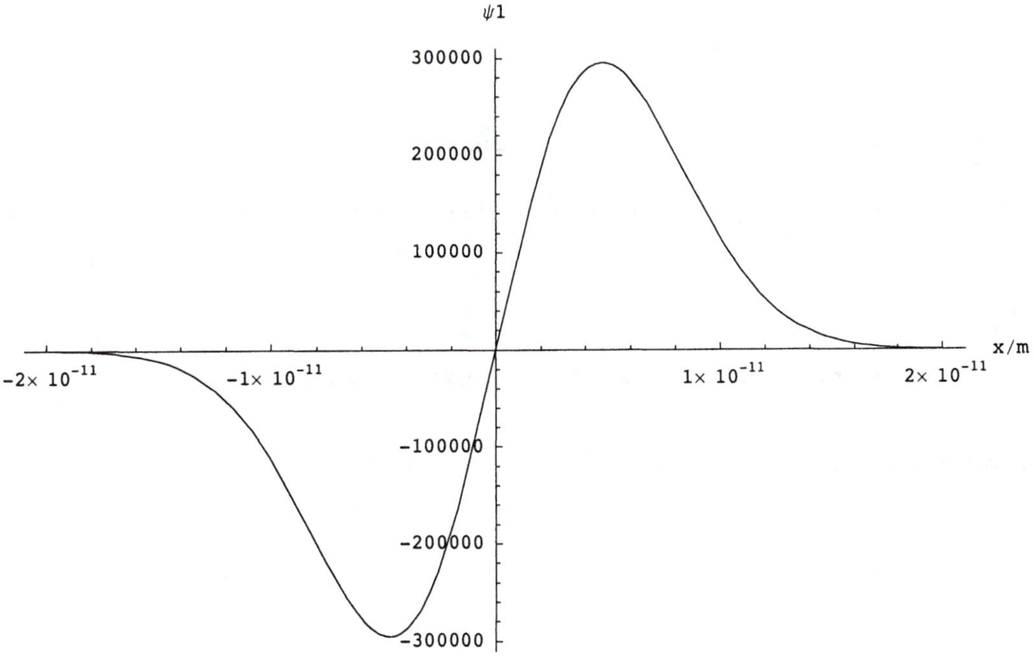

(c) For v = 0

```
xav0=Integrate[x*psihosc[0]^2,{x,-Infinity,Infinity}]
```

0

```
x2av0=Integrate[(x^2)*psihosc[0]^2,{x,-Infinity,Infinity}]
```

1.13791×10^{-23}

According to Example 9.17 $<x^2> = 1/(2*\text{alpha})$

```
1/(2*alpha)
```

```
1.13791×10^-23
```

The standard deviation of x in picometers for v = 0 is given by

```
x2av0^.5/10^-12
```

```
3.3733
```

For v = 1

```
xav1=Integrate[x*psihosc[1]^2,{x,-Infinity,Infinity}]
```

```
0
```

```
x2av1=Integrate[x^2*psihosc[1]^2,{x,-Infinity,Infinity}]
```

```
3.41374×10^-23
```

The standard deviation for x in picometers for v = 1 is given by

```
x2av1^.5/10^-12
```

```
5.84272
```

Note that this is larger than for v = 0 and is about 5% of the mean internuclear distance.

9.E Calculate the probability that a harmonic oscillator is outside of its classical turning points when the vibrational wave number is 4.

SOLUTION

```
psi[n_] := HermiteH[n, y] Exp[-y^2 / 2] / Sqrt[2^n n! Sqrt[Pi]]
```

```
nn = 4;
NIntegrate[Expand[psi[nn]^2], {y, -Infinity, Infinity}]
```

```
1.
```

```
Plot[psi[nn], {y, -5, 5}];
```

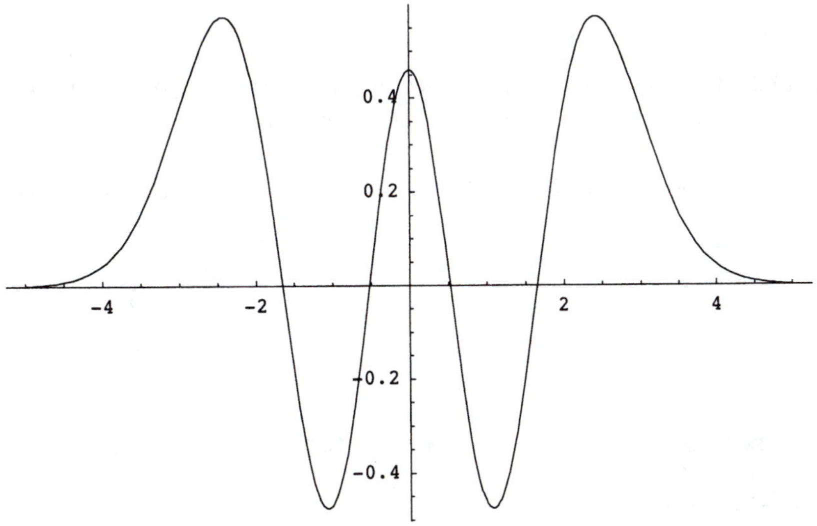


```
Plot[psi[nn]^2, {y, -5, 5}];
```

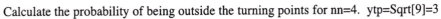

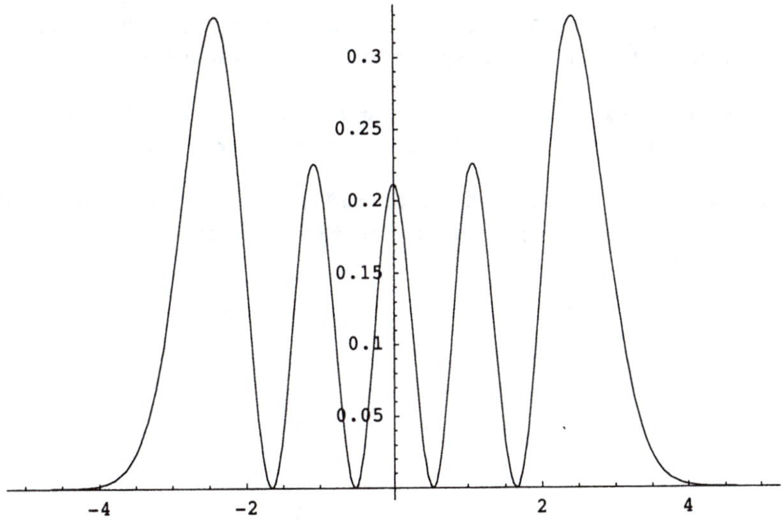

Calculate the probability of being outside the turning points for nn=4. ytp=Sqrt[9]=3

```
pin = NIntegrate[psi[4]^2, {y, -3, 3}]

0.921074
```

The probability of being outside is about 8%. This is in the direction expected from the Bohr correspondence principle. See Noggle, Physical Chemistry Using *Mathematica*, Harper-Collins, New York, 1996, p. 278

9.F Calculate the radiant energy density for a blackbody at 1500, 1700, and 2000 K as a function of (*a*) wavelength in micrometers and (*b*) frequency in Hz.

SOLUTION

(a)

```
ClearAll["Global`*"]

rholambda[lambda_, t_] := Module[{h, c, k}, h = 6.6260755 * 10^-34 (*J s*);
    c = 2.99792458 * 10^8 (*m s^-1*);
        k = 1.380658 * 10^-23 (*J K^-1*);
  ((8 Pi h c) / ((lambda^5) (Exp[(h c) / (lambda k t)] - 1)))]

p2000 = Plot[rholambda[lambda * 10^-6, 2000] / 10^3, {lambda, .1, 6}, PlotRange -> {0, 6},
   AxesLabel -> {"λ/10^-6 m", "ρλ/10^3 J m^-4"}, DisplayFunction -> Identity];

p1500 = Plot[rholambda[lambda * 10^-6, 1500] / 10^3, {lambda, .1, 6}, PlotRange -> {0, 6},
   AxesLabel -> {"λ/10^-6 m", "ρλ/10^3 J m^-4"} , DisplayFunction -> Identity ];

p1700 = Plot[rholambda[lambda * 10^-6, 1700] / 10^3, {lambda, .1, 6}, PlotRange -> {0, 6},
   AxesLabel -> {"λ/10^-6 m", "ρλ/10^3 J m^-4"}, DisplayFunction -> Identity ];
```

```
Show[p1500, p1700, p2000, DisplayFunction -> $DisplayFunction ];
```

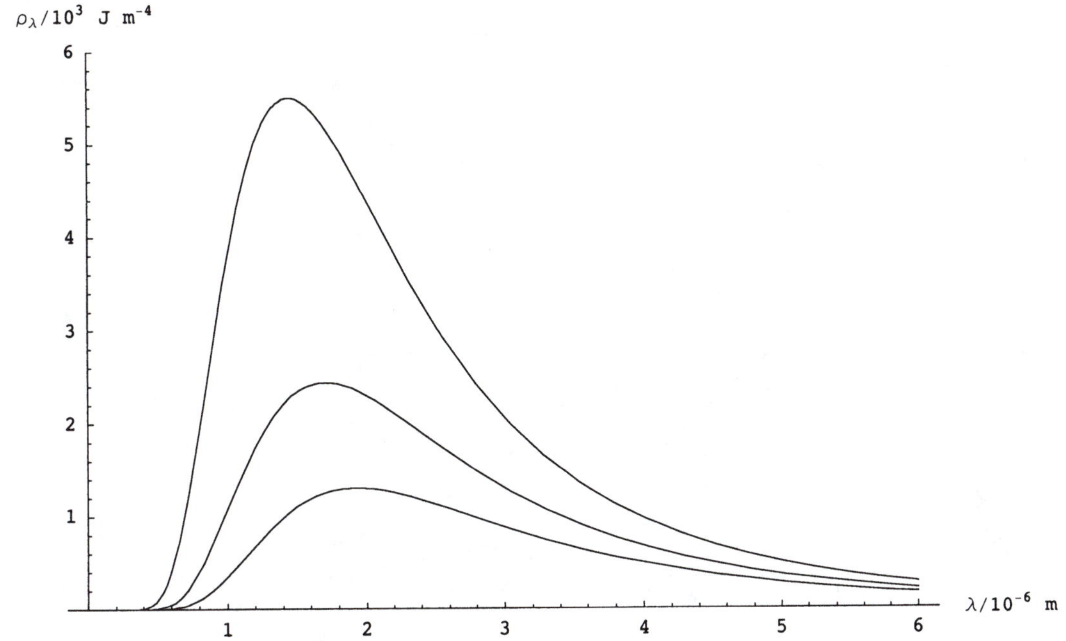

The highest curve is of course for the highest temperature.

(b)

```
rhonu[nu_, t_] := Module[{h, c, k}, h = 6.6260755 * 10^-34 (*J s*);
    c = 2.99792458 * 10^8 (*m s^-1*);
       k = 1.380658 * 10^-23 (*J K^-1*); ((8 Pi h) (nu / c)^3 / (Exp[(h nu) / (k t)] - 1))]

q2000 = Plot[rhonu[nu * 10^14, 2000] / 10^-17, {nu, .1, 4}, PlotRange -> {0, 8},
   AxesLabel -> {"ν/10^14 Hz", "ρν/10^-17 J m^-3 s"} , DisplayFunction -> Identity ];

q1500 = Plot[rhonu[nu * 10^14, 1500] / 10^-17, {nu, .1, 4}, PlotRange -> {0, 8},
   AxesLabel -> {"ν/10^14 Hz", "ρν/10^-17 J m^-3 s"} , DisplayFunction -> Identity ];

q1700 = Plot[rhonu[nu * 10^14, 1700] / 10^-17, {nu, .1, 4}, PlotRange -> {0, 8}, AxesLabel ->
    {"nu" / "10^14 Hz", "ν/10^14 Hz", "ρν/10^-17 J m^-3 s"}, DisplayFunction -> Identity ];
```

```
Show[q1500, q1700, q2000, DisplayFunction -> $DisplayFunction ];
```

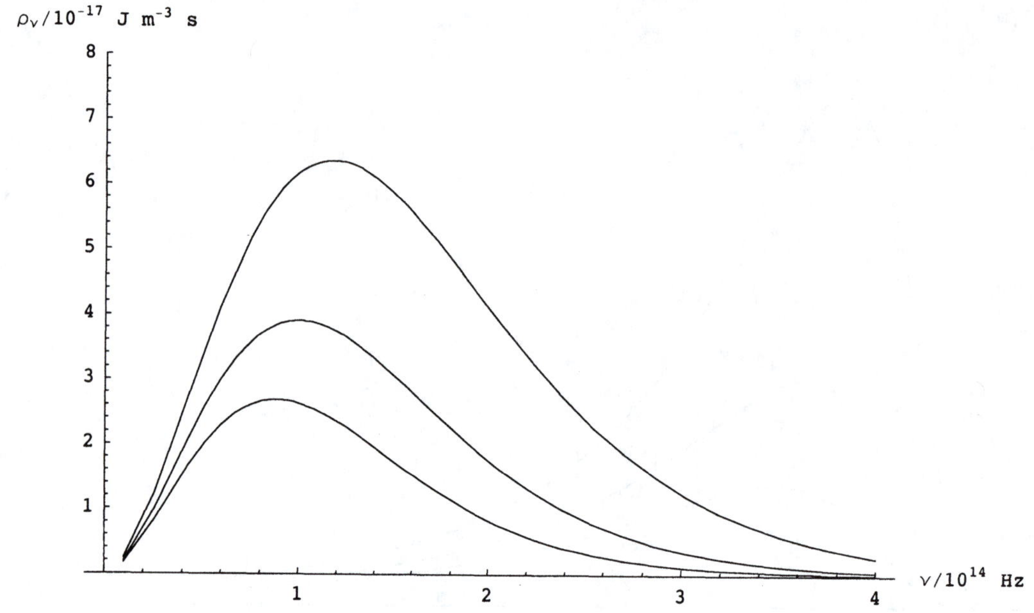

9.G (a) The following differential equation is involved in the discussion of the particle in a box and the harmonic oscillator.

$m\frac{d^2 x}{dt^2} = -kx$

Use *Mathematica* to solve the differential equation and make some plots of solutions. (b) Sunce the potential energy of a classical harmonic oscillator is given by $V = kx^2/2$, plot the potential energy versus x for k = 1 and k = 2.

SOLUTION

(a)

```
ClearAll["Global`*"]

solution = DSolve[x''[t] == -k x[t], x[t], t]

{{x[t] → C[1] Cos[√k t] + C[2] Sin[√k t]}}

x[t] /. solution

{C[1] Cos[√k t] + C[2] Sin[√k t]}

fn1 = x[t] /. solution /. k → 1 /. C[1] → 1 /. C[2] → 1

{Cos[t] + Sin[t]}

Plot[fn1, {t, 0, 10}];
```

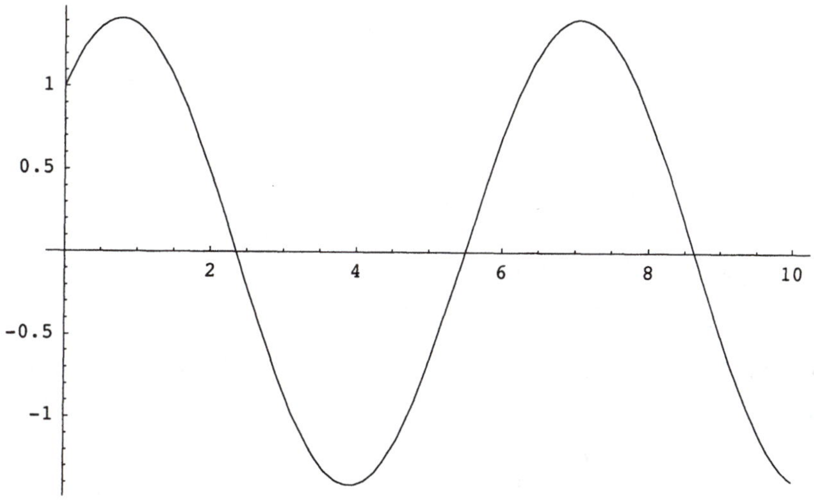


```
fn2 = x[t] /. solution /. k → 1 /. C[1] → 1 /. C[2] → 0

{Cos[t]}
```

```
Plot[fn2, {t, 0, 10}];
```

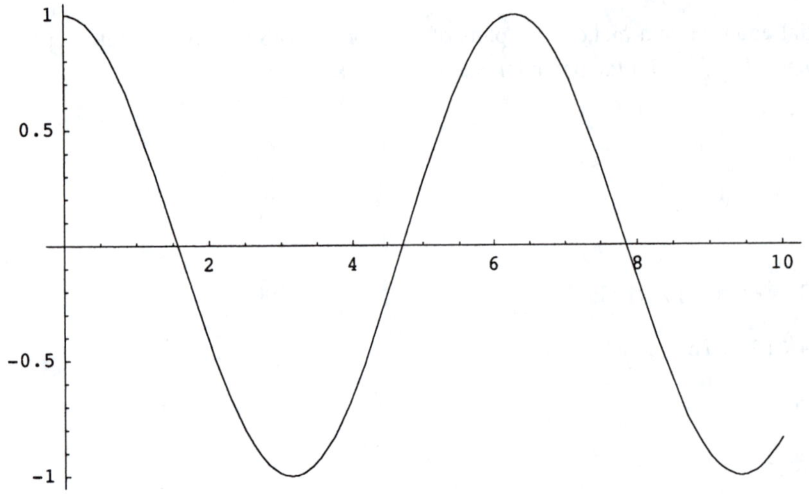

(b)

```
plot1 = Plot[x^2 / 2, {x, -1, 1}, PlotRange → {0, 1}];
```

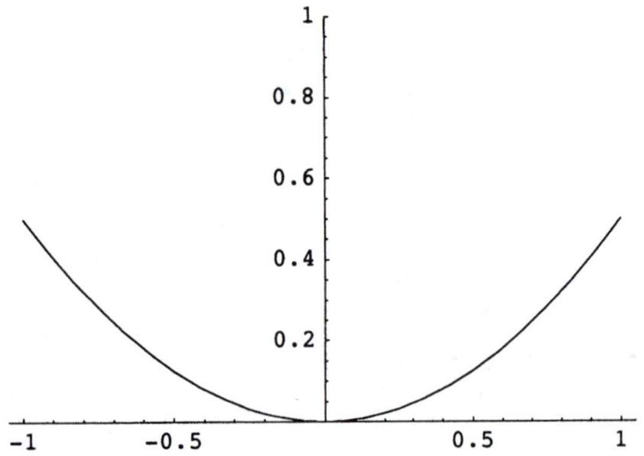


```
plot2 = Plot[x^2, {x, -1, 1}];
```

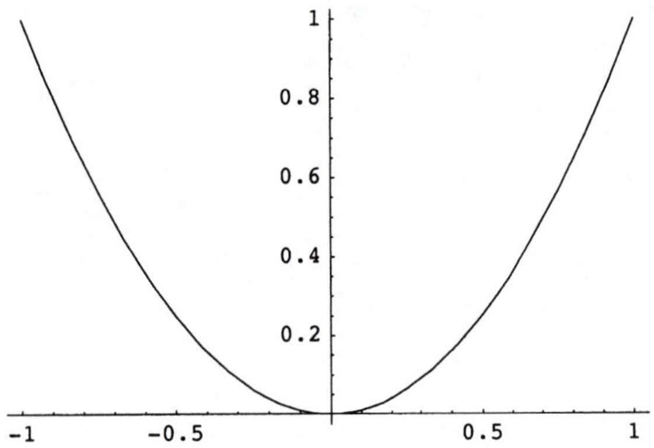

```
Show[plot1, plot2];
```

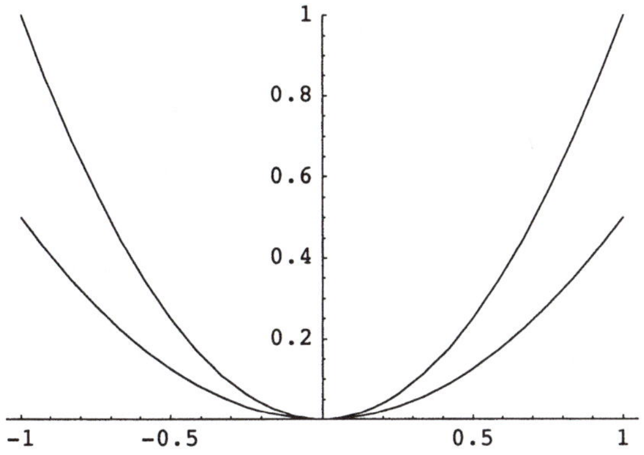

Note that with a bigger force constant the potential energy rises more rapidly with displacement from origin.

9.H Compare the wave functions and probability densities for the harmonic oscillator with the classical potential energy function for the harmonic oscillator. (a) Calculate the energy levels for a harmonic oscillator in units of $\hbar$ for a fundamental vibrational frequency of $1/2\pi\ s^{-1}$. (b) Assuming that $\alpha = 1$, write the expression for the vibrational wave function psi in *Mathematica*. (c) Plot the wave functions for the first five levels in comparison with the classical expression for the potential energy. (d) Plot the probability densities for the first five levels in comparison with the classical expression for the potential energy.

SOLUTION

(a) The fundamental vibrational frequency is $1/2\pi\ s^{-1}$. $E_v = (v + 1/2)\hbar$

```
ClearAll["Global`*"]

e = (v + 1 / 2) /. v → {0, 1, 2, 3, 4}
```

$$\left\{\frac{1}{2}, \frac{3}{2}, \frac{5}{2}, \frac{7}{2}, \frac{9}{2}\right\}$$

(b) The wave function psi is given by

```
psi = ((2^v * v!) ^ - (1 / 2)) * Pi^(-1 / 4) * HermiteH[v, x] * Exp[-x^2 / 2]
```

$$\frac{e^{-\frac{x^2}{2}}\ \text{HermiteH}[v, x]}{\pi^{1/4}\ \sqrt{2^v\ v!}}$$

(c)

```
plot0 = Plot[1 + psi /. v → 0, {x, -5, 5}, PlotRange → {0, 15}, DisplayFunction -> Identity];

plot1 = Plot[3 + psi /. v → 1, {x, -5, 5}, PlotRange → {0, 10}, DisplayFunction -> Identity];

plot2 = Plot[5 + psi /. v → 2, {x, -5, 5}, PlotRange → {0, 15}, DisplayFunction -> Identity];

plot3 = Plot[7 + psi /. v → 3, {x, -5, 5}, PlotRange → {0, 15}, DisplayFunction -> Identity];

plot4 = Plot[9 + psi /. v → 4, {x, -5, 5}, PlotRange → {0, 15}, DisplayFunction -> Identity];
```

We want the classical expression for the potential energy to give values proportional to 1, 3, 5, 7, and 9, and so we take k = 1.

```
plotc = Plot[x^2, {x, -5, 5}, PlotRange → {0, 15}];
```

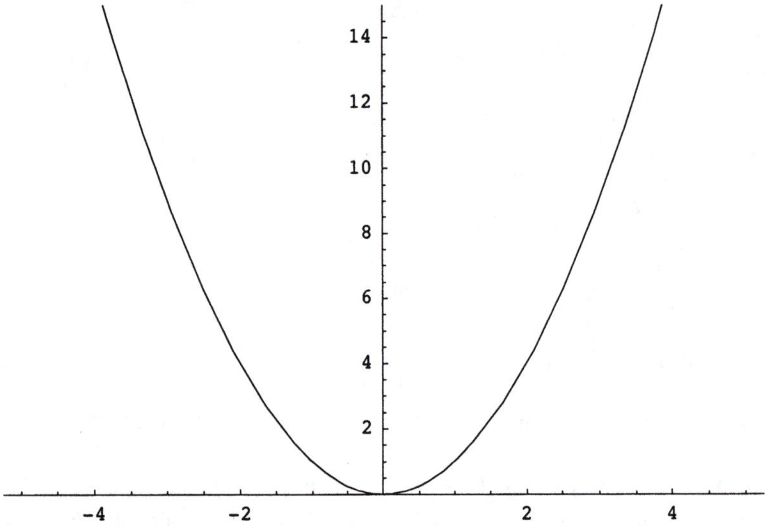

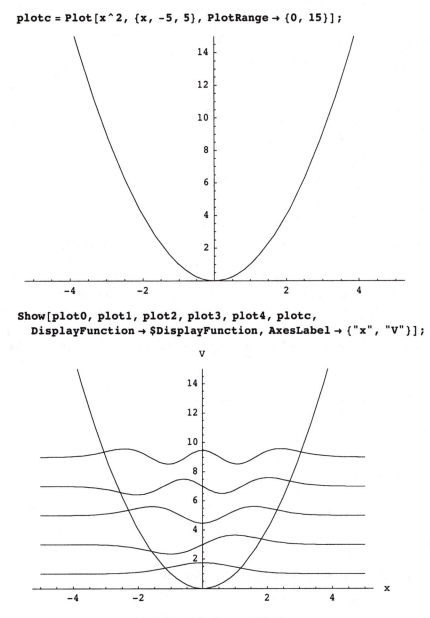

```
Show[plot0, plot1, plot2, plot3, plot4, plotc,
  DisplayFunction → $DisplayFunction, AxesLabel → {"x", "V"}];
```

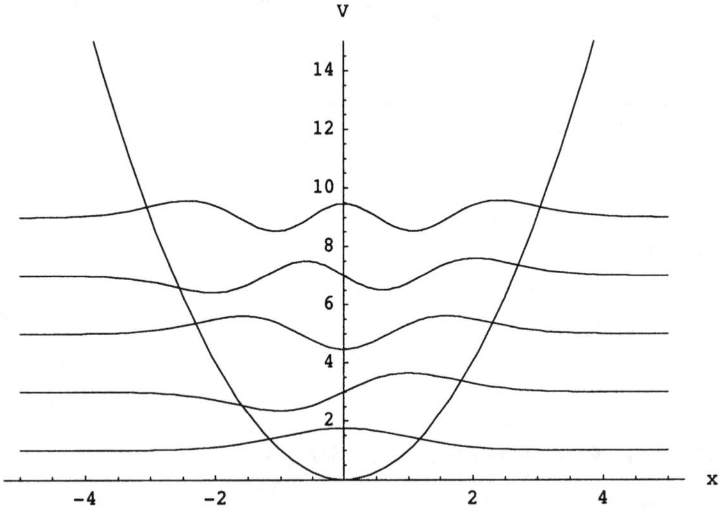

(d) Plot probability densities multiplied by 4 for better visibility.

```
plot0pd =
  Plot[1 + 4 * psi^2 /. v → 0, {x, -5, 5}, PlotRange → {0, 15}, DisplayFunction -> Identity];

plot1pd =
  Plot[3 + 4 * psi^2 /. v → 1, {x, -5, 5}, PlotRange → {0, 10}, DisplayFunction -> Identity];

plot2pd =
  Plot[5 + 4 * psi^2 /. v → 2, {x, -5, 5}, PlotRange → {0, 15}, DisplayFunction -> Identity];

plot3pd =
  Plot[7 + 4 * psi^2 /. v → 3, {x, -5, 5}, PlotRange → {0, 15}, DisplayFunction -> Identity];

plot4pd =
  Plot[9 + 4 * psi^2 /. v → 4, {x, -5, 5}, PlotRange → {0, 15}, DisplayFunction -> Identity];
```

```
Show[plot0pd, plot1pd, plot2pd, plot3pd, plot4pd,
  plotc, DisplayFunction → $DisplayFunction, AxesLabel → {"x", "V"}];
```

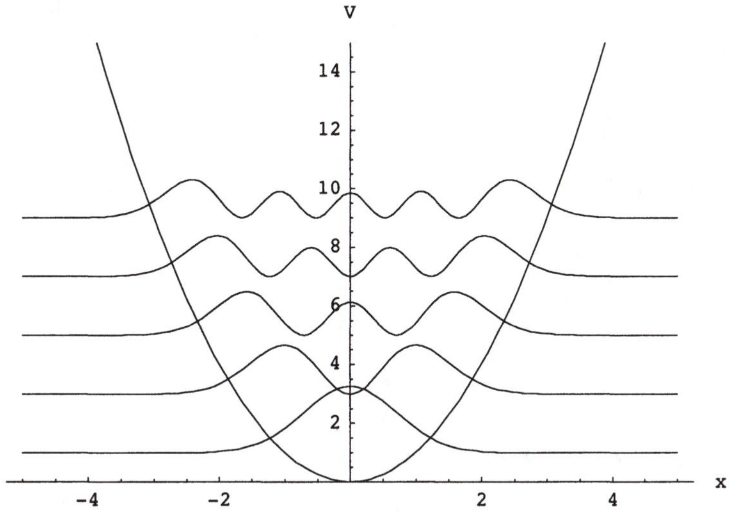

I am not sure the parabola is in the right place. I would like to get rid of the numbers on the two axes.

9.I The amplitude for a travelling wave is given by
A(x.t) = A_0 cos(kx - ωt)
where k = $2\pi/\lambda$ and $\omega = 2\pi/\tau$ so that
A(x,t) = A_0 cos(2πx/λ - 2πt/τ) = A_0 cos(2πx/λ - ωt)
(a) Plot A(x,t)/A_0 versus x for constant t for λ = 2 m and 4 m. (b) Plot A(x,t)/A_0 versus t for constant x for τ = 0.1 s and 0.2 s. (c) In general for wave motion the phase velocity is $\nu\lambda$. For light the phase velocity is c so that at constant t
A(x) = A_0 cos(2πxν/c)
Plot A(x)/A_0 versus x for $\nu = 380\text{x}10^{12}\ s^{-1}$ that is about the highest frequency the human eye can detect..

SOLUTION
(a)

```
Plot[Cos[2 * Pi * x / 2], {x, 0, 10}, AxesLabel → {"x", "A"}];
```

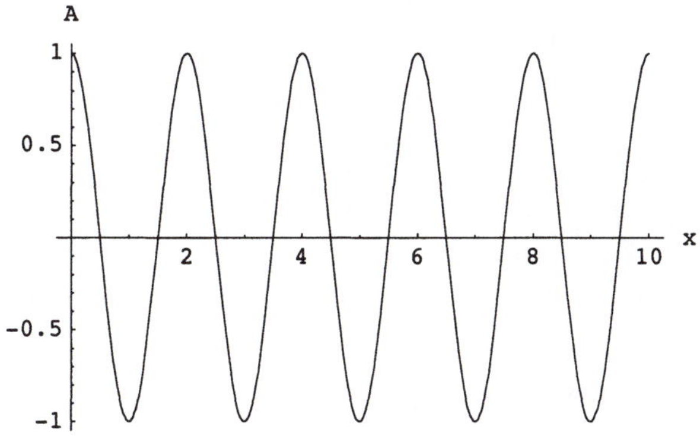

```
Plot[Cos[2*Pi*x/4], {x, 0, 10}, AxesLabel → {"x", "A"}];
```

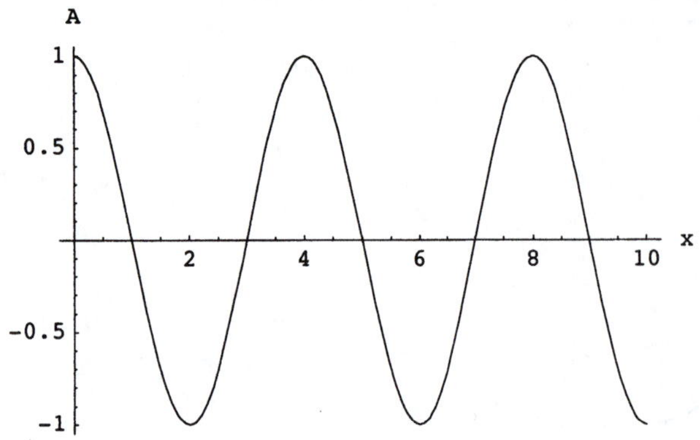

(b)

```
Plot[Cos[-2*Pi*t/.1], {t, 0, 1}, AxesLabel → {"t", "A"}];
```

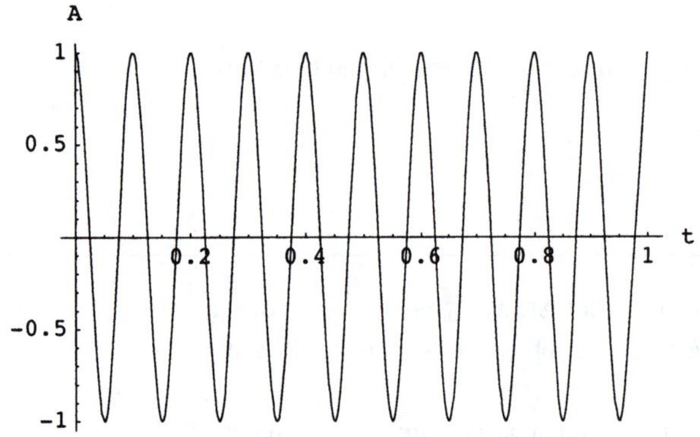


```
Plot[Cos[-2*Pi*t/.2], {t, 0, 1}, AxesLabel → {"t", "A"}];
```

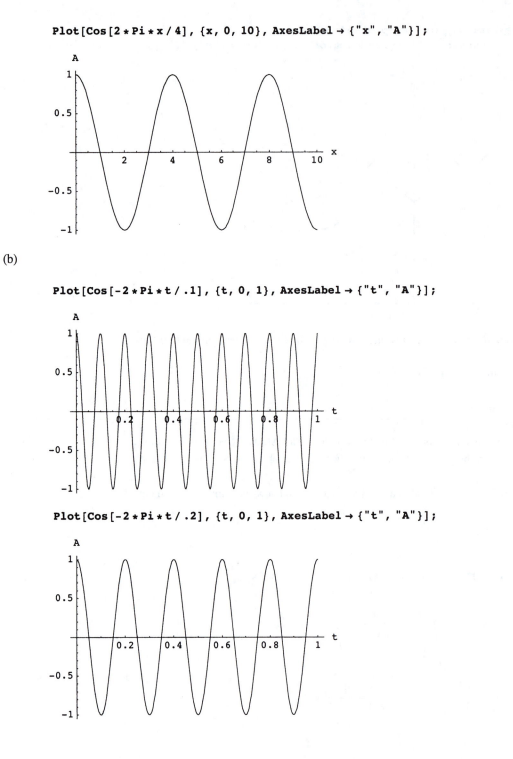

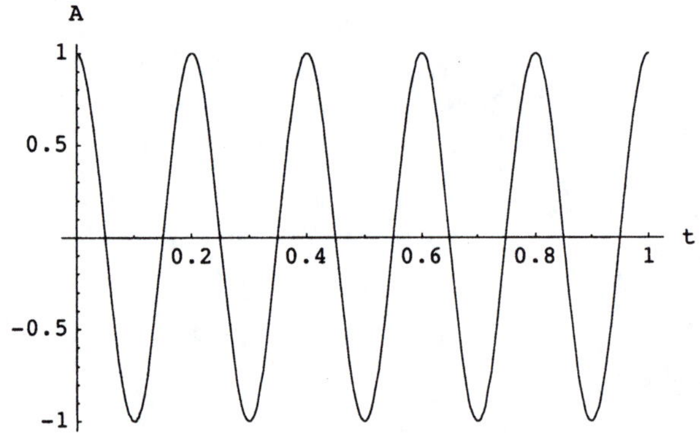

```
Plot[Cos[x], {x, 0, 10}];
```

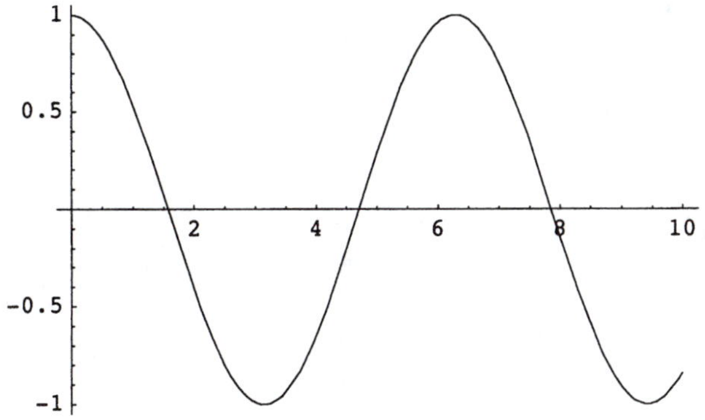


```
Plot[Cos[2 x], {x, 0, 10}];
```

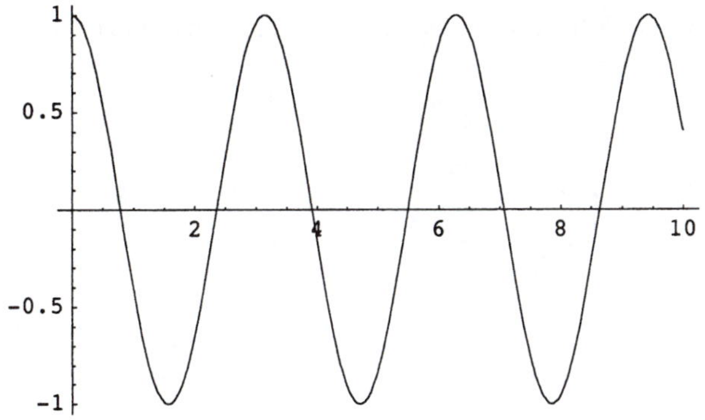

(c)

```
Plot[Cos[2 * Pi * x * (380 * 10^12) / (3 * 10^8)], {x, 0, 10^-6}];
```

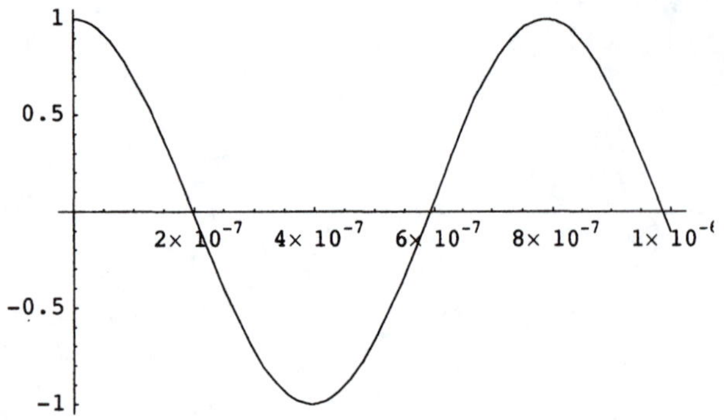

9.J The amplitude of a travelling wave is given by
$A = A_0 \cos(kx - \omega t)$
Show that this equation for the amplitude satisfies the following partial differential equation and derive the expression for the phase velocity v_p.
$$\frac{\partial^2 A}{\partial x^2} = \frac{1}{v_p{}^2}\frac{\partial^2 A}{\partial t^2}$$
How is the phase velocity related to ν and λ?

SOLUTION

```
D[Cos[k * x - ω * t], {x, 2}]

-k² Cos[k x - t ω]
```

```
D[Cos[k*x - ω*t], {t, 2}]
```

$-\omega^2 \text{Cos}[k\,x - t\,\omega]$

Thus the phase velocity is given by
$v_p = \omega/k$
Since $k = 2\pi/\lambda$ and $\omega = 2\pi\nu$, $v_p = \nu\lambda$.

9.K As discussed in the text, when a quantum mechanical harmonic oscillator is excited with a very short pulse of radiation, which necessarily contains a distribution of wave lengths, a number of vibrational levels are excited coherently (that is in phase). (a) Type in the expression for the stationary wavefunction for a single level as a function of the quantum number v and the bond distance x utilizing HermiteH. Plot this function for a couple of levels. (b) Add up the wavefunctions for v = 14, 15, 16, 17, 18, 19, 20, 21 and 22 with relative contributions of these wavefunctions to the superposition that correspond with a Gaussian distribution centered at v = 18. (c) Plot this wavefunction and the probability density function versus x to show that this wave packet is something like a particle. (d) Introduce weighting factors of exp(-ivt) into each of these terms so that the movement of the wave packet can be calculated. As a simplification, the energies of the levels are taken to be proportional to the vibrational quantum numbers v. To construct the probability density, this wave function has to be multiplied with the corresponding wave function with factors of exp(ivt). Plot the probability density function versus x at t = 0. (e) Increase t in steps to show that the system can be thought of as a particle oscillating back and forth. This behavoir makes it possible to study the kinetics of ultrafact reactions. More information is provided by J. S. Baskin and A. H. Zewail, J. Chem. Ed. 78, 737 (2001).

SOLUTION
(a)

```
ClearAll["Global`*"]

psi = ((2^v*v!)^-(1/2))*Pi^(-1/4)*HermiteH[v, x]*Exp[-x^2/2]
```

$$\frac{e^{-\frac{x^2}{2}}\,\text{HermiteH}[v, x]}{\pi^{1/4}\sqrt{2^v\, v!}}$$

```
Plot[psi /. v→14, {x, -10, 10}, PlotRange→{-.8, .8}];
```

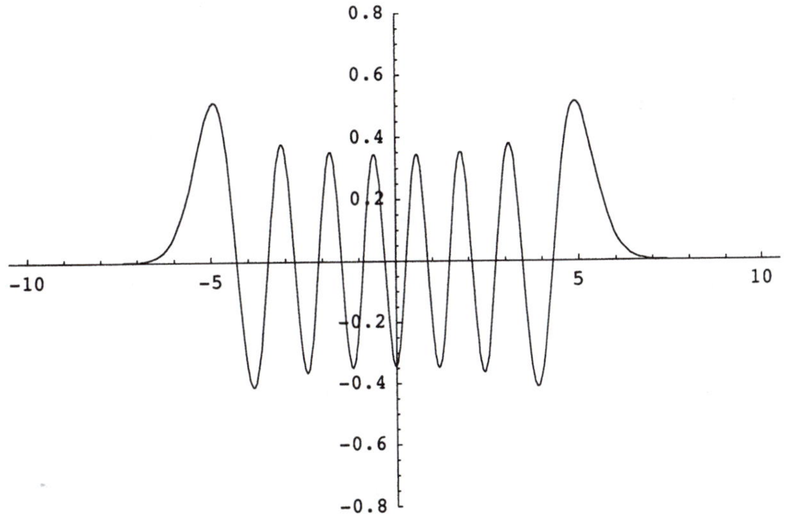

```
Plot[psi /. v → 22, {x, -10, 10}, PlotRange → {-.8, .8}];
```

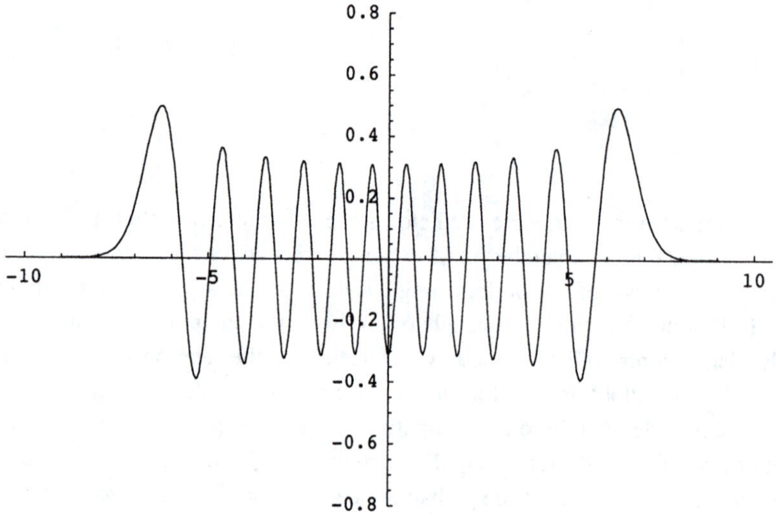

(b)

```
psi14 = psi /. v → 14;

psi15 = psi /. v → 15;

psi16 = psi /. v → 16;

psi17 = psi /. v → 17;

psi18 = psi /. v → 18;

psi19 = psi /. v → 19;

psi20 = psi /. v → 20;

psi21 = psi /. v → 21;

psi22 = psi /. v → 22;

pdf = (1 / Sqrt[2 * Pi]) * Exp[-x^2 / 2]
```

$$\frac{e^{-\frac{x^2}{2}}}{\sqrt{2\pi}}$$

```
pdf /. x → {-2, -1.5, -1, -.5, 0, .5, 1, 1.5, 2} // N

{0.053991, 0.129518, 0.241971, 0.352065, 0.398942, 0.352065, 0.241971, 0.129518, 0.053991}

wf = .054 * psi14 + .130 * psi15 + .242 * psi16 + .352 * psi17 +
   .399 * psi18 + .352 * psi19 + .242 * psi20 + .130 * psi21 + .054 * psi22;
```

(c)

```
plot1 = Plot[wf, {x, -10, 10}, PlotRange → {-2, 2}, AxesLabel → {"x", "ψ"}];
```

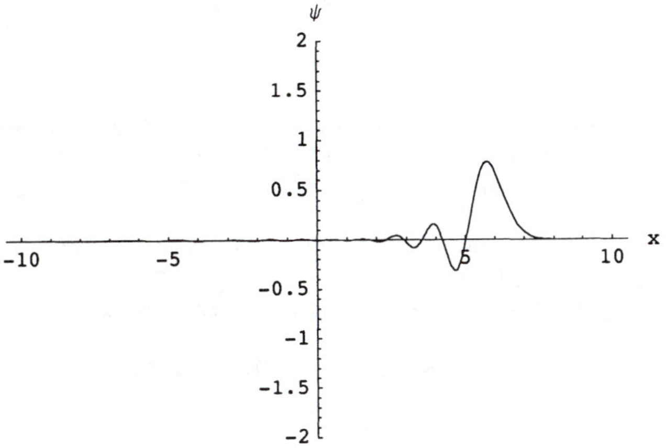


```
plot2 = Plot[wf^2, {x, -10, 10}, PlotRange → {0, .75}, AxesLabel → {"x", "ψ²"}];
```

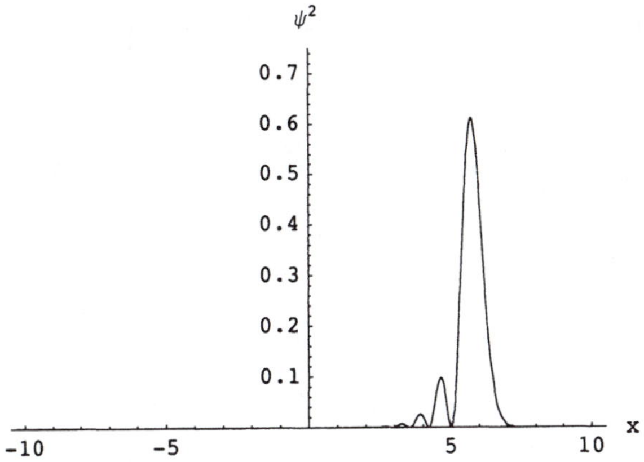

```
Show[GraphicsArray[{plot1,plot2}]];
```

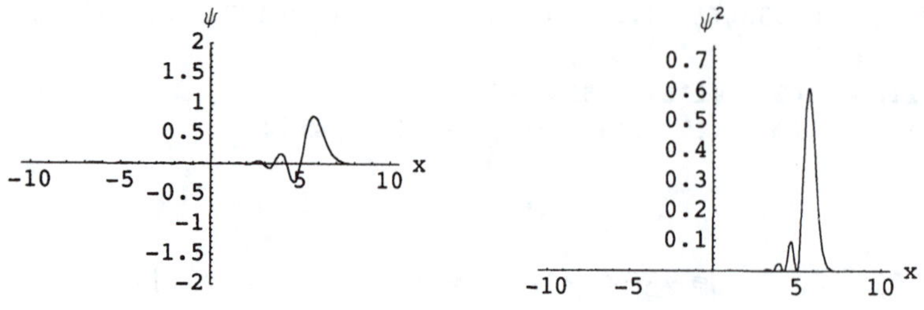

(d)

```
wfminus =
  .054 * psi14 * Exp[-I * 14 * t] + .130 * psi15 * Exp[-I * 15 * t] + .242 * psi16 * Exp[-I * 16 * t] +
   .352 * psi17 * Exp[-I * 17 * t] + .399 * psi18 * Exp[-I * 18 * t] + .352 * psi19 * Exp[-I * 19 * t] +
   .242 * psi20 * Exp[-I * 20 * t] + .130 * psi21 * Exp[-I * 21 * t] + .054 * psi22 * Exp[-I * 22 * t];

wfplus = .054 * psi14 * Exp[I * 14 * t] + .130 * psi15 * Exp[I * 15 * t] + .242 * psi16 * Exp[I * 16 * t] +
   .352 * psi17 * Exp[I * 17 * t] + .399 * psi18 * Exp[I * 18 * t] + .352 * psi19 * Exp[I * 19 * t] +
   .242 * psi20 * Exp[I * 20 * t] + .130 * psi21 * Exp[I * 21 * t] + .054 * psi22 * Exp[I * 22 * t];

t0 = Plot[wfminus * wfplus /. t → 0, {x, -10, 10}, PlotRange → {0, .75}, PlotLabel -> "t = 0"];
```

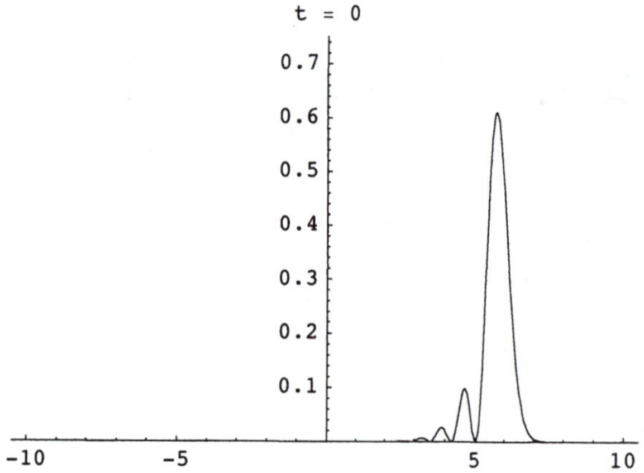

```
t0 = Plot[wfminus * wfplus /. t → 0, {x, -10, 10},
   PlotRange → {0, .75}, Axes → False, PlotLabel -> "t = 0"];
```

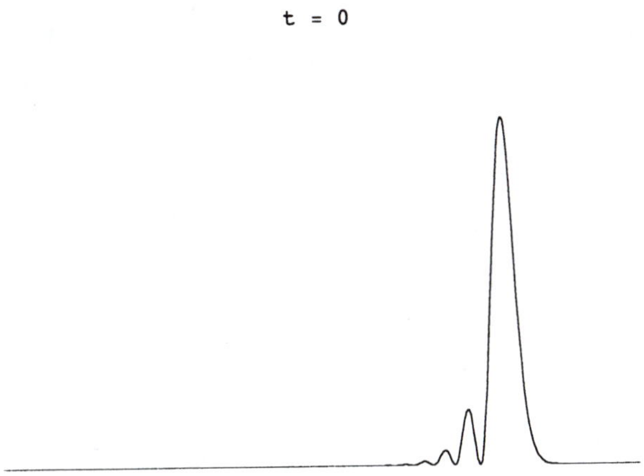

(e)

```
t1 = Plot[wfminus * wfplus /. t → 1, {x, -10, 10}, PlotRange → {0, .75},
   DisplayFunction → Identity, Axes → False, PlotLabel -> "t = 1"];

t2 = Plot[wfminus * wfplus /. t → 2, {x, -10, 10}, PlotRange → {0, .75},
   DisplayFunction → Identity, Axes → False, PlotLabel -> "t = 2"];

t3 = Plot[wfminus * wfplus /. t → 3, {x, -10, 10}, PlotRange → {0, .75},
   DisplayFunction → Identity, Axes → False, PlotLabel -> "t = 3"];

t4 = Plot[wfminus * wfplus /. t → 4, {x, -10, 10}, PlotRange → {0, .75},
   DisplayFunction → Identity, Axes → False, PlotLabel -> "t = 4"];

t5 = Plot[wfminus * wfplus /. t → 5, {x, -10, 10}, PlotRange → {0, .75},
   DisplayFunction → Identity, Axes → False, PlotLabel -> "t = 5"];

t6 = Plot[wfminus * wfplus /. t → 6, {x, -10, 10}, PlotRange → {0, .75},
   DisplayFunction → Identity, Axes → False, PlotLabel -> "t = 6"];

t7 = Plot[wfminus * wfplus /. t → 7, {x, -10, 10}, PlotRange → {0, .75},
   DisplayFunction → Identity, Axes → False, PlotLabel -> "t = 7"];

t8 = Plot[wfminus * wfplus /. t → 8, {x, -10, 10}, PlotRange → {0, .75},
   DisplayFunction → Identity, Axes → False, PlotLabel -> "t = 8"];
```

```
Show[GraphicsArray[{{t0, t1, t2}, {t3, t4, t5}, {t6, t7, t8}}, GraphicsSpacing → .5]];
```

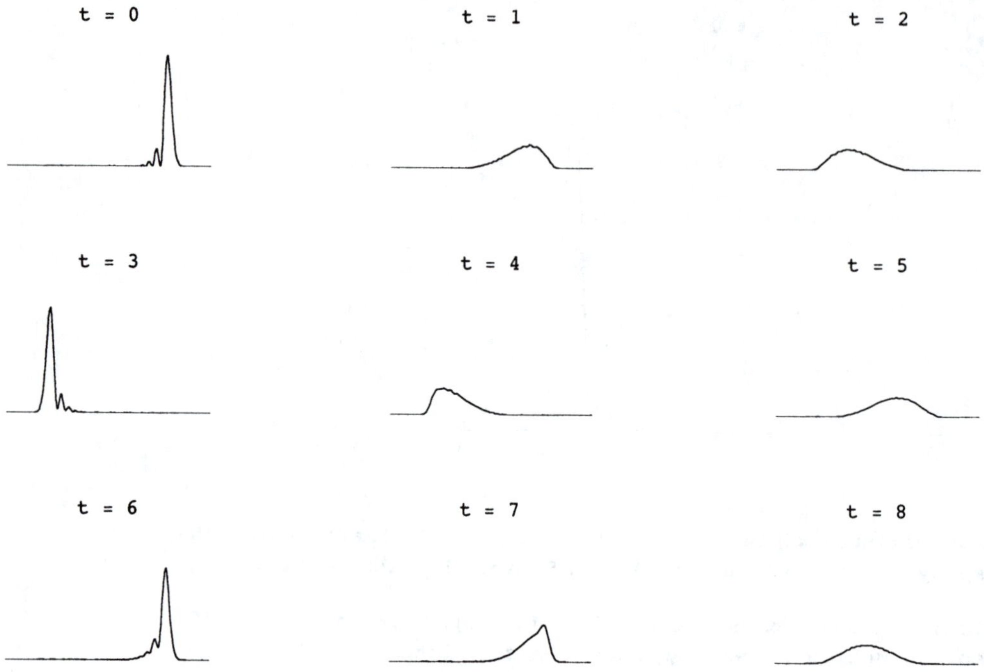

Chapter 10 Atomic Structure

10.A Make three-dimensional plots to show the electron density in the *x*-*y* plane for atomic hydrogen in 1s, 2s, $2p_x$, and $2p_y$ orbitals. Restrict the plots to $x = -0.5$ to 0.5 nm and $y = -0.5$ to 0.5 nm.

SOLUTION

```
ClearAll["Global`*"]

Off[General::spell1];
Off[General::spell];
```

The wavefunction for a 1s orbital for a hydrogen atom is given by

```
1/(Pi^.5)*(1/a0)^1.5*Exp[-r/a0]
```

$$0.56419 \left(\frac{1}{a0}\right)^{1.5} e^{-\frac{r}{a0}}$$

In order to make plots versus x and y, the following assignment is used for the radius:

```
r[x_, y_] := (x^2 + y^2)^.5
```

Thus the 1s wave function is given by

```
f1s[ x_, y_] := 1/ (Pi^.5) (1/a0)^1.5*
                         Exp[-r[x, y]/ a0]
```

The electron density is given by the square of the wave function. The following information is needed to make all of the plots.

```
xMin = -.5;
xMax = .5;
yMin = -.5;
yMax = .5;
a0 = .052918 (*nanometers*);
```

```
fun1s = Plot3D[f1s[ x, y] ^2, {x, xMin, xMax}, {y, yMin, yMax}, PlotRange -> {0, 50},
   PlotPoints -> {25, 25}, AxesLabel -> {"x/nm", "y/nm", "ψ(1s)²"}];
```

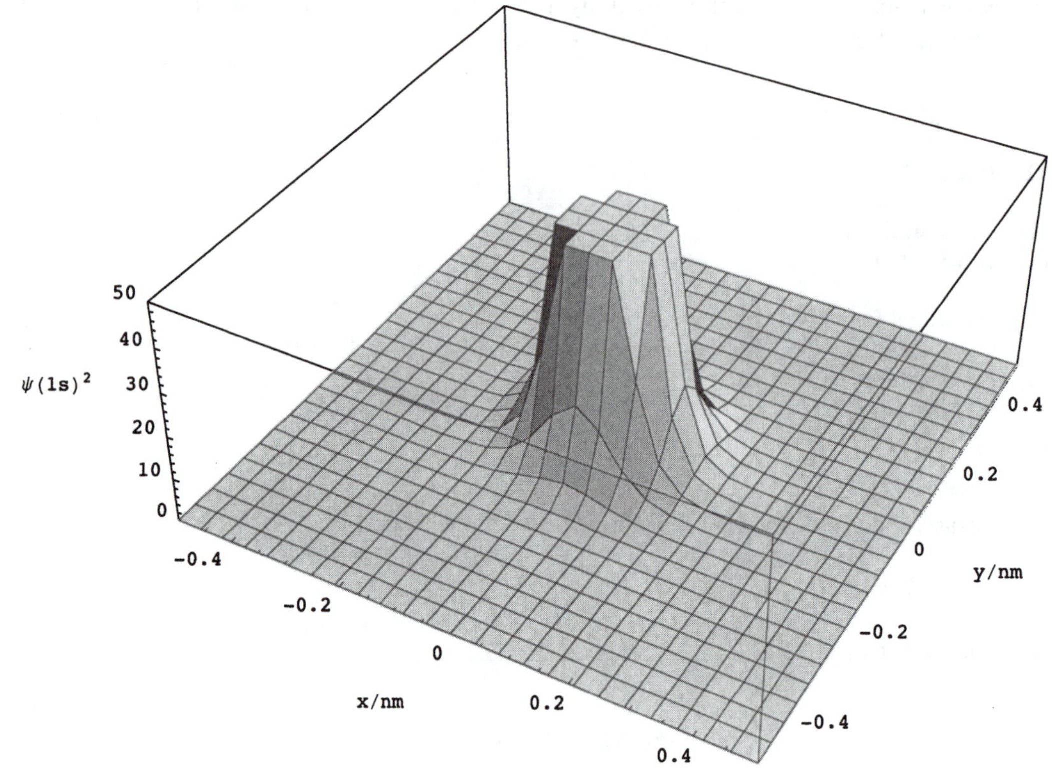

The wavefunction for the 2s orbital for a hydrogen atom is given by

```
f2s[ x_, y_] := 1/ (4 (2 Pi)^.5) (1/a0)^1.5* (2 - r[x, y] /a0) Exp[-r[x, y] / (2 a0)] ;
```

The electron density is given by the square of the wave function.

```
fun2s = Plot3D[f2s[ x, y] ^2, {x, xMin, xMax}, {y, yMin, yMax}, PlotRange -> {0, 50},
   PlotPoints -> {25, 25}, AxesLabel -> {"x/nm", "y/nm", "ψ(2s)² "}];
```

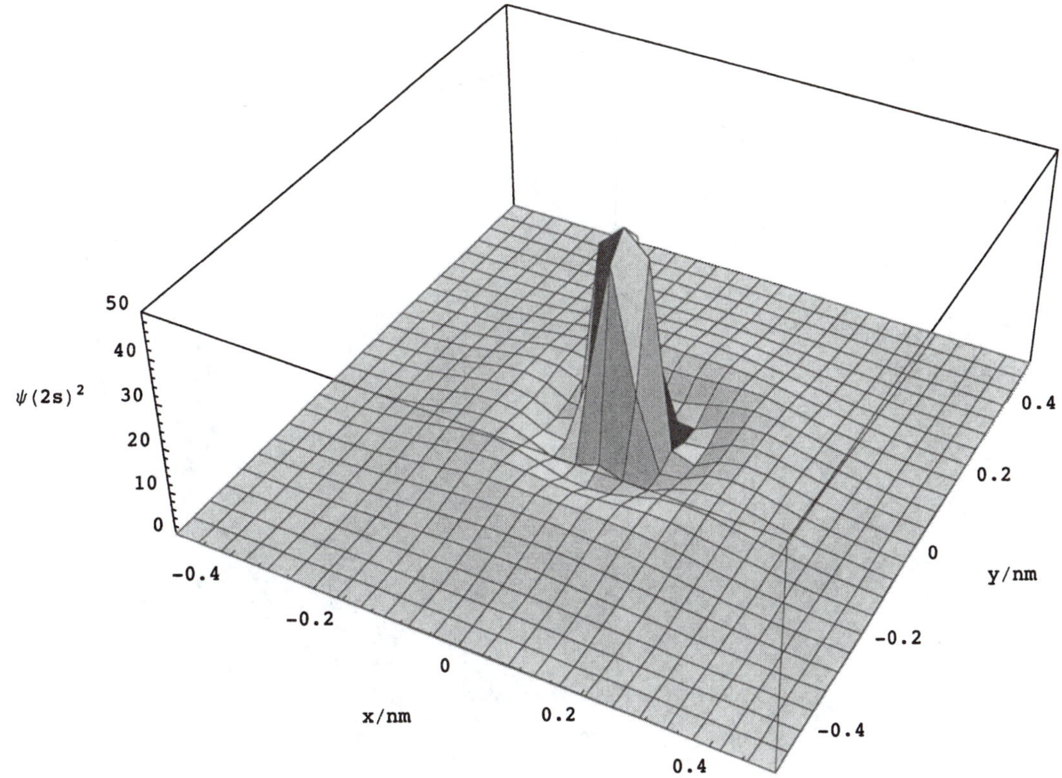

The wavefunction for the $2p_x$ orbital is given by

```
f2px[ x_, y_] := 1 / (4 (2 Pi) ^.5) (1 / a0) ^2.5 *
                              Exp[-r[x, y] / (2 a0)] x ;
```

Note that in spherical coordinates r sin(theta)cos(phi) = x. The electron density is given by the square of the wave function.

```
fun2px = Plot3D[f2px[ x, y]^2,
                        {x, xMin, xMax}, {y, yMin, yMax}, PlotRange -> {0, 50},
   PlotPoints -> {25, 25}, AxesLabel -> {"x/nm", "y/nm", "ψ(2px)²"}];
```

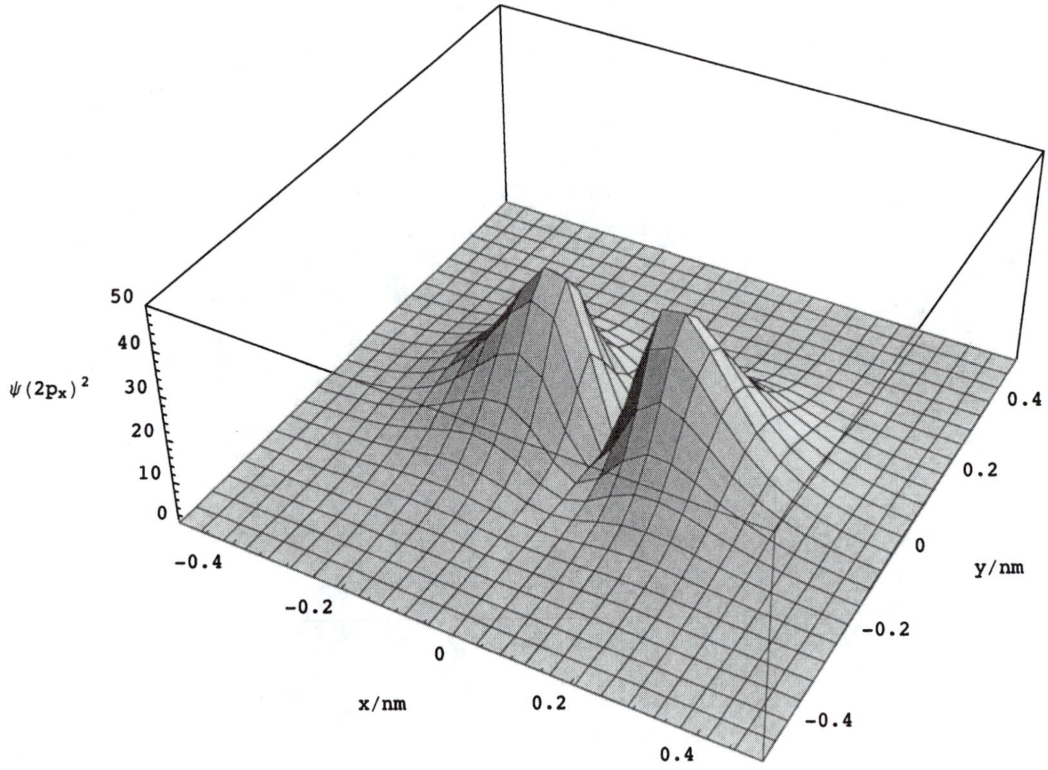

The wavefunction for the $2p_y$ orbital is given by

```
f2py[ x_, y_] := 1/(4 (2 Pi)^.5) (1/a0)^2.5*
                              Exp[-r[x, y]/(2 a0)] y;
```

Note that in spherical coordinates r sin(theta)sin(phi) = y. The electron density is given by the square of the wave function.

```
fun2py = Plot3D[f2py[ x, y]^2,
                        {x, xMin, xMax}, {y, yMin, yMax}, PlotRange -> {0, 50},
   PlotPoints -> {25, 25}, AxesLabel -> {"x/nm", "y/nm", "ψ(2py)²"}];
```

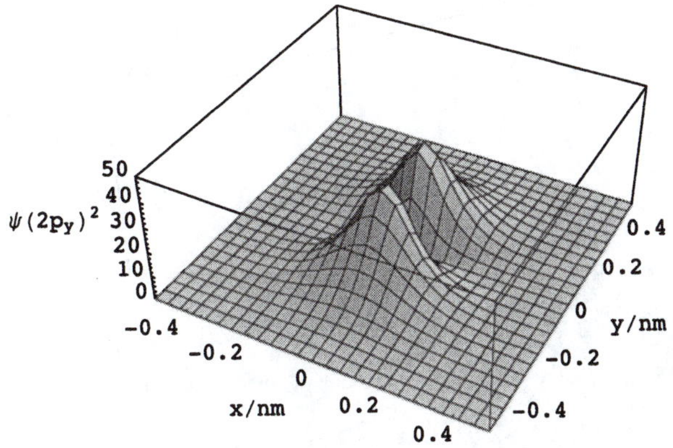

```
Show[fun2py];
```

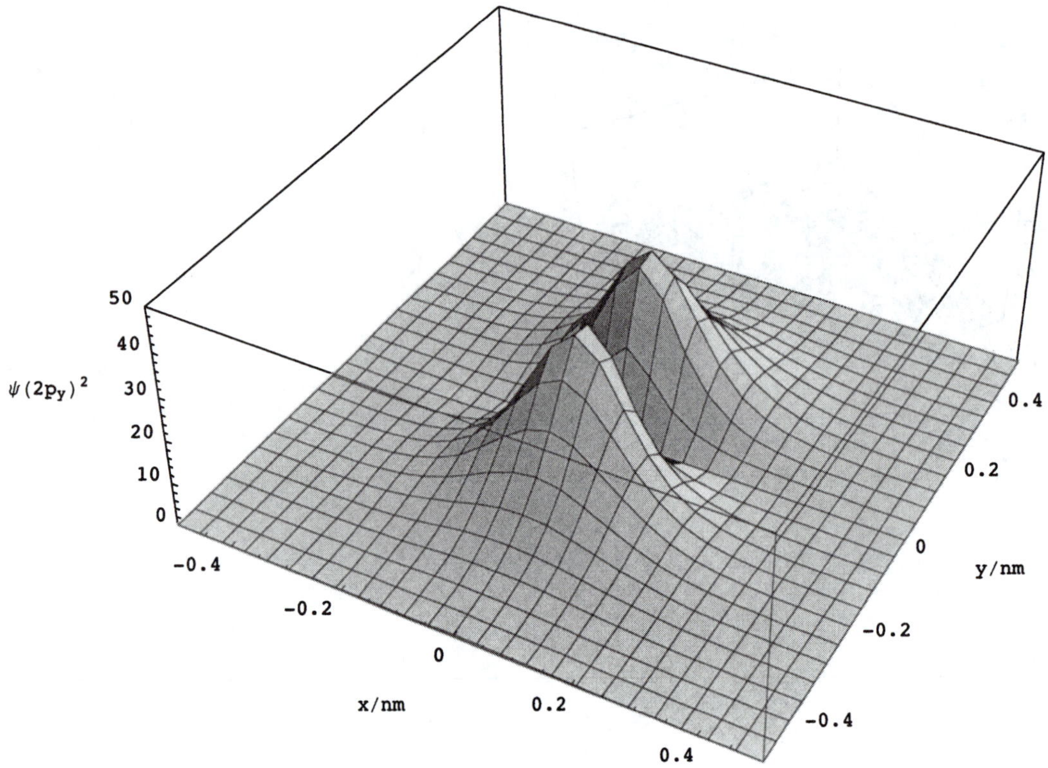

Further calculations can be madewith 3s, 3px,3py,3dx, and 3dx2-y2 orbitals.

10.B The Pauli spin operators for the three directions can be represented by matrices:
$S_x = \begin{pmatrix} 0 & .5 \\ .5 & 0 \end{pmatrix}$, $S_y = \begin{pmatrix} 0 & -.5\,i \\ .5\,i & 0 \end{pmatrix}$, and $S_z = \begin{pmatrix} .5 & 0 \\ 0 & -.5 \end{pmatrix}$.
(*a*) Calculate the commutator for S_x and S_y. (*b*) Calculate the sum of squares of the representations of the operators to obtain the spin matrix S^2. (*c*) Show that the square of the operator of spin angular momentum commutes with the operator for the x-component .

SOLUTION

(a) The three spin operators are given by

```
sx = {{0, 1}, {1, 0}} / 2.

{{0, 0.5}, {0.5, 0}}

sy = {{0, -I}, {I, 0}} / 2.

{{0, -0.5 i}, {0.5 i, 0}}
```

```
sz = {{1, 0}, {0, -1}} / 2.
```

{{0.5, 0}, {0, -0.5}}

```
Print[MatrixForm[sx], "            ", MatrixForm[sy],
    "             ", MatrixForm[sz]]
```

$\begin{pmatrix} 0 & 0.5 \\ 0.5 & 0 \end{pmatrix}$ $\begin{pmatrix} 0 & -0.5\,i \\ 0.5\,i & 0 \end{pmatrix}$ $\begin{pmatrix} 0.5 & 0 \\ 0 & -0.5 \end{pmatrix}$

Calculate the commutator for S_x and S_y.

```
sx.sy - sy.sx // Chop
```

{{0.5 i, 0}, {0, -0.5 i}}

Thus we see that the commutator is equal to I S_z. The other two commutators can be calculated in the same way. Chop replaces approximate real numbers that are close to zero by the exact integer 0.

(b) The sum of the squares of representations of the three operators yields the spin matrix S^2 for the square of the spin, just as the L^2 operator is given by the sum of the squares of the angular momentum operators in the three directions .

```
s2 = sx.sx + sy.sy + sz.sz // Chop
```

{{0.75, 0}, {0, 0.75}}

Note this is as expected from S(S+1)=3/4. These tests show that the Pauli matrices have the properties expected for spin functions.
(c) The square of the operator of spin angular momentum commutes with the operator for the x-component, as shown by.

```
s2.sx - sx.s2
```

{{0., 0.}, {0., 0.}}

10.C (*a*) Calculate the frequencies (in cm^{-1}) for the first 13 lines of the Balmer spectrum of hydrogen atoms. Plot the line spectrum and label the frequency axis. (*b*) Do the same for the Paschen spectrum. (*c*) Do the same for the Brackett spectrum. (*d*) Combine the spectra to show the experimental spectrum.

SOLUTION
(a)

```
ClearAll["Global`*"]

nubarBalmer = 1.09677 * 10^5 * (1 / 4 - 1 / n^2)
```

$109677.\left(\frac{1}{4}-\frac{1}{n^2}\right)$

```
nuB = Table[nubarBalmer, {n, 3, 15}]

{15232.9, 20564.4, 23032.2, 24372.7, 25180.9, 25705.5,
 26065.2, 26322.5, 26512.8, 26657.6, 26770.3, 26859.7, 26931.8}

levelsB = Table[Line[
    {{1.09677 * 10^5 * (1 / 4 - 1 / n^2), 0}, {1.09677 * 10^5 * (1 / 4 - 1 / n^2), 1}}], {n, 3, 14}];

spB = Show[Graphics[{levelsB}], Axes -> {True, False}, PlotRange -> {{0, 30000}, {0, 1.}},
   AxesLabel -> {"λ(cm⁻¹)", " "}, ImageSize -> {500, 100}, AspectRatio -> .2];
```

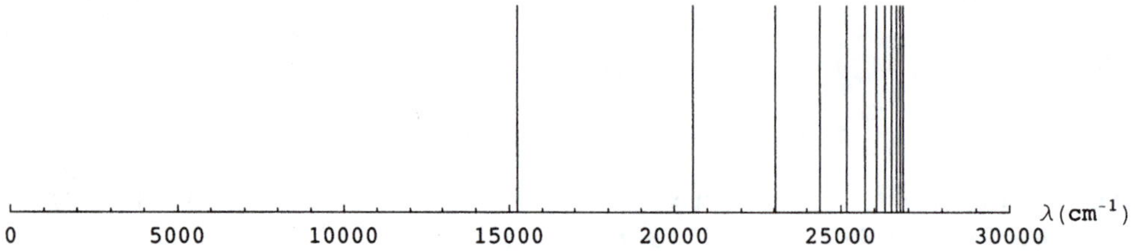

(b)

```
levelsPa = Table[Line[
    {{1.09677 * 10^5 * (1 / 9 - 1 / n^2), 0}, {1.09677 * 10^5 * (1 / 9 - 1 / n^2), 1}}], {n, 4, 15}];

spPa = Show[Graphics[{levelsPa}], Axes -> {True, False}, PlotRange -> {{0, 30000}, {0, 1.}},
   AxesLabel -> {"λ(cm⁻¹)", " "}, ImageSize -> {500, 100}, AspectRatio -> .2];
```

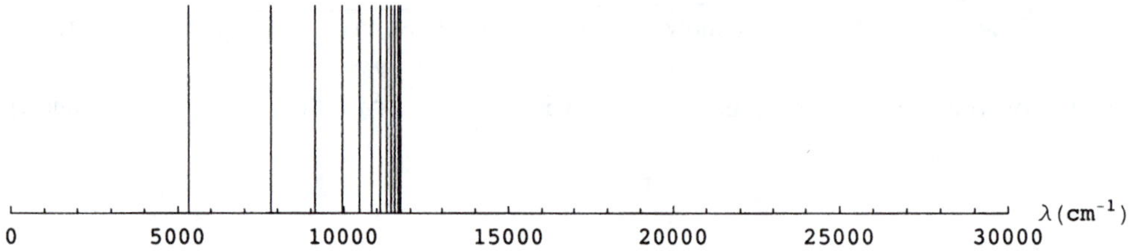

(c)

```
levelsBr = Table[Line[
    {{1.09677 * 10^5 * (1 / 16 - 1 / n^2), 0}, {1.09677 * 10^5 * (1 / 16 - 1 / n^2), 1}}], {n, 5, 16}]

spBr = Show[Graphics[{levelsBr}], Axes -> {True, False}, PlotRange -> {{0, 30000}, {0, 1.}},
   AxesLabel -> {"λ(cm⁻¹)", " "}, ImageSize -> {500, 100}, AspectRatio -> .2];
```

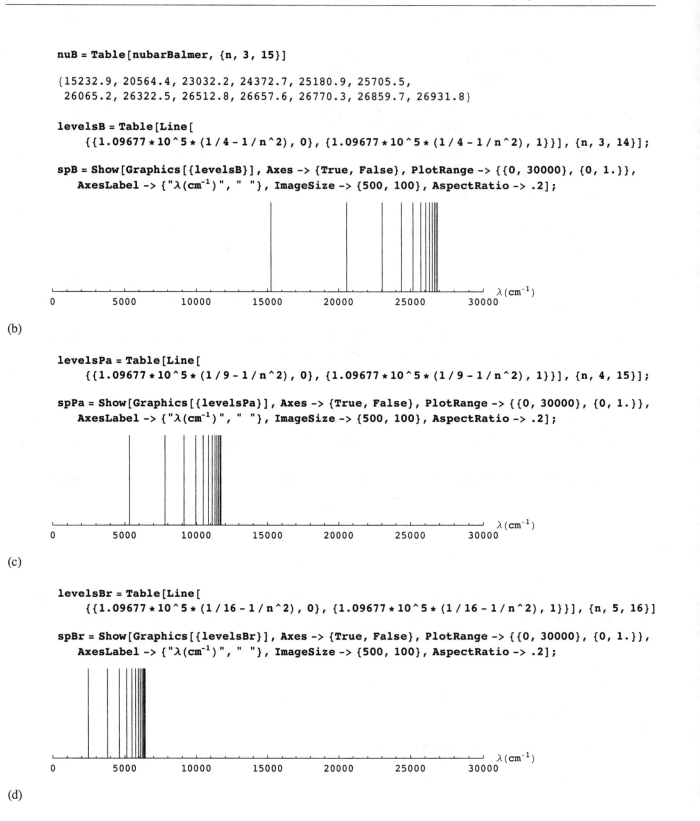

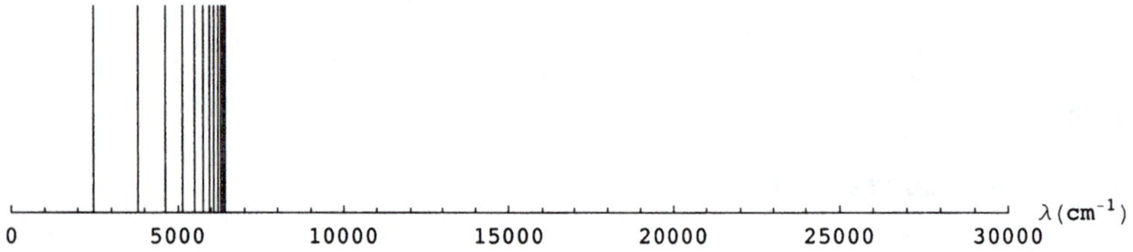

(d)

```
Show[spB, spPa, spBr];
```

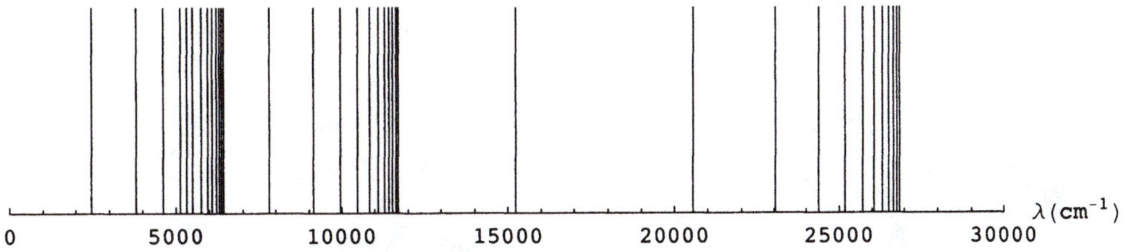

10.D Plot the angular parts of the wavefunctions for the atomic orbitals 2s, $2p_X$, $2p_Y$, and $2p_Z$ using computer software that provides spherical harmonics *Y(ℓ,m)*. In order to plot $2p_x$ and $2p_y$ the sums and differences of *Y*(1,1) and *Y*(1,-1) must be used as explained in the text.

SOLUTION

```
<< Graphics`ParametricPlot3D`

Y[el_, em_] := SphericalHarmonicY[el, em, theta, phi]

Y[0, 0]
```

$$\frac{1}{2\sqrt{\pi}}$$

```
Y[1, 1]
```

$$-\frac{1}{2} e^{i\,phi} \sqrt{\frac{3}{2\pi}}\ \text{Sin[theta]}$$

```
Y[1, -1]
```

$$\frac{1}{2} e^{-i\,phi} \sqrt{\frac{3}{2\pi}}\ \text{Sin[theta]}$$

```
Y[1, 0]
```

$$\frac{1}{2} \sqrt{\frac{3}{\pi}}\ \text{Cos[theta]}$$

```
psi2s = SphericalPlot3D[Abs[Y[0, 0]], {theta, 0, Pi}, {phi, 0, 2*Pi},
   AxesLabel -> {"x", "y", "z"}, DisplayFunction -> Identity, PlotLabel -> "psi2s"];

psi2py = SphericalPlot3D[Abs[Y[1, 1] + Y[1, -1]], {theta, 0, Pi}, {phi, 0, 2*Pi},
   AxesLabel -> {"x", "y", "z"}, DisplayFunction -> Identity, PlotLabel -> "psi2py"];

psi2px = SphericalPlot3D[Abs[Y[1, 1] - Y[1, -1]], {theta, 0, Pi}, {phi, 0, 2*Pi},
   AxesLabel -> {"x", "y", "z"}, DisplayFunction -> Identity, PlotLabel -> "psi2px"];

psi2pz = SphericalPlot3D[Abs[Y[1, 0]],
   {theta, 0, Pi}, {phi, 0, 2*Pi}, AxesLabel -> {"x", "y", "z"},
   DisplayFunction -> Identity, PlotLabel -> "               psi2pz"];
```

```
Show[GraphicsArray[{{psi2s,psi2px},{psi2py,psi2pz}},GraphicsSpacing->.5]];
```

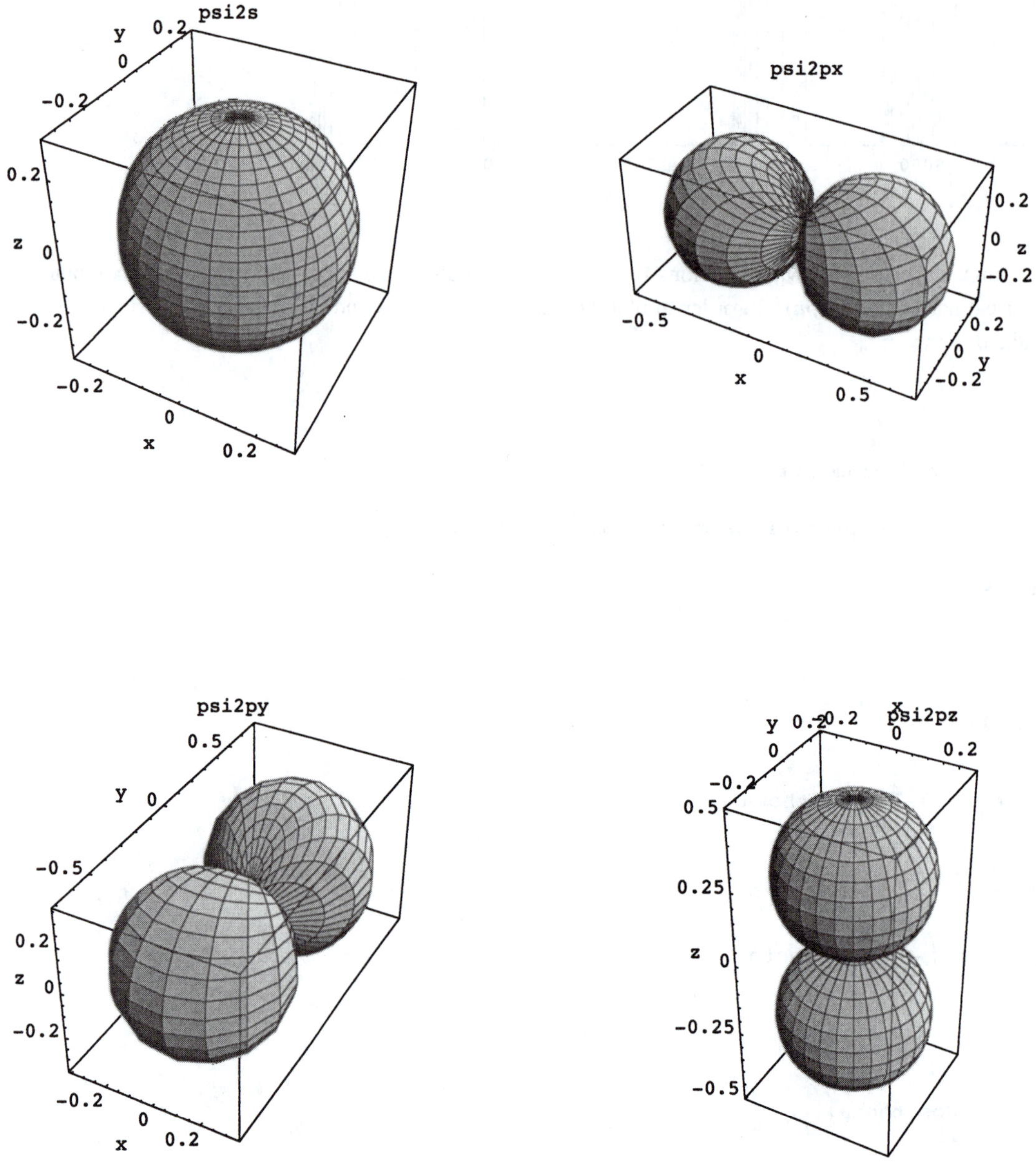

10.E Plot the squares of the angular factors of the atomic orbitals $2p_x$, $2p_y$, and $2p_z$ in three dimensions and compare them with Fig. 10.6.

SOLUTION

```
ClearAll["Global`*"]

<< Graphics`ParametricPlot3D`
```

```
SphericalPlot3D[(Sin[theta] * Cos[phi])^2, {theta, 0, Pi}, {phi, 0, 2*Pi},
  PlotRange -> All, PlotLabel -> "2pₓ", AxesLabel → {"x", "y", "z"}];
```

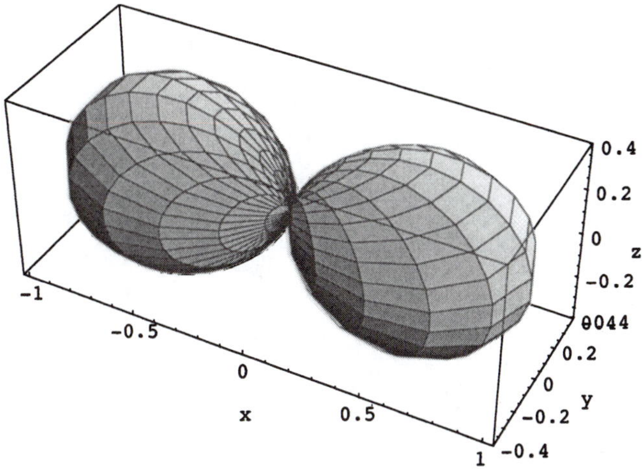


```
SphericalPlot3D[(Sin[theta] * Sin[phi])^2, {theta, 0, Pi}, {phi, 0, 2*Pi},
  PlotRange -> All, PlotLabel -> "2pᵧ", AxesLabel → {"x", "y", "z"}];
```

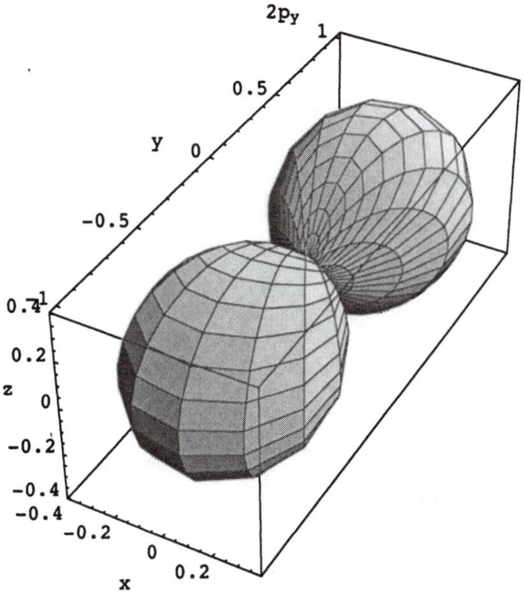

```
SphericalPlot3D[Cos[theta]^2, {theta, 0, Pi}, {phi, 0, 2*Pi}, PlotRange -> All,
  PlotLabel -> "                              2pz", AxesLabel → {"x", "y", "z"}];
```

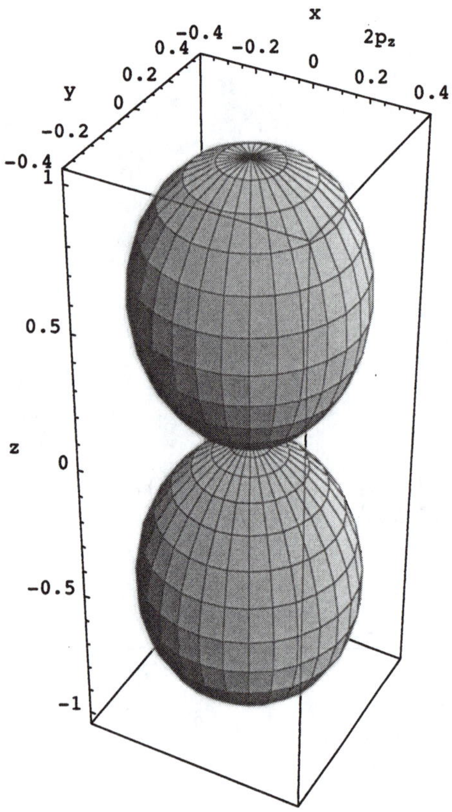

10.F Plot the squares of the angular factors of 3 d_{z^2} , 3 d_{xz} , 3 d_{yz} , 3 d_{xy} , and 3 d_{x2-y2} atomic orbitals in three dimensions. The angular factors are given in Table 10.1. Compare these plots with Fig. 10.6.

SOLUTION

```
<< Graphics`ParametricPlot3D`
```

```
dz2 =
  SphericalPlot3D[(3*Cos[theta]^2 - 1)^2, {theta, 0, Pi}, {phi, 0, 2 Pi}, PlotRange -> All,
        PlotLabel -> "3dz2"];
```

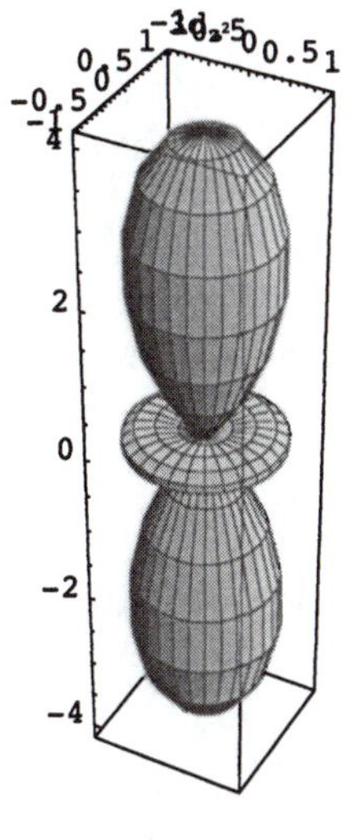


```
dxz = SphericalPlot3D[Sin[theta]^2 Cos[theta]^2 Cos[phi]^2,
   {theta, 0, Pi}, {phi, 0, 2 Pi}, PlotRange -> All,
        PlotLabel -> "3dxz"];
```

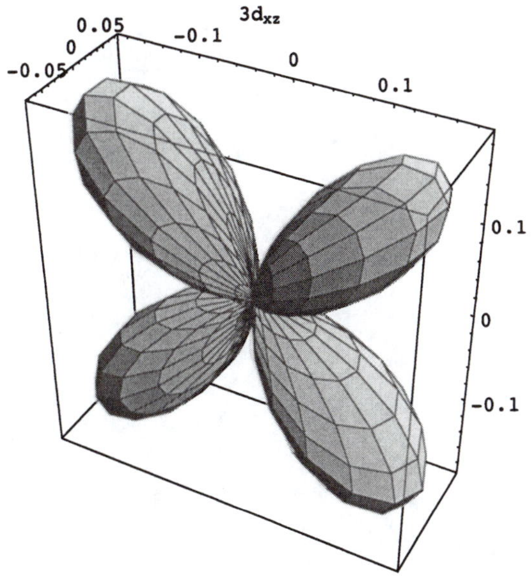

```
dyz = SphericalPlot3D[Sin[theta]^2 Cos[theta]^2 Sin[phi]^2,
   {theta, 0, Pi}, {phi, 0, 2 Pi}, PlotRange -> All,
      PlotLabel -> "3dyz"];
```

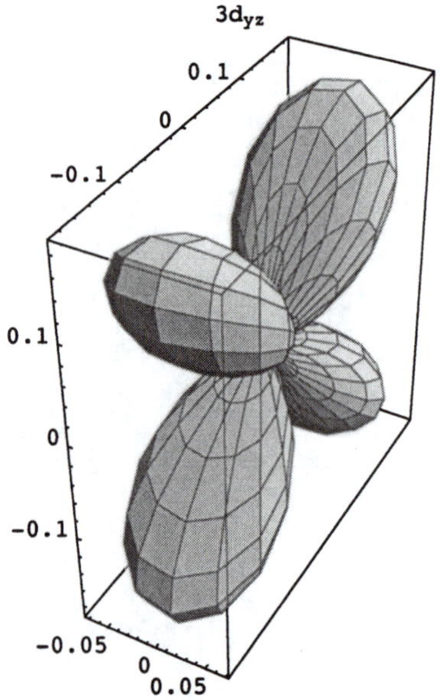


```
dxy = SphericalPlot3D[Sin[theta]^4 Sin[2*phi]^2,
   {theta, 0, Pi}, {phi, 0, 2 Pi}, PlotRange -> All,
      PlotLabel -> "3dxy"];
```

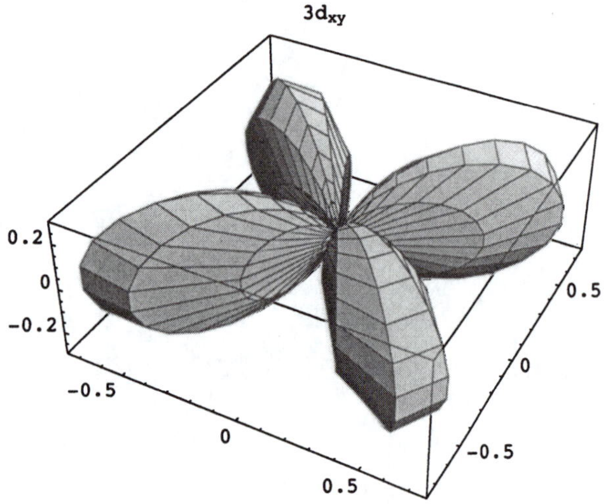

```
dx2y2 = SphericalPlot3D[Sin[theta]^4 Cos[2*phi]^2,
   {theta, 0, Pi}, {phi, 0, 2 Pi}, PlotRange -> All,
      PlotLabel -> "3dx2-y2"];
```

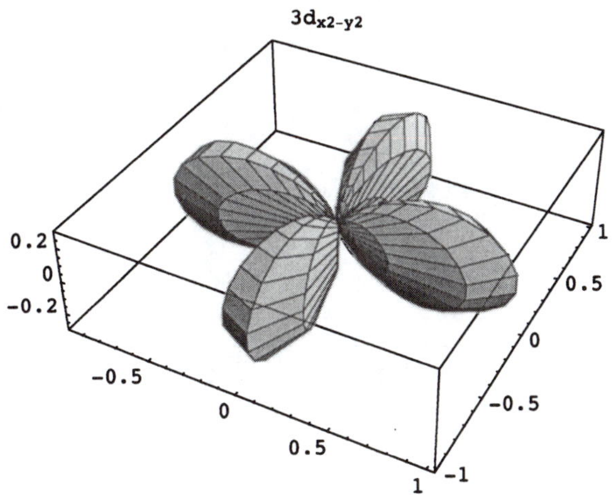

10.G Plot the Coulomb potential, the centrifugal potential, and the sum (effective potential) for a hydrogen atom with $\ell = 1$. Locate the minimum in the effective potential by taking the derivative and plotting it. Why doesn't the radius for the minimum potential correspond with the expectation value for r for the 1s orbital, which is $(3/2)a_0$?

SOLUTION

```
e=1.60218*10^-19(*C*);
epsil0=8.854187*10^-12(*C^2 N^-1 m^-2*);
hbar=1.054572*10^-34(*J s*);
mp=1.672623*10^-27(*kg*);
me=9.109390*10^-31(*kg*);
```

```
mu=(mp^-1+me^-1)^-1
```

9.10443×10^{-31}

```
coulpot=-e^2/(4*Pi*epsil0*r)
```

$-\frac{2.30709\times10^{-28}}{r}$

```
Plot[coulpot,{r,0,2.5*10^-10},PlotRange->{-5*10^-18,5*10^-18},AxesLabel->{"r/m","V/J"}]
;
```

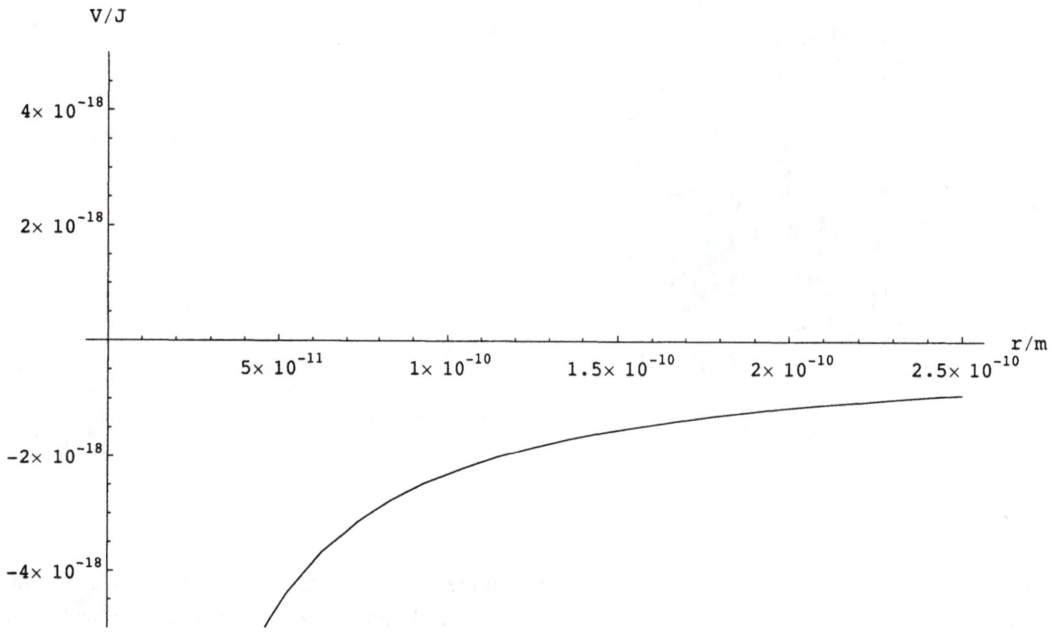


```
centpot=hbar^2/(mu*r^2)
```

$$\frac{1.22152\times10^{-38}}{r^2}$$

```
Plot[centpot,{r,0,2.5*10^-10},PlotRange->{-5*10^-18,5*10^-18},AxesLabel->{"r/m","V/J"}]
;
```

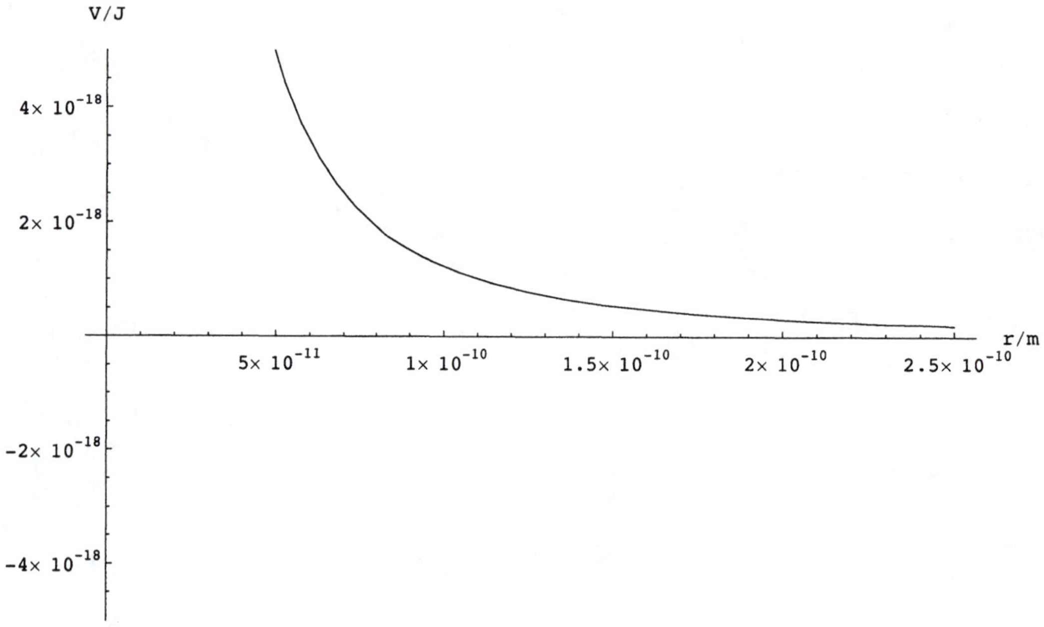

```
Plot[centpot+coulpot,{r,0,2.5*10^-10},PlotRange->{-5*10^-18,5*10^-18},AxesLabel->{"r/m"
,"V/J"}];
```

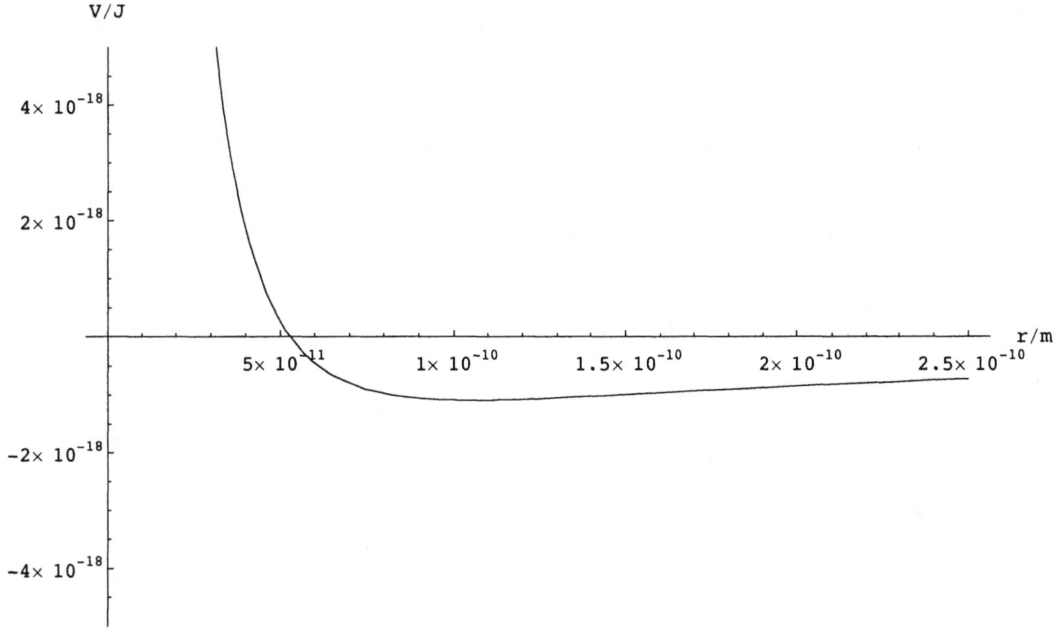


```
Plot[{coulpot,centpot,coulpot+centpot},{r,0,2.5*10^-10},PlotRange->{-5*10^-18,5*10^-18}
,AxesLabel->{"r/m","V/J"}];
```

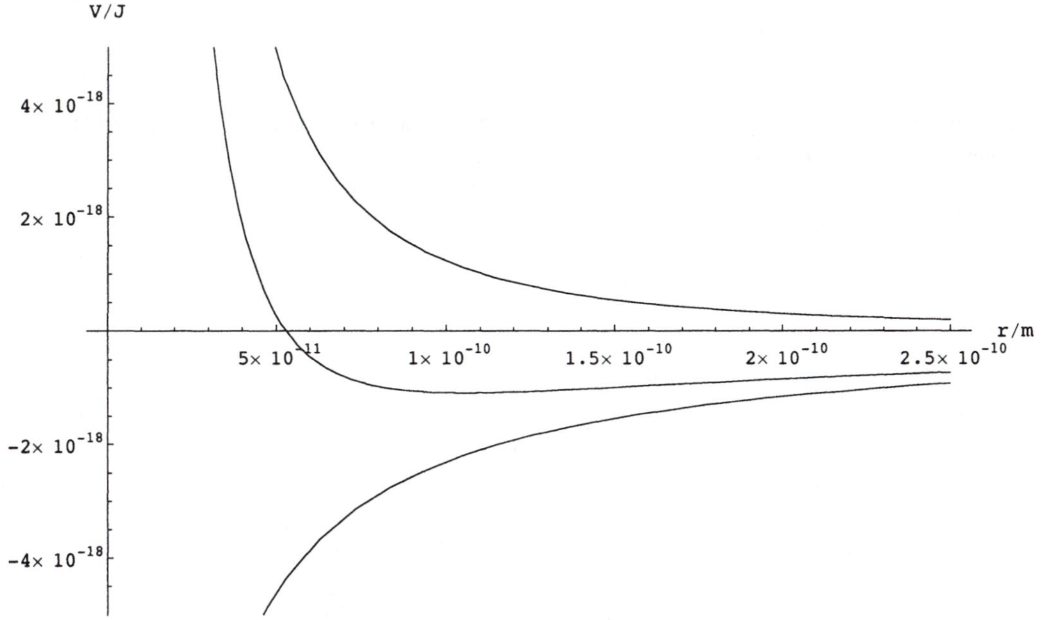

dvt = D[coulpot + centpot, r]

$$-\frac{2.44303\times10^{-38}}{r^3}+\frac{2.30709\times10^{-28}}{r^2}$$

```
Plot[dvt, {r, 0, 2.5 * 10^-10}, PlotRange → {-10^-8, 10^-8}, AxesLabel → {"r/m", "dvt"}];
```

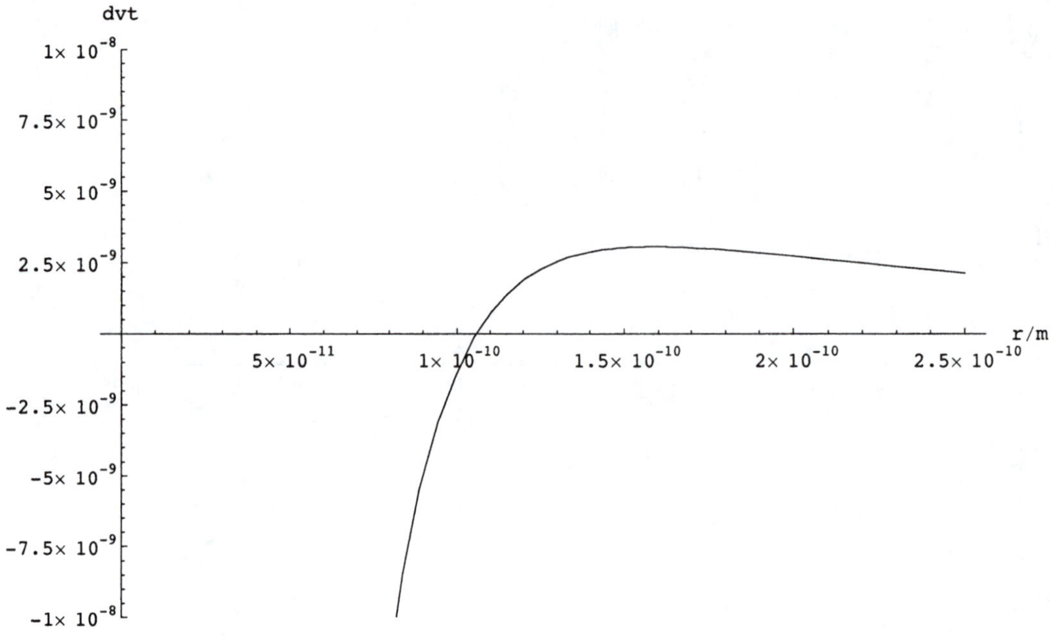

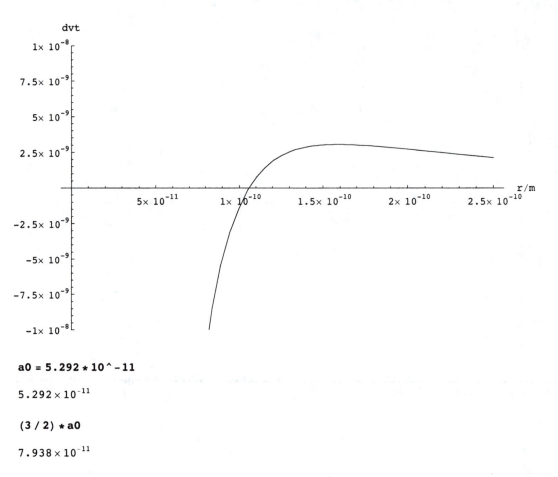

```
a0 = 5.292 * 10^-11
```

5.292×10^{-11}

```
(3 / 2) * a0
```

7.938×10^{-11}

These plots have been concerned with the potential energy, but kinetic energy is also involved in determining the expectation value for r. Further calculations could be done for l = 2.

Chapter 11 Molecular Electronic Structure

11.A Make three-dimensional plots to show the electron density in the *x*-*y* plane for the hydrogen molecule ion $H_2{}^+$. The parameters from a variational treatment (I. N. Levine, *Quantum Chemistry*, 5th ed., Prentice -Hall, 2000, pp. 389-390 are R_e = 2.01 bohrs, k = 1.246 $bohr^{-1}$, and z = 2.965 $bohr^{-1}$. Restrict the plots to x = -0.1 to 0.1 and y = -0.1 to 0.1.

SOLUTION

The molecular orbitals proportional to $\psi(1s)$ in the *x*-*y* plane are

$\psi(1s) = k^{3/2}\ \pi^{-1/2}\,(e^{-kr_A} \pm e^{-kr_B})$

where the plus sign yields the bonding orbital and the minus sign yields the antibonding orbital. The molecular orbitals proportional to $\psi(2p)$ in the *x*-*y* plane are

$\psi(2p) = \frac{\zeta^{5/2}}{4\,(2\,\pi)^{1/2}}\ y\ (e^{-\zeta r_A/2} \pm e^{-\zeta r_A/2})$

```
ClearAll["Global`*"]
xMin = -.1;
xMax = .1;
yMin = -.1;
yMax = .1;
a0 = .052918 (*In nanometers.*);
(*Enter values of R1, k, and zeta in nm.*)
R1 = 2.01 a0;
k = 1.246 / a0;
zeta = 2.965 / a0;

(*In order to make plots of electron density versus x and y, the following
  assignment is used for the radius from the center of mass of the molecule.*)
rA[x_, y_] := ((x + R1 / 2)^2 + y^2)^.5
rB[x_, y_] := ((x - R1 / 2)^2 + y^2)^.5
```

```
(*Type in the wavefunction for the 1s sigma bonding orbital.*)

f1sSig[x_, y_] :=
   k^1.5 Pi^-.5 (Exp[-k rA[x, y]] + Exp[-k rB[x, y]]);
Plot3D[f1sSig[x, y]^2, {x, xMin, xMax},
                              {y, xMin, xMax},
  PlotRange -> {0, 4000}, AxesLabel -> {"x" / "nm", "y" / "nm", density}];
```

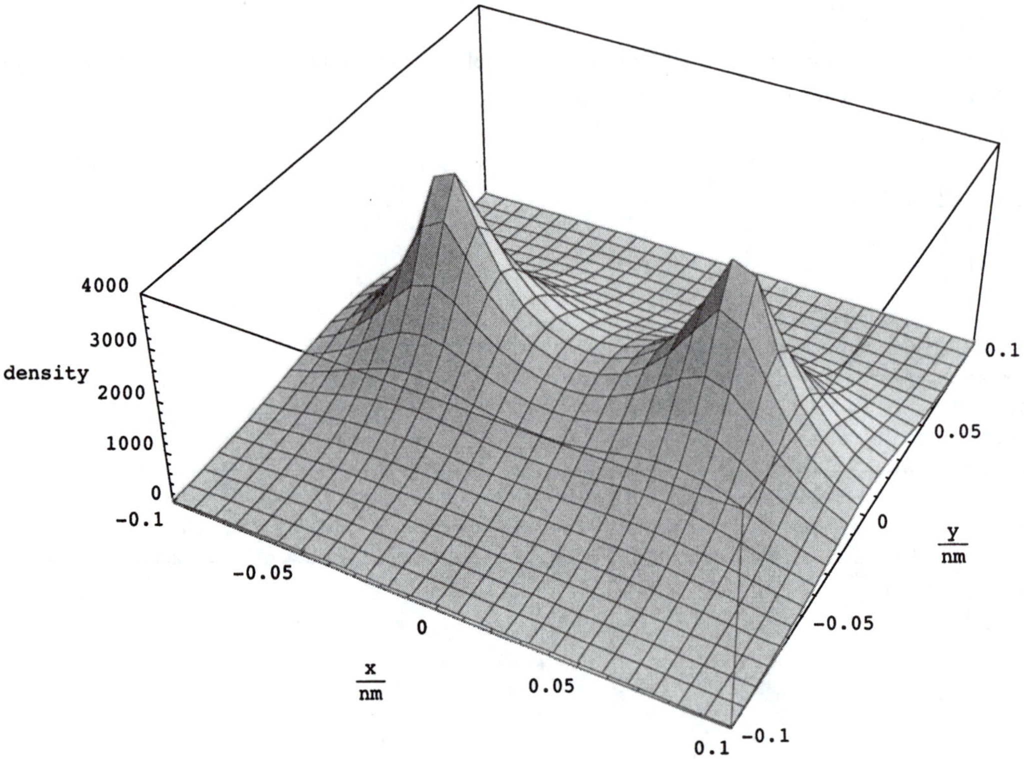

Note the buildup of electron density between the nuclei.

```
(*Type in the wavefunction for the 1s sigma antibonding orbital.*)

f1sSigAnti[ x_, y_] :=
    k^1.5 Pi^-.5 (Exp[-k rA[x, y]] - Exp[-k rB[x, y]]);
Plot3D[f1sSigAnti[x, y]^2, {x, xMin, xMax},
                                        {y, xMin, xMax},
  PlotRange -> {0, 3000}, AxesLabel -> {"x" / "nm", "y" / "nm", density}];
```

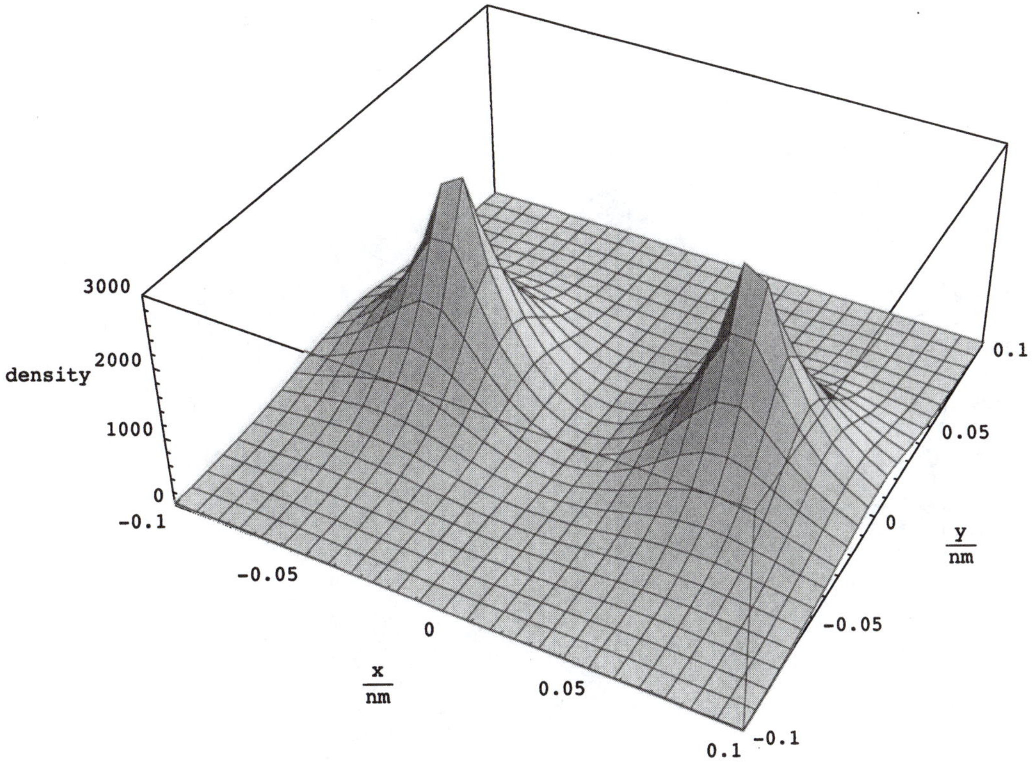

Note the electron density is zero along a line perpendicular to the line between the nuclei and halfway between the nuclei.

```
(*Type in the wavefunction for the 2 p pi bonding orbital.*)

f2pPi[ x_, y_] :=
    zeta^2.5 / (4 (2 Pi) ^.5) * y * (Exp[-zeta rA[x, y] / 2] + Exp[-zeta rB[x, y] / 2]);
Plot3D[f2pPi[x, y]^2, {x, xMin, xMax},
                                        {y, xMin, xMax},
  PlotRange -> {0, 2000}, AxesLabel -> {"x" / "nm", "y" / "nm", density}];
```

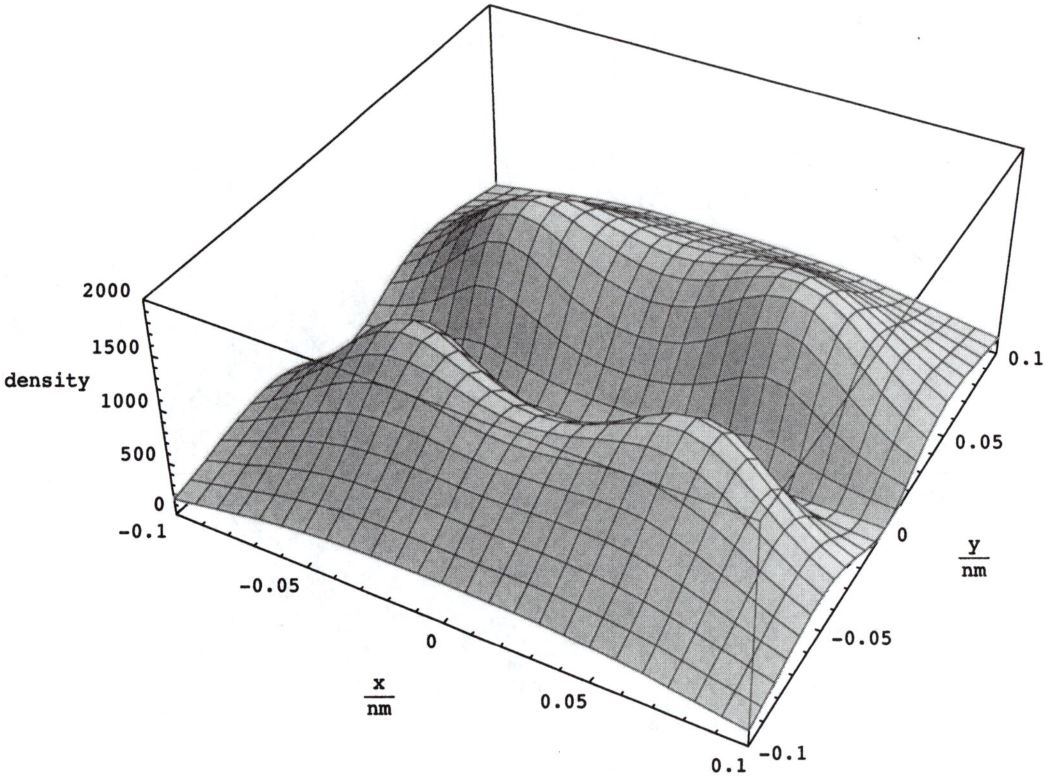

Note the electron density is zero along the line joining the nuclei and that there is electron buildup between the lobes of the p orbitals on the two nuclei.

```
(*Type in the wavefunction for the 2 p pi antibonding orbital.*)

f2pPiAnti[ x_, y_] :=
    zeta^2.5 / (4 (2 Pi) ^.5) * y * (Exp[-zeta rA[x, y] / 2] - Exp[-zeta rB[x, y] / 2]);
Plot3D[f2pPiAnti[x, y] ^2, {x, xMin, xMax},
                                          {y, xMin, xMax},
  PlotRange -> {0, 1000}, AxesLabel -> {"x" / "nm", "y" / "nm", density}];
```

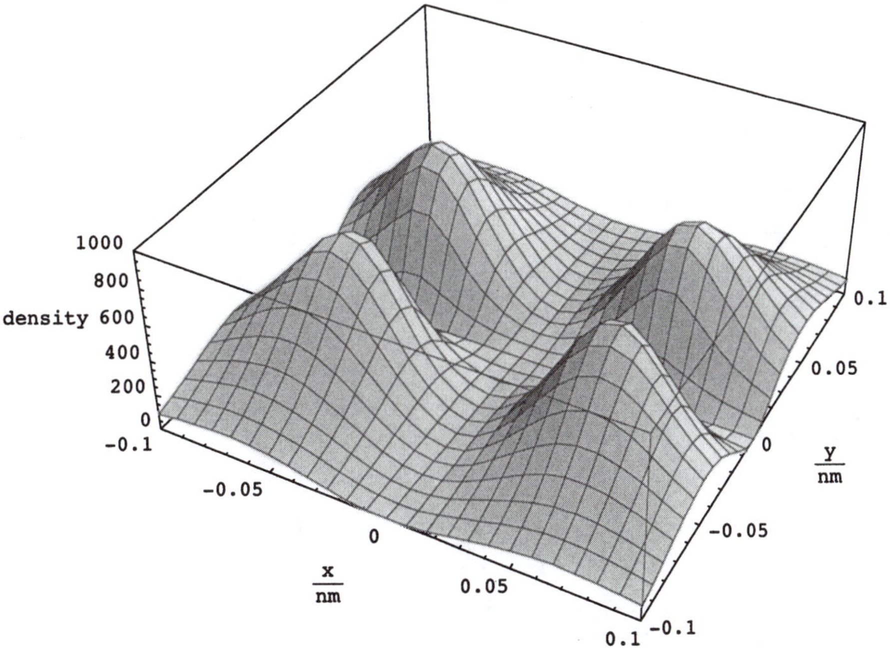

Note that the electron density along the line joining the nuclei is zero and is also zero between the lobes of the p orbitals.

11.B A trial wavefunction for the hydrogen molecule ion is

$\psi_{\pm} = 1s_A \pm 1s_B$

The energies in units of hartrees corresponding with these two wavefunctions can be calculated using

$E_+ = \frac{J+K}{1+S}$

$E_- = \frac{J-K}{1-S}$

where

$S = e^{-R}(1 + R + \frac{R^2}{3})$ [overlap integral]

$J = e^{-2R}(1 + \frac{1}{R})$ [Coulomb integral]

$K = \frac{S}{R} - e^{-R}(1 + R)$ [exchange integral]

where R is the distance between the protons in units of a_0. Plot S, J, K, E_+, and E_- versus R.

SOLUTION

```
ClearAll["Global`*"]
```

```
s = Exp[-r] * (1 + r + (r^2) / 3)
```

$e^{-r}\left(1+r+\frac{r^2}{3}\right)$

```
Plot[s, {r, 0, 10}, AxesLabel → {"R/a₀", "S"}];
```

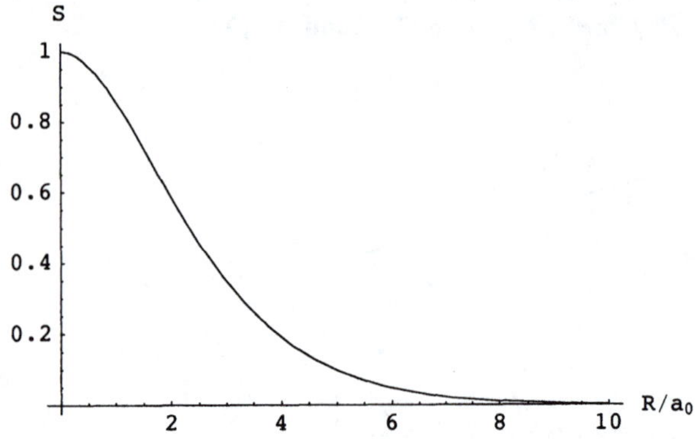


```
j = Exp[-2 * r] * (1 + 1 / r)
```

$e^{-2r}\left(1+\frac{1}{r}\right)$

```
Plot[j, {r, 0, 10}, AxesLabel -> {"R/a₀", "J"}];
```

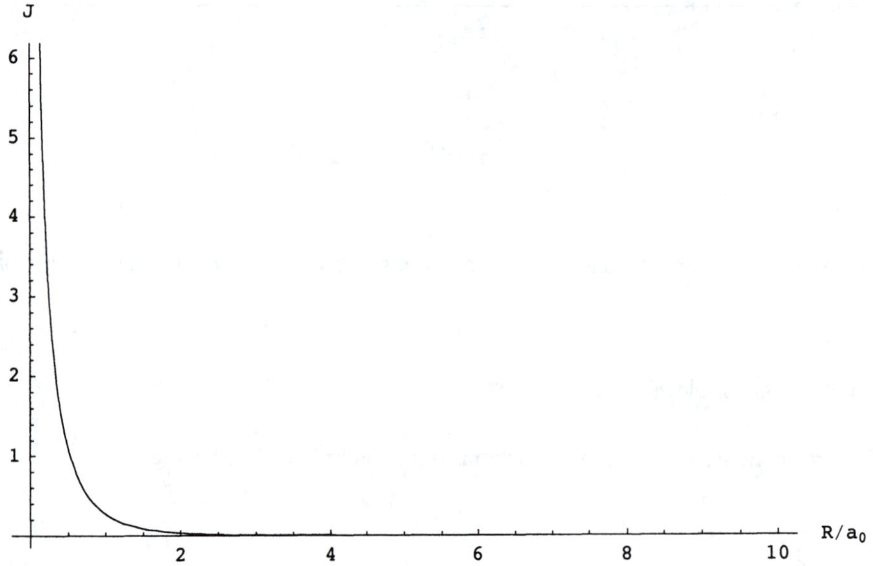


```
k = (s / r) - Exp[-r] * (1 + r)
```

$-e^{-r}(1+r)+\frac{e^{-r}\left(1+r+\frac{r^2}{3}\right)}{r}$

```
Plot[k, {r, 0, 10}, AxesLabel -> {"R/a0", "K"}];
```

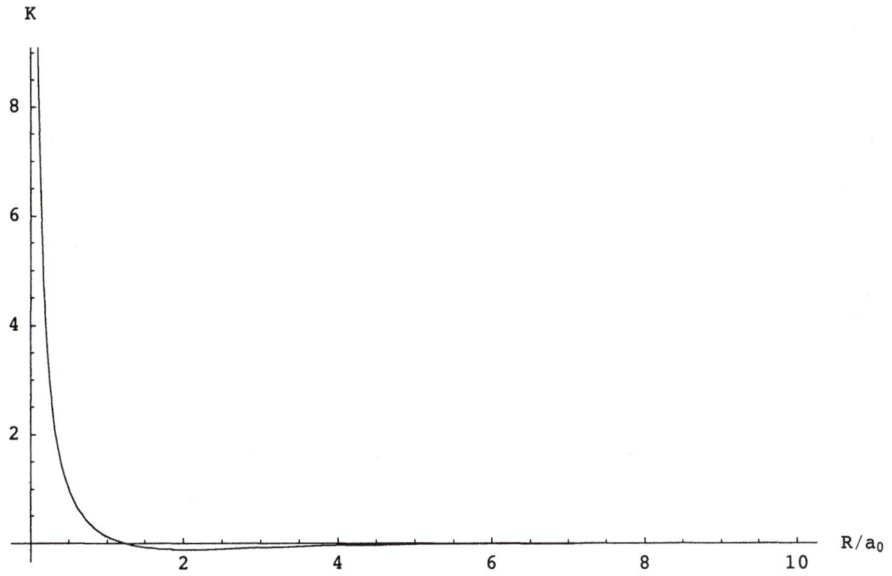


```
eplus = (j + k) / (1 + s)
```

$$\frac{e^{-2r}\left(1+\frac{1}{r}\right)-e^{-r}(1+r)+\frac{e^{-r}\left(1+r+\frac{r^2}{3}\right)}{r}}{1+e^{-r}\left(1+r+\frac{r^2}{3}\right)}$$

```
Plot[eplus, {r, 0, 10}, PlotRange -> {-.1, .5}, AxesLabel -> {"R/a0", "E+/Eh"}];
```

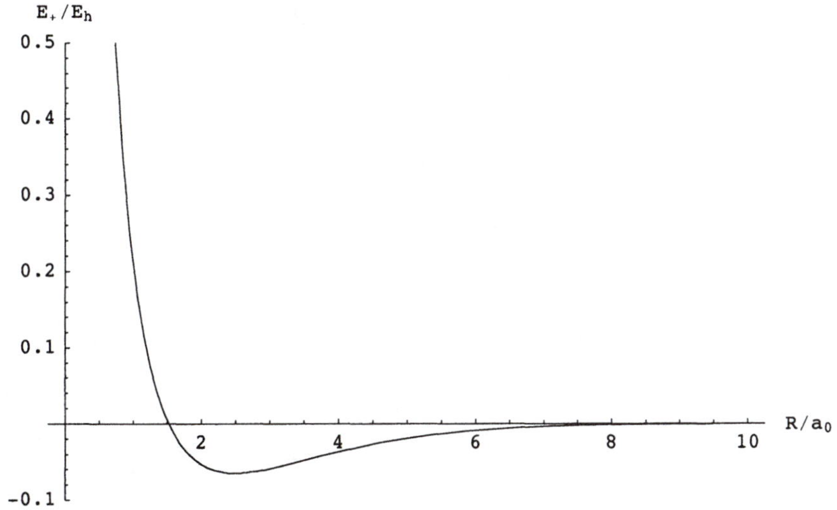


```
eminus = (j - k) / (1 - s)
```

$$\frac{e^{-2r}\left(1+\frac{1}{r}\right)+e^{-r}(1+r)-\frac{e^{-r}\left(1+r+\frac{r^2}{3}\right)}{r}}{1-e^{-r}\left(1+r+\frac{r^2}{3}\right)}$$

```
Plot[eminus, {r, 0, 10}, PlotRange -> {-.1, .5}, AxesLabel -> {"R/a0", "E-/Eh"}];
```

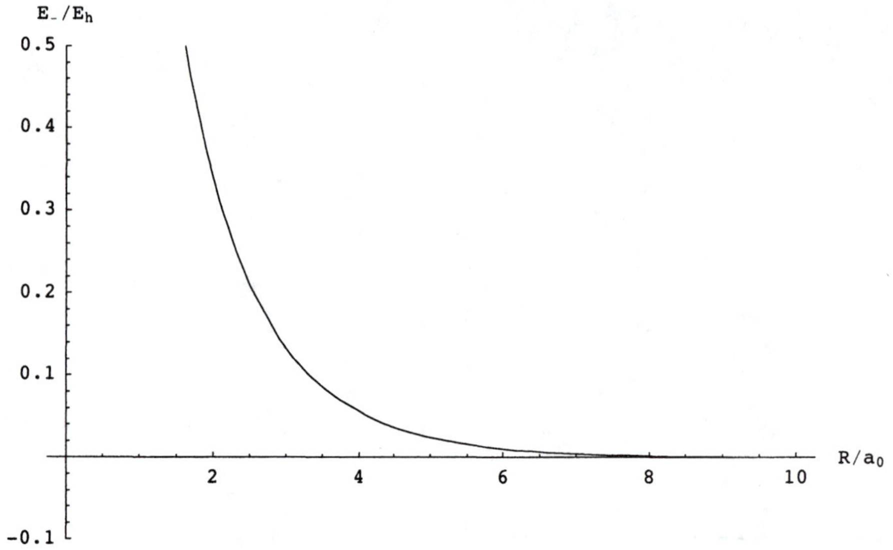

```
Plot[{eplus, eminus}, {r, 0, 10}, PlotRange -> {-.1, .5}, AxesLabel -> {"R/a₀", "E/Eₕ"}];
```

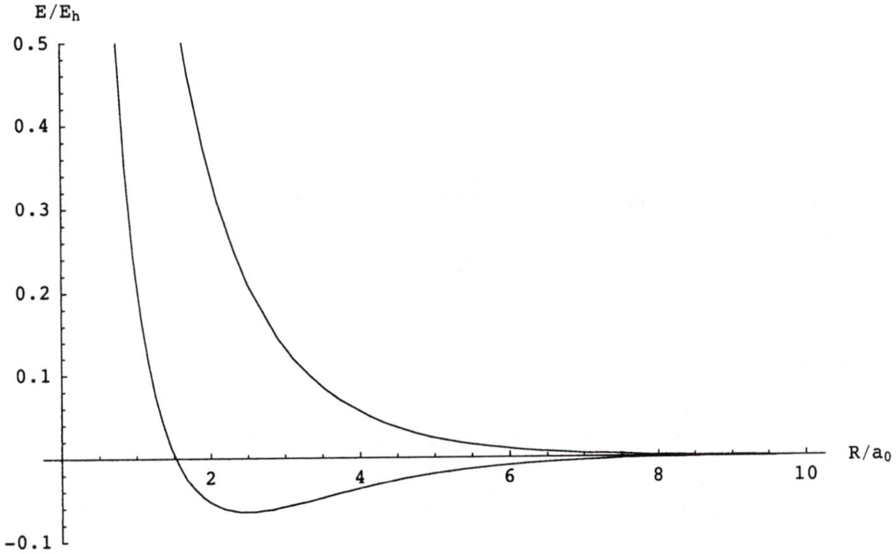

11.C Calculate the energy eigenvalues and eigenvectors for 1,3-butadiene using the Hückel method.

SOLUTION

The eigenvalues of a matrix m are the values of λ_i for which there are vectors x_i such that m . $x_i = \lambda_i x_i$. In *Mathematica*, the eigenvalues and eigenvectors can be calculated at the same time by using the program Eigensystem. The results can be sorted in ascending order of eigenvalues by applying Transpose and Sort, as described by J. E. Feagin, Quantum Methods with *Mathematica*, New York: Springer-Verlag, 1994. As explained in Cropper, the Hückel calculation can be expressed by the matrix equation FC = CE, where F is the Fock matrix, C is the orbital matrix, and E is a diagonal matrix of energies. Since the Hückel method provides no evaluation of α and β, the matrix equation can be written F'C = Cx, in which α is replaced by zero and β is replaced by -1 in the Fock matrix. The entries in the diagonal matrix x are related to the energies of the orbitals by x = (α - E)/β. The Fock matrix for 1,3-butadiene is

```
f = {{0, -1., 0, 0}, {-1., 0, -1., 0}, {0, -1., 0, -1.}, {0, 0, -1., 0}}

{{0, -1., 0, 0}, {-1., 0, -1., 0}, {0, -1., 0, -1.}, {0, 0, -1., 0}}

MatrixForm[f]
```

$$\begin{pmatrix} 0 & -1. & 0 & 0 \\ -1. & 0 & -1. & 0 \\ 0 & -1. & 0 & -1. \\ 0 & 0 & -1. & 0 \end{pmatrix}$$

```
{energies, coeff} = Eigensystem[f] // Transpose // Sort // Transpose

{{-1.61803, -0.618034, 0.618034, 1.61803},
 {{0.371748, 0.601501, 0.601501, 0.371748}, {0.601501, 0.371748, -0.371748, -0.601501},
  {0.601501, -0.371748, -0.371748, 0.601501}, {0.371748, -0.601501, 0.601501, -0.371748}}}
```

```
TableForm[energies]
```

```
-1.61803
-0.618034
0.618034
1.61803
```

```
TableForm[coeff]
```

0.371748	0.601501	0.601501	0.371748
0.601501	0.371748	-0.371748	-0.601501
0.601501	-0.371748	-0.371748	0.601501
0.371748	-0.601501	0.601501	-0.371748

```
TableForm[{energies, coeff}]
```

-1.61803	-0.618034	0.618034	1.61803
0.371748	0.601501	0.601501	0.371748
0.601501	0.371748	-0.371748	-0.601501
0.601501	-0.371748	-0.371748	0.601501
0.371748	-0.601501	0.601501	-0.371748

The first row gives energies in ascending order, and the last four entries in each columngive the coefficients in the wave function. The values of the eigenvalues (energies) and eigenvectors (coeff) can be verified by testing whether the following statement is true.

```
Table[f.coeff[[n]] == energies[[n]] * coeff[[n]], {n, 1, 4}]
```

```
{True, True, True, True}
```

11.D Calculate the energy eigenvalues and eigenvectors for benzene using the Hückel method. (See Computer Problem 11.C.)

SOLUTION

The Fock matrix for benzene is

```
f = {{0, -1., 0, 0, 0, -1.}, {-1., 0, -1., 0, 0, 0}, {0, -1., 0, -1., 0, 0},
   {0, 0, -1., 0, -1., 0}, {0, 0, 0, -1., 0, -1.}, {-1., 0, 0, 0, -1., 0}}
```

```
{{0, -1., 0, 0, 0, -1.}, {-1., 0, -1., 0, 0, 0}, {0, -1., 0, -1., 0, 0},
 {0, 0, -1., 0, -1., 0}, {0, 0, 0, -1., 0, -1.}, {-1., 0, 0, 0, -1., 0}}
```

```
MatrixForm[f]
```

$$\begin{pmatrix} 0 & -1. & 0 & 0 & 0 & -1. \\ -1. & 0 & -1. & 0 & 0 & 0 \\ 0 & -1. & 0 & -1. & 0 & 0 \\ 0 & 0 & -1. & 0 & -1. & 0 \\ 0 & 0 & 0 & -1. & 0 & -1. \\ -1. & 0 & 0 & 0 & -1. & 0 \end{pmatrix}$$

These results can be sorted in ascending order of eigenvalues by applying Transpose and Sort, as described by J. E. Feagin, Quantum Methods with *Mathematica*, New York: Springer-Verlag, 1994.

```
{energies, coeff} = Eigensystem[f] // Transpose // Sort // Transpose

{{-2., -1., -1., 1., 1., 2.},
 {{-0.408248, -0.408248, -0.408248, -0.408248, -0.408248, -0.408248},
  {0.288675, -0.288675, -0.57735, -0.288675, 0.288675, 0.57735},
  {0.5, 0.5, -1.11022×10^-16, -0.5, -0.5, 0.},
  {0.288675, 0.288675, -0.57735, 0.288675, 0.288675, -0.57735},
  {0.5, -0.5, -1.11022×10^-16, 0.5, -0.5, 0.},
  {-0.408248, 0.408248, -0.408248, 0.408248, -0.408248, 0.408248}}}

TableForm[energies]

-2.
-1.
-1.
1.
1.
2.

TableForm[coeff]
```

-0.408248	-0.408248	-0.408248	-0.408248	-0.408248	-0.408248
0.288675	-0.288675	-0.57735	-0.288675	0.288675	0.57735
0.5	0.5	-1.11022×10^{-16}	-0.5	-0.5	0.
0.288675	0.288675	-0.57735	0.288675	0.288675	-0.57735
0.5	-0.5	-1.11022×10^{-16}	0.5	-0.5	0.
-0.408248	0.408248	-0.408248	0.408248	-0.408248	0.408248

The values of the eigenvalues (energies) and eigenvectors (coeff) can be verified by testing whether the following statement is true.

```
Table[f.coeff[[n]] == energies[[n]] * coeff[[n]], {n, 1, 6}]

{True, True, True, True, True, True}
```

This test failed for two of the wave functions, but all True is obtained on some other computers..

11.E Solve the Hückel secular equation for 1,3-butadiene and write the equations for the four energy levels.

SOLUTION

The matrix in equation 11.64 is

```
m = {{x, 1, 0, 0}, {1, x, 1, 0}, {0, 1, x, 1}, {0, 0, 1, x}}

{{x, 1, 0, 0}, {1, x, 1, 0}, {0, 1, x, 1}, {0, 0, 1, x}}
```

```
MatrixForm[m]
```

$$\begin{pmatrix} x & 1 & 0 & 0 \\ 1 & x & 1 & 0 \\ 0 & 1 & x & 1 \\ 0 & 0 & 1 & x \end{pmatrix}$$

The determinant and its solutions are given by

```
Det[m]
```

$1 - 3 x^2 + x^4$

```
Solve[Det[m] == 0, x] // N
```

{{x → -1.61803}, {x → -0.618034}, {x → 0.618034}, {x → 1.61803}}

This leads to equations 11.65.

11.F Plot the $1s_A$ and $1s_B$ wave functions with a separation of $3a_0$ between the protons in the hydrogen molecule ion.

SOLUTION

```
sa = (π^-.5) * Exp[-Abs[x]]
```

$0.56419\, e^{-\text{Abs}[x]}$

```
sb = (π^-.5) * Exp[-Abs[x - 3]]
```

$0.56419\, e^{-\text{Abs}[-3+x]}$

```
Plot[{sa, sb}, {x, -5, 8}, AxesLabel → {"x/a0", "ψ1\\\\\\s"}];
```

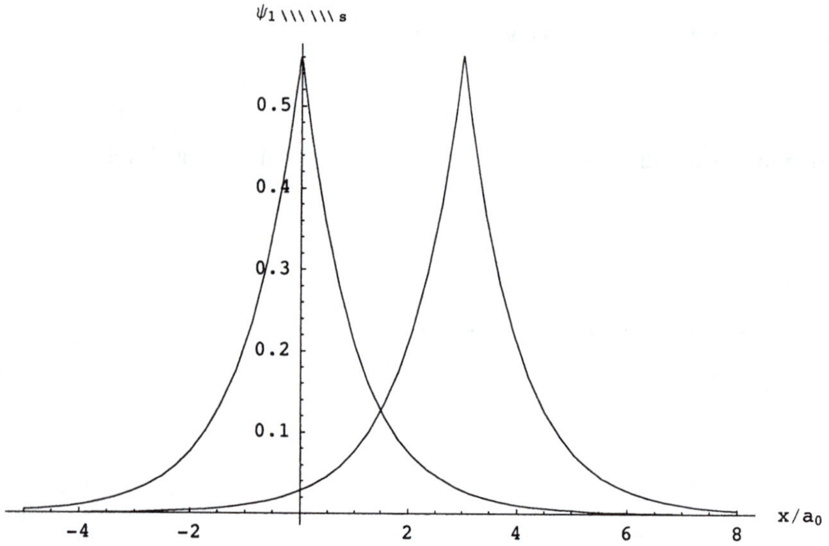

```
Plot[sa * sb, {x, -5, 8}, AxesLabel → {"x/a0", "ψ1 \\\ \\\ sAψ1 \\\ \\\ sB"}];
```

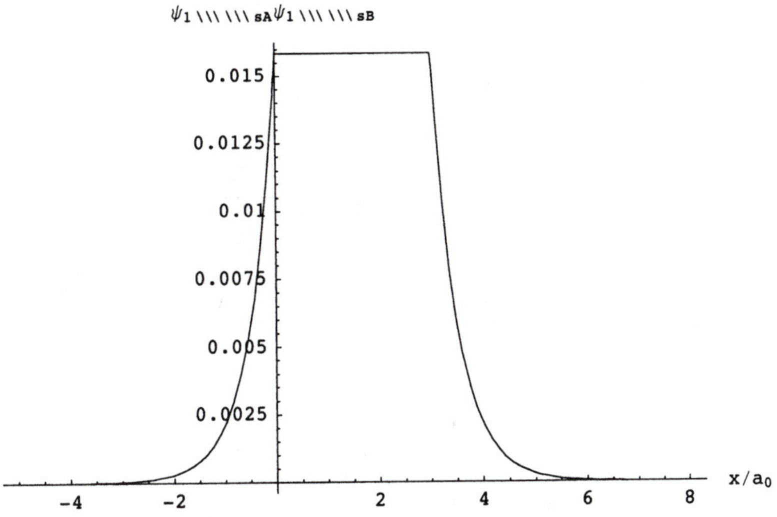

11.G Plot the following normalized molecular orbitals of the hydrogen molecule ion with a separation of $3a_0$ between the protons in the hydrogen molecule ion.

$\psi_g = \frac{1}{\sqrt{2(1+S)}}(1s_A + 1s_B)$

$\psi_u = \frac{1}{\sqrt{2(1-S)}}(1s_A - 1s_B)$

SOLUTION

```
ClearAll["Global`*"]

sa = (π^-.5) * Exp[-Abs[x]];

sb = (π^-.5) * Exp[-Abs[x - 3]];

s = Exp[-r] * (1 + r + (r^2) / 3) /. r -> 3;

psig = (1 / (2 * (1 + s)) ^ .5) * (sa + sb)

0.608917 (0.56419 e^-Abs[-3+x] + 0.56419 e^-Abs[x])

plotg = Plot[psig, {x, -5, 8}, AxesLabel → {"x/a0", "ψg"},
   AxesOrigin → {-5, 0}, DisplayFunction → Identity];

plotg2 = Plot[psig^2, {x, -5, 8}, AxesLabel → {"x/a0", "ψg^2"},
   AxesOrigin → {-5, 0}, DisplayFunction → Identity];

psiu = (1 / (2 * (1 - s)) ^ .5) * (sa - sb)

0.876054 (-0.56419 e^-Abs[-3+x] + 0.56419 e^-Abs[x])

plotu = Plot[psiu, {x, -5, 8}, AxesLabel → {"x/a0", "ψu"},
   AxesOrigin → {-5, 0}, DisplayFunction → Identity];

plotu2 = Plot[psiu^2, {x, -5, 8}, AxesLabel → {"x/a0", "ψu^2"},
   AxesOrigin → {-5, 0}, DisplayFunction → Identity];
```

```
Show[GraphicsArray[{{plotg, plotu}, {plotg2, plotu2}}]];
```

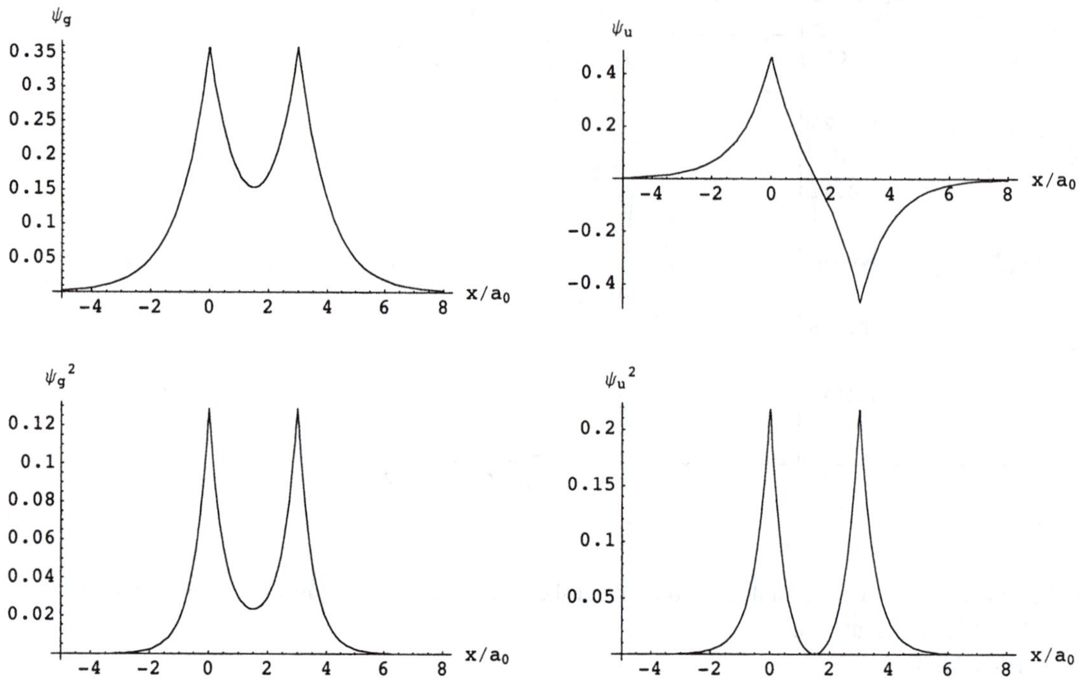

11.H (*a*) Plot the Lennard-Jones potential for the interaction of two nitrogen molecules versus the distance between the molecules. When the energy is expressed as an equivalent temperature, $\epsilon/k = 98.1$ K and $\sigma = 0.37$ nm. (*b*) Calculate the value of R at the minimum by setting the derivative of the potential equal to zero. (*c*) Calculate the value of R at which the potential energy is equal to zero.

SOLUTION

```
pot = 4 * 95.1 * ((.37 / r) ^12 - (.37 / r) ^ 6)
```

$$380.4\left(\frac{6.58295\times 10^{-6}}{r^{12}} - \frac{0.00256573}{r^{6}}\right)$$

(a)

```
Plot[pot, {r, .2, .8}, PlotRange → {-100, 100}, AxesLabel → {"R/nm", "(V/k)/K"}];
```

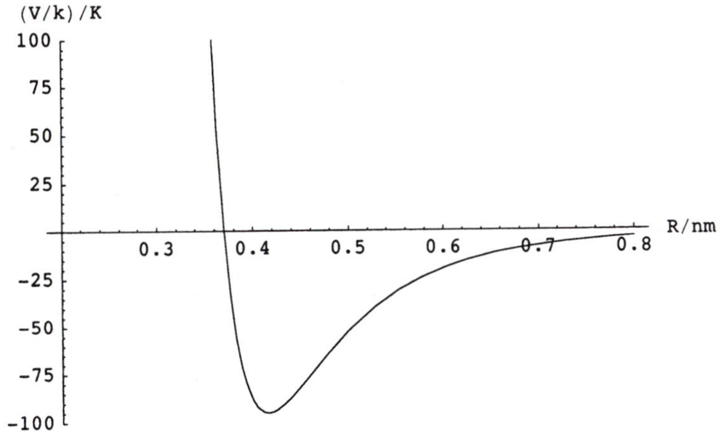

(b)

```
dvt = D[pot, r]
```

$$380.4\left(-\frac{0.0000789954}{r^{13}}+\frac{0.0153944}{r^{7}}\right)$$

```
Solve[dvt == 0, r]
```

```
{{r → -0.415311}, {r → -0.207655 - 0.35967 i}, {r → -0.207655 + 0.35967 i},
 {r → 0.207655 - 0.35967 i}, {r → 0.207655 + 0.35967 i}, {r → 0.415311}}
```

It was shown in the text that at the minimum of the potential energy curve, $r = 2^{1/6}\sigma$.

```
(2^(1/6)) * .37
```

```
0.415311
```

This agrees with the only solution that is real and positive.

(c)

```
Solve[pot == 0, r]
```

```
{{r → -0.37}, {r → -0.185 - 0.320429 i}, {r → -0.185 + 0.320429 i},
 {r → 0.185 - 0.320429 i}, {r → 0.185 + 0.320429 i}, {r → 0.37}}
```

Since R is necessarily positive and real, the only solution that is satisfactory is $R = 0.37$ nm, which is the input value of sigma.

Chapter 12 Symmetry

12.A Show that the matrix product of the $\hat{C}_3$ operation and the $\hat{C}_3{}^2$ operation is equal to the identity operation $\hat{E}$. A matrix for the $\hat{C}_3$ operation is given in Example 125.

SOLUTION

```
c3 = {{-1 / 2, -Sqrt[3] / 2}, {Sqrt[3] / 2, -1 / 2}}
```

$$\{\{-\frac{1}{2}, -\frac{\sqrt{3}}{2}\}, \{\frac{\sqrt{3}}{2}, -\frac{1}{2}\}\}$$

```
MatrixForm[c3]
```

$$\begin{pmatrix} -\frac{1}{2} & -\frac{\sqrt{3}}{2} \\ \frac{\sqrt{3}}{2} & -\frac{1}{2} \end{pmatrix}$$

The $\hat{C}_3{}^2$ operation is the matrix product of the $\hat{C}_3$ operation with itself.

```
c32=c3.c3
```

$$\{\{-\frac{1}{2}, \frac{\sqrt{3}}{2}\}, \{-\frac{\sqrt{3}}{2}, -\frac{1}{2}\}\}$$

```
MatrixForm[c32]
```

$$\begin{pmatrix} -\frac{1}{2} & \frac{\sqrt{3}}{2} \\ -\frac{\sqrt{3}}{2} & -\frac{1}{2} \end{pmatrix}$$

The matrix product of the $\hat{C}_3$ operation and the $\hat{C}_3{}^2$ operation is given by

```
c3.c32
```

```
{{1, 0}, {0, 1}}
```

```
MatrixForm[c3.c32]
```

$$\begin{pmatrix} 1 & 0 \\ 0 & 1 \end{pmatrix}$$

This is the identity matrix.

12.B The multiplication table for the group $\hat{C}_{2v}$ is given in Table 12.2, and the application of these operations to the water molecule is discussed in connection with this table. Since matrices for these operations are given in equations 12.16-12.19, verify that $\hat{\sigma}_v\,'.\,\hat{\sigma}_v = \hat{C}_2$, $\hat{\sigma}_v.\,\hat{C}_2 = \hat{\sigma}_v\,'$, and $\hat{C}_2.\hat{C}_2 = \hat{E}$, where these symbols refer to operations.

SOLUTION

Matrices 12.17, 12.18, and 12.19 are

```
c2 = {{-1, 0, 0}, {0, -1, 0}, {0, 0, 1}}

{{-1, 0, 0}, {0, -1, 0}, {0, 0, 1}}

MatrixForm[c2]
```

$$\begin{pmatrix} -1 & 0 & 0 \\ 0 & -1 & 0 \\ 0 & 0 & 1 \end{pmatrix}$$

```
sigv = {{1, 0, 0}, {0, -1, 0}, {0, 0, 1}}

{{1, 0, 0}, {0, -1, 0}, {0, 0, 1}}

MatrixForm[sigv]
```

$$\begin{pmatrix} 1 & 0 & 0 \\ 0 & -1 & 0 \\ 0 & 0 & 1 \end{pmatrix}$$

```
sigv' = {{-1, 0, 0}, {0, 1, 0}, {0, 0, 1}}

{{-1, 0, 0}, {0, 1, 0}, {0, 0, 1}}

MatrixForm[sigv']
```

$$\begin{pmatrix} -1 & 0 & 0 \\ 0 & 1 & 0 \\ 0 & 0 & 1 \end{pmatrix}$$

Show that $\sigma_v\,'.\,\sigma_v = C_2$.

```
sigv'.sigv

{{-1, 0, 0}, {0, -1, 0}, {0, 0, 1}}
```

```
MatrixForm[sigv'.sigv]
```

$$\begin{pmatrix} -1 & 0 & 0 \\ 0 & -1 & 0 \\ 0 & 0 & 1 \end{pmatrix}$$

This is the matrix for C_2 as expected.
Show that $\sigma_v . C_2 = \sigma_v{}'$.

```
sigv.c2

{{-1, 0, 0}, {0, 1, 0}, {0, 0, 1}}

MatrixForm[sigv.c2]
```

$$\begin{pmatrix} -1 & 0 & 0 \\ 0 & 1 & 0 \\ 0 & 0 & 1 \end{pmatrix}$$

This is $\sigma_v{}'$ as expected.
Show that $C_2 . C_2 = E$.

```
c2.c2

{{1, 0, 0}, {0, 1, 0}, {0, 0, 1}}

MatrixForm[c2.c2]
```

$$\begin{pmatrix} 1 & 0 & 0 \\ 0 & 1 & 0 \\ 0 & 0 & 1 \end{pmatrix}$$

This gives the identity matrix as expected.

12.C Use *Mathematica* to to show the shapes of a tetrahedron, octahedron, dodecahedron, icosohedron, and a bucky ball.

SOLUTION

```
<< Graphics`Graphics`

<< Graphics`Polyhedra`

Show[Polyhedron[Tetrahedron]];
```

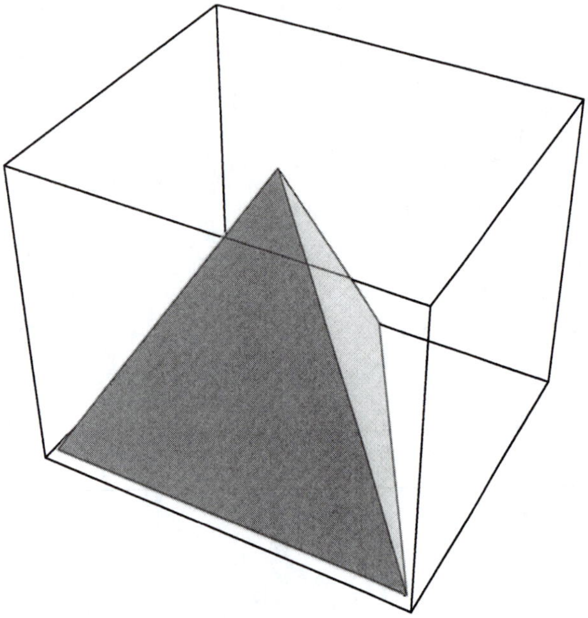

```
Show[Polyhedron[Octahedron]];
```

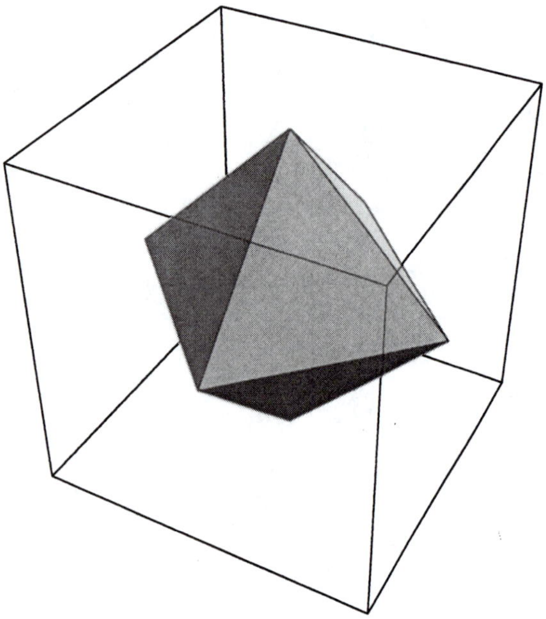

```
Show[Polyhedron[Dodecahedron]];
```

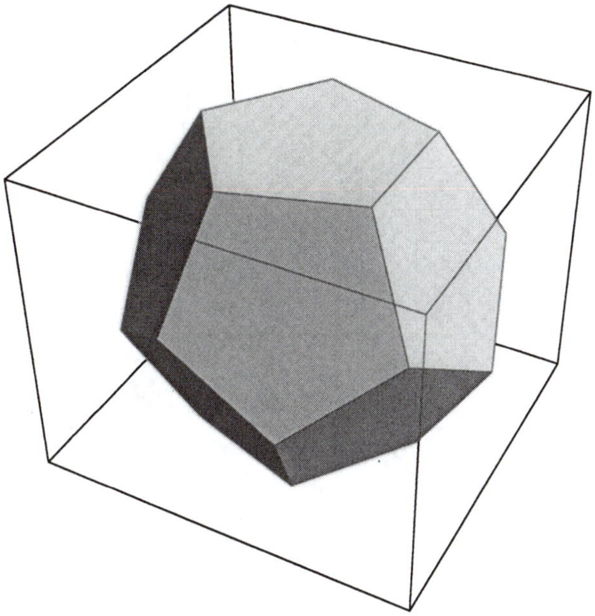

```
Show[Polyhedron[Icosahedron]];
```

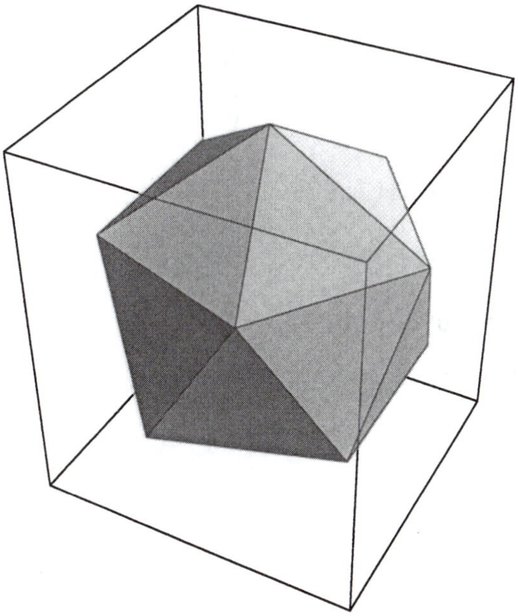

```
ball = First[Truncate[Polyhedron[Icosahedron], 1/3]];
```

```
Show[Graphics3D[ball]];
```

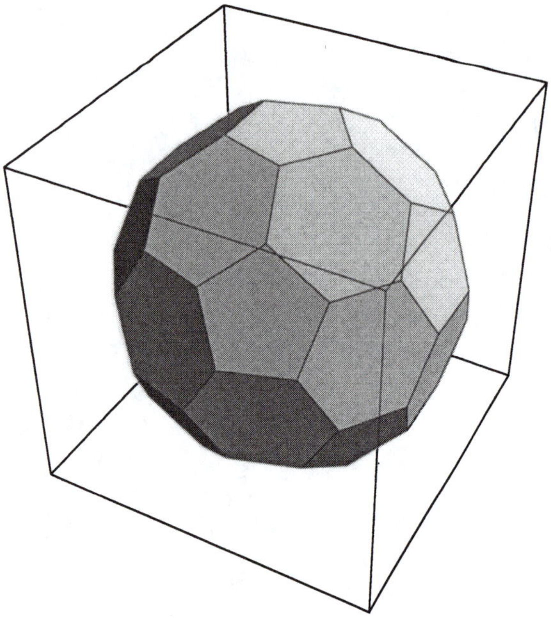

12.D The following matrix transforms a point (x,y,z) through a counterclockwise rotation about the z axis through an angle of θ

```
Cos[θ]     -Sin[θ]     0
Sin[θ]     Cos[θ]      0
0          0           1
```

Apply this matrix to a point at (1,1,1) for the angles 0, $\pi/4$, $\pi/2$, $3\pi/4$, π, $3\pi/2$, and 2π.

SOLUTION

```
trmatrix = {{Cos[θ], -Sin[θ], 0}, {Sin[θ], Cos[θ], 0}, {0, 0, 1}}

{{Cos[θ], -Sin[θ], 0}, {Sin[θ], Cos[θ], 0}, {0, 0, 1}}

TableForm[trmatrix]

Cos[θ]     -Sin[θ]     0
Sin[θ]     Cos[θ]      0
0          0           1

trmatrix /. θ → 0

{{1, 0, 0}, {0, 1, 0}, {0, 0, 1}}

trmatrix.{1, 1, 1} /. θ → 0

{1, 1, 1}

trmatrix.{1, 1, 1} /. θ → π / 4

{0, √2, 1}
```

```
trmatrix.{1, 1, 1} /. θ → π / 2
```

```
{-1, 1, 1}
```

```
trmatrix.{1, 1, 1} /. θ → 3 π / 4
```

```
{-√2 , 0, 1}
```

```
trmatrix.{1, 1, 1} /. θ → π
```

```
{-1, -1, 1}
```

```
trmatrix.{1, 1, 1} /. θ → 3 π / 2
```

```
{1, -1, 1}
```

```
trmatrix.{1, 1, 1} /. θ → 2 π
```

```
{1, 1, 1}
```

Looking down the z-axis

```
ListPlot[{{1, 1}, {0, Sqrt[2]}, {-1, 1}, {-Sqrt[2], 0}, {-1, -1}, {1, -1}},
  AxesLabel -> {"x", "y"}, Prolog → AbsolutePointSize[4], AspectRatio -> 1];
```

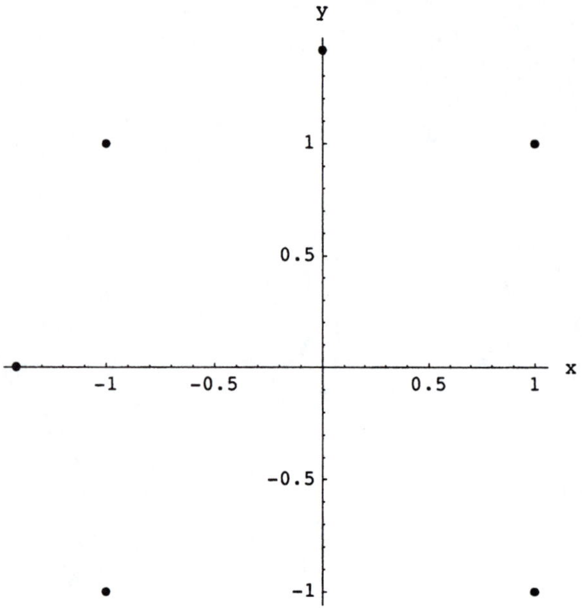

Chapter 13 Rotational and Vibrational Spectroscopy

13.A Plot the Morse potentials for molecular hydrogen in the ground state and the first excited state (see Table 13.4 for parameters), and put the plots on the same graph so that they can be compared with Fig. 11.12.

SOLUTION

The following values of physical constants are needed:

```
ClearAll["Global`*"]
Off[General::spell, General::spell1]

hbar = 1.054572 * 10^-34 (*J s*);

e = 1.6021177 * 10^-19 (*coulomb*);

c = 2.99792458 * 10^8 (*m s^-1*);
```

The following parameters for molecular hydrogen in its ground state are needed:

```
weH2 = 440121 (*harmonic vibration wave number in m^-1*);

wexeH2 = 12134 (*anharmonicity constant in m^-1*);

muH2 = ((1.007825^2) * 1.660540 * 10^-27) /
        (2 * 1.007825) (*kg*)
```

8.36767×10^{-28}

```
reH2 = 74.144 * 10^-3 (*nm*);
```

The equilibrium internuclear distance is given in Table 13.4. The equilibrium dissociation energy (in m^{-1}) for molecular hydrogen in the ground state (1 sigma g) is given by

```
deH2 = 4.47797 / (1.2398425 * 10^-6) + (weH2 / 2) - (wexeH2 / 4)
```

3.82875×10^{6}

where the 4.47797 comes from Table 13.3. For the excited state (ES),

```
weH2ES = 135809 (*harmonic vibration wave number in m^-1*);

wexeH2ES = 2088.8 (*anharmonicity constant in m^-1*);

reH2ES = 129.28 * 10^-3 (*nm*);
```

See Table 13,4. The calculations for molecular hydrogen follow the example in the text.

```
aH2 = weH2 * ((Pi * muH2 * c) / (hbar * deH2)) ^ .5
```

1.94443×10^{10}

```
morsepotentialH2[r_] := (2 * π * hbar * c / e) * deH2 (1 - Exp[-1 * aH2 * 10^-9 * (r - reH2)]) ^ 2;
```

```
mpH2 = Plot[morsepotentialH2[r], {r, .03, .5},
   AxesOrigin -> {0, 0}, AxesLabel -> {"R" / "nm", "V" / "eV"},
    PlotRange -> {{0, 0.5}, {0, 20}}];
```

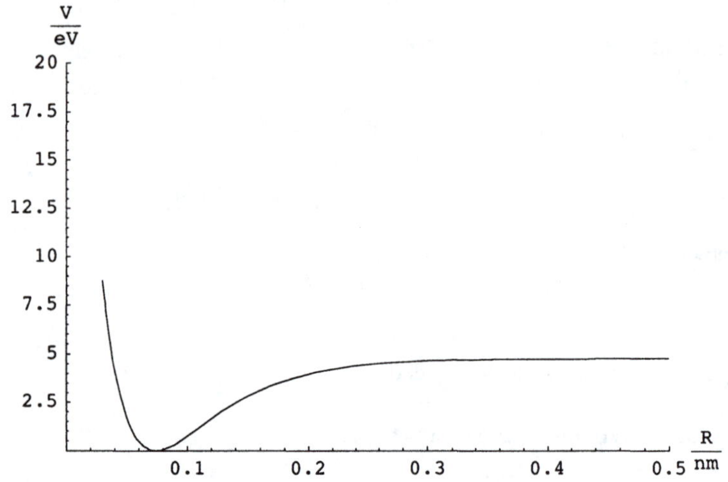

Since the equilibrium dissociation energy of this excited molecule is not known, we can estimate it using the equations in the textbook. The calculations for molecular hydrogen in its first excited state (1 sigma u) are

```
xeH2ES = wexeH2ES / weH2ES (*dimensionless*)
```

0.0153804

```
deH2ES = weH2ES / (4 * xeH2ES) (*m^-1*)
```

2.2075×10^{6}

```
aH2ES = weH2ES * ((Pi * muH2 * c) / (hbar * deH2ES)) ^ .5
```

7.90183×10^{9}

For the excited state, we have to add (91700 cm^{-1})($1.2398x10^{-4}$ eV cm) (see Table 13.4 for the 91 700 and the back cover for conversion factors) to obtain the energy of the excited state relative to the ground state. Since the dissociation energy is given in m^{-1} we have multiplied it by 2*π*hbar*c/e=$1.23989x10^{-6}$ to convert it to eV.

```
morsepotentialH2ES[r_] := 91700 * 1.2398 * 10^-4
+ (2 * π * hbar * c / e) * deH2ES (1 - Exp[-1 * aH2ES * 10^-9 * (r - reH2ES)]) ^ 2;
```

```
mpH2ES = Plot[morsepotentialH2ES[r], {r, .03, .5},
   AxesOrigin -> {0, 0}, AxesLabel -> {"R" / "nm", "V" / "eV"},
    PlotRange -> {{0, 0.5}, {0, 20}}];
```

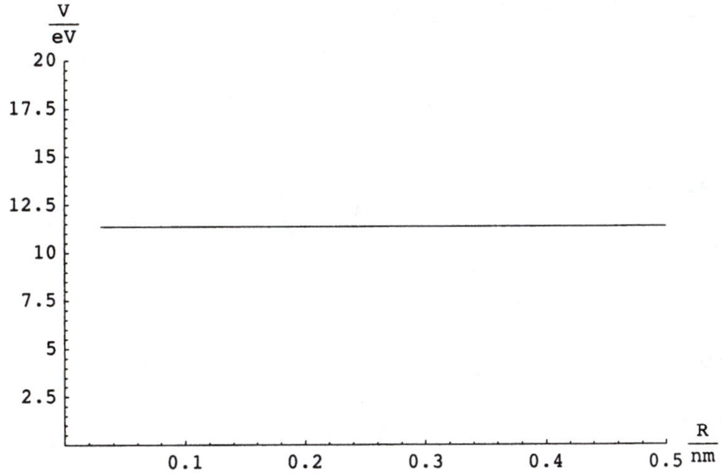


```
Show[mpH2, mpH2ES];
```

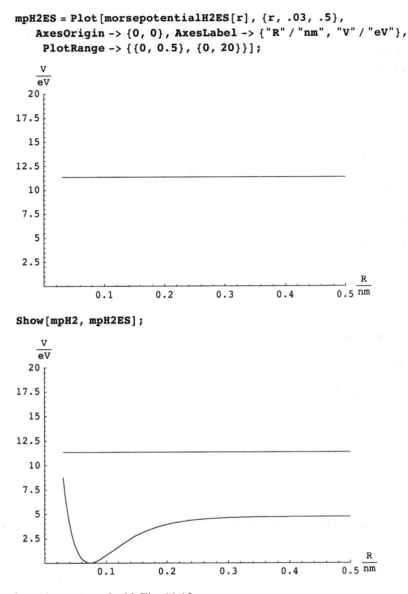

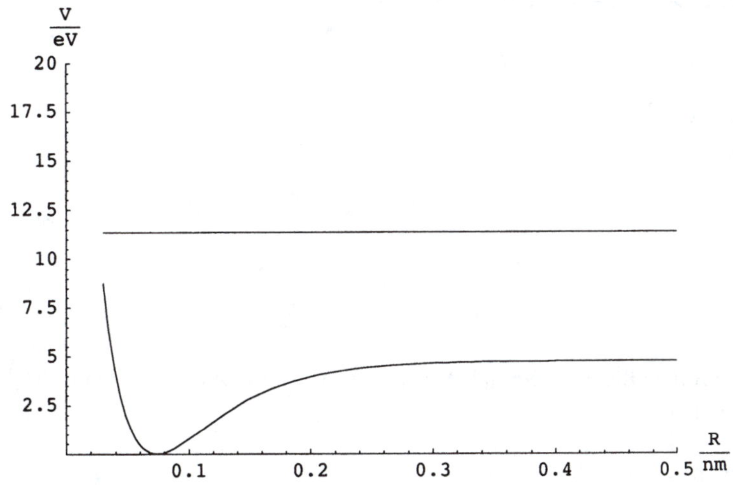

This graph can be compared with Fig. 11.12.

13.B Plot the Morse potential for molecular chlorine in the ground state (see Table 13.4 for parameters) and put in lines for every fifth vibrational level from $v = 0$ to $v = 40$.

SOLUTION

```
ClearAll["Global`*"]
Off[General::spell, General::spell1]
```

The following values of physical constants are needed:

```
hbar = 1.054572 * 10^-34(*J s*);

e = 1.6021177 * 10^-19(*coulomb*);

c = 2.99792458 * 10^8(*m s^-1*);
```

The following parameters for molecular chlorine in its ground state are needed:

```
weCl2 = 55970(*harmonic vibration wave number in m^-1*);

wexeCl2 = 267(*anharmonicity constant in m^-1*);

muCl2 = ((34.968852^2) * 1.660540 * 10^-27) /
        (2 * 34.968852) (*kg*)
```

2.90336×10^{-26}

```
reCl2 = 198.8 * 10^-3(*nm*);

d0Cl2 = 2.47937(*eV*);
```

The calculations for molecular chlorine follow the example in the text. The equilibrium dissociation energy deCl2 is needed, rather than the spectroscopic dissociation energy d0Cl2.

```
deCl2 = d0Cl2 / (1.239892 * 10^-6) + weCl2 / 2 - wexeCl2 / 4
```

2.02758×10^{6}

where the conversion factor converts energies in m^{-1} to eV.

```
aCl2 = weCl2 * ((Pi * muCl2 * c) / (hbar * deCl2))^.5
```

2.00154×10^{10}

```
morsepotentialCl2[r_] := (1.239892 * 10^-6) * deCl2 *
        (1 - Exp[-1 * aCl2 * 10^-9 * (r - reCl2)])^2;
```

```
mpCl2 = Plot[morsepotentialCl2[r],
   {r, 0, .5}, AxesOrigin -> {0, 0}, AxesLabel -> {"R/nm", "V/eV"},
    PlotRange -> {{0, 0.5}, {0, 3}}];
```

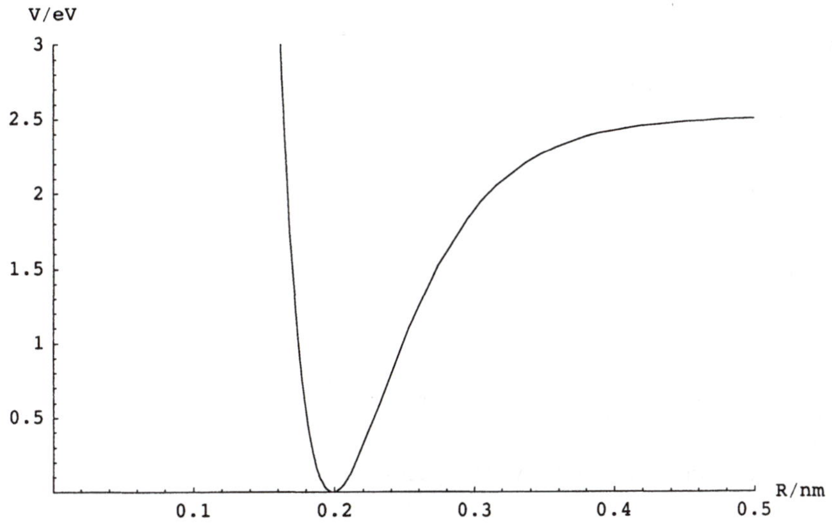

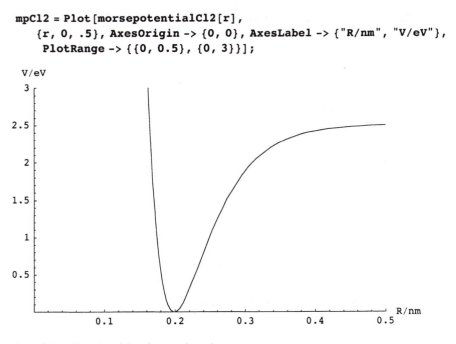

The energies of the vibrational levels are given by

```
evCl2[v_] := weCl2 * (v + 1 / 2) - wexeCl2 * (v + 1 / 2) ^2

evCl2[{0., 5., 10., 15., 20.}]
```

{27918.3, 299758., 558248., 803388., 1.03518×10^6}

In order to draw in horizonal lines for the vibrational levels, it is necessary to calculate the turning points for the classical vibrational motion. Note that there are two values of the radius r that correspond to a certain energy less than deCl2. For the Morse function, these values are readily calculated because the Morse function can be solved for the r.
r = re - (1/a) ln [1 - (morsepotential/deCl2)^.5]
The ratio morsepotential/dCl2 is a positive number, but the square root can be either positive or negative. In *Mathematica* a line is created by Line[{{x1,y1},{x2,y2}}], where are the coordinates are for the two ends of the line . Thus the equations for the horizontal lines for the vibrational levels are given by

```
levels[v_] :=
  Line[{{reCl2 - (10^9 / aCl2) Log[1 - (evCl2[v] / deCl2) ^ .5], evCl2[v] * 1.239892 * 10^-6},
    {reCl2 - (10^9 / aCl2) Log[1 + (evCl2[v] / deCl2) ^ .5], evCl2[v] * 1.239892 * 10^-6}}];
```

Put in every fifth level from v = 0 to 40.

```
lines = Table[Graphics[levels[v]], {v, 0, 40, 5}];
```

```
Show[mpCl2, lines, AxesOrigin -> {0, -.5}, AxesLabel -> {"R/nm", "V/eV"},
    PlotRange -> {{0, 0.5}, {-.5, 3}}];
```

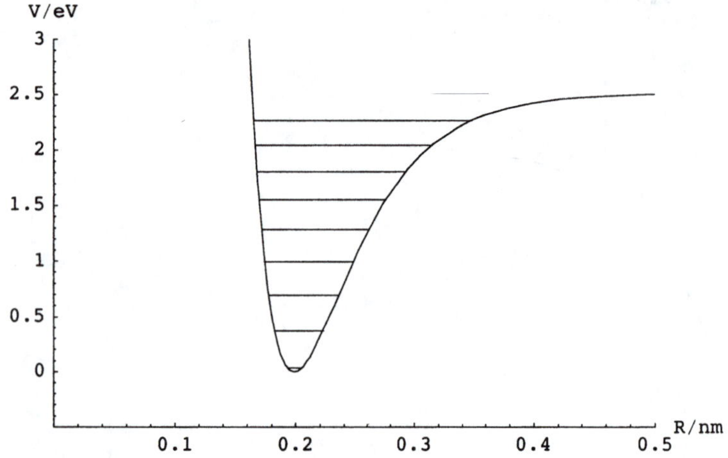

Try a higher upper limit for the vibrational quantum number to see what happens.

13. C (*a*) Calculate the energies (in cm^{-1} of the rotational levels of $H^{35}Cl$ for $J = 0$ to $J = 15$, taking into account the centrifugal distortion ($D = 4.4 \times 10^{-4}\ cm^{-1}$). (*b*) Calculate the frequencies of absorption due to $J \to J + 1$ transitions in cm^{-1} (*c*) Calculate the fractions *f* of the molecules in each of these 15 levels using equation 13.43 at 300 K and plot *f* versus frequencies of absorption in wave numbers (cm^{-1}). (*d*) Calculate the fractions *f* of the molecules in each of these 15 levels using equation 13.43 at 500 K and plot these fractions on the same graph.

SOLUTION

```
ClearAll["Global`*"]
```

(a) The energies of the rotational levels for $H^{35}Cl$ are given in cm^{-1} by
energy = $BJ(J+1) - DJ^2(J+1)^2$
where $B = 10.5934\ cm^{-1}$, and the centrifugal distortion constant $D = 4.4 \times 10^{-4}\ cm^{-1}$
where we is the fundamental vibration frequency.

```
energy[j_] := 10.5934 * j * (j + 1) - (4.4 * 10^-4) * j^2 * (j + 1)^2;
ej = Table[energy[j], {j, 0, 15, 1}]

{0, 21.185, 63.5446, 127.057, 211.692, 317.406, 444.147, 591.851,
 760.444, 949.842, 1159.95, 1390.66, 1641.86, 1913.42, 2205.21, 2517.07}
```

(b) The frequencies of the lines in wave numbers (cm^{-1}) are given by
nubar = $2B(J + 1) - 4D(J + 1)^3$

```
waveno[j_] := 2 * 10.5934 * (j + 1) - 4 * (4.4 * 10^-4) * (j + 1)^3;
nubar = Table[waveno[j], {j, 0, 15, 1}]

{21.185, 42.3595, 63.5129, 84.6346, 105.714, 126.741, 147.704, 168.593,
 189.398, 210.108, 230.712, 251.2, 271.562, 291.786, 311.862, 331.78}
```

(c) To calculate the fractions of the molecules in these energy levels, it is better to use SI units. The energies of the initial states are given by h*c*ej, where c is in cm s^{-1}. The relative populations of the ground states are given by
pop=(2*J+1)*Exp(-h*c*ej/kT)
These populations can be normalized by dividing by the sum of the populations of all levels. As an approximation we will use the sum of the first 15 levels here. The constants involved in the calculation are

```
h = 6.6260755 * 10^-34 (*J s*);

c = 2.99792458 * 10^10 (*cm/s*);

k = 1.380658 * 10^-23 (*J/K*);

pop[j_] := (2 * j + 1) * Exp[-1 * (h * c * (10.5934 * j * (j + 1)
                 - (5.31555 * 10^-4) * j^2 * (j + 1)^2)) / (k * 300)]

Table[pop[j], {j, 0, 15, 1}]

{1, 2.71017, 3.68659, 3.80615, 3.26137, 2.40138, 1.54596, 0.878951, 0.444183,
 0.200419, 0.0810115, 0.0294128, 0.00961299, 0.00283362, 0.000754619, 0.000181849}
```

```
fraction = Table[pop[j], {j, 0, 15, 1}] / (Apply[Plus, Table[pop[j], {j, 0, 15, 1}]])

{0.049853, 0.13511, 0.183787, 0.189748, 0.162589,
 0.119716, 0.0770706, 0.0438183, 0.0221438, 0.00999146, 0.00403867,
 0.00146631, 0.000479236, 0.000141264, 0.00003762, 9.0657×10^-6}

plot1 = ListPlot[Transpose[{nubar, fraction}],
   AxesLabel -> {"cm^-1", "f"}, AxesOrigin -> {0, 0}, Prolog -> AbsolutePointSize[3]];
```

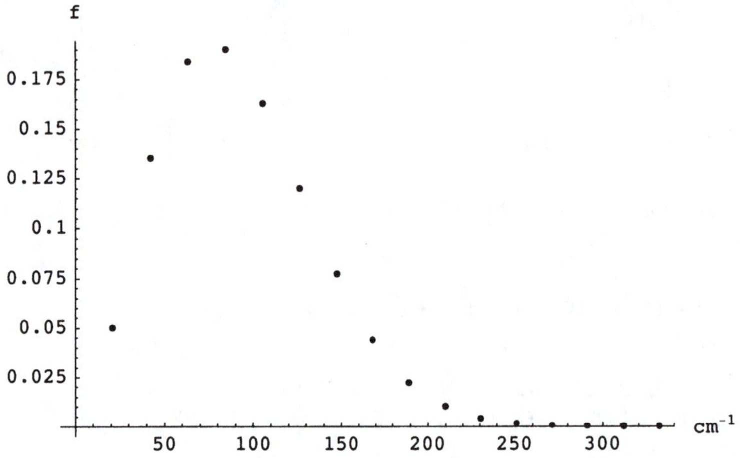

(d) The following calculations produce the corresponding plot at 500 K.

```
pop500[j_] := (2*j + 1) *Exp[-1* (h*c* (10.5934*j* (j + 1)
                        - (5.31555*10^-4) *j^2* (j + 1)^2)) / (k*500)]

Table[pop500[j], {j, 0, 15, 1}]

{1, 2.82258, 4.1645, 4.85659, 4.89482, 4.41402, 3.62323, 2.73411, 1.90861,
 1.23784, 0.748197, 0.422486, 0.223301, 0.11065, 0.051475, 0.0225098}

fraction500 = Table[pop500[j], {j, 0, 15, 1}] / (Apply[Plus, Table[pop500[j], {j, 0, 15, 1}]])

{0.0300888, 0.0849283, 0.125305, 0.146129, 0.14728,
 0.132813, 0.109019, 0.0822661, 0.0574278, 0.0372452, 0.0225124,
 0.0127121, 0.00671888, 0.00332932, 0.00154882, 0.000677293}
```

```
plot2 = ListPlot[Transpose[{nubar, fraction500}],
  AxesLabel -> {"cm-1", "f"}, AxesOrigin -> {0, 0}, Prolog -> AbsolutePointSize[3]];
```

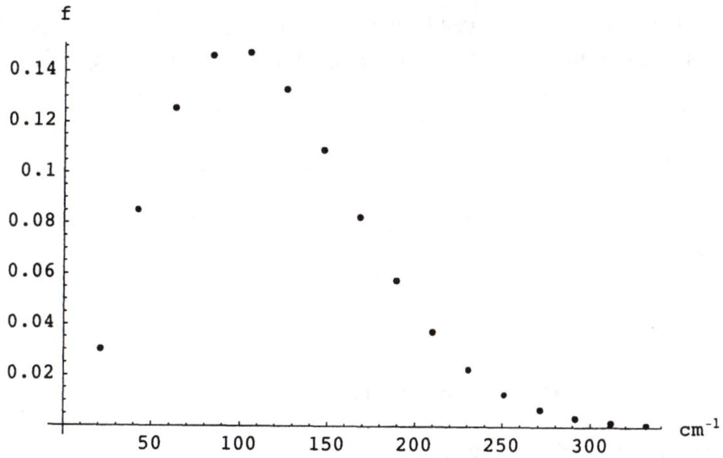

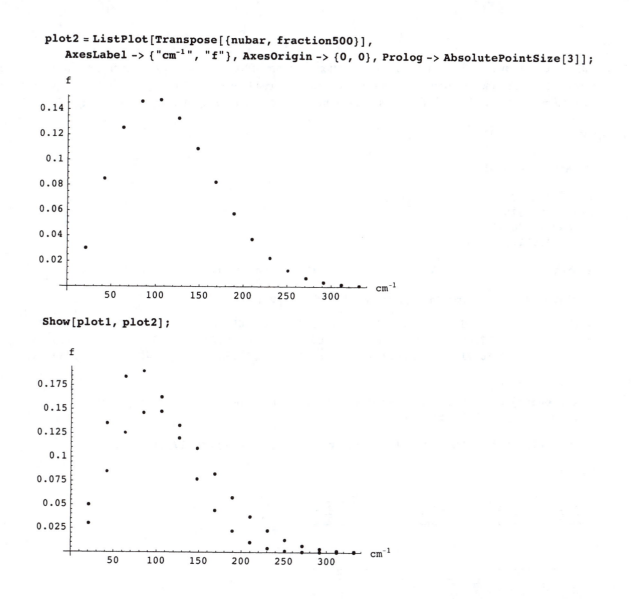

```
Show[plot1, plot2];
```

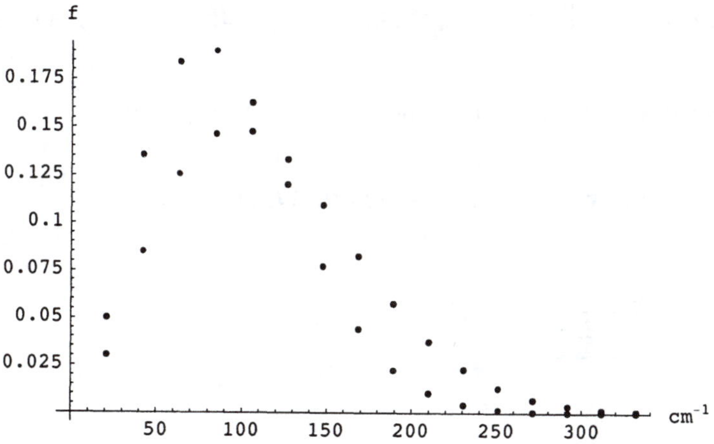

13.D Calculate the frequencies of lines in the vibration-rotation spectrum of HCl for $v = 0$ to $v = 1$, including the dependence of the rotational constant on the vibrational quantum number. Energies can be expressed in cm^{-1}. Plot the line spectra for $v = 0$ to $v = 2$ and $J = 0$ to $J = 10$. Make further plots to show the effect of a larger $\tilde{\omega}_e \tilde{x}_e$, a larger $\tilde{B}_e$, and a larger $\tilde{\alpha}_e$.

SOLUTION

```
ClearAll["Global`*"]

we = 2990.95;
wexe = 52.819;
alphae = .30718;
be = 10.5934;

energy[j_, v_] := we (v + 1 / 2) - wexe (v + 1 / 2) ^2 + (be - alphae (v + 1 / 2)) j (j + 1)

TableForm[Table[energy[j, v], {v, 0, 2}, {j, 0, 10}]]
```

1482.27	1503.15	1544.91	1607.55	1691.07	1795.46	1920.74
4367.58	4387.85	4428.38	4489.17	4570.23	4671.56	4793.15
7147.26	7166.91	7206.21	7265.16	7343.77	7442.02	7559.93

```
levels[j_, v_] :=
 Line[energy = we (v + 1 / 2) - wexe (v + 1 / 2) ^2 + (be - alphae (v + 1 / 2)) j (j + 1);
  {{30, energy}, {70, energy}}]

plotlevels = Table[Graphics[{levels[j, v]}], {v, 0, 2}, {j, 0, 10}];

Show[plotlevels];
```

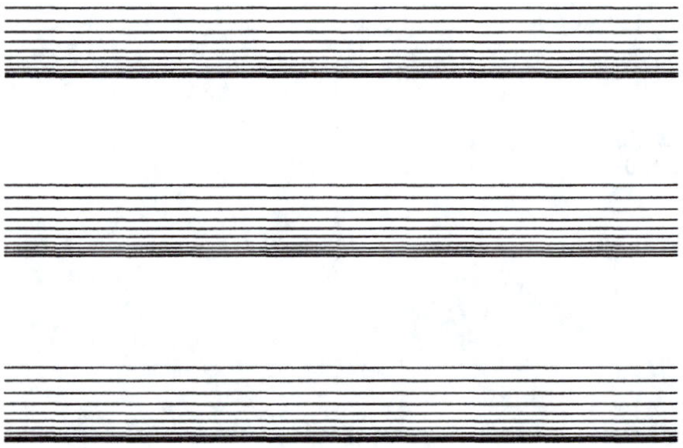

Try the effect of a larger wexe. a larger be, or a larger alphae.

```
wexe = 200;
```

```
Show[Table[Graphics[{levels[j, v]}], {v, 0, 2}, {j, 0, 10}]];
```

```
wexe = 52.819;
be = 20;
```

```
Show[Table[Graphics[{levels[j, v]}], {v, 0, 2}, {j, 0, 10}]];
```

```
alphae = 2;
be = 10.5934;
```

```
Show[Table[Graphics[{levels[j, v]}], {v, 0, 2}, {j, 0, 10}]];
```

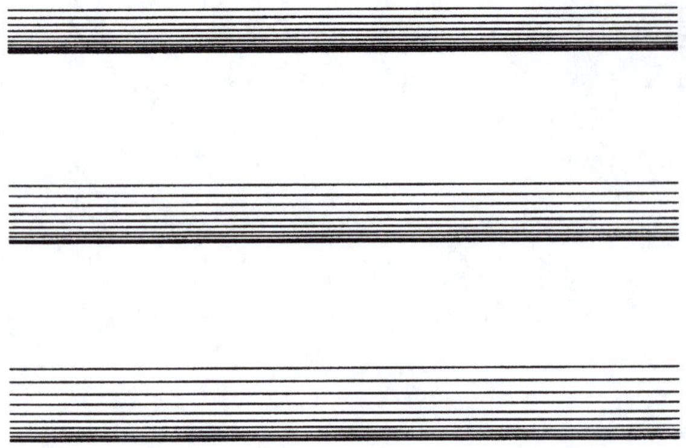

13.E (*a*) Given the spectroscopic constants for $H^{35}Cl$ in Table 13.4 calculate the energy levels (in cm^{-1}) for $J = 0, 1, 2, 3,$ and 4 in the vibrational states $v = 0$ and $v = 1$. (*b*) Calculate the frequencies (in cm^{-1}) of the first two lines in the R branch of $v = 0 \rightarrow 1$ and the first two lines of the P branch of the absorption spectrum. Compare your calculations with Fig. 13.12.

SOLUTION

```
ClearAll["Global`*"]

we = 2990.95;
wexe = 52.819;
alphae = .30718;
be = 10.5934;
```

(a)

```
energy[v_, j_] := we (v + 1 / 2) - wexe (v + 1 / 2) ^ 2 + (be - alphae (v + 1 / 2)) j (j + 1)

tab = Table[energy[v, j], {v, 0, 1}, {j, 0, 4}]

{{1482.27, 1503.15, 1544.91, 1607.55, 1691.07},
 {4367.58, 4387.85, 4428.38, 4489.17, 4570.23}}

TableForm[Join[{Table[x, {x, 0, 4}]}, tab]]
```

0	1	2	3	4
1482.27	1503.15	1544.91	1607.55	1691.07
4367.58	4387.85	4428.38	4489.17	4570.23

This table gives the energies of the H35Cl levels for J = 0, 1 , 2, 3, and 4 in cm^-1. The second line gives the energies for v = 0 and the third line gives the energies for v = 1.

(b) In the R branch of the v = 0 -> 1 spectrum, J increases by 1. Therefore, the frequencies of the first two lines in the R branch are given by

```
4387.85 - 1482.27

2905.58
```

```
4428.38 - 1503.15
```

```
2925.23
```

In the P branch of the v = 0 -> 1 spectrum, J decreases by 1. Therefore, the frequencies of the first two lines in the P branch are given by

```
4367.58 - 1503.15
```

```
2864.43
```

```
4387.85 - 1544.91
```

```
2842.94
```

13.F The observed vibrational frequencies for an anharmonic oscillator are given by
$\tilde{\nu}_{obs} = \tilde{G}(v) - \tilde{G}(0) = \tilde{\nu}_e v - \tilde{\nu}_e x_e\ v(v+1)$
Calculate the fundamental vibration frequency $\tilde{\nu}_e$ and the anharmonicity constant $\tilde{\nu}_e x_e$ for $H^{35}Cl$ for which the frequencies for the 0 -> v transitions are 2885.9, 5668.0, 8347.0, 10 923.1, and 13 396.5 cm^{-1} for v = 1, 2, 3, 4, and 5.

SOLUTION
The parameters can be obtained from a plot of $\tilde{\nu}_{obs}$ /v versus (v + 1).

```
ListPlot[{{2, 2885.9}, {3, 5668 / 2}, {4, 8347 / 3}, {5, 10923.1 / 4}, {6, 13396.5 / 5}},
  Prolog -> AbsolutePointSize[3], PlotRange → {2600, 3000},
  AxesLabel → {"v + 1", "ν̃obs/v"}, AxesOrigin → {2, 2600}];
```

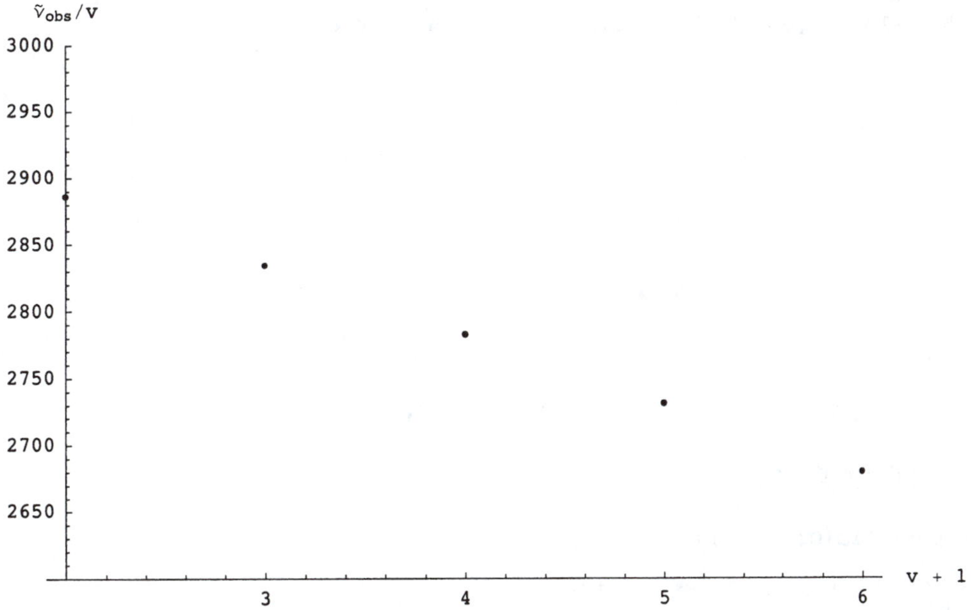


```
Fit[{{2, 2885.9}, {3, 5668 / 2}, {4, 8347 / 3}, {5, 10923.1 / 4}, {6, 13396.5 / 5}}, {1, x}, x]
```

```
2989.03 - 51.6425 x
```

The value for $\tilde{\nu}_e$ is 2990.95 cm^{-1} and the value for $\tilde{\nu}_e x_e$ is 52.819 cm^{-1} in Table 13.4.

```
nucalc = Table[(2989.03 * v - 51.6425 * v * (v + 1)), {v, 5}]

{2885.75, 5668.21, 8347.38, 10923.3, 13395.9}
```

These values are calculated from the empirical equation.

```
nuexpt = {2885.9, 5668.0, 8347.0, 10923.1, 13396.5}

{2885.9, 5668., 8347., 10923.1, 13396.5}

nucalc - nuexpt

{-0.155, 0.205, 0.38, 0.17, -0.625}
```

13.G Plot the Morse potential for the chlorine molecule and the parabolic curve, $V = k(R - R_e)^2/2$, with a force constant k that corresponds with that for the Morse potential at the minimum of the potential energy plot.

SOLUTION
The parameters for the Morse potential for Cl_2 are given in Computer Problem 13.B.

```
deCl2 = 2.02758 * 10^6;
aCl2 = 2.00154 * 10^10;
reCl2 = .1988;

morsepotentialCl2[r_] := (1.239892 * 10^-6) * deCl2 *
        (1 - Exp[-1 * aCl2 * 10^-9 * (r - reCl2)])^2;

plot1 = Plot[morsepotentialCl2[r],
   {r, .1, .5}, PlotRange → {0, 3}, AxesLabel → {"R/nm", "V/eV"}];
```

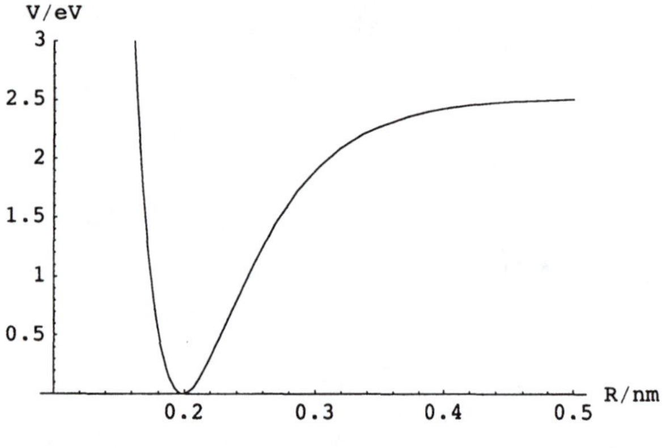

First we examine the slope of this plot:

```
dvt = D[morsepotentialCl2[r], r]
```

$100.637\, e^{-20.0154\,(-0.1988+r)}\,(1 - e^{-20.0154\,(-0.1988+r)})$

```
Plot[dvt, {r, .1, .5}, PlotRange → {-100, 100}, AxesLabel → {"R/nm", "dvt"}];
```

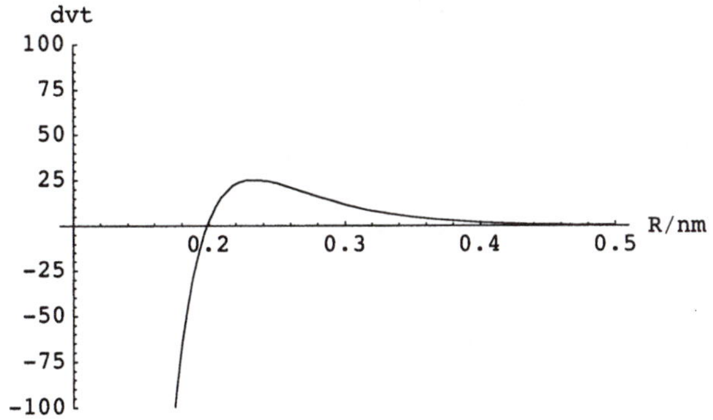

Note that the slope is zero at $R_e = 0.1988$ nm. Second we examine the slope of this plot in the vicinity of R_e.

```
seconddvt = D[morsepotentialCl2[r], {r, 2}]
```

$$2.51398\,(801.232\,e^{-40.0308\,(-0.1988+r)} - 801.232\,e^{-20.0154\,(-0.1988+r)}\,(1 - e^{-20.0154\,(-0.1988+r)}))$$

```
Plot[seconddvt, {r, 0.15, .25}, PlotRange → {0, 3000},
  AxesOrigin → {.1988, 0}, AxesLabel → {"R/nm", "seconddvt"}];
```

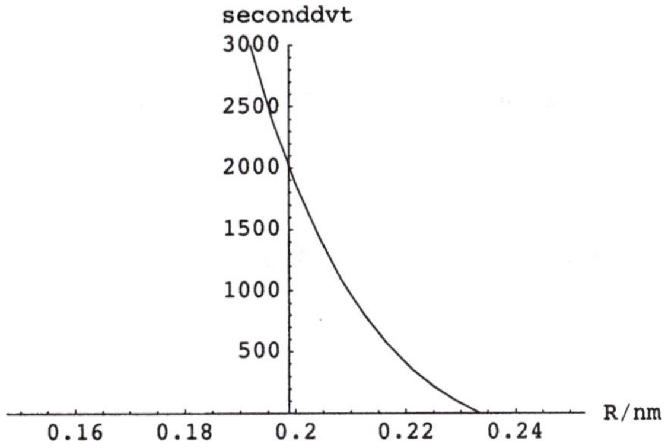

Thus k = 2000 at the minimum.

```
plot2 = Plot[2000 * (r - reCl2) ^2 / 2, {r, .1, .5}, PlotRange → {0, 3}];
```

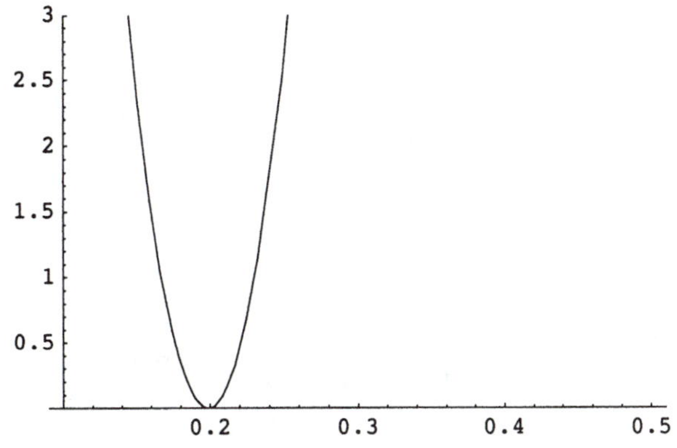

```
Show[plot1, plot2];
```

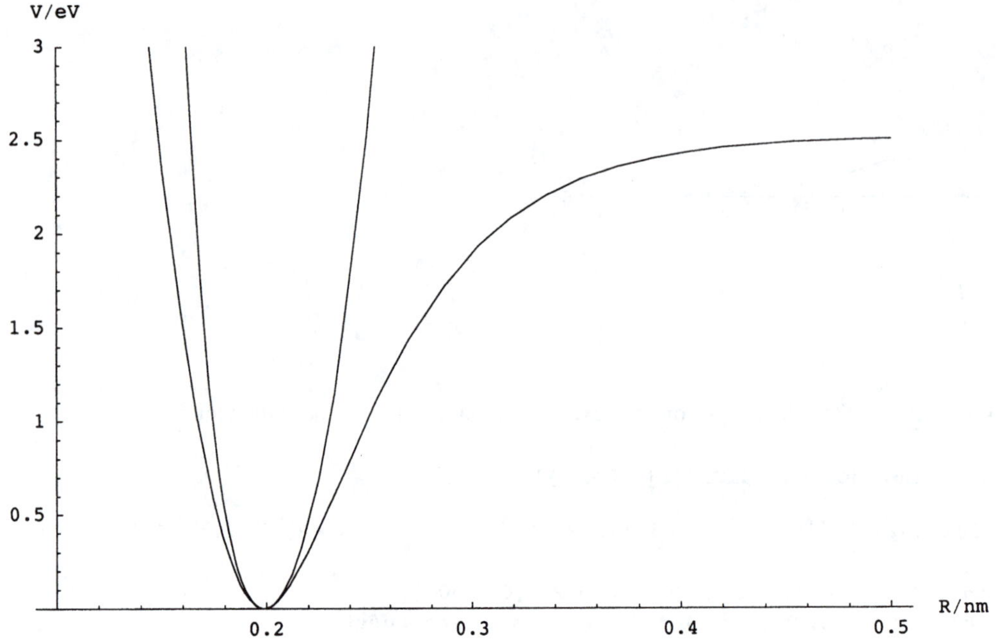

13.H Plot the Morse potential for $H^{35}Cl$ using the parameters calculated in Example 13.7 and the fact that $R_e = 0.1275$ nm.

SOLUTION

```
pot = ((3.725 * 10^4) * (1.2398 * 10^-4)) * (1 - Exp[-18.68 (r - .1275)])^2
```

$$4.61826\,(1 - e^{-18.68\,(-0.1275+r)})^2$$

```
Plot[pot, {r, 0, .5}, AxesLabel → {"R/nm", "V/eV"}, PlotRange → {0, 6}];
```

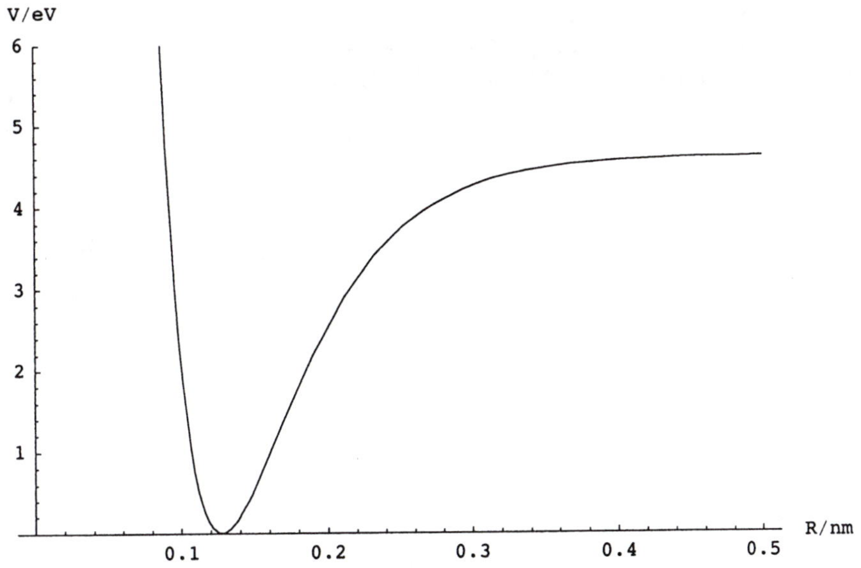

13.I Calculate the vibrational frequencies in wavenumbers for the fundamental absorption band of $H^{35}Cl$ and the first four overtones for (*a*) the harmonic oscillator approximation and (*b*) the anharmonic oscillator. The spectroscopic constants are given in Table 13.4.

SOLUTION

(a) For the harmonic oscillator approximation the frequency in wave numbers is given by
nu = we v v = 1,...,5

```
nuHO = 2990.95 v

2990.95 v

spectHO = Table[nuHO, {v, 1, 5}]

{2990.95, 5981.9, 8972.85, 11963.8, 14954.8}
```

(b) For the anharmonic oscillator

```
nuAO = 2990.95 v - 52.819 v (v + 1)

2990.95 v - 52.819 v (1 + v)

spectAO = Table[nuAO, {v, 1, 5}]

{2885.31, 5664.99, 8339.02, 10907.4, 13370.2}

nos = Table[v, {v, 1, 5}]

{1, 2, 3, 4, 5}
```

```
TableForm[{nos, spectHO, spectAO}]

1            2            3            4            5
2990.95      5981.9       8972.85      11963.8      14954.8
2885.31      5664.99      8339.02      10907.4      13370.2
```

Chapter 14 Electronic Spectroscopy of Molecules

14.A Rhodopsin (see Section 19.11) has a system of four conjugated double bonds, and so we can think of it as having the structure $H(CH = CH)_4 H$. Since each of the eight carbon atoms contributes one electron to the π bond that extends the length of the molecule, the $n = 1$, 2, 3, and 4 levels in the free electron model are filled, and absorption of a photon excites an electron from the $n = 4$ level to the $n = 5$ level. Assuming that the molecule is linear and that a C=C bond contributes 135 pm, a C-C bond contributes 154 pm, and a CH bond at each end contributes 77 pm, calculate the wavelength of maximum absorption.

SOLUTION

```
ClearAll["Global`*"]

h = 6.626 10^-34; c = 3 10^8; me = 9.11 10^-31;

Solve[
 h c / wavelength == h^2 (5^2 - 4^2) / (8 me ((4 135 + 3 154 + 2 77) (10^-12)) ^2), wavelength]

{{wavelength → 4.8995 × 10^-7}}
```

Thus the wavelength expected from the free electron theory is about 490 nm, in agreement with experiment.

14.B (*a*) Calculate the first several lines in the electronic absorption spectrum of $^{12}C^{16}O$, for which $\tilde{T}_e = 6.508\,043 \times 10^4$ cm^{-1}, $\tilde{\nu}_e' = 1514.10$ cm^{-1}, $\tilde{\nu}_e'' = 2169.81$ cm^{-1}, $\tilde{\nu}_e' x_e' = 17.40$ cm^{-1}, and $\tilde{\nu}_e'' x_e'' = 13.29$ cm^{-1}. (*b*) Make a plot of $\tilde{\nu}$ versus the vibrational quantum number v' of the excited state that goes to high enough quantum numbers to show that a convergence limit is reached.

SOLUTION

(a) The frequency of the 0->0 vibronic transition is given by

```
nu00 = 6.508043 * 10^4 + (1514.1 / 2 - 17.4 / 4) - (2169.81 / 2 - 13.29 / 4)

64751.5

nuobs = 64751.5 + 1514.1 * v - 17.40 * v * (v + 1)

64751.5 + 1514.1 v - 17.4 v (1 + v)

Table[nuobs, {v, 0, 5}]

{64751.5, 66230.8, 67675.3, 69085., 70459.9, 71800.}
```

The wavelengths in cm are

```
Table[1 / nuobs, {v, 0, 5}]

{0.0000154437, 0.0000150987, 0.0000147764, 0.0000144749, 0.0000141925, 0.0000139276}
```

(b)

```
ListPlot[Table[nuobs, {v, 0, 60}], AxesLabel -> {"v", "ṽ"}];
```

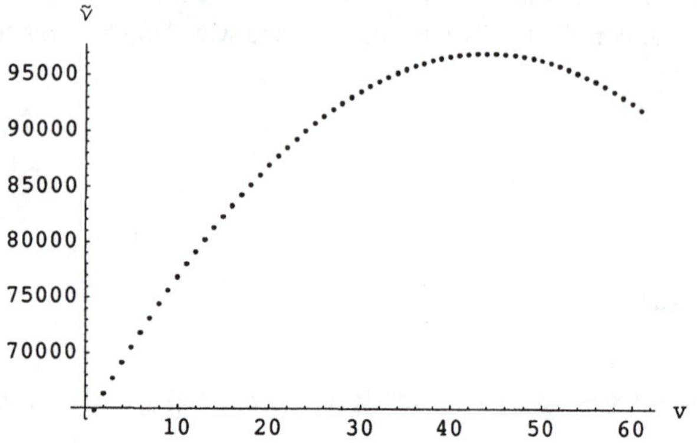

14.C A solution of A and B has an absorbance of 0.800 at λ_1 and 0.500 at λ_2 in a 1-cm cell. At λ_1 the molar absorption coefficient of A is 1.5x10^3 L mol^{-1} cm^{-1} and the molar absorption coefficient of B is 4.0x10^3 L mol^{-1} cm^{-1}. At λ_2 the molar absorption coefficient of A is 3.0x10^3 L mol^{-1} cm^{-1} and the molar absorption coefficient of B is 2.0x10^3 L mol^{-1} cm^{-1}. What is the composition of the solution?

SOLUTION

There are at least two ways to solve simultaneous linear equations in *Mathematica*; Solve is convenient for a small number of equations. LinearSolve is convenient for a large number of equations. The equations are solved using millimolar concentrations

```
Solve[{0.8 == 1.5 * ca + 4 cb, .5 == 3 * ca + 2 cb}, {ca, cb}]

{{ca → 0.0444444, cb → 0.183333}}

m = {{1.5, 4}, {3, 2}}

{{1.5, 4}, {3, 2}}

b = {.8, .5}

{0.8, 0.5}

LinearSolve[m, b]

{0.0444444, 0.183333}
```

Chapter 15 Magnetic Resonance Spectroscopy

15.A Plot the energy for a system containing 1 mol of hydrogen atoms (*a*) aligned with the magnetic field and (*b*) opposed to the magnetic field up to 25 T.

SOLUTION

$E(m_I = +1/2) = -(\text{hbar})\gamma m_I B_z N_A$

```
Clear[b]

eplus = -1 * (1.055 * 10^-34) * (26.7533 * 10^7) * (6.022 * 10^23) * b / 2

-0.008498466654649999*b
```

Since hbar has the units J s, the magnetogyric ratio has the units T^{-1}, the Avogadro constant has units of mol^{-1}, and the magnetic field strength is in T, the energy is given in J mol^{-1}.

```
eminus = 1 * (1.055 * 10^-34) * (26.7533 * 10^7) * (6.022 * 10^23) * b / 2

0.008498466654649999*b
```

```
Plot[{eplus, eminus}, {b, 0, 25}, AxesLabel -> {"B/Hz", "E/J mol^-1"}];
```

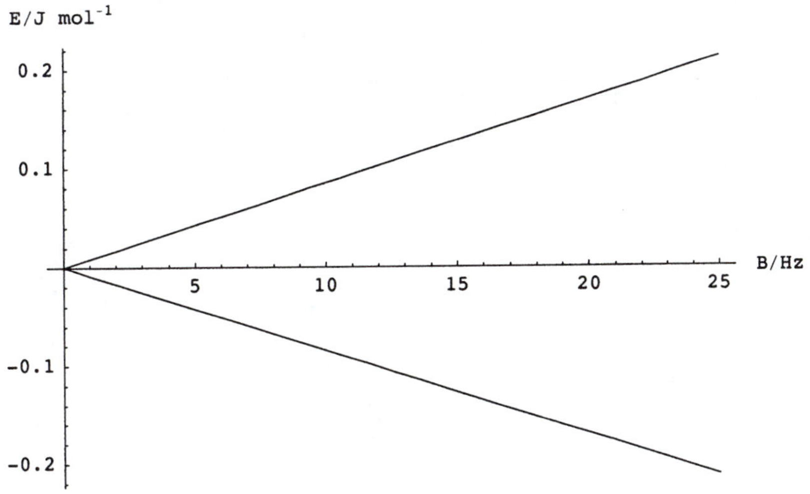

15.B Plot the frequency in MHz versus magnetic field strength up to 25 T for NMR with hydrogen atoms. This is essentially the range available with commercial spectrometers.

SOLUTION

$\nu = \gamma B_z / 2\pi$

```
ClearAll["Global`*"]

nu = (26.7522 * 10^7) * b / ((2 * 10^6) * Pi)

42.57744868583003*b


Plot[nu, {b, 0, 25}, AxesLabel -> {"B/T", "ν/MHz"}];
```

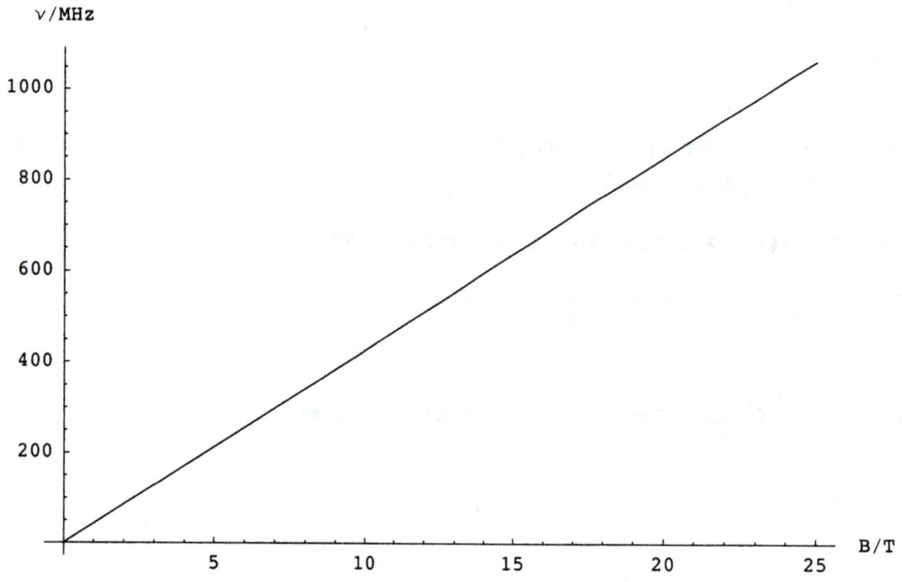

15.C When a system involves a single type of spin, the intensity of magnetization after a 90° pulse is a function of time that we will represent by $f(t)$. This signal, which is referred to as the free induction decay (FID), has the form

$$f(t) = a_1 \cos(2\pi\nu_1 t) \exp(-t/T_2)$$

where a_1 is an amplitude, ν_1 is the Larmor frequency, and T_2 is the transverse relaxation time for the spin. The signal $f(t)$ oscillates and damps down to zero at $t >> T_2$. The Fourier transform of $f(t)$ is a plot of the corresponding spectrum $F(\nu)$ versus frequency. The position of the peak depends on the Larmor frequency ν, and the width of the peak corresponds with the transverse relaxation time T_2. A plot of $f(t)$ versus t is said to be in the time domain, and plot of $F(\nu)$ versus ν is said to be in the frequency domain. Both are useful. A spectrum can be converted to a plot of $f(t)$ versus t with an inverse Fourier transform. When a system involves more than one type of spin, $f(t)$ is made up of a sum of terms of this type, each with its own amplitude, frequency, and transverse relaxation time. As shown in Fig. 15.5, the width of a spectral line at half the peak height is given by $\Delta\nu_{1/2} = 1/\pi T_2$. Perform the following calculations to show the relation between the FID and the spectrum for several simple cases:

(*a*) There is one spin with $\nu_1 = 20\ s^{-1}$ and $T_2 = 0.10$ s.

(*b*) There is one spin with $\nu_1 = 40\ s^{-1}$ and $T_2 = 0.10$ s.

(*c*) There are two spins with $\nu_1 = 40\ s^{-1}$, $\nu_2 = 80\ s^{-1}$, $T_{21} = 0.1$ s, and $T_{22} = 0.1$ s.

SOLUTION

(a)

```
ClearAll["Global`*"]

ft = N[Cos[2 Pi 20 x / 256] * Exp[-x / (256 * .1)]];

datatimedomain = Table[N[{x / 256, ft}], {x, 0, 256}];

plot1=ListPlot[ datatimedomain, PlotJoined ->
True,PlotRange->{{0,.5},{-1,1}},AxesLabel->{"t/s","f(t)"},DisplayFunction->Identity ];

datafreqdomain = Table[ft, {x, 0, 256}];

plot2=ListPlot[Re[Fourier[datafreqdomain]], PlotJoined -> True,

PlotRange->{{0,64},{0,1}},AxesLabel->{"ν/\!\(s\^\(-1\)\)","F(ν)"},DisplayFunction->Iden
tity ];

Show[GraphicsArray[{plot1,plot2}]];
```

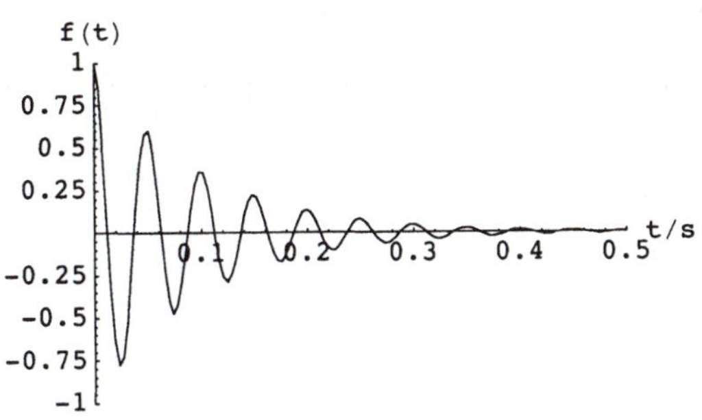

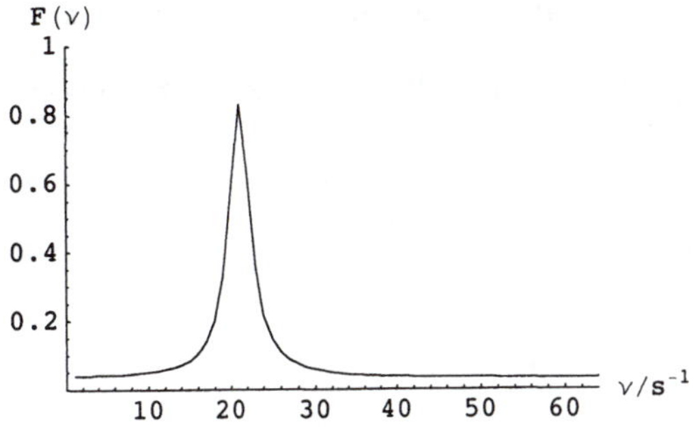

(b)

```
ft = N[Cos[2 Pi 40 x / 256] * Exp[-x / (256 * .1)]];

datatimedomain = Table[N[{x / 256, ft}], {x, 0, 256}];

plot3=ListPlot[ datatimedomain, PlotJoined ->
True,PlotRange->{{0,.5},{-1,1}},AxesLabel->{"t/s","f(t)"},DisplayFunction->Identity ];

datafreqdomain = Table[ft, {x, 0, 256}];

plot4=ListPlot[Re[Fourier[datafreqdomain]], PlotJoined -> True,

PlotRange->{{0,64},{0,1.5}},AxesLabel->{"ν/\!\(s\^\(-1\)\)","F(ν)"},DisplayFunction->Id
entity ];

Show[GraphicsArray[{plot3,plot4}]];
```

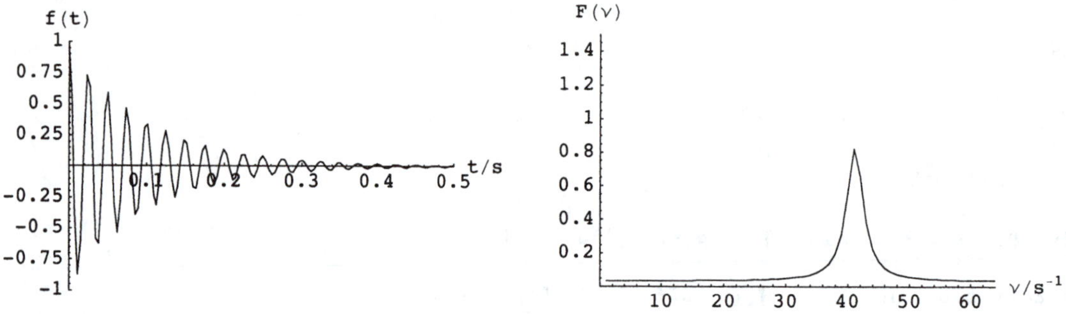

(c)

```
ft = N[(Cos[2 Pi 40 x / 256] + Cos[2 Pi 80 x / 256]) * Exp[-x / (256 * .1)]];

datatimedomain = Table[N[{x / 256, ft}], {x, 0, 256}];

plot5=ListPlot[ datatimedomain, PlotJoined ->
True,PlotRange->{{0,.5},{-1,1}},AxesLabel->{"t/s","f(t)"},DisplayFunction->Identity ];

datafreqdomain = Table[ft, {x, 0, 256}];

plot6=ListPlot[Re[Fourier[datafreqdomain]], PlotJoined -> True,

PlotRange->{{0,128},{0,1}},AxesLabel->{"ν/\!\(s\^\(-1\)\)","F(ν)"},DisplayFunction->Ide
ntity ];
```

```
Show[GraphicsArray[{plot5,plot6}]];
```

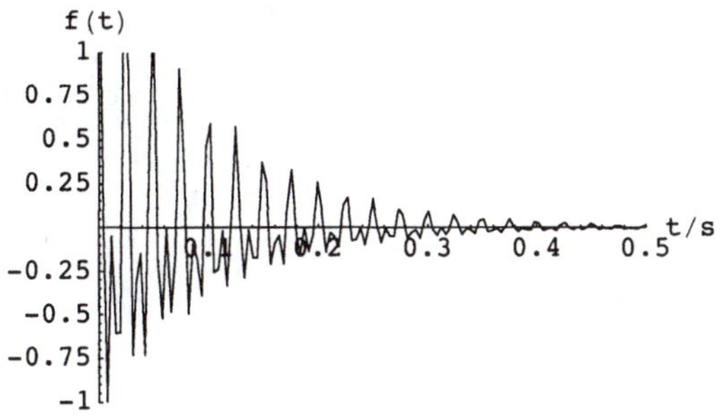

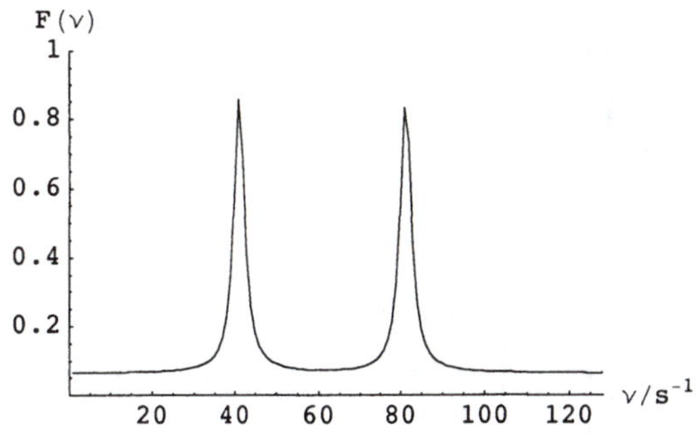

The availability of Fourier in *Mathematica* makes it possible to see the effect on f(t) and F(ν) of changes in Larmor frequencies, transverse relaxation times, and numbers of different molecular species. The built-in object InverseFourier makes it possible to go from the frequency domain to the time domain.

15.D (*a*) Write a program to calculate the free induction decay (FID) for a system involving one spin with $\nu_1 = 20\ s^{-1}$ and $T_2 = 0.10$ s. (*b*) Plot the FID for $t = 0$ to $t = 0.5$ s. (*c*) Make a Fourier transform of this FID to obtain the NMR spectrum and plot the intensity F(ν) from $\nu = 0$ to $\nu = 64\ s^{-1}$.

SOLUTION

```
ClearAll["Global`*"]
```

(a) The following function calculates the function of time (x/256) at 256 data points.

```
ft = N[Cos[2 Pi 20 x / 256] * Exp[-x / (256 * .1)]];

datatimedomain = Table[N[{x / 256, ft}], {x, 0, 256}];
```

(b) Plot the time domain data up to 0.5 s.

```
ListPlot[ datatimedomain, PlotJoined ->
True,PlotRange->{{0,.5},{-1,1}},AxesLabel->{"t/s","f(t)"} ];
```

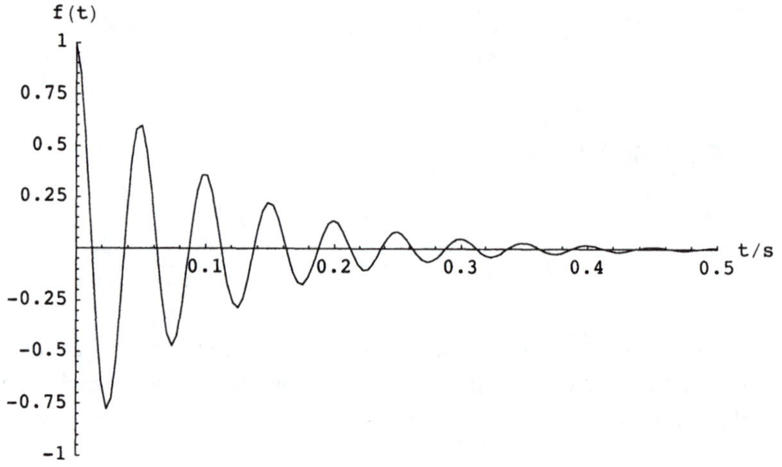

(c) The data required for the Fourier transform is given by

```
datafreqdomain = Table[ft, {x, 0, 256}];

ListPlot[Re[Fourier[datafreqdomain]], PlotJoined -> True,
            PlotRange->{{0,64},{0,1.5}},AxesLabel->{"ν/\!\(s\^-1\)","F(ν)"} ];
```

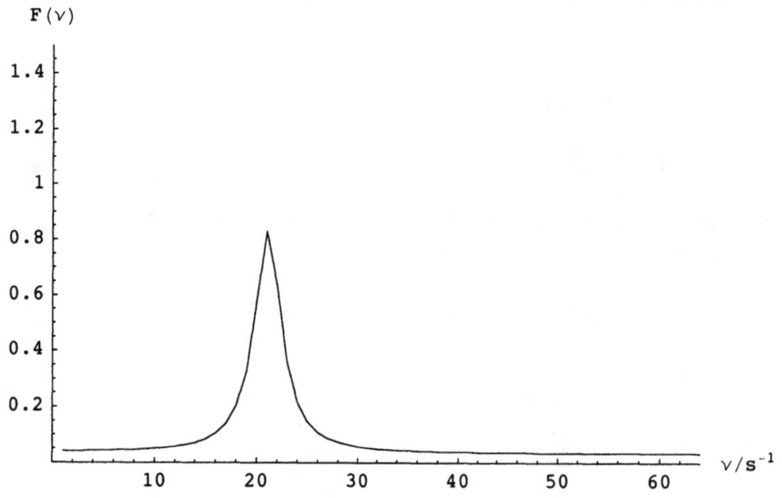

15.E This problem is like Computer Problem 15.C part (*b*) except that the transverse relaxation time is longer. (*a*) Calculate the free induction decay (FID) for a system involving one spin with $\nu_1 = 40\ \mathrm{s}^{-1}$ and $T_2 = 0.30$ s. (*b*) Plot the FID for $t = 0$ to $t = 0.50$ s. (*c*) Make a Fourier transform of this FID to obtain the NMR spectrum and plot the intensity F(ν) from $\nu = 0$ to $\nu = 64\ \mathrm{s}^{-1}$. Is the spectrum broader or narrower? Why?

SOLUTION

```
ClearAll["Global`*"]
```

(a) The following function calculates the function of time (x/256) at 256 data points.

```
ft = N[Cos[2 Pi 40 x / 256] * Exp[-x / (256 * .3)]];

datatimedomain = Table[N[{x / 256, ft}], {x, 0, 256}];
```

(b) Plot the time domain data up to 0.5 s.

```
plot1=ListPlot[ datatimedomain, PlotJoined ->
True,PlotRange->{{0,.5},{-1,1}},AxesLabel->{"t/s","f(t)"},DisplayFunction->Identity ];
```

(c) The data for calculating the spectrum is given by

```
datafreqdomain = Table[ft, {x, 0, 256}];

plot2=ListPlot[Re[Fourier[datafreqdomain]], PlotJoined ->
True,PlotRange->{{0,64},{0,3}},AxesLabel->{"ν/\!\(s\^-1\)","F(ν)"},DisplayFunction->Ide
ntity ];

Show[GraphicsArray[{plot1,plot2}]];
```

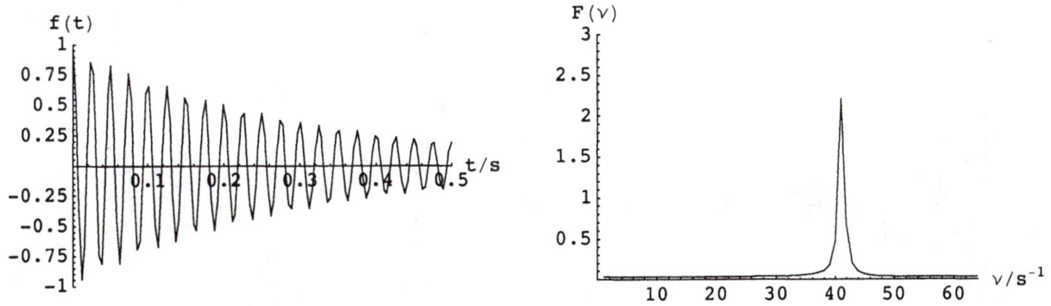

The line is sharper because the FID decays more slowly so that the frequency can be determined more accurately.

15.F (*a*) Write a program to calculate the free induction decay (FID) for a system involving two spins with $a_1 = 0.50$, $a_2 = 1$, $\nu_1 = 40\ s^{-1}$, $\nu_2 = 80\ s^{-1}$, $T_{21} = 0.10$ s, and $T_{22} = 0.10$ s. (*b*) Plot the FID for $t = 0$ to $t = 0.50$ s. (*c*) Make a Fourier transform of this FID to obtain the NMR spectrum and plot the intensity $F(\nu)$ from $\nu = 0$ to $\nu = 128\ s^{-1}$.

SOLUTION

```
ClearAll["Global`*"]
```

(a) The function of time (x/256) now has two terms with different coefficients.

```
ft = N[
   ((.5 Cos[2 Pi 40 x / 256] * Exp[-x / (256 * .1)]) + (Cos[2 Pi 80 x / 256] * Exp[-x / (256 * .1)]))];

datatimedomain = Table[N[{x / 256, ft}], {x, 0, 256}];
```

(b) The FID is plotted using

```
plot1=ListPlot[ datatimedomain, PlotJoined ->
True,PlotRange->{{0,.5},{-1,1}},AxesLabel->{"t/s","f(t)"},DisplayFunction->Identity ];
```

(c) The spectrum is plotted as follows:

```
datafreqdomain = Table[ft, {x, 0, 256}];

plot2=ListPlot[Re[Fourier[datafreqdomain]], PlotJoined -> True,

PlotRange->{{0,128},{0,1}},AxesLabel->{"ν/\!\(s\^-1\)","F(ν)"},DisplayFunction->Identit
y ];

Show[GraphicsArray[{plot1, plot2}]];
```

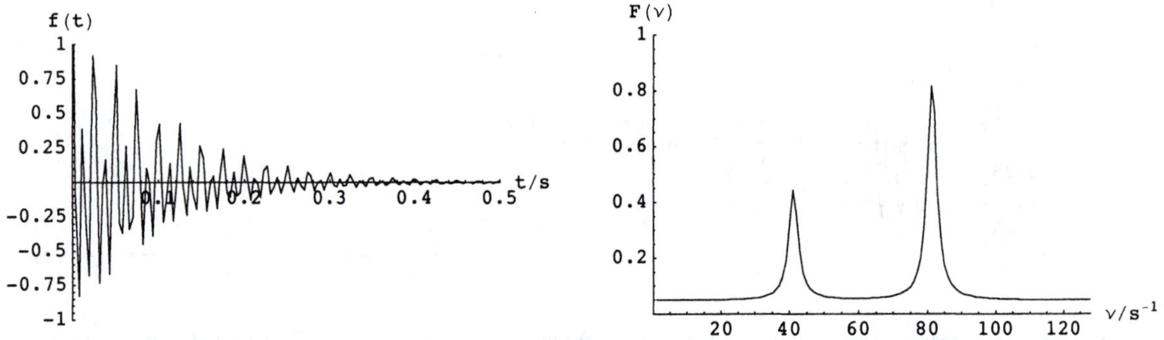

15.G (*a*) Write a program to calculate the free induction decay FID for a system involving two spins with $a_1 = 0.5$, $a_2 = 1$, $\nu_1 = 40\ s^{-1}$, $\nu_2 = 80\ s^{-1}$, $T_{21} = 0.10$ s, and $T_{22} = 0.30$ s. (*b*) Plot the FID for $t = 0$ to $t = 0.50$ s. (*c*) Make a Fourier transform of this FID to obtain the NMR spectrum and plot the intensity $F(\nu)$ from $\nu = 0$ to $\nu = 128\ s^{-1}$.

SOLUTION

```
ClearAll["Global`*"]
```

(a) The function of time (x/256) now has two terms with different coefficients and different values of T_2.

```
ft = N[
   ((.5 Cos[2 Pi 40 x / 256] * Exp[-x / (256 * .1)]) + (Cos[2 Pi 80 x / 256] * Exp[-x / (256 * .3)]))];

datatimedomain = Table[N[{x / 256, ft}], {x, 0, 256}];
```

(b) The FID is plotted as follows:

```
plot1=ListPlot[ datatimedomain, PlotJoined ->
True,PlotRange->{{0,.5},{-1,1}},AxesLabel->{"t/s","f(t)"},DisplayFunction->Identity ];
```

(c) The spectrum is plotted as follows:

```
datafreqdomain = Table[ft, {x, 0, 256}];

plot2=ListPlot[Re[Fourier[datafreqdomain]], PlotJoined -> True,

PlotRange->{{0,128},{0,2}},AxesLabel->{"ν/\!\(s\^-1\)","F(ν)"},DisplayFunction->Identit
y ];

Show[GraphicsArray[{plot1, plot2}]];
```

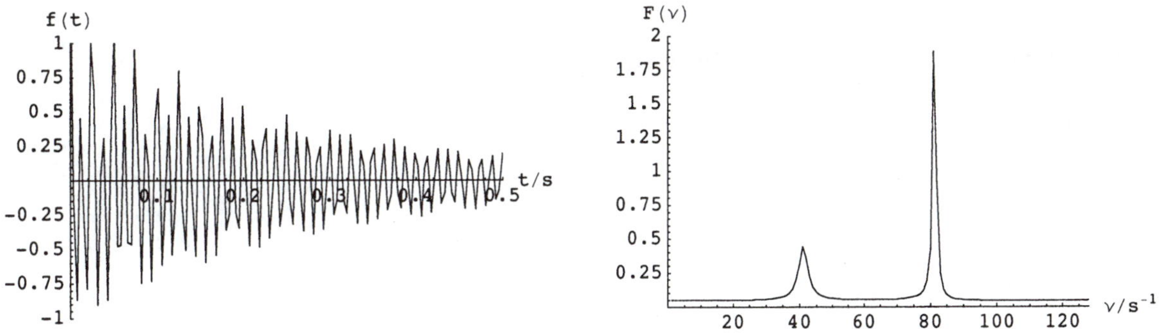

Note that the scale has been changed because the larger peak is so much sharper.

15.H Plot the Lorentzian line shape for absorption when $\nu_0 = \omega/2\pi = 20\ s^{-1}$ and (*a*) $T_2 = 0.10$ s and (*b*) $T_2 = 0.30$ s.

SOLUTION

(a)

```
intenfnu = .1 / (1 + (.1 2 Pi (ν - 20)) ^2)

0.1/(1 + 0.3947841760435743*(-20 + ν)^2)


plot1 =
  Plot[intenfnu, {ν, 0, 64}, AxesLabel -> {"ν/s^-1", "F(ν)"}, DisplayFunction → Identity];
```

(b)

```
intenfnu2 = .3 / (1 + (.3 2 Pi (ν - 20)) ^2)

0.3/(1 + 3.553057584392169*(-20 + ν)^2)


plot2 =
  Plot[intenfnu2, {ν, 0, 64}, AxesLabel -> {"ν/s^-1", "F(ν)"}, DisplayFunction → Identity];

Show[GraphicsArray[{plot1, plot2}]];
```

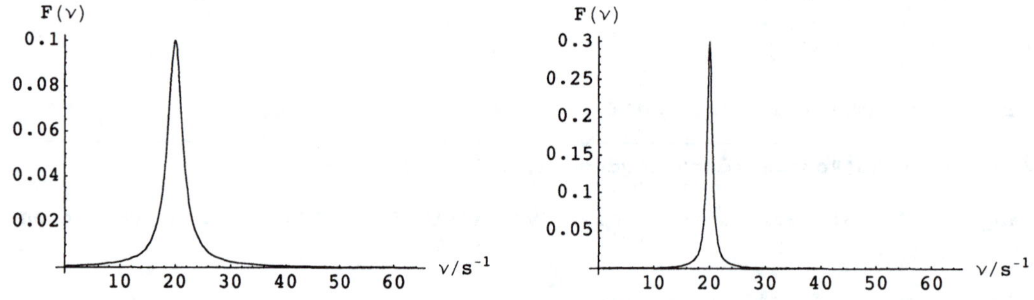

15.I Calculate Pascal's triangle for up to 10 protons.

SOLUTION

```
TableForm[Table[Binomial[n, k], {n, 0, 10}, {k, 0, n}]]
```

1										
1	1									
1	2	1								
1	3	3	1							
1	4	6	4	1						
1	5	10	10	5	1					
1	6	15	20	15	6	1				
1	7	21	35	35	21	7	1			
1	8	28	56	70	56	28	8	1		
1	9	36	84	126	126	84	36	9	1	
1	10	45	120	210	252	210	120	45	10	1

15.J Plot the energy levels of an electron in a magnetic field in J/mol up to 1.0 T. (Compare with Fig. 15.1.)

SOLUTION
The energy levels are given by $E = g_e \mu_B m_s B$.

```
Clear[b]

energyplus = (2.0023) (9.2741 * 10^-24) * (6.022 * 10^23) * b / 2

5.591285612473*b
```

where the energy is expressed in terms of J mol^-1.

```
energyminus = - (2.0023) (9.2741 * 10^-24) * (6.022 * 10^23) * b / 2

-5.591285612473*b
```

```
Plot[{energyplus, energyminus}, {b, 0, 1}, AxesLabel -> {"B/Hz", "E/J mol-1"}];
```

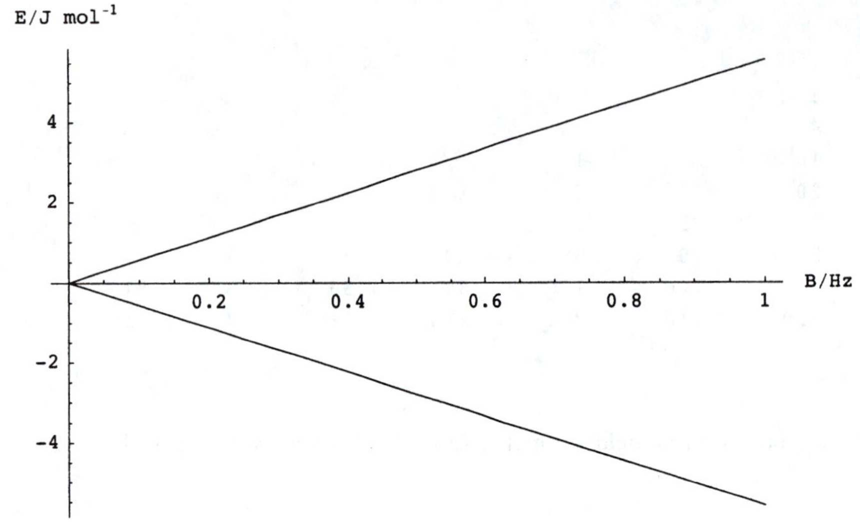

Chapter 16 Statistical Mechanics

16.A Consider a molecule that has two nondegenerate energy levels separated by ϵ. (*a*) Plot the partition function versus kT/ϵ. (*b*) Plot the fractional populations of the two states versus kT/ϵ. (*c*) Plot the ratio of the internal energy U to the energy if all the molecules were in the excited state versus kT/ϵ.

SOLUTION

(a) In order to avoid having to specify ϵ, we will represent kT/ϵ by t. Thus the partition function is given by

```
q = 1 + Exp[-1 / t];
plot1 = Plot[q, {t, 0, 6}, AxesOrigin -> {0, 0},
   PlotRange -> {0, 2}, AxesLabel -> {"kT/ε", "q"}, DisplayFunction → Identity];
```

(b) The fraction N_0/N of the molecules in the ground state is $1/q$. The fraction N_1/N of molecules in the excited state is $\exp(-1/t)/q$.

```
fractgd = 1 / q;
plot2 = Plot[fractgd, {t, 0, 6}, AxesOrigin -> {0, 0},
  PlotRange -> {0, 1}, AxesLabel -> {"kT/ε", "N₀/N"}, DisplayFunction → Identity];

fractex = Exp[-1 / t] / q;
plot3 = Plot[fractex, {t, 0, 6}, AxesOrigin -> {0, 0},
  PlotRange -> {0, 1}, AxesLabel -> {"kT/ε", "N₁/N"}, DisplayFunction → Identity];
```

(c) The ratio of the internal energy U to the energy if all the molecules were in the excited state is given by

```
u = Exp[-1 / t] / q;
plot4 = Plot[u, {t, 0, 6}, AxesOrigin -> {0, 0}, PlotRange -> {0, 1},
   AxesLabel -> {"kT/ε", "U/N₀"}, DisplayFunction → Identity];
```

```
Show[GraphicsArray[{{plot1, plot2}, {plot3, plot4}}, DisplayFunction → $DisplayFunction]];
```

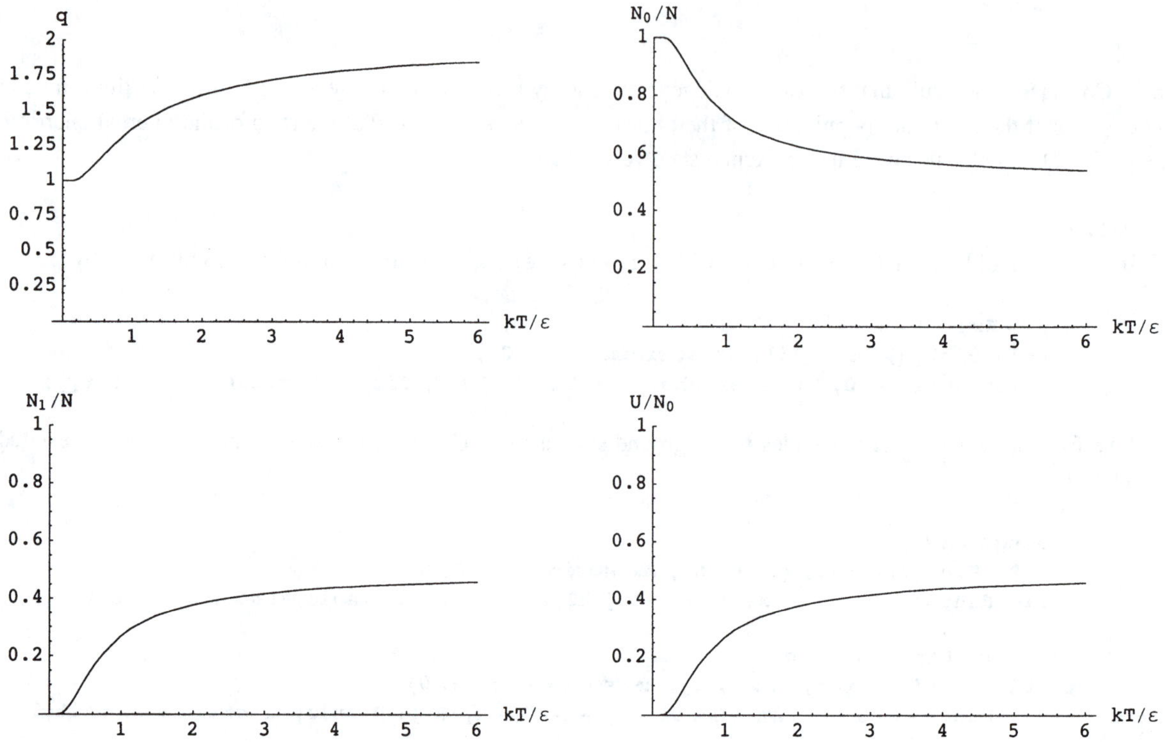

16.B Calculate the fractional populations of $^{16}O_2$ gas molecules in vibrational levels up to $v = 6$ at (*a*) 1000 K and (*b*) 2000 K.

SOLUTION

(a) At 2000 K, $h\nu/kT = 1.136$. The fractional populations N_v/N at 2000 K are given by

```
pop2 = (1 - Exp[-1.136]) Exp[-v 1.136];
plot2 = ListPlot[Transpose[{Table[v, {v, 0, 6}], Table[pop2, {v, 0, 6}]}],
   Prolog -> AbsolutePointSize[4], AxesLabel -> {"v", "Nv/N0"},
   PlotRange -> {0, 1}, DisplayFunction → Identity, PlotLabel → "2000 K"];
```

(b) At 500 K, $h\nu/kT = 1.136(2000)/500$.

```
pop1 = (1 - Exp[-1.136 * 2000 / 1000]) Exp[-v * 1.136 * 2000 / 1000];

plot1 = ListPlot[Transpose[{Table[v, {v, 0, 6}], Table[pop1, {v, 0, 6}]}],
   Prolog -> AbsolutePointSize[4], AxesLabel -> {"v", "Nv/N0"},
   PlotRange -> {0, 1}, DisplayFunction → Identity, PlotLabel → "1000 K"];
```

```
Show[GraphicsArray[{plot1, plot2}], DisplayFunction → $DisplayFunction];
```

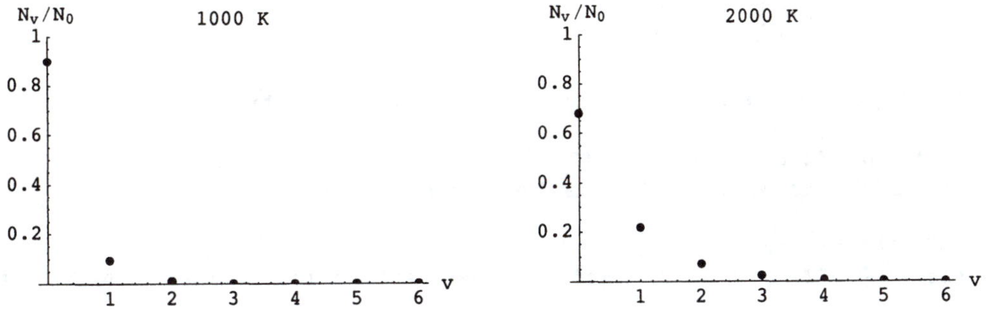

16.C Calculate the molecular partition function for translational motion of a hydrogen atom at 300 K in a volume of 0.2494 m^3. Calculate the thermal wavelength.

SOLUTION
The translational patition function is given by equation 16.76:

```
kB = 1.380658 * 10^-23 (*J K^-1*);
h = 6.6260755 * 10^-34 (*J s*);
nA = 6.022137 * 10^23 (*mol^-1*);
mH = 1.007825 / (nA * 1000) (*kg*);
v = .2494 (*m^3*);
t = 300 (*K*);
qH = ((2 * Pi * mH * kB * t / h^2) ^ (3 / 2)) * v
```

2.46411×10^{29}

The partition function is dimensionless. The thermal wavelength is given by equation 16.77:

```
(v / qH) ^ (1 / 3)
```

1.00403×10^{-10}

The thermal wavelength, which is expressed in meters, is much smaller than the side of the container of the gas, as required by the derivation of equation 16.39.

16.D Plot the molar heat capacity at constant pressure of $^{14}N_2$ (g) versus T from 298.15 to 2000 K. The rigid rotor-harmonic oscillator approximation can be used. Plot values from Table C.3 on the same graph. What do you think is responsible for the differences?

SOLUTION
The various contributions are listed in Table 16.1 , and the vibrational temperature is listed in Table 16.2.

```
Clear[t]
```

```
cpN2 = 8.31451 (7 / 2 + (3392.01 / t) ^2 Exp[3392.01 / t] / (Exp[3392.01 / t] - 1) ^2)
```

$$8.31451\left(\frac{7}{2}+\frac{1.15057\times 10^{7}\, e^{3392.01/t}}{(-1+e^{3392.01/t})^{2}\, t^{2}}\right)$$

```
plot1 = Plot[cpN2, {t, 298.15, 2000}, DisplayFunction -> Identity];

plot2 = ListPlot[{{298.15, 29.124}, {500, 29.58}, {1000, 32.697}, {2000, 35.971}},
   Prolog -> AbsolutePointSize[5], DisplayFunction → Identity];

Show[plot2, plot1,
   AxesLabel -> {"T/K", "C/J K^-1  mol^-1 "}, DisplayFunction → $DisplayFunction];
```

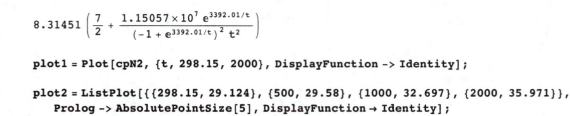

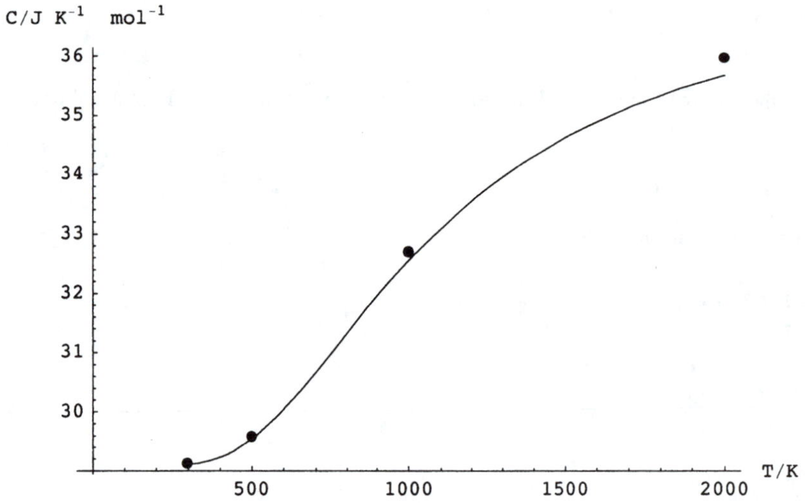

As the temperature increases the molecule deviates more from being a rigid-rotor and a harmonic-oscillator. These deviations bring in additional terms that increase the heat capacity. Also note that atmospheric nitrogen contains 0.366% $^{15}N_2$.

16.E Calculate the relative fractions N_J / N_0 of $^{12}C^{16}O$ molecules in rotational levels up to $J = 20$ at 300 and 500 K. The rotational constant is 2.7771 K.

SOLUTION

```
Clear["Global`*"]

fraction[thetar_, t_, n_] :=
 Module[{ratio, j}, (*Calculates the fraction of diatomic molecules
   with a rotational quantum number J at a specified temperature.*)
   ratio = thetar / t;
      j = Table[x, {x, 0, n}];
      (2 * j + 1) * ratio * Exp[-ratio * j * (j + 1)]]

input300 = Transpose[{Table[x, {x, 0, 20}], fraction[2.7771, 300, 20]}];

input500 = Transpose[{Table[x, {x, 0, 20}], fraction[2.7771, 500, 20]}];

plot300 = ListPlot[input300, Prolog -> AbsolutePointSize[4],
   AxesOrigin -> {0, 0}, AxesLabel -> {"J", "fract."},
   DisplayFunction → Identity, PlotRange → {0, .1}, PlotLabel → "300 K"];

plot500 = ListPlot[input500, Prolog -> AbsolutePointSize[4],
   AxesOrigin -> {0, 0}, AxesLabel -> {"J", "fract."},
   DisplayFunction → Identity, PlotRange -> {0, .1}, PlotLabel → "500 K"];

Show[GraphicsArray[{plot300, plot500}], DisplayFunction → $DisplayFunction];
```

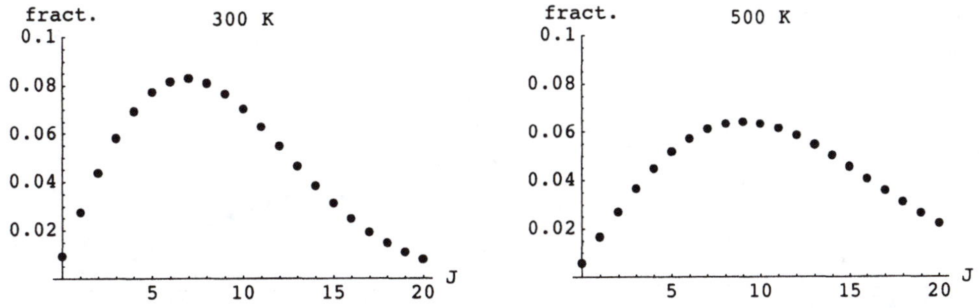

16.F Calculate the relative fractions N_J/N_0 of $^1H^{35}Cl$ molecules in rotational levels up to $J = 20$ at 300 and 1000 K. The rotational constant is 15.2344 K.

SOLUTION

```
Clear["Global`*"]

fraction[thetar_, t_, n_] :=
 Module[{ratio, j}, (*Calculates the fraction of diatomic molecules
    with a rotational quantum number J at a specified temperature.*)
    ratio = thetar / t;
        j = Table[x, {x, 0, n}];
        (2 * j + 1) * ratio * Exp[-ratio * j * (j + 1)]]

input300 = Transpose[{Table[x, {x, 0, 20}], fraction[15.2344, 300, 20]}];

plot300 = ListPlot[input300, Prolog -> AbsolutePointSize[4],
   AxesOrigin -> {0, 0}, AxesLabel -> {"J", "fract."},
   PlotLabel → "300 K", DisplayFunction -> Identity, PlotRange → {0, .2}];

input2000 = Transpose[{Table[x, {x, 0, 20}], fraction[15.2344, 2000, 20]}];

plot2000 = ListPlot[input2000, Prolog -> AbsolutePointSize[4],
   AxesOrigin -> {0, 0}, AxesLabel -> {"J", "fract."},
   PlotLabel → "2000 K", DisplayFunction -> Identity, PlotRange → {0, .2}];

Show[GraphicsArray[{plot300, plot2000}], DisplayFunction -> $DisplayFunction];
```

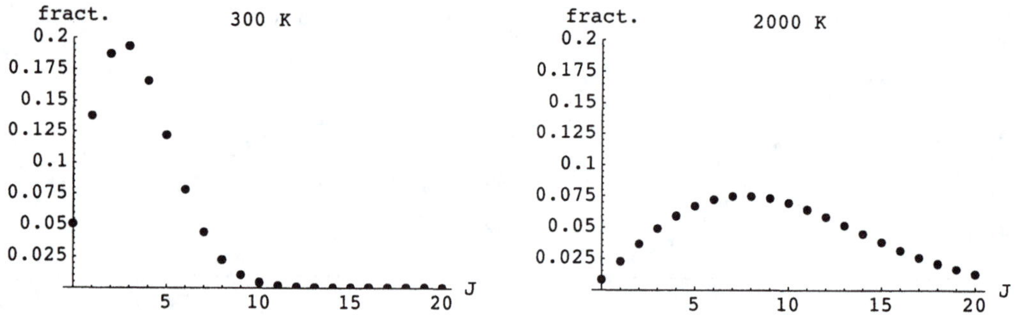

You can try some other diatomic molecules and other temperatures.

16.G Calculate the equilibrium constant for the reaction H_2 (g) = 2H(g) at 500, 1000, and 2000 K using the harmonic oscillator,-rigid rotor approximation. Compare these results with values calculated using Appendix C.3.

SOLUTION

```
ClearAll["Global`*"]

kH2diss[t_] := Module[{qHdV, qH2dV, na, mH, mH2, kB,
   h, thetarotH2, thetavibH2, doH2, rjoulesgoH, goH2, sigmaH2, k},
      (*This program calculates the equilibrium constant (KP) for H2 (g) =
    2 H (g) at any desired temperature,
   in the harmonic oscillator-rigid rotor approximation.  The values of all the
      constants involved are protected.  The spectroscopic dissociation energy is in
      kJ mol^-1.  The standard pressure is 10^5 pascals.  This program can readily
      be modified to calculate KP for any other dissociation X2 (g)  = 2 X (g).*)
       na = 6.022137 * 10^23;
       kB = 1.380658 * 10^-23;
       h = 6.6260755 * 10^-34;
       rjoules = 8.31451;
       mH = 1.007825 / (na * 1000);
       mH2 = 2 * mH;
       thetarotH2 = 87.547;
       thetavibH2 = 6338.2;
       doH2 = 432.073;
       goH = 2;
       goH2 = 1;
       sigmaH2 = 2;
       (*The molecular partition
    function divided by V for the hydrogen atom is given by*)
qHdV = goH * (2 * N[Pi] * mH * kB * t / h^2) ^ (3 / 2);
       (*The molecular partition function divided by V for H2 is given by*)
       qH2dV = ((2 * N[Pi] * mH2 * kB * t / h^2) ^ (3 / 2)) * (t / (sigmaH2 * thetarotH2)) *
    ((1 - Exp[-thetavibH2 / t]) ^-1) * goH2 * Exp[doH2 / (rjoules * t / 1000)];
       (*The numerical value of the equilibrium constant is given by*)
    k = (kB * t / 10^5) * (qHdV^2) / qH2dV]

kH2diss[{500, 1000, 2000}]

{5.29896 × 10^-41, 5.54086 × 10^-18, 2.89429 × 10^-6}
```

The corresponding values calculated in the thermodynamics chapter are 4.84×10^{-41}, 5.15×10^{-18}, and 2.65×10^{-6}.

16.H Calculate the equilibrium constant for the reaction I_2 (g) = 2I(g) at 500, 1000, and 2000 K using the harmonic oscillator-rigid rotor approximation. Compare these results with values calculated using Appendix C.3 in Computer Problem 5.D.

SOLUTION

```
Clear["Global`*"]

kI2diss[t_] := Module[{qIdV, qI2dV, na, mI, mI2, kB,
   h, thetarotI2, thetavibI2, doI2, rjoules, goI, goI2, sigmaI2, k},
      (*This program calculates the equilibrium constant (KP) for I2 (g) =
    2 I (g) at any desired temperature,
   in the harmonic oscillator-rigid rotor approximation. The values of all the
      constants involved are protected.  The spectroscopic dissociation energy is in
      kJ mol^-1.  The standard pressure is 10^5 pascals.  This program can readily
      be modified to calculate KP for any other dissociation X2 (g) = 2 X (g).*)
        na = 6.022137 * 10^23;
        kB = 1.380658 * 10^-23;
        h = 6.6260755 * 10^-34;
        rjoules = 8.31451;
        mI = 126.904477 / (na * 1000);
        mI2 = 2 * mI;
        thetarotI2 = .05376;
        thetavibI2 = 309.65;
        doI2 = 148.81;
        goI = 4;
        goI2 = 1;
        sigmaI2 = 2;
        (*The molecular partition
    function divided by V for the iodine atom is given by*)
qIdV = goI * (2 * N[Pi] * mI * kB * t / h^2) ^ (3 / 2);
        (*The molecular partition function divided by V for I2 is given by*)
        qI2dV = ((2 * N[Pi] * mI2 * kB * t / h^2) ^ (3 / 2)) * (t / (sigmaI2 * thetarotI2)) *
    ((1 - Exp[-thetavibI2 / t]) ^ -1) * goI2 * Exp[doI2 / (rjoules * t / 1000)];
        (*The numerical value of the equilibrium constant is given by*)
   k = (kB * t / 10^5) * (qIdV^2) / qI2dV]

kI2diss[{500, 1000, 2000}]

{3.315 × 10^-11, 0.00320549, 37.5967}
```

The corresponding values calculated in the thermodynamics chapter are 3.24×10^{-11}, 0.00308, and 34.37.

16.I Calculate the equilibrium constant for the reaction HI(g) = H(g) + I(g) at 500, 1000, and 2000 K using the harmonic oscillator-rigid rotor approximation. Compare these results with values calculated using Appendix C.3.

SOLUTION

```
Clear["Global`*"]

kHIdiss[t_] := Module[{qIdV, qHdV, qHIdV, na, mI, mH, mHI, kB,
   h, thetarotHI, thetavibHI, doHI, rjoules, goI, goH, goHI, sigmaHI, k},
      (*This program calculates the equilibrium constant (KP) for HI (g) =
    H (g) + I (g) at any desired temperature,
   in the harmonic oscillator-rigid rotor approximation.  The values of all the
     constants involved are protected.  The spectroscopic dissociation energy is in
     kJ mol^-1.  The standard pressure is 10^5 pascals.  This program can readily
     be modified to calculate KP for any other dissociation of HX (g) to atoms.*)
        na = 6.022137 * 10^23;
        kB = 1.380658 * 10^-23;
        h = 6.6260755 * 10^-34;
        rjoules = 8.31451;
        mI = 126.904477 / (na * 1000);
        mH = 1.007825 / (na * 1000);
        mHI = (1.007825 + 126.904477) / (na * 1000);
        thetarotHI = 9.369;
        thetavibHI = 3322.24;
        doHI = 294.67;
        goI = 4;
        goH = 2;
        goHI = 1;
        sigmaHI = 1;
        (*The molecular partition
    function divided by V for the iodine atom is given by*)
qIdV = goI * (2 * N[Pi] * mI * kB * t / h^2) ^ (3 / 2);
        (*The molecular partition
    function divided by V for the hydrogen atom is given by*)
qHdV = goH * (2 * N[Pi] * mH * kB * t / h^2) ^ (3 / 2);
        (*The molecular partition function divided by V for HI is given by*)
        qHIdV = ((2 * N[Pi] * mHI * kB * t / h^2) ^ (3 / 2)) * (t / (sigmaHI * thetarotHI)) *
    ((1 - Exp[-thetavibHI / t]) ^-1) * goHI * Exp[doHI / (rjoules * t / 1000)];
        (*The numerical value of the equilibrium constant is given by*)
    k = (kB * t / 10^5) * (qIdV * qHdV) / qHIdV]

kHIdiss[{500, 1000, 2000}]

{3.57685 × 10^-27, 2.40574 × 10^-11, 0.00283838}
```

The corresponding values calculated in Chapter 5 are $3.50\text{x}10^{-27}$, $2.34\text{x}10^{-11}$, and 0.605.

16.J Calculate the equilibrium constant for the reaction $2HI(g) = H_2(g) + I_2(g)$ at 500, 1000, and 2000 K using the harmonic oscillator-rigid rotor approximation. Compare these results with values calculated using Appendix C.3.

SOLUTION

```
Clear["Global`*"]

k2HItoH2andI2[t_] := Module[{na, kB, h, rjoules, mHI, thetarotHI, thetavibHI,
   doHI, goHI, sigmaHI, mH2, thetarotH2, thetavibH2, doH2, goH2, sigmaH2, mI2,
   thetarotI2, thetavibI2, doI2, goI2, sigmaI2, qHIdV, qH2dV, qI2dV, k},
      (*This program calculates the equilibrium constant (KP) for 2 HI (g)=
    H2 (g) + I2 (g) at any desired temperature, in the harmonic oscillator-
    rigid rotor approximation. The values of all the constants involved are
     protected. The spectroscopic dissociation energies are in kJ mol^-1.*)
    (*The general constants are*)
       na = 6.022137 * 10^23;
       kB = 1.380658 * 10^-23;
       h = 6.6260755 * 10^-34;
       rjoules = 8.31451;
       (*The properties of HI are*)
       mHI = (1.007825 + 126.904477) / (na * 1000);
       thetarotHI = 9.369;
       thetavibHI = 3322.24;
       doHI = 294.67;
       goHI = 1;
       sigmaHI = 1;
       (*The properties of H2 are*)
    mH2 = 2 * 1.007825 / (na * 1000);
       thetarotH2 = 87.547;
       thetavibH2 = 6338.2;
       doH2 = 432.073;
       goH2 = 1;
       sigmaH2 = 2;
       (*The properties of I2 are*)
    mI2 = 2 * 126.904477 / (na * 1000);
       thetarotI2 = .05376;
       thetavibI2 = 309.65;
       doI2 = 148.81;
       goI2 = 1;
       sigmaI2 = 2;
       (*The molecular partition function divided by V for HI is given by*)
       qHIdV = ((2 * N[Pi] * mHI * kB * t / h^2) ^ (3 / 2)) * (t / (sigmaHI * thetarotHI)) *
    ((1 - Exp[-thetavibHI / t]) ^-1) * goHI * Exp[doHI / (rjoules * t / 1000)];
       (*The molecular partition function divided by V for H2 is given by*)
       qH2dV = ((2 * N[Pi] * mH2 * kB * t / h^2) ^ (3 / 2)) * (t / (sigmaH2 * thetarotH2)) *
    ((1 - Exp[-thetavibH2 / t]) ^-1) * goH2 * Exp[doH2 / (rjoules * t / 1000)];
       (*The molecular partition function divided by V for I2 is given by*)
       qI2dV = ((2 * N[Pi] * mI2 * kB * t / h^2) ^ (3 / 2)) * (t / (sigmaI2 * thetarotI2)) *
    ((1 - Exp[-thetavibI2 / t]) ^-1) * goI2 * Exp[doI2 / (rjoules * t / 1000)];
       (*The numerical value of the equilibrium constant is given by*)
    k = (qH2dV * qI2dV) / (qHIdV^2)]
```

```
k2HItoH2andI2[{500, 1000, 2000}]

{0.0072833, 0.0325857, 0.0740374}
```

The corresponding values calculated in Chapter 5 are 0.0078, 0.034, and 0.080.

16.K (*a*) Make a three-dimensional plot of the molar entropy of O(g) from 298 to 1000 K and 1 to 100 bar using the Sackur-Tetrode equation. (*b*) Make a two-dimensional plot of the molar entropy at 1 bar for 298 K to 1000 K.

SOLUTION

(a)

```
ClearAll["Global`*"]

sm[t_, p_] := 8.31451 * (-1.151693 + 1.5 * Log[16] - Log[p] + 2.5 * Log[t])

Plot3D[sm[t, p], {p, 1, 100}, {t, 298, 1000},
  AxesLabel -> {"P/bar", "  T/K", "S̄/J K^-1 mol^-1"}, PlotRange -> {100, 180}];
```

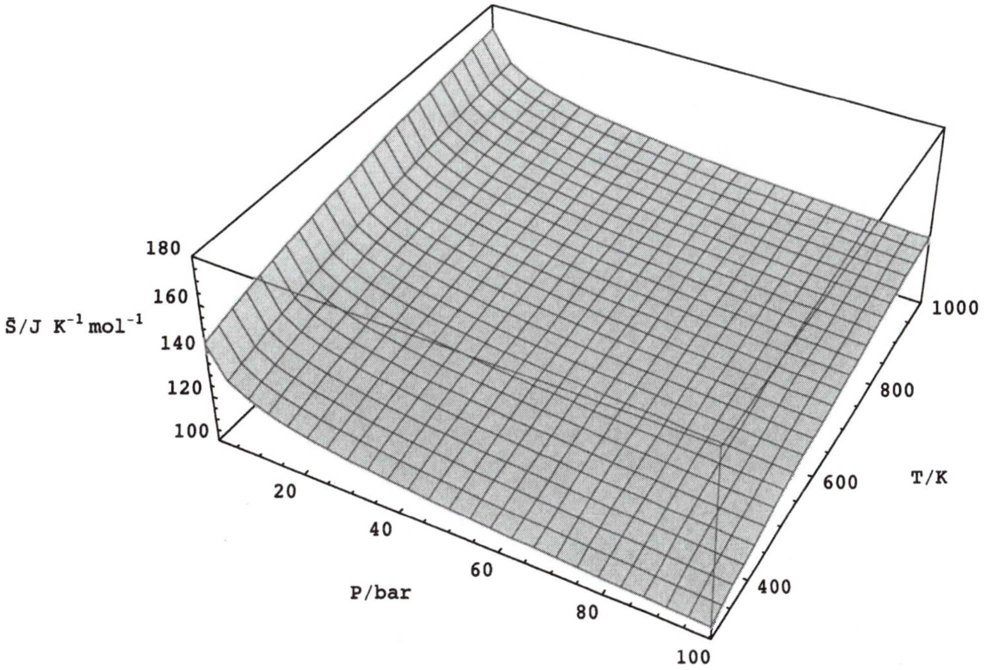

(b)

```
Plot[sm[t, p] /. p → 1, {t, 298, 1000}, AxesOrigin -> {298, 140},
  PlotRange -> {140, 180}, AxesLabel -> {"T/K", "S̄/J K^-1 mol^-1"}];
```

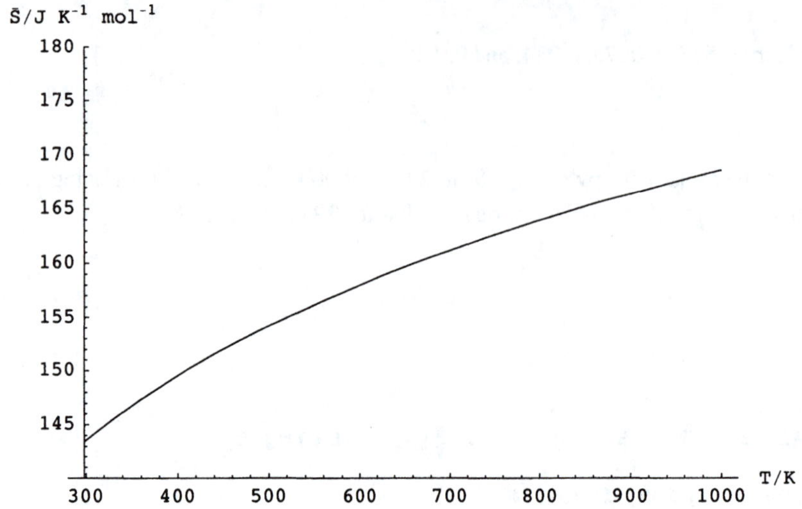

These molar entropies are lower than in Appendix Table C.3 because of the neglect of electronic excitation.

Chapter 17 Kinetic Theory of Gases

17.A Plot the probability density for the molecular speed of nitrogen molecules at 200, 500, and 800 K.

SOLUTION

```
ClearAll["Global`*"]

4 Pi v^2 (m/ ((2 Pi 1.381 10^-23) t)) ^ (3/2) Exp[-m v^2/ (2 (1.381 10^-23) t)] * 1000
```

$$1.55471\times10^{37}\, e^{-\frac{3.62056\times10^{22}\, m\, v^2}{t}} \left(\frac{m}{t}\right)^{3/2} v^2$$

```
fv[v_, t_] :=
 4 Pi v^2 (m/ ((2 Pi 1.381 10^-23) t)) ^ (3/2) Exp[-m v^2/ (2 (1.381 10^-23) t)] * 1000

m = 2 (14.007 10^-3) / (6.022 10^23)
```

4.65194×10^{-26}

The factor of 1000 is used so that the ordinate of the plot will be of the order of magnitude of unity.

```
Plot[{fv[v, 200], fv[v, 500], fv[v, 800]},
  {v, 0, 2000}, AxesLabel -> {"v" / "m s^-1", "10^3F(v)" / "s m^-1"}];
```

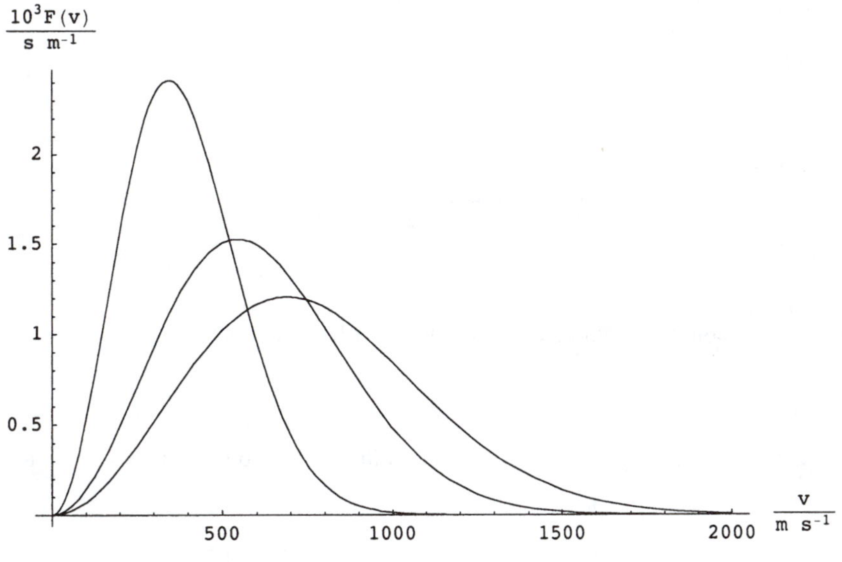

You can plot probability densities for other molecules at other temperatures.

17.B Plot the probability density of the x component of the velocity of nitrogen molecules at 200, 500, and 800 K.

SOLUTION

```
ClearAll["Global`*"]
```

```
(m / ((2 Pi 1.381 10^-23) t))^(1 / 2) Exp[-m (vx^2) / ((2 1.381 10^-23) t)] * 1000
```

$$1.07353\times10^{14}\, e^{-\frac{3.62056\times10^{22}\, m\, vx^2}{t}} \sqrt{\frac{m}{t}}$$

```
fvx[vx_, t_] := (m / ((2 Pi 1.381 10^-23) t))^(1 / 2) Exp[-m (vx^2) / ((2 1.381 10^-23) t)] * 100(
```

The factor of 1000 is used so that the ordinate of the plot will be of the order of magnitude of unity.

```
m = 2 (14.007 10^-3) / (6.022 10^23)
```

4.65194×10^{-26}

```
Plot[{fvx[vx, 200], fvx[vx, 500], fvx[vx, 800]}, {vx, -2000, 2000},
  AxesLabel -> {"vx" / "m s-1", "10^3 f(vx)" / "s m-1"}, AxesOrigin -> {-2000, 0}];
```

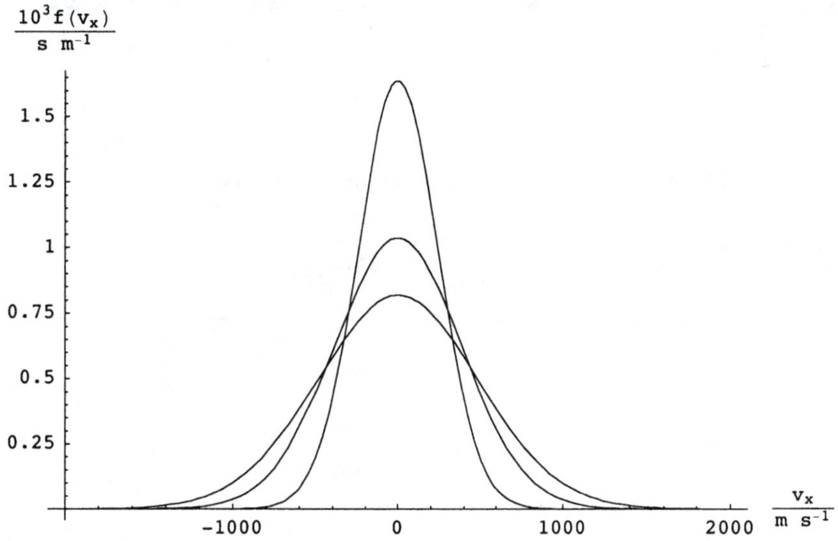

You can plot probability densities of the x-component for other molecules and other temperatures.

17.C Plot the probability density for the velocity of oxygen molecules in an arbitrary direction at 100, 300, 500, and 1000 K.

SOLUTION

```
Clear["Global`*"]
```

```
(((32*10^-3) / (2*π*8.314*t))^.5)*Exp[(-1*(32*10^-3)*v^2) / (2*8.314*t)]*1000
```

$$24.7503\, e^{-\frac{0.00192446\, v^2}{t}} \left(\frac{1}{t}\right)^{0.5}$$

```
pd[v_, t_] :=
  (((32*10^-3) / (2*N[Pi]*8.314*t))^.5)*Exp[(-1*(32*10^-3)*v^2) / (2*8.314*t)]*1000
```

The factor of 1000 is used so that the ordinate of the plot will be of the order of magnitude of unity.

```
Plot[{pd[v, 100], pd[v, 300], pd[v, 500], pd[v, 1000]}, {v, -1000, 1000},
  AxesOrigin -> {-1000, 0}, AxesLabel -> {"vx" / "m s-1", "f(vx)" / "103"}];
```

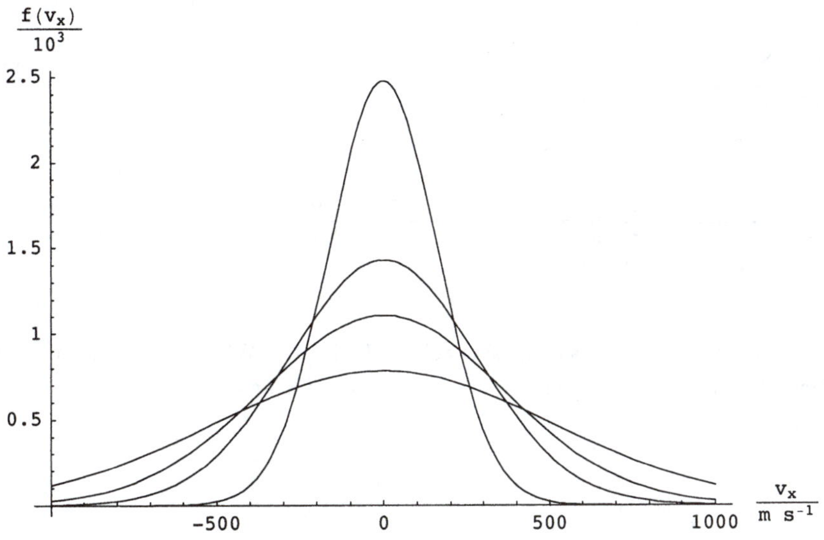

You can plot densities of the x-component for other molecules and other temperatures.

17.D Calculate the probability density of various speeds v for molecular oxygen at 100, 300, 500, and 1000 K.

SOLUTION

```
ClearAll["Global`*"]
```

```
(4*N[Pi]*(v^2))*(((32*10^-3) / (2*N[Pi]*8.314*t))^1.5)*
 Exp[(-1*(32*10^-3)*v^2) / (2*8.314*t)]*1000
```

$$0.190524\, e^{-\frac{0.00192446\, v^2}{t}} \left(\frac{1}{t}\right)^{1.5} v^2$$

```
f[v_, t_] := (4*N[Pi]*(v^2))*(((32*10^-3) / (2*N[Pi]*8.314*t))^1.5)*
  Exp[(-1*(32*10^-3)*v^2) / (2*8.314*t)]*1000
```

```
Plot[{f[v, 100], f[v, 300], f[v, 500], f[v, 1000]},
  {v, 0, 1500}, AxesLabel -> {"v" / "m s^-1", "10^3F(v)" / "s m^-1"}];
```

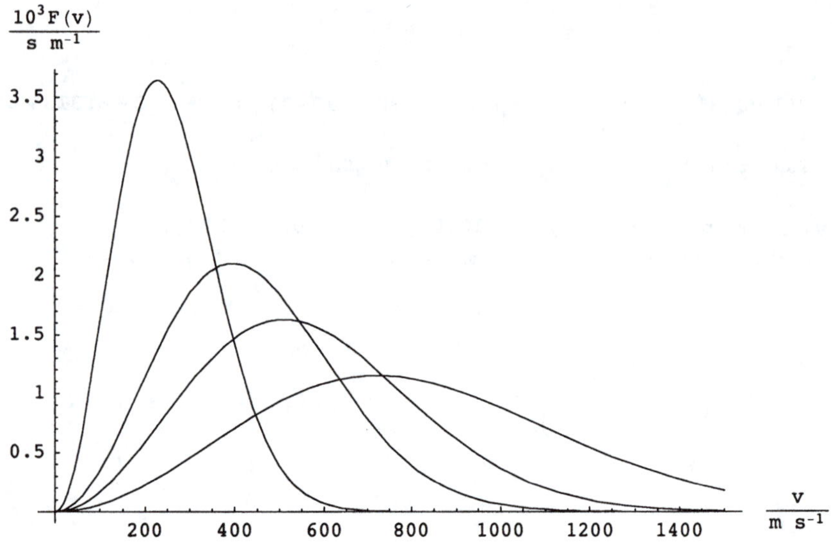

17.E A sodium atom emits a frequency of 5×10^8 MHz from an emission cell at 500 K. (*a*) What is the standard deviation of the spectral line due to Doppler broadening? (*b*) Plot the shape of the spectral lines at 500 and 1000 K.

SOLUTION

(a)

```
ClearAll["Global`*"]
```

The standard deviations at the two temperatures in megahertz are given by

```
stddev1 = (10^-6) * ((((5*10^14)^2)*8.314*500) / ((.02299)*((3*10^8)^2)))^.5

708.711

stddev2 = (10^-6) * ((((5*10^14)^2)*8.314*1000) / ((.02299)*((3*10^8)^2)))^.5

1002.27
```

(b)

```
pnu1 = ((2*Pi*stddev1)^-.5)*Exp[-(nu^2) / (2*stddev1^2)]
```

$0.0149856\, e^{-9.95478\times10^{-7}\, nu^2}$

```
pnu2 = ((2*Pi*stddev2)^-.5)*Exp[-(nu^2) / (2*stddev2^2)]
```

$0.0126014\, e^{-4.97739\times10^{-7}\, nu^2}$

```
Plot[{pnu1, pnu2}, {nu, -5*10^3, 5*10^3},
  AxesOrigin -> {-5000, 0}, AxesLabel -> {"ν/MHz", "Prob."}];
```

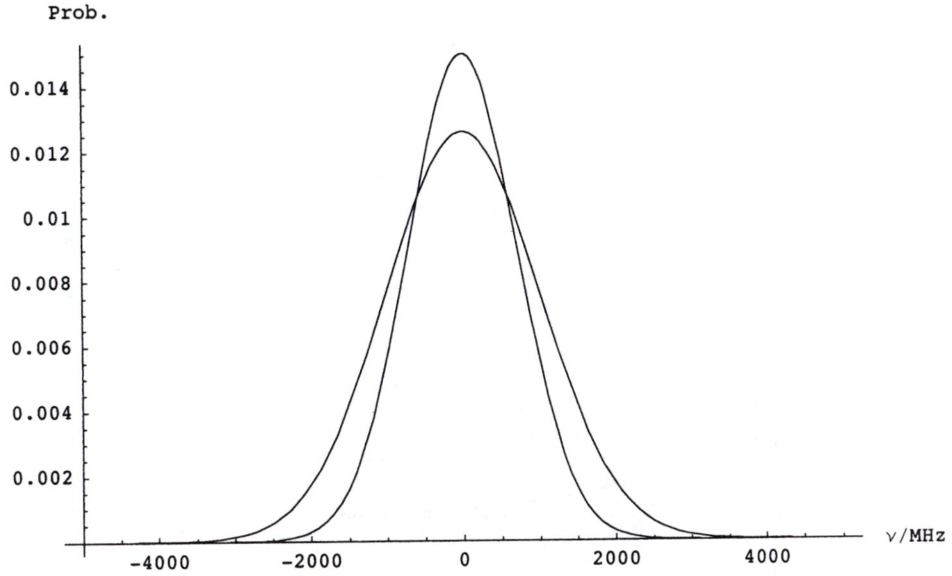

Note that the plots of intensity versus frequency in MHz are normalized. You can plot line shapes for other atoms at other temperatures.

17.F (*a*) Calculate the probability density for molecular energy of a molecule of ideal gas at 100, 300, and 500 K. Note that the probability density is independent of the molar mass. (*b*) Calculate the average energies of an ideal gas molecule at these temperatures. If it bothers you that the energy at the maximum of the probability density for molecular energy differs from the average energy, see B. A. Morrow and D. F. Tessler, J. Chem. Educ. **59**, 193 (1982).

SOLUTION

```
ClearAll["Global`*"]
```

(a)

```
k = 1.381*10^-23 (*J T^-1*)
```

1.381×10^{-23}

```
2 Pi ((Pi k t)^-1.5) (e^.5) Exp[-e/(k t)]
```

$$\frac{2.19869\times10^{34}\, e^{0.5}\, e^{-\frac{7.24113\times10^{22}\, e}{t}}}{t^{1.5}}$$

```
feden[t_, e_] := 2 Pi ((Pi k t)^-1.5) (e^.5) Exp[-e/(k t)]
```

```
Plot[{feden[100, e], feden[300, e], feden[500, e]},
  {e, 0, 10^-20}, AxesLabel -> { " ε"/ " J", "F(ε)"/"J^-1"}];
```

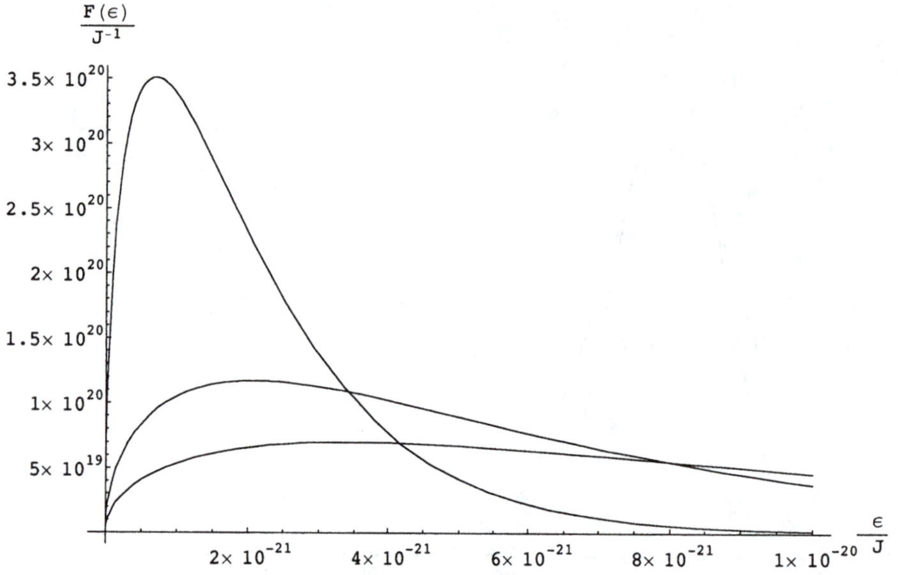

(b) Calculate the average energies at these temperatures.

```
Table[1.5*k*t, {t, 100, 500, 200}] (*J*)
```

$\{2.0715 \times 10^{-21},\ 6.2145 \times 10^{-21},\ 1.03575 \times 10^{-20}\}$

17.G Plot points at the ends of random vectors with a normal distribution in a plane.

SOLUTION

The Standard Add-On Packages of *Mathematica* contain the needed function Normal Distribution in the Statistics Package.

```
Needs["Statistics`NormalDistribution`"]

NormalDistribution[0, 1];
```

The first parameter is the mean and the second is the standard deviation. The following command makes a table of random numbers corresponding to the normal distribution with a mean of 0 and a standard deviation of 1. Note that different numbers are obtained each time that this is run.

```
Table[Random[NormalDistribution[0, 1]], {5}]

{-0.494045, -0.230128, 0.40791, -0.285283, 1.31971}

Table[Random[NormalDistribution[0, 1]], {5}]

{-0.0484889, 1.76306, -1.48368, -0.687512, -1.11485}
```

A set of numbers like this is taken as the radial coordinate in the command

```
makeRadial[r_, ang_] := r {Cos[ang], Sin[ang]}
```

```
Table[makeRadial[Random[NormalDistribution[0, 1]], Random[Real, {0, 2 Pi}]], {5}]
```

```
{{0.171449, -0.30661}, {1.31254, 0.0974495},
 {0.0349564, -0.58058}, {-1.06306, 0.237885}, {0.682472, -0.0391913}}
```

These are the coordinates for points in a plane that have random locations but are distributed about the origin with normal distribution. The following two commands calculate 1000 points and plots them.

```
Table[makeRadial[Random[NormalDistribution[0, 1]], Random[Real, {0, 2 Pi}]], {1000}];

rplot1 = ListPlot[%, AspectRatio -> 1, DisplayFunction → Identity];

Table[makeRadial[Random[NormalDistribution[0, 1]], Random[Real, {0, 2 Pi}]], {1000}];

rplot2 = ListPlot[%, AspectRatio -> 1, DisplayFunction → Identity];

Show[GraphicsArray[{rplot1,rplot2}]];
```

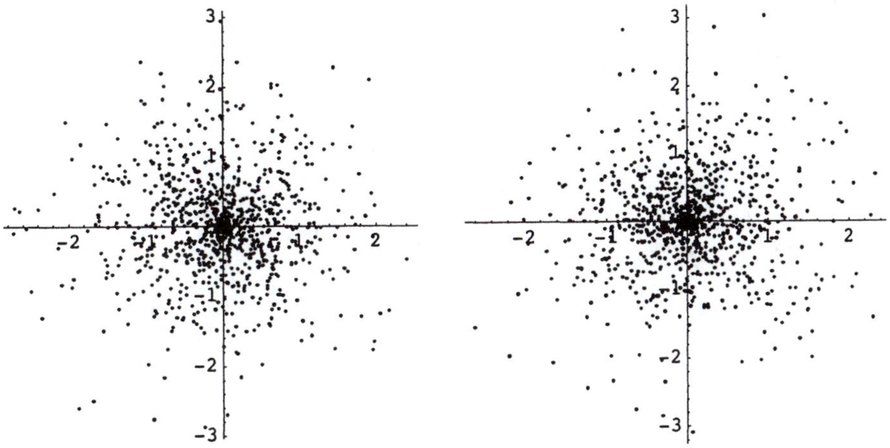

17.H Plot the most probable speed, the mean speed, and the root-mean-square speed of ideal gas molecules in m/s at 273.15 K versus molar mass M in g mol^{-1}.

SOLUTION

```
veemp=(2*8.31451*273.15/(m*10^-3))^.5
```

$2131.25 \left(\frac{1}{m}\right)^{0.5}$

```
veemp/.m->2.016
```

```
1501.03
```

```
vmean=(8*8.31451*273.15/(π*m*10^-3))^.5
```

$2404.86 \left(\frac{1}{m}\right)^{0.5}$

```
vmean/.m->2.016
```

```
1693.73
```

```
vrms=(3*8.31451*273.15/(m*10^-3))^.5
```

$2610.23 \left(\frac{1}{m}\right)^{0.5}$

```
vrms/.m->2.016
```

1838.38

```
Plot[{veemp,vmean,vrms},{m,1,100},AxesLabel->{"M/g mo\!\(l\^-1\)","v/m \!\(s\^-1\)"}];
```

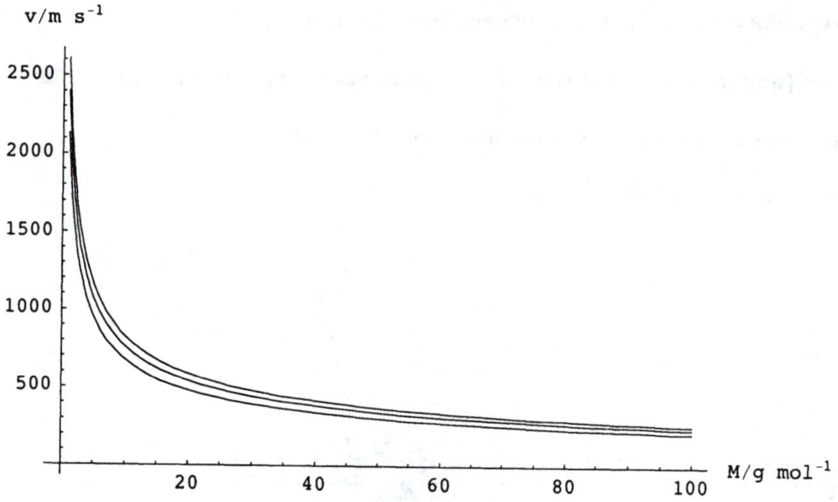

Chapter 18 Experimental Kinetics and Gas Reactions

18.A Two consecutive first order reactions are represented by

$$A_1 \xrightarrow{k_1} A_2 \xrightarrow{k_2} A_3$$

for which we have seen the general solution in equations 18.47-18.50. Use a mathematical application for solving differential equations to calculate the concentrations of A_1, A_2, and A_3 as a function of time without using equations 18.47-18.50. Assume that initially A_1 is at unit concentration and treat three cases: (*a*) $k_1 = 1\ s^{-1}$, $k_2 = 1\ s^{-1}$; (*b*) $k_1 = 1\ s^{-1}$, $k_2 = 5\ s^{-1}$; (*c*) $k_1 = 1\ s^{-1}$, $k_2 = 25\ s^{-1}$.

SOLUTION

(a) In Mathematica, a derivative is indicated by an apostrophe, and so the rate equations and initial conditions are given by

```
ClearAll["Global`*"]

eqns = {c1'[t] == -c1[t], c2'[t] == c1[t] - c2[t],
        c3'[t] == c2[t], c1[0] == 1, c2[0] == 0, c3[0] == 0};
```

We need a list of variables to be calculated, and that is given by

```
vars = {c1[t], c2[t], c3[t]};
```

NDSolve is used to calculate interpolating functions that give the concentrations of the three reactants at times in the specified range.

```
concs = NDSolve[eqns, vars, {t, 0, 3}];
```

A plot can be obtained by evaluating these interpolation functions. The ReplaceAll operation (/.) applies the interpolating functions to the three concentration variables.

```
Plot[Evaluate[vars /. concs], {t, 0, 3}, AxesLabel -> {"t" / "s", "[Ai]"}];
```

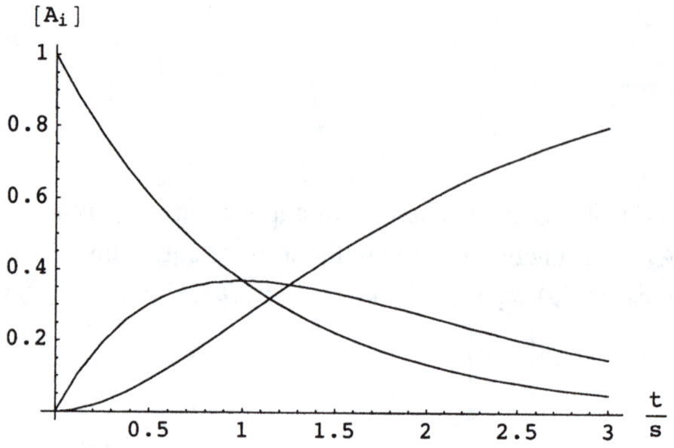

(b)

```
ClearAll["Global`*"]

eqns = {c1'[t] == -c1[t], c2'[t] == c1[t] - 5 c2[t],
        c3'[t] == 5 c2[t], c1[0] == 1, c2[0] == 0, c3[0] == 0};

vars = {c1[t], c2[t], c3[t]};

concs = NDSolve[eqns, vars, {t, 0, 3}];

Plot[Evaluate[vars /. concs], {t, 0, 3}, AxesLabel -> {"t" / "s", "[Ai]"}];
```

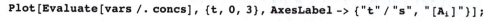

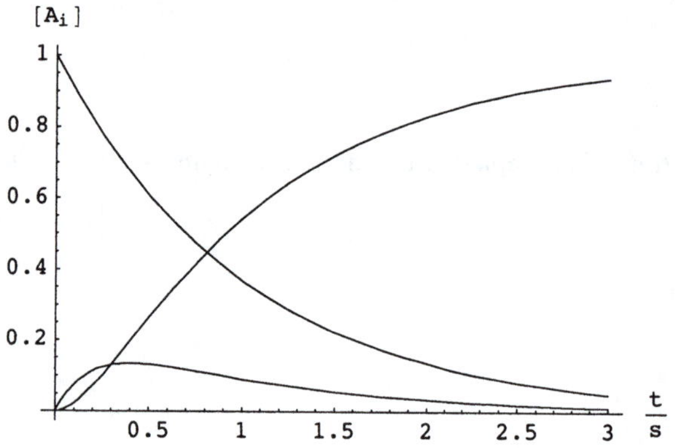

(c)

```
ClearAll["Global`*"]

eqns = {c1'[t] == -c1[t], c2'[t] == c1[t] - 25 c2[t],
        c3'[t] == 25 c2[t], c1[0] == 1, c2[0] == 0, c3[0] == 0};

vars = {c1[t], c2[t], c3[t]};

concs = NDSolve[eqns, vars, {t, 0, 3}];
```

```
Plot[Evaluate[vars /. concs], {t, 0, 3}, AxesLabel -> {"t" / "s", "[Ai]"}];
```

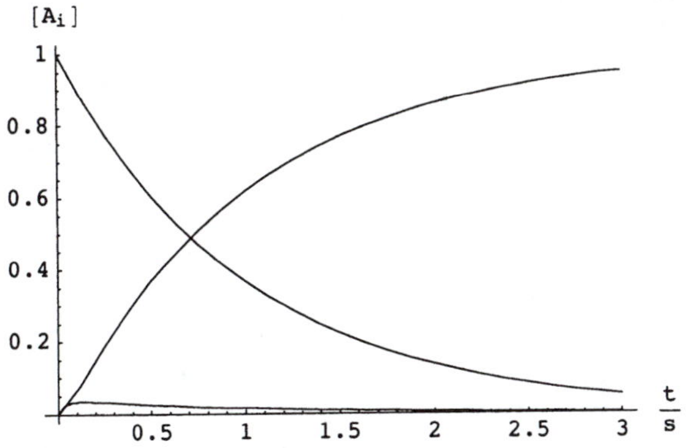

18.B Two consecutive reversible first order reactions are represented by

$$A_1 \underset{k_2}{\overset{k_1}{\rightleftharpoons}} A_2 \underset{k_4}{\overset{k_3}{\rightleftharpoons}} A_3$$

Use a mathematical application for solving differential equations to calculate the concentrations of A_1, A_2, and A_3 as a function of time. Assume that initially A_1 is at unit concentration and treat three cases: (*a*) $k_1 = k_2 = k_3 = k_4 = 1\ s^{-1}$, (*b*) $k_1 = k_2 = k_4 = 1\ s^{-1}$, $k_3 = 3\ s^{-1}$, (*c*) $k_1 = k_2 = k_4 = 1\ s^{-1}$, $k_3 = 9\ s^{-1}$,

SOLUTION

(a)

```
ClearAll["Global`*"]

eqns1 = {c1'[t] == -c1[t] + c2[t],
         c2'[t] == c1[t] - 2 c2[t] + c3[t],
         c3'[t] == c2[t] - c3[t],
c1[0] == 1, c2[0] == 0, c3[0] == 0};

vars1 = {c1[t], c2[t], c3[t]};

conc1 = NDSolve[eqns1, vars1, {t, 3}];
```

```
Plot[Evaluate[vars1 /. conc1], {t, 0, 3},
  AxesLabel -> {"t" / "s", "[Ai]"}, PlotRange -> {0, 1}];
```

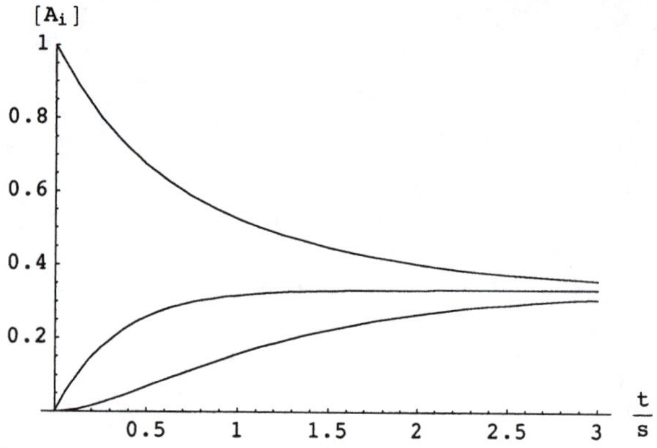

(b)

```
ClearAll["Global`*"]

eqns1 = {c1'[t] == -c1[t] + c2[t],
         c2'[t] == c1[t] - 4 c2[t] + c3[t],
         c3'[t] == 3 c2[t] - c3[t],
c1[0] == 1, c2[0] == 0, c3[0] == 0};

vars1 = {c1[t], c2[t], c3[t]};

conc1 = NDSolve[eqns1, vars1, {t, 3}];
```

```
Plot[Evaluate[vars1 /. conc1], {t, 0, 3},
  AxesLabel -> {"t" / "s", "[Ai]"}, PlotRange -> {0, 1}];
```

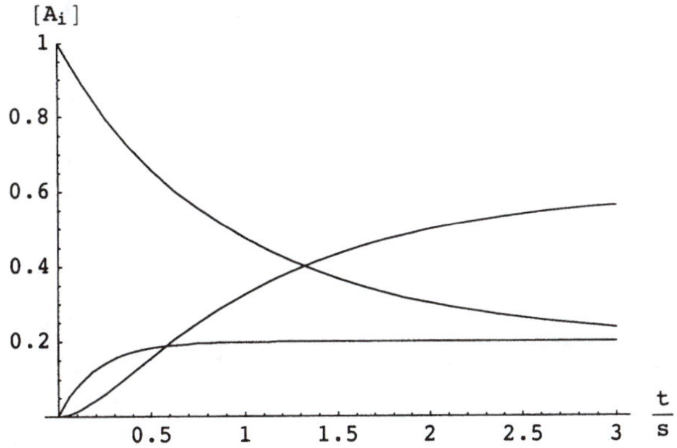

(c)

```
ClearAll["Global`*"]

eqns1 = {c1'[t] == -c1[t] + c2[t],
        c2'[t] == c1[t] - 10 c2[t] + c3[t],
        c3'[t] == 9 c2[t] - c3[t],
   c1[0] == 1, c2[0] == 0, c3[0] == 0};

vars1 = {c1[t], c2[t], c3[t]};

conc1 = NDSolve[eqns1, vars1, {t, 3}];
```

```
Plot[Evaluate[vars1 /. conc1], {t, 0, 3},
  AxesLabel -> {"t" / "s", "[Ai]"}, PlotRange -> {0, 1}];
```

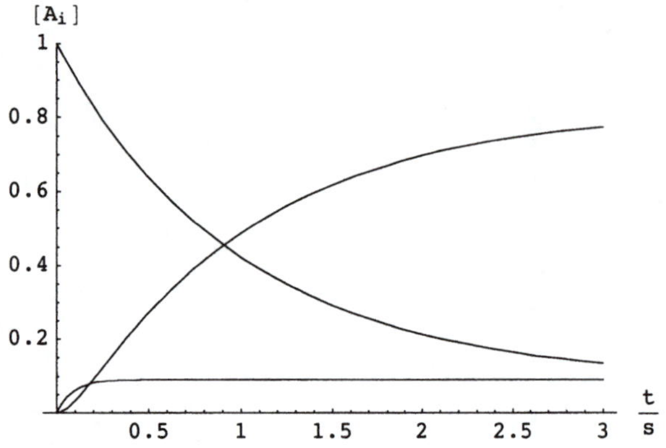

18.C A reaction A + B = C + D is reversible and has an equilibrium constant equal to 2. (*a*) If the initial concentrations of A and B are 1 and 0.5 mol L^{-1}, respectively, plot the four concentrations as a function of time, assuming the rate constant for the forward reaction is 1 L mol^{-1} s^{-1}. (*b*) To confirm the equilibrium concentrations of C and D, calculate these concentrations using the equilibrium constant expression.

SOLUTION

(a) The four rate equations and the initial conditions are given by

```
ClearAll["Global`*"]

eqns1 = {cA'[t] == -cA[t] * cB[t] + .5 cC[t] * cD[t],
         cB'[t] == -cA[t] * cB[t] + .5 cC[t] * cD[t],
         cC'[t] == cA[t] * cB[t] - .5 cC[t] * cD[t],
         cD'[t] == cA[t] * cB[t] - .5 cC[t] * cD[t],
cA[0] == 1, cB[0] == .5, cC[0] == 0, cD[0] == 0};

vars1 = {cA[t], cB[t], cC[t], cD[t]};

conc1 = NDSolve[eqns1, vars1, {t, 3}];

Plot[Evaluate[vars1 /. conc1], {t, 0, 3}, AxesLabel -> {"t" / "s", "conc"}];
```

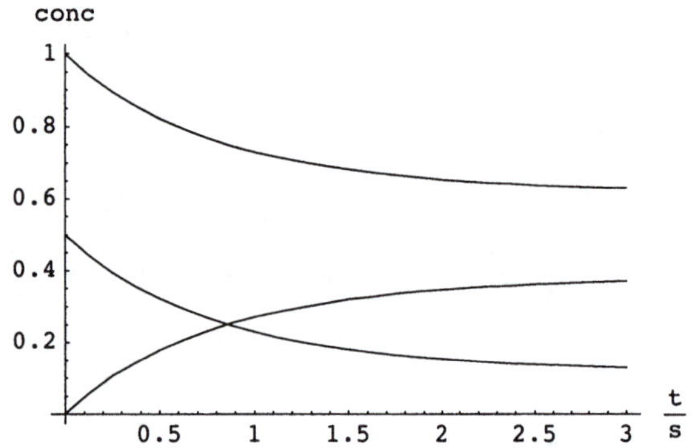

(b) Calculation of equilibrium concentrations of C and D:

```
FindRoot[2 * (1 - x) * (.5 - x) == x^2, {x, .5}]

{x → 0.381966}
```

18.D In the case of a reaction like that in the preceding problem, there may be a question as to whether there is an intermediate X:

$$A + B \underset{10\ s^{-1}}{\overset{1\ L\ mol^{-1}\ s^{-1}}{\rightleftarrows}} X \underset{0.5\ L\ mol^{-1}\ s^{-1}}{\overset{10\ s^{-1}}{\rightleftarrows}} C + D$$

For the indicated values of the rate constants, explore the effects of intermediate X on the plots of concentration versus time.

SOLUTION

(a) The five rate equations and the initial conditions are given by

```
ClearAll["Global`*"]

eqns1 = {cA'[t] == -cA[t] * cB[t] + 10 cX[t],
         cB'[t] == -cA[t] * cB[t] + 10 cX[t],
         cC'[t] == 10 cX[t] - .5 cC[t] * cD[t],
         cD'[t] == 10 cX[t] - .5 cC[t] * cD[t],
         cX'[t] == cA[t] * cB[t] + .5 cC[t] * cD[t] - 20 cX[t],
cA[0] == 1, cB[0] == .5, cC[0] == 0, cD[0] == 0, cX[0] == 0};

vars1 = {cA[t], cB[t], cC[t], cD[t], cX[t]};

conc1 = NDSolve[eqns1, vars1, {t, 3}];

Plot[Evaluate[vars1 /. conc1], {t, 0, 3}, AxesLabel -> {"t" / "s", "conc"}];
```

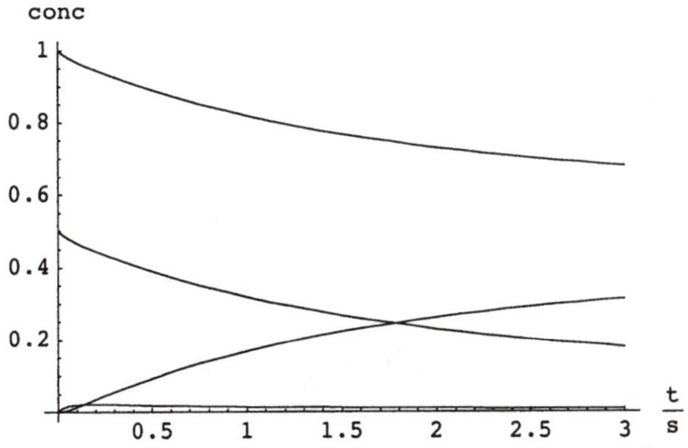

Note that C and D have equal concentrations at each time. The concentration of X barely shows on this plot, but its concentration goes through a maximum. This plot shows that intermediate X is in a nearly steady state after the induction period (about 0.1 s) of its formation. Note that there is an induction period in the formation of C and D, by comparing this plot with that in the preceding problem. Thus even if X cannot be detected spectroscopically, the induction period in the formation of C and D will reveal the existence of the intermediate. This induction period can be seen more clearly by looking at the first 0.5 s.

```
Plot[Evaluate[vars1 /. conc1], {t, 0, .5},
  AxesLabel -> {"t" / "s", "conc"}, PlotRange → {0, .1}];
```

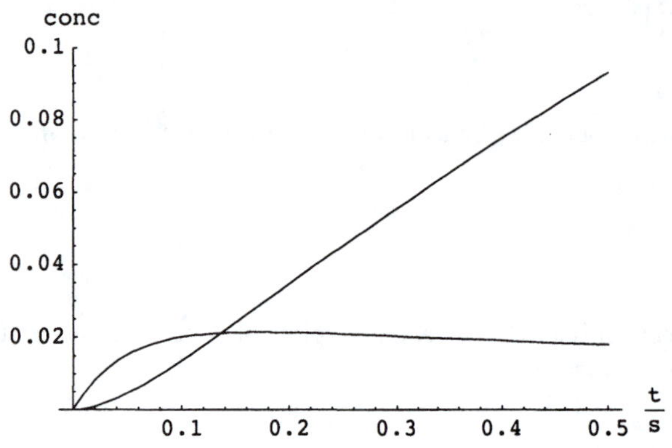

18.E The simplest example of an autocatalytic reaction is A + B -> 2B. Assuming that the rate constant is unity and that $[A]_0 = 1$, plot [A] and [B] versus time and explore the effect of varying the initial concentration of B from 0.01 to 0.2. Note that [B] levels off at $[A]_0 + [B]_0$.

SOLUTION

When $[B]_0 = 0.01$

```
eqns = {ca'[t] == -ca[t] * cb[t], cb'[t] == ca[t] * cb[t], ca[0] == 1, cb[0] == .01};
vars = {ca[t], cb[t]};
concs = NDSolve[eqns, vars, {t, 0, 10}];

Plot[Evaluate[vars /. concs], {t, 0, 10}, AxesLabel -> {"t", "conc"}];
```

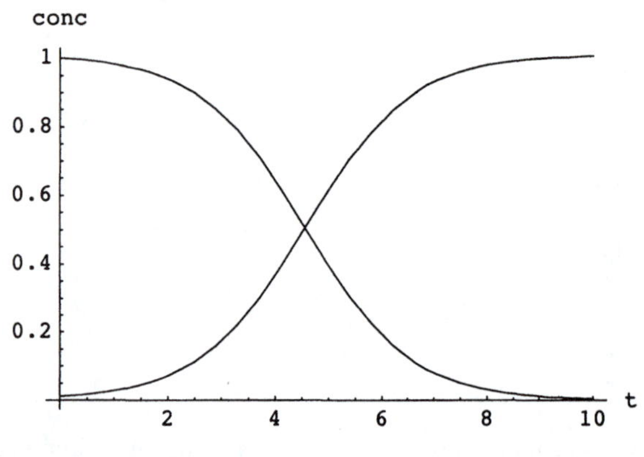

When $[B]_0 = 0.2$

```
eqns = {ca'[t] == -ca[t] * cb[t], cb'[t] == ca[t] * cb[t], ca[0] == 1, cb[0] == .2};
vars = {ca[t], cb[t]};
concs = NDSolve[eqns, vars, {t, 0, 10}];
```

```
Plot[Evaluate[vars /. concs], {t, 0, 10}, AxesLabel -> {"t", "conc"}];
```

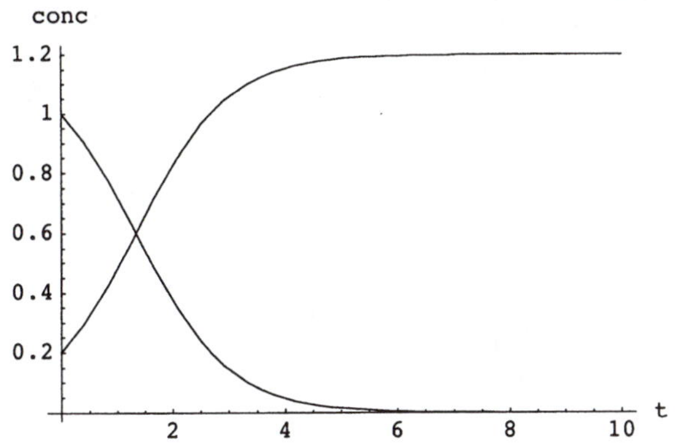

18.F Calculate the activation energy and the pre-exponential factor for the gas reaction
$N_2O_5 = 2NO_2 + (1/2)O_2$

T/K	273	298	308	318	328	338
$k \times 10^5 / s^{-1}$	.0787	3.46	13.5	49.8	150	487

SOLUTION

```
ClearAll["Global`*"]

recipT = 1 / {273, 298, 308, 318, 328, 338};

lnk = Log[{.0787 * 10^-5, 3.46 * 10^-5, 13.5 * 10^-5, 49.8 * 10^-5, 150 * 10^-5, 487 * 10^-5}];

data = Transpose[Join[{recipT, lnk}]];

pointplot = ListPlot[data, AxesOrigin -> {.0029, -15},
   Prolog -> AbsolutePointSize[3], DisplayFunction → Identity];

eq = Fit[data, {1, x}, x]

31.2729 - 12375.8 x

lineplot = Plot[31.2729 - 12375 x, {x, .0029, .0037},
   AxesOrigin -> {0.0029, -15}, DisplayFunction → Identity];
```

```
Show[pointplot, lineplot,
  DisplayFunction → $DisplayFunction, AxesLabel -> {"K" / "T", "ln k"}];
```

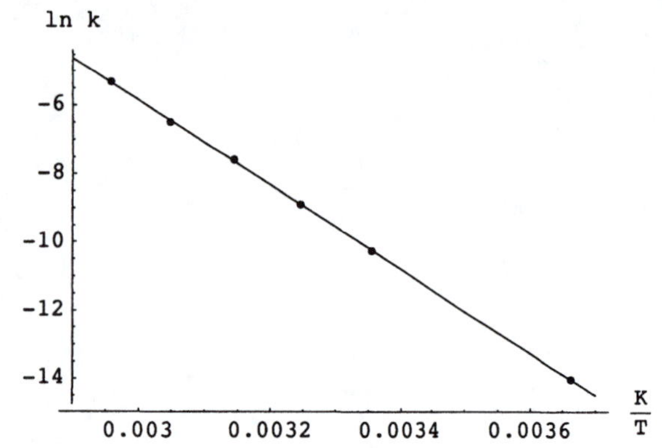

The activation energy in J mol^{-1} is given by

```
12375 * 8.3145
```

```
102892.
```

The preexponential factor in s^{-1} is given by

```
Exp[31.27]
```

3.8053×10^{13}

Data for other reactions can be treated in this way.

18.G (*a*) Calculate the activation energy and the pre-exponential factor for the hydrolysis of 2-chlorooctane. (*b*) Calculate k at 50 ^{o} C.

t/^{o} C	0	25	35	45
k /s^{-1}	1.06 x10^{-5}	3.19 x10^{-4}	9.86 x10^{-4}	2.92 x10^{-3}

SOLUTION

```
ClearAll["Global`*"]
```

(a)

```
recipT = 1 / {273, (273 + 25), (273 + 35), (273 + 45)};

lnk = Log[{1.06 * 10^-5, 3.19 * 10^-4, 9.86 * 10^-4, 2.92 * 10^-3}];

data = Transpose[Join[{recipT, lnk}]];

pointplot = ListPlot[data, AxesOrigin -> {.0031, -12},
   Prolog -> AbsolutePointSize[3], DisplayFunction → Identity];

eq = Fit[data, {1, x}, x]

28.3171 - 10852.4 x

lineplot = Plot[28.3171 - 10852 x, {x, .0031, .0037},
   AxesOrigin -> {0.0031, -12}, DisplayFunction → Identity];

Show[pointplot, lineplot,
  AxesLabel -> {"K" / "T", "ln k"}, DisplayFunction → $DisplayFunction];
```

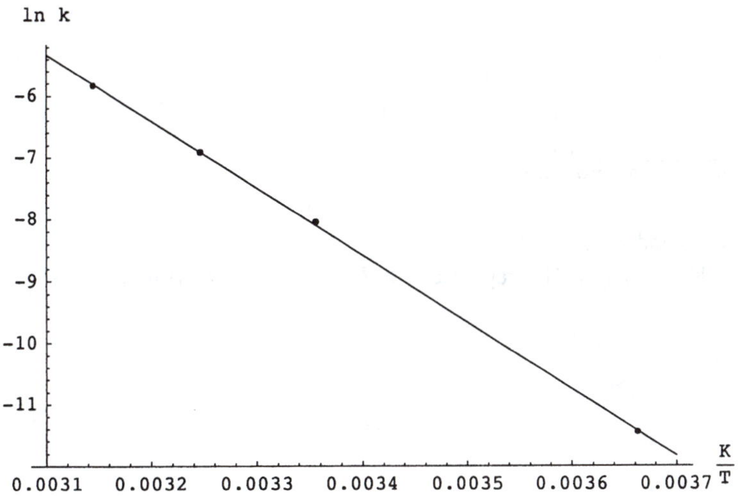

The activation energy in joules mol^{-1} is given by

```
10852 * 8.3145

90229.
```

The pre-exponential factor in s^{-1} is given by

```
A = Exp[28.32]

1.99168 × 10^12
```

(b) The value of the first order rate constant at 50 ^{o} C in s^{-1} is given by

```
Exp[28.32] * Exp[-10852 / (273.15 + 50)]

0.0051853
```

18.H (a) Plot k_{uni} versus [M] according to the Lindemann equation with $k_1 = 1$, $k_2 = 3$, and $k_3 = 1$. (b) Plot $\log_{10}(k_{uni}/k^{\infty})$ versus $\log_{10}$ (P), where the pressure is expressed in pascals. Compare this plot with Fig. 18.11.

SOLUTION

(a) $k_{uni} = k_1 k_3 [M]/(k_2 [M] + k_3)$

```
Plot[m / (1 + 3 * m), {m, 0, 2}, AxesLabel -> {"[M]" / "mol L^-1", "k_uni" / "s^-1"}];
```

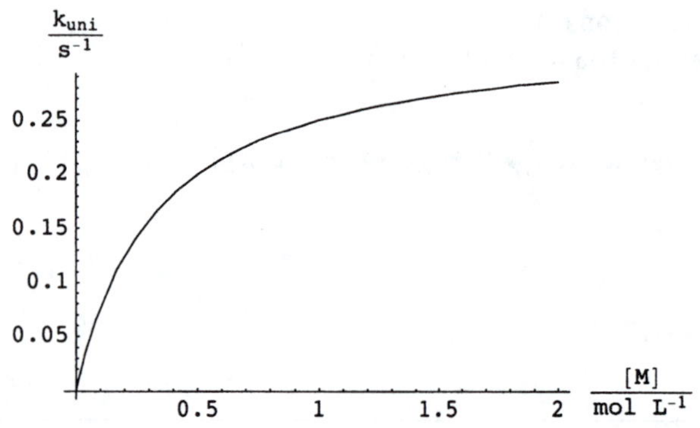

(b) $k_{uni} = k^{\infty} [M]/((k_3/k_2) + [M])$ where $k^{\infty} = 1/3$ and $k_3/k_2 = 1/3$

Since $P = [M]RT$, where $R = 8.314*10^3$ L Pa K^{-1} mol^{-1}, [M] is replaced with P/RT. Replace P with 10^x.

```
Plot[Log[10, (1 / 3) * 10^x / ((8.314 * 3) * (298)) / ((1 / 3) + 10^x / ((8.314 * 3) * (298)))],
  {x, 0, 6}, AxesLabel -> {"log(P/Pa)", "log(k_uni/k_∞)"}, AxesOrigin -> {0, -4}];
```

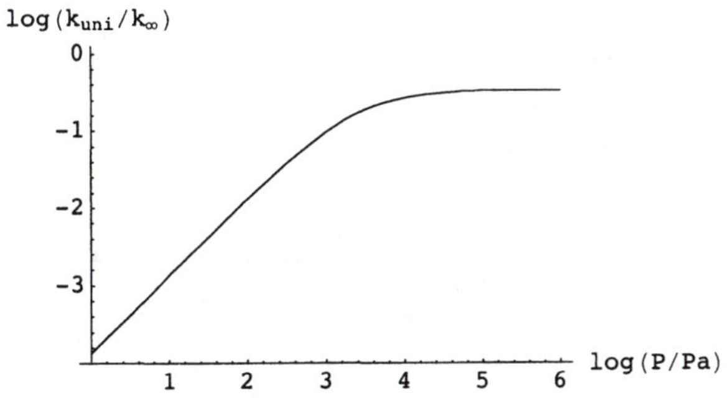

18.I For the monomolecular triangle reaction (see equations 18.60 and 18.61) calculate concentrations as a function of time for the following three cases and discuss the results in terms of the principle of detailed balance: (a) The rate constants in mechanism 18.60 are all unity. (b) The rate constants in mechanism 18.61 are all unity. (c) The rate constants in mechanism 18.61 are all unity except for $k_1 = 1.1$. The initial concentration of A can be taken as 1 M.

SOLUTION

(a) In Mathematica, a derivative is indicated by an apostrophe, and so the rate equations and initial conditions are given by

```
ClearAll["Global`*"]

eqns = {c1'[t] == c3[t] - c1[t], c2'[t] == c1[t] - c2[t],
        c3'[t] == c2[t] - c3[t], c1[0] == 1, c2[0] == 0, c3[0] == 0};
```

The list of variables to be calculated is given by

```
vars = {c1[t], c2[t], c3[t]};
```

NDSolve is used to calculate interpolating functions that give the concentrations of the three reactants at times in the specified range.

```
concs = NDSolve[eqns, vars, {t, 0, 20}];
```

A plot can be obtained by evaluating these interpolation functions. The ReplaceAll operation (/.) applies the interpolating functions to the three concentration variables.

```
plot1a = Plot[Evaluate[vars /. concs],
   {t, 0, 10}, AxesLabel -> {"t" / "s", "[A_i]"}, PlotRange → {0, 1},
   PlotStyle → {{Dashing[{1, 0}]}, {Dashing[{.01, .01}]}, {Dashing[{0.02, 0.02}]}},
   DisplayFunction → Identity];


plot1b = Plot[Evaluate[vars /. concs], {t, 0, 10},
   AxesLabel -> {"t" / "s", "[A_i]"}, PlotRange → {.32, .4}, AxesOrigin → {0, .32},
   PlotStyle → {{Dashing[{1, 0}]}, {Dashing[{.01, .01}]}, {Dashing[{0.02, 0.02}]}},
   DisplayFunction → Identity];
```

```
Show[GraphicsArray[{plot1a, plot1b}]];
```

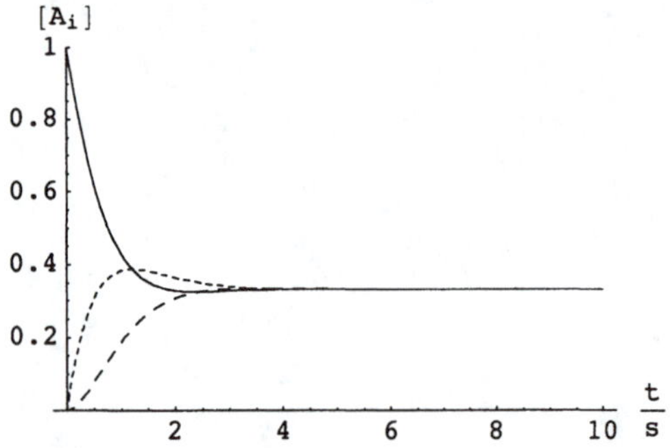

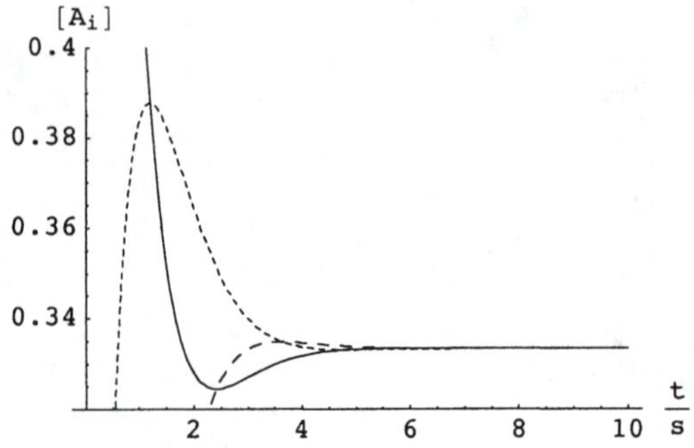

Notice the overshoots and undershoots. The system continues to oscillate at longer times. The oscillations at longer times have lower amplidudes. The system reaches a steady state, but no real system would behave this way.

(b) Now consider the reversible mechanism with all six rate constants equal to unity.

```
eqns2 = {c1'[t] == c2[t] + c3[t] - 2 * c1[t], c2'[t] == c1[t] + c3[t] - 2 * c2[t],
        c3'[t] == c2[t] + c1[t] - 2 * c3[t], c1[0] == 1, c2[0] == 0, c3[0] == 0};

vars2 = {c1[t], c2[t], c3[t]};

concs2 = NDSolve[eqns2, vars2, {t, 0, 10}];

plot2a = Plot[Evaluate[vars2 /. concs2],
   {t, 0, 10}, AxesLabel -> {"t" / "s", "[Ai]"}, PlotRange → {0, 1},
   PlotStyle → {{Dashing[{1, 0}]}, {Dashing[{.01, .01}]}, {Dashing[{0.02, 0.02}]}},
   DisplayFunction → Identity];

plot2b = Plot[Evaluate[vars2 /. concs2], {t, 0, 10},
   AxesLabel -> {"t" / "s", "[Ai]"}, PlotRange → {.32, .4}, AxesOrigin → {0, .32},
   PlotStyle → {{Dashing[{1, 0}]}, {Dashing[{.01, .01}]}, {Dashing[{0.02, 0.02}]}},
   DisplayFunction → Identity];
```

```
Show[GraphicsArray[{plot2a, plot2b}]];
```

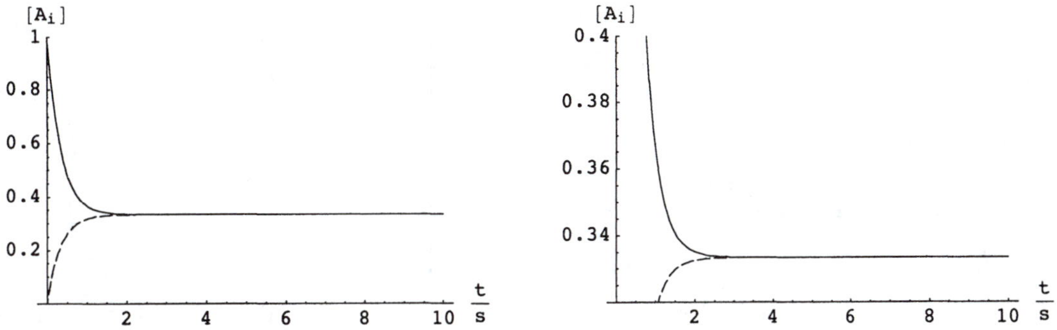

There are no overshoots in the approach to equilibrium. The concentrations of A_2 and A_3 are equal at all times.

(c) Now make k_1 = 1.1, while the other rate constans are unity.

```
eqns3 = {c1'[t] == c2[t] + c3[t] - 2.1*c1[t], c2'[t] == 1.1*c1[t] + c3[t] - 2*c2[t],
        c3'[t] == c2[t] + c1[t] - 2*c3[t], c1[0] == 1, c2[0] == 0, c3[0] == 0};

vars3 = {c1[t], c2[t], c3[t]};

concs3 = NDSolve[eqns3, vars3, {t, 0, 10}];

plot3a = Plot[Evaluate[vars3 /. concs3],
   {t, 0, 10}, AxesLabel -> {"t" / "s", "[Ai]"}, PlotRange → {0, 1},
   PlotStyle → {{Dashing[{1, 0}]}, {Dashing[{.01, .01}]}, {Dashing[{0.02, 0.02}]}},
   DisplayFunction → Identity];

plot3b = Plot[Evaluate[vars3 /. concs3], {t, 0, 10},
   AxesLabel -> {"t" / "s", "[Ai]"}, PlotRange → {.32, .4}, AxesOrigin → {0, .32},
   PlotStyle → {{Dashing[{1, 0}]}, {Dashing[{.01, .01}]}, {Dashing[{0.02, 0.02}]}},
   DisplayFunction → Identity];
```

```
Show[GraphicsArray[{plot3a, plot3b}]];
```

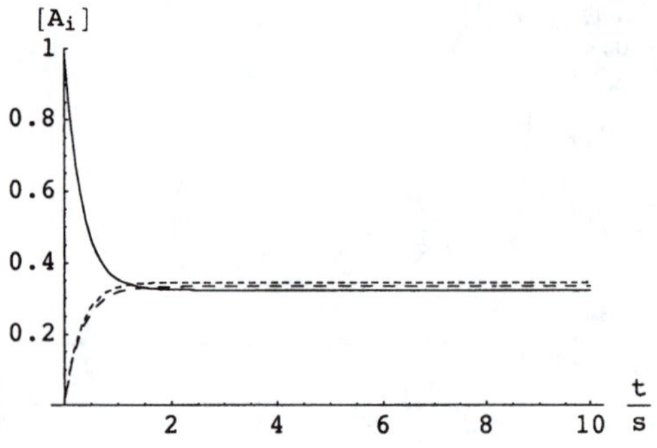

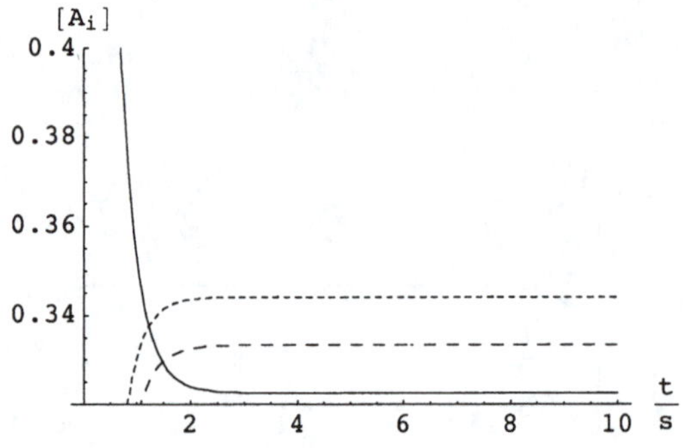

The concentrations at long times do not correspond with the equilibrium constants for the three reactions. Thus this mechanism reaches a steady state rather than equilibrium. A real system would not behave like this. The rate constants used here do not satisfy the requirement that $k_1 \, k_3 \, k_5 = k_2 \, k_4 \, k_6$.

18.J Assume that a reaction A = B goes through an intermediate I, but that $k_2 << k_{-1}$.

$$A \underset{k_{-1}}{\overset{k_1}{\rightleftharpoons}} I \overset{k_2}{\longrightarrow} B$$

Under these conditions, the first step remains at equilibrium.

$$\frac{d[B]}{dt} = k[A] = \frac{k_1 k_2}{k_{-1}} [A] = K k_2 [A]$$

Assume that for the first step, K = exp(-ΔH/RT), and $k_2 = 1 \text{ s}^{-1}$ at 298 K and E_a = 20 kJ mol^{-1}. Plot lnk versus 1/T for ΔH = 0, -10, -20, and -30 kJ mol^{-1}.

SOLUTION

When ΔH = 0 and T = 298 K,
k=a*Exp[-20*10^3/(8.314*298)]
When ΔH ≠ 0,
k=a*Exp[-(20*10^3+ΔH)/(8.314*298)]

```
a = 1 / Exp[-20 * 10^3 / (8.314 * 298)]

3204.85

ka = 3.20 * 10^3 * Exp[-20 * 10^3 / (8.3145 * (1 / x))]

3200. e^(-2405.44 x)

kb = 3.20 * 10^3 * Exp[-10 * 10^3 / (8.3145 * (1 / x))]

3200. e^(-1202.72 x)

kc = 3.20 * 10^3 * Exp[0 * 10^3 / (8.3145 * (1 / x))]

3200.
```

```
kd = 3.20 * 10^3 * Exp[10 * 10^3 / (8.3145 * (1 / x))]
```

$3200.\ e^{1202.72\ x}$

Note that when ΔH for the first step is sufficiently negative, the rate constant decreases with increasing temperature.

```
Plot[{Log[ka], Log[kb], Log[kc], Log[kd]}, {x, .002, .003}, AxesLabel → {"1/T", "lnk"}];
```

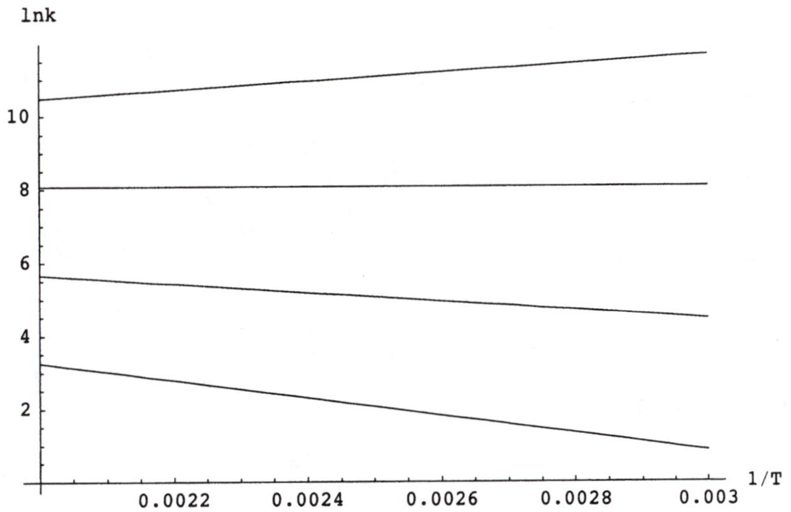

18.K Sometimes Arrhenius plots are curved and can be represented by
$k = aT^n\ e^{-E/RT}$
(a) Plot logk versus 1/T for n = -2, -1, 0, 1, 2 when a = 10^8 and E = 10 kJ mol $^{-1}$. Use the temperature range 300 K to 1000 K.
(b) Calclate the activation energies at temperatures 333, 500, and 1000 K and for n = -2, -1, 0, 1, 2 and make a table..

SOLUTION

(a)

```
k1 = 10^8 * (1 / x) ^-2 * Exp[-10000 / (8.314 * (1 / x))]
```

$100000000\ e^{-1202.79\ x}\ x^2$

where T = 1/x.

```
k2 = 10^8 * (1 / x) ^-1 * Exp[-10000 / (8.314 * (1 / x))]
```

$100000000\ e^{-1202.79\ x}\ x$

```
k3 = 10^8 * (1 / x) ^0 * Exp[-10000 / (8.314 * (1 / x))]
```

$100000000\ e^{-1202.79\ x}$

```
k4 = 10^8 * (1 / x) ^1 * Exp[-10000 / (8.314 * (1 / x))]
```

$$\frac{100000000\ e^{-1202.79\ x}}{x}$$

```
k5 = 10^8 * (1 / x) ^2 * Exp[-10000 / (8.314 * (1 / x))]
```

$$\frac{100000000\ e^{-1202.79\ x}}{x^2}$$

```
Plot[{Log[10, k1], Log[10, k2], Log[10, k3], Log[10, k4], Log[10, k5]}, {x, .001, .003},
  AxesOrigin → {.001, -1}, PlotRange → {-1, 14}, AxesLabel → {"K/T", "logk"}];
```

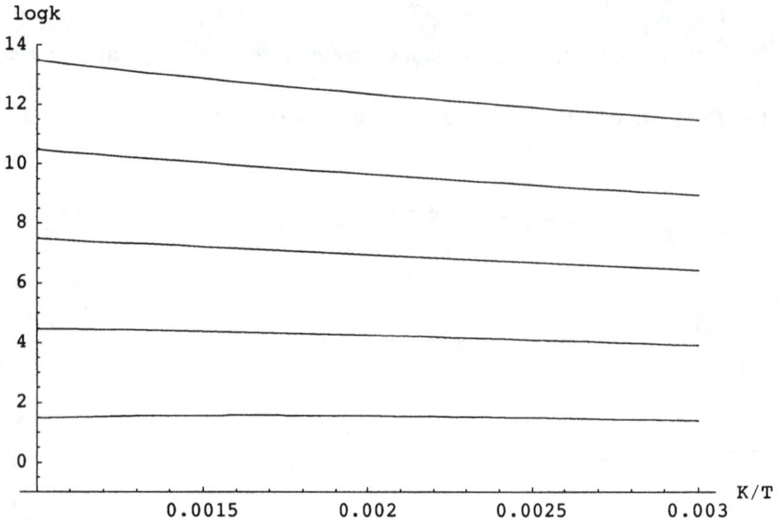

The lowest plot is for n = -2 and the highest is for n = 2.

(b)

```
k1t = 10^8 * t^-2 * Exp[-10000 / (8.314 * t)]

100000000 e^-1202.79/t
----------------------
         t^2

line1 = 8.314 * t^2 * D[Log[k1t], t] /. t → {333, 500, 1000}

{4462.88, 1686., -6628.}

k2t = 10^8 * t^-1 * Exp[-10000 / (8.314 * t)]

100000000 e^-1202.79/t
----------------------
          t

line2 = 8.314 * t^2 * D[Log[k2t], t] /. t → {333, 500, 1000}

{7231.44, 5843., 1686.}

k3t = 10^8 * t^0 * Exp[-10000 / (8.314 * t)]

100000000 e^-1202.79/t

line3 = 8.314 * t^2 * D[Log[k3t], t] /. t → {333, 500, 1000}

{10000., 10000., 10000.}

k4t = 10^8 * t^1 * Exp[-10000 / (8.314 * t)]

100000000 e^-1202.79/t t

line4 = 8.314 * t^2 * D[Log[k4t], t] /. t → {333, 500, 1000}

{12768.6, 14157., 18314.}
```

```
k5t = 10^8 * t^2 * Exp[-10000 / (8.314 * t)]
```

$100000000\, e^{-1202.79/t}\, t^2$

```
line5 = 8.314 * t^2 * D[Log[k5t], t] /. t → {333, 500, 1000}
```

{15537.1, 18314., 26628.}

```
TableForm[{line1, line2, line3, line4, line5}, TableHeadings →
  {{"n=-2", "n=-1", "n=0", "n=1", "n=2"}, {"333 K", "500 K", "1000 K"}}]
```

	333 K	500 K	1000 K
n=-2	4462.88	1686.	-6628.
n=-1	7231.44	5843.	1686.
n=0	10000.	10000.	10000.
n=1	12768.6	14157.	18314.
n=2	15537.1	18314.	26628.

Note the negative activation energy at n = -2 and 1000 K.

Chapter 19 Chemical Dynamics and Photochemistry

19.A In molecular beam experiments, reactant molecules can be accelerated to supersonic velocities by allowing a dilute mixture of the reactant in an inert carrier gas to expand through a pinhole into a vacuum. (*a*) Use equation 19.40 to calculate the peak velocity of ethane molecules in a carrier gas of helium that expands from a source chamber at 298.15 K. (*b*) Calculate the temperature at which ethane molecules have this root-mean-square velocity.

SOLUTION

(a) The approximate equation for the peak velocity is given by

$v_p = (\frac{2RT}{M_{He}})^{1/2}(\frac{\gamma}{\gamma-1})^{1/2}$

For helium, the molar heat capacity at constant volume is (3/2)R and the molar heat capacity at constant pressure is (5/2)R so that $\gamma = 5/3$. Therefore, the peak velocity of ethane in the supersonic beam is given by

```
vpeak = (2 8.314 298.15 / .00400) ^ .5 ((5 / 3) / (5 / 3 - 1)) ^ .5

1760.26
```

```
Solve[1760 == (3 8.3145 t / .030) ^ .5, t]

Solve::ifun : Inverse functions are being used by Solve, so some
   solutions may not be found; use Reduce for complete solution information. More...

{{t → 3725.54}}
```

Thus the peak velocity of ethane molecules in the supersonic beam corresponds with a temperature of 3700 K.

19.B The simplest equation that gives a saddle-shaped surface is $z = y^2 - x^2$. Plot this surface in three dimensions and think about the path that a vibrating molecule would take across it with the minimum energy. Consider different angles of approach.

SOLUTION

```
Plot3D[z = y^2 - x^2, {x, -1, 1}, {y, -1, 1}, AxesLabel → {"x", "y", "z"}];
```

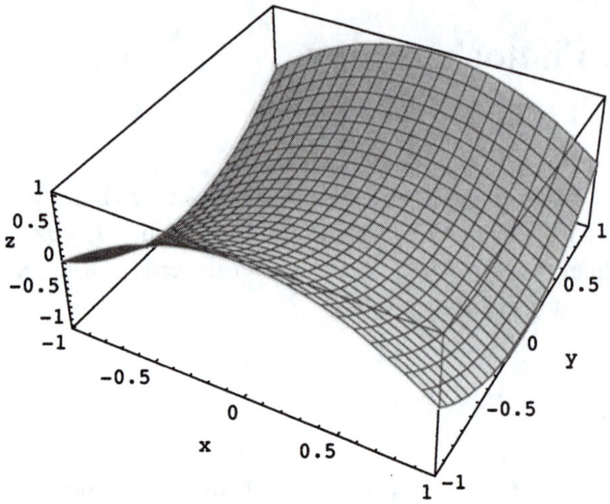

19.C Plot the ratio of the reaction cross section S to the collision cross section πd_{12}^2 as a function of the ratio of the collision energy ϵ to the minimum energy ϵ_c along the line of centers to cause reaction.

SOLUTION

```
Plot[1 - 1 / r, {r, 0, 10}, PlotRange → {0, 1}, AxesLabel → {"ε/εc", "S/πd12^2"}];
```

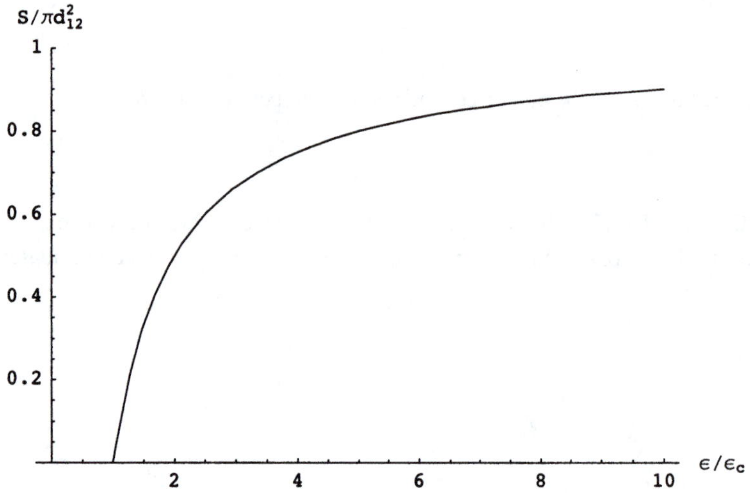

19.D If the average energy of a probe photon in a femtosecond transition state experiment is 5362 cm^{-1}, what is the average speed <v> of the wave packet in the dissociation of NaI?

SOLUTION

$E = (1/2)\,\mu < v >^2$

```
μ = (23 * 10^-3) * (126.9 * 10^-3) / ((149.9 * 10^-3) * (6.02 * 10^23))
```

3.23438×10^{-26}

This reduced mass is in kg.
The energy in joules is

```
5362 * (1.986 * 10^-5) * (10^-18)
```

1.06489×10^{-19}

The average velocity in m/s is given by

```
(2 * (1.065 * 10^-19) / μ) ^.5

2566.22
```

This agrees with Baskin and Zewail.

Chapter 20 Kinetics in the Liquid Phase

20.A In Example 20.9 the following four rate constants were deduced from steady-state rate measurements at 25 ^oC and pH 8.

$$E + \text{fumarate} \underset{0.6\times10^3}{\overset{1.14\times10^8}{\rightleftharpoons}} X \underset{8\times10^6}{\overset{0.2\times10^8}{\rightleftharpoons}} E + \text{L-malate}$$

where the unimolecular rate constants are in s^{-1} and the bimolecular constants are in mol L^{-1} s^{-1}. Plot the concentrations of the the four reactants when the initial concentrations of fumarate and free enzymatic sites, [E], are 10^{-4} M and 10^{-8} M, respectively.

SOLUTION

(a) The four rate equations and the intial conditions are given by

```
ClearAll["Global`*"]

Off[General::spell1];
Off[General::spell];

eqns1 = {cfum'[t] == -(1.14*10^8) cE[t]*cfum[t] + (.6*10^3) cX[t],
        cX'[t] ==
     (1.14*10^8)*cE[t]*cfum[t] + (8*10^6)*cE[t]*cmal[t] - (.8*10^3)*cX[t],
        cmal'[t] == (.2*10^3)*cX[t] - (8*10^6)*cE[t]*cmal[t],
        cE'[t] ==
     (.8*10^3)*cX[t] - (1.14*10^8)*cE[t]*cfum[t] - (8*10^6) cE[t]*cmal[t],
cfum[0] == 10^-4, cE[0] == 10^-8, cmal[0] == 0, cX[0] == 0};

vars1 = {cfum[t], cE[t], cX[t], cmal[t]};

conc1 = NDSolve[eqns1, vars1, {t, 100}];

Plot[Evaluate[vars1 /. conc1], {t, 0, 100}, AxesLabel -> {"t" / "s", "conc"}];
```

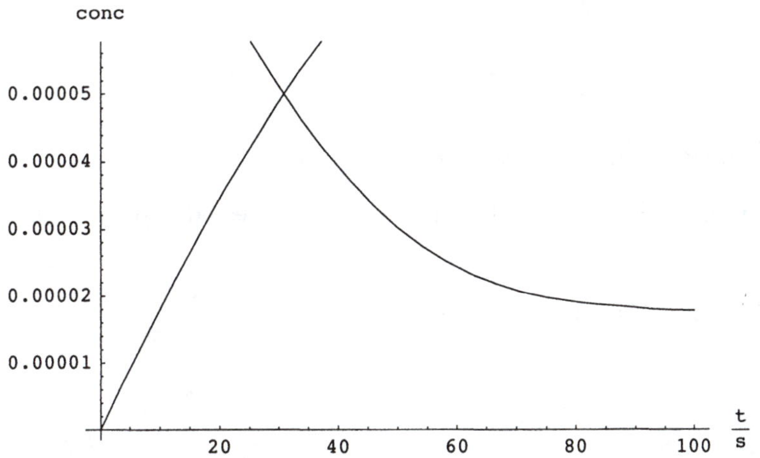

This plot does not show [E]and [X] because they are so small, and so we use a log plot.

```
Plot[Evaluate[Log[10, vars1] /. conc1], {t, 0, 100},
  AxesOrigin -> {0, -12}, AxesLabel -> {"t" / "s", "Log[10,conc]"}];
```

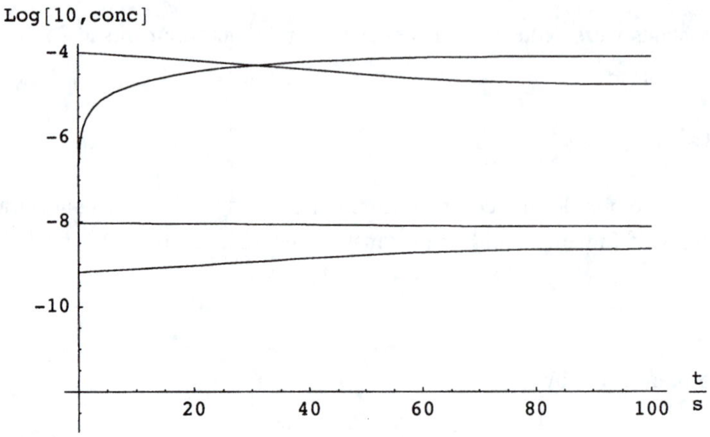

The next to the lowest plot is that for log[X], and it shows that X decreases a little during the steady state. The initial formation of X is not shown because it occurs so rapidly. However, we can see it if we expand the initial time scale by 10^4.

```
conc1 = NDSolve[eqns1, vars1, {t, .01}];
```

```
Plot[Evaluate[Log[10, vars1] /. conc1], {t, 0, .01}, AxesOrigin -> {0, -20},
  PlotRange -> {-20, -2}, AxesLabel -> {"t" / "s", "Log[10,conc]"}];
```

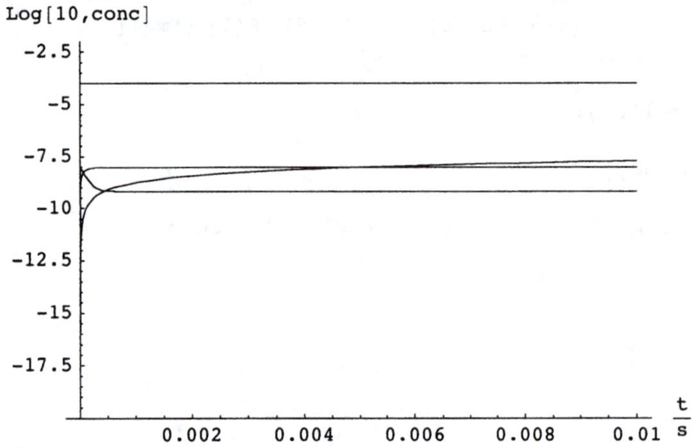

This shows that it takes 0.01 s for malate to reach $10^{-7.5}$ M. To see the induction period in the formation of X, we expand the initial time scale by a factor of 10^2.

```
conc1 = NDSolve[eqns1, vars1, {t, .0001}];
```

```
Plot[Evaluate[Log[10, vars1] /. conc1], {t, 0, .0001},
  AxesOrigin -> {0, -20}, AxesLabel -> {"t" / "s", "Log[10,conc]"}];
```

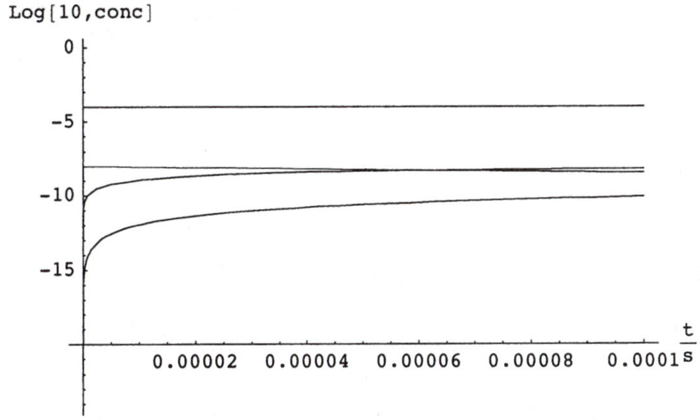

This shows [X] starting at about 10^{-10} M and becoming equal to [E] by 0.00006 s.

The steady state concentration of X is given by

$[X]/[E]_0 = (k_1 [fum] + k_4 [mal])/(k_2 + k_3 + k_1 [fum] + k_4 [mal])$

Therefore, early in the reaction, [X] = 9.3×10^{-9} M. When the reverse reaction was studied, early in the reaction [X] = 5×10^{-9} M. Therefore, the steady state concentration of X decreases during the reaction

```
cX = ((1.14 * 10^8) * (10^-4) / ((.8 * 10^3) + (1.14 * 10^8) * (10^-4))) * 10^-8
```

9.34426×10^{-9}

```
cX = ((8 * 10^6) * (10^-4) / ((.8 * 10^3) + (8 * 10^6) * (10^-4))) * 10^-8
```

$5. \times 10^{-9}$

20.B The diffusion coefficient of sucrose at 25 °C in dilute aqueous solution is 5.1×10^{-6} cm^2 s^{-1}. (*a*) Plot the c/c_0, where c_0 is the initial concentration, of sucrose versus distance at 1, 4, 9, 16, and 25 hours. (*b*) Plot the concentration gradient $-dc/dx$ at these times.

SOLUTION

```
diff = 5.1 * 10^-6(*cm^2 s^-1*);

co = 1.;

conc[x_, t_] := N[.5 * co * (1 - Erf[x / Sqrt[4 * diff * t]])];

sec = 1.; min = 60 sec; hour = 60 min;
```

```
Plot[{conc[x, 1 sec], conc[x, 1 hour], conc[x, 4 hour], conc[x, 9 hour],
    conc[x, 16 hour], conc[x, 25 hour]}, {x, -2, 2},
  PlotRange → {{-2, 2}, {0, 1}}, AxesOrigin -> {-2, 0}, AxesLabel -> {"x/cm", "c/c0"}];
```

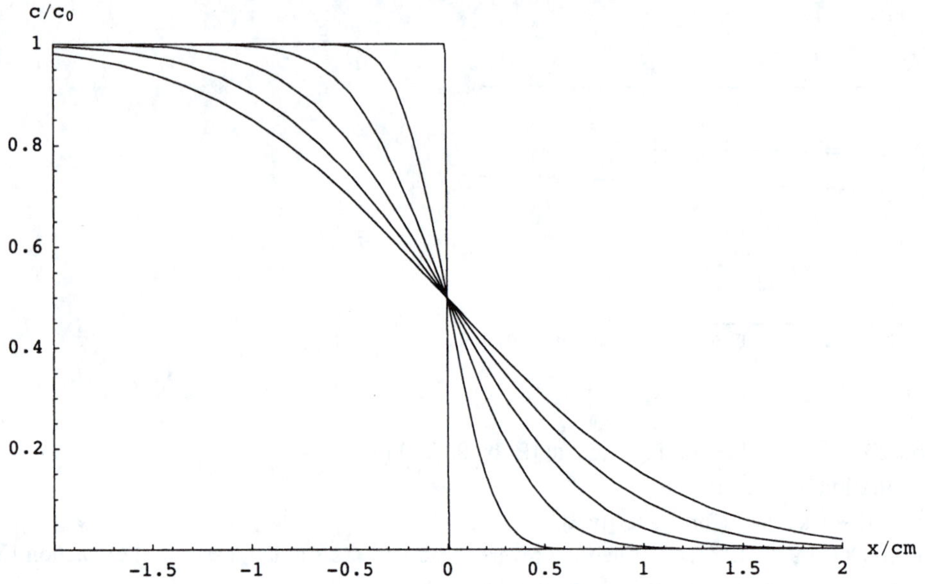

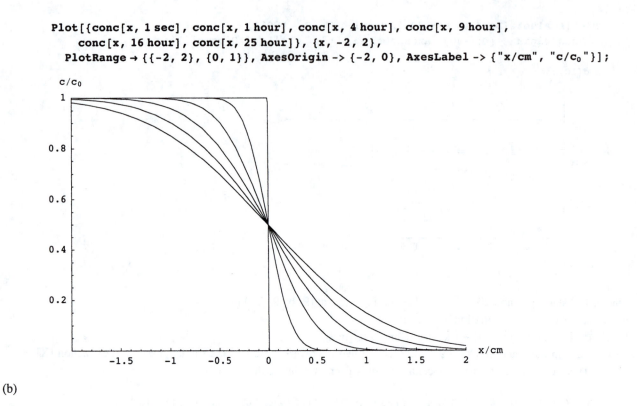

(b)

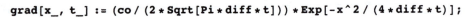

```
grad[x_, t_] := (co / (2 * Sqrt[Pi * diff * t])) * Exp[-x^2 / (4 * diff * t)];
```

```
Plot[{grad[x, 1 hour], grad[x, 4 hour], grad[x, 9 hour], grad[x, 16 hour], grad[x, 25 hour]},
  {x, -2, 2}, PlotRange → {{-2, 2}, {0, 2.5}},
  AxesOrigin -> {-2, 0}, AxesLabel -> {"x/cm", "-dc/dx"}];
```

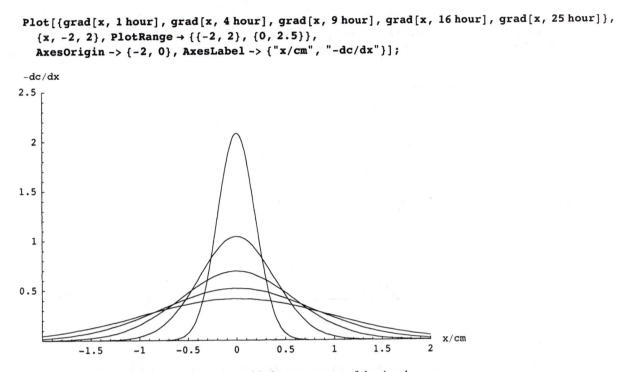

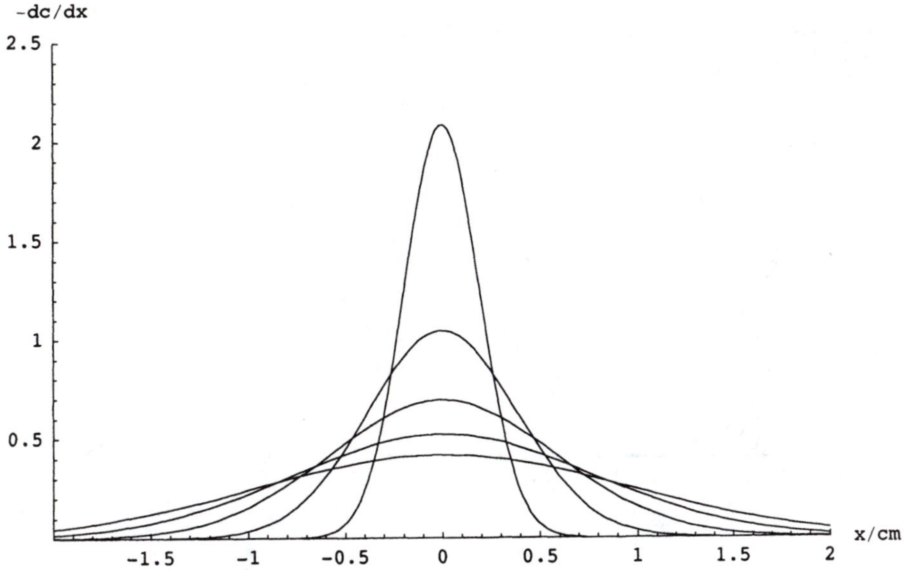

Note that the breadth of the diffuse region increases with the square root of the time because standard deviation = $\sigma = \sqrt{2\mathrm{Dt}}$

20.C The diffusion coefficient of hemoglobin at 25 ^{o}C in dilute aqueous solution is 6.9x10^{-7} cm^{2} s^{-1}. (*a*) Plot the c/c_0 of hemoglobin, where c_0 is the initial concentration, versus distance at 1, 4, 9, 16, and 25 hours. (*b*) Plot the concentration gradient -dc/dx at these times.

SOLUTION

```
ClearAll["Global`*"]

diff = 6.9 * 10^-7 (*cm^2 s^-1*);

co = 1.;

conc[x_, t_] := N[.5 * co * (1 - Erf[x / Sqrt[4 * diff * t]])]

sec = 1.; min = 60 sec; hour = 60 min;
```

```
Plot[{conc[x, 1 sec], conc[x, 1 hour], conc[x, 4 hour], conc[x, 9 hour],
    conc[x, 16 hour], conc[x, 25 hour]}, {x, -2, 2},
  PlotRange -> {{-2, 2}, {0, 1}}, AxesOrigin -> {-2, 0}, AxesLabel -> {"x/cm", "c/c0"}];
```

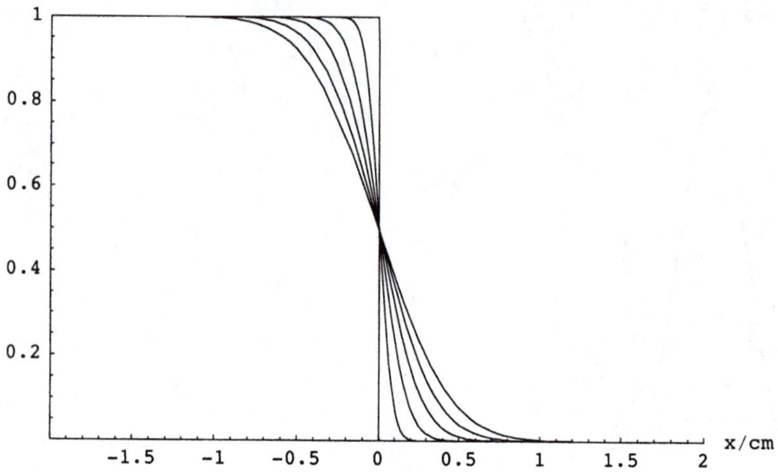

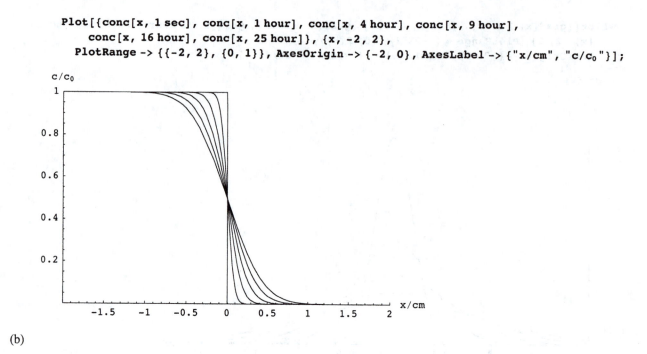

(b)

```
grad[x_, t_] := (co / (2 * Sqrt[Pi * diff * t])) * Exp[-x^2 / (4 * diff * t)]
```

```
Plot[{grad[x, 1 hour], grad[x, 4 hour], grad[x, 9 hour], grad[x, 16 hour], grad[x, 25 hour]},
  {x, -2, 2}, PlotRange -> {{-2, 2}, {0, 6}},
  AxesOrigin -> {-2, 0}, AxesLabel -> {"x/cm", "-dc/dx"}];
```

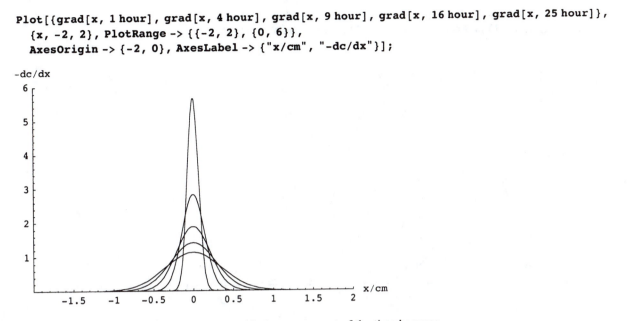

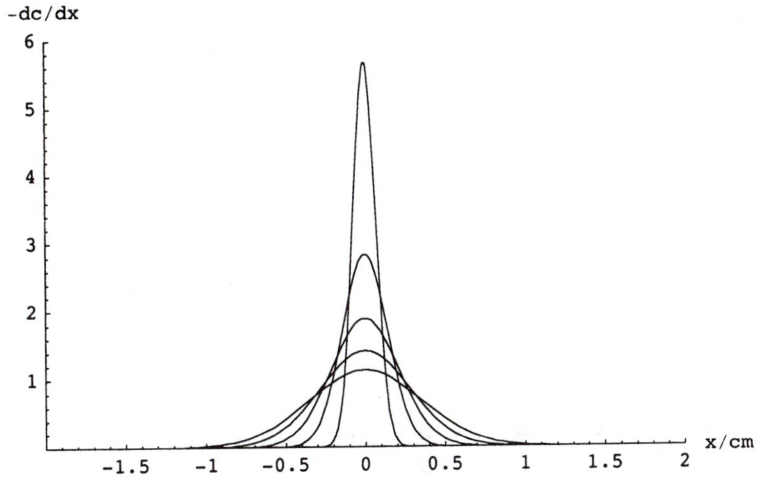

Note that the breadth of the diffuse region increases with the square root of the time because standard deviation = $\sigma = \sqrt{2\,Dt}$

20.D As described in equations 20.86 to 20.88 the maximum rate of an enzyme-catalyzed reaction may go through a maximum as the pH is varied. Plot the apparent rate constant divided by the rate constant for the rate-determining step versus pH for an enzyme-substrate complex that has $K_a = 10^{-4}$ and $K_b = 10^{-5}$, 10^{-6}, 10^{-7}, 10^{-8}, and 10^{-9}.

SOLUTION

```
ClearAll["Global`*"]

relrate[pkb_, pH_] := 1 / (1 + ((10^-pH) / (10^-4)) + ((10^-pkb) / (10^-pH)))
```

```
Plot[Evaluate[Table[relrate[pkb, pH], {pkb, 5, 9}]],
  {pH, 3, 11}, AxesOrigin → {3, 0}, AxesLabel -> {"pH", "k'/k"}];
```

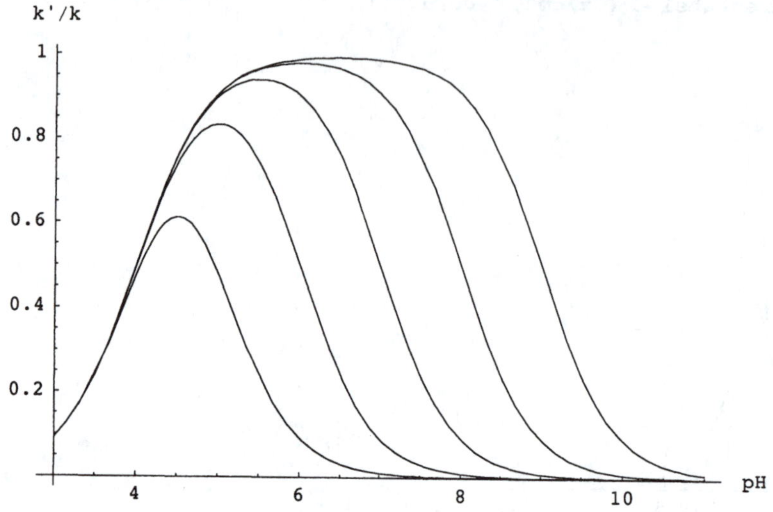

20.E Calculate the Michaelis constant K_M and maximum velocity V_M for fumarase when L-malate is the substrate using the Eadie-Hofstee method and make a plot to show how well the Michaelis equation fits the data.

$[M]/10^{-6}$ M	0.1	0.333	1.0	3.33	10	33.3	100
v	1.9	4.2	6.1	6.5	7.2	7.4	6.9

SOLUTION

When we multiply equation 20.73 by $[E]_0$ we obtain the equation

$v/[M] = v_{max}/K_M - v/K_M$

```
data = {{1.9, 4.2, 6.1, 6.5, 7.2, 7.4, 6.9},
  {1.9, 4.2, 6.1, 6.5, 7.2, 7.4, 6.9} / {.1, .333, 1, 3.33, 10, 33.3, 100}};

plot1 = ListPlot[Transpose[data], AxesOrigin -> {0, 0},
  Prolog -> AbsolutePointSize[4], DisplayFunction → Identity];

eqn = Fit[Transpose[data], {1, v}, v]

26.6556 - 3.62969 v

plot2 = Plot[eqn, {v, 0, 8}, DisplayFunction → Identity];
```

```
Show[plot1, plot2, AxesLabel -> {"v", "v/[M]"},
  PlotRange -> {{0, 8}, {0, 27}}, DisplayFunction -> $DisplayFunction];
```

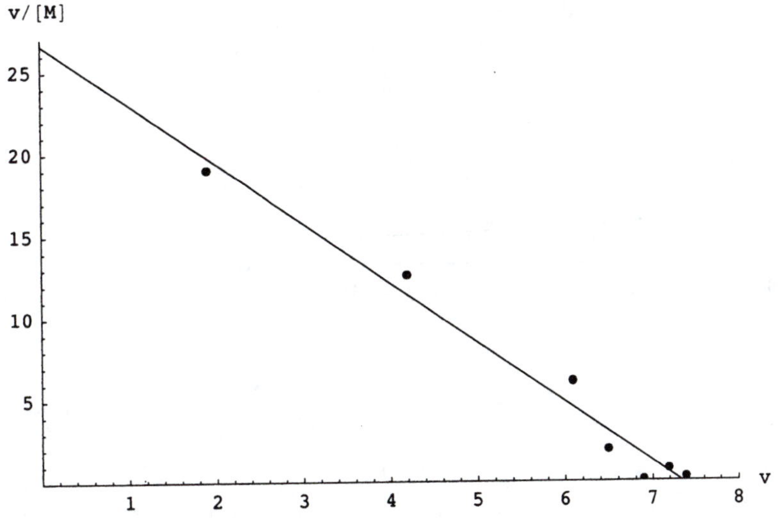

The Michaelis constant K_M is the negative reciprocal of the slope.

```
km = -1/(-3.63)

0.275482
```

where this has the units of micromolar. The maximum velocity is K_M times the intercept

```
.275 * 26.7

7.3425
```

20.F Plot the electrostatic factor f for the second order rate constant at 25 ^oC in water for the reaction of two ions as a function of R_{12} when there are (*a*) opposite unit charges and (*b*) like unit charges.

SOLUTION

(a) As shown in Example 20.6, the equation for opposite charges is

$$f = (-7.12 * 10\wedge-10 / R_{12}) / (e^{-7.12*10\wedge-10/R12} - 1)$$

where R_{12} is expressed in meters.

```
fopp = (-7.12 * 10^-10 / r12) / (Exp[-7.12 * 10^-10 / r12] - 1)
```

$$-\frac{7.12\times10^{-10}}{(-1+e^{-7.12\times10^{-10}/r12})\ r12}$$

```
Plot[Log[fopp], {r12, 0, 10^-9}, PlotRange → {-3, 3}, AxesLabel → {"  R12 /m", "ln f"}];
```

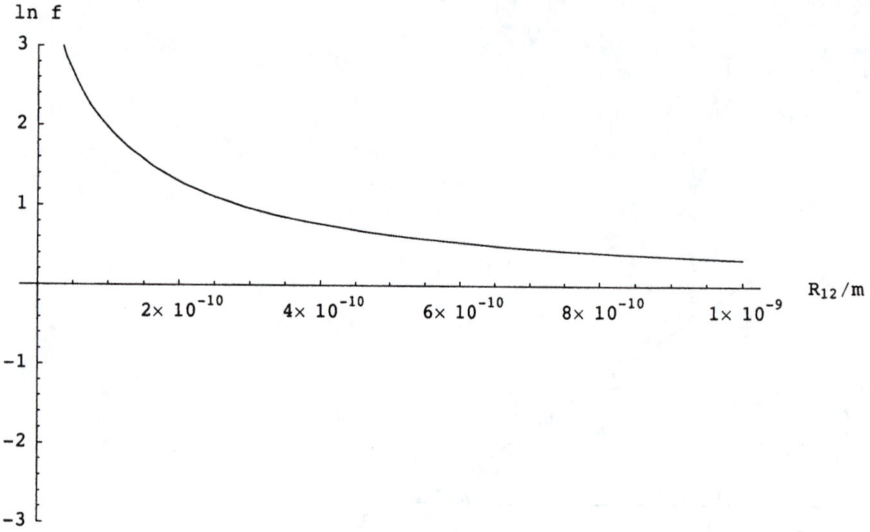

(b) When the unit charges have the same sign, the electrostatic factor is given by

```
flike = (7.12 * 10^-10 / r12) / (Exp[7.12 * 10^-10 / r12] - 1)
```

$$\frac{7.12 \times 10^{-10}}{(-1 + e^{7.12\times10^{-10}/r12})\ r12}$$

```
Plot[{Log[fopp], Log[flike]}, {r12, 0, 10^-9},
  PlotRange → {3, -3}, AxesLabel → {"  R12/m", "ln f"}];
```

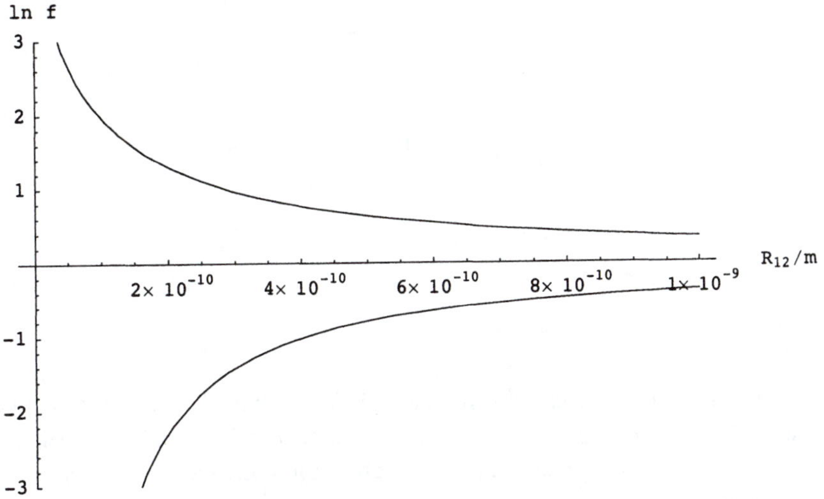

Note that the second-order rate constant is changed by a larger factor when the ions have the same charge.

20.G Solve the simultaneous rate equations for the Lotka mechanism for an autocatalytic reaction described in Section 20.x. Assume the concentration of A is held at 1 M and the rate constants are $k_1 = 0.3\ s^{-1}$, $k_2 = 0.6\ M^{-1}\ s^{-1}$, and $k_3 = 0.8\ s^{-1}$. The initial concentrations of X, Y, and P are to be taken as 0.2 M, 0.1 M, and 0 M. (a) Plot the concentrations of X, Y, and P at short times and discuss why the plots have these shapes. (b) Plot the concentrations of X, Y, and P at longer times and discuss what happens. The *Mathematica* programs for making these plots are given in Ferreira, et al., J. Chem. Educ. 76, 861 (1999).

SOLUTION

```
a = 1;

k1 = .3;

k2 = .6;

k3 = .8;
```

(a) t = 0 to 10 s

```
sol = NDSolve[{x'[t] == k1 * a - k2 * x[t] * y[t],
    y'[t] == k2 * x[t] * y[t] - k3 * y[t],
    p'[t] == k3 * y[t],
    x[0] == 0.2, p[0] == 0.0, y[0] == 0.1},
   {x, y, p}, {t, 0, 10}, MaxSteps → Infinity]

{{x → InterpolatingFunction[{{0., 10.}}, <>],
  y → InterpolatingFunction[{{0., 10.}}, <>], p → InterpolatingFunction[{{0., 10.}}, <>]}}
```

```
Plot[Evaluate[{x[t], y[t], p[t]} /. sol],
  {t, 0, 10}, PlotRange → All, AxesLabel → {"t/s", "c/M"}];
```

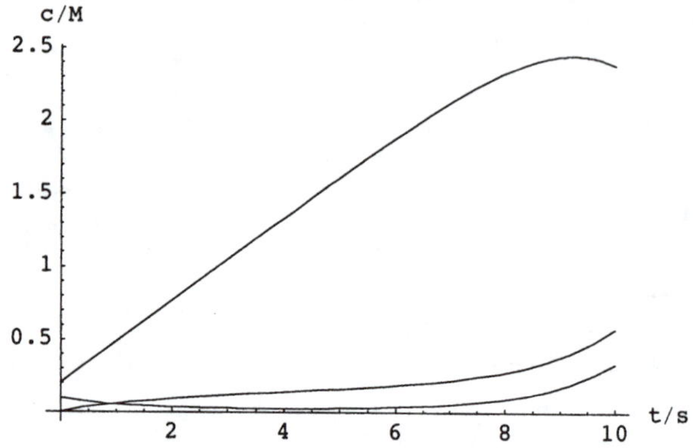

At 10 s the top curve is for X, the middle curve is for P, and and the lowest curve is for Y. Initially X and Y react to produce P, and Y decreases. X increases because it is being produced at a constant rate by the first reaction and is not being used up very fast by the second reaction. As X increases, Y is soon produced more rapidly because the second reaction is autocatalytic. After 9 seconds the second reaction begins pulling down the concentration of X.

(b) t = 0 to t = 100 s

```
sol2 = NDSolve[{x'[t] == k1 * a - k2 * x[t] * y[t],
   y'[t] == k2 * x[t] * y[t] - k3 * y[t],
   p'[t] == k3 * y[t],
   x[0] == 0.2, p[0] == 0.0, y[0] == 0.1},
  {x, y, p}, {t, 0, 100}, MaxSteps → Infinity]

{{x → InterpolatingFunction[{{0., 100.}}, <>],
  y → InterpolatingFunction[{{0., 100.}}, <>],
  p → InterpolatingFunction[{{0., 100.}}, <>]}}
```

This time we will plot X and Y together and P separately.

```
plota = Plot[Evaluate[{x[t], y[t]} /. sol2],
   {t, 0, 100}, PlotRange → All, AxesLabel → {"t/s", "c/M"}];
```

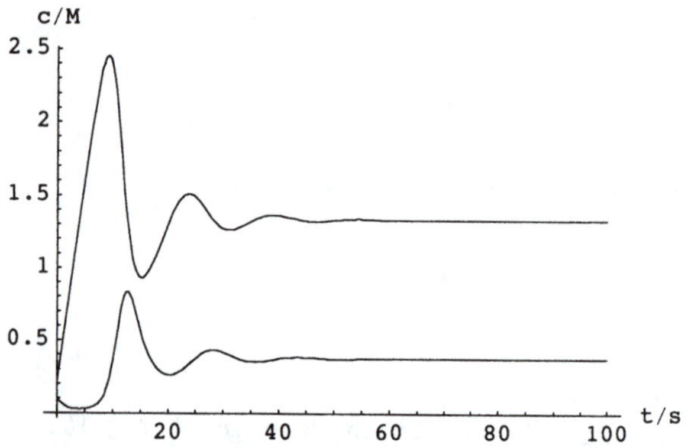

As the concentration of Y gets higher it pulls down the concentration of X, and this causes an oscillation. The next oscillation is smaller, and finally a steady state is reached.

```
plotb =
  Plot[Evaluate[p[t] /. sol2], {t, 0, 100}, PlotRange → All, AxesLabel → {"t/s", "c/M"}];
```

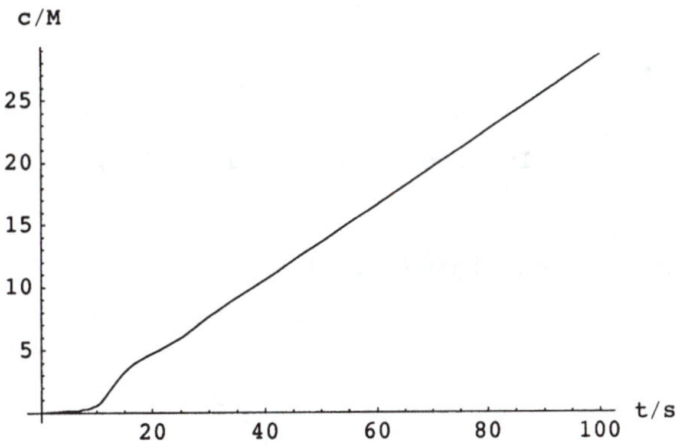

The production of P is slow at first because there is an induction period, but then it is produced most rapidly when Y goes through a maximum. When X and Y get into steady states, their concentrations are no longer changing and P is produced at a steady rate equal to the rate X is being produced.

```
Show[GraphicsArray[{{plota, plotb}}, GraphicsSpacing → 0]];
```

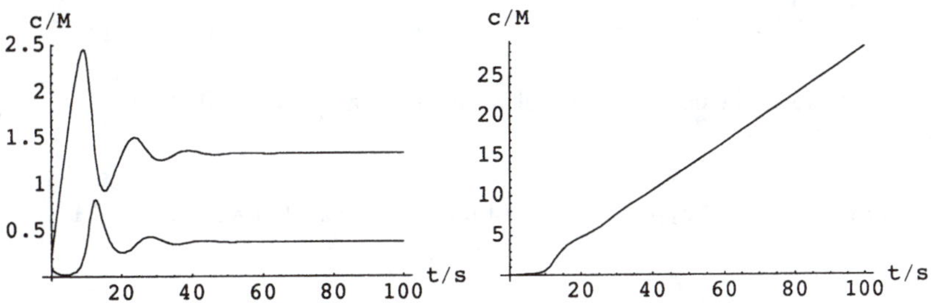

20.H The Lotka-Volterra mechanism for an autocatalytic reaction gives more striking results than the Lotka mechanism.

A + X -> 2X k_1

X + Y = 2Y k_2

Y -> P k_3

Assume the concentration of A is held at 1 M and the rate constants are $k_1 = 1\ s^{-1}$, $k_2 = 1.7\ M^{-1} s^{-1}$, and $k_3 = 1.6\ s^{-1}$. The initial concentrations of X and Y are to be taken as 0.2 M, 0.1 M, and 0 M. Plot the concentrations of X, Y, and P at short times. (b) Plot the concentration of product P as a function of time. Try varying the rate constants and initial concentrations to see what happens. *Mathematica* programs for making related plots are given in Ferreira, et al., J. Chem. Educ. 76, 861 (1999).

SOLUTION

```
a = 1;

k1 = 1;

k2 = 1.7;

k3 = 1.6;
```

(a) Plot the concentrations of X and Y at t = 0 to 30 s

```
sol = NDSolve[{x'[t] == k1 * a * x[t] - k2 * x[t] * y[t],
   y'[t] == k2 * x[t] * y[t] - k3 * y[t],
   p'[t] == k3 * y[t],
   x[0] == 0.2, p[0] == 0.0, y[0] == 0.1},
  {x, y, p}, {t, 0, 30}, MaxSteps → Infinity]

{{x → InterpolatingFunction[{{0., 30.}}, <>],
  y → InterpolatingFunction[{{0., 30.}}, <>], p → InterpolatingFunction[{{0., 30.}}, <>]}}

Plot[Evaluate[{x[t], y[t]} /. sol], {t, 0, 30},
  PlotRange → All, AxesLabel → {"t/s", "c/M"}, AxesOrigin → {0, 0}];
```

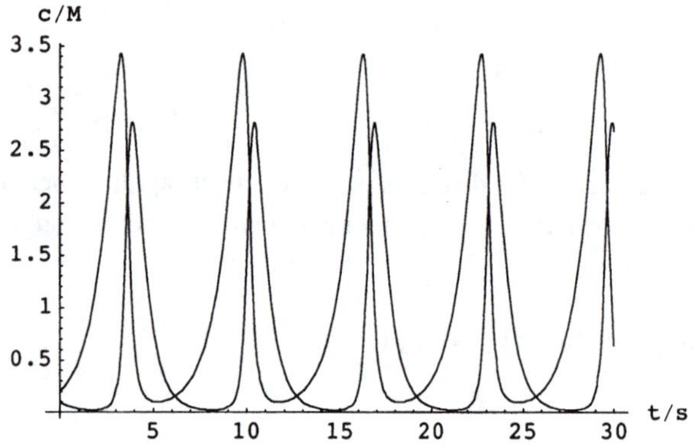

As the concentration of Y gets higher it pulls down the concentration of X, and this causes an oscillation.

(b) Plot the concentration of product versus time.

```
plotb = Plot[Evaluate[p[t] /. sol], {t, 0, 30}, PlotRange → All, AxesLabel → {"t/s", "c/M"}];
```

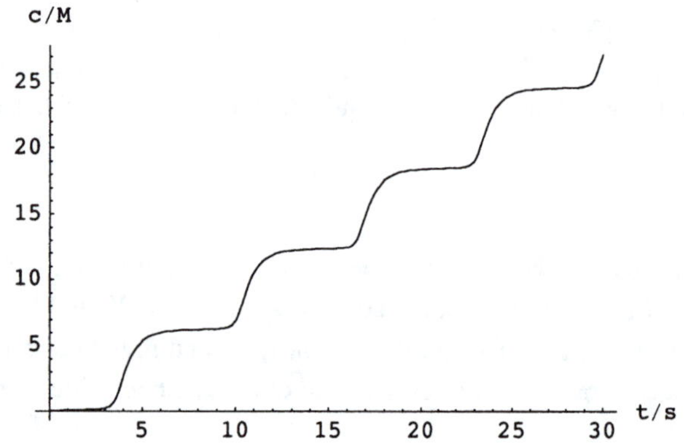

The production of P is slow at first because there is an induction period, but then it is produced most rapidly when Y goes through a maximum.

Chapter 21 Macromolecules

21.A (*a*) Plot the probability density $W(\xi)$ that a random walker along a line is at the distance ξ to $\xi + d\xi$ from the origin after 100 steps, 1000 steps and 10,000 steps. Assume that the step length is $h = 1$ m. (*b*) Check that the standard deviation in each case is given by $\sigma = n^{1/2}\ h$.

SOLUTION

(a)

```
ClearAll["Global`*"]

wx[n_, x_] := (1 / ((2 Pi n) ^ .5)) Exp[-x ^ 2 / (2 n)]

Plot[{wx[100, x], wx[1000, x], wx[10000, x]}, {x, -200, 200},
  PlotRange -> {0, .05}, AxesLabel -> {"ξ/m", "W(ξ)"}, AxesOrigin -> {-200, 0}];
```

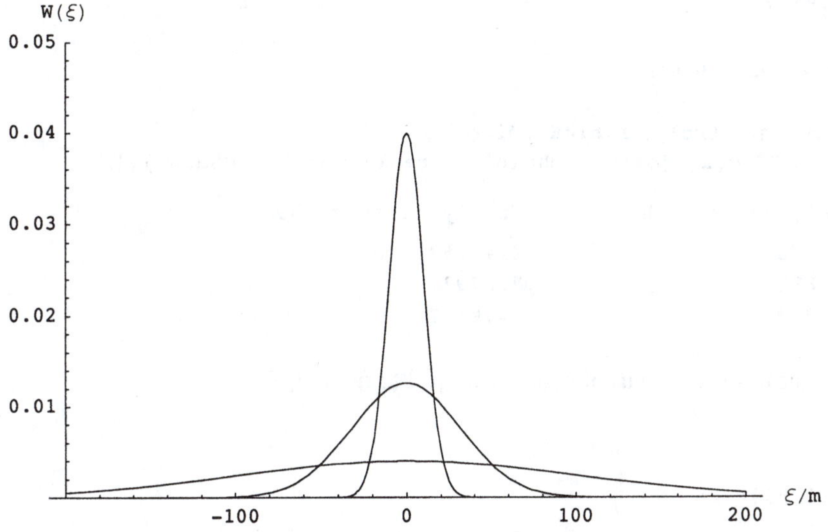

(b)

```
{100, 1000, 10000} ^ .5 * 1

{10., 31.6228, 100.}
```

These agree with the standard deviations of the above plots.

21.B For polyethylene $H(CH_2\ CH_2)_n H$ calculate (*a*) the root-mean-square end-to-end distance for a freely jointed chain and a freely rotated chain for molar masses for 10^3, 10^4, and 10^5 g mol^{-1} and make a table. (*b*) Calculate the radius of gyration for a freely jointed chain and a freely rotated chain for molar masses of 10^3, 10^4, and 10^5 g mol^{-1} and make a table. The bond length is 0.15 nm, and there are tetrahedral bond angles (109 °).

SOLUTION

(a) The number N of bonds is given by $N = M/(14\ \text{g mol}^{-1})$:

```
m = {10^3, 10^4, 10^5}
```

```
{1000, 10000, 100000}
```

```
n = m / 14 // N
```

```
{71.4286, 714.286, 7142.86}
```

For a freely jointed chain, $< r^2 >^{1/2}$ is given by $(N^{1/2})$0.15 nm:

```
fjc = (n^.5) * .15
```

```
{1.26773, 4.00892, 12.6773}
```

For a freely rotated chain, $< r^2 >^{1/2}$ is given by $(N^{1/2})$0.15 nm (1+cos71°)/(1-cos71°) since theta = 180° - 109° = 71°.

```
frc = (n^.5) * .15 * (1 + Cos[71 * 2 * Pi / 360]) / (1 - Cos[71 * 2 * Pi / 360])
```

```
{2.49167, 7.87937, 24.9167}
```

Table of root-mean-square end-to -end distances in nm

```
TableForm[Transpose[{m, fjc, frc}], TableHeadings ->
  {{}, {"M" / "g mol^-1", "Freely jointed chain", "Freely rotated chain"}}]
```

M / g mol^-1	Freely jointed chain	Freely rotated chain
1000	1.26773	2.49167
10000	4.00892	7.87937
100000	12.6773	24.9167

(b) For a freely jointed chain, the root-mean-square radius of gyration is equal to fjc/Sqrt[6]:

```
rgfjc = fjc / Sqrt[6]
```

```
{0.517549, 1.63663, 5.17549}
```

For a freely rotated chain the root-mean-square radius of gyrationis equal to frc/Sqrt[6]:

```
rgfrc = frc / Sqrt[6]
```

```
{1.01722, 3.21674, 10.1722}
```

```
{m, fjc, frc}
```

```
{{1000, 10000, 100000}, {1.26773, 4.00892, 12.6773}, {2.49167, 7.87937, 24.9167}}
```

Table of radius of gyration in nm

```
TableForm[Transpose[{m, rgfjc, rgfrc}], TableHeadings ->
  {{}, {"M" / "g mol^-1", "Freely jointed chain", "Freely rotated chain"}}]
```

$\frac{M}{\text{g mol}^{\wedge}\text{-1}}$	Freely jointed chain	Freely rotated chain
1000	0.517549	1.01722
10000	1.63663	3.21674
100000	5.17549	10.1722

21.C (*a*) Plot the mole fraction distribution of condensation polymer versus chain length *i* for extents of reaction from 0.95 to 0.59. (*b*) Plot the weight fraction distribution of condensation polymer versus chain length *i* for extents of reaction from 0.95 to 0.99.

SOLUTION

(a)

```
molefract[p_, i_] := (1 - p) * p^ (i - 1)

Plot[Evaluate[Table[molefract[p, i], {p, .95, .99, .01}]],
  {i, 0, 250}, AxesLabel -> {"i", "πi"}, PlotRange -> {0, .04}];
```

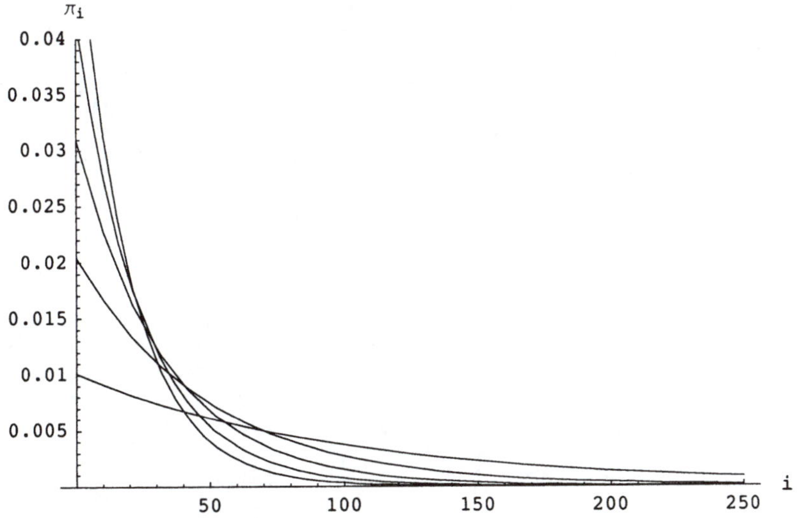

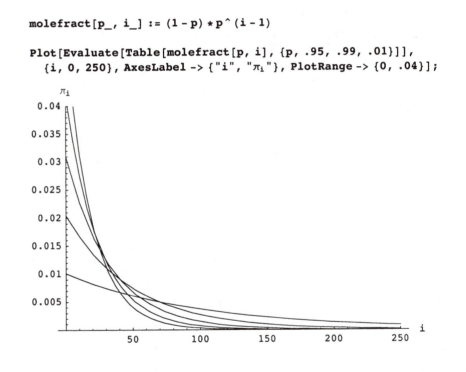

(b)

```
weightfract[p_, i_] = molefract[p, i] * i * (1 - p)
```

$i\,(1-p)^2\,p^{-1+i}$

```
Plot[Evaluate[Table[weightfract[p, i], {p, .95, .99, .01}]],
  {i, 0, 250}, AxesLabel -> {"i", "wᵢ"}, PlotRange -> {0, .03}];
```

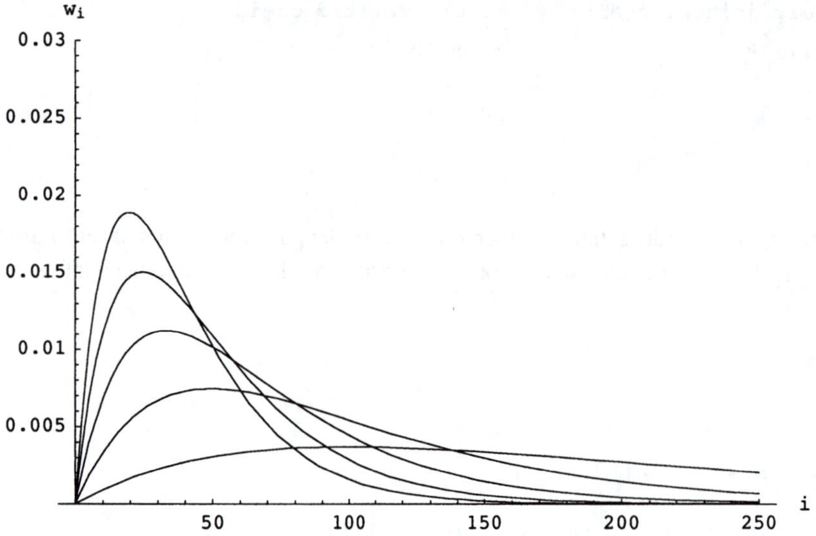

21.D Plot the viscosity data in problem 21.10 as $(1/c) \ln (\eta/\eta_0)$ to see whether the same intrinsic viscosity is obtained as in problem 21.12.

SOLUTION

```
c = {.00249, .00499, .00999, .01998}

{0.00249, 0.00499, 0.00999, 0.01998}

relvis = {1.355, 1.782, 2.879, 6.09}

{1.355, 1.782, 2.879, 6.09}

data = {c, (1 / c) * Log[relvis]}

{{0.00249, 0.00499, 0.00999, 0.01998}, {122.009, 115.779, 105.85, 90.4228}}

plot1 = ListPlot[Transpose[data], AxesOrigin -> {0, 80}, PlotRange -> {{0, .03}, {80, 260}},
   Prolog -> AbsolutePointSize[3], AxesLabel → {"c", "ln (η/η₀)"}];
```

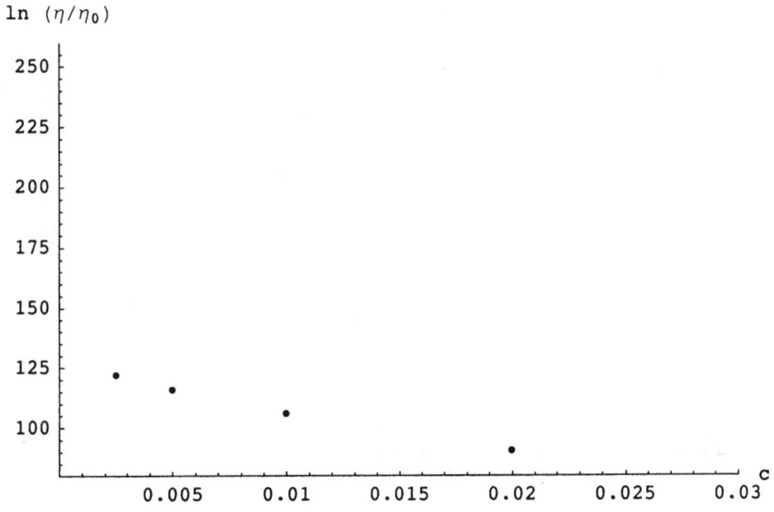


```
data2 = {c, (relvis - 1) / c}

{{0.00249, 0.00499, 0.00999, 0.01998}, {142.57, 156.713, 188.088, 254.755}}
```

```
plot2 = ListPlot[Transpose[data2], PlotRange -> {{0, .03}, {80, 260}},
   Prolog -> AbsolutePointSize[4], AxesLabel → {"c", "(ηrel-1)/c"}];
```

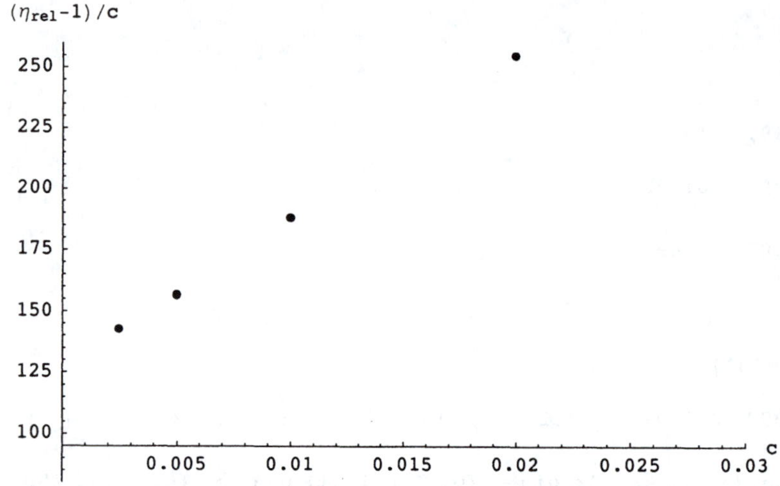


```
Show[plot1, plot2, AxesOrigin -> {0, 80}, AxesLabel -> {"c/g cm^-3", ""}];
```

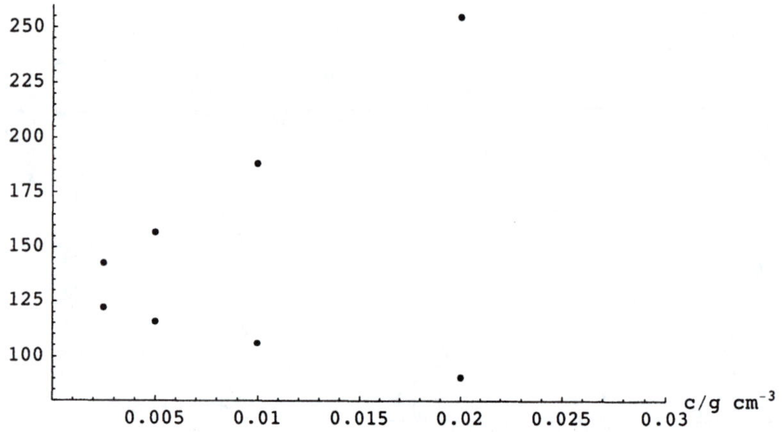

Since the plot of $(1/c)\ln(\eta/\eta_0)$ has a lower slope, it may be preferable.

21.E A hydroxy acid with $M_0 = 100$ g/mol is polymerized to the point that 99% of the monomers have reacted. (a) Plot the weight fraction w_i versus the chain size i. (b) Plot the weight fraction w_i versus the molar mass if the polymer. (c) Calculate the number average molar mass and the mass average molaar mass, and indicate them on the plot in (b).

SOLUTION

(a)

```
Plot[i * (1 - p) ^2 * p^ (i - 1) /. p → .99, {i, 0, 350},
  AxesOrigin -> {0, 0}, PlotRange → {0, .004}, AxesLabel → {"i", "wi"}];
```

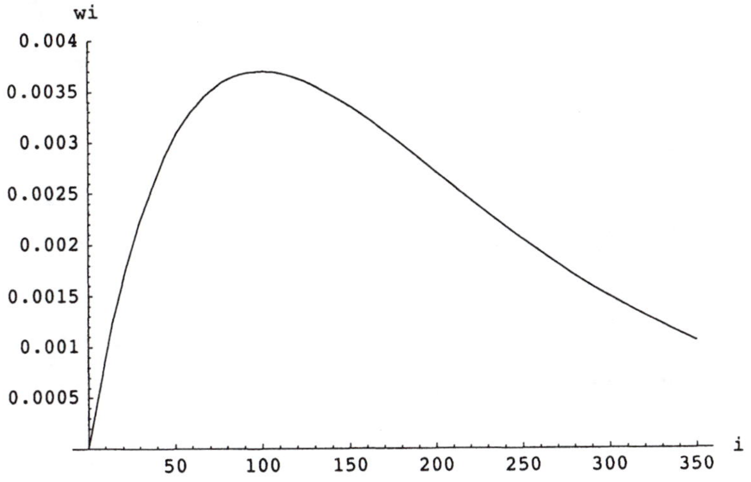

(b) Replace i with m/mo=m/100

```
plot2 = Plot[(m / 100) * (1 - p) ^2 * p^ ((m / 100) - 1) /. p → .99, {m, 0, 35000},
   AxesOrigin -> {0, 0}, PlotRange → {0, .004}, AxesLabel → {"M/(g/mol)", "wi"}];
```

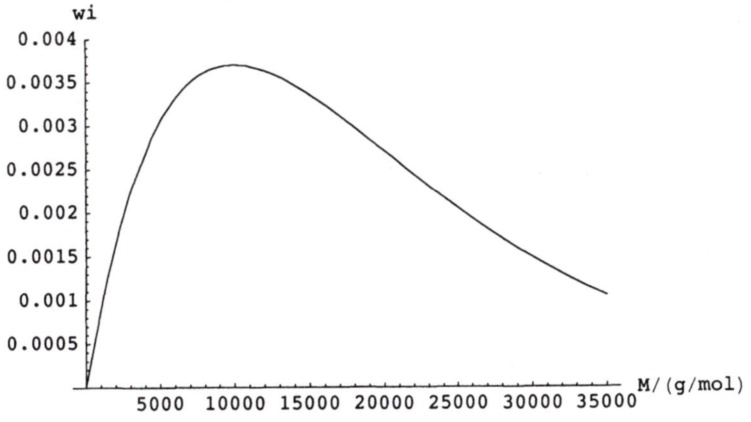

(c)

```
mn = 100 / (1 - .99)

10000.

mm = 100 * 1.99 / (1 - .99)

19900.

noavg = Line[{{10000, 0}, {10000, .004}}];

massavg = Line[{{19900, 0}, {19900, 0.004}}]

Line[{{19900, 0}, {19900, 0.004}}]
```

```
plot3 = Show[Graphics[noavg], Graphics[massavg]];
```

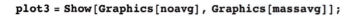

```
Show[plot2, plot3, AxesLabel → {"M/(g/mol)", "wi"}];
```

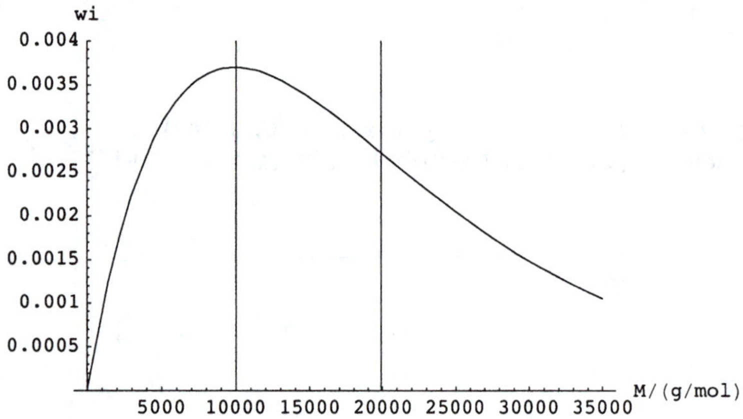

Chapter 22 Electric and Magnetic Properties of Molecules

22.A Calculate the dipole moment of fluorobenzene, using a plot of P_m versus $1/T$, from the following molar polarizations for gaseous samples.

T/K	343.6	371.4	414.1	453.2	507.0
P_m / cm^3 mol^{-1}	69.9	66.8	62.5	59.3	55.8

SOLUTION

```
t = {343.6, 371.4, 414.1, 453.2, 507.};
```

```
p = {69.9, 66.8, 62.5, 59.3, 55.8} * 10^-6;
```

Note the molar polarizations have been converted to m^3 mol^{-1} .

```
data = Transpose[{1 / t, p}]
```

```
{{0.00291036, 0.0000699}, {0.00269251, 0.0000668},
 {0.00241488, 0.0000625}, {0.00220653, 0.0000593}, {0.00197239, 0.0000558}}
```

```
plot1 = ListPlot[data, AxesOrigin -> {.0018, .000052},
   Prolog -> AbsolutePointSize[3], DisplayFunction → Identity];
```

```
Fit[data, {1, x}, x]
```

```
0.0000259829 + 0.0151177 x
```

```
plot2 = Plot[.00002598 + .01512 * x, {x, .0018, .003}, DisplayFunction → Identity];
```

```
Show[plot1, plot2, AxesLabel -> {"K/T", "Pm/m^3 mol^-1"},
  DisplayFunction → $DisplayFunction];
```

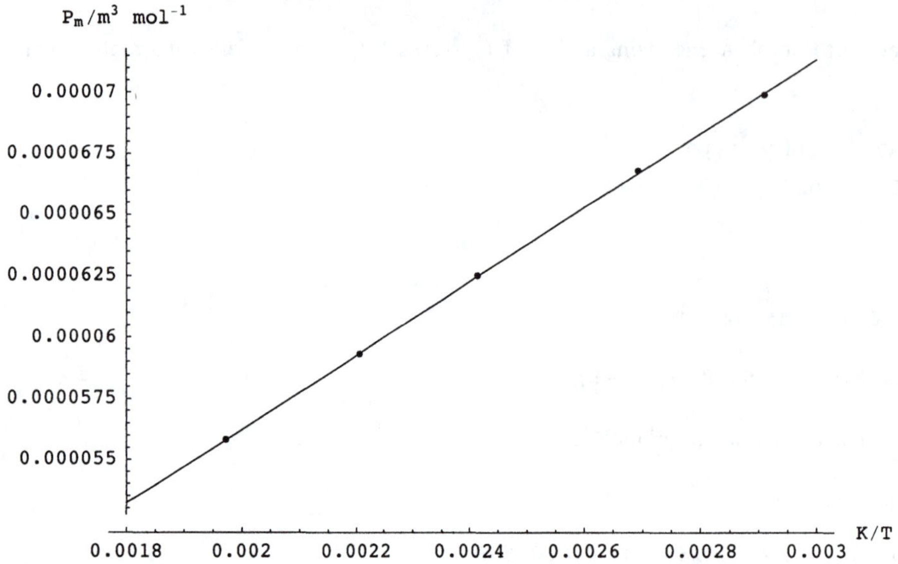

The intercept yields the value of α.

```
Solve[.00002598 == (6.022 * 10^23) * α / (3 * (8.854 * 10^-12)), α]
```

$\{\{\alpha \to 1.14593 \times 10^{-39}\}\}$

The molecular polarizability has units of $C^2\ m^2\ J^{-1}$.
The slope yields the value of the dipole moment.

```
Solve[.015118 == ((6.022 * 10^23) * μ^2) / (9 * (8.854 * 10^-12) * (1.381 * 10^-23)), μ]
```

$\{\{\mu \to -5.25611 \times 10^{-30}\}, \{\mu \to 5.25611 \times 10^{-30}\}\}$

The dipole moment has units of C m. Since 1 debye = 3.336×10^{-30} C m, the dipole moment in units of D is

```
(5.256 * 10^-30) / (3.336 * 10^-30)
```

1.57554

22.B Plot the frequency dependence of the real part ε' and the imaginary part ϵ'' of the relative permittivity for a process with a single relaxation time τ with $\epsilon_l = 15$ and $\epsilon_h = 5$, as shown in Fig. 22.5. (*a*) Do this by using equation 22.29 with equations 22.30 and 22.31. (*b*) Do this using equation 22.28. Obtaining the same result in two ways will confirm these two ways of expressing the relative permittivity.

SOLUTION

(a)

```
e = 5 + (15. - 5.) / (1. + (10^x)^2) - I (15. - 5.) (10^x) / (1 + (10^x)^2)
```

$$5 - \frac{10.\ i\ 10^x}{1 + 10^{2x}} + \frac{10.}{1. + 10^{2x}}$$

```
re = Re[Table[e, {x, -3, 3, .1}]];

ωτ = Table[x, {x, -3, 3, .1}];

plot1 = ListPlot[Transpose[{ωτ, re}], PlotJoined -> True,
   AxesOrigin -> {-3, 0}, AxesLabel -> {"ωτ", "ϵ"}, PlotRange → {0, 15}];
```

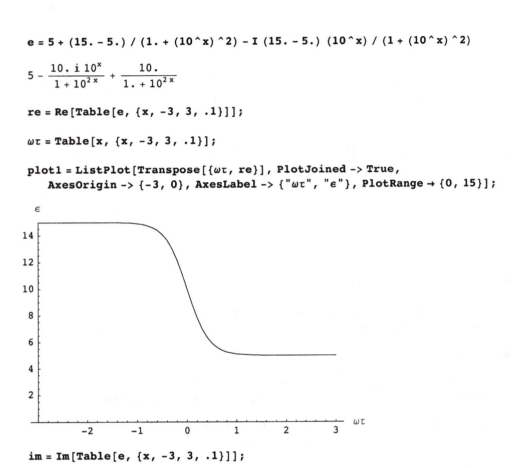

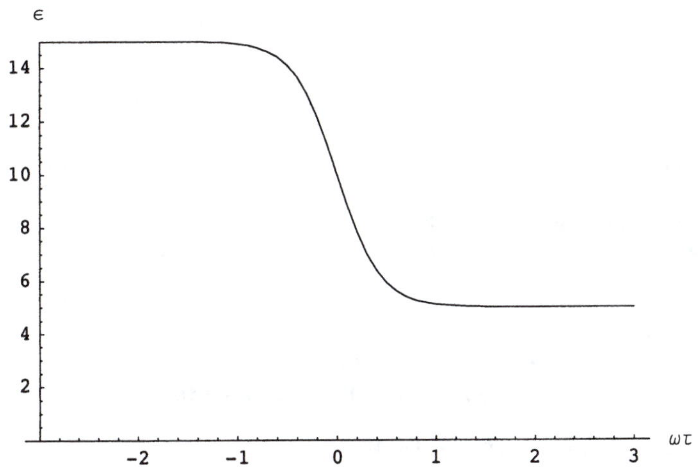


```
im = Im[Table[e, {x, -3, 3, .1}]];

plot2 = ListPlot[Transpose[{ωτ, -1 * im}], PlotJoined -> True, DisplayFunction -> Identity];
```

```
Show[plot1, plot2];
```

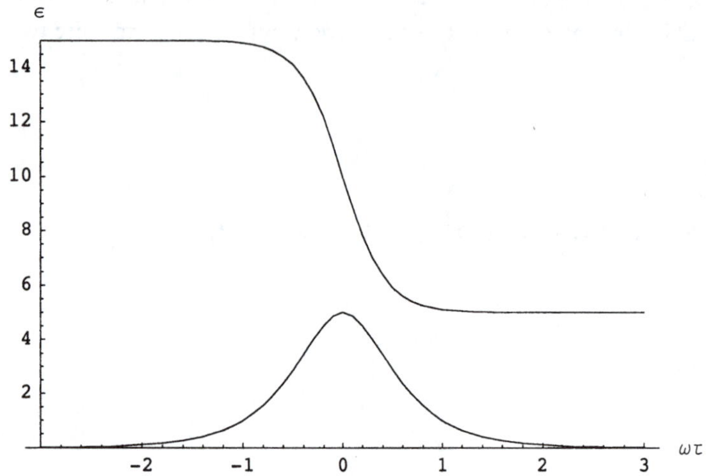

(b)

```
ee = 5 + (15 - 5) / (1 + I 10^x)
```

$$5 + \frac{10}{1 + i\,10^x}$$

```
ree = Re[Table[ee, {x, -3, 3, .1}]];

plot12 = ListPlot[Transpose[{ωτ, ree}], PlotJoined -> True,
   AxesOrigin -> {-3, 0}, AxesLabel -> {"ωτ", "ϵ"}, DisplayFunction → Identity];

ime = Im[Table[ee, {x, -3, 3, .1}]];

plot22 = ListPlot[Transpose[{ωτ, -1 * ime}], PlotJoined -> True,
   AxesOrigin -> {-3, 0}, AxesLabel -> {"ωτ", "ϵ"}, DisplayFunction → Identity];

Show[plot12, plot22, DisplayFunction → $DisplayFunction];
```

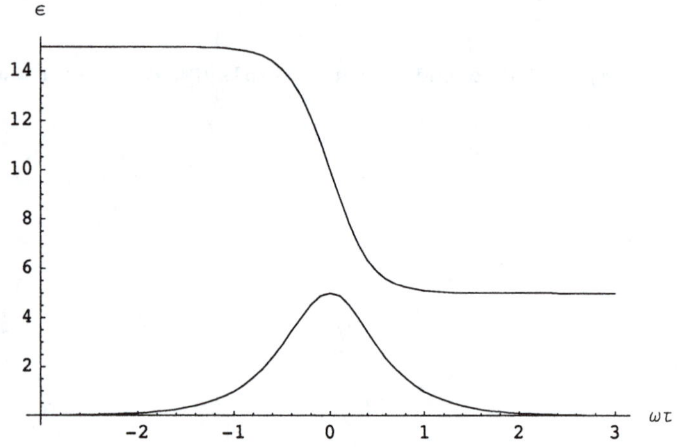

Chapter 23 Solid-State Chemistry

23.A Calculate the angles theta for the first order Bragg reflections for the 100, 110, and 111 planes of an orthorhombic unit cell with $a = 488.2$ pm, $b = 665.7$ pm, and $c = 831.6$ pm. The wavelength of the monochromatic X-rays is 154.433 pm..

SOLUTION

```
ClearAll["Global`*"]
```

The following equation gives the Bragg angles (theta) in radians, and so the angles in degees are obtained by dividing by Degrees, which is a constant equal to Pi/180.

```
theta[h_, k_, l_] := ArcSin[lambda / (2 Sqrt[(1 / (h^2 / a^2 + k^2 / b^2 + l^2 / c^2))])]

a = 488.2 (*pm*) ; b = 665.7; c = 831.6; lambda = 154.433;

theta[1, 0, 0] / Degree

9.10044

theta[1, 1, 0] / Degree

11.3113

theta[1, 1, 1] / Degree

12.5334
```

23.B Copper forms face-centered cubic crystals with a 361.6 pm unit cell at 25 °C. Calculate the first five Bragg angles obtained with 154.05 pm X-rays.

SOLUTION

```
ClearAll["Global`*"]

theta[h_, k_, l_] := ArcSin[lambda Sqrt[h^2 + k^2 + l^2] / (2 a)]

a = 361.6 (*pm*) ; lambda = 154.05;

theta[1, 1, 1] / Degree

21.6507

theta[2, 0, 0] / Degree

25.2154

theta[2, 2, 0] / Degree

37.0483
```

```
theta[3, 1, 1] / Degree
```

```
44.9492
```

```
theta[2, 2, 2] / Degree
```

```
47.5523
```

23.C Calculate the Fourier transform of the function
$\sin 2\pi\nu_1 t + 2\sin 2\pi\nu_2 t$, where $\nu_1 = 30\ s^{-1}$ and $\nu_2 = 20\ s^{-1}$. (*a*) Plot the function in the time domain over a period of one second. (*b*) Plot the corresponding spectrum in the frequency domain.

SOLUTION

(a)

```
fn = N[Sin[30 2 Pi n / 256] + 2 Sin[20 2 Pi n / 256]];

datatimedomain = Table[N[{n / 256, fn}], {n, 0, 256}];

ListPlot[datatimedomain, PlotJoined → True, AxesLabel → {"t/s", "fn"}];
```

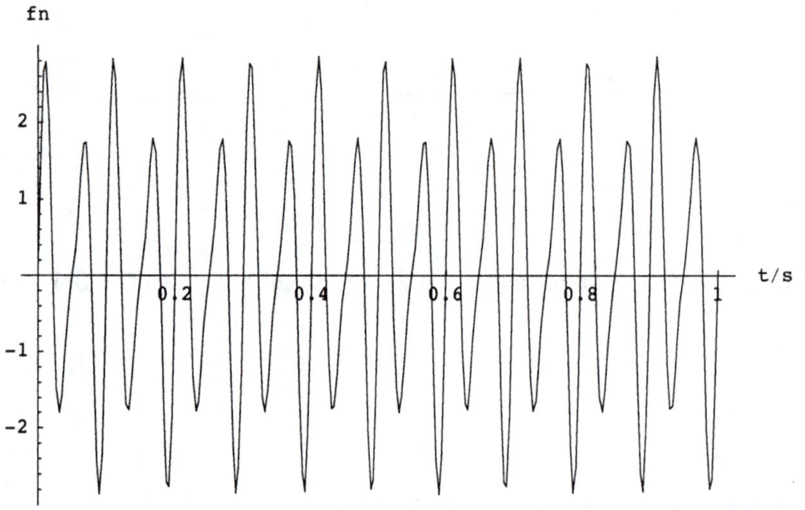

(b)

```
tab = Table[fn, {n, 256}];
```

```
ListPlot[Re[Fourier[tab]], PlotJoined → True, PlotRange → All, AxesLabel → {"ν/s⁻¹", ""}];
```

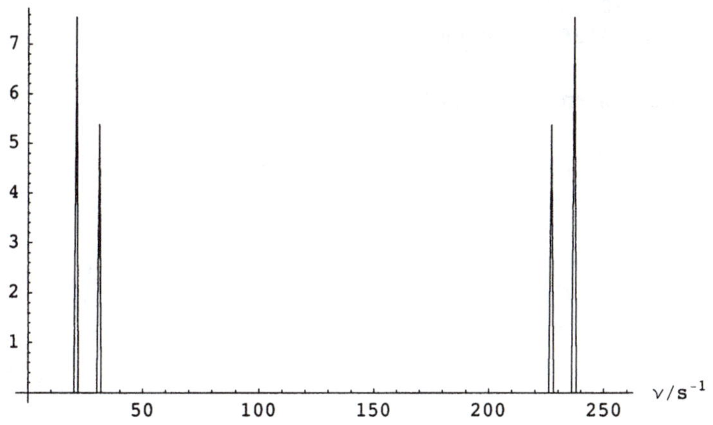

The Fourier transform spectrum shows strong peaks at 20 and 30 cycles per second, and symmetric peaks at 257 – 20 and 256 - 30.

23.D Calculate the angles for the first eleven first-order Bragg reflections for a face-centered cubic crystal with $\lambda/a = 0.289$.

SOLUTION

```
millerindices = {{1, 1, 1}, {2, 0, 0}, {2, 2, 0}, {3, 1, 1},
   {2, 2, 2}, {4, 0, 0}, {3, 3, 1}, {4, 2, 0}, {4, 2, 2}, {3, 3, 3}, {4, 4, 0}};
```

```
MatrixForm[millerindices]
```

$$\begin{pmatrix} 1 & 1 & 1 \\ 2 & 0 & 0 \\ 2 & 2 & 0 \\ 3 & 1 & 1 \\ 2 & 2 & 2 \\ 4 & 0 & 0 \\ 3 & 3 & 1 \\ 4 & 2 & 0 \\ 4 & 2 & 2 \\ 3 & 3 & 3 \\ 4 & 4 & 0 \end{pmatrix}$$

```
N[Apply[Plus, Transpose[millerindices^2]]]
```

```
{3., 4., 8., 11., 12., 16., 19., 20., 24., 27., 32.}
```

```
ArcSin[(.289 / 2) * Sqrt[N[Apply[Plus, Transpose[millerindices^2]]]]] / Degree

{14.4942, 16.7981, 24.1237, 28.6366, 30.0372,
 35.31, 39.0399, 40.2575, 45.0645, 48.6635, 54.8269}
```

You can also calculate the angles for primitive and body-centered cubic crystals.

Chapter 24 Surface Dynamics

24.A According to the BET theory, plot V / V_m, where V_m is the volume of gas required to form a monolayer, versus $P/P°$, where $P°$ is the saturation vapor pressure of the gas. Show the effect of changing the constant c from 0.10 to 200.

SOLUTION

```
vol[c_, x_] := c x / ((1 - x) (1 + (c - 1) x))

Plot[{vol[.1, x], vol[1, x], vol[3, x], vol[10, x], vol[30, x], vol[200, x]},
  {x, 0, .9}, PlotRange -> {0, 3}, AxesLabel -> {"P/P°", "V/Vm"}];
```

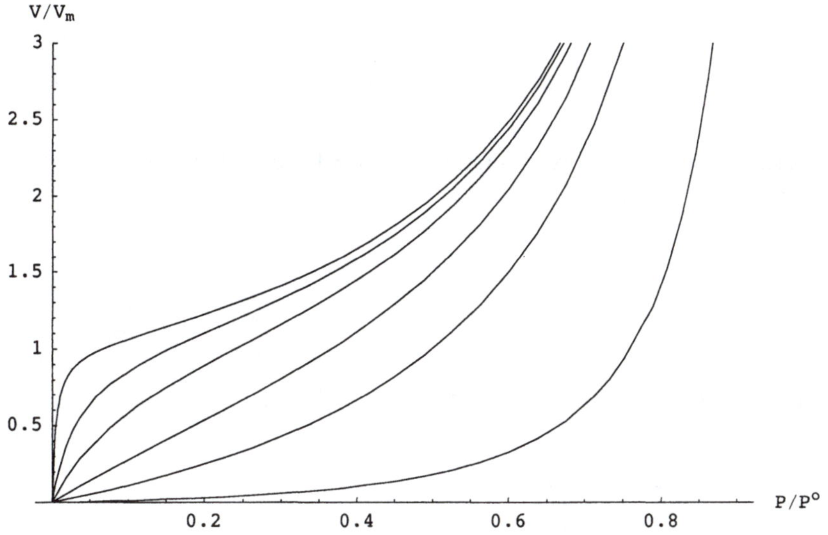

24.B Suppose that at 298.15 K, the adsorption isotherm for a gas on a solid is given by
$\theta = KP/(1 + KP)$
where θ is the fractional coverage of the surface, K(298.15 K) = 3 bar, and ΔH_{ads} = -20 kJ mol^{-1}. Plot θ versus P at 298.15 K and 350 K.

SOLUTION
At 350 K the constant K is given by

```
k350 = 3 * ((298.15 / 350) ^ .5) * Exp[((20 * 10^3) / 8.3145) * ((1 / 350) - (1 / 298.15))]

0.837986
```

```
Plot[{3*p/(1+3*p), .837*p/(1+.837*p)}, {p, 0, 2}, AxesLabel -> {"P/bar", "θ"}];
```

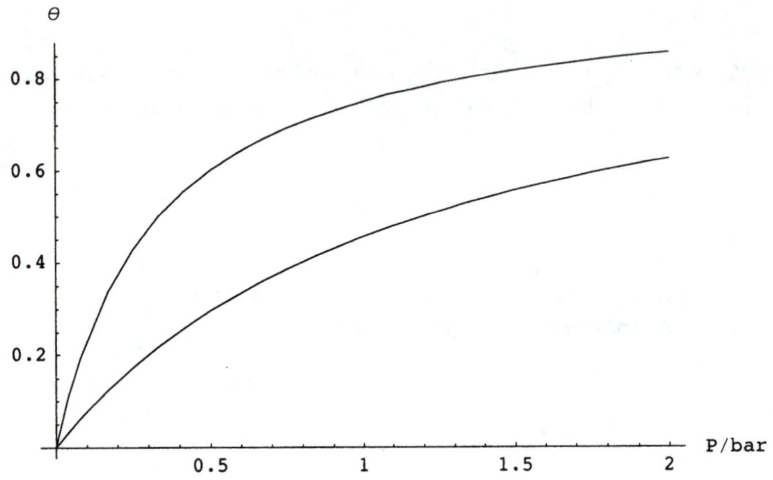

24.C Plot the fractional coverage Θ of a surface for dissociative adsorption with $K = 1$ and for Langmuir adsorption with $K = 1$. Superimpose these plots so that the differences will be clearer.

SOLUTION

Dissociative adsorption

```
plot1 = Plot[p^.5/(1+p^.5), {p, 0, 10}, AxesLabel → {"P", "θ"}];
```

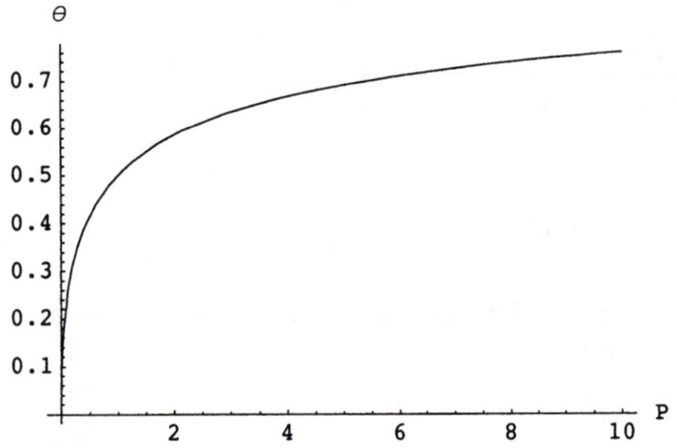

Lagmuir adsorption

```
plot2 = Plot[p / (1 + p), {p, 0, 10}, AxesLabel → {"P", "θ"}];
```

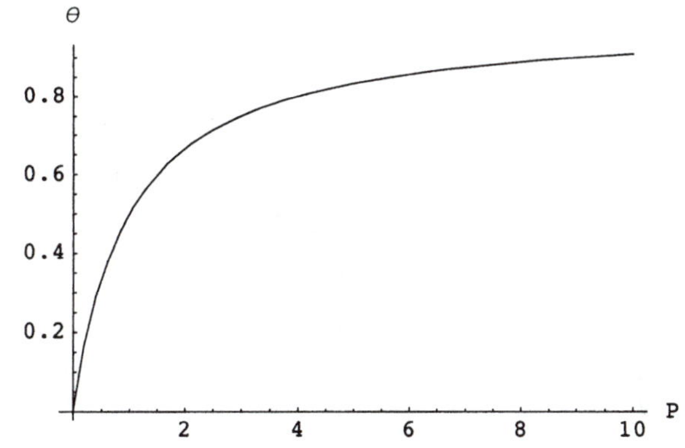


```
Show[plot1, plot2];
```

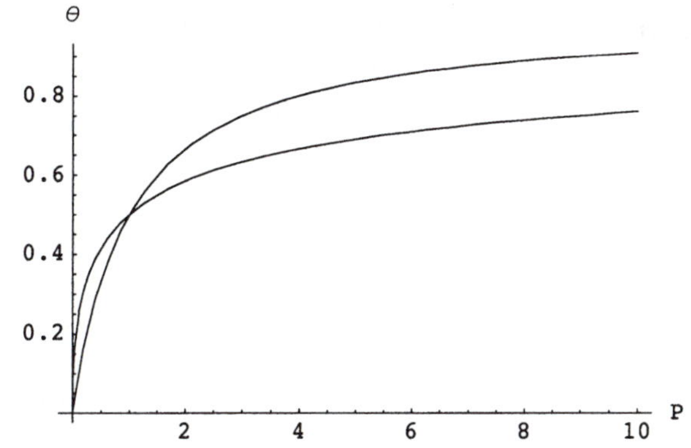

24.D Plot the de Broglie wavelength λ in nanometers as a function of the accelerating potential difference ϕ in volts from zero to 100 volts. Check that the result in Example 24.4 is confirmed.

SOLUTION

```
Plot[(1.504 / ϕ) ^ .5, {ϕ, 0, 100}, AxesLabel → {"ϕ/V", "λ/nm"}, PlotRange → {0, 2}];
```

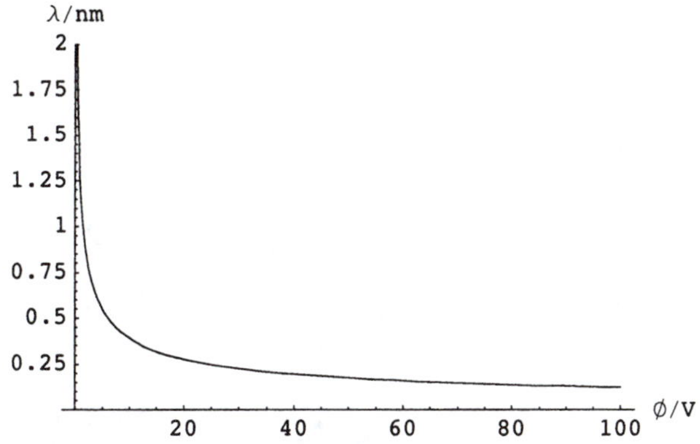

Since λ increases so much at low voltages, it may be convenient to use a logarithmic scale on the ordinate. This is accomplished with LogPlot.

```
<< Graphics`Graphics`
```

```
LogPlot[(1.504/ϕ)^.5, {ϕ, 0, 100}, AxesLabel → {"ϕ/V", "λ/nm"}];
```

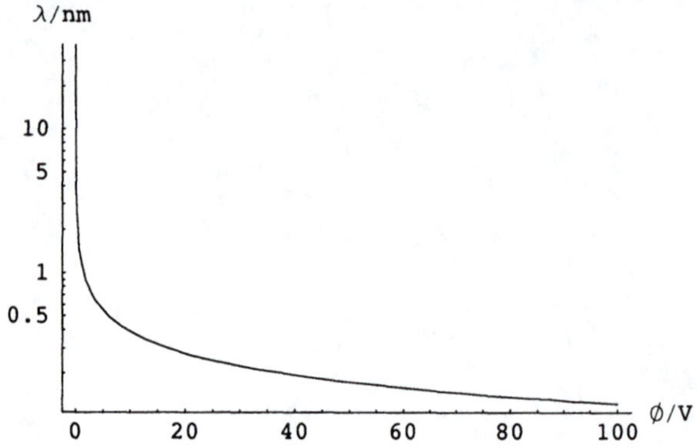

A plot with logarithmic scales on both axes is more convenient for interpolation.

```
LogLogPlot[(1.504/ϕ)^.5, {ϕ, 0, 100}, AxesLabel → {"ϕ/V", "λ/nm"}];
```

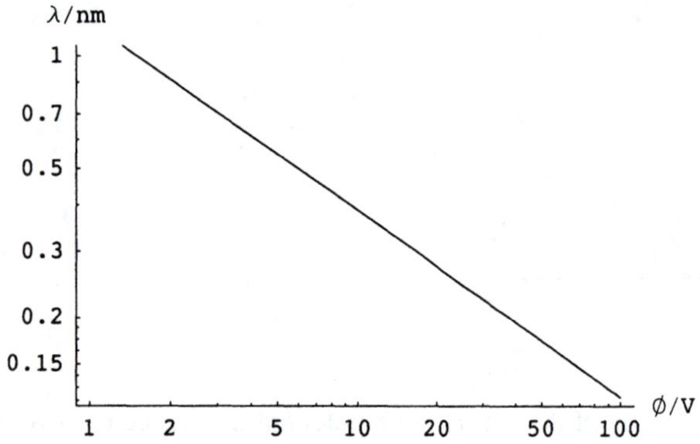

ReadMe

Robert J. Silbey, Robert A. Alberty, and Moungi G. Bawendi, SOLUTION MANUAL TO ACCOMPANY PHYSICAL CHEMISTRY, Fourth Edition, New York: Wiley.

The availability of mathematical programs for personal computers has provided new opportunities for solving physical chemistry problems and the possibility of going more deeply into topics requiring longer calculations. In each chapter in the fourth edition, there are some Computer Problems that require a personal computer with a mathematical application. These problems go more deeply into certain topics and show how to obtain solutions in calculations that are too long for a hand-held calculator. In this edition more computer problems have been introduced throughout the chapters, and an icon has been used to indicate that certain problems in the first two sets may be more conveniently solved on a person computer with a mathematical application. The Computer Problems and their complete solutions in *Mathematica* 5.0 are provided in the Solutions Manual (4th ed.) and on the Web at http://www.wiley.com/college/silbey.

The basic reference on *Mathematica* is

S. Wolfram, The *Mathematica* Book, Fourth Edition, Cambridge University Press, 1999.

There are two other sources of *Mathematica* programs of interest in learning physical chemistry:

J. H. Noggle, Physical Chemistry Using *Mathematica*, HarperCollins, New York, 1996.
W. H. Cropper, *Mathematica* Computer Programs for Physical Chemistry, Springer-Verlag, New York, 1998.

You do not need to be a *Mathematica* programmer to use the programs given in the Solutions Manual. The solutions manual on the Web can be used in any personal computer with *Mathematica* 5.0 installed. A student version is available at a reduced price. In the material on the Web there is a folder for each chapter in the textbook, and each problem is in a *Mathematica* notebook. Of course, you do not have to have access to the Web to use these problems; you can type them in your computer. If you use the Web there are two ways to use these notebooks: (1) You can run the program from the Web and even make changes, but the changes will not be retained in your computer. (2) You can transfer a notebook to a file on your hard disk, and make changes, which will be retained when the notebook is saved.

When using a PC:

To start *Mathematica* from a hard drive, click Start, point to Programs, and select the *Mathematica* program group. This will put a new *Mathematica* notebook on the screen. Type 2+2, hold down the Shift key, and press Enter. This first calculation will take longer than subsequent calculations because the kernel has to start up. Note that your input is labelled In[1]= , and the output is labelled Out[2]= . The input and output cells are indicated on the right edge of the screen are in a larger cell. A notebook can be copied from the Web and be opened with File/Open. Navigate in the dialog box until you see the name of the file you want. Select the notebook and click Open. The whole notebook can be evaluated. When you start typing, it will be taken as *Mathematica* input. If you want to type text, type Text from the Style submenu, or use the keyboard shortcut Alt-[7]. To run a cell put the cursor anywhere in the cell and hold down Shift and press Enter. To close a notebook without quitting *Mathematica*, select File/Close. You can also drag a notebook into your hard disk, and use it there. Then when changes are made in the notebook and saved, the altered notebook remains in the file on the hard drive.

When using a Mac:

To start *Mathematica*, click on its icon. This will put a new *Mathematica* notebook on the screen. Type 2+2, and press Enter. This first calculation will take longer than subsequent calculations because the kernel has to start up. Note that your input is labelled In[1]= , and the output is labelled Out[2]= . The input and output cells are indicated on the right edge of the screen are in a larger cell. A notebook can be copied from the Web and be opened with File/Open. Select the notebook and click Open. A whole notebook can be evaluated by using Kernel/Evaluate/EvaluateNotebook from the pull-down menu.. To create a new cell, move the cursor until it becomes horizontal. When you start typing, it will be taken as *Mathematica* input. If you want to type text, use the keyboard shortcut Command-[7]. To run a cell put the cursor anywhere in the cell and press Enter. To close a notebook without quitting *Mathematica*, select File/Close. You can also drag a notebook onto your hard disk, and use it there. When changes are made in the notebook and saved, the altered notebook remains in the file on the hard drive.

In *Mathematica* the input is in darker type that the output. *Mathematica* recognizes upper-case and lower-case letters as different. In *Mathematica*, physical quantities are represented by lower case letters or abbreviations beginning with a lowercase letter because all built-in *Mathematica* objects start with capital letters. The same calculation can be made with other values of the parameters by simply blocking the values, typing in different numbers, and re-evaluating the notebook or re-running the necessary lines. When other problems are worked on in the same *Mathematica* session, it is often a good idea to use ClearAll["Global`*"] to eliminate objects from the previous problem or problems.

A great deal can be learned about programming in *Mathematica* by observing how these problems are solved. The pull-down Help in *Mathematica* is very useful becuse it shows exacly what has to be specified in using built-in *Mathematica* objects (commands). Help can also be used to access material in the The *Mathematica* Book.

The Introduction to *Mathematica* that is included in the Solutions Manual should be helpful for people who have not had previous experience with *Mathematica*. An index is provided in the Solutions Manual to show where various *Mathematica* commands are used in the Computer Problems.

Introduction to Mathematica

Robert J. Silbey, Robert A. Alberty, and Moungi G. Bawendi, SOLUTION MANUAL TO ACCOMPANY PHYSICAL CHEMISTRY, Fourth Edition, New York: Wiley.

1. In order to run the Computer Problems described in the Solutions Manual or the Web, *Mathematica* 5.0 must be installed in the computer. There is a *Mathematica* notebook for each problem. The opening of a notebook in a PC or a Mac is described in the ReadMe. A *Mathematica* Notebook is a stuctured interactive document that is organized into a sequence of cells, as indicated along the right margin. There are various ways to use the notebooks, but the recommended way is to transfer a notebook to a folder on your hard disk. The whole notebook can be run in *Mathematica* using Kernel/Evaluation/-Evaluate Notebook. *Mathematica* has two parts, the Front End and the Kernel; the Front End handles the screen and the Kernel does the calculations. You can also experiment with the program, and if you mess up, you can always get another copy of the problem from the Web. The calculations in a Notebook can also be run cell by cell in the order in which they are in the Notebook by putting the curson anywhere within a cell and pressing the Enter key on a Mac or Shift-Enter on a PC.

2. Note that the Commands in *Mathematica* start with capital letters, while variables and constants start with lowercase letters. *Mathematica* is a high level language with hundreds of commands, which means it can be used to write programs without dealing with as much detail as in lower level languages. If you want to find out what a command does, simply type, for example,

```
? Plot

Plot[f, {x, xmin, xmax}] generates a plot of f as a function of x from
  xmin to xmax. Plot[{f1, f2, ... }, {x, xmin, xmax}] plots several functions fi.
```

This shows the input for the command. More information can be obtained by using two question marks.

```
??Plot

Plot[f, {x, xmin, xmax}] generates a plot of f as a function of x from
  xmin to xmax. Plot[{f1, f2, ... }, {x, xmin, xmax}] plots several functions fi.

Attributes[Plot] = {HoldAll, Protected}

Options[Plot] = {AspectRatio → 1/GoldenRatio, Axes → Automatic, AxesLabel → None,
  AxesOrigin → Automatic, AxesStyle → Automatic, Background → Automatic,
  ColorOutput → Automatic, Compiled → True, DefaultColor → Automatic, Epilog → {},
  Frame → False, FrameLabel → None, FrameStyle → Automatic, FrameTicks → Automatic,
  GridLines → None, ImageSize → Automatic, MaxBend → 10., PlotDivision → 30., PlotLabel → None,
  PlotPoints → 25, PlotRange → Automatic, PlotRegion → Automatic, PlotStyle → Automatic,
  Prolog → {}, RotateLabel → True, Ticks → Automatic, DefaultFont :→ $DefaultFont,
  DisplayFunction :→ $DisplayFunction, FormatType :→ $FormatType, TextStyle :→ $TextStyle}
```

Plots can be altered in many ways, but the defaults are used as much as possible in the Solutions Manual. More information and examples can be obtained from Help.

3. Each Computer Problem has a statement of the problem, which is just like that in the textbook, and a SOLUTION in *Mathematica*. An attempt has been made to keep the programs as simple as possible for beginners with *Mathematica*. Note that in *Mathematica* a symbol is assigned a value by simply typing, for example,

```
m = 10

10
```

and pressing the enter key. Now typing m and pressing the enter key yields 10.

```
m

10
```

Several assignements can be made on a single line by putting semicolons between them, and if a semicolon is placed at the end of the line, the values will not be repeated in an Out[ ].

```
a = 1; b = 2;
```

In *Mathematica* an integer means an exact number, but the presence of a decimal point indicates an approximate number.

```
a / b
```

$\frac{1}{2}$

```
a = 1.; b = 2.;

a / b

0.5
```

If you want to change the value of m to 20, you can type m=20. If you do not want m to have a value, use

```
Clear[m]

m

m
```

When you work several problems in a session with *Mathematica*, symbols tend to accumulate and may interfere in other calculations. A workspace can be cleared of sysmbols by use of

```
ClearAll["Global`*"]
```

The programs in the Solutions Manual show the key strokes actually used. Note that a product can be written a*b or a b, where there is a space between a and b. Raising a to the power b is entered as a^b. *Mathematica* offers the possiblity of entering equations so that they look just like equations in textbooks; this is done with Palettes that are found under File. In a few problems both modes are used to show that computatins can be made either way.

4. *Mathematica* has some built-in constants, which are represented by Pi, E, I, Infinity, and Degree:

```
Pi
```

π

You can obtain the numerical value as follows:

```
Pi // N

3.14159

E

e

E // N

2.71828

I

i

I // N

0. + 1. i

Degree

°

Degree // N

0.0174533
```

Note that a degree is equal to $2\pi/360$ radian.

5. *Mathematica* can be used like a hand-held calculator, and this is very useful for calculations that involve a number of steps because the expression can be proof read and errors can be corrected. *Mathematica* has the advantage that you can express the quantity you want to calculate in terms of symbols like m, a, and b, which are assigned values separate from the equation. Another advantage over a hand-held calculator is that *Mathematica* can handle units, but the Computer Problems given here concentrate on getting numerical answers without units. Of course, you have to use physical quantities with internally consistent units and to know the units of the quantity you calculate.

As an example, let us consider the ideal gas law: $PV = nRT$. Capitol letters are used to mame *Mathematica* commands and options, and so we follow the convention of using lower case letters so that P, V, n, R, and T are represented by p, v, n, r, and t. The ideal gas law is given by

```
p * v == n * r * t

p v == n r t
```

Note that == is used for equals because = is used for assignment, as illustrated above. When pressures are are expressed in bars and volumes are expressed in liters, the gas constant is 0.083145 L bar K^{-1} mol^{-1}. The pressure in bars of 2 moles of ideal gas in a volume of 10 liters at 298.15 K is obtained when the following assignments are made :

```
v = 10.; n = 2.; r = .0831451; t = 298.15;
```

```
p == n * r * t / v
```

```
p == 4.95794
```

We can avoid rearranging the equation by use of Solve.

```
Solve[p * v == n * r * t, p]
```

```
{{p → 4.95794}}
```

The form of this answer is called a "rule," and it can be used as follows in another calculation by use of ReplaceAll (/.). For example, we can check the volume that corresponds with this pressure by use of

```
ClearAll["Global`*"]
```

```
n = 2.; r = .0831451; t = 298.15;
v = n * r * t / p /. p → 4.95794
```

```
10.
```

Solve deals primarily with linear and polynomial equations. When other kinds of functions are involved, FindRoot can be used, but an approximate solution has to be given.

6. *Mathematica* has many built-in functions like Sin[ ], Log[ ], and Exp[ ], but any desired function can be written. For example, the function that calculates the square of x is

```
f[x_] := x^2
```

The symbol x_ means that the value of x is going to be supplied later. In defining a function, the underline is referred to as a blank.

```
f[9]
```

```
81
```

Note that Mathematica works with symbols as well as numbers.

```
f[1 + a]
```

```
(1 + a)^2
```

Functions in *Mathematica* are essentially procedures that execute commands. Note that := is used rather than = in writing functions. The symbol := is used for a delayed assignment (that is assignment given later in brackets), whereas = is used for an immediate assignment. In writing equations in *Mathematica* == is used. Note that *Mathematica* distinguishes between three kinds of brackets: [ ] is used for arguments in functions, ( ) is used in matheematical expressions, and { } is used in lists. Note that x^2y means (x^2) y, not x^(2y). Some calculations can be made with lists, which are put in as {3,5,1}. For example,

```
f[{3, 5, 1}]
```

```
{9, 25, 1}
```

7. In writing longer programs it is a good idea to define a function using Module so that the values of variables in the function can be protected.

```
?Module
```

```
Module[{x, y, ... }, expr] specifies that occurrences of the symbols x, y, ... in expr should
  be treated as local. Module[{x = x0, ... }, expr] defines initial values for x, ... .
```

As an example, let us return to the calculation of the volume of an ideal gas. The values of various variables are kept local (inside the operation of the program) by use of

```
volume[n_, t_, p_] := Module[{r}, r = .0831451; n*r*t/p]
```

The volume (in m^3) of 2 moles of gas at 298.15 K and 4.95794 bar is given by

```
volume[2, 298.15, 4.95794]

10.
```

This program can be extended to the calculation of the volume for a mixture of n_1 moles of gas 1, n_2 moles of gas 2, ... by use of

```
volume[n_, t_, p_] :=
 Module[{r, amount}, r = .0831451; amount = Apply[Plus, n]; amount*r*t/p]
```

The volume (in m^3) of a mixture of 2 moles of gas 1, 1.6 moles of gas 2, and 0.75 moles of gas 3 at 298.15 K and 1.5 bar can be calculated as follows starting with a list:

```
n = {2., 1.6, .75}

{2., 1.6, 0.75}

volume[n, 298, 1.5]

71.854
```

An advantage of *Mathematica* is that it accepts lists as input quantities.

8. It is easy to make plots in *Mathematica,* as shown by

```
Plot[Sin[x], {x, 0, 2 Pi}]
```

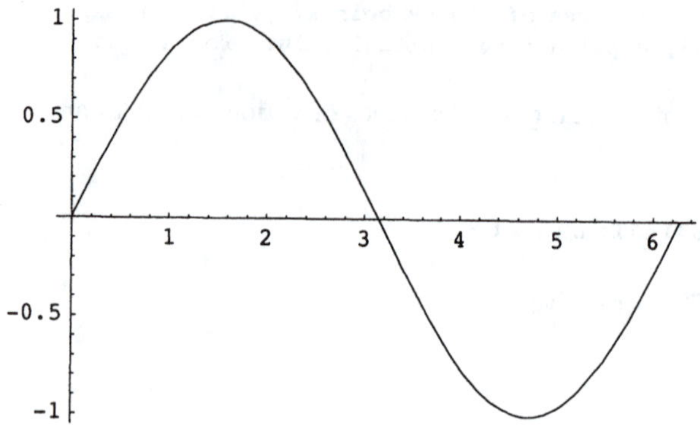


```
- Graphics -
```

Various options can be used to label the axes, etc.

9. The partial derivative of a function f with respect to x is obtained with D[f,x]. The nth derivative is obtained with D[f,{x,n}]. The indefinite integral of a function f with respect to x is obtained with Integrate[f,x]. A large book of integrals will list a few thousand indefinite integrals, but *Mathematica* can actually do a vastly wider range of integrals. Definite integrals are evaluated using Integrate[f, {x, x_{min}, x_{max} }]. Differential equations are solved by use of DSolve.

10. Some of the Computer Problems suggest further calculations that can be made using the programs given. For example, the calculation may be repeated for other substances, or a different temperature range, etc. Rather than retyping functions, you can copy and paste them. In making plots a program may be equivalent to a whole series of figures, in the sense that some variable, or variables can be readily changed.

11. One of the best ways to learn how to use new commands in *Mathematica* is to see them in action. For this reason an index of commands used in the Computer Problem is included in the Solutions Manual and on the Web.

12. References

Some very nice books have been written specifically to help people getting started with *Mathematica.*

K. R. Coombes, B. R. Hunt, R. L. Lipsman, J. E. Osborn, and G. J. Stuck, *The Mathematica Primer*, Cambridge University Press, 1998.

C-K. Cheung, G. E. Keough, C. Landraitis, and R. H. Gross, *Getting Started with Mathematica*, Wiley, New York, 1998.

H. F. W. Hoft and M. H. Hoft, *Computing with Mathematica*, Academic Press, San Diego, 1998.

B. F. Torrence and E. A. Torrence, The Student's Introduction to *Mathematica*, Cambridge University Press, 1999.

Index of Mathematica Commands

Graphics options are omitted.

Notes

Notes

Notes

Notes

Notes